西南林业大学馆藏蚂蚁模式标本

西南林业大学生物多样性保护学院
云南省森林灾害预警与控制重点实验室
云南生物多样性研究院

徐正会　刘　霞　张新民　熊忠平　钱昱含　和秋菊 / 著

内容简介

中国西南山地地貌和气候复杂，拥有东喜马拉雅和横断山2个全球生物多样性热点地区。蚂蚁是地球陆地上分布最广泛、种类最丰富的社会性昆虫，与人类生产和生活关系紧密。本书以中英文图鉴形式记录主要在西南山地发现的保存于西南林业大学的136个蚂蚁新种的模式标本，占全球已知蚂蚁物种总数的1%，隶属于10亚科39属。总论部分介绍了蚂蚁的社会性、生物学特性、生态功能、国内外对中国蚂蚁的分类研究历程、西南林业大学的分类研究历程及分类研究修订、蚂蚁的形态特征及分亚科检索表。分论部分由每个新种的英文原始描述、中文对照、原始描述中的黑白点线图和模式标本的高清照片组成。书后附有参考文献、中文名和学名索引、136个新种的名录及其分类地位和地理分布。该书系统总结了西南林业大学在西南山地的蚂蚁分类研究成果，对学术交流、生物学教学、生物多样性保护具有重要参考价值。本书适合大专院校生命科学师生、科研人员、自然和生物多样性保护工作者、蚂蚁爱好者和青少年读者阅读。

About This Book

The mountainous landform and climate in Southwest China are complex. It has two global biodiversity hotspot regions, the eastern Himalayas and the Hengduan Mountains. Ants are the most widespread and diverse social insects on the Earth, and closely related to human production and life. This book is written in both Chinese and English to record the type specimens of 136 new species of ants housed in the Southwest Forestry University, occupying 1% of the total known ant species in the world, which were mainly discovered in the southwestern mountains. The 136 species belong to 10 subfamilies and 39 genera of Formicidae. The general theory section introduces the sociality, biological characters, ecological function of ants, the taxonomic history on Chinese ants at home and abroad, the taxonomic course of Southwest Forestry University and related revision, morphological characters of ants and key to subfamilies. The individual theory section consists of the original English descriptions of each new species, corresponding Chinese descriptions, the black and white illustrations in the original descriptions and the high resolution colorful images. References, indexes of Chinese and scientific names, list of 136 new species including taxonomic status and geographical distribution are attached to the book. The book is a systematic summary of the research achievements of Southwest Forestry University on ant taxonomy in the southwestern mountains. It has important reference value for academic exchange, biology teaching and biodiversity conservation. This book is suitable for college students and teachers of life sciences, scientific research personnels, nature and biodiversity conservationists, ant lovers and young readers.

图书在版编目（CIP）数据

西南林业大学馆藏蚂蚁模式标本 / 徐正会等著. --
北京：中国林业出版社, 2024.5
　　ISBN 978-7-5219-2732-0
　　Ⅰ.①西… Ⅱ.①徐… Ⅲ.①蚁科—标本—图集
Ⅳ.①Q969.54-64
　　中国国家版本馆CIP数据核字(2024)第111645号

策划与责任编辑：葛宝庆
装帧设计：刘临川　张丽

出版发行：中国林业出版社
（100009，北京市西城区刘海胡同7号，电话83143612）
电子邮箱：cfphzbs@163.com
网址：www.cfph.net
印刷：北京雅昌艺术印刷有限公司
版次：2024年5月第1版
印次：2024年5月第1次
开本：889mm×1194mm　1/16
印张：22
字数：687千字
定价：280.00元

Type Specimens of Ants Housed in Southwest Forestry University

College of Biodiversity Conservation, Southwest Forestry University
Key Laboratory of Forest Disaster Warning and Control in Yunnan Province
Yunnan Academy of Biodiversity

by

Zheng-hui Xu, Xia Liu, Xin-min Zhang, Zhong-ping Xiong,
Yu-han Qian, Qiu-ju He

China Forestry Publishing House
2024

前 言
PREFACE

中国幅员辽阔，地貌复杂，气候多样，孕育了多彩的生物多样性，是全球12个生物多样性最丰富的国家之一。因受到喜马拉雅造山运动的影响，中国的西南地区被强烈挤压褶皱，形成了西北东南走向的喜马拉雅山区和南北走向的横断山区，这里的地貌和气候更加复杂多样，生物物种高度富集，因而成为全球30个生物多样性热点地区中的东喜马拉雅和横断山2个热点地区，是具有国际意义的陆地生物多样性保护关键地区。

蚂蚁是地球陆地上分布最广泛、种类和数量最多的社会性昆虫，估计全球有2万种。蚂蚁的数量很大，估计地球上蚂蚁个体的总数在10^{15}头以上。蚂蚁起源于距今8千万年前的白垩纪中后期，与被子植物同步进化繁荣，能够改良土壤、为植物授粉，可以为地球上1.1万种植物传播种子，还能捕食约10万种其他昆虫，控制害虫数量，在生态系统中具有重要功能。黄猄蚁 *Oecophylla smaragdina*（Fabricius）等种类具有食用和药用价值，红火蚁 *Solenopsis invicta* Buren等种类危害人类健康、干扰农事活动。所以，研究蚂蚁的区系及多样性对认识利用其生态功能和防控有害蚂蚁十分必要。

China is a vast country with complex landforms and diverse climates, and has given birth to a rich diversity of life. It is one of the 12 countries with the highest biodiversity in the world. Affected by the Himalayan orogeny, the southwest region of China was strongly extruded and folded, forming the northwest-southeast trending Himalayan Mountains and the north-south trending Hengduan Mountains. The landscape and climate are more complex and diverse, and the biological species are highly enriched in these areas, so it has become two hotspots, the eastern Himalayas and Hengduan Mountains, among the 30 biodiversity hotspots in the world. It is also a key area of international significance for the conservation of terrestrial biodiversity.

Ants are the most widespread, diverse, and numerous social insects on Earth's land with an estimated 20,000 species worldwide. The individual amount of ants is very large, and it is estimated that there are more than 10^{15} individuals of ants on the earth. Ants originated about 80 million years ago in the middle and late Cretaceous period, evolving and flourishing in sync with angiosperms. They can improve soil, pollinate plants, spread seeds for 11,000 species of plants on the Earth, prey on about 100,000 other insects, control the number of insect pests, therefore they play an important role in the ecosystem. The weaver ant, *Oecophylla smaragdina* (Fabricius), *etc*. has edible and medicinal value. However, the fire ant, *Solenopsis invicta* Buren, *etc*. endangers human health and interferes with agricultural activities. Therefore, it is necessary to study the fauna and diversity of ants to understand and utilize their ecological functions and prevent and control harmful ants.

早在1858年西方人就开启了对中国蚂蚁的分类研究。研究范围从香港开始，一路北上，先后报道广东、澳门、福建、台湾、浙江、江西、江苏、上海、山东、北京等沿海地区的种类。后来又有西藏、甘肃、湖南、重庆、四川、云南、湖北、陕西、内蒙古、黑龙江、新疆、广西等内地蚂蚁的研究报道。直至1982年，我国分类学家才开始研究报道我国的蚂蚁物种。至1995年，我国蚂蚁分类先驱唐觉等在《中国经济昆虫志第四十七册：膜翅目·蚁科（一）》中记载中国已知蚂蚁7亚科48属123个种和亚种。同年，我国著名蚂蚁分类学家吴坚和王常禄在《中国蚂蚁》中记载我国已知蚂蚁9亚科67属230种。

为了完善我国蚂蚁区系研究，由徐正会带领的西南林业大学蚂蚁课题组从1990年起开启了中国西南山地生物多样性热点地区及青藏高原、祁连山、新疆天山、四川大凉山、川西高原等地的蚂蚁多样性研究。通过34年的持续探索，初步查清了上述地区的蚂蚁种类、区系特征、群落结构、多样性特点和物种分布格局，积累研究标本10万余号，共100万余头。通过分类研究，发表蚂蚁新种139个、新属4个，发现中国新记录属9个、新记录种80余个。为了方便蚂蚁分类领域的学术交流、生物学教学与研究、生物多样性保护与利用、生命科学领域学生和蚂蚁爱好者学习参考，我们系统总结了西南林业大学在蚂蚁分类领域的研究成果，写成《西南林业大学馆藏蚂蚁模式标本》一书。

本书以中英文图鉴形式记录西南山地发现的保存于西南林业大学的136个蚂蚁新种的模式标本，隶属于10亚科39属。总论部分介绍了蚂蚁的社会性、生物学特性、生态功能、国内外对中国蚂蚁的分类研究历程、西南林业大学的分类研究历程及分类研究修订、蚂蚁的形态特征及分亚科检索表。分论部分由每个新种的英文原始描述、中文对照、原始描述中的黑白点线图和模式标本的高清照片组成。书后附有参考文献、中文名和学名索引、136个新种的名录及其分类地位和地理分布。

Early in 1858, the western scholars began to study the taxonomy of Chinese ants. The research scope started from Hong Kong and proceeded north, and ant species of coastal areas were successively reported from Guangdong, Macau, Fujian, Taiwan, Zhejiang, Jiangxi, Jiangsu, Shanghai, Shandong, Beijing, *etc*. Later, ant species of inland areas were also reported from Tibet, Gansu, Hunan, Chongqing, Sichuan, Yunnan, Hubei, Shaanxi, Inner Mongolia, Heilongjiang, Xinjiang, Guangxi, *etc*. Until 1982, our taxonomists have only just begun to study and report ant species of our country. To 1995, our country's ant classification pioneer Jue Tang et al. recorded 7 subfamilies, 48 genera, 123 species and subspecies of ants known in China in the book "*Economic Insect Fauna of China. Fasc. 47. Hymenoptera: Formicidae (1)* ". In the same year, our country's famous ant taxonomists Jian Wu and Chang-lu Wang recorded 9 subfamilies, 67 genera and 230 species of ants known in our country in the book "*The Ants of China*".

In order to improve the ant fauna research in China, the ant team of Southwest Forestry University led by Zheng-hui Xu has begun the researches on ant diversity in the biodiversity hotspots of mountainous areas in southwest China, Qinghai-Tibet Plateau, Qilian Mountains, Tianshan Mountains of Xinjiang, Daliangshan of Sichuan, Western Sichuan Plateau, *etc*. since 1990. Through 34 years of continuous exploration, the species, faunal characteristics, community structure, diversity feature and species distribution pattern of ants in the above mentioned areas have been preliminarily revealed. More than 100,000 numbered specimens and more than 1 million ant individuals have been accumulated. After classification study, 139 new species and 4 new genera of ants were published, and 9 new record genera and more than 80 new record species were found in China. To facilitate the academic exchange in ant taxonomy, biology teaching and research, biodiversity conservation and utilization, life science students and ant enthusiasts learning reference, we have systematically summarized the research achievements of Southwest Forestry University in the field of ant taxonomy and have written this book.

The book recorded the type specimens of 136 new species belonging to 10 subfamilies and 39 genera of ants in the form of Chinese and English pictorial book, all the new species were found in the mountainous area of southwest China with the type specimens housed in Southwest Forestry University. The general theory section introduces the sociality, biological characteristics, ecological functions of ants, the taxonomic study history on Chinese ants at home and abroad, the study course of Southwest Forestry University and related taxonomic revision, morphology of ants and the key to subfamilies. The individual theory section consists of the original English descriptions of each new species, corresponding Chinese descriptions, the black and white illustrations in the original descriptions and the high resolution colorful images. References, indexes of Chinese and scientific names, list of 136 new species including taxonomic status and geographical distribution are attached to the book.

本书由西南林业大学生物多样性保护学院、云南省森林灾害预警与控制重点实验室和云南生物多样性研究院组织编写。由云南省"增A去D"森林保护国际一流专业建设经费和云南省一流学科林学学科建设经费资助出版。该书研究成果在下列基金项目经费资助下获得：国家自然科学基金地区项目"青藏高原及邻近地区蚂蚁区系与物种多样性研究"（31860615，2019/01-2022/12）、"新疆天山地区蚁科昆虫多样性及系统发育研究"（31860166，2019/01-2022/12）、"川西高原地区蚂蚁区系与物种多样性研究"（32060122，2021/01-2024/12）、"四川大凉山地区蚂蚁区系与物种多样性研究"（31760633，2018/01-2021/12）、"喜马拉雅地区蚂蚁多样性研究"（31260521，2013/01-2016/12），国家自然科学基金面上项目"藏东南地区蚂蚁区系与物种多样性研究"（30870333，2009/01-2011/12），国家自然科学基金经典分类项目"滇西北地区蚂蚁区系分类与物种多样性研究"（30260016，2003/01-2005/12），国家自然科学基金青年项目"滇南热带雨林蚁科昆虫物种与生态系统多样性研究"（39500118，1996/01-1998/12），国家自然科学基金委员会应急管理项目"《中国动物志》的编研"（31750002）子课题"中国动物志 昆虫纲 膜翅目 蚁科（二）"（2018/01-2022/12）；云南省应用基础研究基金面上项目"利用捕食性蚂蚁防治热带亚热带林木害虫基础研究"（2004C0041M，2004/07-2007/12）、"哀牢山自然保护区蚂蚁生物多样性及其利用研究"（2001C0042M，2001/07-2004/07）、"高黎贡山自然保护区蚁科昆虫生物多样性研究"（97C006G，1997/07-2000/07）、"西双版纳自然保护区蚁科昆虫生物多样性研究"（95C067Q，1995/07-1998/07）。

This book was organized to write by the College of Biodiversity Conservation, Southwest Forestry University, Key Laboratory of Forest Disaster Warning and Control in Yunnan Province and Yunnan Academy of Biodiversity. The book was published with the support of both Yunnan Provincial "Increasing A and Eliminating D" Construction Fund for International First-class Speciality Forest Protection and Yunnan Provincial First-class Discipline Construction Fund for Forestry. The research achivements of this book were supported by the following fundations: National Natural Science Foundation of China (NSFC): regional projects "A Research on the Ant Fauna and Species Diversity of Qinghai-Tibet Plateau and Adjacent Area" (31860615, 2019/01-2022/12), "Diversity and Phylogeny of Formicidae of Mt. Tianshan in Xinjiang" (31860166, 2019/01-2022/12), "Study on the Fauna and Species Diversity of Ants from Western Sichuan Plateau" (32060122, 2021/01-2024/12), "Study on the Fauna and Species Diversity of Ants from Daliang Mountain in Sichuan" (31760633, 2018/01-2021/12) and "A Research on the Ant Diversity of Himalaya Area" (31260521, 2013/01-2016/12); general project "A Study on the Ant Fauna and Species Diversity in Southeast Tibet" (30870333, 2009/01-2011/12); classical classification project "Studies on the Fauna, Taxonomy and Species Diversity of Ants of Northwestern Yunnan" (30260016, 2003/01-2005/12); youth project "A Study on the Species and Ecosystem Diversity of Formicidae Insects in Tropical Rain Forest of Southern Yunnan" (39500118, 1996/01-1998/12); National Natural Science Foundation of China Emergency Management Program "The Editing and Research of Fauna Sinica" (31750002) Subproject "Fauna Sinica Insecta Hymenoptera Formicidae (II)" (2018/01-2022/12); Yunnan Provincial Applied and Basic Research Foundation general projects "Basic Studies on the Control of Tree Pests in Tropical and Subtropical Forests by Predatory Ants" (2004C0041M, 2004/07-2007/12), "A Study on Ant Biodiversity and Its Utilization in Ailao Mountains Nature Reserve" (2001C0042M, 2001/07-2004/07), "A Study on the Biodiversity of Formicidae in Gaoligong Mountains Nature Reserve" (97C006G, 1997/07-2000/07) and "A Study on the Biodiversity of Formicidae in Xishuangbanna Nature Reserve" (95C067Q, 1995/07-1998/07).

本书的前期研究是在郑哲民先生（陕西师范大学）指导下开始的。唐觉先生（原浙江农业大学）和吴坚先生（中国林业科学研究院森林保护研究所）对我们的研究提出了宝贵的意见和建议；赵清山先生（原中国林业部森林病虫害防治总站）把他的中文译著《印度动物志：膜翅目第2卷：蚂蚁和青蜂》寄给我们。下列专家学者分别为我们查询复制了部分分类文献：夏凯龄教授（原中国科学院上海昆虫研究所），梁铬球教授（中山大学昆虫研究所），吴铱先生（原西南林学院），马恩波博士（山西大学），博士生欧晓红、李保平和任国栋（陕西师范大学），杨兵（中国科学院昆明动物研究所），王莉萍（云南省林业科学研究院），王吉斌（原中国林业部西北林业调查规划设计院），孙波（原西北林学院经济林系）。

在前期研究中，下列国外蚁类学家为我们赠送了他们的分类文献：金兵珍博士（韩国汉城圆光大学）；寺山守先生（日本东京桐朋教育研究所），小野山敬一博士（日本北海道带广农业和兽医大学），绪方一夫博士（日本福冈九州大学）；威廉·布朗教授（美国康奈尔大学），菲利普·沃德博士（美国加利福尼亚大学），查尔斯·库格勒博士（美国瑞德福大学），马克·杜布瓦先生（美国伊利诺州橡树圈），罗伊·斯奈林先生（美国洛杉基县自然历史博物馆）；安德烈·弗朗索瓦教授（加拿大希库蒂米魁北克大学）；巴瑞·博尔顿先生（英国自然历史博物馆），塞德里克·科灵伍德先生（英国利兹城市博物馆）；贝恩哈德·塞费尔特博士（德国哥利兹自然科学研究所国家博物馆），艾尔弗雷德·布申格教授（德国达姆斯塔德生物-动物-技术学院）；西萨尔·巴洛尼乌尔巴尼先生（瑞士巴塞尔大学动物学研究所）；亨利·卡尼昂先生（法国保尔萨巴蒂尔大学）；泽维尔·艾斯巴达勒先生（西班牙巴塞罗那自治大学）；亚历山大·拉德申科先生（乌克兰科学院）；安德烈·普林斯先生和哈米什·罗伯逊博士（南非博物馆）。

The initial research for this book was conducted under the guidance of Prof. Zhe-min Zheng (Shaanxi Normal University). Prof. Jue Tang (Former Zhejiang Agricultural University) and Prof. Jian Wu (Institute of Forest Protection, Chinese Academy of Forestry) who put forward valuable opinions and suggestions for our study. Mr. Qing-shan Zhao (Former General Station of Forest Pests Prevention and Control, China Ministry of Forestry) sent us his Chinese translation of *The Fauna of British India including Ceylon and Burma, Hymenoptera, Vol. II. Ants and Cuckoo-Wasps*. The following experts and scholars reproduced some of the classification literatures for our query: Prof. Kai-ling Xia (Former Shanghai Institute of Entomology, Chinese Academy of Sciences), Prof. Ge-qiu Liang (Institute of Entomology, Sun Yat-sen University), Prof. Yi Wu (Former Southwest Forestry College), Dr. En-bo Ma (Shanxi University), doctoral candidates Xiao-hong Ou, Bao-ping Li and Guo-dong Ren (Shaanxi Normal University), Bing Yang (Kunming Institute of Zoology, Chinese Academy of Sciences), Li-ping Wang (Yunnan Academy of Forestry Sciences), Ji-bin Wang (Northwest Forestry Survey Planning and Design Institute, Former Ministry of Forestry of China) and Bo Sun (Department of Economic Forestry, Former Northwest Forestry College).

In our earlier study, the following foreign myrmecologists donated their taxonomic literatures to us: Dr. Byungjin Kim (Won Kwang University, Seoul, South Korea), Mr. Mamoru Terayama (Toho Institute of Education, Tokyo, Japan), Dr. Keiichi Onoyama (Obihiro University of Agriculture and Veterinary Medicine, Hokkaido, Japan), Dr. Kasuo Ogata (Kyushu University, Fukuoka, Japan), Prof. William L. Brown, Jr. (Cornell University, USA), Dr. Philip S. Ward (University of California, USA), Dr. Charles Kugler (Radford University, USA), Mr. Mark B. DuBois (Oakwood Circle, Illinois, USA), Mr. Roy R. Snelling (Natural History Museum of Los Angeles County, USA), Prof. André Francoeur (Universitè du Québec à Chicoutimi, Canada), Mr. Barry Bolton (British Museum Natural History, United Kingdom), Mr. Cedric A. Collingwood (Leeds City Museum, United Kingdom), Dr. Bernhard Seifert (Staatliches Museum Für Naturkunde-Forshungsstelle-Görlitz, Germany), Prof. Alfred Buschinger (Fachbererch Biologie-Zoologie-Technische Hochschule Darmstad, Germany), Mr. Cesare Baroni Urbani (Zoologische Institut Universität Basel, Switzerland), Mr. Henri Cagniant (Universite Paul Sabatir, France), Mr. Xavier Espadaler (Universitat Autónoma de Bacelona, Spain), Mr. Alexander G. Radchenko (Ukrainian Academy of Sciences, Ukraine), Mrs. André J. Prins and Hamish G. Robertson (South African Museum, South Africa).

在蚂蚁课题组的研究过程中，周善义教授和陈志林博士（广西师范大学）为我们提供了馆藏蚂蚁正模标本照片和研究文献，并赠送部分馆藏蚂蚁副模标本；吴卫教授（新疆大学）帮助我们查看馆藏蚂蚁标本；贺达汉教授和辛明博士（宁夏大学）帮助我们查看和拍摄馆藏蚂蚁模式标本，并赠送部分馆藏蚂蚁副模标本；马丽滨教授（陕西师范大学）帮助拍摄了馆藏蚂蚁模式标本照片；王小艺研究员、王鸿斌研究员、王梅博士、杨忠岐研究员、曹亮明博士（中国林业科学研究院）帮助我们查看和拍摄馆藏蚂蚁模式标本；乔格侠研究员、陈军研究员、张魁艳博士（中国科学院动物研究所）为我们查看和拍摄馆藏蚂蚁模式标本提供了帮助；蒋国芳（原广西壮族自治区林业科学研究所）、牛瑶（河南师范大学）、任国栋（河北大学）、杨效东（中国科学院西双版纳热带植物园）、魏琮（西北农林科技大学）、山根正气教授（日本鹿儿岛大学）和江口克之博士（日本长崎大学）、刘聪博士（日本冲绳科学技术大学院研究生大学）向我们赠送了蚂蚁标本。中村彰宏博士（中国科学院西双版纳热带植物园）、克里斯·伯威尔博士（澳大利亚昆士兰博物馆）、迈克尔·斯泰伯博士（德国弗莱堡大学）、满沛（新西兰奥克兰理工大学）、刘冠临（北京师范大学附属实验中学）、张智英（云南大学）采集并向我们赠送了部分蚂蚁新种的模式标本。

In the study process of our ant team, Prof. Shan-yi Zhou and Dr. Zhi-lin Chen (Guangxi Normal University) provided us images of holotype specimens of ants in their collection and literatures, and donated part of the paratype specimens to us. Prof. Wei Wu (Xinjiang University) help us observe the ant specimens in their collection. Prof. Da-han He and Dr. Ming Xin (Ningxia University) help us observe and take images of the type specimens of ants in their collection and donated part paratype specimens to us. Prof. Li-bin Ma (Shaanxi Normal University) took images of ant type specimens in their collection for us. Prof. Xiao-yi Wang, Prof. Hong-bin Wang, Dr. Mei Wang, Prof. Zhong-qi Yang and Dr. Liang-ming Cao (Chinese Academy of Forestry) help us observe and take images of ant type specimens in their collection. Prof. Ge-xia Qiao, Prof. Jun Chen and Dr. Kui-yan Zhang (Institute of Zoology, Chinese Academy of Sciences) provided help to observe and take images of ant type specimens in their collection. Guo-fang Jiang (Former Forestry Science Institute of Guangxi Zhuang Autonomous Region), Yao Niu (Henan Normal University), Guo-dong Ren (Hebei University), Xiao-dong Yang (Xishuangbanna Tropical Botanical Garden, Chinese Academy of Sciences), Cong Wei (Northwest Sci-Tech University of Agriculture and Forestry), Prof. Seiki Yamane (Kagoshima University, Japan), Dr. Katsuyuki Eguchi (Nagasaki University, Japan) and Dr. Cong Liu (Okinawa Institute of Science and Technology Graduate University, Japan) donated ant specimens to us. Dr. Akihiro Nakamura (Xishuangbanna Tropical Botanical Garden, Chinese Academy of Sciences), Dr. Chris J. Burwell (Queensland Museum, Australia), Dr. Michael Staab (University of Freiburg, Germany), Pei Man (Auckland University of Technology, New Zealand, Guan-lin Liu (The Experimental High School Attached to Beijing Normal University, China) and Zhi-ying Zhang (Yunnan University) collected and donated some type specimens of new ant species to us.

下列人员先后邀请我们参与国内自然保护区、国家公园和生物多样性关键地区的蚂蚁调查，从而丰富了我们的蚂蚁研究成果和标本收藏。李安·阿隆索博士（美国保护国际）邀请蚂蚁课题组徐正会参加了四川甘孜州蚂蚁调查，吕植教授和王昊博士（北京大学生命科学学院和山水自然保护中心）邀请徐正会参加藏东南地区、三江源国家级自然保护区、内蒙古额尔古纳、黑龙江大兴安岭和小兴安岭蚂蚁调查，刘少英研究员（四川省林业科学研究院）邀请徐正会参加西藏工布自然保护区蚂蚁调查，陈振宁教授（青海师范大学）邀请蚂蚁课题组参加祁连山国家公园青海片区蚂蚁调查，石福明教授（河北大学）邀请徐正会参加四川鞍子河国家级自然保护区和王朗国家级自然保护区蚂蚁调查，魏琮教授（西北农林科技大学）邀请蚂蚁课题组参加湖北神农架国家级自然保护区蚂蚁调查，肖文教授和黄志旁博士（大理大学）邀请蚂蚁课题组参加云南云岭自然保护区蚂蚁调查，杨宇明教授（原西南林学院）邀请蚂蚁课题组参加云南铜壁关自然保护区和大围山自然保护区蚂蚁调查，和世钧先生和王娟教授（西南林业大学）邀请蚂蚁课题组参加云南乌蒙山自然保护区蚂蚁调查，杜凡教授（西南林业大学）邀请蚂蚁课题组参加云南南滚河国家级自然保护区蚂蚁调查。在野外调查工作中，蚂蚁课题组得到了赵远潮（新疆维吾尔自治区林业和草原局）、马学林（武警日喀则中队）、陈平（西藏自治区林业和草原局）、王海宏（西藏自治区林芝市林业和草原局）、张伟（西藏自治区朗县）、靳代缨（三江源国家级自然保护区）、杨佐忠和李斌（四川省林业和草原局）、赵晓东（云南省林业和草原局）的大力帮助。

巴瑞·博尔顿先生（英国自然历史博物馆）允许使用其全球蚂蚁分类目录（AntCat），布莱恩·费歇尔博士及加州科学院（美国）允许使用其世界蚂蚁图像数据库（AntWeb），AntWiki团队允许使用其世界蚂蚁知识网（AntWiki），AntMaps团队允许使用其全球蚂蚁地理分布信息网（AntMaps），马丁·普费弗博士（蒙古国国立大学）允许使用其亚欧蚂蚁分类图像数据库（AntBase）。

The following people have invited us to participate ant survey in the domestic nature reserves, national parks and key biodiversity areas in China, thus enriched our ant research results and specimens collection. Dr. Leeanne E. Alonso (Conservation International, USA) invited Zheng-hui Xu of the ant team to participate in the survey of ants in Ganzi Prefecture, Sichuan Province. Prof. Zhi Lv and Dr. Hao Wang (School of Life Sciences, Peking University and Shanshui Nature Conservation Center) invited Zheng-hui Xu to join the survey of ants in southeastern Tibet, Sanjiangyuan National Nature Reserve, Erguna of Inner Mongolia, Greater Khingan Mountains and Lesser Khingan Mountains of Heilongjiang Province. Prof. Shao-ying Liu (Sichuan Academy of Forestry Sciences) invited Zheng-hui Xu to take part in the survey of ants in Gongbo Nature Reserve of Tibet. Prof. Zhen-ning Chen (Qinghai Normal University) invited our ant team to engage in the survey of ants in Qinghai section of Qilian Mountains National Park. Prof. Fu-ming Shi (Hebei University) invited Zheng-hui Xu to participate in the survey of ants in Anzihe National Nature Reserve and Wanglang National Nature Reserve of Sichuan Province. Prof. Cong Wei (Northwest Sci-Tech University of Agriculture and Forestry) invited ant team to join the survey of ants in Shennongjia National Nature Reserve of Hubei Province. Prof. Wen Xiao and Dr. Zhi-pang Huang (Dali University) invited ant team to take part in the survey of ants in Yunling Nature Reserve of Yunnan Province. Prof. Yu-ming Yang (Former Southwest Forestry College) invited ant team to engage in the survey of ants in Tongbiguan Nature Reserve and Daweishan Nature Reserve of Yunnan Province. Mr. Shi-jun He and Prof. Juan Wang (Southwest Forestry University) invited ant team to join the survey of ants in Wumeng Mountain Nature Reserve of Yunnan Province. Prof. Fan Du (Southwest Forestry University) invited ant team to take a hand in the survey of ants in Nangunhe National Nature Reserve of Yunnan Province. In field work, our ant team get a lot of help from Yuan-chao Zhao (Forestry and Grassland Bureau of Xinjiang Uygur Autonomous Region), Xue-lin Ma (Shigatse Squadron of the Armed Police), Ping Chen (Forestry and Grassland Bureau of Tibet Autonomous Region), Hai-hong Wang (Forestry and Grassland Bureau of Nyingchi City, Tibet Autonomous Region), Wei Zhang (Nang County, Tibet Autonomous Region), Dai-ying Jin (Sanjiangyuan National Nature Reserve), Zuo-zhong Yang and Bin Li (Forestry and Grassland Bureau of Sichuan Province), and Xiao-dong Zhao (Forestry and Grassland Bureau of Yunnan Province).

Mr. Barry Bolton (British Natural History Museum, United Kingdom) allows us to use An Online Catalog of the Ants of the World (AntCat), Dr. Brian L. Fisher and California Academy of Sciences (USA) allow us to utilize the world's online database of ant images (AntWeb), AntWiki team allows us to make use of the wealth of information on the world's ants (AntWiki), AntMaps team allow us to use the global ant geographic distribution information network (AntMaps), Dr. Martin Pfeiffer (National University of Mongolia, Mongolia) allow us to make use of A Taxonomic Ant Picturebase of Asia and Europe (AntBase).

下列人员参与了蚂蚁课题组不同时间的研究工作。西南林业大学的蚂蚁课题组成员：杨元昌、王仁师、杨比伦、胡刚、于新文、张庆、吴伟、李巧、周雪英、许国莲。西南林业大学硕士研究生：赵宇翔、梅象信、张继玲、傅美招、杨俊伍、陈友、史胜利、郭萧、杨忠文、陈龙官、姜海波、段艳、褚姣娇、张成林、于娜娜、李春良、宋扬、莫福燕、李文琼、李安娜、张翔、甘田、窦报坤、诸慧琴、罗成龙、和玉成、黄钊、赵梦乔、祁彪、李彪、翟奖、钱怡顺、陈超、郭宁妍、崔文夏、杨蕊、杨林、冷雪艳、邓方超、赵元昊、王先会、付毫、杨龙、韩秀、李婷、段加焕、尹晓丹、李朝义、尤诗佳、苏海舸、陈欣康。西南林业大学的本专科学生：王丽琼、陈志萍、杜永超、代色平、李天生、赖玉初、曾光、柳太勇、何云峰、龙启珍、李继乖、付磊、吴定敏、陈志强、蒋兴成、柴正群、周兴国、王文华、徐文川、何宗辉、赵忠良、王玲、和作萍、唐金奎、史生朝、王亚丽、石云、李丽梅、陈志峰、何娟、张扬、张鹏、王亮、张宁、张力、蒋华、郝永强、李海斌、罗丛凤、陈鑫、普顺荣、杨文涛、郑莹、庄江旭、李斌、李晓艳、黄雪蓉、袁定宇、冉茂君、王辉、刘兰兰、都红、武必念。蚂蚁课题组科研助理：马学云。

本书是在庄翔麟和杨松（西南林业大学生物多样性保护学院）亲切关怀下写作的。杨斌、罗旭、付建生、唐甜甜、王雷光、刘朝茂（西南林业大学生物多样性保护学院），张大才、苏小冰、王宏虬、施蕊、王曙光、李法营、朱家颖（西南林业大学林学院）对研究和出版工作给以全力支持。葛宝庆对本书精心编辑，张丽和刘临川对本书装帧作悉心设计。

我们谨在此对上述提供各种帮助的中外人士，以及因遗忘或疏漏未能提及的友好人士表示衷心的感谢！限于作者的研究深度和写作能力，书中错误在所难免，期待广大读者和研究人员不吝指正。

著者
2024年4月

The following personnel participated in the research work of the ant team at different times. Members of the ant team of Southwest Forestry University: Yuan-chang Yang, Ren-shi Wang, Bin-lun Yang, Gang Hu, Xin-wen Yu, Qing Zhang, Wei Wu, Qiao Li, Xue-ying Zhou and Guo-lian Xu. Postgraduates of Southwest Forestry University: Yu-xiang Zhao, Xiang-xin Mei, Ji-ling Zhang, Mei-zhao Fu, Jun-wu Yang, You Chen, Sheng-li Shi, Xiao Guo, Zhong-wen Yang, Long-guan Chen, Hai-bo Jiang, Yan Duan, Jiao-jiao Chu, Cheng-lin Zhang, Na-na Yu, Chun-liang Li, Yang Song, Fu-yan Mo, Wen-qiong Li, An-na Li, Xiang Zhang, Tian Gan, Bao-kun Dou, Hui-qin Zhu, Cheng-long Luo, Yu-cheng He, Zhao Huang, Meng-qiao Zhao, Biao Qi, Biao Li, Jiang Zhai, Yi-shun Qian, Chao Chen, Ning-yan Guo, Wen-xia Cui, Rui Yang, Lin Yang, Xue-yan Leng, Fang-chao Deng, Yuan-hao Zhao, Xian-hui Wang, Hao Fu, Long Yang, Xiu Han, Ting Li, Jia-huan Duan, Xiao-dan Yin, Zhao-yi Li, Shi-jia You, Hai-ge Su and Xin-kang Chen. Students of Southwest Forestry University: Li-qiong Wang, Zhi-ping Chen, Yong-chao Du, Se-ping Dai, Tian-sheng Li, Yu-chu Lai, Guang Zeng, Tai-yong Liu, Yun-feng He, Qi-zhen Long, Ji-guai Li, Lei Fu, Ding-min Wu, Zhi-qiang Chen, Xing-cheng Jiang, Zheng-qun Chai, Xing-guo Zhou, Wen-hua Wang, Wen-chuan Xu, Zong-hui He, Zhong-liang Zhao, Ling Wang, Zuo-ping He, Jin-kui Tang, Sheng-zhao Shi, Ya-li Wang, Yun Shi, Li-mei Li, Zhi-feng Chen, Juan He, Yang Zhang, Peng Zhang, Liang Wang, Ning Zhang, Li Zhang, Hua Jiang, Yong-qiang Hao, Hai-bin Li, Cong-feng Luo, Xin Chen, Shun-rong Pu, Wen-tao Yang, Ying Zheng, Jiang-xu Zhuang, Bin Li, Xiao-yan Li, Xue-rong Huang, Ding-yu Yuan, Mao-jun Ran, Hui Wang, Lan-lan Liu, Hong Du and Bi-nian Wu. Research assistant of the ant team: Xue-yun Ma.

This book was written under the tender care of Xiang-lin Zhuang and Yang Song (College of Biodiversity Conservation, Southwest Forestry University). Bin Yang, Xu Luo, Jian-sheng Fu, Tian-tian Tang, Lei-guang Wang and Chao-mao Liu (College of Biodiversity Conservation, Southwest Forestry University) and Da-cai Zhang, Xiao-bing Su, Hong-Qiu Wang, Rui Shi, Shu-guang Wang, Fa-ying Li and Jia-ying Zhu (College of Forestry, Southwest Forestry University) provided full support to the research and publication. Bao-qing Ge edited the book carefully. Li Zhang and Lin-chuan Liu designed binding of the book carefully.

We would like to express our heartfelt thanks to the above-mentioned Chinese and foreign people, and the friendly people who have not been mentioned because of forgetting or omission, for their various assistance! Limited to the depth of the author's research and writing ability, mistakes are inevitable in this book, so we look forward to readers and researchers to correct generously.

Authors
April 2024

西南林業大學
馆藏蚂蚁模式标本

Type Specimens of Ants Housed in Southwest Forestry University

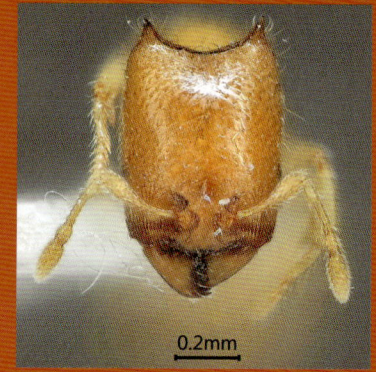

目 录

前 言

总 论 ... 001

    1 中国蚁科昆虫的国外分类研究历程 ... 007
    2 中国蚁科昆虫的国内分类研究历程 ... 011
    3 西南林业大学的蚁科昆虫分类研究历程 ... 015
    4 西南林业大学蚁科昆虫分类研究的修订 ... 019
    5 蚁科 Formicidae 昆虫形态特征 ... 025
    6 测量指标和比例及其缩写 ... 028
    7 蚁科分亚科检索表 ... 030

分 论 ... 033

阿佤钝猛蚁 ... 036	昌平卷尾猛蚁 ... 068	门巴小眼猛蚁 ... 100
细齿钝猛蚁 ... 038	赵氏卷尾猛蚁 ... 070	片突厚结猛蚁 ... 102
康巴钝猛蚁 ... 040	短背短猛蚁 ... 072	郑氏厚结猛蚁 ... 104
梅里钝猛蚁 ... 042	直唇隐猛蚁 ... 074	巴卡猛蚁 ... 106
八齿钝猛蚁 ... 044	黑色埃猛蚁 ... 076	坝湾猛蚁 ... 108
三叶钝猛蚁 ... 046	黄帝细颚猛蚁 ... 078	二齿猛蚁 ... 110
卓玛钝猛蚁 ... 048	老子细颚猛蚁 ... 080	龙林猛蚁 ... 112
木兰版纳猛蚁 ... 050	孟子细颚猛蚁 ... 082	勐腊猛蚁 ... 114
小眼迷猛蚁 ... 052	盘古细颚猛蚁 ... 084	南贡山猛蚁 ... 116
版纳曲颊猛蚁 ... 054	孙子细颚猛蚁 ... 086	五齿猛蚁 ... 118
版纳盘猛蚁 ... 056	炎帝细颚猛蚁 ... 088	片马猛蚁 ... 120
滇盘猛蚁 ... 058	庄子细颚猛蚁 ... 090	黄色猛蚁 ... 122
布氏卷尾猛蚁 ... 060	锥头小眼猛蚁 ... 092	长柄小盲猛蚁 ... 124
克平卷尾猛蚁 ... 062	傣小眼猛蚁 ... 094	六刺云行军蚁 ... 126
龙门卷尾猛蚁 ... 064	哈尼小眼猛蚁 ... 096	北京细蚁 ... 128
怒江卷尾猛蚁 ... 066	珞巴小眼猛蚁 ... 098	德宏细蚁 ... 130

昆明细蚁 132	墨脱奇蚁 192	霜降切胸蚁 252
秦岭细蚁 134	弯钩棱胸蚁 194	夏切胸蚁 254
云南细蚁 136	哀牢山塔蚁 196	小寒切胸蚁 256
北京原细蚁 138	弄巴塔蚁 198	小满切胸蚁 258
双色原细蚁 140	杨氏塔蚁 200	小暑切胸蚁 260
单色原细蚁 142	女娲角腹蚁 202	小雪切胸蚁 262
叉颚原细蚁 144	暗首棒角蚁 204	夏至切胸蚁 264
耿马原细蚁 146	哀牢窄结蚁 206	雨水切胸蚁 266
西藏原细蚁 148	乌蒙窄结蚁 208	心头铺道蚁 268
无缘细长蚁 150	雅鲁藏布窄结蚁 210	圆叶铺道蚁 270
凹唇细长蚁 152	白露切胸蚁 212	玉龙铺道蚁 272
隆背细长蚁 154	春切胸蚁 214	西藏犁沟蚁 274
叉唇细长蚁 156	春分切胸蚁 216	凹头臭蚁 276
尖唇细长蚁 158	处暑切胸蚁 218	鞍背臭蚁 278
大禹圆鳞蚁 160	大寒切胸蚁 220	鳞结臭蚁 280
平背高黎贡蚁 162	大暑切胸蚁 222	尖齿刺结蚁 282
阿诗玛无刺蚁 164	大雪切胸蚁 224	网纹刺结蚁 284
疏毛无刺蚁 166	冬切胸蚁 226	楔结长齿蚁 286
景颇弯蚁 168	冬至切胸蚁 228	巴卡多刺蚁 288
尼玛弯蚁 170	谷雨切胸蚁 230	短胸多刺蚁 290
尖刺稀切叶蚁 172	寒露切胸蚁 232	方肩多刺蚁 292
高结稀切叶蚁 174	惊蛰切胸蚁 234	驼背多刺蚁 294
双角稀切叶蚁 176	立春切胸蚁 236	齿肩多刺蚁 296
弯刺稀切叶蚁 178	立冬切胸蚁 238	圆肩多刺蚁 298
钝齿稀切叶蚁 180	立秋切胸蚁 240	圆顶多刺蚁 300
直背稀切叶蚁 182	立夏切胸蚁 242	大眼前结蚁 302
纹头稀切叶蚁 184	芒种切胸蚁 244	黑角前结蚁 304
条纹稀切叶蚁 186	清明切胸蚁 246	双齿唇拟毛蚁 306
郑氏华丽蚁 188	秋切胸蚁 248	
裂唇奇蚁 190	秋分切胸蚁 250	

参考文献 308

附 录 319

模式标本馆藏在西南林业大学的蚂蚁新种名录 319

中文名索引 330

学名索引 332

CONTENTS

PREFACE

General Theory 001

    1 Abroad Taxonomic Study History on Chinese Formicidae Insects 007
    2 Domestic Taxonomic Study History on Chinese Formicidae Insects 011
    3 Taxonomic Study Course on Formicidae Insects in Southwest Forestry University 015
    4 Revision of the Taxonomy of Formicidae Insects in Southwest Forestry University 019
    5 Morphology of Formicidae Insects 025
    6 Measurements and Indices and Their Abbreviations 028
    7 Key to Subfamilies of Formicidae 030

Individual Theory 033

Amblyopone awa Xu & Chu, 2012 036
Amblyopone crenata Xu, 2001 038
Amblyopone kangba Xu & Chu, 2012 040
Amblyopone meiliana Xu & Chu, 2012 042
Amblyopone octodentata Xu, 2006 044
Amblyopone triloba Xu, 2001 046
Amblyopone zoma Xu & Chu, 2012 048
Bannapone mulanae Xu, 2000 050
Mystrium oculatum Xu, 1998 052
Gnamptogenys bannana Xu & Zhang, 1996 054
Discothyrea banna Xu, Burwell & Nakamura, 2014 056
Discothyrea diana Xu, Burwell & Nakamura, 2014 058
Proceratium bruelheidei Staab, Xu & Hita Garcia, 2018 060
Proceratium kepingmai Staab, Xu & Hita Garcia, 2018 062
Proceratium longmenense Xu, 2006 064
Proceratium nujiangense Xu, 2006 066
Proceratium shohei Staab, Xu & Hita Garcia, 2018 068
Proceratium zhaoi Xu, 2000 070
Brachyponera brevidorsa Xu, 1994 072
Cryptopone recticlypea Xu, 1998 074
Emeryopone melaina Xu, 1998 076
Leptogenys huangdii Xu, 2000 078
Leptogenys laozii Xu, 2000 080
Leptogenys mengzii Xu, 2000 082
Leptogenys pangui Xu, 2000 084
Leptogenys sunzii Xu & He, 2015 086
Leptogenys yandii Xu & He, 2015 088

Leptogenys zhuangzii Xu, 2000 090

Myopias conicara Xu, 1998 092

Myopias daia Xu, Burwell & Nakamura, 2014 094

Myopias hania Xu & Liu, 2012 096

Myopias luoba Xu & Liu, 2012 098

Myopias menba Xu & Liu, 2012 100

Pachycondyla lobocarena Xu, 1995 102

Pachycondyla zhengi Xu, 1995 104

Ponera baka Xu, 2001 106

Ponera bawana Xu, 2001 108

Ponera diodonta Xu, 2001 110

Ponera longlina Xu, 2001 112

Ponera menglana Xu, 2001 114

Ponera nangongshana Xu, 2001 116

Ponera pentodontos Xu, 2001 118

Ponera pianmana Xu, 2001 120

Ponera xantha Xu, 2001 122

Probolomyrmex longiscapus Xu & Zeng, 2000 124

Yunodorylus sexspinus Xu, 2000 126

Leptanilla beijingensis Qian, Xu, Man & Liu, 2024 128

Leptanilla dehongensis Qian, Xu, Man & Liu, 2024 130

Leptanilla kunmingensis Xu & Zhang, 2002 132

Leptanilla qinlingensis Qian, Xu, Man & Liu, 2024 134

Leptanilla yunnanensis Xu, 2002 136

Protanilla beijingensis Man, Ran, Chen & Xu, 2017 ... 138

Protanilla bicolor Xu, 2002 140

Protanilla concolor Xu, 2002 142

Protanilla furcomandibula Xu & Zhang, 2002 144

Protanilla gengma Xu, 2012 146

Protanilla tibeta Xu, 2012 148

Tetraponera amargina Xu & Chai, 2004 150

Tetraponera concava Xu & Chai, 2004 152

Tetraponera convexa Xu & Chai, 2004 154

Tetraponera furcata Xu & Chai, 2004 156

Tetraponera protensa Xu & Chai, 2004 158

Epitritus dayui Xu, 2000 160

Gaoligongidris planodorsa Xu, 2012 162

Kartidris ashima Xu & Zheng, 1995 164

Kartidris sparsipila Xu, 1999 166

Lordomyrma jingpo Liu, Xu & Hita Garcia, 2021 .. 168

Lordomyrma nima Liu, Xu & Hita Garcia, 2021 170

Oligomyrmex acutispinus Xu, 2003 172

Oligomyrmex altinodus Xu, 2003 174

Oligomyrmex bihornatus Xu, 2003 176

Oligomyrmex curvispinus Xu, 2003 178

Oligomyrmex obtusidentus Xu, 2003 180

Oligomyrmex rectidorsus Xu, 2003 182

Oligomyrmex reticapitus Xu, 2003 184

Oligomyrmex striatus Xu, 2003 186

Paratopula zhengi Xu & Xu, 2011 188

Perissomyrmex fissus Xu & Wang, 2004 190

Perissomyrmex medogensis Xu & Zhang, 2012 192

Pristomyrmex hamatus Xu & Zhang, 2002 194

Pyramica ailaoshana Xu & Zhou, 2004 196

Pyramica nongba Xu & Zhou, 2004 198

Pyramica yangi Xu & Zhou, 2004 200

Recurvidris nuwa Xu & Zheng, 1995 202

Rhopalomastix umbracapita Xu, 1999 204

Stenamma ailaoense Liu & Xu, 2011 206

Stenamma wumengense Liu & Xu, 2011 208

Stenamma yaluzangbum Liu & Xu, 2011 210

Temnothorax bailu Qian & Xu, 2024 212

Temnothorax chun Qian & Xu, 2024 214

Temnothorax chunfen Qian & Xu, 2024 216

Temnothorax chushu Qian & Xu, 2024 218

Temnothorax dahan Qian & Xu, 2024 220

Temnothorax dashu Qian & Xu, 2024 222

Temnothorax daxue Qian & Xu, 2024 224

Temnothorax dong Qian & Xu, 2024 226

Temnothorax dongzhi Qian & Xu, 2024 228

Temnothorax guyu Qian & Xu, 2024 230

Temnothorax hanlu Qian & Xu, 2024 232
Temnothorax jingzhe Qian & Xu, 2024 234
Temnothorax lichun Qian & Xu, 2024 236
Temnothorax lidong Qian & Xu, 2024 238
Temnothorax liqiu Qian & Xu, 2024 240
Temnothorax lixia Qian & Xu, 2024 242
Temnothorax mangzhong Qian & Xu, 2024 244
Temnothorax qingming Qian & Xu, 2024 246
Temnothorax qiu Qian & Xu, 2024 248
Temnothorax qiufen Qian & Xu, 2024 250
Temnothorax shuangjiang Qian & Xu, 2024 252
Temnothorax xia Qian & Xu, 2024 254
Temnothorax xiaohan Qian & Xu, 2024 256
Temnothorax xiaoman Qian & Xu, 2024 258
Temnothorax xiaoshu Qian & Xu, 2024 260
Temnothorax xiaoxue Qian & Xu, 2024 262
Temnothorax xiazhi Qian & Xu, 2024 264
Temnothorax yushui Qian & Xu, 2024 266
Tetramorium cardiocarenum Xu & Zheng, 1994 268
Tetramorium cyclolobium Xu & Zheng, 1994 270
Tetramorium yulongense Xu & Zheng, 1994 272
Vombisidris tibeta Xu & Yu, 2012 274
Dolichoderus incisus Xu, 1995 276
Dolichoderus sagmanotus Xu, 2001 278
Dolichoderus squamanodus Xu, 2001 280
Lepisiota acuta Xu, 1994 282
Lepisiota reticulata Xu, 1994 284
Myrmoteras cuneonodum Xu, 1998 286
Polyrhachis bakana Xu, 1998 288
Polyrhachis brevicorpa Xu, 2002 290
Polyrhachis cornihumera Xu, 2002 292
Polyrhachis cyphonota Xu, 1998 294
Polyrhachis dentihumera Xu, 2002 296
Polyrhachis orbihumera Xu, 2002 298
Polyrhachis rotoccipita Xu, 2002 300
Prenolepis magnocula Xu, 1995 302
Prenolepis nigriflagella Xu, 1995 304
Pseudolasius bidenticlypeus Xu, 1997 306

References 308

Appendix 319

Checklist of New Species of Ants with Type Specimens Housed in Southwest Forestry University 319

Index of Chinese Names 330

Index of Scientific Names 332

西南林业大学
馆藏蚂蚁模式标本

Type Specimens of Ants Housed in
Southwest Forestry University

总论
General Theory

蚂蚁是我们身边常见的小动物，它们经常出现在房前屋后、道路两旁、公园绿地或森林草地之中，要么成群结队搬运食物，要么单个活动四处觅食。它们是人类童年较早结识的自然观察对象，是我们非常熟悉的社会性昆虫。蚂蚁给人类的印象是力大无比、集体合作、辛勤劳作。人类也经常拿蚂蚁与大象作对比，以凸显大象的巨大和蚂蚁的渺小。蚂蚁虽小，然而数量惊人，其群体力量不可小觑。根据徐正会等（1999a）对我国西双版纳热带雨林的蚂蚁数量、生物量和生态功能的测定，西双版纳的蚂蚁总数为 5.2×10^{13} 头（52万亿头），相应的生物量为 1.04×10^5 吨（10.4万吨），相当于34,667头成年亚洲象的体重之和；这些蚂蚁1年内搬运的土壤和有机质为 1.5×10^9 吨（15亿吨）。这样大的数量、生物量和生态功能是不是很惊人呢？

蚂蚁隶属于动物界Animalia节肢动物门Arthropoda昆虫纲Insecta膜翅目Hymenoptera蚁科Formicidae，是自然界分布最广泛的昆虫类群之一。除了地球的两极和高山的雪线以上极寒冷区域，陆地上几乎到处都有蚂蚁的踪迹。蚂蚁的种类很多，估计全球有20000种。蚂蚁的数量很大，估计地球上蚂蚁个体的总数在 10^{15} 头以上（Hölldobler & Wilson, 1990）。目前全球已经记载16亚科346属14172种，化石蚂蚁4亚科160属769种（Bolton, 2023）。中国已经记载12亚科119属1113种（AntWiki, 2023），物种数仅占全球已知种类的7.9%，说明我国还有很多蚂蚁有待发现。云南是全国蚂蚁种类最丰富的省份，已经记载605种，占全国已知种类的54.4%。

Ants are common small animals around us. They often appear in front and behind houses, on both sides of roads, in parks or forest meadows, either moving their food in groups, or moving around individually to look for food. They are the observed objects in nature that humans meet early in childhood, and also a kind of social insects which we are very familiar with. Ants give the impression of being extremely powerful, cooperative and hard-working. Humans also often compare ants with elephants to highlight the elephant's size and the ant's smallness. Ants are small, but their numbers are staggering, and their collective power can not to be underestimated. According to the determination of population, biomass and ecological function of ants in Xishuangbanna tropical rain forest of our country by Xu et al. (1999a), the total number of ants in Xishuangbanna is 5.2×10^{13} heads (52 trillion heads), and the corresponding biomass is 1.04×10^5 tons (104,000 tons), which is the weight of 34,667 adult Asian elephants combined. These ants can move 1.5×10^9 tons (1.5 billion tons) of soil and organic matter in one year. Isn't it amazing how large the amount, the biomass and the ecological function ants own?

Ants belong to Formicidae, Hymenoptera, Insecta, Arthropoda, Animalia. They are one of the most widely distributed insect groups in nature. Ants are almost everywhere on the land except for the poles of the Earth and high mountains above the snow line extremely cold areas. There are many kinds of ants, with estimated 20000 species worldwide. The number of ant individuals is huge, and it is estimated that there are more than 10^{15} individuals of ants on the earth (Hölldobler & Wilson, 1990). At present, 16 subfamilies, 346 genera and 14172 species of living ants and four subfamilies, 160 genera and 769 species of fossil ants have been recorded in the world (Bolton, 2023). And 12 subfamilies, 119 genera and 1113 species have been recorded in China (AntWiki, 2023), only 7.9% of the world's known species. It shows that there are still many ants to be found in our country. Yunnan is the province with the richest ant species in China, with 605 species recorded, which accounts for 54.4% of the known species in our country.

所有蚂蚁种类均属于社会性昆虫。社会性昆虫有3个特征：一是同种个体间相互合作，照顾幼体；二是同种个体间有明确的分工和形态的分化，各司其职；三是同种蚁巢内至少有2个重叠的世代，上一代在一段时间内照顾子代。蚂蚁社会中通常包含蚁后、雄蚁、工蚁3个类型，一些类群的工蚁进一步分化为体形较大的大型工蚁（兵蚁）和体形较小的工蚁。蚁后为雌性，早期有翅，交配后翅脱落，负责繁殖后代，寿命最长，可以生活几年至十几年，蚁巢的寿命通常取决于蚁后的寿命。雄蚁为雄性，繁殖季节才出现，有翅，寿命短暂，交配后不久死亡。工蚁为雌性，终生无翅，不交配也不生育，其职能复杂，包括筑巢、觅食、哺育幼蚁、照顾蚁后、清洁蚁巢、保卫蚁巢等，工蚁的寿命较短，半年至1年以上不等。兵蚁为大型的工蚁，雌性，终生无翅，不交配也不生育，体形比小型工蚁大得多，职能侧重保卫，也包括搬运大型食物和咬碎坚硬食物等，寿命与小型工蚁相仿。

蚂蚁通常在春夏季繁殖。在天气晴好的时候，巢内出现大量有翅雌蚁和雄蚁，它们在空中飞舞或在地面聚集，寻找配偶，组建新巢。但是大多数繁殖蚁被鸟类、兽类、蛙类或其他昆虫捕食，只有少数个体成功建巢。蚂蚁的巢包括游动巢、土壤巢、地表巢、木质巢、层纸巢、丝质巢六类。蚂蚁的食物很复杂，归纳起来有小型动物、植物、蜜露、真菌四大类（徐正会，2002a）。蚂蚁在亲缘关系上与胡蜂和蜜蜂比较接近，均属于比较进化的完全变态类昆虫，一生经历卵、幼虫、蛹、成虫4个虫态，与白蚁的亲缘关系很远。白蚁属于相对原始的不全变态类昆虫，与蜚蠊、螳螂等亲缘关系较近，一生只经历卵、若虫、成虫3个虫态。蚂蚁与白蚁的共同之处是分别进化出了社会性，均为社会性昆虫。

All ant species are social insects. The sociali insects have three characteristics: ① individuals of the same species cooperate and care for the young; ② there is a clear division of labor and form of differentiation between individuals of the same species, each with its own role; ③ there are at least two overlapping generations in the same nest, with the previous generation taking care of the offspring for a period of time. Ant societies usually consist of three castes: queens, males, and workers. The workers of some groups are further differentiated into larger workers (the soldiers) and smaller workers. The queen is a female, early winged, after mating the wings fall off, who is responsible for the reproduction of offspring. Its life span is the longest, and it can live a few years to more than a decade. The lifespan of a nest usually depends on the lifespan of the queen. Male ants are male caste, only appear in breeding season and winged, short-lived and die soon after mating. Workers are female, wingless for life, who do not mate and do not procreate. Their functions are complex, include nesting, foraging, feeding the young, caring the queen, cleaning and defending the nest, *etc*. Workers' lives are shorter, ranging from six months to more than one year. Soldiers are large workers, female, wingless for life, neither mate nor procreate, they are much larger than the small workers, and their function focuses on security, also includes moving large foods and crushing hard foods. Their life span is similar to that of small workers.

Ants usually breed in spring and summer. On sunny days, the nest is teeming with winged females and males. They fly in the air or gather on the ground, looking for mates and building new nests. But most breeding ants are eaten by birds, mammals, frogs or other insects, and only a few individuals can successfully nest. Ant nests include six types: moveable nest, soil nest, ground nest, wooden nest, layer paper nest and silk nest. The food of ant is complicated, which can be divided into four categories: small animals, plants, honeydew and fungi (Xu, 2022a). Ants are closely related to wasps and bees, all of which belong to complete metamorphic insects of comparative evolution, and experience 4 stages during their lifetime: eggs, larvae, pupae and adults. They are distantly related to termites. Termites belong to the relatively primitive incomplete metamorphic insects, closely related to cockroaches and mantids, and experience only 3 states during their lifetime: eggs, nymphs and adults. What ants and termites have in common is that each has evolved sociality and both are social insects.

我们的祖先很早就对蚂蚁产生兴趣，并对蚂蚁进行观察记录，他们认为蚂蚁的活动依节令和气候而变，观察蚂蚁的行为可以预知节气更迭和天气变化。蚂蚁可以作为食物或药物，食用后可以强身健体，治疗疾病。知名中医吴志成（1991）通过多年临床研究，发现鼎突多刺蚁 Polyrhachis vicina Roger（今称双齿多刺蚁 Polyrhachis dives Smith）制剂可用于治疗类风湿性关节炎，并著有《蚂蚁与类风湿性关节炎》专著。今天，分布于热带和南亚热带的黄猄蚁 Oecophylla smaragdina（Fabricius）是云南西双版纳、德宏、临沧、保山等地傣族等少数民族钟爱的食用昆虫，可以凉拌，还可以做成火锅。木重头蚁 Carebara lignata Westwood 分布于我国云南及东南亚、南亚地区，其工蚁微小，但是蚁后的蛹个体硕大，在云南德宏傣族景颇族自治州民间是美味食用昆虫。当地少数民族能通过工蚁的活动轨迹找到蚁巢，在蚁后的蛹长成的时节，他们打开蚁巢，掏出蚁后的蛹来食用，并把有生殖能力的蚁后小心放回巢中，来年继续繁殖。

黄猄蚁除了作为食物，还是著名的天敌昆虫。早在西晋（公元304年）时期，广东果农已经利用黄猄蚁防治柑橘害虫了，成为人类最早以虫治虫的生物防治成功案例（陈守坚，1962；唐觉等，1995）。在广东、福建一带，山区的农民将树上的黄猄蚁蚁巢整个取下，用布袋装上，运到集市出售。种植柑橘的果农买回蚁巢，将其安放在果树之上，再用绳索将果树连起来，这样黄猄蚁就可以从蚁巢到达各棵果树捕食害虫，达到以虫治虫的目的。今天，在西双版纳巴西橡胶树 Hevea brasiliensis 种植园内，黄猄蚁仍然是得力的捕虫能手。据徐正会等（1999a）调查，黄猄蚁可以在45.5%的橡胶树树冠上筑巢，很好地保护了这些橡胶树，促进了橡胶丰产。

Our ancestors took an early interest in ants and made observations about them. They thought that ants' activities vary according to season and climate, observing ants' behavior can predict solar terms and weather changes. Ants can be used as food or medicine, after eating one can strengthen one's body and cure one's disease. Through many years of clinical research, a well-known Chinese medicine doctor Wu (1991) found that the preparation of *Polyrhachis vicina* Roger (now called *Polyrhachis dives* Smith) was useful in the treatment of rheumatoid arthritis, and has written the monograph "Ants and Rheumatoid Arthritis". Today, the weaver ant *Oecophylla smaragdina* (Fabricius) which distributed in the tropics and south subtropics is a favorite food insect of the Dai and other minority nationalities in Xishuangbanna, Dehong, Lincang, Baoshan, *etc.* of Yunnan Province, which can be made into cold ant salad or hot pot. The wood heavy head ant *Carebara lignata* Westwood distributes in Yunnan of our country, southeast Asia and south Asia. Its workers are tiny, but the queen's pupae are huge, and the pupae are delicious edible insects in the folk of Dehong Dai and Jingpo Autonomous Prefecture, Yunnan Province. Local minorities can trace the activity of workers to the nest. At the time when the Queen's pupae are fully formed, they open the nest and took out the queen's pupae to eat, and carefully returned the reproductive queen to her nest to breed in the next year.

In addition to being a food source, the weaver ant *Oecophylla smaragdina* (Fabricius) is also known as a famous natural enemy insect. As early as the Western Jin Dynasty (AD 304) period, fruit farmers in Guangdong have been using the ant to control citrus pests. It has become the earliest successful case of biological control using insects to control pests (Chen, 1962; Tang et al., 1995). In Guangdong and Fujian Provinces, farmers in the mountains removed the entire nest from the tree, put it in a cloth bag and sold it to the market. Citrus growers bought back the ant nests and secured them on the trees. Then the fruit trees were connected by ropes, so the ant can traveled from the nest to the fruit trees to hunt for pests, thus the aim of pest control was achieved. Today, in the Brazilian rubber tree *Hevea brasiliensis* plantation, in Xishuangbanna, *Oecophylla smaragdina* (Fabricius) is still a skilled insect pests catcher. According to the investigation of Xu et al. (1999a), the weaver ant can nest on 45.5% of the rubber tree canopy, the rubber trees were well protected, and the ant promoted the high yield of rubber.

相关研究表明，地球上的蚂蚁能捕食大约10万种其他昆虫，在食物链中具有重要地位（Thomas & Settele，2004）。分子生物学研究揭示，蚂蚁与被子植物是同步进化的（Moreau et al.，2006），今天地球上的蚂蚁能够帮助1.1万种植物传播种子（Lengyel et al.，2009）。在云南金沙江干热河谷和昆明植物园等地，已经观察到盘腹蚁属 *Aphaenogaster* 物种可以帮助百部科 Stemonaceae 植物传播种子。因为该科内的某些植物种子上演化出了蚂蚁喜欢取食的油脂体。这些油脂体能散发出特殊气味，吸引蚂蚁前来搬运。蚂蚁取食完油脂体后将种子丢弃，从而帮助百部科种子实现传播扩散。虽然蚂蚁在授粉方面不如蜜蜂和蝴蝶那样高效，也可以帮助许多植物授粉。许多蚂蚁喜欢甜食，植物花朵中的蜜露经常吸引大量蚂蚁个体前来取食。在青藏高原等高海拔地区，经常观察到红蚁属 *Myrmica* 和蚁属 *Formica* 物种的个体钻进很深的管状或喇叭状花冠内吸食花蜜，帮助植物授粉，促进了高原植物的繁荣。

蚂蚁在生态系统中最重要的功能是改良土壤。已有研究表明，大多数蚂蚁物种在土壤内筑巢，在地表觅食，它们将食物运到地下食用后降解为有机肥料，提高了土壤肥力。蚂蚁在地下筑巢的过程中，将土壤切割为小颗粒运出蚁巢，散布或堆积于蚁巢之外。堆积的土壤颗粒很疏松，非常适合植物种子萌发生长。蚁巢内被掏空的土壤空间有利于透气、透水，很适合植物根系的生长。所以，蚂蚁的活动促进了植物的繁荣，植物的繁荣又促成了蚂蚁社会的兴旺，二者相得益彰。

Related studies show that ants on the Earth can prey on approximately 100000 other insects (Thomas & Settele, 2004) and play an important role in the food chain. Molecular biology studies have revealed that ants and angiosperms evolved synchronously (Moreau et al., 2006). Ants on the Earth today help spread seeds among 11,000 plant species (Lengyel et al., 2009). In places like the dry-hot valley of Jinsha River and Kunming Botanical Garden in Yunnan Province, it has been observed that *Aphaenogaster* species can help Stemonaceae plants spread their seeds. Because the seeds of some plants in this family evolved the fat body that ants like to eat, and fat body can give off a special smell, attracting ants to come to carry. The ants discard the seeds after eating the fat body, thus help Stemonaceae seeds achieve the spread of diffusion. Although ants are not as efficient at pollination as bees and butterflies, they can also help pollinate many plants. Many ants have a sweet tooth, so nectar in plant flowers often attracts large numbers of ant individuals to feed. In the Qinghai-Tibet Plateau and other high-altitude areas, it is often observed that individuals of *Myrmica* and *Formica* species burrowing deep into tubular or trumpet-shaped corolla to suck nectar, help pollinate plants and promote the prosperity of plateau plants.

The most important function of ants in ecosystem is to improve the soil. Studies have shown that most ant species nest in the soil and forage on the ground, and they transport food underground for consumption and then degrade it into organic fertilizer, improving soil fertility. In the process of building a nest underground, they cut the soil into small particles and carried them out of the nest, spreading or stacking them outside. The accumulated soil particles are very loose and suitable for the germination and growth of plant seeds. The hollowed soil space in the nest is conducive to air and water permeability, and very suitable for the growth of plant roots. Therefore, the activities of ants promote the prosperity of plants, and the prosperity of plants in turn has contributed to the prosperity of ant society, both plants and ants complement each other.

蚂蚁与其他昆虫之间也存在许多相互依存关系，蚂蚁与蚜虫、介壳虫等分泌蜜露的昆虫之间存在非常紧密的关系。蚜虫和介壳虫分泌的蜜露是蚂蚁喜欢的甜食，作为回报，蚂蚁会充当这些昆虫的"保镖"，为它们驱赶寄生蜂、瓢虫等天敌。高原上的蚂蚁甚至会把蚜虫、介壳虫等搬进蚁巢内过冬，有些蚂蚁物种的雌蚁在繁殖季节建巢时甚至会从"娘家"带上介壳虫，作为将来蜜糖的供应源。蚂蚁饲养蚜虫、介壳虫的行为，好比人类饲养奶牛。不过，蚂蚁饲养这些泌露昆虫的历史比人类饲养奶牛要早多了，要知道蚂蚁起源于距今8000万年的白垩纪中后期（Hölldobler & Wilson，1990），而人类的起源是距今大约1400万年。除了饲养蚜虫、介壳虫等，蚂蚁还会种蘑菇。产于南美洲热带雨林的芭切叶蚁属 Atta 物种用上颚将树叶切成小片，运回蚁巢内培养真菌作为食物，是地球上会种蘑菇、以真菌为食的蚂蚁，就像白蚁培养鸡㙮菌作为食物一样。不过到目前为止我国尚未发现此类会种蘑菇的切叶蚁。

原产南美洲的红火蚁 Solenopsis invicta Buren，自2004年入侵我国广东省吴川县以来（曾玲等，2005），至今已经快速扩散至我国华南、华东、华中、西南地区。在城市、农地、林地等环境中广泛定殖，繁殖力强大，食物繁杂，竞争力突出，适应性极强，防治难度很大。其螯针排出的毒液能导致皮肤红肿、疼痛，体质过敏者被蜇伤后长时间发痒，甚至休克。红火蚁的入侵已经严重干扰人类的农事活动，甚至危及公共卫生和生态系统安全，需要加强防控工作。

可见蚂蚁与植物之间、蚂蚁与其他昆虫和动物之间存在千丝万缕的联系。正是这些看似不经意的联系，构筑成了自然界的食物链和食物网关系，让地球上的生物多样性生生不息，绵延不断，永久造福地球生物圈和人类社会。

There are also many interdependent relationships between ants and other insects. Ants have a very close relationship with aphids, scale insects and other honeydew-secreting insects. The honeydew secreted by aphids and scale insects is a sweet food for ants. In return, the ants act as bodyguards for the insects, for them to drive parasitic wasps, ladybugs and other natural enemies. The ants on the plateau even move aphids, scale insects, *etc*. into their nests to overwinter. In some ant species, females even bring scale insects from their parents' homes when they build their nests during the breeding season, as a future source of honey. The behavior of ants raising aphids and scale insects is just like humans raising cows. But ants raised these dew-producing insects long before humans raised cows. Remember that ants originated about 80 million years ago in the mid-to-late Cretaceous period (Hölldobler & Wilson, 1990), however the origin of man is about 14 million years ago. In addition to raising aphids and scale insects, ants also grow mushrooms. The leaf-cutting ants of the genus *Atta* found in tropical rainforest of South America use their jaws to cut leaves into small pieces, and transport them back to the nest to cultivate fungi as food. They are the Earth's mushroom-growing, fungus-eating ants, and just like termites growing termitomyces for food. However, so far our country has not found this kind of mushroom-growing leaf-cutting ants.

So far, the red fire ant native to South America, *Solenopsis invicta* Buren, has rapidly spread to the south, east, central and southwest of our country since invading Wuchuan County of Guangdong Province in 2004 (Zeng et al., 2005). The pest ant is widely colonized in urban, agricultural and woodland environments, which has strong fecundity, complicated food, outstanding competitiveness and strong adaptability, and it is very difficult to prevent and control it. Its sting expels venom that can cause skin swelling and pain. After being stung, the allergic person itches for a long time and even goes into shock. The invasion of red fire ant has seriously disrupted human agricultural activities, even endanger public health and ecosystem security, so we need to strengthen the prevention and control work.

It can be seen that there are countless links between ants and plants, as well as between ants and other insects and animals. It is these seemingly casual links that make up the natural food chain and food webs, maintain the circle of biodiversity on the Earth endlessly, so as to permanently benefit the Earth's biosphere and human society.

1 中国蚁科昆虫的国外分类研究历程
1 Abroad Taxonomic Study History on Chinese Formicidae Insects

Linnaeus（1758）发表《自然系统》，建立了生物命名的双名法，提出了物种定义，科学分类学由此诞生。在双名法指导下，欧美发达国家致力于地方性动物区系研究，蚂蚁的分类研究也不例外。直至西方探险家从东方国家带回大量动植物标本，Smith（1858）首次在中国香港和华南地区的标本中发现并命名了2个中国的蚂蚁新种：警觉多刺蚁 Polyrhachis vigilans Smith 和暴躁多刺蚁 P. tyrannica Smith，标志着国外研究人员对中国蚂蚁研究的开始。

Darwin（1859）发表《物种起源》，首次提出进化论。在进化论指导下，西方国家的探险活动遍及全球，包括中国。Mayr（1866，1870，1889）分别报道中国香港、西藏等地的蚂蚁7种。Forel 于1879—1922年在13篇论文中记载中国香港、台湾、江西、上海、甘肃、西藏等地蚂蚁27种。Emery 于1894—1925年在4篇论文中记载中国上海、青岛、满洲里、香港的蚂蚁5种。Ruzsky（1905，1914，1915）报道中国东北地区和西藏的蚂蚁21种。Wheeler 于1909—1933年先后在13篇论文中报道中国台湾、重庆、浙江莫干山、苏州、广东、青岛、北京、厦门、江西九江、长沙、澳门、上海、云南、福州、香港、湖南岳麓山等地的蚂蚁180个种、亚种和变种。Yano（1911）报道中国台湾的多

Linnaeus (1758) published "Systema Naturae," and established the binomial nomenclature for naming organisms and proposed the concept of species, marking the birth of scientific taxonomy. Under the guidance of binomial nomenclature, developed countries in Europe and the United States were committed to the study of local fauna, the taxonomic study of ants was no exception. Until western explorers brought back from the east a large number of specimens of animals and plants, Smith (1858) first discovered and named two new ant species, Polyrhachis vigilans Smith and P. tyrannica Smith, from China in specimens from Hong Kong and southern China. This marked the beginning of the study of Chinese ants by foreign researchers.

Darwin (1859) published "On the Origin of Species," first proposed the theory of evolution. Under the guidance of the theory of evolution, the exploration activities of Western countries were all over the world, including China. Mayr (1866, 1870, 1889) reported seven species of ants in Hong Kong, Tibet and other places of China separately. Between 1879 and 1922, Forel recorded 27 species of ants in Hong Kong, Taiwan, Jiangxi, Shanghai, Gansu, Tibet and other places of China in 13 papers. Between 1894 and 1925, Emery accounted five species of ants in Shanghai, Qingdao, Manzhouli and Hong Kong of China in four papers. Ruzsky (1905, 1914, 1915) recognized 21 ant species in Manchuria and Tibet of China separately. Between 1909 and 1933, Wheeler published 13 papers and reported 180 species, subspecies and varieties of ants in Taiwan, Chongqing, Moganshan of Zhejiang, Suzhou, Guangdong, Qingdao, Beijing, Xiamen, Jiujiang of Jiangxi, Changsha, Macau,

刺蚁 Polyrhachis 4种。Viehmeyer（1912，1922）记载中国华南和福建、四川的蚂蚁9种。Stitz（1923，1934）记载中国海南和西北地区蚂蚁28种。Santschi（1925，1928）报道中国北京、上海等地蚂蚁11种。Matsumura和Uchida（1926）报道中国台湾的蚂蚁2种。Donisthorpe（1929，1947）分别记载中国西藏珠穆朗玛峰和广西北部湾的蚂蚁7种。Teranishi（1936，1940）分别记录中国东北和台湾的蚂蚁114种。Menozzi（1939）报道中国西藏喜马拉雅山和喀喇昆仑山的蚂蚁9种。Wheeler（1930）发表《中国已知蚂蚁名录》，系统总结这个时期西方人研究中国蚂蚁的成果，记录中国已知蚂蚁7亚科58属138种54亚种53变种。

Huxley（1940）发表《新系统学》与Mayr（1942）发表《系统学与物种起源》标志着动物分类学种群研究阶段的开始，对以往发表的分类阶元进行修订成为分类学研究的重点工作。此后，Eidmann（1941）记载中国四川、西藏和青海的蚂蚁43种。Yasumatsu（1941a，1941b，1962）报道中国东北、华东和四川、北京、湖北、台湾、广东、广西、香港等地蚂蚁5种。Weber（1947，1950）报道中国西藏蚂蚁3种。Brown于1948—1978年先后在8篇论文中报道中国四川、陕西、福州、台湾、香港和华南、东南等地蚂蚁17种。Wilson（1955，1964）记载中国哈尔滨、新疆天山、陕西、西藏喀喇昆仑山、四川新津、昆明、浙江莫干山、苏州、台湾和华南等地蚂蚁7种。Collingwood（1962，1970，1982）记录中国东北和内蒙古、西藏、四川、江苏等地蚂蚁54种。Dlussky（1965）报道中国西藏、青海和内蒙古的蚂蚁8种。Bolton于1977—2023年先后在10余篇修订性论文和《世界蚂蚁线上目录》中记载中国大陆蚂蚁10亚科117属1042种，记载中国台湾省蚂蚁10亚科61属180种。Terayama等于1984—2012年先后在13篇论文中报道中国台湾地区蚂蚁11亚科69属264种。

Shanghai, Yunnan, Fuzhou, Hong Kong, Yuelu Mountain of Hunan and other places of China. Yano (1911) reported four species of *Polyrhachis* in Taiwan of China. Viehmeyer (1912, 1922) recorded nine species of ants in southern China, Fujian and Sichuan of China. Stitz (1923, 1934) accounted 28 species of ants in Hainan and northwest region of China. Santschi (1925, 1928) recognized 11 species of ants in Beijing, Shanghai and other places of China. Matsumura & Uchida (1926) reported two species of ants in Taiwan of China. Donisthorpe (1929, 1947) recorded seven species of ants on Mount Everest of Tibet and Gulf of Tonkin of Guangxi in China. Teranishi (1936, 1940) accounted 114 ant species in northeastern China and Taiwan of China seperately. Menozzi (1939) reported nine species of ants in the Himalayas and Karakoram Mountains of Tibet, China. Wheeler (1930) published "*A List of the Known Chinese Ants*", systematically summarized the achievements of western studies on Chinese ants during this period, and recorded 7 subfamilies, 58 genera, 138 species, 54 subspecies and 53 varieties of ants known in China.

Huxley (1940) published "*New Systematics*", and Mayr (1943) published "*Systematics and the Origin of Species*", marking the beginning of population study period of animal taxonomy, the revision of previously published taxonomic categories became the key tasks. Since then, Eidmann (1941) recorded 43 species of ants in Sichuan, Tibet and Qinghai of China. Yasumatsu (1941a, 1941b, 1962) reported five species of ants in northeastern China, Sichuan, Beijing, Hubei, eastern China, Taiwan, Guangdong, Guangxi and Hong Kong of China. Weber (1947, 1950) recognized three species of ants in Tibet, China. Between 1948 and 1978, Brown published eight papers reporting 17 species of ants in Sichuan, Shaanxi, Fuzhou, Taiwan, Hong Kong, southern China and southeastern China of China. Wilson (1955, 1964) accounted seven species of ants in Harbin, Tianshan of Xinjiang, Shaanxi, Karakoram Mountains of Tibet, Xinjin of Sichuan, Kunming, Moganshan of Zhejiang, Suzhou, Taiwan and southern China of China. Collingwood (1962, 1970, 1982) recorded 54 species of ants in northeastern China, Inner Mongolia, Tibet, Sichuan, Jiangsu and other places of China. Dlussky (1965) recognized eight species of ants in Tibet, Qinghai and Inner Mongolia of China. Between 1977 and 2023, Bolton recorded 10 subfamilies, 117 genera and 1,042 species of ants in mainland China and 10 sabfamilie, 61 genera and 180 species of ants in Taiwan Province of China in ten revisionary papers and "*An Online Catalog of the Ants of the World*". Between 1984 and 2012, Terayama et al. published 13 papers reporting 11 subfamilies, 69 genera and 264 species of ants in Taiwan, China.

Mullis（1990）发明聚合酶链式反应（PCR）技术以后，促进了动物分类学中系统发育问题的解决，关于分类阶元之间亲缘关系的探讨方兴未艾，极大地促进了分类学的发展，标志着动物分类学分子技术应用阶段的开始。在这个时期，Moreau等（2006）、Brady等（2006，2014）的蚂蚁系统发育研究成果被应用于蚂蚁分类研究中，修订和完善了全球蚂蚁分类系统。寺山守（1990）记载中国台湾的蚂蚁41种。Kohout（1994）报道中国西藏蚂蚁1种。Ogata等（1995）记录中国台湾蚂蚁1种。DuBois（1998）报道中国蚂蚁1种。Elmes和Radchenko（1998）报道中国台湾蚂蚁4种。Seifert（2003，2020，2023a，2023b）记录中国蚂蚁6种。Lattke（2004）报道中国四川、云南、台湾等地蚂蚁5种。LaPolla（2004）报道中国四川、云南、江西、台湾、广西、澳门等地蚂蚁3种。Radchenko等（1995，2010）于1995—2010年记录中国西藏、四川、云南、陕西、湖南、广西等地蚂蚁40种。Guénard和Dunn（2012）记载中国大陆蚂蚁12亚科103属939种，记载中国台湾蚂蚁11亚科69属294种。Guénard等（2013）报道中国云南西双版纳蚂蚁1种。Staab（2014，2018）报道中国云南、湖南、江西、浙江、台湾等地蚂蚁9种。Williams和LaPolla（2016）记录中国云南、湖北、浙江、台湾、广西等地蚂蚁10种。Yamane等（2018）报道中国台湾蚂蚁4种。Pierce等（2019）报道香港蚂蚁2种。最近，Brassard等（2020）报道中国澳门蚂蚁1新种。Okido等（2020）发表中国蚂蚁2新种。Seifert（2023a，2023b）发现中国新疆、西藏、青海的蚂蚁2新种。Hamer等（2023a，2023b）描述中国香港蚂蚁4新种。Griebenow（2024）报道中国香港蚂蚁1新种。

After the invention of Polymerase Chain Reaction (PCR) technology by Mullis (1990), it promoted the solution of phylogenetic problems in animal taxonomy. The study on the relationship among taxa was in the ascendant. It has greatly promoted the development of taxonomy, and marked the beginning of the period of molecular techniques application in animal taxonomy. In this period, the results of the studies on ant phylogeny by Moreau et al. (2006) and Brady et al. (2006, 2014), were applied to the study of ant taxonomy, and the global ant classification system was revised and improved. Terayama (1990) recorded 41 species of ants in Taiwan, China. Kohout (1994) reported one species of ants in Tibet, China. Ogata et al. (1995) accounted one species of ants in Taiwan, China. DuBois (1998) recognized one species of ants in China. Elmes & Radchenko (1998) reported four species of ants in Taiwan, China. Seifert (2003, 2020, 2023a, 2023b) accounted six new species of ants in China. Lattke (2004) reported five species of ants in Sichuan, Yunnan, Taiwan and other places of China. LaPolla (2004) recognized three species of ants in Sichuan, Yunnan, Jiangxi, Taiwan, Guangxi, Macau and other places of China. Between 1995 and 2010, Radchenko et al. (1995, 2010) recorded 40 species of ants in Tibet, Sichuan, Yunnan, Shaanxi, Hunan, Guangxi and other places of China. Guénard & Dunn (2012) recorded 12 subfamilies, 103 genera and 939 species of ants in mainland China and 11 subfamilies, 69 genera and 294 species of ants in Taiwan, China. Guénard et al. (2013) reported one species of ants in Xishuangbanna, Yunnan, China. Staab (2014, 2018) recognized nine species of ants in Yunnan, Hunan, Jiangxi, Zhejiang and Taiwan of China. Williams & LaPolla (2016) accounted ten species of ants in Yunnan, Hubei, Zhejiang, Taiwan and Guangxi of China. Yamane et al. (2018) reported four species of ants in Taiwan, China. Pierce et al. (2019) recognized two species of ants in Hong Kong. Recently Brassard et al. (2020) reported one new species of ants in Macau, China. Okido et al. (2020) published two new species of ants in China. Seifert (2023a, 2023b) discovered three new species of ants in Xinjiang, Tibet and Qinghai of China. Hamer et al. (2023a, 2023b) described four new species of ants from Hong Kong, China. Griebenow (2024) reported one new species of ants in Hong Kong, China.

国外对中国蚂蚁的研究，主要集中于分类学领域。作为东方大国，中国地域辽阔，地貌丰富，气候多样，孕育了众多的蚂蚁种类，对西方蚁类学家富有吸引力。从1858年至今，一直有外国学者和旅行者来华采集、研究中国的蚂蚁物种。然而，在科学分类学诞生后的100年里，中国的蚁科昆虫研究是空白的。1858—1939年，西方人对中国蚂蚁的分类研究是片段化的，局限于记录中国沿海、东北、西藏的零散蚂蚁物种。相比之下，Wheeler（1930）的分类研究比较系统，首次整理出中国已知蚂蚁名录。1940—1984年，分类学的一个鲜明特点是开始对以往记录的物种进行修订研究，Wilson（1955）、Yasumatsu（1962）、Brown（1978）等相继对中国蚂蚁种类开展修订研究。1990年以来，随着分子生物学、超微形态学和互联网技术的普及，全球性的修订和系统发育研究趋向成熟。Lattke（2004）、LaPolla（2004）、Bolton（2007）、Radchenko和Elmes（2010）、Williams和LaPolla（2016）、Staab等（2018）等人完成了中国部分物种的修订工作。Guénard和Dunn（2012）系统整理了中国的蚂蚁名录。

The studies on Chinese ants abroad mainly focus on the field of taxonomy. As a big country in the east, China has a vast territory with rich landforms and diverse climates, and it breeds numerous ant species and is attractive to the western myrmecologists. Ever since 1858, foreign scholars and travelers have been coming to China to collect and study Chinese ant species. However, in the 100 years after the birth of scientific taxonomy, studies on Formicidae in China were blank. From 1858 to 1939, the taxonomic study of Chinese ants in the west was fragmented, limited to record scattered ant species along China's coast, northeast and Tibet of China. By contrast, the taxonomic study of Wheeler (1930) is relatively systematic, and he compiled a list of known Chinese ants for the first time. From 1940 to 1984, a distinctive feature of taxonomy was the beginning of revision of previously recorded species. Wilson (1955), Yasumatsu (1962), Brown (1978) and others successively carried out revisionary studies on ant species of China. Since 1990, with the popularization of molecular biology, ultramorphology and internet technology, global revision and phylogeny research has become mature. Lattke (2004), LaPolla (2004), Bolton (2007), Radchenko & Elmes (2010), Williams & LaPolla (2016), Staab et al. (2018) and others completed the revision of some species of China. Guénard & Dunn (2012) systematically compiled a list of the Chinese ants.

2 中国蚁科昆虫的国内分类研究历程
2 Domestic Taxonomic Study History on Chinese Formicidae Insects

我国早期著名昆虫学家胡经甫发表了《中国昆虫目录》（Wu，1941），记载我国蚂蚁7亚科58属184种56亚种56变种，拉开了中国蚁科昆虫国内研究的序幕。1982—1995年，我国蚂蚁分类学先驱唐觉及其同事先后在6篇论文中报道了浙江舟山群岛、湖南、西南横断山区的蚂蚁种类。唐觉等（1995）出版国内第一部蚁类学专著《中国经济昆虫志第四十七册：膜翅目·蚁科（一）》，记载中国已知蚂蚁7亚科48属123个种和亚种。李参和陈益（1992）报道浙江、福建的蚂蚁2种。随后，吴坚、王常禄、王敏生等于1987—1994年在7篇论文中报道了中国6个属的分类研究及湖南省蚂蚁分类研究成果。吴坚和王常禄（1995）出版国内第二部蚁类学专著《中国蚂蚁》，记载中国已知蚂蚁9亚科67属230种。周樑镒和寺山守（1991）记录我国台湾省蚂蚁198种。程量等（1992）报道中国上海蚂蚁1种。

1994年以后，徐正会等于1994—2024年先后在76篇论文中报道了中国蚂蚁45个属的分类研究成果，发表新属4个、新种139个。徐正会（2002a）出版专著《西双版纳自然保护区蚁科昆虫生物多样性研究》，记载云南西双版纳的蚂蚁9亚科76属286种。徐正会等（2022）出版专著《高黎贡山蚂蚁图鉴》，记载云南高黎贡山蚂蚁11亚科67属245种。张玮等（1995）报道四川蚂蚁1种。刘红等

Wu, C.F., a famous early entomologist of our country, published "*Catalogus Insectorum Sinensium*", recording 7 subfamilies 58 genera 184 species 56 subspecies and 56 varieties of ants in our country, and the domestic study of Formicidae insects of China began. From 1982 to 1995, the pioneer of ant taxonomy in our country Tang and his colleagues reported the species of ants in Zhoushan Islands of Zhejiang, Hunan and Hengduan Mountains of southwestern China in six papers. Tang et al. (1995) published the first domestic myrmecological monograph "*Economic Insect Fauna of China. Fasc. 47. Hymenoptera: Formicidae (1)*", recording 7 subfamilies, 48 genera, 123 known species and subspecies of ants in China. Li & Chen (1992) recognized two species of ants in Zhejiang and Fujian Provinces. Subsequently, Wu, Wang, Wang and others reported the taxonomic achievements of 6 genera in China and the taxonomic study of ants in Hunan Province from 1987 to 1994. Wu & Wang (1995) published the second domestic myrmecological monograph "*The Ants of China*", recording 9 subfamilies, 67 genera and 230 known species of ants in China. Chou & Terayama (1991) accounted 198 species of ants in Taiwan Province of our country. Cheng (1992) reported one species of ants in Shanghai, China.

After 1994, Xu et al. reported the taxonomic results of 45 ant genera in China in 76 papers from 1994 to 2024 and published 4 new genera and 139 new species. Xu (2002a) published the monograph "*A Study on the Biodiversity of Formicidae Ants of Xishuangbanna Nature Reserve*", recording 9 subfamilies, 76 genera and 286 species of ants in Xishuangbanna of Yunnan. Xu et al. (2022) published the monograph "*Pictorial Book of Ants of Mt. Gaoligong*",

（1995）报道吉林东部山区蚂蚁11种。夏永娟和郑哲民（1995，1997a，1997b）记载新疆蚂蚁45种。周善义等于1996—2020年在38篇论文中报道了中国蚂蚁26个属的分类研究成果，发表新种105个。周善义（2001）出版专著《广西蚂蚁》，记载广西壮族自治区蚂蚁8亚科64属204种。周善义等（2020）出版《中国习见蚂蚁生态图鉴》，记载蚂蚁11亚科66属161种。Lin和Wu（1996，1998，2001）记录中国台湾蚂蚁30种。王维等于1997—2007年在11篇论文报道了中国6个属的分类研究成果，发表新种11个。王维等（2009）出版专著《湖北省蚁科昆虫分类研究》，记载湖北省蚂蚁9亚科50属150种。张明伟等（1997）报道辽宁蚂蚁21种。长有德和贺达汉于1998—2002年先后在9篇论文中报道了中国西北地区7个属的分类研究成果，记载西北地区蚂蚁75种。魏琮等（1999）报道陕西等地蚂蚁13种。

2000年以来，Wei等（2001a，2001b）报道陕西蚂蚁2新种。张玮和郑哲民（2002）记载四川蚂蚁78种。Fellowes（2003）报道了中国海南蚂蚁的属。黄人鑫等（2004）记载新疆蚂蚁42种。Huang等（2004，2006，2008）报道中国蚂蚁3新种。刘福林等（2005）报道河南商丘地区蚂蚁42种。李淑萍等（2005）记录河南蚂蚁81种。黄建华和周善义（2006，2007a，2007b）整理了中国切叶蚁亚科Myrmicinae名录，合计记录392个种、亚种和变种。马永林等（2008）记载宁夏蚂蚁68种。Liu和Xu（2011）报道中国蚂蚁3新种。冉浩和周善义（2011，2012，2013）整理了中国蚁型亚科群名录，合计记录285种。王思忠等（2011）报道四川成都市蚂蚁29种。陈志林等于2011—2023年先后在12篇论文中报道了中国蚂蚁9个属的分类研究成果，发表新种28个。陈志林等（2021）出版《广西花坪蚂蚁图鉴》，记载广西花坪国家级自然保护区蚂蚁9亚科54属147种。张玮和周善义（2016）记录广东南岭国家森林公园蚂蚁38种。Liu等（2015a，2015b，2022）报道中国蚂蚁4新种和17个新记录种。Luo和Guénard（2016）报道香港蚂蚁2种。Man等（2017）记录北

recording 11 subfamilies, 67 genera and 245 species of ants in Mt. Gaoligong of Yunnan. Zhang et al. (1995) reported one species of ants in Sichuan. Liu et al. (1995) recognized 11 species of ants in the eastern mountains of Jilin. Xia & Zheng (1995, 1997 a, 1997 b) recorded 45 species of ants in Xinjiang. Zhou et al., reported the taxonomic results of 26 ant genera in China in 38 papers from 1996 to 2020 and published 105 new species. Zhou (2001) published the monograph "*Ants of Guangxi*", recording 8 subfamilies, 64 genera and 204 species of ants in Guangxi. Zhou et al. (2020) published the book "*Ecological Atlas of Ants Commonly Seen in China*", recording 11 subfamilies, 66 genera and 161 species. Lin & Wu (1996, 1998, 2001) accounted 30 species of ants in Taiwan of China. Wang et al. reported the taxonomic results of 6 ant genera in China in 11 papers from 1997 to 2007 and published 11 new species. Wang et al. (2009) published the monograph "*A Taxonomic Study on the Family Formicidae from Hubei Province*", recording 9 subfamilies, 50 genera and 150 species of ants in Hubei. Zhang et al. (1997) recognized 21 species of ants in Liaoning. From 1998 to 2002, Chang & He reported the taxonomic results of 7 ant genera in northwestern China in 9 papers, and recorded 75 species of ants in northwestern China. Wei et al. (1999) accounted 13 species of ants in Shaanxi and other places.

Since 2000, Wei et al. (2001a, 2001b) reported two new species of ants in Shaanxi. Zhang & Zheng (2002) recorded 78 species of ants in Sichuan. Fellowes (2003) accounted the genera of ants in Hainan. Huang et al. (2004) recorded 42 species of ants in Xinjiang. Huang et al. (2004, 2006, 2008) recognized 3 new species of ants in China. Liu et al. (2005) reported 42 species of ants in Shangqiu region of Henan. Li et al. (2005) accounted 81 species of ants in Henan. Huang & Zhou (2006, 2007a, 2007b) compiled a checklist of Myrmicinae of China, recorded 392 species, subspecies and varieties. Ma et al. (2008) recorded 68 species of ants in Ningxia. Liu & Xu reported three new species of ants in China. Ran & Zhou (2011, 2012, 2013) compiled a checklist the formicomorph subfamilies of China, recorded 285 species. Wang et al. (2011) recognized 29 species of ants in Chengdu, Sichuan. Chen et al. reported the taxonomic results of 8 ant genera in China in 12 papers from 2011 to 2022 and published 28 new species. Chen et al. (2021) published the monograph "*Illustrated Handbook of Ants of Huaping, Guangxi*", recording 9 subfamilies, 54 genera and 147 species of Huaping Nature Reserve of Guangxi. Zhang & Zhou (2016) accounted 38 species of ants in Nanling National Forest Park of Guangdong. Liu et al. (2015a, 2015b, 2022) reported four new species and 17 new record species

京蚂蚁2种。Hsu等（2017a，2017b）记载中国台湾蚂蚁2种。Leong等（2017，2018，2019）报道中国台湾蚂蚁14种。Leong等（2018）报道澳门蚂蚁1新种。Tang等（2019，2023，2024）记录中国蚂蚁13新种。梅象信等（2019）出版《河南蚁科昆虫》，记载河南省蚂蚁7亚科43属150种。Wong和Guénard（2020）报道香港和澳门的蚂蚁4种。Liu等（2021）记录中国蚂蚁2种。骆春璇等（2021）报道中国蚂蚁1种。马丽滨等（2022）出版《云南树栖型蚂蚁高清图鉴》，记载云南省蚂蚁7亚科31属61种。Huang和Zhong（2023）报道云南蚂蚁2种。Tang和Guénard（2023）记载中国香港蚂蚁10种。Liu等（2024）报道中国蚂蚁1种。Qian等（2024a，2024b）记载中国蚂蚁31种。

国内蚁科昆虫的分类研究始于唐觉和李参先生，他们开拓了国内的蚂蚁分类研究，首次将蚂蚁分类知识介绍给国内读者，并系统总结了外国学者对我国蚂蚁的分类研究成果。后续有吴坚、王常禄、王敏生等人的系统分类研究，他们借助全国林业系统的基层力量，将蚂蚁标本采集范围首次扩大到全国范围，系统记载我国各地的代表性蚂蚁类群，写成了至今仍有重要参考价值的《中国蚂蚁》专著。之后有徐正会、周善义、王维、张玮、黄建华、长有德、黄人鑫、吴卫等人的更广泛研究，在国家自然科学基金和地方基金的有力支持下，他们将研究范围扩大到西双版纳热带雨林、西南山地、横断山区、喜马拉雅山、青藏高原、新疆天山、祁连山、大凉山、梵净山、十万大山等边疆生物多样性热点和关键地区，并对诸多蚂蚁类群开展了深入研究，成果丰硕。当今有陈志林、梁志文、Tang、刘聪、刘霞、张新民、辛明、马丽滨等后起之秀，他们在逐渐成长并将继承前人的分类学事业，逐渐将研究范围扩展到华南地区、大湾区、海南岛、川西高原、四川盆地、塔里木盆地、贺兰山、六盘山、秦岭等众多生物多样性高度富集的地方。通过四代人的努力，加上外国学者在中国的研究成果，目前我国已经记载蚁科昆虫12亚科119属1113种，分别

of ants in China. Luo & Guénard (2016) recognized two species of ants in Hongkong. Man et al. (2017) recorded two species of ants in Beijing. Hsu et al. (2017a, 2017b) accounted 2 species of ants in Taiwan, China. Leong et al. (2017, 2018, 2019) reported 14 species of ants in Taiwan, China. Leong et al. (2018) recognized one new species of ants in Macau. Tang et al. (2019, 2023, 2024) recorded 13 species of ants in China. Mei et al. (2019) published the book *Formicidae Insects of Henan*, recording 7 subfamilies, 43 genera and 150 species of ants in Henan. Wong & Guénard （2020）reported four species of ants in Hong Kong and Macau. Liu et al. (2021) recorded two species of ants in China. Luo et al. (2021) recognized one species of ants in China. Ma et al. (2022) published the book *Arboreal Ants of Yunnan: A Field Guide with High-resolution Photographs*, recording 7 subfamilies, 31 genera and 61 species of ants in Yunnan. Huang & Zhong (2023) reported two species of ants in Yunnan. Tang & Guénard (2023) recorded ten species of ants in Hong Kong, China. Liu et al. (2024) reported one new species of ants in China. Qian et al. (2024a, 2024b) recorded 31 species of ants in China.

The taxonomic study of Formicidae insects in China began with Mr. Jue Tang and Mr. Shen Li. They pioneered the study of ant taxonomy in China, introduced the taxonomic knowledge to domestic readers for the first time, and systematically summarized the classification results of ants in our country by foreign scholars. Follow-up was the systematic classification by Jian Wu, Chang-lu Wang, Min-sheng Wang and others. For the first time, they expanded their ant collection to a national scale with the help of the national forestry system, and systematically recorded representative ant groups in various parts of our country. They wrote a monograph *The Ants of China* which still has important reference value today. Afterwards, there were more extensive researches by Zheng-hui Xu, Shan-yi Zhou, Wei Wang, Wei Zhang, Jian-hua Huang, You-de Chang, Ren-xin Huang, Wei Wu and others. Under the strong support of the National Natural Science Foundation and local foundations, they expanded the study scope to the hotspots and key areas of biodiversity conservation in frontier regions like Xishuangbanna tropical rainforest, Southwestern Mountains, Hengduan Mountains, Himalayas, Qinghai-Tibet Plateau, Tianshan of Xinjiang, Qilian Mountains, Daliangshan Mountains, Fanjing Mountains and Shiwandashan Mountains, etc., and conducted in-depth studies on many ant groups with fruitful results. Today, there are rising stars such as Zhi-lin Chen, Chi-man Leong, Kit-lam Tang, Cong Liu, Xia Liu, Xin-min Zhang, Ming Xin, Li-bin Ma and others. They are growing up and will do well to follow in the footsteps of their

占全球蚁科昆虫已知亚科（16亚科）的75%，已知属（346属）的34.4%，已知种（14,172种）的7.9%。

predecessors, and gradually expand the study scope to where biodiversity highly concentrated like South China, Greater Bay Area, Hainan Island, Western Sichuan Plateau, Sichuan Basin, Tarim Basin, Helan Mountain, Liupan Mountain, Qinling Mountains and other places. Through the efforts of four generations, in addition to the research achievements of foreign scholars in China, 12 subfamilies, 119 genera and 1113 species of Formicidae have been recorded in our country up to date, respectively accounting for 75% of the known subfamilies (16 subfamilies), 34.4% of the known genera (346 genera) and 7.9% of the known species (14,172 species) of Formicidae worldwide.

3 西南林业大学的蚁科昆虫分类研究历程
3 Taxonomic Study Course on Formicidae Insects in Southwest Forestry University

1990—1993年，徐正会前往陕西师范大学动物学专业攻读博士学位，在我国著名昆虫分类学家郑哲民教授指导下开展"中国滇桂黔三省区蚂蚁区系研究"，开启了西南林业大学在蚂蚁分类领域的探索。1993年，徐正会获得理学博士学位，成为国内第一位蚁类学博士，同年返回西南林业大学的前身西南林学院，带领蚂蚁课题组在中国西南山地生物多样性热点地区和西部边疆地区开展了一系列蚂蚁分类研究工作。

1994年起，徐正会（1994a）报道了中国西南地区短猛蚁属 *Brachyponera* 分类研究并发表1新种。徐正会（1994b）完成了中国西南地区刺结蚁属 *Lepisiota* 分类研究并描述2新种。Xu 和 Zheng（1994）发表中国西南地区铺道蚁属 *Tetramorium* 3新种。徐正会（1995a）完成中国臭蚁属 *Dolichoderus* 分类研究并发表1新种。Xu 和 Zheng（1995）发现中国西南地区角腹蚁属 *Recurvidris* 和无刺蚁属 *Kartidris* 2新种。Xu（1995b）描述云南前结蚁属 *Prenolepis* 2新种。Xu（1995c, 1996）报道中国厚结猛蚁属 *Pachycondyla* 2新种。Xu 和 Zhang（1996）发表中国西南地区曲颊猛蚁属 *Gnamptogenys* 1新种。Xu（1997）完成中国拟毛蚁属 *Pseudolasius* 分类研究并发表1新种。Xu（1998a）首次报道迷猛蚁属 *Mystrium* 在中国的分布，并描述云南迷猛蚁属和隐

Between 1990 and 1993, Zheng-hui Xu went to Shaanxi Normal University study for a doctor's degree in zoology. Under the guidance of Professor Zhe-min Zheng, a famous insect taxonomist in China, he carried out "A Faunistic Study of the Chinese Ants from the Provinces of Yunnan, Guangxi and Guizhou (Insecta: Hymenoptera: Formicidae)", and opened the exploration of ant classification in Southwest Forestry University. In 1993, Zheng-hui Xu was awarded a doctorate in science, and became the first PhD of myrmecology in China. In the same year, he returned to Southwest Forestry College, the predecessor of Southwest Forestry University, and he led the ant team carrying out a series taxonomic studies on ants in the biodiversity hotspot mountains of southwest China and the western frontier regions.

Since 1994, Xu (1994a) has reported the taxonomy of *Brachyponera* in southwestern China and published a new species. Xu (1994b) carried out the taxonomy of *Lepisiota* in southwestern China and described two new species. Xu & Zheng (1994) published three new species of *Tetramorium* in southwestern China. Xu (1995a) accomplished the taxonomy of *Dolichoderus* in China and published a new species. Xu & Zheng (1995) discovered two new species of the genera *Recurvidris* and *Kartidris* in southwestern China. Xu (1995b) described two new species of *Prenolepis* from Yunnan. Xu (1995c, 1996) reported two new species of *Pachycondyla* in China. Xu & Zhang (1996) published a new species of *Gnamptogenys* in southwestern China. Xu (1997) completed the taxonomy of *Pseudolasius* in China and published a new species. Xu (1998a) reported the distribution of *Mystrium* in China for the first time and described two new species of

猛蚁属 Cryptopone 2 新种。Xu（1998b）发现云南多刺蚁属 Polyrhachis 2 新种。Xu（1998c）首次记录埃猛蚁属 Emeryopone 和长齿蚁属 Myrmoteras 在中国的分布，并发表云南埃猛蚁属、小眼猛蚁属 Myopias 和长齿蚁属 3 新种。Xu（1999）完成中国盲切叶蚁属 Carebara、棒角蚁属 Rhopalomastix 和无刺蚁属 Kartidris 的系统分类研究，首次记录棒角蚁属在中国的分布，并描述云南 2 新种。

2000 年之后，Xu（2000a）发表云南猛蚁亚科 Ponerinae 和行军蚁亚科 Dorylinae 2 新属（版纳猛蚁属 Bannapone 和云行军蚁属 Yunodorylus）和 2 新种。Xu（2000b）报道了中国卷尾猛蚁属 Proceratium 系统分类研究并描述 1 新种。Xu（2000c）发现中国细颚猛蚁属 Leptogenys 5 新种和 1 个新记录种。Xu 和 Zeng（2000）首次记录宽猛蚁属 Platythyrea 在中国的分布并描述云南小盲猛蚁属 Probolomyrmex 1 新种。Xu（2000d）发表圆鳞蚁属 Epitritus 1 新种。Xu（2001a）完成了中国猛蚁属 Ponera 系统分类研究并描述云南西双版纳 5 新种。Xu（2001b）发现中国云南臭蚁属 2 新种。Xu（2001c）发表云南高黎贡山猛蚁属 4 新种。Xu（2001d）完成了中国钝猛蚁属 Amblyopone 系统分类研究并描述 2 新种。Wei 等（2001a）报道陕西圆颚蚁属 Strongylognathus 1 新种。Xu（2002b）报道了中国细蚁亚科 Leptanillinae 系统分类研究，首次报道原细蚁属 Protanilla 在中国的分布并发表云南西双版纳细蚁属 Leptanilla 和原细蚁属 3 新种。Xu 和 Zhang（2002a）发现昆明西山细蚁属和原细蚁属 2 新种。Xu（2002c）完成了中国多刺蚁属驼多刺蚁亚属 Polyrhachis（Cyrtomyrma）的系统分类研究并描述 5 新种。徐正会（2002a）首次记录了海胆蚁属 Echinopla 在中国的分布。Xu 和 Zhang（2002b）完成中国棱胸蚁属 Pristomyrmex 系统分类研究并发表云南 1 新种。Xu（2003）报道了中国稀切叶蚁属 Oligomyrmex 系统分类研究并发现 8 新种。Zhou 和 Xu（2003）完成了中国瘤颚蚁属 Strumigenys 的系统分类研究并发表江西 1 新种。Xu 和 Chai（2004）报道了中国细长蚁属 Tetraponera 系统分类研究并描述云南 5 新种。Xu 和 Zhou（2004）完成了中国塔蚁

Mystrium and Cryptopone in Yunnan. Xu (1998b) discovered two new species of Polyrhachis in Yunnan. Xu (1998c) recorded the distribution of Emeryopone and Myrmoteras in China for the first time and published three new species of the genera Emeryopias, Myopias and Myrmoteras in Yunnan. Xu (1999) carried out the systematic taxonomy of the genera Carebara, Rhopalomastix and Kartidris of China, recorded the distribution of Rhopalomastix in China for the first time, and described two new species from Yunnan.

After 2000, Xu (2000a) published two new genera, Bannapone and Yunodorylus, of the subfamilies Ponerinae and Dorylinae and two new species in Yunnan. Xu (2000b) reported the taxonomy of Proceratium and described a new species from Yunnan. Xu (2000c) discovered five new species and a new record species of Leptogenys in China. Xu & Zeng (2000) recorded the distribution of Platythyrea in China for the first time and described a new species of Probolomyrmex in Yunnan. Xu (2000d) published a new species of Epitritus in Yunnan. Xu (2001a) carried out the taxonomy of Ponera in China and described five new species from Xishuangbanna, Yunnan. Xu (2001b) discovered two new species of Dolichoderus in Yunnan. Xu (2001c) published four new species of Ponera in Mt. Gaoligong, Yunnan. Xu (2001d) accomplished the taxonomy of Amblyopone in China and described two new species. Wei et al. (2001a) reported a new species of Strongylognathus in Shaanxi. Xu (2002b) reported the systematic taxonomy of the subfamily Leptanillinae, the distribution of Protanilla in China for the first time and published three new species of Leptanilla and Protanilla in Xishuangbanna, Yunnan. Xu & Zhang (2002a) discovered two new species of Leptanilla and Protanilla in Xishan Mountain, Kunming, Yunnan. Xu (2002c) completed the taxonomy Polyrhachis (Cyrtomyrma) in China and described five new species. Xu (2002a) recorded the distribution of Echinopla in China for the first time. Xu & Zhang (2002b) finished the taxonomy of Pristomyrmex in China and described a new species from Yunnan. Xu (2003) reported the taxonomy of Oligomyrmex in China and discovered eight new species. Zhou & Xu (2003) carried out the taxonomy of Strumigenys in China and published a new species from Jiangxi. Xu & Chai (2004) reported the taxonomy of Tetraponera in China and described five new species in Yunnan. Xu & Zhou (2004) accomplished the taxonomy of Pyramica in China and discovered three new species in Yunnan. Xu & Wang (2004) reported the distribution of Perissomyrmex in China for the first time and published a new species from Ailao Mountain, Yunnan. Xu (2006) recorded three new species of the genera Amblyopone and Proceratium in Yunnan.

属 *Pyramica* 系统分类研究并发现云南 3 新种。Xu 和 Wang（2004）首次报道奇蚁属 *Perissomyrmex* 在中国的分布并发表云南哀牢山 1 新种。Xu（2006）记录云南钝猛蚁属和卷尾猛蚁属 3 新种。

2010 年以来，Xu 和 Xu（2011）发表西藏华丽蚁属 *Paratopula* 1 新种。Xu 和 He（2011）报道了红矛猛蚁 *Myopopone castanea* (Smith) 在西藏喜马拉雅地区的分布。Liu 和 Xu（2011）描述喜马拉雅山和横断山区窄结蚁属 *Stenamma* 3 新种。Xu 和 Liu（2012）发现西藏和云南小眼猛蚁属 3 新种。Xu（2012a）首次报道了盗蚁属 *Harpagoxenus* 在中国的分布及 1 个中国新记录种。Xu（2012b）描述切叶蚁亚科 Myrmicinae 1 新属（高黎贡蚁属 *Gaoligongidris*）及云南 1 新种。Xu（2012c）发表细蚁亚科 1 新属（叉细蚁属 *Furcotanilla*）及西藏和云南原细蚁属 2 新种。Xu 和 Zhang（2012）修订了世界奇蚁属物种并描述西藏 1 新种。Xu 和 Chu（2012）发现中国西南地区钝猛蚁属 4 新种。Xu 和 Yu（2012）发表西藏犁沟蚁属 *Vombisidris* 1 新种。Xu 等（2014a）描述云南盘猛蚁属 *Discothyrea* 2 新种。Xu 等（2014b）发现云南小眼猛蚁属 1 新种。Xu 和 He（2015）修订了东洋界细颚猛蚁属种类并发表云南 2 新种。Xu 和 Zhou（2015）编制了全球海胆蚁属分种检索表并报道了中国的已知种类。Man 等（2017）描述北京原细蚁属 1 新种。Staab 等（2018）完成了中国卷尾猛蚁属的系统分类研究并发表 3 新种。Liu 等（2021）修订了中国弯蚁属 *Lordomyrma* 物种并描述西藏和云南 2 新种。Jaitrong 等（2022）发现亚洲大陆宽猛蚁属 1 新种。Qian 等（2024）编制了全球细蚁属已知种检索表并描述云南、陕西和北京 3 新种。Qian 和 Xu（2024）完成了中国细胸蚁属 *Leptothorax* 和切胸蚁属 *Temnothorax* 的分类研究并发现中国切胸蚁属 28 新种。

通过以上研究，西南林业大学蚂蚁课题组合计发表中国蚂蚁新种 139 种，占中国已知蚂蚁物种总数（1113 种）的 12.5%，占全球已知蚂蚁物种总数（14,172 种）的 1.0%。在 139 个新种之中，有 136 个新种的模式标本保存在西南林业大学标本馆

Since 2010, Xu & Xu (2011) have published a new species of *Paratopula* in Tibet. Xu & He (2011) reported the distribution of *Myopopone castanea* (Smith) in the Himalayas, Tibet. Liu & Xu (2011) described three new species of *Stenamma* from the Himalayas and Hengduan Mountains. Xu & Liu (2012) discovered three new species of *Myopias* in Tibet and Yunnan. Xu (2012a) reported the distribution of *Harpagoxenus* in China for the first time and a new record species in China. Xu (2012b) described a new genus of the subfamily Myrmicinae, *Gaoligongidris*, and a new species from Yunnan. Xu (2012c) published a new genus of the subfamily Leptanillinae, *Furcotanilla*, and two new species of *Protanilla* in Tibet and Yunnan. Xu & Zhang (2012) revised the species of *Perissomyrmex* in the world and described a new species from Tibet. Xu & Chu (2012) discovered four new species of *Amblyopone* in southwestern China. Xu & Yu (2012) published a new species of *Vombisidris* in Tibet. Xu et al. (2014a) described two new species of *Discothyrea* from Yunnan. Xu et al. (2014b) discovered a new species of *Myopias* in Yunnan. Xu & He (2015) revised the species of *Leptogenys* in the Oriental realm and published two new species in Yunnan. Xu & Zhou (2015) compiled a key to the world species of *Echinopla* and reported the known species in China. Man et al. (2017) described a new species of *Protanilla* from Beijing. Staab et al. (2018) carried out the taxonomy of *Proceratium* in China and published three new species. Liu et al. (2021) revised the Chinese species of *Lordomyrma* and described two new species from Tibet and Yunnan. Jaitrong et al. (2022) discovered a new species of *Platythyrea* from continental Asia. Qian et al. (2024) compiled a key to the global known species of *Leptanilla* and described three new species from Yunnan, Shaanxi and Beijing. Qian & Xu (2024) accomplished the taxonomy of *Leptothorax* and *Temnothorax* in China and discovered 28 new species of *Temnothorax* in China.

Through the above researches, the ant team of Southwest Forestry University published a total of 139 new species of ants in China, accounting for 12.5 % of the total known ant species (1113 species) in China and 1.0 % of the total known ant species in the world (14,172 species). Among the 139 new species, the type specimens of 136 species were housed in the Ant Subcollection, Forest Insect Collection, Specimen Museum of Southwest Forestry University. For the other three new species, the type specimens of *Strumigenys jiangxiensis* Zhou & Xu, 2003 were deposited in the Insect Collection, College of Life Sciences, Guangxi Normal University; the type specimens of *Strongylognathus tylonum* Wei, Xu & He, 2001 were deposited in the Insect Collection,

森林昆虫标本室蚂蚁标本分室。另外3个新种，江西瘤颚蚁 *Strumigenys jiangxiensis* Zhou & Xu, 2003 的模式标本保存在广西师范大学生命科学学院昆虫标本室，瘤点圆颚蚁 *Strongylognathus tylonum* Wei, Xu & He, 2001 的模式标本保存在西北农林科技大学林学院昆虫标本室，霍氏宽猛蚁 *Platythyrea homasawini* Jaitrong, Xu & Khachonpisitsak, 2022 的模式标本保存在泰国国家科学博物馆自然历史博物馆。

College of Forestry, Northwest Sci-Tech University of Agriculture and Forestry; and the type specimens of *Platythyrea homasawini* Jaitrong, Xu & Khachonpisitsak, 2022 were deposited in the Natural History Museum of the National Science Museum, Thailand.

4 西南林业大学蚁科昆虫分类研究的修订
4 Revision of the Taxonomy of Formicidae Insects in Southwest Forestry University

分类学和其他科学一样,随着人类对自然的探索不断深入,分类知识的积累也在不断更新和修正。今天,随着分子生物学知识不断被应用到分类学之中,蚁科昆虫的分类系统也在不断被完善。由于分类系统的更新,西南林业大学蚂蚁课题组发表的139个新种之中,有23个种已经发生了属的转移,被组合到了其他属之中。此外,在前期研究过程中,因受到分类文献不足或记载不详、缺乏详尽和正确的插图及研究人员分类研究经历不足等因素制约,所发表新种之中,有6个新种是不成立的,已经被修订为其他物种的次异名。这里对组合到其他属的物种和修订为次异名的物种进行系统梳理,以便读者了解最新的分类学信息。

Like other sciences, taxonomy deepens along with human's constant exploration of nature, and the accumulation of classification knowledge is also constantly updated and revised. Today, as the knowledge of molecular biology applied to taxonomy, the classification system of Formicidae is also improving. Due to the update of the classification system, of the 139 new species published by the ant team of Southwest Forestry University, 23 species have been transferred, and combined into other genera. Besides, in the prophase research process, due to the constraints of lack of taxonomic literature, not detailed record, lack of detailed and correct illustrations, and limited experience of researchers in taxonomy, six of the new species published are not valid and have been revised as junior synonyms of other species. A systematic combing of species combined into other genera and revised as junior synonyms is made here to keep the readers abreast of the latest taxonomic information.

4.1 组合到其他属的物种

(1)阿佤钝猛蚁 *Amblyopone awa* Xu & Chu, 2012组合为阿佤点眼猛蚁 *Stigmatomma awa* (Xu & Chu, 2012)

Amblyopone awa Xu & Chu, 2012: 1192, figs. 51-56 (w.q.) CHINA (Yunnan).

Stigmatomma awa (Xu & Chu, 2012): Yoshimura & Fisher, 2014: 15.

(2)细齿钝猛蚁 *Amblyopone crenata* Xu, 2001组合为细齿点眼猛蚁 *Stigmatomma crenatum* (Xu, 2001)

Amblyopone crenata Xu, 2001d: 553, figs. 18-20 (w.) CHINA (Yunnan).

Stigmatomma crenatum (Xu, 2001): Yoshimura & Fisher, 2012a: 19.

4.1 Species Combined into Other Genera

(1) *Amblyopone awa* Xu & Chu, 2012 combined as *Stigmatomma awa* (Xu & Chu, 2012)

Amblyopone awa Xu & Chu, 2012: 1192, figs. 51-56 (w.q.) CHINA (Yunnan).

Stigmatomma awa (Xu & Chu, 2012): Yoshimura & Fisher, 2014: 15.

(2) *Amblyopone crenata* Xu, 2001 combined as *Stigmatomma crenatum* (Xu, 2001)

Amblyopone crenata Xu, 2001d: 553, figs. 18-20 (w.) CHINA (Yunnan).

Stigmatomma crenatum (Xu, 2001): Yoshimura & Fisher, 2012a: 19.

（3）康巴钝猛蚁 *Amblyopone kangba* Xu & Chu, 2012 组合为康巴点眼猛蚁 *Stigmatomma kangba* (Xu & Chu, 2012)

Amblyopone kangba Xu & Chu, 2012: 1187, figs. 13-15 (w.) CHINA (Tibet).

Stigmatomma kangba (Xu & Chu, 2012): Yoshimura & Fisher, 2014: 15.

（4）梅里钝猛蚁 *Amblyopone meiliana* Xu & Chu, 2012 组合为梅里点眼猛蚁 *Stigmatomma meilianum* (Xu & Chu, 2012)

Amblyopone meiliana Xu & Chu, 2012: 1190, figs. 33-35 (w.) CHINA (Yunnan).

Stigmatomma meilianum (Xu & Chu, 2012): Yoshimura & Fisher, 2014: 15.

（5）八齿钝猛蚁 *Amblyopone octodentata* Xu, 2006 组合为八齿点眼猛蚁 *Stigmatomma octodentatum* (Xu, 2006)

Amblyopone octodentata Xu, 2006: 152, figs. 1-7 (w.q.) CHINA (Yunnan).

Stigmatomma octodentatum (Xu, 2006): Yoshimura & Fisher, 2012a: 19.

（6）三叶钝猛蚁 *Amblyopone triloba* Xu, 2001 组合为三叶点眼猛蚁 *Stigmatomma trilobum* (Xu, 2001)

Amblyopone triloba Xu, 2001d: 552, figs. 11-13 (w.) CHINA (Yunnan).

Stigmatomma trilobum (Xu, 2001): Yoshimura & Fisher, 2012a: 19.

（7）卓玛钝猛蚁 *Amblyopone zoma* Xu & Chu, 2012 组合为卓玛点眼猛蚁 *Stigmatomma zoma* (Xu & Chu, 2012)

Amblyopone zoma Xu & Chu, 2012: 1189, figs. 28-30 (w.) CHINA (Tibet).

Stigmatomma zoma (Xu & Chu, 2012): Yoshimura & Fisher, 2014: 15.

（8）木兰版纳猛蚁 *Bannapone mulanae* Xu, 2000 组合为木兰点眼猛蚁 *Stigmatomma mulanae* (Xu, 2000)

Bannapone mulanae Xu, 2000a: 301, figs. 10-12 (q.) CHINA (Yunnan).

(3) *Amblyopone kangba* Xu & Chu, 2012 combined as *Stigmatomma kangba* (Xu & Chu, 2012)

Amblyopone kangba Xu & Chu, 2012: 1187, figs. 13-15 (w.) CHINA (Tibet).

Stigmatomma kangba (Xu & Chu, 2012): Yoshimura & Fisher, 2014: 15.

(4) *Amblyopone meiliana* Xu & Chu, 2012 combined as *Stigmatomma meilianum* (Xu & Chu, 2012)

Amblyopone meiliana Xu & Chu, 2012: 1190, figs. 33-35 (w.) CHINA (Yunnan).

Stigmatomma meilianum (Xu & Chu, 2012): Yoshimura & Fisher, 2014: 15.

(5) *Amblyopone octodentata* Xu, 2006 combined as *Stigmatomma octodentatum* (Xu, 2006)

Amblyopone octodentata Xu, 2006: 152, figs. 1-7 (w.q.) CHINA (Yunnan).

Stigmatomma octodentatum (Xu, 2006): Yoshimura & Fisher, 2012a: 19.

(6) *Amblyopone triloba* Xu, 2001 combined as *Stigmatomma trilobum* (Xu, 2001)

Amblyopone triloba Xu, 2001d: 552, figs. 11-13 (w.) CHINA (Yunnan).

Stigmatomma trilobum (Xu, 2001): Yoshimura & Fisher, 2012a: 19.

(7) *Amblyopone zoma* Xu & Chu, 2012 combined as *Stigmatomma zoma* (Xu & Chu, 2012)

Amblyopone zoma Xu & Chu, 2012: 1189, figs. 28-30 (w.) CHINA (Tibet).

Stigmatomma zoma (Xu & Chu, 2012): Yoshimura & Fisher, 2014: 15.

(8) *Bannapone mulanae* Xu, 2000 combined as *Stigmatomma mulanae* (Xu, 2000)

Bannapone mulanae Xu, 2000a: 301, figs. 10-12 (q.) CHINA (Yunnan).

Stigmatomma mulanae (Xu, 2000): Ward & Fisher, 2016: 691.

（9）版纳曲颊猛蚁 *Gnamptogenys bannana* Xu & Zhang, 1996组合为版纳雕猛蚁 *Stictoponera bannana* (Xu & Zhang, 1996)

Gnamptogenys bannana Xu & Zhang, 1996: 55, figs. 1-7 (w.) CHINA (Yunnan).

Stictoponera bannana (Xu & Zhang, 1996): Camacho et al., 2022: 12.

（10）片突厚结猛蚁 *Pachycondyla lobocarena* Xu, 1995组合为片突扁头猛蚁 *Ectomomyrmex lobocarenus* (Xu, 1995)

Pachycondyla lobocarena Xu, 1995c: 109, figs. 15, 16 (w.) CHINA (Yunnan).

Ectomomyrmex lobocarenus (Xu, 1995): Schmidt & Shattuck, 2014: 193.

（11）郑氏厚结猛蚁 *Pachycondyla zhengi* Xu, 1995组合为郑氏扁头猛蚁 *Ectomomyrmex zhengi* (Xu, 1995)

Pachycondyla zhengi Xu, 1995c: 110, figs. 19, 20 (w.) CHINA (Yunnan).

Ectomomyrmex zhengi (Xu, 1995): Schmidt & Shattuck, 2014: 194.

（12）大禹圆鳞蚁 *Epitritus dayui* Xu, 2000组合为大禹瘤颚蚁 *Strumigenys dayui* (Xu, 2000)

Epitritus dayui Xu, 2000d: 299, figs. 9-12 (w.) CHINA (Yunnan).

Strumigenys dayui (Xu, 2000): Baroni Urbani & De Andrade, 2007: 118.

（13）尖刺稀切叶蚁 *Oligomyrmex acutispinus* Xu, 2003组合为尖刺重头蚁 *Carebara acutispina* (Xu, 2003)

Oligomyrmex acutispinus Xu, 2003: 315, figs. 16-19 (s.w.) CHINA (Yunnan).

Carebara acutispina (Xu, 2003) : Guénard & Dunn, 2012: 41.

（14）高结稀切叶蚁 *Oligomyrmex altinodus* Xu, 2003组合为高结重头蚁 *Carebara altinodus* (Xu, 2003)

Oligomyrmex altinodus Xu, 2003: 312, figs. 5-8 (s.w.) CHINA (Yunnan).

Stigmatomma mulanae (Xu, 2000): Ward & Fisher, 2016: 691.

(9) *Gnamptogenys bannana* Xu & Zhang, 1996 combined as *Stictoponera bannana* (Xu & Zhang, 1996)

Gnamptogenys bannana Xu & Zhang, 1996: 55, figs. 1-7 (w.) CHINA (Yunnan).

Stictoponera bannana (Xu & Zhang, 1996): Camacho et al., 2022: 12.

(10) *Pachycondyla lobocarena* Xu, 1995 combined as *Ectomomyrmex lobocarenus* (Xu, 1995)

Pachycondyla lobocarena Xu, 1995c: 109, figs. 15, 16 (w.) CHINA (Yunnan).

Ectomomyrmex lobocarenus (Xu, 1995): Schmidt & Shattuck, 2014: 193.

(11) *Pachycondyla zhengi* Xu, 1995 combined as *Ectomomyrmex zhengi* (Xu, 1995)

Pachycondyla zhengi Xu, 1995c: 110, figs. 19, 20 (w.) CHINA (Yunnan).

Ectomomyrmex zhengi (Xu, 1995): Schmidt & Shattuck, 2014: 194.

(12) *Epitritus dayui* Xu, 2000 combined as *Strumigenys dayui* (Xu, 2000)

Epitritus dayui Xu, 2000d: 299, figs. 9-12 (w.) CHINA (Yunnan).

Strumigenys dayui (Xu, 2000): Baroni Urbani & De Andrade, 2007: 118.

(13) *Oligomyrmex acutispinus* Xu, 2003 combined as *Carebara acutispina* (Xu, 2003)

Oligomyrmex acutispinus Xu, 2003: 315, figs. 16-19 (s.w.) CHINA (Yunnan).

Carebara acutispina (Xu, 2003) : Guénard & Dunn, 2012: 41.

(14) *Oligomyrmex altinodus* Xu, 2003 combined as *Carebara altinodus* (Xu, 2003)

Oligomyrmex altinodus Xu, 2003: 312, figs. 5-8 (s.w.) CHINA (Yunnan).

Carebara altinodus (Xu, 2003): Guénard & Dunn, 2012: 41.

（15）双角稀切叶蚁 *Oligomyrmex bihornatus* Xu, 2003 组合为双角重头蚁 *Carebara bihornata* (Xu, 2003)

Oligomyrmex bihornatus Xu, 2003: 317, figs. 24-27 (s.w.) CHINA (Yunnan).

Carebara bihornata (Xu, 2003): Guénard & Dunn, 2012: 41.

（16）弯刺稀切叶蚁 *Oligomyrmex curvispinus* Xu, 2003 组合为弯刺重头蚁 *Carebara curvispina* (Xu, 2003)

Oligomyrmex curvispinus Xu, 2003: 313, figs. 9-12 (s.w.) CHINA (Yunnan).

Carebara curvispina (Xu, 2003): Guénard & Dunn, 2012: 41.

（17）钝齿稀切叶蚁 *Oligomyrmex obtusidentus* Xu, 2003 组合为钝齿重头蚁 *Carebara obtusidenta* (Xu, 2003)

Oligomyrmex obtusidentus Xu, 2003: 316, figs. 20-23 (s.w.) CHINA (Yunnan).

Carebara obtusidenta (Xu, 2003): Guénard & Dunn, 2012: 41.

（18）直背稀切叶蚁 *Oligomyrmex rectidorsus* Xu, 2003 组合为直背重头蚁 *Carebara rectidorsa* (Xu, 2003)

Oligomyrmex rectidorsus Xu, 2003: 319, figs. 32-35 (s.w.) CHINA (Yunnan).

Carebara rectidorsa (Xu, 2003): Guénard & Dunn, 2012: 41.

（19）纹头稀切叶蚁 *Oligomyrmex reticapitus* Xu, 2003 组合为纹头重头蚁 *Carebara reticapita* (Xu, 2003)

Oligomyrmex reticapitus Xu, 2003: 319, figs. 38-41 (s.w.) CHINA (Yunnan).

Carebara reticapita (Xu, 2003): Guénard & Dunn, 2012: 41.

（20）条纹稀切叶蚁 *Oligomyrmex striatus* Xu, 2003 组合为条纹重头蚁 *Carebara striata* (Xu, 2003)

Oligomyrmex striatus Xu, 2003: 314, figs. 13-15 (s.) CHINA (Yunnan).

Carebara striata (Xu, 2003): Fernández, 2010: 202.

Carebara altinodus (Xu, 2003): Guénard & Dunn, 2012: 41.

(15) *Oligomyrmex bihornatus* Xu, 2003 combined as *Carebara bihornata* (Xu, 2003)

Oligomyrmex bihornatus Xu, 2003: 317, figs. 24-27 (s.w.) CHINA (Yunnan).

Carebara bihornata (Xu, 2003): Guénard & Dunn, 2012: 41.

(16) *Oligomyrmex curvispinus* Xu, 2003 combined as *Carebara curvispina* (Xu, 2003)

Oligomyrmex curvispinus Xu, 2003: 313, figs. 9-12 (s.w.) CHINA (Yunnan).

Carebara curvispina (Xu, 2003): Guénard & Dunn, 2012: 41.

(17) *Oligomyrmex obtusidentus* Xu, 2003 combined as *Carebara obtusidenta* (Xu, 2003)

Oligomyrmex obtusidentus Xu, 2003: 316, figs. 20-23 (s.w.) CHINA (Yunnan).

Carebara obtusidenta (Xu, 2003): Guénard & Dunn, 2012: 41.

(18) *Oligomyrmex rectidorsus* Xu, 2003 combined as *Carebara rectidorsa* (Xu, 2003)

Oligomyrmex rectidorsus Xu, 2003: 319, figs. 32-35 (s.w.) CHINA (Yunnan).

Carebara rectidorsa (Xu, 2003): Guénard & Dunn, 2012: 41.

(19) *Oligomyrmex reticapitus* Xu, 2003 combined as *Carebara reticapita* (Xu, 2003)

Oligomyrmex reticapitus Xu, 2003: 319, figs. 38-41 (s.w.) CHINA (Yunnan).

Carebara reticapita (Xu, 2003): Guénard & Dunn, 2012: 41.

(20) *Oligomyrmex striatus* Xu, 2003 combined as *Carebara striata* (Xu, 2003)

Oligomyrmex striatus Xu, 2003: 314, figs. 13-15 (s.) CHINA (Yunnan).

Carebara striata (Xu, 2003): Fernández, 2010: 202.

（21）哀牢山塔蚁 *Pyramica ailaoshana* Xu & Zhou, 2004 组合为哀牢山瘤颚蚁 *Strumigenys ailaoshana* (Xu & Zhou, 2004)

Pyramica ailaoshana Xu & Zhou, 2004: 445, figs. 19, 20 (w.q.) CHINA (Yunnan).

Strumigenys ailaoshana (Xu & Zhou, 2004)：Baroni Urbani & De Andrade, 2007: 115.

（22）弄巴塔蚁 *Pyramica nongba* Xu & Zhou, 2004 组合为弄巴瘤颚蚁 *Strumigenys nongba* (Xu & Zhou, 2004)

Pyramica nongba Xu & Zhou, 2004: 443, figs. 7, 8 (w.) CHINA (Yunnan).

Strumigenys nongba (Xu & Zhou, 2004): Baroni Urbani & De Andrade, 2007: 125.

（23）杨氏塔蚁 *Pyramica yangi* Xu & Zhou, 2004 组合为杨氏瘤颚蚁 *Strumigenys yangi* (Xu & Zhou, 2004)

Pyramica yangi Xu & Zhou, 2004: 447, figs. 25, 26 (w.) CHINA (Yunnan).

Strumigenys yangi (Xu & Zhou, 2004): Baroni Urbani & De Andrade, 2007: 130.

4.2 修订为次异名的物种

（1）小眼迷猛蚁 *Mystrium oculatum* Xu, 1998 是卡氏迷猛蚁 *Mystrium camillae* Emery, 1889 的次异名

Mystrium camillae Emery, 1889: 491, pl. 10, figs. 1-3 (w.q.) MYANMAR.

Mystrium oculatum Xu, 1998a: 161, figs. 1, 2 (w.) CHINA (Yunnan). Junior synonym of *Mystrium camillae*: Bihn & Verhaagh, 2007: 3.

（2）版纳曲颊猛蚁 *Gnamptogenys bannana* Xu & Zhang, 1996 是双色雕猛蚁 *Stictoponera bicolor* (Emery, 1889) 的次异名

Ectatomma (Stictoponera) bicolor Emery, 1889: 493 (w.) MYANMAR.

Stictoponera bicolor (Emery, 1889): Emery, 1911e: 48.

Gnamptogenys bicolor (Emery, 1889): Brown, 1958g: 227.

(21) *Pyramica ailaoshana* Xu & Zhou, 2004 combined as *Strumigenys ailaoshana* (Xu & Zhou, 2004)

Pyramica ailaoshana Xu & Zhou, 2004: 445, figs. 19, 20 (w.q.) CHINA (Yunnan).

Strumigenys ailaoshana (Xu & Zhou, 2004): Baroni Urbani & De Andrade, 2007: 115.

(22) *Pyramica nongba* Xu & Zhou, 2004 combined as *Strumigenys nongba* (Xu & Zhou, 2004)

Pyramica nongba Xu & Zhou, 2004: 443, figs. 7, 8 (w.) CHINA (Yunnan).

Strumigenys nongba (Xu & Zhou, 2004): Baroni Urbani & De Andrade, 2007: 125.

(23) *Pyramica yangi* Xu & Zhou, 2004 combined as *Strumigenys yangi* (Xu & Zhou, 2004)

Pyramica yangi Xu & Zhou, 2004: 447, figs. 25, 26 (w.) CHINA (Yunnan).

Strumigenys yangi (Xu & Zhou, 2004): Baroni Urbani & De Andrade, 2007: 130.

4.2 Species Revised as Junior Synonyms

(1) *Mystrium oculatum* Xu, 1998 is a junior synonym of *Mystrium camillae* Emery, 1889

Mystrium camillae Emery, 1889: 491, pl. 10, figs. 1-3 (w.q.) MYANMAR.

Mystrium oculatum Xu, 1998a: 161, figs. 1, 2 (w.) CHINA (Yunnan). Junior synonym of *Mystrium camillae*: Bihn & Verhaagh, 2007: 3.

(2) *Gnamptogenys bannana* Xu & Zhang, 1996 is a junior synonym of *Stictoponera bicolor* (Emery, 1889)

Ectatomma (Stictoponera) bicolor Emery, 1889: 493 (w.) MYANMAR.

Stictoponera bicolor (Emery, 1889): Emery, 1911e: 48.

Gnamptogenys bicolor (Emery, 1889): Brown, 1958g: 227.

Gnamptogenys bannana Xu & Zhang, 1996: 55, figs. 1-7 (w.) CHINA (Yunnan).

Stictoponera bannana (Xu & Zhang, 1996): Camacho et al., 2022: 12.

Stictoponera bannana (Xu & Zhang, 1996): Junior synonym of *Stictoponera bicolor*: Lattke, 2004: 81.

（3）怒江卷尾猛蚁*Proceratium nujiangense* Xu, 2006是赵氏卷尾猛蚁*Proceratium zhaoi* Xu, 2000的次异名

Proceratium zhaoi Xu, 2000b: 435, figs. 3-10 (w.q.) CHINA (Yunnan).

Proceratium nujiangense Xu, 2006: 153, figs. 8-13 (w.q.) CHINA (Yunnan). Junior synonym of *Proceratium zhaoi*: Staab et al., 2018: 164.

（4）黄帝细颚猛蚁*Leptogenys huangdii* Xu, 2000是光亮细颚猛蚁*Leptogenys lucidula* Emery, 1895的次异名

Leptogenys lucidula Emery, 1895m: 462 (w.) MYANMAR.

Leptogenys huangdii Xu, 2000c: 119, figs. 5-8 (w.) CHINA (Yunnan). Junior synonym of *Leptogenys lucidula*: Xu & He, 2015: 143.

（5）大眼前结蚁*Prenolepis magnocula* Xu, 1995是那氏前结蚁*Prenolepis naoroji* Forel, 1902的次异名

Prenolepis naoroji Forel, 1902d: 290 (w.) INDIA.

Prenolepis magnocula Xu, 1995b: 339, figs. 4-6 (w.) CHINA (Yunnan). Junior synonym of *Prenolepis naoroji*: Williams & LaPolla, 2016: 234.

（6）黑角前结蚁*Prenolepis nigriflagella* Xu, 1995是黑腹前结蚁*Prenolepis melanogaster* Emery, 1893的次异名

Prenolepis melanogaster Emery, 1893g: 223 (in text) (w.) MYANMAR.

Prenolepis nigriflagella Xu, 1995b: 338, figs. 1-3 (w.) CHINA (Yunnan). Junior synonym of *Prenolepis melanogaster*: Williams & LaPolla, 2016: 234.

Gnamptogenys bannana Xu & Zhang, 1996: 55, figs. 1-7 (w.) CHINA (Yunnan).

Stictoponera bannana (Xu & Zhang, 1996): Camacho et al., 2022: 12.

Stictoponera bannana (Xu & Zhang, 1996): Junior synonym of *Stictoponera bicolor*: Lattke, 2004: 81.

(3) *Proceratium nujiangense* Xu, 2006 is a junior synonym of *Proceratium zhaoi* Xu, 2000

Proceratium zhaoi Xu, 2000b: 435, figs. 3-10 (w.q.) CHINA (Yunnan).

Proceratium nujiangense Xu, 2006: 153, figs. 8-13 (w.q.) CHINA (Yunnan). Junior synonym of *Proceratium zhaoi*: Staab et al., 2018: 164.

(4) *Leptogenys huangdii* Xu, 2000 is a junior synonym of *Leptogenys lucidula* Emery, 1895

Leptogenys lucidula Emery, 1895m: 462 (w.) MYANMAR.

Leptogenys huangdii Xu, 2000c: 119, figs. 5-8 (w.) CHINA (Yunnan). Junior synonym of *Leptogenys lucidula*: Xu & He, 2015: 143.

(5) *Prenolepis magnocula* Xu, 1995 is a junior synonym of *Prenolepis naoroji* Forel, 1902

Prenolepis naoroji Forel, 1902d: 290 (w.) INDIA.

Prenolepis magnocula Xu, 1995b: 339, figs. 4-6 (w.) CHINA (Yunnan). Junior synonym of *Prenolepis naoroji*: Williams & LaPolla, 2016: 234.

(6) *Prenolepis nigriflagella* Xu, 1995 is a junior synonym of *Prenolepis melanogaster* Emery, 1893

Prenolepis melanogaster Emery, 1893g: 223 (in text) (w.) MYANMAR.

Prenolepis nigriflagella Xu, 1995b: 338, figs. 1-3 (w.) CHINA (Yunnan). Junior synonym of *Prenolepis melanogaster*: Williams & LaPolla, 2016: 234.

5 蚁科 Formicidae 昆虫形态特征
5 Morphology of Formicidae Insects

蚂蚁均为真社会性昆虫，蚁巢多年生，存在无翅的工蚁，有性个体具有同步婚飞习性。头部前口式，下唇和下咽之间具下口腔囊。工蚁和雌蚁触角在长的柄节和鞭节间呈膝状弯曲。足的基节端孔向侧面开口，完全包围转节基部，包括转节前关节，因而所有基节端膜隐藏。中足和后足基节窝小，圆形，单关节，向腹面开口，基节端部强烈弯向侧面。后胸侧板腺存在。并胸腹节气门位于侧面，远离并胸腹节前上角，通常位于并胸腹节中部。雌蚁的翅可脱落，在交尾后脱去。前翅缺 3rs-m 和 2m-cu 脉，后翅 C 脉未延伸到翅前缘，后翅基室不向端部延伸。腹部第 2 节缩小形成独立的结状或鳞片状的腹柄，通常第 3 节也缩小形成独立的后腹柄（AntWiki，2024）。

蚁科 Formicidae 的主要特征为胸腹部之间有细腰，细腰由 1~2 个独立的体节组成，分别称为腹柄和后腹柄。形态学上的腹部第 1 节并入胸部成为并胸腹节；第 2 节两端缢缩形成独立的腹柄，或者第 2~3 节均在两端缢缩分别形成腹柄和后腹柄；其余腹节组成后腹部。极少数腹柄前端缢缩，后面不缢缩，与后腹部宽阔连接。

The ants are eusocial insects, colonies perennial, wingless worker caste present, sexual individuals with synchronous nuptial flights. Head capsule prognathous, infrabuccal sac present between labium and hypopharynx. Antennae of workers and gynes geniculate between long scape and funiculus. Disticoxal foramen directed laterally and completely enclosing protrochanteral base, including protrochanteral condyles, such that all disticoxal membrane concealed. All meso- and metacoxal cavities small, circular, monocondylic, ventrally-directed, and disticoxae strongly produced laterally. Metapleural gland present. Propodeal spiracle located on lateral propodeal face distant from the anterodorsal propodeal corner, often near propodeum midlength. Wings of alate gyne deciduous, being shed after copulation. Forewing 3rs-m and 2m-cu absent, hindwing C not extending along anterior margin, even spectrally. Hindwing basal cell not produced distally. Second abdominal segment reduced, forming an isolated node-like or scale-like petiole, frequently the third abdominal segment also reduced and formed the isolated postpetiole (AntWiki, 2024).

The ant family Formicidae is characterized by a thin waist between the mesosoma and abdomen, the waist consists of 1-2 isolated segments, called petiole and postpetiole respectively. The morphological first segment of abdomen merges into the thorax and becomes the propodeum. The second abdominal segment constricted at both ends to form the petiole, or the second and third abdominal segments constricted at both ends to form the petiole and postpetiole respectively. The rest abdominal segments formed the gaster. In rare cases, the petiole constricted anteriorly, but not constricted posteriorly, and broadly attached to the gaster.

蚁科物种全部为社会性昆虫,蚁巢内个体有形态和职能的分工。雌蚁体形较大,具翅,交配后翅脱落,后腹部膨大变成蚁后,专司生殖,但是个体稀少。雄蚁体形瘦长,具翅,与雌蚁一起负责生殖,其数量众多,交配后不久死亡,但是只有繁殖季节可以见到。工蚁是性腺不发育的雌性,体形较小,无翅,在蚁巢内数量最多。一些类群的工蚁分化为体形显著较大的大型工蚁(又称兵蚁)和体形明显较小的小型工蚁。在少数类群中,工蚁有多型现象,其个体大小呈现梯度变化。因为工蚁在蚁巢中数量最多,在自然环境中最常见,所以蚂蚁的形态分类主要依据工蚁的特征;如果巢内个体分化出大型工蚁和小型工蚁,则同时依据各型工蚁的特征进行分类,但以大型工蚁为主(图1、图2)。

All the species of Formicidae are social insects, and the individuals in the nest have a division of form and function. The females are larger, winged, and the wings shed after mating, with its gaster gradually expanding and becoming a queen. The queen specialized in reproduction, but its individuals are rare. The males are lanky and winged, responsible for reproduction together with the females, numerous in individuals and die soon after mating, but only seen during the breeding season. The workers are small, wingless females with underdeveloped gonads, and have the largest number in the nest. The workers of some groups are differentiated into larger workers significantly larger in size (also known as soldiers) and small workers obviously smaller in size. In a few groups, workers are polymorphic, and their individual size varies in a gradient. Because workers are the most abundant in the nest and the most common in the natural environment, the morphological classification of ants is mainly based on the characteristics of workers. If the individuals in the nest differentiate into large and small workers, then the classification is based on all types of workers, but mainly large worker (figures 1-2).

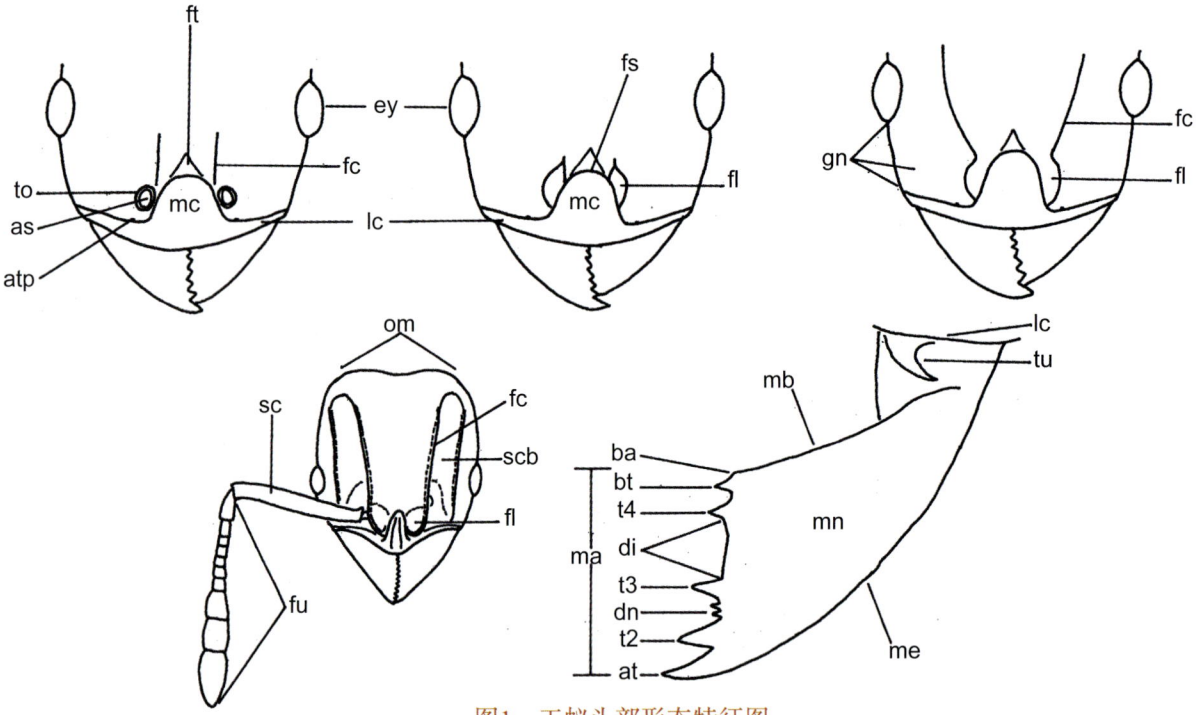

图1 工蚁头部形态特征图

as-触角槽;at-上颚端齿;atp-前幕骨陷;ba-上颚基角;bt-上颚基齿;di-齿间隙;dn-细齿;ey-复眼;fc-额脊;fl-额叶;fs-额唇基缝;ft-额三角区;fu-触角鞭节;gn-颊区;lc-唇基侧区;ma-上颚咀嚼缘;mb-上颚基缘;mc-唇基中区;me-上颚外缘;mn-上颚;om-头后缘;sc-触角柄节;scb-触角沟;t-上颚齿数目;to-触角窝;tu-上颚凹陷(引自Bolton, 1994)

Figure 1 Morphological features of the heads of workers

as-antennal socket; at-apical tooth of mandible; atp-anterior tentorial pit; ba-basal angle of mandible; bt-basal tooth of mandible; di-diastema; dn-denticle; ey-eye; fc-frontal carina; fl-frontal lobe; fs-fronto-clypeal suture; ft-frontal triangle; fu-funiculus of antenna; gn-gena; lc-lateral portion of clypeus; ma-apical (masticatory) margin of mandible; mb-basal margin of mandible; mc-median portion of clypeus; me-external margin of mandible; mn-mandible; om-occipital margin of head; sc-scape of antenna; scb-antennal scrobe; t-tooth number; to-torulus; tu-trulleum (Cited from Bolton, 1994)

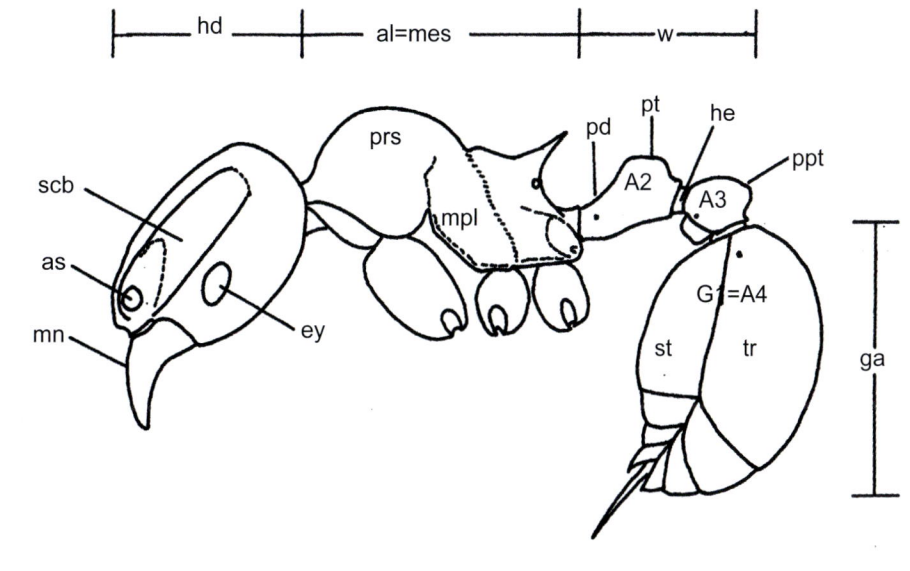

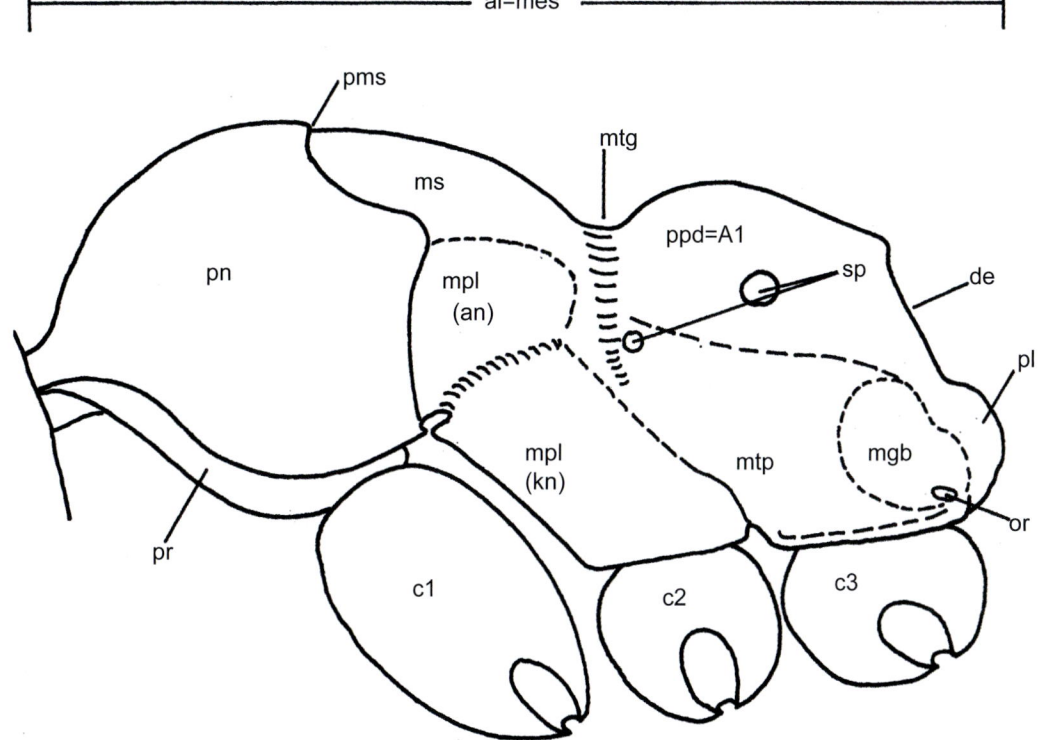

图2 工蚁身体分类特征图

A-腹节数目；al-胸部；an-前上侧片；as-触角槽；c-基节数目；de-并胸腹节斜面；ey-复眼；G-后腹部节数目；ga-后腹部；hd-头部；he-后腹柄收缩部；kn-前下侧片；mes-胸部；mgb-后胸侧板腺泡；mn-上颚；mpl-中胸侧板；ms-中胸背板；mtg-后胸沟；mtp-后胸侧板；or-后胸侧板腺口；pd-腹柄小柄；pl-并胸腹节侧叶；pms-前中胸背板缝；pn-前胸背板；ppd-并胸腹节；ppt-后腹柄；pr-前胸侧板；prs-前中胸背板；pt-腹柄；scb-触角沟；sp-气门；st-腹板；tr-背板；w-腰部（引自Bolton，1994）

Figure 2 Morphological features of the bodies of workers

A-abdominal segment number; al-alitrunk; an-anepisternum; as-antennal socket; c-coxa number; de-declivity of propodeum; ey-eye; G-gastral segment number; ga-gaster; hd head; he-helcium; kn-katepistemum; mes-mesosoma; mgb-metapleural gland bulla; mn-mandible; mpl-mesopleuron; ms-mesonotum; mtg-metanotal groove; mtp-metapleuron; or-orifice of metapleural gland; pd-peduncle of petiole; pl-propodeal lobe; pms-promesonotal suture; pn-pronotum; ppd-propodeum; ppt-postpetiole; pr-propleuron; prs-promesonotum; pt-petiole; scb-antennal scrobe; sp-spiracle; st-sternite; tr-tergite; w-waist (Cited from Bolton, 1994)

6 测量指标和比例及其缩写
6 Measurements and Indices and Their Abbreviations

本书中的主要标准测量指标和比例及其缩写参照 Bolton（1975，1976）制定的指标，附加上颚长（ML）、上颚头长比（MI）、复眼最大直径（ED）、腹柄长高比（LPI）、腹柄长宽比（DPI）、腹柄结长（PNL）、腹柄结高（PNH）、腹柄结宽（PNW）、后腹柄长（PPL）、后腹柄高（PPH）、后腹柄宽（PPW）、后腹柄结长（PPNL）、后腹柄结高（PPNH）、后腹柄结宽（PPNW）。

体长（TL）：身体伸展状态下，上颚顶端至腹末端的长度。

头长（HL）：头部正面观从唇基前缘中点至头后缘中点的直线长度，如果唇基前缘或头后缘凹陷，则从凹陷处两端的连线中点测量。

头宽（HW）：正面观头部的最大宽度，不包括复眼。

头比（CI）：$CI = HW \times 100 / HL$。

柄节长（SL）：触角柄节的直线长度，不包括基部狭缩处和球形的插入部。

柄节比（SI）：$SI = SL \times 100 / HW$。

上颚长（ML）：上颚顶端至基部关节处的直线长度。

上颚头长比（MI）：$MI = ML \times 100 / HL$。

复眼最大直径（ED）：复眼长径的长度。

Main standard measurements and indices and their abbreviations used in this book are as defined in Bolton (1975, 1976) with supplementary of ML, MI, ED, LPI, DPI, PNL, PNH, PNW, PPL, PPH, PPW, PPNL, PPNH, PPNW.

Total Length (TL): total outstretched length of the individual, from the mandibular apex to the gastral apex.

Head Length (HL): straight-line length of head in perfect full-face view, measured from the mid-point of the anterior clypeal margin to the midpoint of the posterior margin. In species where one or both of these margins are concave, the measurement is taken from the mid-point of a transverse line that spans the apices of the projecting portions.

Head Width (HW): maximum width of head in full-face view, excluding the eyes.

Cephalic Index (CI) = $HW \times 100 / HL$.

Scape Length (SL): straight-line length of the antennal scape, excluding the basal constriction or neck.

Scape Index (SI) = $SL \times 100 / HW$.

Mandible Length (ML): straight-line length of mandible measured from apex to basal articulation.

Mandibulo-cephalic index (MI) = $ML \times 100 / HL$.

Eye Diameter (ED): the longest diameter of eye.

前胸背板宽（PW）：背面观前胸背板的最大宽度。

胸长（WL = AL = MSL）：胸部侧面观从前胸背板与颈片交界点至后胸侧板后下角之间的对角线长度。

腹柄长（PL）：侧面观腹柄前关节至后关节之间的长度。

腹柄高（PH）：侧面观腹柄结顶端至腹柄下突顶端之间的垂直高度。

腹柄宽（DPW）：背面观腹柄的最大宽度。

腹柄长高比（LPI）：LPI = PH × 100 / PL。

腹柄长宽比（DPI）：DPI = DPW × 100 / PL。

腹柄结长（PNL = NL）：侧面观腹柄结的最大长度，不包括前面和后面收缩的柄状部。

腹柄结高（PNH = NH）：侧面观腹柄结顶端至腹柄下突下端之间的垂直高度。

腹柄结宽（PNW = NW）：背面观腹柄结的最大宽度。

后腹柄长（PPL）：侧面观后腹柄前关节至后关节之间的长度。

后腹柄高（PPH）：侧面观后腹柄结顶端至后腹柄下突顶端之间的垂直高度。

后腹柄宽（PPW）：背面观后腹柄的最大宽度。

后腹柄结长（PPNL）：侧面观后腹柄结的最大长度，不包括前面和后面收缩的柄状部。

后腹柄结高（PPNH）：侧面观后腹柄结顶端至后腹柄下突下端之间的垂直高度。

后腹柄结宽（PPNW）：背面观后腹柄结的最大宽度。

所有测量指标单位均为毫米（mm）。

Pronotal Width (PW): maximum width of pronotum measured in dorsal view.

Weber's Length (WL = AL, alitrunk length = MSL, mesosoma length): the diagonal length of the mesosoma in lateral view from the point at which the pronotum meets the cervical shield to the posterior base of the metapleuron.

Petiole Length (PL): length of petiole measured in lateral view from the anterior articulation to the posterior articulation of petiole.

Petiole Height (PH): height of petiole measured in lateral view from the apex of the ventral (subpetiolar) process vertically to a line intersecting the dorsal most point of the node.

Dorsal Petiole Width (DPW): maximum width of petiole in dorsal view.

Lateral Petiole Index (LPI) = PH × 100 / PL.

Dorsal Petiole Index (DPI) = DPW × 100 / PL.

Petiolar Node Length (PNL): maximum length of petiolar node in lateral view, excluding the constricted anterior and posterior peduncle portions.

Petiolar Node Height (PNH): height of petiolar node measured in lateral view from the apex of the ventral (subpetiolar) process vertically to a line intersecting the dorsal most point of the node.

Petiolar Node Width (PNW): maximum width of petiolar node measured in dorsal view.

Postpetiole Length (PPL): length of postpetiole measured in lateral view from the anterior articulation to the posterior articulation of the postpetiole.

Postpetiole Height (PPH): height of postpetiole measured in lateral view from the apex of the ventral (subpostpetiolar) process vertically to a line intersecting the dorsal most point of the node.

Postpetiole Width (PPW): maximum width of postpetiole in dorsal view.

Postpetiolar Node Length (PPNL): maximum length of petiolar node in lateral view, excluding the constricted anterior and posterior peduncle portions.

Postpetiolar Node Height (PPNH): height of postpetiolar node measured in lateral view from the apex of the ventral (subpetiolar) process vertically to a line intersecting the dorsal most point of the node.

Postpetiolar Node Width (PPNW): maximum width of postpetiolar node measured in dorsal view.

All measurements are expressed in millimeters.

7 蚁科分亚科检索表
7 Key to Subfamilies of Formicidae

依据Bolton（1995）分类系统，中国已知蚁科昆虫12亚科，本书记载模式标本的136个蚂蚁新种隶属于其中的10个亚科。此处，提供中国已知12个亚科的检索表供读者参考使用。

According to Bolton (1995) classification system, 12 subfamilies of Formicidae are known in China. The 136 new species of ants with type specimens recorded in this book belong to 10 subfamilies of them. A key to the 12 subfamilies known in China is provided for readers' reference.

中国蚁科工蚁分亚科检索表

1. 后腹部末节臀板背面具1列或许多木钉状刺 ················· 粗角蚁亚科 Cerapachyinae
   后腹部末节臀板背面缺木钉状刺 ··· 2
2. 胸部和后腹部之间有2个缩小独立的体节，即腹柄和后腹柄 ······················· 3
   胸部和后腹部之间只有1个缩小独立的体节，即腹柄 ································ 6
3. 前中胸背板缝存在，可以自由活动 ··· 4
   前中胸背板缝消失或愈合，仅在背面呈1条弱的横沟，不能自由活动 ············ 5
4. 复眼缺失，或至多具1个至少数几个小眼 ······················· 细蚁亚科 Leptanillinae
   复眼发达，占据头侧缘的1/3以上 ··························· 伪切叶蚁亚科 Pseudomyrmecinae
5. 正面观触角窝靠近头部前缘。缺复眼 ···························· 盲蚁亚科 Aenictinae
   正面观触角窝远离头部前缘。通常具复眼，少数缺复眼 ·········· 切叶蚁亚科 Myrmicinae
6. 后腹部末端缺螯针，开口圆孔状或横缝状 ·· 7
   后腹部末端具螯针，经常伸出，可缩入体内，通过体壁可见 ······················· 8
7. 后腹部末端开口呈圆孔状，孔口边缘通常具1圈放射状毛 ·········· 蚁亚科 Formicinae
   后腹部末端开口横缝状，孔口边缘缺放射状毛 ·················· 臭蚁亚科 Dolichoderinae
8. 后腹部末端臀板背面凹陷，具1~2对指向后侧方的刺 ············ 行军蚁亚科 Dorylinae
   后腹部末端臀板背面不凹陷，缺指向后侧方的刺 ······································· 9
9. 腹柄与后腹部宽阔连接，节间不缢缩 ·························· 钝猛蚁亚科 Amblyoponinae

腹柄与后腹部狭窄连接，节间强烈缢缩 ………………………………………………………… 10
10. 后腹部第2节背面具粗糙纵脊纹。后足基节背面具1个背刺………… 刺猛蚁亚科Ectatomminae
    后腹部第2节背面缺条纹或脊纹。后足基节缺背刺 ………………………………………… 11
11. 胸部背面缺前中胸背板缝。后腹部第2节背板显著扩大，末端向前弯曲 …… 卷尾猛蚁亚科Proceratiinae
    胸部背面有前中胸背板缝。后腹部第2节背板不显著扩大，末端指向后方或下方 …… 猛蚁亚科Ponerinae

Key to Subfamilies of Formicidae in China Based on Worker Caste

1. Dorsum of pygidium of the last gastral segment with a row of or many peg–like spines ……… Cerapachyinae
   Dorsum of pygidium of the last gastral segment without peg–like spines ……………………………… 2
2. Body with two reduced isolated segments, the petiole and postpetiole, between mesosoma and gaster ……… 3
   Body with a single reduced isolated segment, the petiole, between mesosoma and gaster ……………… 6
3. Promesonotal suture present, freely flexible ………………………………………………………… 4
   Promesonotal suture absent or fused and appearing as a weak transverse groove on the dorsum, inflexible … 5
4. Eyes absent, or at most with one or a few ommatidia ……………………………………… Leptanillinae
   Eyes developed, occupying more than 1/3 of the lateral margin of head ……………………… Pseudomyrmecinae
5. In full–face view, antennal sockets close to anterior margin of head. Eyes absent ……………… Aenictinae
   In full–face view, antennal sockets far away from anterior margin of head. Eyes usually present, rarely absent ……………………………………………………………………………………… Myrmicinae
6. Sting absent, pex of gaster with a circular opening or a transverse slit–like orifice ……………………… 7
   Sting present, often prominent, can be retracted into the body, visible through the body wall ……………… 8
7. Apex of gaster with a circular opening, usually surrounded by a fringe of hairs …………………… Formicinae
   Apex of gaster with a transverse slit–like orifice, not surrounded by a fringe of hairs ………… Dolichoderinae
8. Dorsum of pygidium of the last gastral segment concave, with 1–2 pairs of posterolaterally pointed spines ……………………………………………………………………………………… Dorylinae
   Dorsum of pygidium of the last gastral segment not concave, without 1–2 pairs of posterolaterally pointed spines ……………………………………………………………………………………… 9
9. Petiole broadly attached to first gastral segment, without constriction between them ………… Amblyoponinae
   Petiole narrowly attached to first gastral segment, with strong constriction between them ……………… 10
10. Dorsum of second gastral segment coarsely and longitudinally costate. Hind coxa with dorsal spine ……………………………………………………………………………………… Ectatomminae
    Dorsum of second gastral segment lacking striate or costate sculpture. Hind coxa lacking dorsal spine … 11
11. Dorsum of mesosoma lacking promesonotal suture. Tergite of second gastral segment strongly enlarged, apex of gaster directed anteriorly ……………………………………………………………… Proceratiinae
    Dorsum of mesosoma with promesonotal suture. Tergite of second gastral segment not strongly enlarged, apex of gaster directed rearwards or ventrally ……………………………………………………… Ponerinae

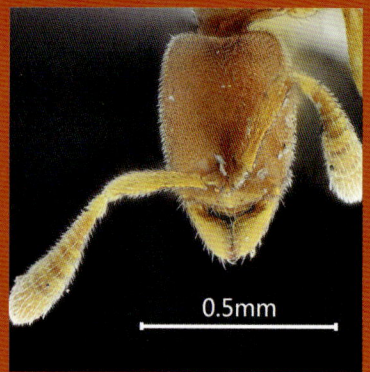

西南林業大學
馆藏蚂蚁模式标本

Type Specimens of Ants Housed in
Southwest Forestry University

分 论
Individual Theory

本书分论部分以图鉴形式记录过去30年里西南林业大学蚂蚁课题组主要在中国西南山地发现和发表的136个蚂蚁新种的模式标本，这些物种隶属于蚁科的10亚科39属，其模式标本均保存在中国云南省昆明市西南林业大学标本馆森林昆虫标本室蚂蚁标本分室。每个物种的记录包括中文名、学名、文献引证、英文原始描述、对应的中文描述、原始描述中的黑白插图和模式标本的高清照片。其中，短背短猛蚁 *Brachyponera brevidorsa* Xu, 1994、尖齿刺结蚁 *Lepisiota acuta* Xu, 1994、网纹刺结蚁 *Lepisiota reticulata* Xu, 1994、凹头臭蚁 *Dolichoderus incisus* Xu, 1995这4个新种以中文发表，本书提供了对应的英文描述。分论中保持了新种发表时学名的原貌，以方便读者查询，因分类地位变更或修订而发生的种名变动，详见总论的相关部分。

书中每个物种的英文和中文描述占1个版面，彩色插图和黑白插图占1个版面，2个版面对页排版，方便读者阅读。因为不同时间对新种的描述详略不等，前期的描述比较简短，后期的描述比较详细，为了保持全书体例一致，统一省略了原始描述中副模标本的采集信息和保存机构信息，仅保留正模标本的采集信息和保存机构信息。因为不同年份发表的新种，其地理分布的行政区划名称中的一些地名迄今已经被修改，书中正模标本的地理分布一律以新种发表时的地名为准。原始描述中记录了副模蚁后、副模雌蚁或副模雄蚁的物种，其相关描述也在本书中予以省略。这些被省略的信息，可以阅读书后列出的相关参考文献。唯一依据脱翅雌蚁命名的新种——木兰版纳猛蚁 *Bannapone mulanae* Xu, 2000，书中记录了其脱翅雌蚁的原始描述。前期发表的新种，其原始描述中缺少词源，本书进行了补充，以便读者了解种名的含义。

In the individual theory section of this book, the type specimens of 136 new species of ants mainly discovered in the southwest mountains of China and published by the ant team of Southwest Forestry University in the past 30 years were recorded in the form of pictorial book. These species belong to 39 genera of 10 subfamilies of Formicidae. The type specimens of these species were housed in the Ant Specimen Sub-collection, Forest Insect Collection, Specimen Museum of Southwest Forestry University, Kunming, Yunnan Province, China. Records of each species include Chinese name, Latin name, literature citation, original description in English, the corresponding Chinese description, the black and white illustrations in the original description and the high-resolution images of the type specimens. Among them, the four new species, *Brachyponera brevidorsa* Xu, 1994, *Lepisiota acuta* Xu, 1994, *Lepisiota reticulata* Xu, 1994 and *Dolichoderus incisus* Xu, 1995, were published in Chinese, and this book provided a corresponding description in English for each species. The original Latin names of all the new species are preserved in the individual theory section for searching convenience of readers. A change in the Latin name of a species resulting from a genera change or revision in the status of the classification is detailed in the relevant section of the general theory.

In the book, for the convenience of readers, the English and Chinese descriptions of each species take up one page, the color images and black-and-white illustrations take up one page, and the two pages are opposite for page layout typeset. Because the description of the new species varied slightly at different times, the initial descriptions are brief, while the later descriptions are more detailed. In order to keep the style consistent throughout the book, the collection information and preservation institution information of paratype specimens in the original description are uniformly omitted, and only the collection information and preservation institution information of holotype specimen are preserved. Because some place names in the administrative divisin in geographical distribution of the new species published in different years have been revised today, the geographical distribution of the holotypes is uniformly keeping the original place names when the new species published in the book. The species whose paratype queen, paratype female and paratype male were recorded in the original description, its related description is also omitted in this book. For the information that has been omitted, the readers can read the relevant references listed in the book. The only new species named after the dealate female, *Bannapone mulanae* Xu, 2000, the original description of whom is recorded in the book. Previously published new species lacking etymology in their original descriptions, the etymology was supplemented in this book so that the readers can understand the meaning of the species names.

书中记录的131个新种的模式标本采用徐正会等（1999b）和徐正会（2002a）制定的样地调查法及徐正会等（2011）制定的搜索调查法采集。此外，布氏卷尾猛蚁 *Proceratium bruelheidei* Staab, Xu & Hita Garcia, 2018、克平卷尾猛蚁 *P. kepingmai* Staab, Xu & Hita Garcia, 2018、萧氏卷尾猛蚁 *P. shohei* Staab, Xu & Hita Garcia, 2018这3个新种的模式标本分别由梅尔·诺亚克、迈克尔·斯泰伯、贝努瓦·盖纳德、本杰明·布兰查德和刘聪采用Winkler地被物分离法采集；北京细蚁 *Leptanilla beijingensis* Qian, Xu, Man & Liu, 2024新种的模式标本由满沛采用地下陷阱法采集；秦岭细蚁 *Leptanilla qinlingensis* Qian, Xu, Man & Liu, 2024新种的模式标本由刘冠临采用搜索调查法采集。

新种的模式标本用胶水粘在插针的三角纸顶端制作成干制标本，然后在带有测微尺的江南体视显微镜（Jiangnan XTB-1，江南光学仪器厂，中国南京）和舜宇体视显微镜（SDPTOP SZM，宁波舜宇仪器有限公司，中国宁波）下观察、测量和描述，在莫提克体视显微镜（Motic-700Z，莫提克光学技术有限公司，中国香港）下绘制黑白插图。采用励扬显微照相系统（LY-WN-YH，成都励扬精密机电有限公司，中国成都）拍摄显微高清彩色照片，采用Zerene Stacker version 1.04软件（Zerene Systems LLC，美国）堆叠合成彩色照片。其中，木兰版纳猛蚁 *Bannapone mulanae* Xu, 2000的彩色照片由美国加州科学院的艾普利尔·诺贝尔采用莱卡照相系统（德国）拍摄。

The type specimens of the 131 new species recorded in this book were collected using the plot-sampling method developed by Xu et al. (1999b) and Xu (2002a) and the search method developed by Xu et al. (2011). In addition, the type specimens of the three new species, *Proceratium bruelheidei* Staab, Xu & Hita Garcia, 2018, *P. kepingmai* Staab, Xu & Hita Garcia, 2018 and *P. shohei* Staab, Xu & Hita Garcia, 2018, were collected using Winkler litter extractor by Merle Noack, Michael Staab, Benoit Guénard, Benjamin Blanchard and Cong Liu respectively. The type specimens of the new species, *Leptanilla beijingensis* Qian, Xu, Man & Liu, 2024, were collected using a subterranean pitfall trap by Pei Man. The type specimens of the new species, *Leptanilla qinlingensis* Qian, Xu, Man & Liu, 2024, were collected using search method by Guan-Lin Liu.

The type specimens of the new species were made into dried specimens sticked with glue at the tip of the pinned triangular paper, and then observed, measured and described under the stereo-microscope with a micrometer of Jiangnan (Jiangnan XTB-1, Jiangnan Optical Instrument Factory, Nanjing, China) and Sunny (SDPTOP SZM, Ningbo Sunny Instruments Co. Ltd, Ningbo, China). The black and white illustrations were drawn under a Motic stereo-microscope (Motic-700Z, Motic Optical Technology Co. Ltd, Hong Kong, China). The high-resolution images of the type specimens were obtained using a Liyang Super Resolution System (LY-WN-YH, Chengdu Liyang Precision Machinery Co., Ltd, Chengdu, China). The stacking color images were composed using Zerene Stacker version 1.04 software (Zerene Systems LLC, USA). In addition, the color images of *Bannapone mulanae* Xu, 2000 were took under a Leica Camera System (Germany) by April Nobile (California Academy of Sciences, USA).

001 阿佤钝猛蚁
Amblyopone awa Xu & Chu, 2012

Amblyopone awa Xu & Chu, 2012, Sociobiology, 59(4): 1192, figs. 51-56 (w.q.) CHINA (Yunnan).

Holotype worker: TL 3.8, HL 0.80, HW 0.68, CI 84, SL 0.40, SI 59, ED 0.03, ML 0.55, PW 0.45, AL 1.05, PL 0.38, PH 0.43, DPW 0.43, LPI 113, DPI 113. In full-face view, head roughly trapezoidal, widened forward and longer than broad. Occipital margin widely weakly concave, occipital corners bluntly angled. Sides weakly convex, anterolateral corners acutely toothed. Mandibles elongate, masticatory margin with a long apical tooth, a short subapical tooth, and 3 pairs of curved teeth; inner margin about as long as masticatory margin, with a pair of curved teeth, a short subbasal tooth, and a large basal tooth. Anterior clypeal margin with 8 teeth, which combined into 4 pairs. Antennae short, 12-segmented; apices of scapes reached to about 2/3 of the distance from antennal sockets to occipital corners; funiculi incrassate toward apex. Eyes very small, each with 3 facets, and located well behind the midpoints of the sides of head. In lateral view, pronotum weakly convex. Promesonotal suture distinctly notched. Mesonotum short and convex. Metanotal groove absent. Propodeal dorsum straight, about 2 times as long as declivity, posterodorsal corner rounded, declivity weakly convex. Petiole trapezoidal, dorsal and anterior faces nearly straight, anterodorsal corner close to a right angle; ventral face oblique, nearly straight; subpetiolar process roughly rectangular, with a large elliptical sub-transparent fenestra, ventral face straight, posteroventral corner rightly angled. In dorsal view, mesothorax constricted. Propodeum slightly widened backward. Propodeal declivity longitudinally concave. Petiole broader than long, width: length = 1.25 : 1, anterior margin and sides weakly convex. Mandibles longitudinally striate. Head densely punctured, interfaces appear as micro-reticulations. Pronotum densely punctured, the narrow longitudinal middle strip without punctures. Dorsa of mesonotum and propodeum abundantly punctured. Sides of mesothorax and metathorax finely longitudinally striate. Petiole and gaster finely sparsely punctured. Dorsal surfaces of head and body with sparse suberect short hairs and dense decumbent pubescence. Scapes with sparse suberect hairs and dense decumbent pubescence. Tibiae with dense decumbent pubescence, but without suberect hairs. Color reddish brown. Eyes black. Antennae and legs yellowish brown. **Paratype workers:** TL 3.6-4.2, HL 0.78-0.90, HW 0.65-0.75, CI 81-90, SL 0.40-0.48, SI 57-63, ED 0.03-0.04, ML 0.53-0.60, PW 0.43-0.50, AL 1.03-1.20, PL 0.38-0.45, PH 0.43-0.48, DPW 0.41-0.48, LPI 106-113, DPI 103-113 (*n*=5). As holotype, but eyes with 3-6 facets, posteroventral corner of subpetiolar process toothed or bluntly angled, head and alitrunk reddish brown to blakish brown.

Holotype: Worker, China: Yunnan Province, Cangyuan County, Banlao Town, Huguang Village, 1720 m, 2011.Ⅲ.17, collected by Yong-qiang Hao from a soil sample in the monsoon evergreen broad-leaf forest, No. A11- 855.

Etymology: The new species is named after an intimate call "Awa" of the minority nationality "Wa" people who commonly live in the area of the holotype locality.

正模工蚁： 正面观，头部近梯形，向前变宽，长大于宽；后缘宽形轻度凹陷，后角钝角状；侧缘轻度隆起，前侧角具锐齿。上颚伸长，咀嚼缘具1个长端齿、1个短的亚端齿和3对弯齿；内缘约与咀嚼缘等长，具1对弯齿、1个短的亚基齿和1个大基齿。唇基前缘具8齿，合并为4对。触角短，12节；柄节顶端到达触角窝至头后角间距约2/3处，鞭节向端部变粗。复眼很小，每个复眼具3个小眼，位于头侧缘中点之后。侧面观，前胸背板轻度隆起。前中胸背板缝明显凹陷。中胸背板短而隆起。缺后胸沟。并胸腹节背面直，约为斜面长的2倍，后上角钝圆，斜面轻度隆起。腹柄梯形，背面和前面近平直，前上角近直角；腹面倾斜，近平直；腹柄下突近长方形，具1个大的椭圆形半透明窗斑，腹面平直，后下角直角形。背面观，中胸收缩。并胸腹节向后轻度变宽，斜面纵向凹陷。腹柄宽大于长，宽：长=1.25：1，前面和侧面轻度隆起。上颚具纵条纹。头部具密集刻点，界面呈微网纹。前胸背板具密集刻点，狭窄的中央纵带无刻点。中胸背板和并胸腹节背面具丰富刻点。中胸和后胸侧面具细纵条纹。腹柄和后腹部具稀疏细刻点。头部和身体背面具稀疏亚直立短毛和密集倾斜绒毛被。柄节具稀疏亚直立毛和密集倾斜绒毛被。胫节具密集倾斜绒毛被，缺亚直立毛。身体红棕色，复眼黑色，触角和足黄棕色。**副模工蚁：** 特征同正模，但复眼具3～6个小眼，腹柄下突后下角齿状或钝角状，头胸部红棕色至黑棕色。

正模： 工蚁，中国云南省沧源县班老乡湖广村，1720m，2011.Ⅲ.17，郝永强采于季风常绿阔叶林土壤样中，No. A11-855。

词源： 该新种根据普遍生活在正模标本产地的少数民族"佤族"的昵称"阿佤"命名。

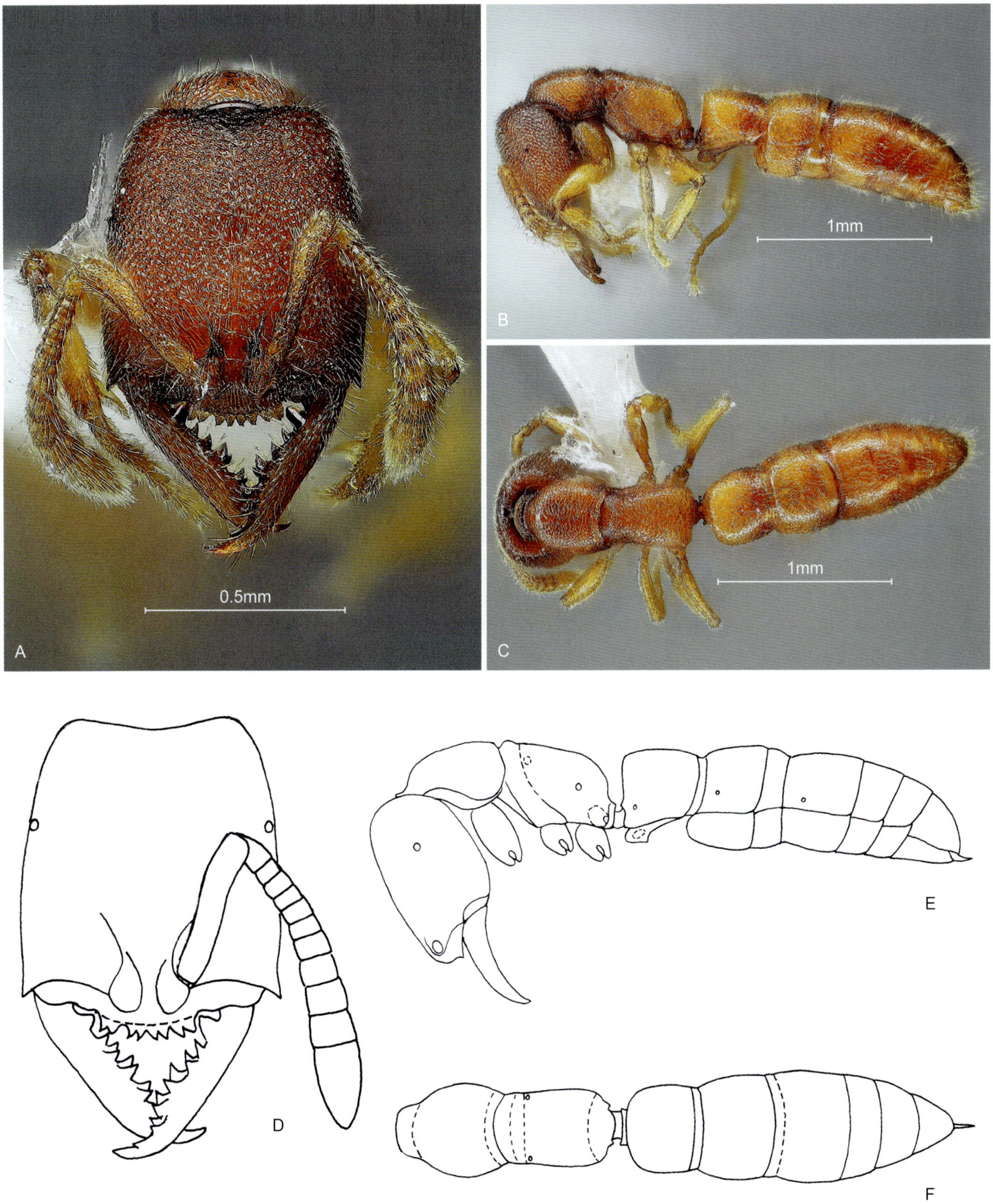

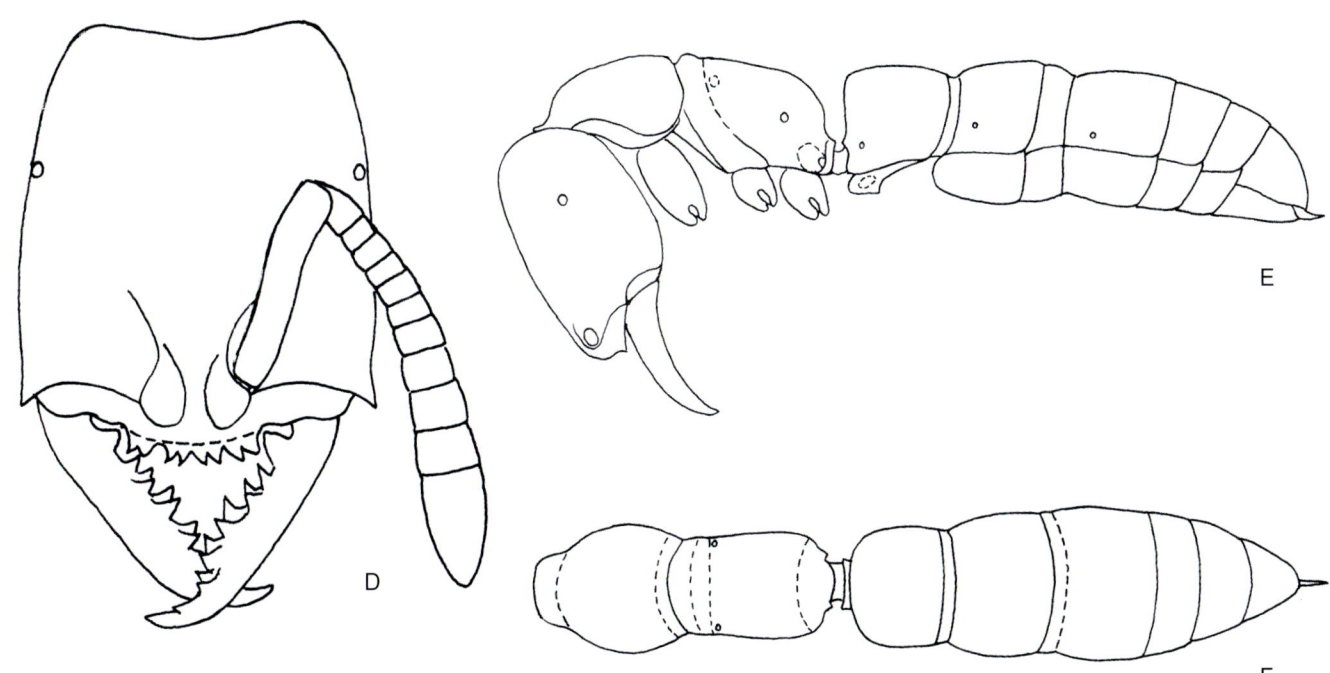

阿佤钝猛蚁 *Amblyopone awa*

A, D. 工蚁头部正面观；B, E. 工蚁身体侧面观；C, F. 工蚁身体背面观（D-F. 引自Xu和Chu, 2012）
A, D. Head of worker in full-face view; B, E. Body of worker in lateral view; C, F. Body of worker in dorsal view (D-F. cited from Xu & Chu, 2012)

细齿钝猛蚁
Amblyopone crenata Xu, 2001

Amblyopone crenata Xu, 2001d, Acta Zootaxonomica Sinica, 26(4): 553, figs. 18-20 (w.) CHINA (Yunnan).

Holotype worker: TL 7.4, HL 1.53, HW 1.28, CI 84, SL 0.90, SI 71, PW 0.85, AL 2.00, ED 0.11, ML 1.18, PL 0.78, PH 0.83, DPW 0.78. In full-face view, head roughly trapezoid, longer than broad, weakly widened forward. Occipital margin weakly concave, occipital corners rounded. Sides weakly convex, anterolateral corner of head with small genal tooth. Mandibles slender, relatively straight, with 2 rows of teeth, the dorsal row with 6 teeth, the ventral row with 5 ones. Anterior margin of clypeus weakly convex, with 12 minute denticles. Median furrow present on the anterior 2/3 dorsum of head. Antennae 12-segmented, apex of scape reached to 3/4 of the distance from antennal socket to occipital corner, segments 4-10 about as broad as long. Eyes relatively large, situated behind the midline of head, each with 19 facets. In lateral view, pronotum weakly convex, both promesonotal suture and metanotal groove distinctly notched. Dorsum of propodeum evenly convex, longer than declivity, the latter truncate, posterodorsal corner of propodeum rounded. In lateral view, petiolar node roughly trapezoid, narrowed backward, anterior face truncate and straight, dorsum evenly convex, anterodorsal corner bluntly prominent and close to a right angle, ventral face oblique. Subpetiolar process nearly rectangular, without fenestra, rounded anteriorly, posteroventral corner toothed. In dorsal view, mesothorax strongly constricted, petiolar node nearly square, as broad as long, anterior and lateral borders weakly convex, posterior border straight. Constriction distinct between the 2 basal segments of gaster. Mandibles longitudinally and obliquely striate, with apex smooth. Clypeus longitudinally striate. Head with dense large punctures, dull. Thorax, petiole and first segment of gaster with dense mid-size punctures, less dull. Segments 2-5 of gaster with sparse fine punctures, relatively shining. Whole body surface and appendages with sparse suberect hairs and dense decumbent pubescence. Body in color black. Mandibles, clypeus, antennae, legs and apex of gaster reddish brown. **Paratype workers:** TL 6.7-7.3, HL 1.43-1.53, HW 1.25-1.30, CI 85-88, SL 0.90, SI 69-72, PW 0.83-0.85, AL 1.88-2.00, ED 0.13, ML 1.00-1.10, PL 0.73-0.78. PH 0.78-0.85, DPW 0.75-0.80 (*n*=3). As holotype, but eye with 18-19 facets, gaster black or brownish black.

Holotype: Worker, China: Yunnan Province, Mengla County, Shangyong Town, Manzhuang Village, 900 m, 1997.Ⅷ.14, collected by Yun-feng He in semi-evergreen monsoon forest of Xishuangbanna National Nature Reserve, No. A97-1680.

Etymology: The name of the new species is descriptive of the "*crenate*" anterior clypeal margin plus the suffix "*-ata*".

正模工蚁： 正面观，头部近梯形，长大于宽，向前轻度变宽；后缘轻度凹陷，后角窄圆；头侧缘轻度隆起，前侧角具小齿。上颚细长，较直，具2列齿，背面1列具6齿，腹面1列具5齿。唇基前缘轻度隆起，具12个小齿。头部背面前部2/3具中央纵沟。触角12节，柄节顶端到达触角窝至头后角间距的3/4处，4～10节长宽约相等。复眼较大，位于头中线之后，每个复眼具19个小眼。侧面观，前胸背板轻度隆起，前中胸背板缝和后胸沟明显凹陷。并胸腹节背面均匀隆起，长于斜面，斜面平截，后上角窄圆。侧面观，腹柄结近梯形，向后变窄，前面平截且直，背面均匀隆起，前上角突出近直角，腹面倾斜。腹柄下突近长方形，无窗斑，前面圆，后下角齿状。背面观，中胸强烈收缩，腹柄结近方形，长宽相等，前缘和侧缘轻度隆起，后缘直。后腹部基部2节间缢缩明显。上颚具倾斜纵条纹，端部光滑。唇基具纵条纹。头部具密集大刻点，暗。胸部、腹柄和后腹部第1节具密集中等大小刻点，较暗。后腹部2～5节具稀疏细刻点，较光亮。身体表面和附肢具稀疏亚直立毛和密集倾斜绒毛被。身体黑色，上颚、唇基、触角、足和后腹部末端红棕色。**副模工蚁：** 特征如正模，但复眼具18～19个小眼，后腹部黑色或棕黑色。

正模： 工蚁，中国云南省勐腊县尚勇乡曼庄村，900m，1997.Ⅷ.14，何云峰采于滇南西双版纳国家级自然保护区半常绿季雨林中，No. A97-1680。

词源： 该新种的名称由描述唇基前缘特征的"锯齿状的"加后缀"具有"组成。

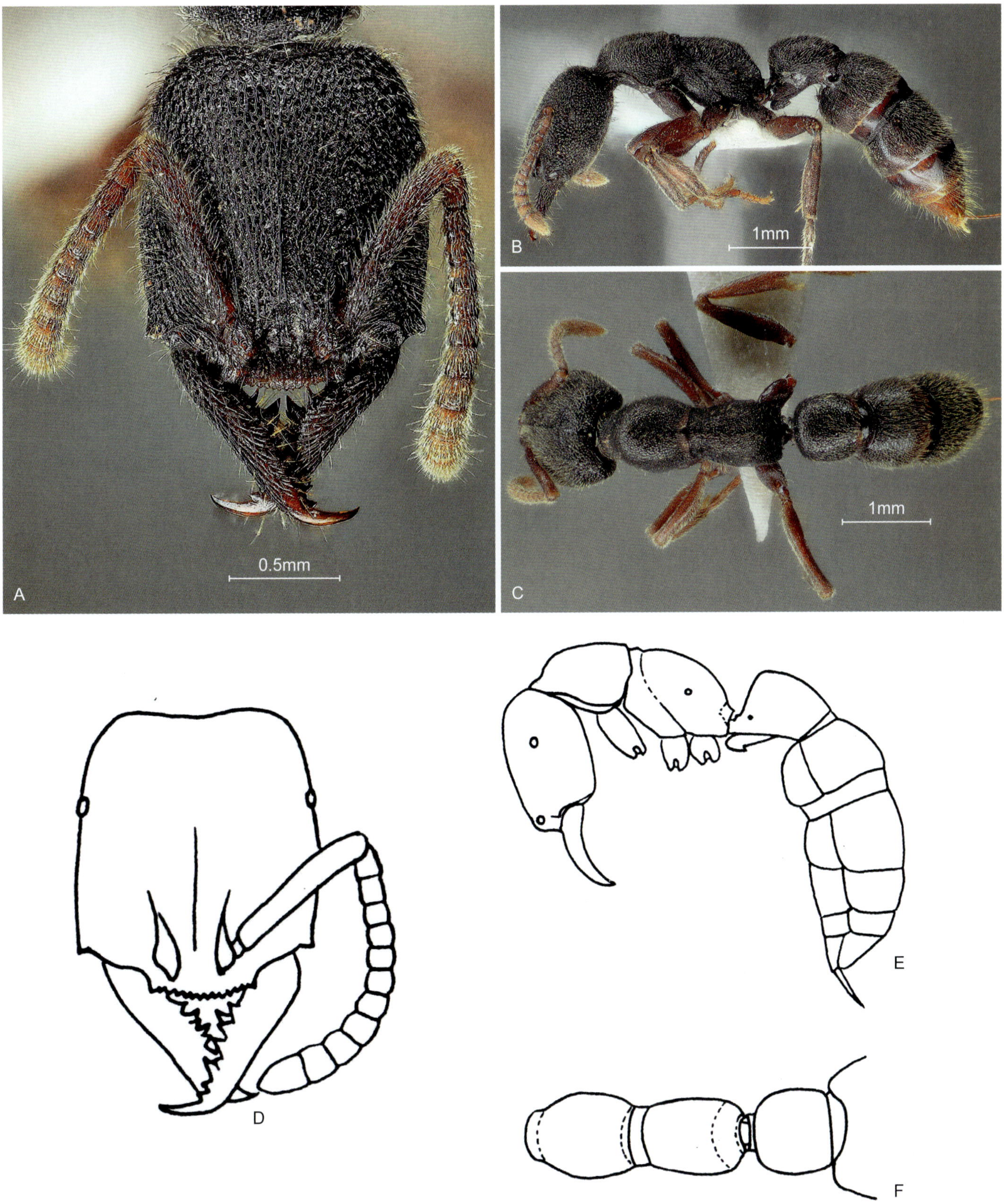

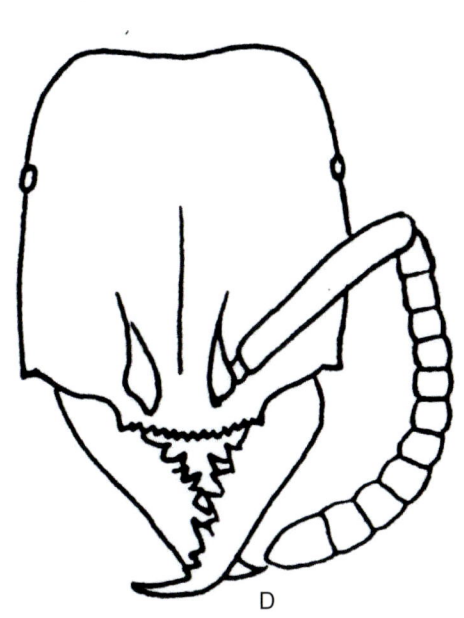
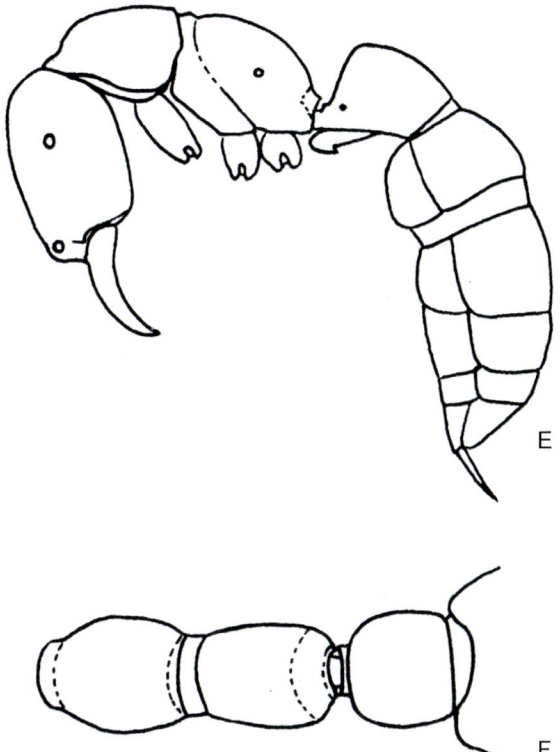

细齿钝猛蚁 *Amblyopone crenata*

A, D. 工蚁头部正面观；B, E. 工蚁身体侧面观；C, F. 工蚁身体背面观（D-F. 引自Xu, 2001d）
A, D. Head of worker in full-face view; B, E. Body of worker in lateral view; C, F. Body of worker in dorsal view (D-F. cited from Xu, 2001d)

康巴钝猛蚁
Amblyopone kangba Xu & Chu, 2012

Amblyopone kangba Xu & Chu, 2012, Sociobiology, 59(4): 1187, figs. 13-15 (w.) CHINA (Tibet).

Holotype worker: TL 6.7, HL 1.40, HW 1.40, CI 100, SL 0.80, SI 57, ED 0.10, ML 1.17, PW 0.93, AL 2.07, PL 0.70, PH 0.87, DPW 0.80, LPI 124, DPI 114. In full-face view, head square, as broad as long, slightly widened forward. Occipital margin weakly widely concave, occipital corners bluntly angled. Sides nearly straight. Anterolateral corners each with a reduced tiny tooth. Mandibles elongate and linear, masticatory margin very short, about 1/4 length of the inner margin, with 3 simple teeth; inner margin with 2 rows of curved teeth, each row with 6 teeth, the basal tooth large and triangular. Middle portion of anterior clypeal margin weakly protruding forward, slightly concave, with 4 tiny short rectangular denticles; anterolateral corners rightly angled. Frontal lobes slightly surpassed anterior clypeal margin. Antennae short, 12-segmented; apices of scapes reached to 2/3 of the distance from antennal sockets to occipital corners; funiculi incrassate toward apices. Eyes small, located behind the midpoints of the sides of head, each with about 9 facets. In lateral view, dorsum of alitrunk weakly convex, promesonotal suture distinctly notched. Mesonotum short and convex. Metanotal groove absent. Dorsum of propodeum nearly straight, about 1.5 times as long as declivity, posterodorsal corner rounded, declivity weakly convex. Dorsal face of petiole weakly convex, anterior face straight, anterodorsal corner bluntly angled; ventral face oblique and weakly concave, subpetiolar process roughly square, with a circular sub-transparent fenestra. Constriction between the two basal gastral segments distinct, sting strong and extruding. In dorsal view, mesothorax constricted, mesonotum very short. Propodeal declivity weakly concave. Petiole broader than long, width : length = 1.2 : 1. Mandibles longitudinally striate. Head with fine elongate reticulations. Dorsal faces of alitrunk, petiole, and gaster sparsely punctured, the punctures decreased in diameter from alitrunk to gaster, interfaces smooth and shining. The longitudinal middle strip of pronotum without punctures. Sides of alitrunk, petiole, and gaster densely punctured; sides of mesothorax and metathorax longitudinally striate. Dorsal surfaces of head and body with abundant suberect short hairs and dense decumbent pubescence. Scapes and hind tibiae with sparse suberect hairs and dense decumbent pubescence. Color reddish brown. Occiput blackish brown. Antennae and legs yellowish brown. **Paratype workers:** TL 6.5-7.0, HL 1.33-1.40, HW 1.33-1.40, CI 100- 102, SL 0.77-0.83, SI 57-60, ED 0.07-0.08, ML 1.17-1.23, PW 0.87-0.97, AL 1.93-2.07, PL 0.67-0.77, PH 0.87-0.90, DPW 0.78-0.83, LPI 117-130, DPI 109-118 (*n*=4). As holotype, but middle portion of anterior clypeal margin with 4-6 tiny rectangular denticles; Color yellowish brown to reddish brown.

Holotype: Worker, China: Tibet Autonomous Region, Zayu County, Zhuwagen Town, Cibaqiao, 1750 m, 2010.Ⅷ. 30, collected by Xia Liu from a soil sample in the forest of *Pinus yunnanensis* (Pinaceae), No.A10-3405.

Etymology: The new species is named after a race of the Tibetan people "Kangba" who live in the southwestern Tibet.

正模工蚁：正面观，头部方形，长宽相等，向前轻度变宽；后缘轻度宽形凹陷，后角钝角状；侧缘近平直，前侧角具1个退化的小齿。上颚伸长，线形；咀嚼缘很短，约为内缘长的1/4，具3个简单齿；内缘具2列弯齿，每列具6齿，基齿大，三角形。唇基前缘中部轻度向前突出，微凹，具4个微小的长方形齿；前侧角直角状。额叶稍超过唇基前缘。触角短，12节，柄节顶端到达触角窝至头后角的2/3处，鞭节向端部变粗。复眼小，位于头侧缘中点之后，每个复眼约具9个小眼。侧面观，胸部背面轻度隆起，前中胸背面缝明显切入，中胸背板短而隆起，缺后胸沟；并胸腹节背面近平直，约为斜面长的1.5倍，后上角宽圆，斜面轻度隆起。腹柄背面轻度隆起，前面平直，前上角钝角状；腹面倾斜，轻度凹陷，腹柄下突近方形，具1个圆形半透明窗斑。后腹部基部2节间缢缩明显，螫针发达，伸出。背面观，中胸收缩，中胸背板很短，并胸腹节斜面轻度凹陷；腹柄宽大于长，宽：长=1.2：1。上颚具纵条纹。头部具伸长的细网纹。胸部、腹柄和后腹部背面具稀疏刻点，从胸部至后腹部刻点直径变小，界面光滑发亮。前胸背板中央纵带无刻点；胸部、腹柄和后腹部侧面具密集刻点；中胸和后胸侧面具纵条纹。头部和身体背面具丰富亚直立短毛和密集倾斜绒毛被。柄节和后足胫节具稀疏亚直立毛和密集倾斜绒毛被。身体红棕色，后头黑棕色，触角和足黄棕色。**副模工蚁**：特征同正模，但是唇基前缘中部具4～6个微小的长方形齿，身体黄棕色至红棕色。

正模：工蚁，中国西藏自治区察隅县竹瓦根镇慈巴桥，1750m，2010.Ⅷ.30，刘霞采于云南松林土壤样中，No.A10-3405。

词源：该新种以居住在藏东南地区的藏族的一个支系"康巴"命名。

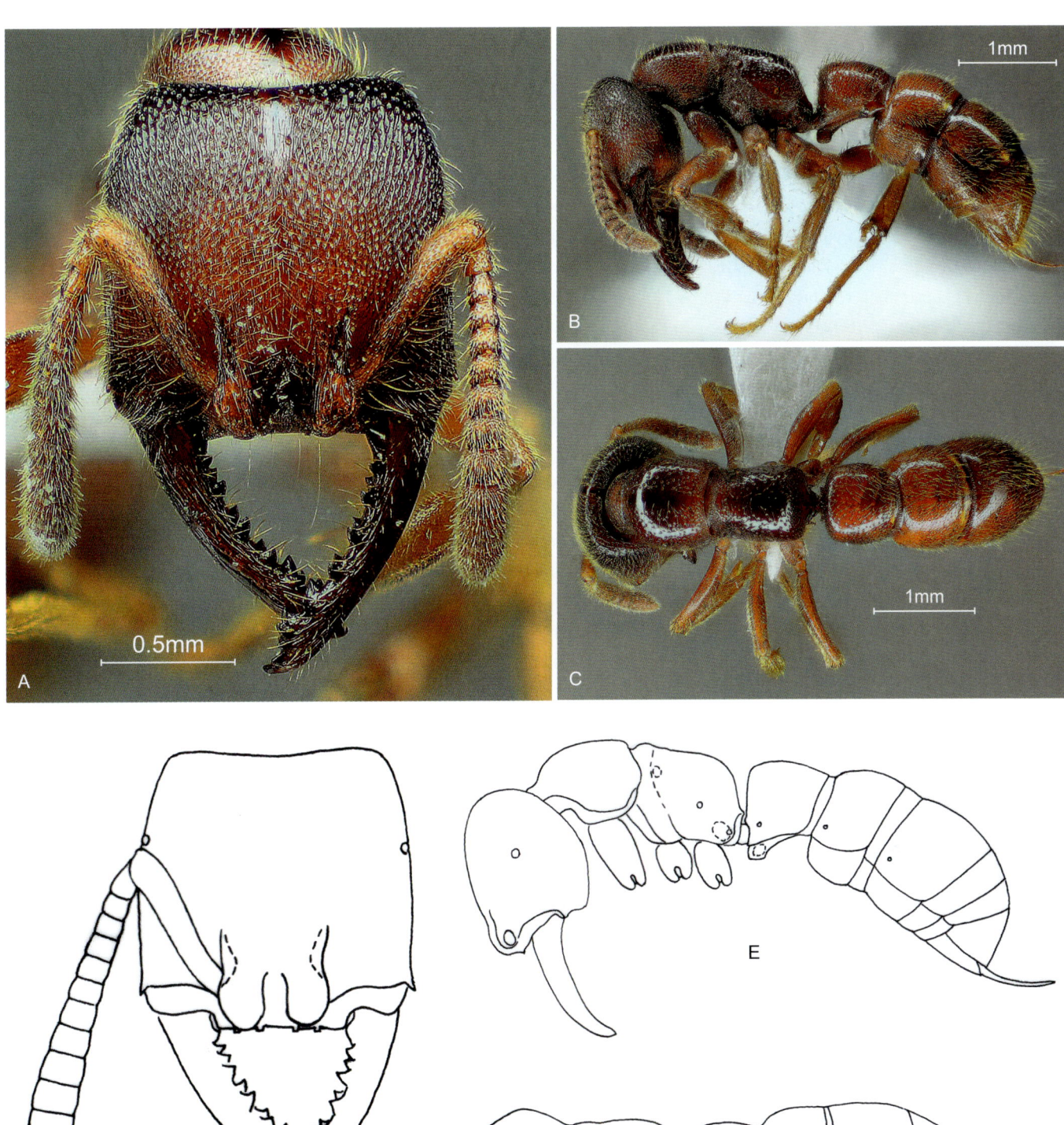

康巴钝猛蚁 *Amblyopone kangba*

A, D. 工蚁头部正面观；B, E. 工蚁身体侧面观；C, F. 工蚁身体背面观（D-F. 引自Xu和Chu, 2012）
A, D. Head of worker in full-face view; B, E. Body of worker in lateral view; C, F. Body of worker in dorsal view (D-F. cited from Xu & Chu, 2012)

梅里钝猛蚁
Amblyopone meiliana Xu & Chu, 2012

Amblyopone meiliana Xu & Chu, 2012, Sociobiology, 59(4): 1190, figs. 33-35 (w.) CHINA (Yunnan).

Holotype worker: TL 4.9, HL 1.10, HW 0.95, CI 86, SL 0.63, SI 66, ED 0.05, ML 0.78, PW 0.63, AL 1.50, PL 0.53, PH 0.60, DPW 0.60, LPI 114, DPI 100. In full-face view, head roughly trapezoidal, widened forward and longer than broad. Occipital margin widely weakly concave, occipital corners rounded. Sides nearly straight, anterolateral corners acutely toothed. Mandibles elongate, masticatory margin about as long as inner margin, with a long apical tooth, a short subapical tooth, and 3 pairs of curved teeth; inner margin with a pair of curved teeth, a short subbasal tooth, and a large basal tooth. Anterior clypeal margin roundly convex, with a broad middle lobe, a narrow lobe on each side, and a simple tooth between the middle and lateral lobes; the broad middle lobe with 4 denticles at apex, the lateral lobes slightly bifid at apices. Antennae short, 12-segmented; apices of scapes reached to 2/3 of the distance from antennal sockets to occipital corners; funiculi incrassate toward apex. Eyes small, each with 5 facets, and located well behind the midpoints of the sides of head. In lateral view, posterior 2/3 of pronotum nearly straight, mesonotum short and convex. Promesonotal suture distinctly notched, metanotal groove weakly depressed. Propodeal dorsum straight, about 1.5 times as long as declivity, posterodorsal corner rounded, declivity nearly straight. Petiole roughly trapezoidal, dorsal and anterior faces nearly straight, anterodorsal corner rounded; ventral face oblique and weakly concave. Subpetiolar process nearly triangular, with a large circular sub-transparent fenestra, anterior face rounded, ventral face straight, posteroventral corner rightly angled. In dorsal view, mesothorax constricted. Propodeum widened backward. Petiole slightly broader than long, width : length = 1.1 : 1, anterior margin and sides weakly convex. Mandibles and clypeus longitudinally striate. Head densely punctured, interfaces appear as micro-reticulations. Dorsal face of alitrunk densely punctured, the longitudinal middle strip on pronotum and propodeum smooth and shining. Sides of pronotum transversely striate. Sides of mesothorax, metathorax, and propodeum longitudinally striate. Declivity nearly smooth. Dorsal faces of petiole and gaster smooth, sides of petiole finely reticulate, sides of gaster finely punctured. Dorsal surfaces of head and body with sparse suberect short hairs and dense decumbent pubescence. Scapes and tibiae with sparse suberect hairs and dense decumbent pubescence. Color reddish brown. Eyes blackish. Legs yellowish brown.

Holotype: Worker, China: Yunnan Province, Deqin County, Yunling Town, Mingyong Village, 3250 m, 2004.Ⅹ.10, collected by Sheng-li Shi from a ground sample in the conifer-broadleaf mixed forest on the east slope of the Snow Mt. Meili, No. A04-536.

Etymology: The new species is named after the type locality "Snow Mt. Meili", the highest mountain in Yunnan Province.

正模工蚁：正面观，头部近梯形，向前变宽，长大于宽；后缘宽形浅凹，后角窄圆；侧面近平直，前侧角尖齿状。上颚伸长，咀嚼缘约与内缘等长，具1个长端齿、1个短的亚端齿和3对弯齿；内缘具1对弯齿、1个短的亚基齿和1个大基齿。唇基前缘圆形隆起，具1个宽的中叶，两侧各具1个窄的侧叶，中叶和侧叶之间具1个简单齿；中叶顶端具4个细齿，侧叶顶端轻度分叉。触角短，12节，柄节顶端到达触角窝至头后角间距的2/3处，鞭节向端部变粗。复眼小，每个复眼具5个小眼，明显位于头侧缘中点之后。侧面观，前胸背板后部2/3近平直，中胸背板短而隆起。前中胸背板缝明显切入，后胸沟轻度凹陷。并胸腹节背面平直，约为斜面长度的1.5倍，后上角宽圆，斜面近平直。腹柄近梯形，背面和前面近平直，前上角窄圆；腹面倾斜，轻度凹陷；腹柄下突近三角形，具1个大的圆形半透明窗斑，前面圆，腹面平直，后下角直角形。背面观，中胸收缩，并胸腹节向后变宽；腹柄宽稍大于长，宽：长=1.1：1，前面和侧面轻度隆起。上颚和唇基具纵条纹。头部具密集刻点，界面具微网纹。胸部背面具密集刻点，前胸背板和并胸腹节背面的中央纵带光滑发亮。前胸背板侧面具横条纹；中胸、后胸和并胸腹节侧面具纵条纹，并胸腹节斜面近光滑。腹柄背面和后腹部光滑，腹柄侧面具细网纹，后腹部侧面具细刻点。头部和身体背面具稀疏亚直立短毛和密集倾斜绒毛被。柄节和胫节具稀疏亚直立毛和密集倾斜绒毛被。身体红棕色，复眼黑色，足黄棕色。

正模：工蚁，中国云南省德钦县云岭乡明永村，3250m，2004.Ⅹ.10，史胜利采于梅里雪山东坡针阔混交林地表样中，No. A04-536。

词源：该新种以云南省最高的山脉、模式产地"梅里雪山"命名。

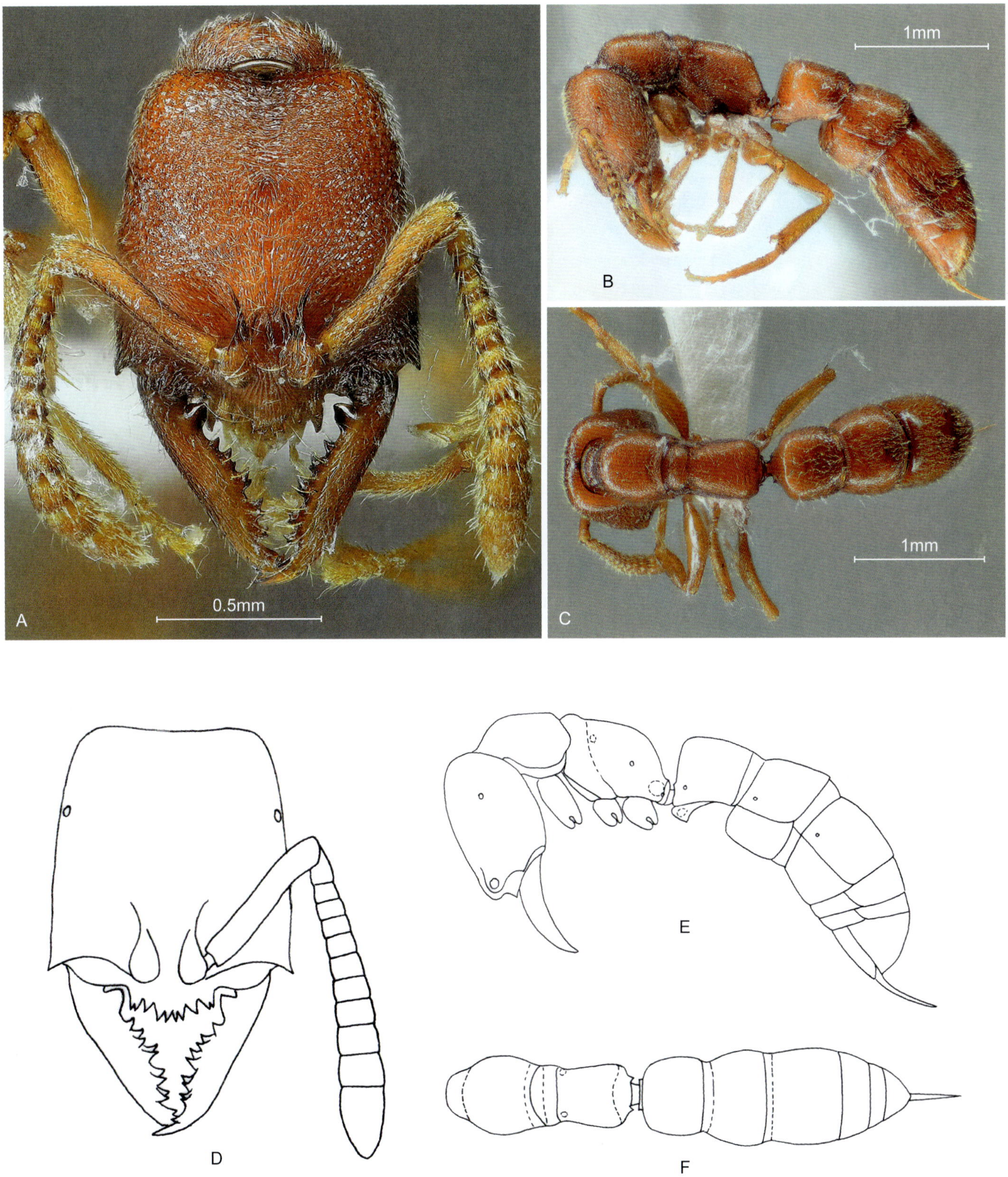

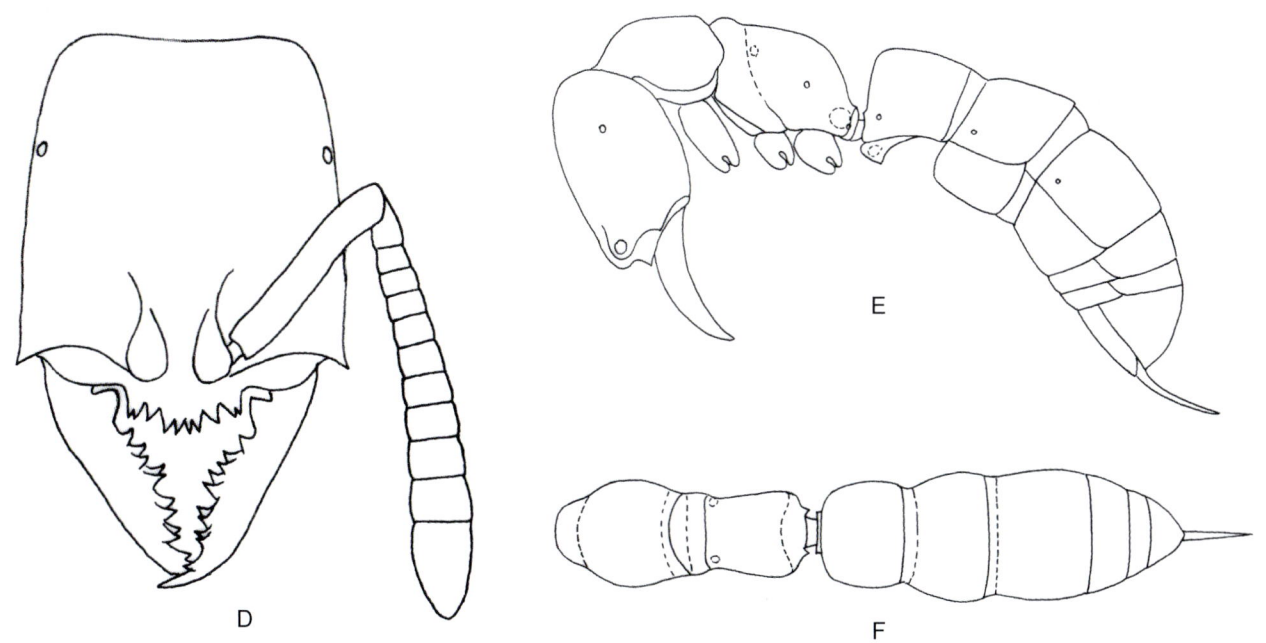

梅里钝猛蚁 *Amblyopone meiliana*

A, D. 工蚁头部正面观；B, E. 工蚁身体侧面观；C, F. 工蚁身体背面观（D-F. 引自Xu和Chu, 2012）
A, D. Head of worker in full-face view; B, E. Body of worker in lateral view; C, F. Body of worker in dorsal view (D-F. cited from Xu & Chu, 2012)

八齿钝猛蚁
Amblyopone octodentata Xu, 2006

Amblyopone octodentata Xu, 2006, Myrmecologische Nachrichten, 8: 152, figs. 1-7 (w.q.) CHINA (Yunnan).

Holotype worker: TL 4.3, HL 0.93, HW 0.80, CI 86, SL 0.47, SI 58, ML 0.67, ED 0.06, PW 0.52, AL 1.20, PL 0.58, PH 0.43, DPW 0.50. In full-face view, head nearly trapezoid and widened forward, longer than broad. Occipital margin weakly concave, occipital corners roundly prominent. Sides of head relatively straight. Mandibles narrow and slender, with 8 teeth, the basal 2 teeth and the apical 2 ones simple, the middle 4 teeth each bifurcated, the apical tooth slender. Anterolateral corner of head elongated into an acute tooth. Anterior margin of clypeus roundly convex, with 8 simple teeth. Antennae 12-segmented, apices of scapes reached to 3/5 of the distance from antennal socket to occipital corner. Eyes small, with 5-6 facets, placed at posterior 2/5 of lateral side of head. In lateral view, pronotum weakly convex. Promesonotal suture depressed. Mesonotum very short, with straight dorsum. Metanotal groove only visible on sides. Dorsum of propodeum very long, straight and sloped posteriad, posterodorsal corner rounded, declivity weakly convex. In lateral view, petiole narrowed backward, anterior face weakly concave, dorsal face nearly straight, anterodorsal corner right-angled. Subpetiolar process cuneiform, with large elliptic translucent fenestra, anterior face roundly convex, ventral face straight, posteroventral corner acutely toothed. In dorsal view, petiole about as broad as long, anterior and lateral borders weakly convex, anterolateral corners rounded, posterior border straight. Mandibles finely longitudinally striate. Head, pronotum, dorsal faces of mesonotum and propodeum densely and coarsely punctate, interstices appear as fine reticulations. Sides of mesothorax, metathorax and propodeum densely, finely and longitudinally striate. Petiole and first gastral segment densely and finely punctate. Second gastral segment sparsely and finely punctate, interstices smooth. Remaining gastral segments smooth and shining. Dorsal faces of head and body with dense erect or suberect short hairs and dense decumbent pubescence. Scapes and tibiae with sparse erect or suberect hairs and dense decumbent pubescence. Body color brown, legs brownish yellow. **Paratype workers:** TL 4.2-4.5, HL 0.87-0.93, HW 0.77-0.78, CI 84-88, SL 0.43-0.47, SI 57-60, ML 0.63-0.67, ED 0.05, PW 0.52-0.53, AL 1.17-1.20, PL 0.47-0.53, PH 0.42-0.47, DPW 0.45-0.48 ($n=4$). As holotype.

Holotype: Worker, China: Yunnan Province, Kunming City, Xishan Mountain Forest Park, Nie'ermu, 2150 m, 2001.V.3, collected by Yu-xiang Zhao in *Pinus armandii* forest, No. A00476.

Etymology: The species name *octodentata* combines Latin *octo-* (eight) + word root *dent* (tooth) + suffix *-atus* (feminine form *-ata*, with), it refers to the eight teeth of the anterior margin of the clypeus.

正模工蚁： 正面观，头部近梯形，向前变宽，长大于宽；后缘浅凹，后角圆突；侧缘较直。上颚窄而细长，具8齿，基部2齿和端部2齿简单，中间4齿均二叉状，端齿细长。头部前侧角延伸呈尖齿。唇基前缘圆形隆起，具8个简单齿。触角12节，柄节顶端到达触角窝至头后角间距的3/5处。复眼小，具5~6个小眼，位于头侧缘后部2/5处。侧面观，前胸背板轻度隆起，前中胸背板缝凹陷；中胸背板很短，背面平直；后胸沟仅在侧面可见。并腹胸节背面很长，平直，向后呈坡形，后上角宽圆，斜面轻度隆起。腹柄向后变窄，前面轻度凹陷，背面近平直，前上角直角形。腹柄下突楔形，具大的椭圆形透明窗斑，前面圆形隆起，腹面平直，后下角具锐齿。背面观，腹柄长宽约相等，前面和侧面轻度隆起，前侧角窄圆，后缘直。上颚具细纵条纹。头部、前胸背板、中胸背板和并胸腹节背面具密集粗糙刻点，界面呈细网纹。中胸侧面、后胸侧面和并胸腹节侧面具密集细纵条纹。腹柄和后腹部第1节具密集细刻点；后腹部第2节具稀疏细刻点，界面光滑；后腹部其余腹节光滑发亮。头部和身体背面具密集直立或亚直立短毛和密集倾斜绒毛被。柄节和胫节具稀疏直立或亚直立毛和密集倾斜绒毛被。身体棕色，足棕黄色。**副模工蚁：** 特征同正模。

正模： 工蚁，中国云南省昆明市西山森林公园聂耳墓，2150m，2001.V.3，赵宇翔采于华山松林中，No. A00476。

词源： 该新种的种名"*octodentata*"由拉丁语"*octo-*"（八）+词根"*dent*"（牙齿）+后缀"*-atus*"（阴性形式-"*ata*"，具有）组成，指唇基前缘具8个齿。

Individual Theory 分　论

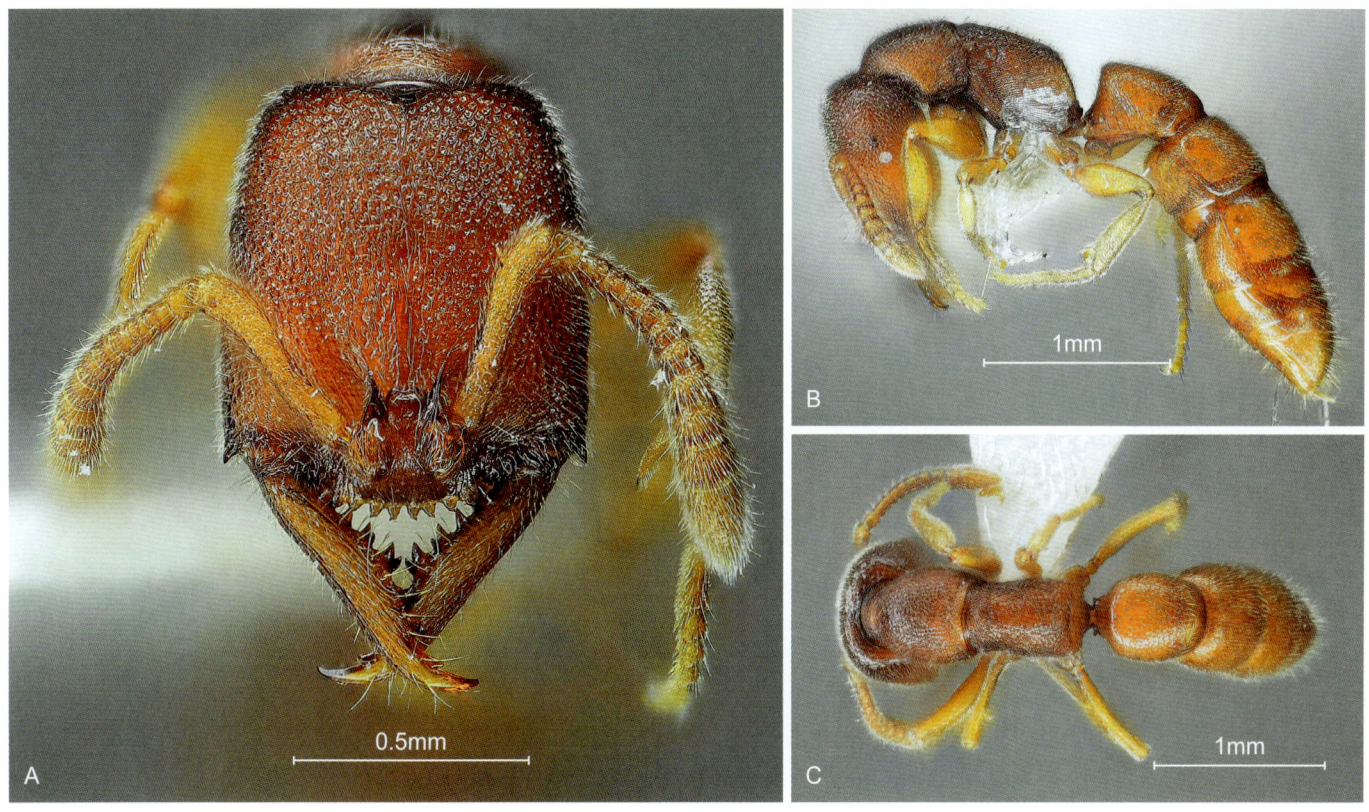

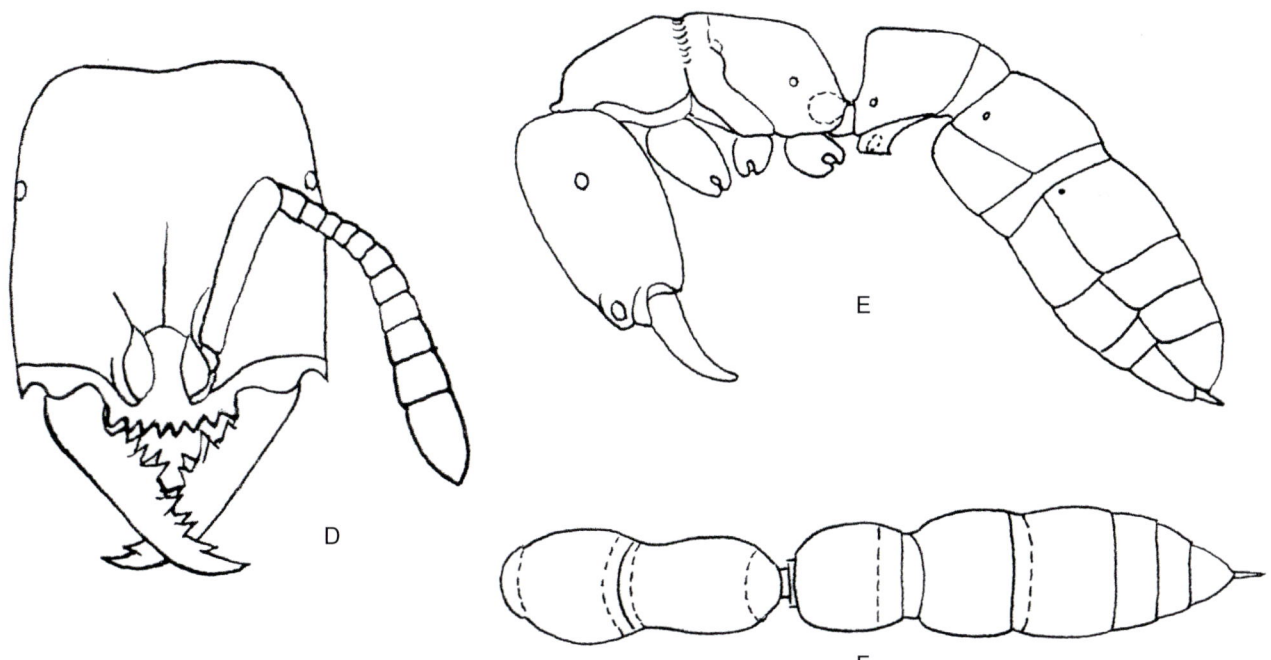

八齿钝猛蚁 *Amblyopone octodentata*
A, D. 工蚁头部正面观；B, E. 工蚁身体侧面观；C, F. 工蚁身体背面观（D-F. 引自Xu, 2006）
A, D. Head of worker in full-face view; B, E. Body of worker in lateral view; C, F. Body of worker in dorsal view (D-F. cited from Xu, 2006)

三叶钝猛蚁
Amblyopone triloba Xu, 2001

Amblyopone triloba Xu, 2001d, Acta Zootaxonomica Sinica, 26(4): 552, figs. 11-13 (w.) CHINA (Yunnan).

Holotype worker: TL 4.5, HL 1.00, HW 0.85, Cl 85, SL 0.43, SI 62, PW 0.55, AL 1.23, ED 0.00, ML 0.65, PL 0.45, PH 0.55, DPW 0.50. In full-face view, head roughly trapezoid, longer than broad, widened forward. Occipital margin weakly concave, occipital corners roundly prominent, sides weakly convex. Anterolateral corner of head with acute genal tooth. Mandibles slender, relatively straight, inner and masticatory margins about equal in length, inner margin with 5-6 teeth, masticatory margin with 4 teeth. Anterior margin of clypeus roundly convex and divided into 3 lobes, middle lobe with 4 denticles, lateral lobes each with 2 denticles, besides an extra tooth present between middle lobe and right lateral lobe. Antennae short, 12-segmented, apex of scape reached to 2/3 of the distance from antennal socket to occipital corner, segments 2-10 each broader than long. Eyes absent. In lateral view, dorsum of thorax weakly convex, promesonotal suture distinctly notched, metanotal groove slightly depressed. Dorsum of propodeum straight, longer than declivity, declivity flat, posterodorsal corner of propodeum bluntly angled. In lateral view, petiolar node roughly trapezoid, narrowed backward, anterior face truncate and straight, dorsum weakly convex, anterodorsal corner nearly a right angle, ventral face oblique. Subpetiolar process large, nearly square, with large elliptic fenestra, anteroventral corner rounded, posteroventral corner bluntly angled. In dorsal view, petiolar node nearly square, length∶width = 9∶10, anterior and lateral borders weakly convex, posterior border nearly straight. Constriction distinct between the 2 basal segments of gaster. Mandibles longitudinally striate, with apex smooth and shining. Clypeus finely and longitudinally striate. Head densely and coarsely punctured, with the longitudinal central stripe of dorsum finely and longitudinally striate, dull. Pronotum with sides densely and coarsely punctured, but smooth on the dorsum. Sides of mesothorax, metathorax and propodeum finely and longitudinally striate, dorsum of propodeum punctured, declivity smooth. Petiole with sides densely punctured but smooth on dorsum. First segment of gaster with sides finely and densely punctured but smooth on dorsum. Segments 2-5 of gaster smooth. Whole body surface and appendages with sparse suberect short hairs and dense decumbent pubescence. Body in color reddish brown. Legs brownish yellow.

Holotype: Worker, China: Yunnan Province, Lushui County, Pianma Town, Pianma Village, 2500 m, 1999.Ⅳ.25, collected by Zheng-hui Xu in the subalpine moist evergreen broadleaf forest of the Gaoligong Mountain National Nature Reserve, No. A99-27.

Etymology: The species name *triloba* combines Greek *tri-* (three) + word root *lob* (lobe) + suffix *-a* (feminine form), it refers to the three lobes of the anterior margin of the clypeus.

正模工蚁： 正面观，头部近梯形，长大于宽，向前变宽；后缘轻度凹陷，后角圆突；侧缘轻度隆起，前侧角具尖锐的颊齿。上颚细长，较直，内缘和咀嚼缘约等长，内缘具5～6个齿，咀嚼缘具4个齿。唇基前缘圆形隆起，分裂成3个叶，中叶顶端具4个小齿，侧叶各具2个小齿；此外，中叶和右侧叶之间还具1个额外的齿。触角短，12节，柄节末端到达触角窝至头后角间距的2/3处，第2～10节宽大于长。缺复眼。侧面观，胸部背面轻度隆起，前中胸背板缝明显切入，后胸沟轻微凹陷。并胸腹节背面直，长于斜面，斜面平坦，后上角钝角状。腹柄结近梯形，向后变窄，前面平截且直，背面轻度隆起，前上角近直角形，腹面倾斜；腹柄下突大，近方形，具大的椭圆形窗斑，前下角圆，后下角钝角状。背面观，腹柄结近方形，长∶宽=9∶10，前缘和侧缘轻度隆起，后缘近平直。后腹部基部2节间收缩明显。上颚具纵条纹，端部光滑发亮。唇基具细纵条纹。头部具密集粗糙刻点，背面中央纵带具细纵条纹，暗。前胸背板侧面具密集粗糙刻点，背面光滑。中胸侧面、后胸侧面和并胸腹节侧面具细纵条纹，并胸腹节背面具刻点，斜面光滑。腹柄侧面具密集刻点，背面光滑。后腹部第1节侧面具细密刻点，背面光滑；后腹部第2～5节光滑。身体表面和附肢具稀疏亚直立短毛和密集倾斜绒毛被。身体红棕色，足棕黄色。

正模： 工蚁，中国云南省泸水县片马镇片马村，2500m，1999.Ⅳ.25，徐正会采于高黎贡山国家级自然保护区中山湿性常绿阔叶林中，No. A99-27。

词源： 该新种的种名"*triloba*"由希腊语"*tri-*"（三）+词根"*lob*"（叶）+后缀"*-a*"（阴性形式）组成，指唇基前缘具3个叶。

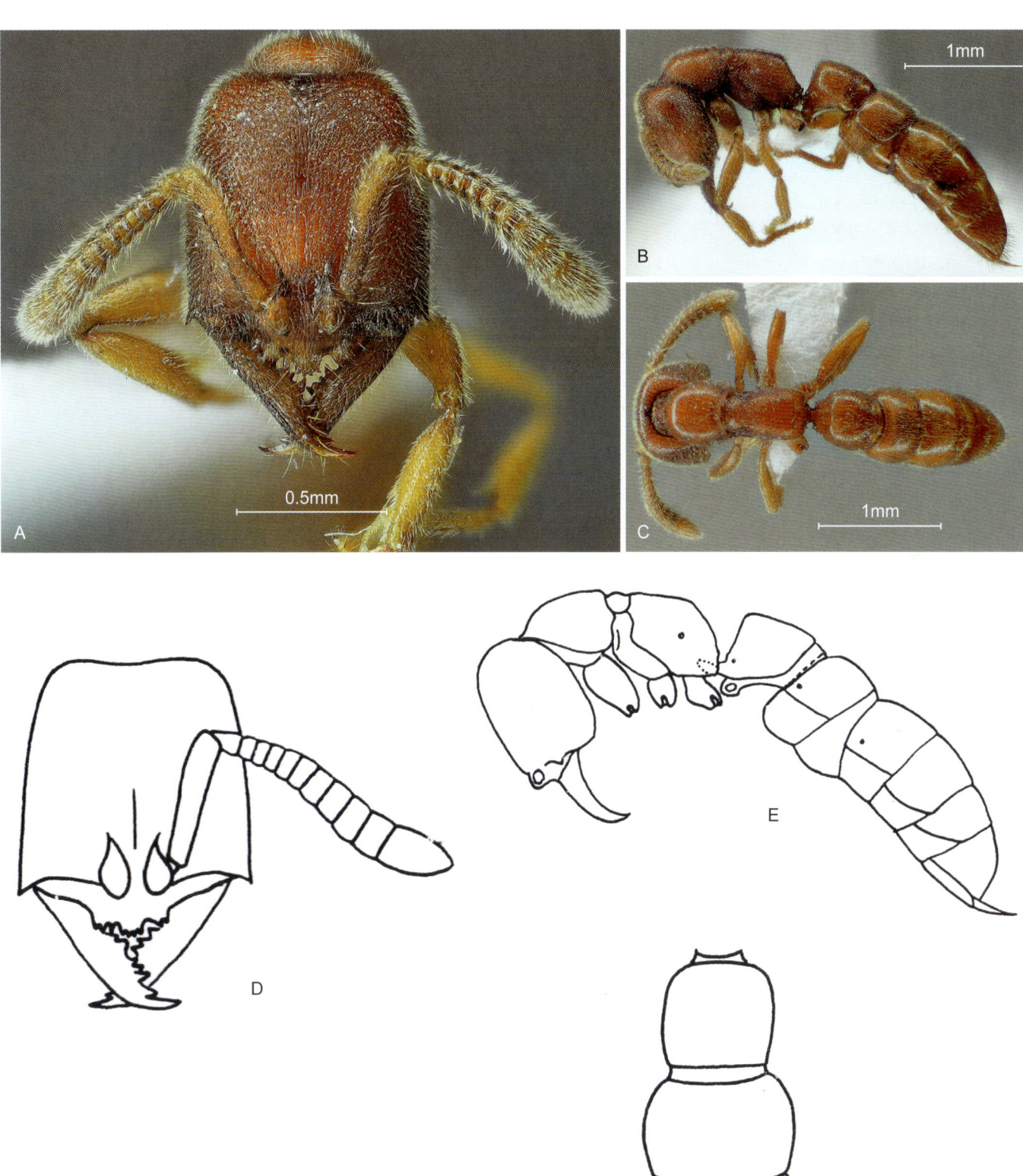

三叶钝猛蚁 *Amblyopone triloba*

A, D. 工蚁头部正面观；B, E. 工蚁身体侧面观；C. 工蚁身体背面观；F. 工蚁腹柄和后腹柄背面观（D-F. 引自Xu, 2001d）

A, D. Head of worker in full-face view; B, E. Body of worker in lateral view; C. Body of worker in dorsal view; F. Petiole and postpetiole of worker in dorsal view (D-F. cited from Xu, 2001d)

卓玛钝猛蚁
Amblyopone zoma Xu & Chu, 2012

Amblyopone zoma Xu & Chu, 2012, Sociobiology, 59(4): 1189, figs. 28-30 (w.) CHINA (Tibet).

Holotype worker: TL 4.5, HL 1.00, HW 0.85, CI 85, SL 0.48, SI 56, ED 0.05, ML 0.68, PW 0.58, AL 1.25, PL 0.48, PH 0.55, DPW 0.55, LPI 116, DPI 100. In full-face view, head roughly trapezoidal, widened forward and longer than broad. Occipital margin widely weakly concave, occipital corners bluntly angled. Sides weakly convex. Anterolateral corners acutely toothed. Mandibles elongate triangular, masticatory margin with a long apical tooth, a short subapical tooth, and 3 pairs of curved teeth; inner margin about as long as masticatory margin, with a pair of curved teeth, a short subbasal tooth, and a large basal tooth. Anterior clypeal margin roughly triangularly protruding, with a large middle lobe, and 3 teeth on each side, the most lateral tooth large and lobe-like; the large middle lobe truncated at apex, with a small denticle on each side. Antennae short, 12-segmented; apices of scapes reached to 4/7 of the distance from antennal sockets to occipital corners; funiculi incrassate toward apex. Eyes small, each with about 6 facets, and located behind the midpoints of the sides of head. In lateral view, pronotum weakly convex. Promesonotal suture deeply notched. Mesonotum short and convex. Metanotal groove absent. Propodeal dorsum straight, about 1.5 times as long as declivity; posterodorsal corner very bluntly angled; declivity nearly straight. Dorsal and anterior faces of petiole nearly straight, anterodorsal corner nearly rightly angled; ventral face oblique and weakly concave; subpetiolar process roughly rectangular, with a large elliptical sub-transparent fenestra, anteroventral corner prominent, ventral face straight, posteroventral corner toothed. Sternite of the first gastral segment ill developed, anteroventral corner toothed, and the segment looks narrower. In dorsal view, mesothorax constricted. Propodeum widened backward. Petiole broader than long, width : length = 1.2 : 1, anterior margin and sides weakly convex. Mandibles longitudinally striate. Head densely punctured, interfaces appear as micro-reticulations. Pronotum densely punctured. Mesonotum, propodeum, petiole, and first gastral segment abundantly punctured. The middle longitudinal narrow strip on alitrunk without punctures. Sides of mesothorax, metathorax, and propodeum longitudinally striate. Gastral segments 2-5 finely sparsely punctured. Dorsal surfaces of head and body with dense subdecumbent pubescence, gaster with sparse suberect hairs and dense subdecumbent pubescence. Scapes with sparse suberect hairs and dense decumbent pubescence. Tibiae with dense decumbent pubescence. Head brown, eyes grey. Body yellowish brown, legs yellow.

Holotype: Worker, China: Tibet Autonomous Region, Medog County, Beibeng Town, Gangouhe, 740 m, 2011.Ⅶ.19, collected by Zheng-hui Xu from a soil sample in the valley tropical rainforest, No. A11-3676.

Etymology: The new species is named after a common female name "Zoma" widely used in Tibet.

正模工蚁： 正面观，头部近梯形，向前变宽，长大于宽；后缘轻度宽形凹陷，后角钝角状；侧缘轻度隆起，前侧角尖齿状。上颚长三角形，咀嚼缘具1个长端齿、1个短的亚端齿和3对弯齿；内缘约与咀嚼缘等长，具1对弯齿、1个短的亚基齿和1个大基齿。唇基前缘大约呈三角形突出，具1个大形中叶，中叶每侧具3个齿，最外侧的齿大且呈叶状；大形中叶顶端平截，每侧具1个小齿。触角短，12节，柄节末端到达触角窝至头后角间距的4/7处，鞭节向顶端变粗。复眼小，每个复眼约具6个小眼，位于头侧缘中点之后。侧面观，前胸背板轻度隆起，前中胸背板缝深切，中胸背板短而隆起，缺后胸沟。并胸腹节背板直，约为斜面长的1.5倍，后上角钝角状，斜面近平直。腹柄背面和前面近平直，前上角近直角形；腹面倾斜，轻度凹陷；腹柄下突近长方形，具1个大的椭圆形窗斑，前下角突出，腹面直，后下角齿状。后腹部第1节腹板不发达，前下角具齿，整个腹节看上去较窄。背面观，中胸收缩，并胸腹节向后变宽。腹柄宽大于长，宽：长=1.2：1，前缘和侧缘轻度隆起。上颚具纵条纹。头部具密集刻点，界面呈微网状。前胸背板具密集刻点。中胸背板、并胸腹节、腹柄和后腹部第1节具丰富刻点。胸部背面的狭窄中央纵带无刻点。中胸侧面、后胸侧面和并胸腹节侧面具纵条纹。后腹部第2~5节具稀疏细刻点。头部和身体背面具密集的亚倾斜绒毛被，后腹部具稀疏的亚直毛和密集的亚倾斜绒毛被，柄节具稀疏的亚直立毛和密集的倾斜绒毛被，胫节具密集的倾斜绒毛被。头部棕色，复眼灰色，身体黄棕色，足黄色。

正模： 工蚁，中国西藏自治区墨脱县背崩乡干沟河，740m，2011.Ⅶ.19，徐正会采于沟谷热带雨林土壤样中，No. A11-3676。

词源： 该新种以西藏广泛使用的一个女性名字"卓玛"命名。

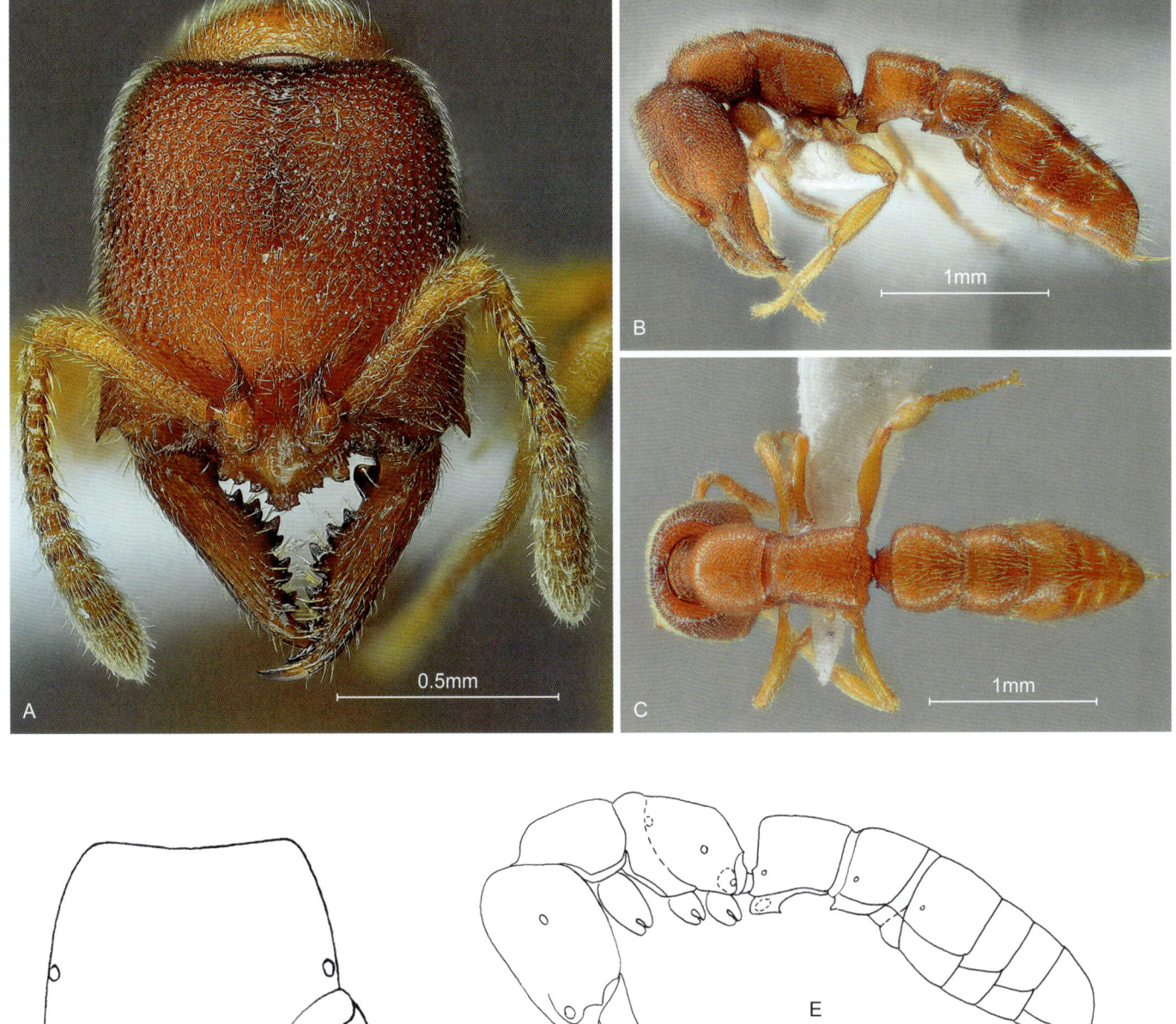

卓玛钝猛蚁 *Amblyopone zoma*

A, D. 工蚁头部正面观; B, E. 工蚁身体侧面观; C, F. 工蚁身体背面观（D-F. 引自Xu和Chu, 2012）
A, D. Head of worker in full-face view; B, E. Body of worker in lateral view; C, F. Body of worker in dorsal view (D-F. cited from Xu & Chu, 2012)

木兰版纳猛蚁
Bannapone mulanae Xu, 2000

Bannapone mulanae Xu, 2000a, Zoological Research, 21(4): 301, figs. 10-12 (q.) CHINA (Yunnan).

Holotype dealate female: TL 2.1, HL 0.38, HW 0.32, CI 84, SL 0.20, SI 63, PW 0.28, AL 0.58, ED 0.08, ML 0.30, PL 0.14, PH 0.19, DPW 0.20, LPI 136, DPI 105. In full-face view, head rectangular and depressed, longer than broad, slightly broader anteriorly. Occipital margin weakly emarginate in the middle. Occipital corners roundly prominent. Sides nearly straight. Mandibles long triangular, inner margin distinctly shorter than masticatory margin. Masticatory margin with 3 teeth, the apical tooth long and acute, bifid at most apex, then followed by 2 blunt finger-like teeth, which slightly expended at apex. Clypeus very narrow and transverse, anterior margin weakly convex. Antennae short, 11-segmented, apex of scape reached to 1/2 of the distance from antennal socket to occipital corner, funiculi 3-9 broader than long. Eyes moderately large, situated at the mid-length of the side of the head. Ocelli present. Alitrunk with full complement of flight sclerites and certainly winged when virgin. In lateral view dorsum of alitrunk evenly convex. Dorsum of propodeum slightly depressed, posterodorsal corner rounded, declivity depressed vertically. In lateral view petiolar node roughly rectangular and higher than long, broadly attached to first gastral segment, anterodorsal corner roundly prominent. Subpetiolar process narrow, cuneiform. Gaster large, roughly cylindrical, segments with the similar length, constriction between the 2 basal segments indistinct. Sting long and functional. Mandibles finely rugulose, less shining. Head, alitrunk, petiole and gaster densely and finely punctured and relatively dull, gaster weakly punctured and more shining. Dorsum of head and body, and appendages with abundant decumbent pubescence, erect hairs very sparse and absent on scapes and tibiae. Body in color yellowish brown. Eyes, ocelli area and tegulae black.

Holotype: Dealate female, China: Yunnan Province, Mengla County, Shangyong Town, Manzhuang Village, 900 m, collected by Tai-yong Liu from a soil sample in semi-evergreen monsoon forest, 1998.Ⅲ.10, No. A98-418.

Etymology: The new species is named after the ancient Chinese legend heroine "Mu-lan Hua" of the Northern and Southern Dynasties (A. D. 420-589).

正模脱翅雌蚁： 正面观，头部长方形，扁平，长大于宽，向前稍变宽；后缘中部轻度凹入，后角圆突；侧缘近平直。上颚长三角形，内缘明显短于咀嚼缘；咀嚼缘具3个齿，端齿长而尖锐，在最顶端分叉；随后具2个钝的指状齿，其顶端轻度膨大。唇基很窄，横向，前缘轻度隆起。触角短，11节，柄节末端到达触角窝至头后角间距的1/2处，鞭节3～9节宽大于长。复眼中等大小，位于头侧缘中点位置；具单眼。胸部具完整的飞行骨片，交尾前具翅。侧面观，胸部背面适度隆起，并胸腹节背面轻度凹陷，后上角圆，斜面垂直凹陷。腹柄结近长方形，高大于长，与后腹部第1节宽阔连接，前上角圆突；腹柄下突狭窄，楔形。后腹部大，近圆柱形，各腹节长度相似，基部2节间收缩不明显。螯针长，具防卫功能。上颚具细皱纹，不太光亮。头部、胸部、腹柄和后腹部具密集细刻点，较暗；后腹部刻点较弱，较光亮。头部背面、身体和附肢具丰富倾斜绒毛被，立毛很稀疏，柄节和胫节缺立毛。身体黄棕色，复眼、单眼区和翅基片黑色。

正模： 脱翅雌蚁，中国云南省勐腊县尚勇乡曼庄村，900m，1998.Ⅲ.10，柳太勇采于半常绿季雨林土壤样中，No. A98-418。

词源： 该新种以中国古代传说中的南北朝时期（公元420—589年）巾帼英雄花木兰命名。

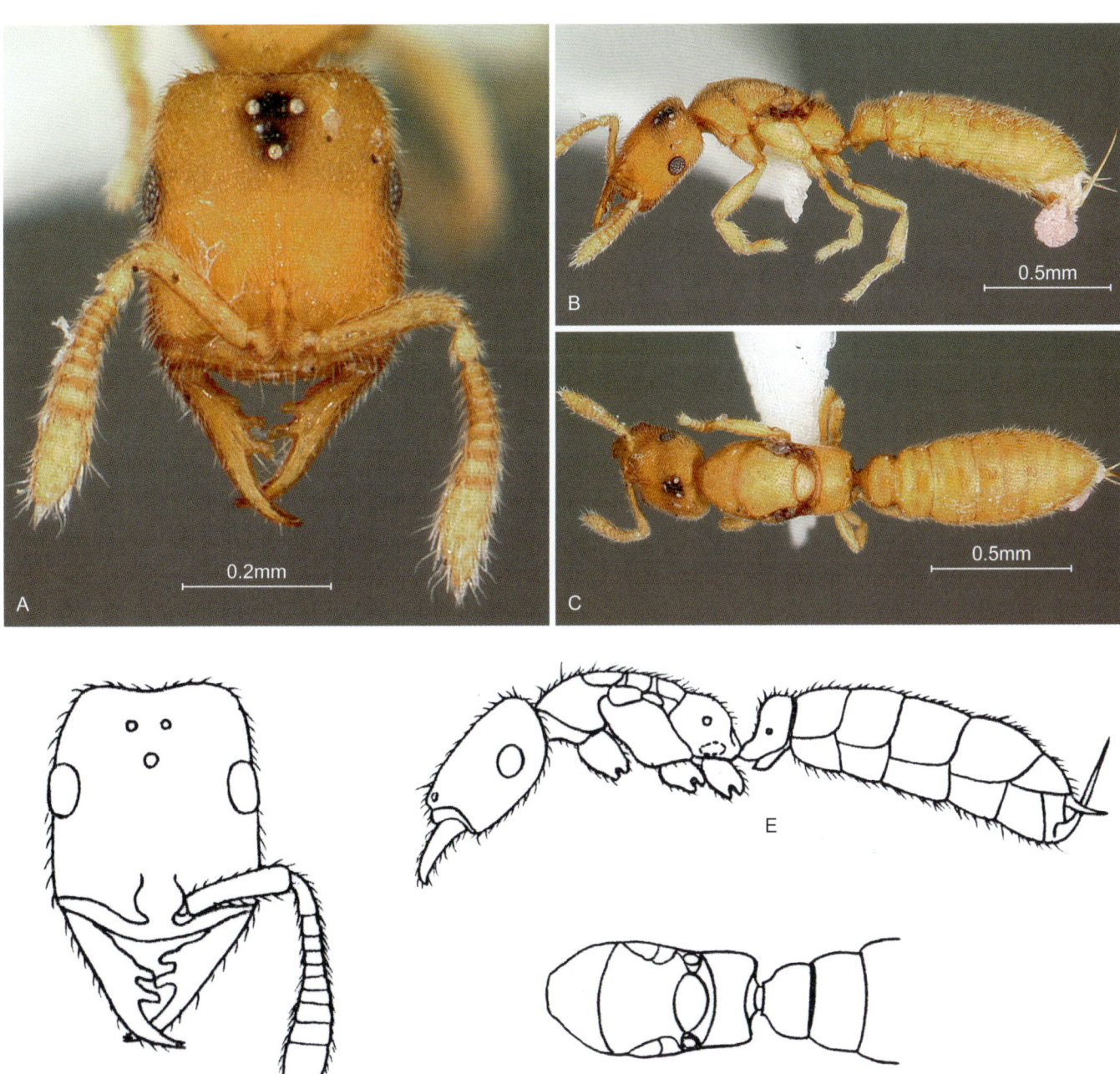

木兰版纳猛蚁 *Bannapone mulanae*

A, D. 脱翅雌蚁头部正面观；B, E. 脱翅雌蚁身体侧面观；C, F. 脱翅雌蚁身体背面观（D-F. 引自Xu, 2000a）
A, D. Head of dealate female in full-face view; B, E. Body of dealate female in lateral view; C, F. Body of dealate female in dorsal view (D-F. cited from Xu, 2000a)

小眼迷猛蚁
Mystrium oculatum Xu, 1998

Mystrium oculatum Xu, 1998a, Zoological Research, 19(2): 161, figs. 1, 2 (w.) CHINA (Yunnan).

Holotype worker: TL 4.7, HL 1.25, HW 1.33, CI 106, SL 0.75, SI 57, PW 0.68, AL 1.25, ED 0.05, ML 1.35. In full-face view, head slightly wider than long, much broader in front than in the post. Occipital margin deeply and roundly emarginate. Occipital corners extruding. Anterior 2/3 of the lateral margin straight, the posterior 1/3 narrowed posteriorly. Anterolateral corner of head produced into an acute spine. Central dorsum of head flat. Mandibles long and linear, the basal 2/3 straight, apical 1/3 incurved, apex slightly expanded; dorsum has a longitudinal carina; inner margin with 2 rows of hamulus-like denticles, each row with about 12 denticles; apical tooth hooked, ventrally curved. Central portion of clypeus flat, depressed; anterior margin roundly extruded in the middle, with a row of teeth; lateral portion of clypeus extended into a blunt angle at the base of mandible. Frontal lobes small, frontal carinae absent. Antennae with 12 joints: scape falling short of the occipital corner by about 1/3 of its length; flagellum incrassate towards apex, the apical 4 joints formed a weak club. Eyes very small, situated in the lateral margin and behind the middle line of the head, each with 5-6 ommatidia. In lateral view, dorsum of alitrunk about at the same level, promesonotal suture wide and deeply depressed; metanotal groove narrow, shallowly depressed. Dorsum of propodeum short, weakly convex, and formed a blunt angle with the declivity; declivity long and truncate, slope, about 2 times as long as dorsum. In lateral view, petiolar node rectangular, higher than long, articulated to the gaster by whole of its posterior face; anterior face truncate, dorsal face weakly convex and constricted near the posterior margin; subpetiolar process narrow and long, anteroventrally pointed, blunt at apex. In dorsal view, pronotum and propodeum broad, mesonotum constricted. Petiolar node transverse, long elliptic, about 2 times as broad as long. Constriction between the 2 basal segments of gaster weak, but obvious. Outer faces of mandibles longitudinally rugose. Head, dorsum of alitrunk, dorsum of petiolar node, and dorsum of first gastral tergum coarsely reticulate; lateral faces of alitrunk, posterior face of propodeum, anterior and lateral faces of petiolar node, and gaster finely and densely punctulate, punctures of gaster relatively weaker. Head, alitrunk, petiole, gaster, mandibles, antennae, and legs with abundant decumbent short clavate setae, apical 4 joints of antenna, ventral or inner faces of legs, and apex of gaster with normal subdecumbent hairs. Colour yellowish brown, eyes and apices of mandibles black. **Paratype workers:** TL 4.1-4.7, HL 1.03-1.25, HW 1.13-1.33, CI 106-112, SL 0.68-0.75, SI 57-60, PW 0.63-0.68, AL 1.13-1.25, ED 0.05-0.06, ML 1.03-1.35 (*n*=5). As holotype.

Holotype: Worker, China: Yunnan Province, Mengla County, Menglun Town, Bakaxiaozhai (21.9° N, 101.2° E), 840 m, 1996.Ⅲ.8, collected by Zheng-hui Xu from a ground sample of seasonal rain forest, No. A96-318.

Etymology: The species name *oculatum* combines Latin *ocul-* (eye) + suffix *-atum* (feminine form *-ata*, with), it refers to the small eyes on the head.

正模工蚁： 正面观，头宽稍大于长，前部明显宽于后部；后缘圆形深凹，后角突出；侧缘前部2/3平直，后部1/3向后变窄；前侧角突出形成锐刺。头部背面中央平坦。上颚长，线形，基部2/3直，顶端1/3弯曲，顶端稍膨大；背面具1条纵脊，内缘具2列钩状小齿，每列约具12个齿；端齿钩状，弯向腹面。唇基中部平坦，凹陷；前缘中部圆形突出，具1列齿；唇基两侧在上颚基部延伸成钝角状。额叶小，缺额脊。触角12节，柄节末端至头后角间距为柄节长的1/3；鞭节向顶端变粗，端部4节形成弱的触角棒。复眼很小，位于头侧缘中线之后，每个复眼具5～6个小眼。侧面观，胸部背面约在同一水平上，前中胸背板缝宽且深凹，后胸沟窄而浅凹。并胸腹节背部短，轻度隆起，与斜面形成钝角；斜面长，平截，坡形，约为背面长的2倍。腹柄结长方形，高大于长，与后腹部宽阔连接，前面平截，背面轻度隆起，在后缘附近收缩；腹柄下突窄而长，指向前下方，顶端钝。背面观，前胸背板和并胸腹节宽，中胸背板收缩。腹柄结横形，长椭圆形，宽约为长的2倍。后腹部基部2节间轻度收缩，但是明显。上颚外面具纵皱纹。头部、胸部背面、腹柄结背面和后腹部第1节背面具粗糙网纹；胸部侧面、并胸腹节后面、腹柄结前面和侧面以及后腹部具细密刻点，后腹部刻点较弱。头部、胸部、腹柄、后腹部、上颚、触角和足具丰富倾斜棒状短毛；触角端部4节、足的腹面或内侧以及后腹部端部具正常的亚倾斜毛。身体黄棕色，复眼和上颚顶端黑色。**副模工蚁：** 特征同正模。

正模： 工蚁，中国云南省勐腊县勐仑镇巴卡小寨（21.9°N，101.2°E），840m，1996.Ⅲ.8，徐正会采于季

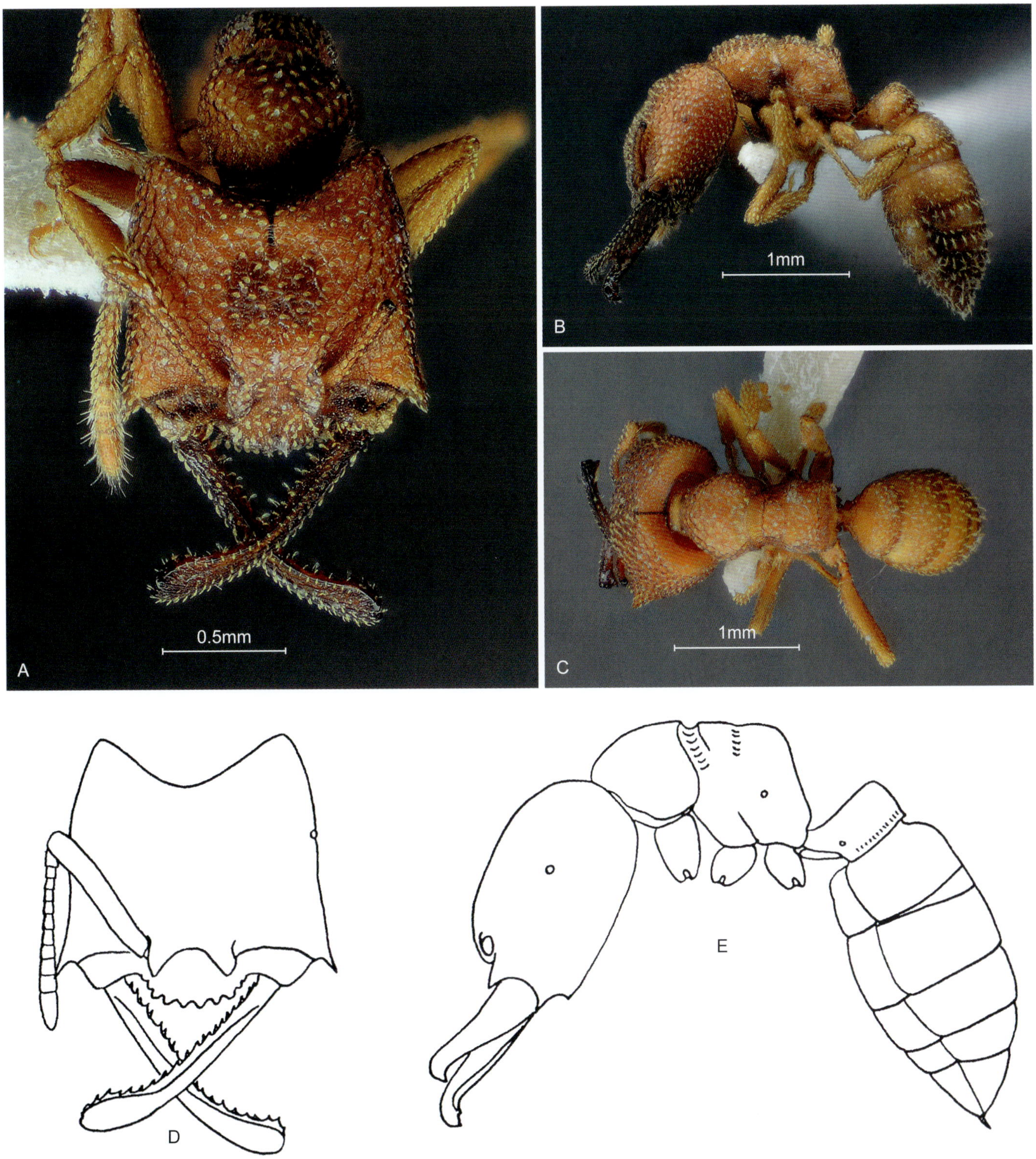

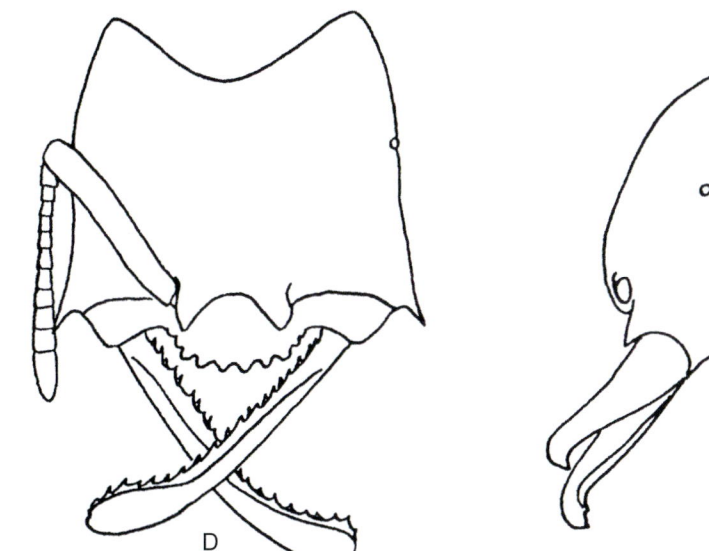
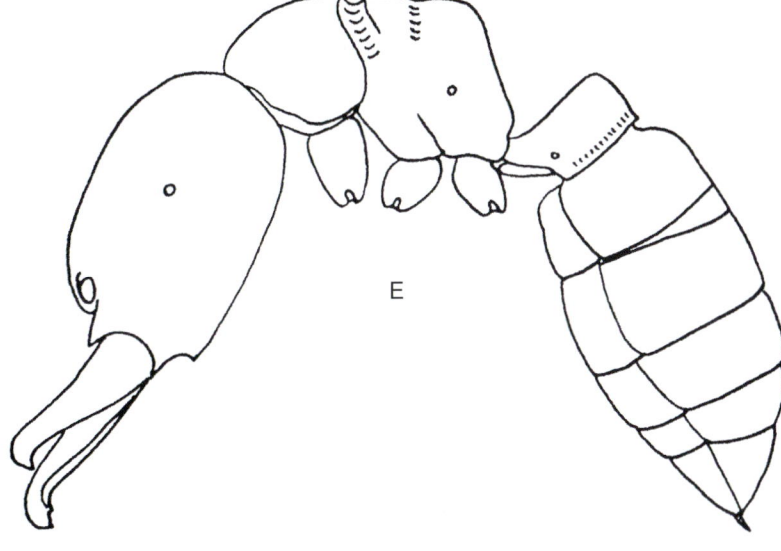

小眼迷猛蚁 Mystrium oculatum

A, D. 工蚁头部正面观；B, E. 工蚁身体侧面观；C. 工蚁身体背面观（D-E. 引自Xu, 1998a）
A, D. Head of worker in full-face view; B, E. Body of worker in lateral view; C. Body of worker in dorsal view (D-E. cited from Xu, 1998a)

节性雨林地表样中，No. A96-318。

词源： 该新种的种名"*oculatum*"由拉丁语"*ocul-*"（眼睛）+后缀"*-atum*"（阴性形式"*-ata*"，具有）组成，指头部具小的复眼。

版纳曲颊猛蚁
Gnamptogenys bannana Xu & Zhang, 1996

Gnamptogenys bannana Xu & Zhang, 1996, Entomotaxonomia, 18(1): 55, figs. 1-7 (w.) CHINA (Yunnan).

Holotype worker: TL 6.0, HL 1.38, HW 1.13, CI 82, SL 1.18, SI 104, PW 1.00, AL 1.98, ED 0.20, ML 0.70, NL 0.60, NW 0.68. In full-face view, head longer than broad, narrowing anteriorly, the occipital corners elongate posteriorly into lobe-like processes, the occipital margin deeply concave. Eyes large and prominent, situated behind the middle line of the head. Mandibles elongately triangular, finely longitudinally rugose, with the apex strongly curving down, the masticatory margin with numerous minute teeth. Clypeus low and flat, the anterior margin bluntly angled. Frontal area weakly defined. Frontal carinae strong, expanding laterally and covering the insertions of the antennae, parallel with each other and extending backward to the anterior margins of the eyes, with the ends incurved. Antennae with 12 segments, scapes extending backward just to the apices of the elongate occipital corners, flagellum incrassate toward apex. In lateral view, the dorsal surface of alitrunk forming a complete arch, lowering down posteriorly. Promesonotal suture and metanotal groove inconspicuous on the dorsal surface. Pronotum forming a groove on each side of its anterior face to accept the elongate lobes of the occipital corners. Propodeum with a pair of blunt teeth at apex, the apical face depressed. Hind coxa with a strong blunt spine on the upper surface. Each tibia with a pectinate spur, claw with a subbasal tooth. Petiolar node low, paniform; subpetiolar process large, lamelliform, with a tooth-like anteroventral process; in dorsal view the node broader than long, narrowing forward. Gaster with a curved transverse ridge anteroventrally; the two basal segments very large, with distinct constriction between them; the second one strongly curved downward; sting exerted. Head, alitrunk, petiole and first gastric segment with coarse foveolae; clypeus longitudinally striate; anterior dorsal surface of the head with coarse striations. Central portion of the first, and the other gastric segments smooth and shining; a few foveolae are present on the lateral surfaces of the second segment. Dorsal surfaces of head and body with abundant erect or suberect hairs. Flagellum of antennae, margins of alitrunk and petiole, coxae, and apex of gaster with dense pubescence. Dorsal surfaces of scapes and hind tibiae with abundant suberect hairs; furthermore, a few suberect long hairs present on the scapes. Head, femurs and tibiae of legs, and gaster black; anterior portion of head, mandibles, antennae, alitrunk, coxae and tarsi of legs, petiole and apex of the gaster reddish brown. **Paratype workers:** TL 4.6-6.3, HL 1.23-1.50, HW 0.96-1.20, CI 77-82, SL 1.03-1.20, SI 100-108, PW 0.83-1.08, AL 1.68-2.03, ED 0.26-0.30, ML 0.65-0.75, NL 0.48-0.60, NW 0.54-0.70 (n=5). As holotype.

Holotype: Worker, China: Yunnan Province, Mengla County, Menglun Town, Menglun (21.9°N, 101.2°E), 560 m, 1989.Ⅵ.15, collected by Yao Niu, No. A89-1.

Etymology: The new species is named after the abbreviation "Banna" of the type locality "Xishuangbanna", where Menglun Town belonging to.

正模工蚁： 正面观，头长大于宽，向前变窄；后角向后伸长成叶状突；后缘深凹。复眼大而突起，位于头中线之后。上颚长三角形，具细纵皱纹，顶端强烈下弯，咀嚼缘具众多细齿。唇基低而平坦，前缘钝角状。额区边界不明显。额脊发达，向侧面扩展，遮盖触角插入部，互相平行，向后延伸至复眼前缘，末端内弯。触角12节，柄节末端刚到达延伸的头后角顶端，鞭节向端部变粗。侧面观，胸部背面形成完整的弓形，向后降低；前中胸背板缝和后胸沟在背面不明显。前胸背板两侧前面各形成1个凹槽，以接纳头后角延伸的叶状部。并胸腹节后端具1对钝齿，斜面凹陷。后足基节背面具1个发达的粗刺，胫节各具1个梳状距，爪具1个亚基齿。腹柄结低，半球形；腹柄下突大，片状，前下角具齿状突。背面观，腹柄结宽大于长，向前变窄。后腹部前下角具弯曲的横脊；基部2节很大，节间明显收缩；第2节强烈下弯，螫针伸出。头部、胸部、腹柄和后腹部第1节具粗糙凹坑；唇基具纵条纹；头部背面前部具粗条纹。后腹部第1节背面中部和其他腹节光滑发亮；第2节侧面具少数凹坑。头部和身体背面具丰富立毛或亚直立毛；触角鞭节、胸部和腹柄边缘、基节和后腹部端部具密集绒毛被；柄节和后足胫节表面具丰富亚直立毛；此外，柄节具少数亚直立长毛。头部、足的腿节和胫节以及后腹部黑色；头前部、上颚、触角、胸部、足的基节和跗节、腹柄和后腹部末端红棕色。**副模工蚁：** 特征同正模。

正模： 工蚁，中国云南省勐腊县勐仑镇勐仑（21.9°N，101.2°E），560m，1989.Ⅵ.15，牛瑶采集，No. A89-1。

词源： 该新种以模式产地"西双版纳"的缩写"版纳"命名，勐仑镇属于该地区。

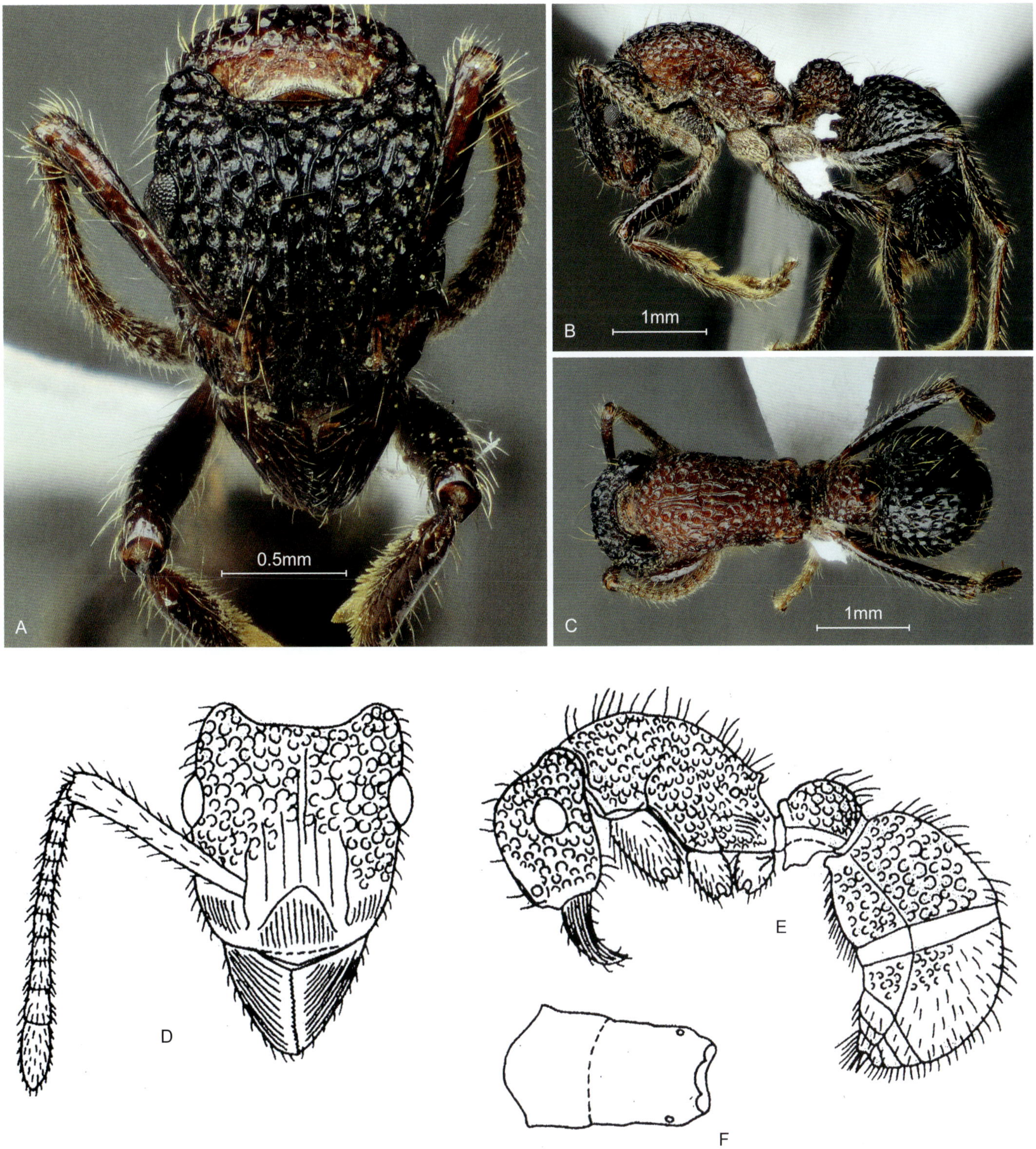

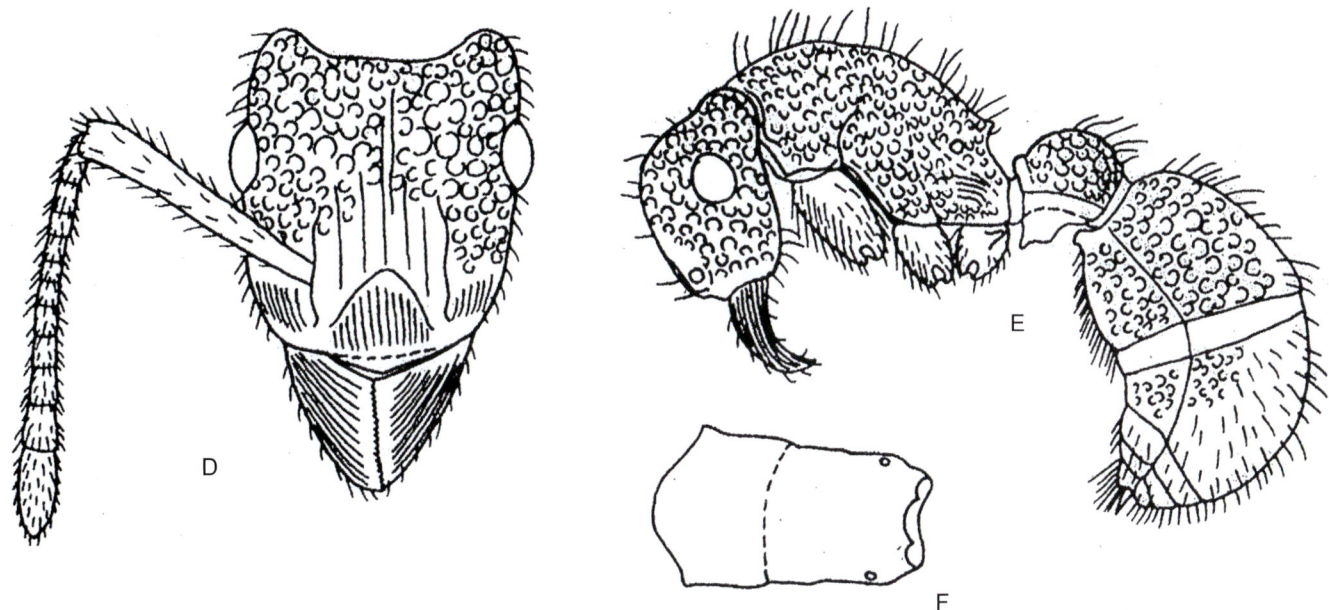

版纳曲颊猛蚁 *Gnamptogenys bannana*

A, D. 工蚁头部正面观；B, E. 工蚁身体侧面观；C. 工蚁身体背面观；F. 工蚁胸部背面观（D-F. 引自Xu和Zhang, 1996）
A, D. Head of worker in full-face view; B, E. Body of worker in lateral view; C. Body of worker in dorsal view; F. Alitrunk of worker in dorsal view (D-F. cited from Xu & Zhang, 1996)

版纳盘猛蚁

Discothyrea banna Xu, Burwell & Nakamura, 2014

Discothyrea banna Xu et al., 2014a, Asian Myrmecology, 6: 35, figs. 1-6 (w.) CHINA (Yunnan).

Holotype worker: TL 2.8, HL 0.78, HW 0.68, CI 87, SL 0.55, SI 81, ED 0.10, PW 0.53, MSL 0.83, PL 0.20, PH 0.38, DPW 0.30, LPI 188, DPI 150. In full-face view, head longer than broad, roughly trapezoidal and narrowed anteriorly. Posterior margin nearly straight, posterolateral corners rounded. Sides evenly convex, deeply notched in front of mandibular insertions. Mandibles triangular, masticatory margin edentate, apical tooth acute. Anterior margin of clypeus weakly convex. Frontal lobes confused each other and formed a large roughly rhombic frontal area, distinctly longer than broad, lateral corners bluntly angled, anterolateral margins straight, posterolateral margins weakly concave. Frontal carinae relatively long and well separated, extending posteriorly to 1/2 of head-length. Antennae 9-segmented, apices of scapes reaching to 2/3 of distance from antennal sockets to posterolateral corners. Eyes moderately large and convex, with 6 ommatidia on maximum diameter, located at mid-length of sides. In posterior view, frontal area weakly elevated and roughly rhombic, with a broad base, anterior margin weakly convex, lateral corners acutely angled. In lateral view, dorsum of mesosoma weakly convex and sloping down posteriorly, promesonotal suture and metanotal groove absent. Posterodorsal corner of propodeum rounded, declivity weakly concave. Propodeal lobes large and rounded. Petiolar node high, roughly triangular and narrowed upward, anterior and dorsal faces forming a single arch, strongly convex, anterodorsal corner indistinct. Subpetiolar process short, very bluntly angled ventrally and translucent. First gastral segment large, about 2/3 of the total length of gaster. Constriction between the two basal gastral segments broad and deep. In dorsal view, mesosoma trapezoidal and narrowed posteriorly, humeral corners bluntly angled, lateral margins nearly straight, posterior margin weakly concave medially. Petiolar node transverse and trapezoidal, length : width = 1 : 1.5, widening posteriorly, anterior margin nearly straight, lateral and posterior margins straight, anterolateral corners bluntly angled. Mandibles finely punctured. Head, mesosoma, petiole, and first gastral segment densely and coarsely punctured, interface much smaller than puncture diameter. Punctures on first gastral segment even larger, and foveolate on its sides. Rest gastral segments finely punctured and relatively shining. Whole body covered with dense, suberect to subdecumbent short pubescence, without standing hairs. Scapes and tibiae with dense decumbent short pubescence, without standing hairs. Colour reddish brown. Apical antennal segments, legs, and gastral apex yellowish brown. Eyes black.

Paratype workers: TL 2.5-2.9, HL 0.78-0.85, HW 0.65-0.70, CI 81-85, SL 0.53-0.58, SI 79-85, ED 0.09-0.10, PW 0.48-0.58, MSL 0.80-0.90, PL 0.20-0.25, PH 0.35-0.40, DPW 0.28-0.35, LPI 156-178, DPI 122-163 (*n*=10). As holotype.

Holotype: Worker, China: Yunnan Province, Mengla County, Shangyong Town, Manzhuang Village (21.449722°N, 101.736389°E), 980 m, 2012.Ⅲ.23, collected by Wen-xia Cui from a ground sample in semi-evergreen monsoon forest, No. A12-311.

Etymology: The specific epithet refers to "Banna", an abbreviation of Xishuangbanna, the type locality of this species.

正模工蚁： 正面观，头长大于宽，近梯形，向前变窄；后缘近平直，后角圆；侧缘均匀隆起，在上颚基部前面深切。上颚三角形，咀嚼缘缺齿，端齿尖锐。唇基前缘轻度隆起。额叶互相融合，形成一个大的近菱形的额区，长明显大于宽，侧角钝角状，前侧缘直，后侧缘轻度凹陷。额脊较长，互相远离，向后延伸至头长的1/2处。触角9节，柄节顶端到达触角窝至头后角间距的2/3处。复眼中等大，隆起，最大直径上具6个小眼，位于头侧缘中部。后面观，额区轻度升高，近菱形，基部宽，前缘轻度隆起，侧角锐角状。侧面观，胸部背面轻度隆起，向后降低，前中胸背板缝和后胸沟缺失。并胸腹节后上角圆，斜面轻度凹陷。后侧叶大，端部圆。腹柄结高，近三角形，向上变窄，前面和背面形成单一弓形，强烈隆起，前上角不明显；腹柄下突短，腹面成极钝的角状，半透明。后腹部第1节大，约占后腹部总长度的2/3；基部2节间缢缩宽且深。背面观，胸部呈梯形，向后变窄，肩角钝角状，侧缘近平直，后缘中央轻度凹陷。腹柄结横形，呈梯形，长：宽=1：1.5，向后变宽，前缘近平直，侧缘和后缘直，前侧角钝角状。上颚具细刻点。头部、胸部、腹柄和后腹部第1节具密集粗刻点，界面远小于刻点直径；后腹部第1节刻点更大，在侧面呈凹坑状；其余腹节具细刻点，较光亮。全身具密集的亚直立或亚倾斜绒毛被，缺立毛。柄节和胫节具密集倾斜短绒毛被，缺立毛。身体红棕色，触角端部几节、足和后腹部末端黄棕色，复眼黑色。**副模工蚁：** 特征同正模。

正模： 工蚁，中国云南省勐腊县尚勇乡曼庄村（21.449722°N，101.736389°E），980m，2012.Ⅲ.23，崔

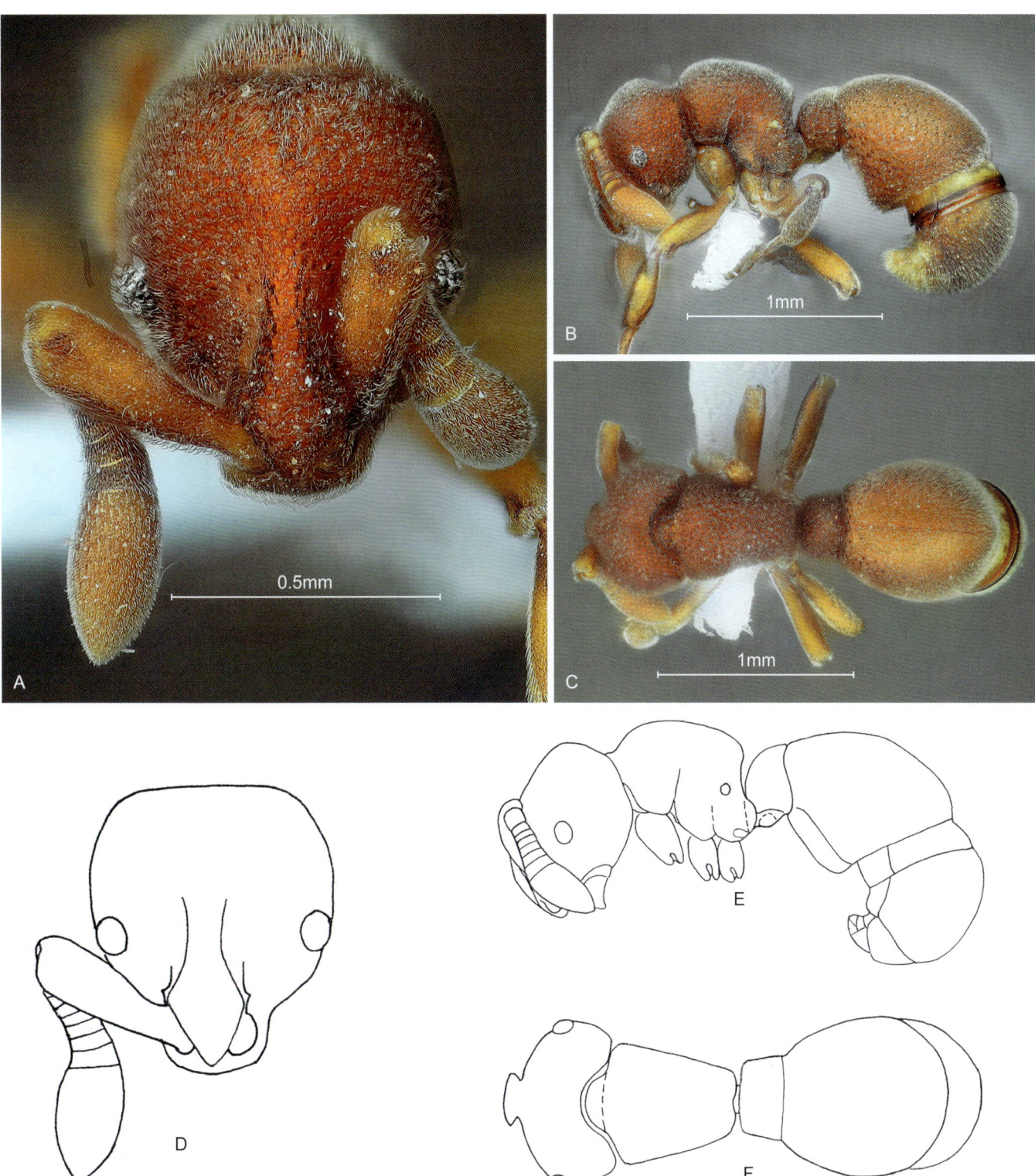

版纳盘猛蚁 *Discothyrea banna*

A, D. 工蚁头部正面观；B, E. 工蚁身体侧面观；C, F. 工蚁身体背面观（D-F. 引自Xu等, 2014a）

A, D. Head of worker in full-face view; B, E. Body of worker in lateral view; C, F. Body of worker in dorsal view (D-F. cited from Xu et al., 2014a)

文夏采于半常绿季雨林地表样中，No. A12-311。

词源： 种加词指"版纳"，是该种的模式产地西双版纳的缩写。

滇盘猛蚁
Discothyrea diana Xu, Burwell & Nakamura, 2014

Discothyrea diana Xu et al., 2014a, Asian Myrmecology, 6: 38, figs. 7-12 (w.) CHINA (Yunnan).

Holotype worker: TL 1.8, HL 0.58, HW 0.50, CI 87, SL 0.33, SI 65, ED 0.05, PW 0.40, MSL 0.58, PL 0.10, PH 0.26, DPW 0.21, LPI 263, DPI 213. In full-face view, head longer than broad, roughly trapezoidal, narrowed anteriorly. Posterior margin weakly concave medially, posterolateral corners rounded. Sides evenly convex, deeply notched in front of mandibular insertions. Mandibles triangular, masticatory margins edentate, apical tooth acute. Anterior margin of clypeus weakly convex, anterolateral corners rounded. Frontal lobes fused each other and formed a roughly triangular area, about as broad as long, lateral corners acutely angled, anterolateral margins weakly convex. Frontal carinae short and close to each other, extending posteriorly to about 1/3 of head length. Antennae 7-segmented, apices of scapes reaching to 3/5 of distance from antennal sockets to posterolateral corners. Eyes small and convex, consisting of about 6 ommatidia, located slightly in front of mid-length of sides. In posterior view, frontal area strongly elevated and rectangular, with a narrow base, anterior and lateral margins nearly straight. In lateral view, dorsum of mesosoma evenly convex, and sloping down posteriorly, promesonotal suture and metanotal groove absent. Posterodorsal corner of propodeum rightly angled, declivity evenly concave. Propodeal lobe large and rounded. Petiolar node short and low, dorsal face sloping down anteriorly and weakly concave, anterodorsal corner bluntly angled. Subpetiolar process large, roughly triangular, ventrally pointed and translucent. First gastral segment about 2/3 of the total length of gaster. Constriction between the two basal gastral segments narrow and weak. In dorsal view, mesosoma roughly trapezoidal, narrowed posteriorly, humeral corners bluntly angled, lateral margins straight, propodeum more strongly narrowed, posterior margin weakly concave, posterolateral corners bluntly angled. Petiolar node transverse and rectangular, length : width = 1 : 3.5, transversely depressed, anterior margin nearly straight, posterior margin weakly concave, anterolateral corners right-angled. Head, mesosoma, petiole, and first gastral segment densely and coarsely punctured, interface much smaller than puncture diameter, punctures on sides of head and first gastral segment even larger. Rest of gastral segments densely finely punctured. Whole body covered with dense subdecumbent to decumbent short pubescence, without standing hairs. Scapes and tibiae with dense decumbent pubescence, without standing hairs. Colour reddish brown. Apical antennal segments, legs, and gastral apex yellowish brown. Eyes black. **Paratype workers:** TL 1.8-1.9, HL 0.55-0.58, HW 0.48-0.50, CI 86-91, SL 0.30-0.33, SI 60-68, ED 0.04-0.05, PW 0.35-0.40, MSL 0.55-0.63, PL 0.09-0.11, PH 0.25-0.28, DPW 0.21-0.23, LPI 244-286, DPI 200-243 (n=8). As holotype, but color reddish brown to yellowish brown, apical antennal segments, legs, and gastral apex yellowish brown to yellowish.

Holotype: Worker, China: Yunnan Province, Mengla County, Mengla Town, Bubang Village, QCAS site 800-2 (21.613°N, 101.578°E), 741 m, 2012.Ⅶ, collected by Akihiro Nakamura & Chris J. Burwell in rainforest using Berlese method, No. A12-1812.

Etymology: The specific epithet refers to an abbreviation for Yunnan Province "Dian", the type locality of this species.

正模工蚁：正面观，头长大于宽，近梯形，向前变窄；后缘中部轻度凹陷，后角圆；侧缘均匀隆起，在上颚基部前面深切。上颚三角形，咀嚼缘缺齿，端齿尖锐。唇基前缘轻度隆起，前侧角圆。额叶互相融合，形成1个近三角形的区域，长宽相当，侧角呈锐角，前侧缘轻度隆起。额脊短，彼此靠近，向后延伸至头长的1/3处。触角7节，柄节末端到达触角窝至头后角间距的3/5处。复眼小而隆起，由约6个小眼组成，位于头侧缘中点稍前处。后面观，额区强烈升高呈长方形，基部狭窄，前缘和侧缘近平直。侧面观，胸部背面均匀隆起，向后降低，前中胸背板缝和后胸沟缺失；并胸腹节后上角直角形，斜面均匀凹陷。后侧叶大，端部圆。腹柄结短而低，背面向前坡形降低且轻度凹陷，前上角钝角状；腹柄下突大，近三角形，指向腹面，半透明。后腹部第1节约占后腹部总长度的2/3，基部2节间缢缩窄且弱。背面观，胸部近梯形，向后变窄，肩角钝角状，侧缘直；并胸腹节更强烈地变窄，后缘轻度凹陷，后侧角钝角状。腹柄结横形，呈长方形，长：宽=1：3.5，背面横向凹陷，前缘近平直，后缘轻度凹陷，前侧角呈直角。头部、胸部、腹柄和后腹部第1节具密集粗糙刻点，界面远小于刻点直径；头侧缘和后腹部第1节侧缘刻点较大，后腹部其余腹节具密集细刻点。全身具密集亚倾斜至倾斜绒毛被，缺立毛；柄节和胫节具密集倾斜绒毛被，缺立毛。身体红棕色，触角端部几节、足和后腹部末端黄棕色，复眼黑色。**副模工蚁：**特征同正模，但身体为红棕色至黄棕色，触角端部几节、足和后腹部末端黄棕色至淡黄色。

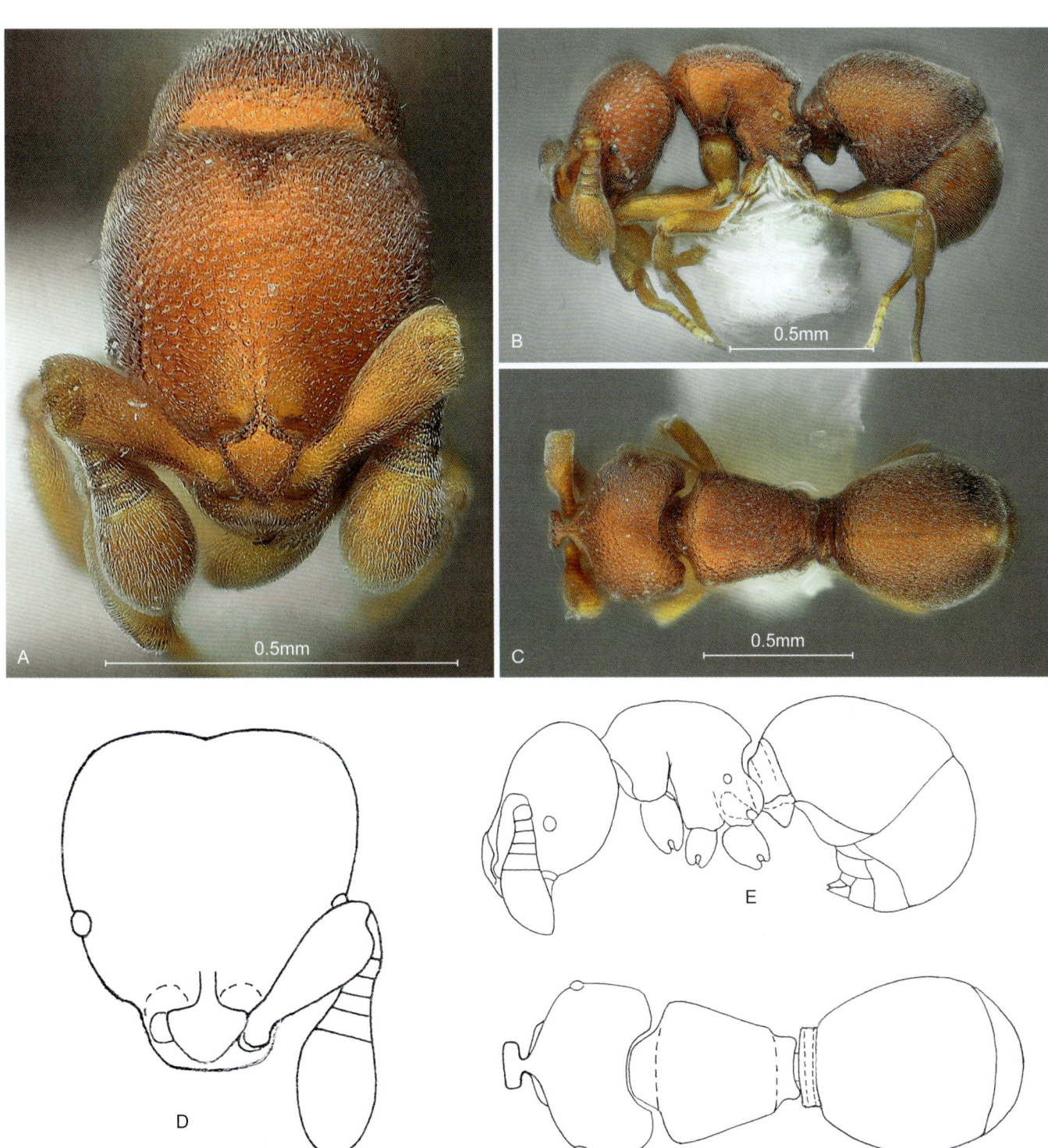

滇盘猛蚁 *Discothyrea diana*

A, D. 工蚁头部正面观；B, E. 工蚁身体侧面观；C, F. 工蚁身体背面观（D-F. 引自Xu等，2014a）
A, D. Head of worker in full-face view; B, E. Body of worker in lateral view; C, F. Body of worker in dorsal view (D-F. cited from Xu et al., 2014a)

正模： 工蚁，中国云南省勐腊县勐腊镇补蚌村，QCAS样点800-2（21.613°N，101.578°E），741m，2012.Ⅶ，Akihiro Nakamura 和 Chris J. Burwell 采用Berlese分离法在雨林中采集，No. A12-1812。

词源： 种加词是云南省的简称"滇"，为该种的模式产地。

布氏卷尾猛蚁
Proceratium bruelheidei Staab, Xu & Hita Garcia, 2018

Proceratium bruelheidei Staab et al., 2018, ZooKeys, 2018, 770: 148, figs. 4B, 6B, 7B, 7D, 8, 9, 24 (w.) CHINA (Jiangxi).

Holotype worker: TL 3.9, HL 0.84, HW 0.79, CI 93, SL 0.50, SI 63, ED 0.04, WL 1.06, PL 0.39, DPW 0.32. In full-face view, head slightly longer than broad, anterior sides straight to very weakly convex, posterior sides narrowing dorsally, vertex convex. Clypeus reduced and narrow, with a broadly triangular median anterior projection. Frontal carinae relatively short, slightly covering antennal insertions, constantly diverging posteriorly, lateral expansions of anterior part of frontal carinae developed as broad lamellae, raised, conspicuously and broadly extending laterally above antennal insertions. Eyes reduced, minute, consisting of one to four ommatidia and located on midline of head. Antennae 12-segmented, scapes short, not reaching posterior head margin and thickening apically. Mandibles elongate and triangular, masticatory margin with four teeth. Mesosoma in lateral view slightly convex. Lower mesopleurae with well-demarcated sutures, upper mesopleurae with inconspicuous sutures, no other sutures developed on lateral and dorsal mesosoma; posterodorsal corner of propodeum broadly angular, propodeal lobes weakly developed as bluntly rounded lamellae; propodeal declivity almost vertical, sides of propodeum separated from declivity by distinct lamellate margins; propodeal spiracle rounded, at mid height. Petiolar node in lateral view high, nodiform, with a straight and sloping anterior face, dorsum of node broadly rounded, posterior face as steep as anterior face and relatively short, less than half as long as anterior face; petiole in dorsal view longer than broad, apex of node almost as long as broad; ventral process of petiole well developed, with a roughly trapezoid projection of varying shape and ventral outline. Constriction between abdominal segments Ⅲ and Ⅳ deep. Abdominal segment Ⅳ very large, strongly recurved and posteriorly rounded, remaining abdominal tergites and sternites inconspicuous and projecting anteriorly. Sting large and extended. Whole body covered with dense short decumbent to suberect pubescent hairs; additionally, dorsal surfaces of body with abundant significantly longer suberect and erect hairs. Mandibles striate; entire body densely punctate; on sides of pronotum punctures aligned in diffuse lines, appearing striate; abdominal segments Ⅴ-Ⅶ very superficially reticulate and shiny. Body color uniformly orange brown to reddish brown, vertex of head slightly darker, legs, antennal funiculus, and abdominal segments V-VII yellowish brown. **Paratype workers:** TL 3.6-4.0, HL 0.79-0.86, HW 0.73-0.79, CI 89-94, SL 0.49-0.53, SI 62-67, ED 0.03-0.04, WL 1.03-1.10, PL 0.36-0.39, DPW 0.30-0.32 (*n*=7). As holotype.

Holotype: Worker, China: Jiangxi Province, ca. 15 km SE of Wuyuan, near the village Xingangshan (29.123333°N, 117.906944°E), 158 m, 2015.Ⅳ.26, collected by Merle Noack in early successional tree plantation of the BEF-China experiment using Winkler leaf litter extraction, label "MN290" (CASENT0790023), deposited in Southwest Forestry University.

Etymology: The species epithet is a patronym in honor of the German botanist Prof. Helge Bruelheide and his efforts in establishing and promoting the BEF-China project. All specimens of this species were collected on BEF-China field sites (cited from Staab et al., 2018).

正模工蚁： 正面观，头长稍大于宽，侧缘前部平直至轻微隆起，侧缘后部向后变窄，头顶隆起。唇基缩小，狭窄，具1个宽三角形的中央前突。额脊较短，局部覆盖触角插入部，向后持续分歧，额脊前部的侧向扩展部发达，呈宽的片状，升高，明显向侧面扩展至触角插入部上方。复眼退化，微小，由1~4个小眼组成，位于头中线上。触角12节，柄节短，未到达头后缘，向端部变粗。上颚长三角形，咀嚼缘具4个齿。侧面观，胸部背面轻度隆起。中胸侧板下部具界线明显的缝，中胸侧板上部具不明显的缝，此外胸部侧面和背面缺明显的沟和缝；并胸腹节后上角钝角状；后侧叶不发达，末端钝圆；斜面几乎垂直，并胸腹节侧面与斜面分界具明显边缘；并胸腹节气门圆形，位于侧面中部。腹柄结高，呈结形，前面平直坡形，背面宽圆，后面与前面一样陡，较短，短于前面长度的一半。背面观，腹柄长大于宽，腹柄结顶端长宽几乎相等；腹柄下突发达，近梯形，形状和腹面轮廓可变。后腹部基部2节间深度缢缩；第2节很大，强烈下弯，后面圆形，其余腹节的背板和腹板不明显，指向前方。螫针发达，伸出。全身具密集的倾斜至亚直立短绒毛被；此外，身体背面具丰富的明显较长的亚直立或直立毛。上颚具条纹；全身具密集刻点；前胸背板侧面刻点聚集呈条纹状；后腹部第3~5节具肤浅的网纹，光亮。身体橙棕色至红棕色，头顶浅黑色，足、触角鞭节和后腹部第3~5节黄棕色。**副模工蚁：** 特征同正模。

正模： 工蚁，中国江西省婺源东南约15km新岗山村附近（29.123333°N，117.906944°E），158m，

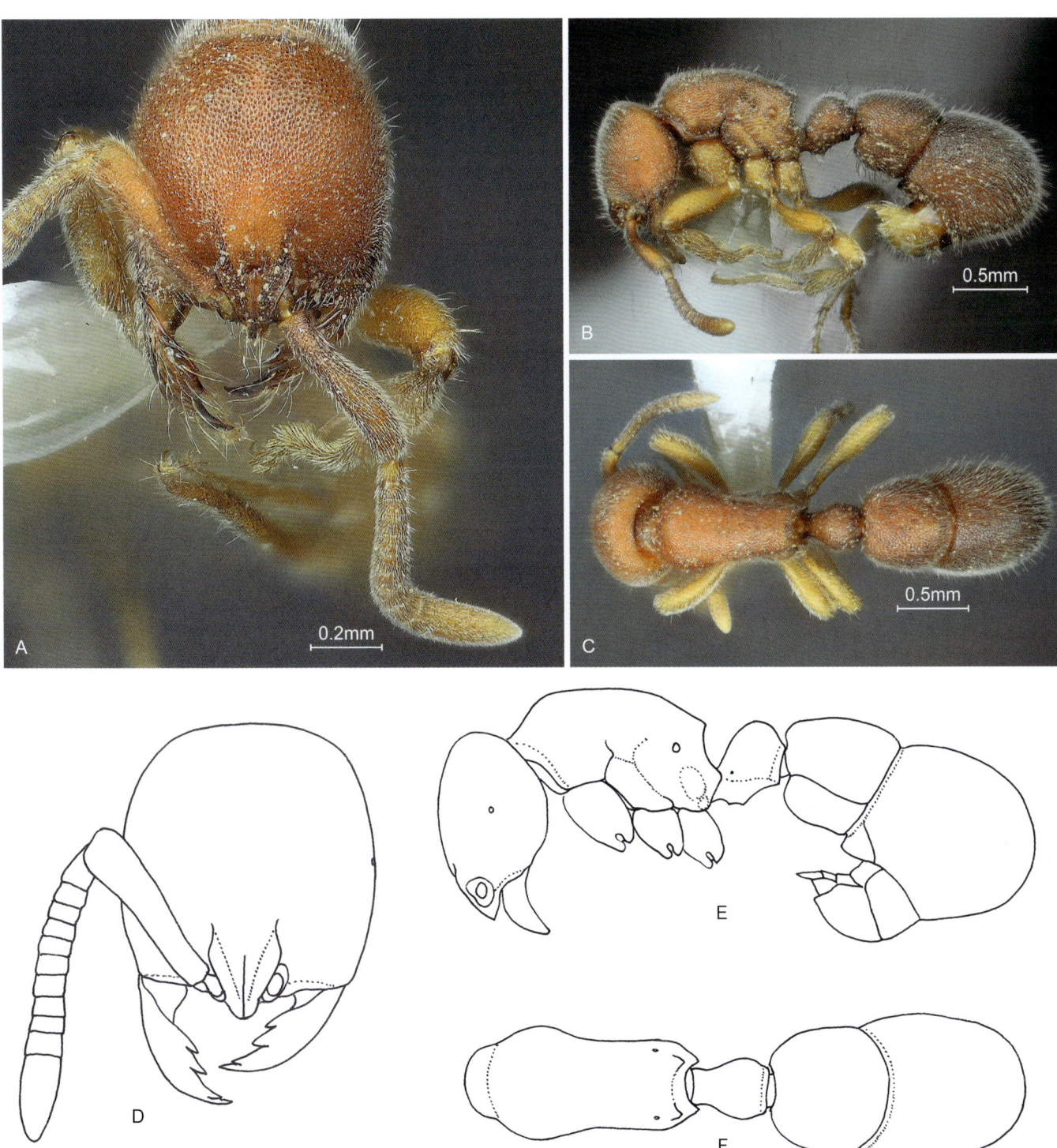

布氏卷尾猛蚁 *Proceratium bruelheidei*

A, D. 工蚁头部正面观；B, E. 工蚁身体侧面观；C, F. 工蚁身体背面观
A, D. Head of worker in full-face view; B, E. Body of worker in lateral view; C, F. Body of worker in dorsal view

2015.Ⅳ.26，Merle Noack采用Winkler地被物分离法采集于BEF-China实验站的早期演替人工林中，标签"MN290"（CASENT0790023），保存于西南林业大学。

词源： 该物种的名称是德国植物学家Helge Bruelheide教授的姓氏，以纪念他为设立和推动BEF-China项目所做的努力。该物种的所有标本均采于BEF-China野外实验点（引自Staab等，2018）。

克平卷尾猛蚁
Proceratium kepingmai Staab, Xu & Hita Garcia, 2018

Proceratium kepingmai Staab et al., 2018, ZooKeys, 770: 157, figs. 5B, 7A, 7C, 12, 13, 24 (w.) CHINA (Jiangxi).

Holotype worker: TL 4.4, HL 0.92, HW 0.86, CI 92, SL 0.57, SI 66, ED 0.02, WL 1.14, PL 0.45, DPW 0.36. In full-face view, head slightly longer than broad, sides weakly convex, broadest at the eye level and gently narrowing anteriorly and posteriorly, vertex weakly convex, almost straight. Clypeus reduced and narrow, with a broadly triangular median anterior projection. Frontal carinae relatively short, slightly covering antennal insertions, constantly diverging posteriorly, lateral expansions of anterior part of frontal carinae developed as broad lamellae, raised, conspicuously and broadly extending laterally above antennal insertions. Eyes reduced, minute, consisting of a single ommatidium and located on midline of head. Antennae 12-segmented, scapes short, not reaching posterior head margin and thickening apically. Mandibles elongate and triangular, masticatory margin with four teeth. Mesosoma in lateral view slightly convex. Lower mesopleurae with well-demarcated sutures; upper mesopleurae with inconspicuous sutures, no other sutures developed on lateral and dorsal mesosoma; posterodorsal corner of propodeum broadly angular, propodeal lobes weakly developed as bluntly rounded lamellae; propodeal declivity almost vertical; sides of propodeum separated from declivity by distinct lamellate margins; propodeal spiracle rounded, at mid height. Petiolar node in lateral view high, nodiform, with a straight and sloping anterior face, dorsum of node broadly rounded, posterior face half as long and steeper than anterior face; petiole in dorsal view longer than broad but apex of node clearly broader than long; ventral process moderately developed on anterior petiole, with a relatively indistinct rectangular projection. Constriction between abdominal segments Ⅲ and Ⅳ deep. Abdominal segment Ⅳ very large, recurved and posteriorly strongly rounded, remaining abdominal tergites and sternites inconspicuous and projecting anteriorly. Sting large and extended. Whole body covered with dense short decumbent to suberect pubescent hairs; additionally, the dorsal surfaces of body interspersed with significantly longer and erect hairs. Mandibles striate; entire body densely punctate; on sides of pronotum punctures aligned in diffuse lines, appearing striate; abdominal segments Ⅴ–Ⅶ very superficially punctured and shiny. Body color uniformly orange brown to reddish brown, vertex of head slightly darker, legs, antennal funiculus, and abdominal segments Ⅴ–Ⅶ yellowish brown. **Paratype worker:** TL 4.5, HL 0.98, HW 0.90, CI 93, SL 0.59, SI 66, ED 0.03, WL 1.24, PL 0.46, DPW 0.37 (n=1). As holotype.

Holotype: Worker, China: Jiangxi Province, ca. 15 km SE of Wuyuan, near the village Xingangshan (29.124444°N, 117.911111°E), 270 m, 2015.Ⅲ.26, collected by Michael Staab in secondary subtropical mixed forest using Winkler leaf litter extraction, label "MS1836" (CASENT0790031), deposited in Southwest Forestry University.

Etymology: The species epithet is a patronym in honor of the Chinese botanist Prof. Keping Ma and his efforts in establishing the BEF-China project and promoting biodiversity research and nature conservation in China (cited from Staab et al., 2018).

正模工蚁：正面观，头长稍大于宽；侧缘轻度隆起，在复眼水平处最宽，向前向后均轻度变窄；头顶轻度隆起，几乎平直。唇基缩小，狭窄，具1个宽三角形的中央前突。额脊较短，局部覆盖触角插入部，向后持续分歧，额脊前部的侧向扩展部发达，呈宽的片状，升高，明显向侧面扩展至触角插入部上方。复眼退化，很小，由1个小眼组成，位于头中线上。触角12节，柄节短，未到达头后缘，向顶端变粗。上颚长三角形，咀嚼缘具4个齿。侧面观，胸部背面轻度隆起。中胸侧板下部具界线明显的缝，中胸侧板上部具不明显的缝，此外胸部侧面和背面缺明显的沟和缝；并胸腹节后上角钝角状；后侧叶不发达，末端钝圆；斜面几乎垂直，并胸腹节侧面与斜面分界具明显边缘；并胸腹节气门圆形，位于侧面中部。腹柄结高，结形，前面平直坡形，背面宽圆，后面长为前面的一半，比前面陡。背面观，腹柄长大于宽，但腹柄结顶端明显宽大于长；腹柄下突在腹柄前部中等发达，近长方形。后腹部基部2节间深度缢缩，第2节很大，强烈下弯，后面圆形，其余腹节的背板和腹板不明显，指向前方。螫针发达，伸出。全身具密集的倾斜至亚直立短绒毛被；此外，身体背面散布着明显更长的立毛。上颚具条纹；全身具密集刻点；前胸背板侧面刻点聚集呈条纹状；后腹部第3~5节具肤浅的刻点，光亮。身体橙棕色至红棕色，头顶浅黑色，足、触角鞭节和后腹部第3~5节黄棕色。**副模工蚁：**特征同正模。

正模：工蚁，中国江西省婺源东南约15km新岗山村附近（29.124444°N，117.911111°E），270m，2015.Ⅲ.26，Michael Staab采用Winkler地被物分离法采于次生亚热带混交林中，标签"MS1836"

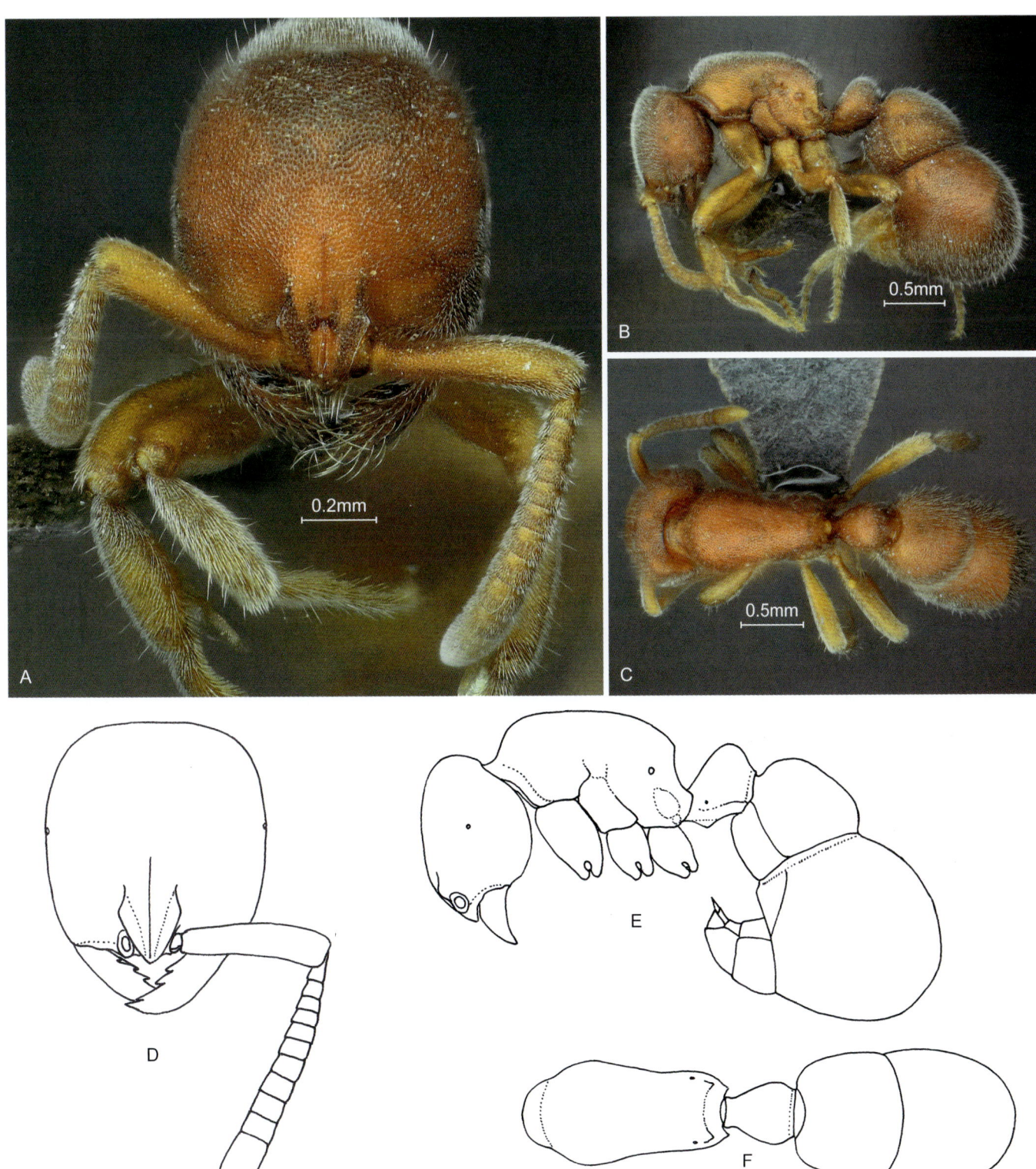

克平卷尾猛蚁 *Proceratium kepingmai*
A, D. 工蚁头部正面观；B, E. 工蚁身体侧面观；C, F. 工蚁身体背面观
A, D. Head of worker in full-face view; B, E. Body of worker in lateral view; C, F. Body of worker in dorsal view

(CASENT0790031)，保存于西南林业大学。

词源： 该物种的名称是中国植物学家马克平教授的姓名，以纪念他在设立BEF-China项目和促进中国生物多样性研究和自然保护中所作的努力（引自Staab等，2018）。

015 龙门卷尾猛蚁
Proceratium longmenense Xu, 2006

Proceratium longmenense Xu, 2006, Myrmecologische Nachrichten, 8: 154, figs. 14-16 (w.) CHINA (Yunnan).

Holotype worker: TL 3.2, HL 0.87, HW 0.73, CI 85, SL 0.50, SI 68, ML 0.40, ED 0.02, PW 0.52, AL 0.97, PL 0.35, PH 0.33, DPW 0.32. In full-face view, head nearly square, slightly longer than broad. Occipital margin straight. Occipital corners rounded. Sides weakly convex. Masticatory margins of mandibles with 4 teeth, which reduced in size from apex to base. Frontal carinae straight, convergent forward. Anterior margin of clypeus with a wide and triangular projection in the middle. Antennae thick, with 12 segments. Apices of scapes reached to 4/5 of the distance from antennal socket to occipital corner. Eyes with only one facet. In lateral view, dorsum of alitrunk complete and weakly convex. Promesonotal suture and metanotal groove vanished on dorsum. Posterodorsal corner of propodeum extrudent into a right-angled tooth. Lateral lobes of propodeum blunt and rounded. Petiole roughly triangular and inclined backward, anterior and dorsal faces convex, anterodorsal corner rounded, posterior face short and straight. Subpetiolar process small and nearly rectangular, ventral face straight, anteroventral corner rightly angled, posteroventral corner bluntly angled. Constriction between the two basal segments of gaster distinct, second segment very large, apical three segments short and under the second one. Mandibles finely longitudinally striate. Head, alitrunk, petiole, and gaster finely densely punctate and dim. Gastral segments 3-5 smooth and shining. Dorsum of whole body with sparse short erect hairs and dense decumbent pubescence. Erect hairs abundant on dorsum of alitrunk. Dorsal faces of scapes and tibiae with dense decumbent pubescence, but without erect hairs. Whole body yellowish brown. Mandibles, antennae, legs, and apex of gaster brownish yellow.

Holotype: Worker, China: Yunnan Province, Kunming City, Xishan Mountain Forest Park, Longmen, 2050 m, 2001.Ⅴ.5, collected by Zheng-hui Xu in subtropical monsoon evergreen broadleaf forest, No. A00514.

Etymology: The new species is named after the type specimen locality Longmen, a place of the Xishan Mountain Forest Park.

正模工蚁： 正面观，头部近方形，长稍大于宽；后缘平直，后角圆；侧缘轻度隆起。上颚咀嚼缘具4个齿，从端部到基部依次变小。额脊直，向前收敛。唇基前缘中部具1个宽三角形前突。触角粗，12节，柄节末端到达触角窝至头后角间距的4/5处。复眼仅具1个小眼。侧面观，胸部背面完整，轻度隆起，前中胸背板缝和后胸缝在背部消失。并胸腹节后上角突出成直角形齿。后侧叶顶端钝圆。腹柄近三角形，后倾，前面和背面隆起，前上角圆，后面短而直；腹柄下突小，近长方形，腹面直，前下角直角形，后下角钝角状。后腹部基部2节间缢缩明显，第2节很大，端部的3节短，位于第2节之下。上颚具细纵条纹。头部、胸部、腹柄和后腹部具细密刻点，暗；后腹部第3～5节光滑发亮。身体背面具稀疏直立短毛和密集倾斜绒毛被，胸部背面直立毛丰富；柄节和胫节背面具密集倾斜绒毛被，缺直立毛。身体黄棕色，上颚、触角、足和后腹部末端棕黄色。

正模： 工蚁，中国云南省昆明市西山森林公园龙门，2050m，2001.Ⅴ.5，徐正会采于亚热带季风常绿阔叶林中，No. A00514。

词源： 该新种以模式标本产地"龙门"，西山森林公园的一个地点命名。

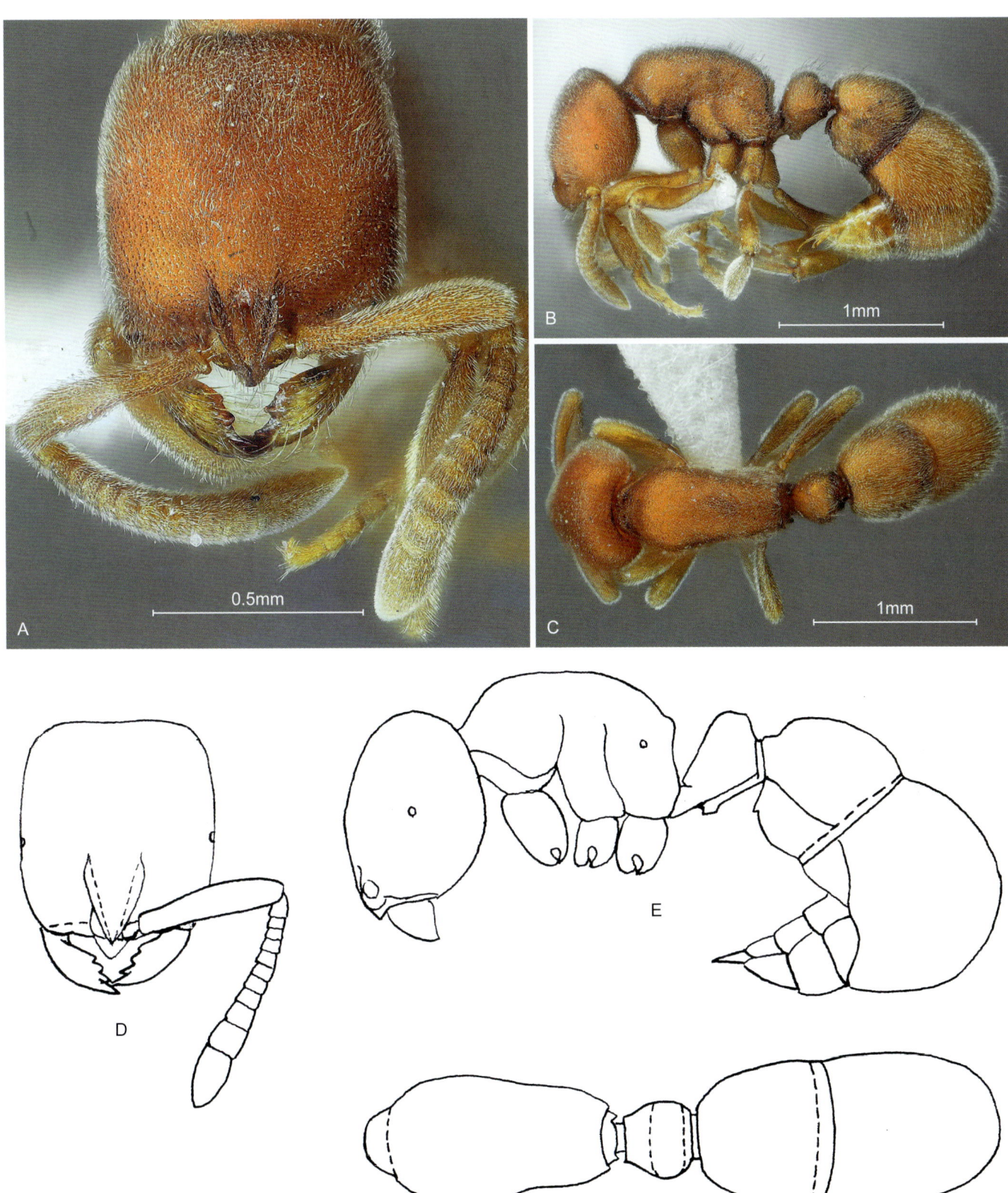

龙门卷尾猛蚁 *Proceratium longmenense*

A, D. 工蚁头部正面观；B, E. 工蚁身体侧面观；C, F. 工蚁身体背面观（D-F. 引自Xu, 2006）

A, D. Head of worker in full-face view; B, E. Body of worker in lateral view; C, F. Body of worker in dorsal view (D-F. cited from Xu, 2006)

怒江卷尾猛蚁
Proceratium nujiangense Xu, 2006

Proceratium nujiangense Xu, 2006, Myrmecologische Nachrichten, 8: 153, figs. 8-13 (w.q.) CHINA (Yunnan).

Holotype worker: TL 2.5, HL 0.62, HW 0.55, CI 89, SL 0.35, SI 64, ML 0.30, ED 0.02, PW 0.38, AL 0.70, PL 0.22, PH 0.25, DPW 0.27. In full-face view, head nearly square, slightly longer than broad. Occipital margin straight, occipital corners roundly prominent, sides of head moderately convex. Masticatory margins of mandibles with 4 teeth, which reduced in size from apex to base. Anterior margin of clypeus with a narrow acute central projection. Antennae with 12 segments. Apices of scapes reach to 3/5 of the distance from antennal socket to occipital corner. Eyes minute, with 1 facet. In lateral view, dorsum of alitrunk complete and straight, pronotum and propodeum convex, promesonotal suture and metanotal groove vanished on dorsal face. Posterodorsal corner of propodeum extrudent and bluntly angled. Lateral lobes of propodeum rounded. Petiolar node nearly triangular, anterior face long, straight and slope-like, dorsal face roundly convex, posterior face short and straight. Subpetiolar process short and small, nearly triangular. Constriction between the two basal segments of gaster distinct, deep and narrow. Second gastral segment very large, gastral segments 3-5 small. Sting extrudent. Mandibles coarsely and longitudinally striate. Head, alitrunk, petiole and gaster densely and finely punctate, dull. Whole body with dense decumbent short pubescence, dorsal face of body without erect hairs. Mandibles, anterior margin of clypeus and ventral face of body with sparse erect hairs. Whole body brown. Eyes grey. Terminal segment of antenna brownish yellow. **Paratype workers:** TL 2.4-2.8, HL 0.62-0.68, HW 0.53-0.62, CI 86-90, SL 0.33-0.38, SI 61-66, ML 0.30-0.37, ED 0.02, PW 0.37-0.43, AL 0.70-0.80, PL 0.23-0.28, PH 0.23-0.30, DPW 0.25-0.30 (n=7). As holotype.

Holotype: Worker, China: Yunnan Province, Baoshan City, Lujiang Town, Bawan Village, 1500 m, 1998.Ⅷ.11, collected by Qi-zhen Long in *Pinus yunnanensis* forest on the east slope of Nujiang River Valley, No. A98-1964.

Etymology: The new species is named after the big river Nujiang, also called Thanlwin. In the valley of the river the new species was discovered.

正模工蚁： 正面观，头部近方形，长稍大于宽；后缘直，后角圆突；侧缘中度隆起。上颚咀嚼缘具4个齿，从端部到基部依次变小。唇基前缘具1个狭窄尖锐的中央突起。触角12节，柄节顶端到达触角窝至头后角间距的3/5处。复眼微小，仅具1个小眼。侧面观，胸部背面完整近平直，前胸背板和并胸腹节隆起，前中胸背板缝和后胸沟在背面消失；并胸腹节后上角突出呈钝角状。并胸腹节侧叶顶端钝圆。腹柄结近三角形，前面长而直，坡形；背面圆突，后面短而直；腹柄下突短而小，近三角形。后腹部基部2节间收缩明显，深而窄；第2节很大，第3~5节小。螯针伸出。上颚具粗糙纵条纹。头部、胸部、腹柄和后腹部具密集细刻点，暗。全身具密集倾斜短绒毛被，身体背面缺立毛；上颚、唇基前缘和身体腹面具稀疏直立毛。身体棕色，复眼灰色，触角端节棕黄色。**副模工蚁：** 特征同正模。

正模： 工蚁，中国云南省保山市潞江镇坝湾村，1500m，1998.Ⅷ.11，龙启珍采于怒江河谷东坡云南松林中，No. A98-1964。

词源： 该新种以怒江命名，也被称为萨尔温江。在怒江河谷发现了该新种。

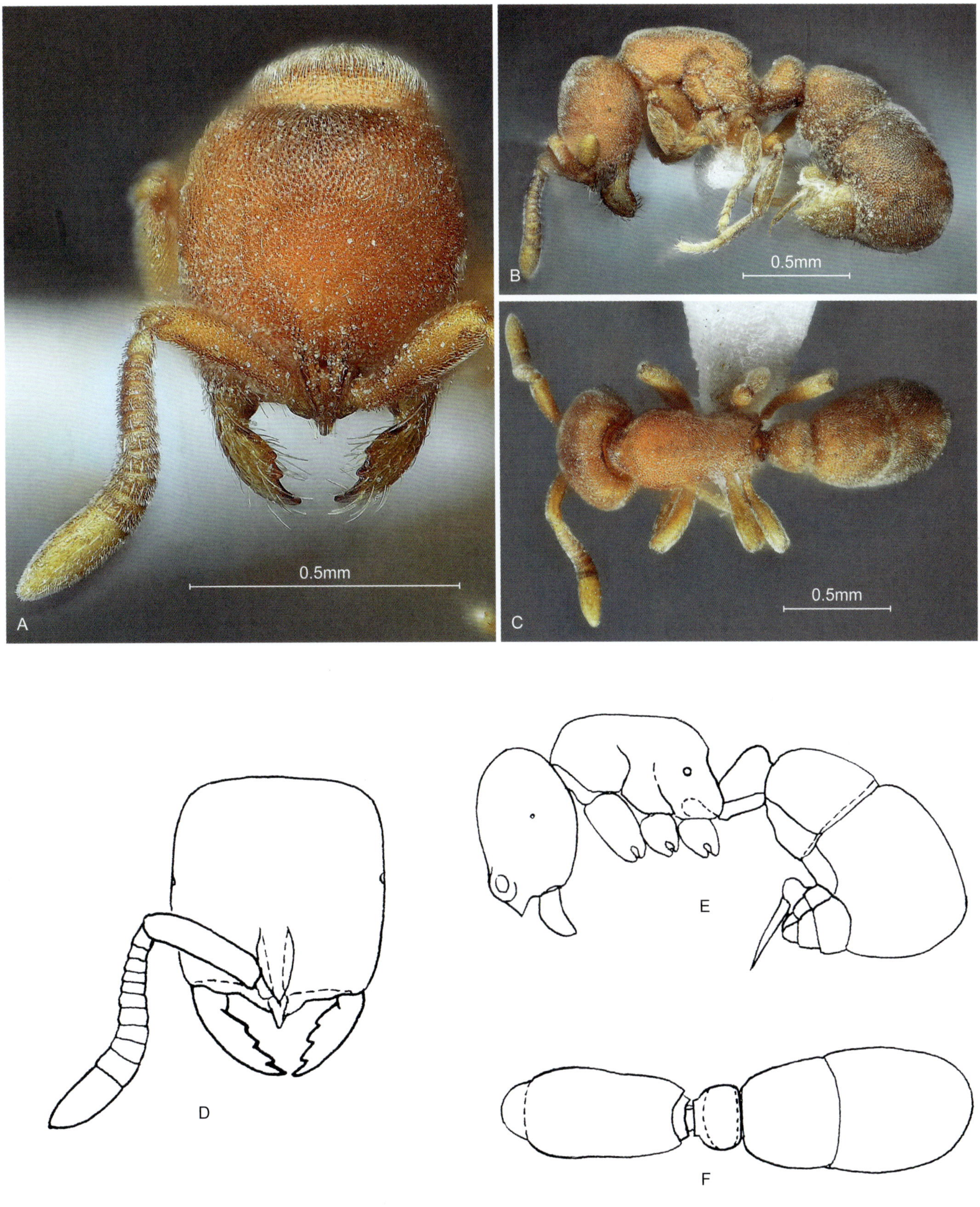

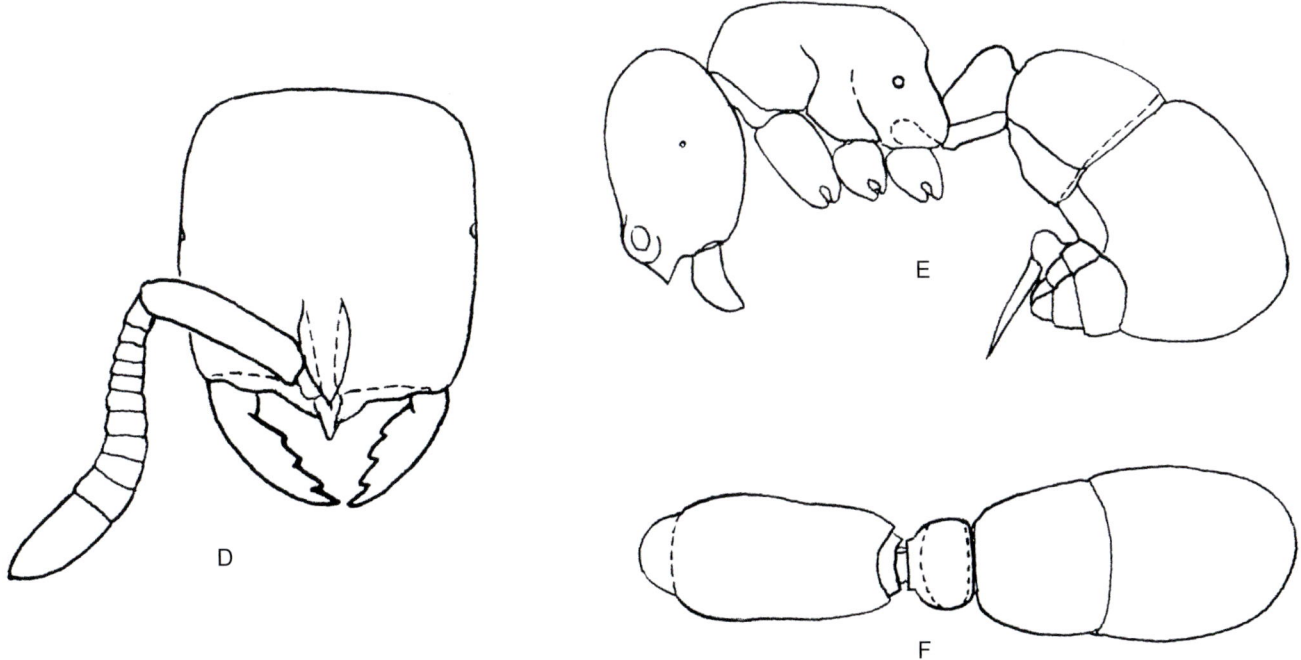

怒江卷尾猛蚁 *Proceratium nujiangense*
A, D. 工蚁头部正面观；B, E. 工蚁身体侧面观；C, F. 工蚁身体背面观（D-F. 引自Xu, 2006）
A, D. Head of worker in full-face view; B, E. Body of worker in lateral view; C, F. Body of worker in dorsal view (D-F. cited from Xu, 2006)

昌平卷尾猛蚁
017 *Proceratium shohei* Staab, Xu & Hita Garcia, 2018

Proceratium shohei Staab et al., 2018, ZooKeys, 770: 176, figs. 3A, 3C, 22, 23, 25 (w.) CHINA (Yunnan).

Holotype worker: TL 4.2, HL 0.99, HW 0.89, CI 90, SL 0.71, SI 80, ED 0.09, WL 1.25, PL 0.47, DPW 0.44. In full-face view, head slightly longer than broad, sides and vertex weakly convex, almost straight. Clypeus relatively broad, surrounding antennal insertions and protruding anteriorly, anterior clypeal margin with a distinct notch. Frontal carinae relatively short, broadly separated from each other, constantly diverging posteriorly and not covering antennal insertions, lateral expansions of frontal carinae slightly concave in full-face view; frontal area convex; frontal furrow absent. Genal carinae strongly developed. Eyes consisting of one convex ommatidium, located slightly anterior to the midline of head. Antennae 12-segmented, scapes comparatively long, not reaching posterior head margin and thickening apically. Mandibles elongate and triangular, masticatory margin with three teeth. Mesosoma in lateral view convex. Lower mesopleurae with demarcated sutures, upper mesopleurae and promesonotum with inconspicuous and very shallow sutures; posterodorsal corners of propodeum with broad teeth that project over less than half of the propodeal lobes in lateral view, propodeal lobes strongly developed as broadly triangular teeth protruding dorsolaterally; propodeal declivity almost vertical, sides of propodeum separated from declivity by lamellate margins; propodeal spiracle located above mid height. Petiole in dorsal view longer than broad, sides consistently diverging posteriorly; in lateral view, petiolar node relatively compressed dorsoventrally, its anterior face slightly sloping; dorsum of node relatively flat, weakly convex; ventral face inconspicuous with a thin lamella and no projection. Constriction between abdominal segments Ⅲ and Ⅳ deep. Abdominal segment Ⅳ very large, very strongly recurved and posteriorly rounded, remaining abdominal tergites and sternites inconspicuous and projecting anteriorly. Sting large and extended. Whole body covered with dense relatively short decumbent to erect hairs; additionally significantly longer suberect to erect hairs abundant on whole body. Mandibles striate; head, mesosoma, petiole, and abdominal segment Ⅲ foveolate with superimposed punctures and granules, the foveae relatively deep, large, and irregular; abdominal segment Ⅳ smooth and shiny, dorsally without sculpture, laterally superficially punctured. Body color uniformly dark ferruginous-brown, antennae, legs, and abdominal segments Ⅴ-Ⅶ orange brown.

Holotype: Worker, China: Yunnan Province, Xishuangbanna, Kilometer 55 station (21.964°N, 101.202°E), 820 m, 2013.Ⅵ.13, collected by Benoit Guénard, Benjamin Blanchard and Cong Liu in rain forest using Winkler leaf litter extraction, label '#05121' (CASENT0717686), deposited in Southwest Forestry University.

Etymology: This species is named in honor of Dr. Shohei Suzuki (1979—2016), a Japanese marine biologist whose life was tragically lost in a diving accident while conducting coral reef research in Okinawa (cited from Staab et al., 2018).

正模工蚁： 正面观，头长稍大于宽；侧缘和头顶轻度隆起，近平直。唇基较宽，围绕触角插入部，并向前伸出，唇基前缘具明显的缺刻。额脊较短，互相远离，向后持续分歧，未遮盖触角插入部，额脊的侧向扩展部轻度凹陷。额区隆起，缺额沟。颊区隆脊很发达。复眼由1个隆起的小眼构成，位于头中线稍前处。触角12节，柄节较长，未到达头后缘，向末端变粗。上颚长三角形，咀嚼缘具3个齿。侧面观，胸部背面隆起；中胸侧板下部具分界缝，中胸侧板上部和前中胸背板具不明显且很浅的缝。并胸腹节后上角具宽齿，侧面观伸出长度短于并胸腹节侧叶长度的一半；并胸腹节侧叶很发达，呈宽三角形齿，向背侧方伸出；并胸腹节斜面近垂直，侧面与斜面之间有片状边缘分隔；并胸腹节气门位于侧面中点之上。背面观，腹柄长大于宽，侧缘向后持续变宽；侧面观，腹柄结相对背腹压扁，前面缓坡形；背面较平，轻度隆起；腹面具不明显的薄脊，缺腹柄下突。后腹部基部2节间收缩很深，第2节很大，强烈下弯，后面圆，其余腹节的背板和腹板不明显，指向前方。螫针发达，伸出。全身具密集较短的倾斜至亚直立毛，此外还具显著较长的丰富亚直立至直立毛。上颚具条纹；头部、胸部、腹柄和后腹部第1节具凹坑，还叠加有刻点和粒纹，凹坑较深，大且不规则；后腹部第2节光滑发亮，背面无刻纹，侧面具肤浅刻点。身体暗红棕色，触角、足和后腹部第3~5节橙棕色。

正模： 工蚁，中国云南省西双版纳55千米观测站（21.964°N，101.202°E），820m，2013.Ⅵ.13，Benoit Guénard、Benjamin Blanchard和刘聪采用Winkler地被物分离法采于雨林中，标签"#05121"（CASENT0717686），保存于西南林业大学。

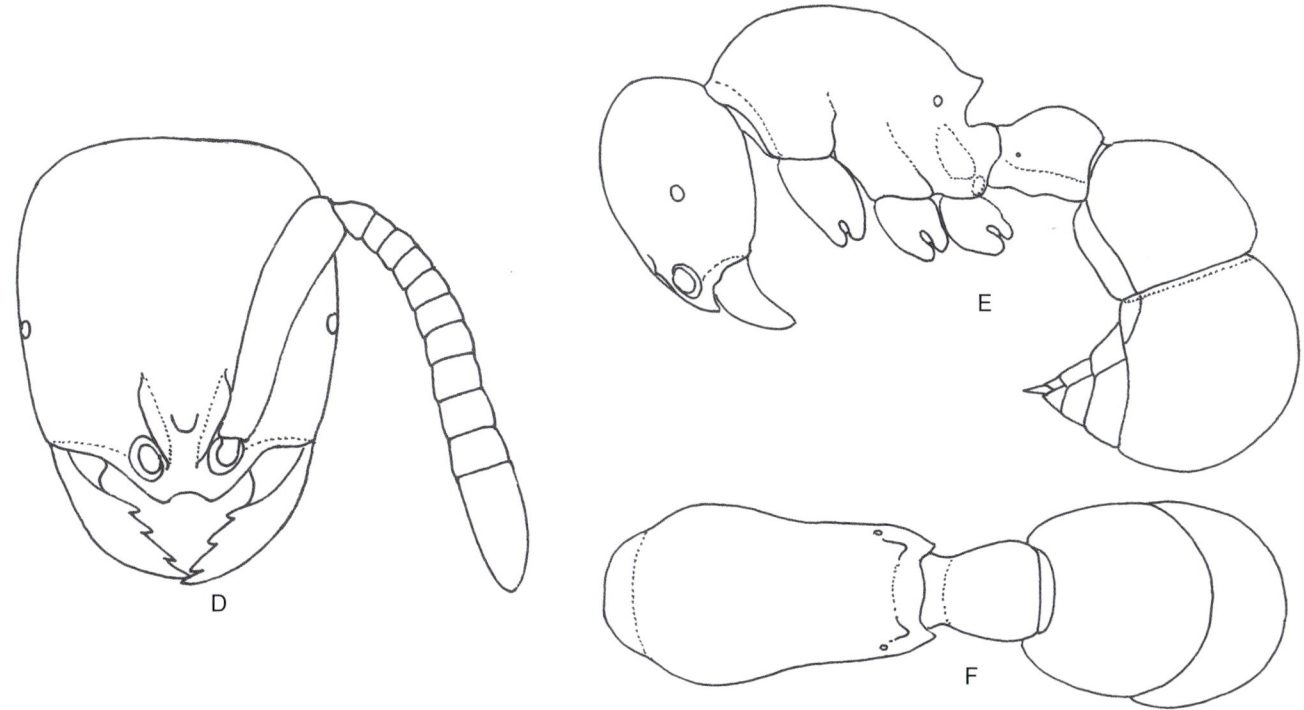

昌平卷尾猛蚁 *Proceratium shohei*

A, D. 工蚁头部正面观;B, E. 工蚁身体侧面观;C, F. 工蚁身体背面观
A, D. Head of worker in full-face view; B, E. Body of worker in lateral view; C, F. Body of worker in dorsal view

词源: 这个物种是为了纪念铃木昌平博士(1979—2016)而命名的,他是一位日本海洋生物学家,在冲绳进行珊瑚礁研究时不幸在一次潜水事故中失去了生命(引自Staab等,2018)。

赵氏卷尾猛蚁
Proceratium zhaoi Xu, 2000

Proceratium zhaoi Xu, 2000b, Acta Zootaxonomica Sinica 25(4): 435, figs. 3-10 (w.q.) CHINA (Yunnan).

Holotype worker: TL 2.4, HL 0.62, HW 0.52, CI 84, SL 0.34, SI 65, PW 0.38, AL 0.70, ED 0.03, ML 0.30, PL 0.22, PH 0.28, DPW 0.23. In full-face view, head roughly rectangular, slightly longer than broad and slightly narrowed forward. Occipital margin straight, occipital corners rounded. Sides weakly convex. Frontal carinae suberect, without laterally expended frontal lobes. Antennal sockets completely exposed. Anterior margin of clypeus with a triangular projection in the middle. Mandible with 4 teeth. Antenna stout, apex of scape reaching to 1/2 of the distance from antennal socket to occipital corner, segments 3-11 broader than long. Eye minute, with only 1 facet. In lateral view, dorsum of alitrunk complete and weakly convex, lowed down backward, without sutures. Posterodorsal corner of propodeum bluntly angled, without tooth, declivity depressed. Metapleural lobe rounded at apex. In lateral view petiolar node thick, narrowed upward and inclined backward, anterior and dorsal faces weakly convex, anterodorsal corner higher than posterodorsal corner. Subpetiolar process large and roughly rectangular, with a posteriorly pointed tooth. In dorsal view petiolar node transverse and broader than long, narrowed forward. Mandibles finely and longitudinally striate. Head, alitrunk, petiole and gaster finely and densely punctured, relatively dim. Head, body and appendages with dense decumbent pubescence, but without hairs. Body in color yellowish brown. **Paratype workers:** TL 2.0-2.5, HL 0.60-0.64, HW 0.52-0.56, CI 86-90, SL 0.32-0.34, SI 61-65, PW 0.34-0.38, AL 0.66-0.74, ED 0.02-0.03, ML 0.28-0.34, PL 0.20-0.24, PH 0.26-0.30, DPW 0.22-0.26 (*n*=6). As holotype, but subpetiolar process rectangular to triangular.

Holotype: Worker, China: Yunnan Province, Menghai County, Meng'a Town, Papo Village, 1280 m, 1997.XII.10, collected by Zheng-hui Xu in a soil sample in deciduous broadleaf forest, No. A97-2338.

Etymology: This new species is named after Mr. Qing-shan Zhao (Former South Institute of Forest Plant Quarantine, Department of Forestry of China, Yiyang, Jiangxi Province) for his contribution in translation *The Fauna of British India Including Ceylon and Burma. Hymenoptera. Volume II. Ants and Cuckoo-wasps* from English into Chinese for the Chinese readers.

正模工蚁： 正面观，头近长方形，长稍大于宽，向前稍变窄；后缘直，后角圆；侧缘轻度隆起。额脊亚直立，缺侧向扩展的额叶。触角窝完全外露。唇基前缘中部具三角形突起。上颚具4个齿。触角粗，柄节末端到达触角窝至头后角间距的1/2处，第3～11节宽大于长。复眼微小，仅具1个小眼。侧面观，胸部背面完整，轻度隆起，向后降低，背面缺沟缝。并胸腹节后上角钝角状，缺齿，斜面凹陷。后侧叶端部钝圆。腹柄结厚，向上变窄，向后倾斜，前面和背面轻度隆起，前上角高于后上角；腹柄下突大，近长方形，具指向后方的齿。背面观，腹柄结横形，宽大于长，向前变窄。上颚具细纵条纹。头部、胸部、腹柄和后腹部具细密刻点，较暗。头部、身体和附肢具密集倾斜绒毛被，缺立毛。身体黄棕色。**副模工蚁：** 特征同正模，但是腹柄下突长方形至三角形。

正模： 工蚁，中国云南省勐海县勐阿乡怕迫村，1280m，1997.XII.10，徐正会采于落叶阔叶林土壤样中，No. A97-2338。

词源： 该新种以赵清山先生（原中国林业部南方森林植物检疫所，江西省弋阳）姓氏命名，以纪念他为中国读者将《印度动物志：膜翅目第2卷：蚂蚁和青蜂》由英文翻译成中文作出的贡献。

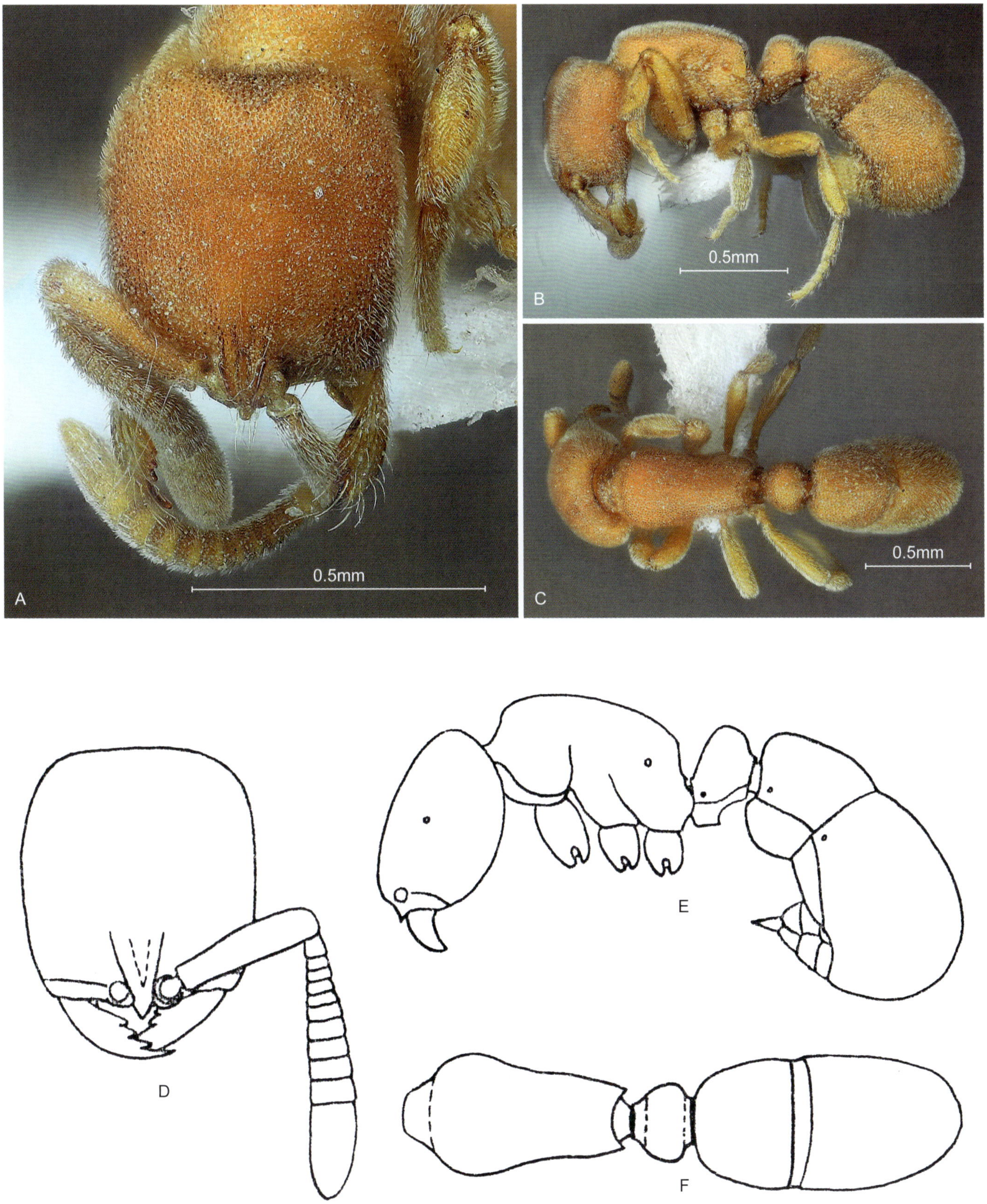

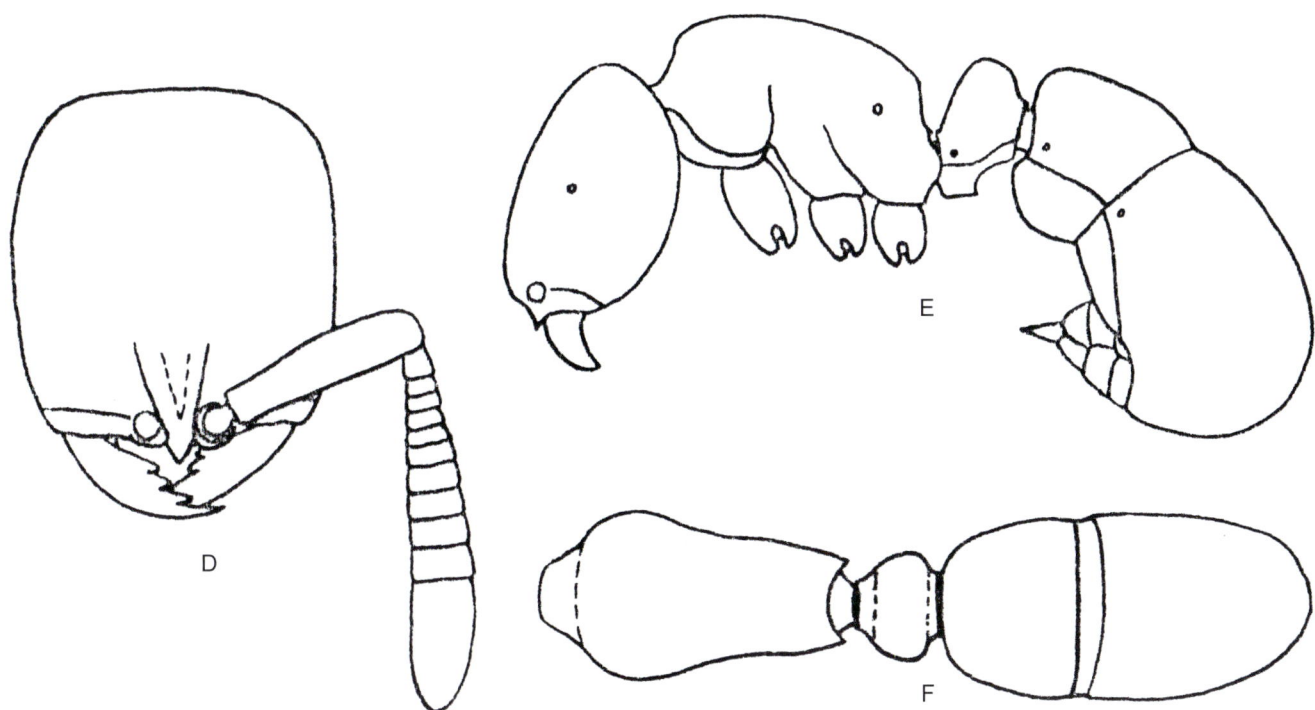

赵氏卷尾猛蚁 *Proceratium zhaoi*

A, D. 工蚁头部正面观；B, E. 工蚁身体侧面观；C, F. 工蚁身体背面观（D-F. 引自Xu, 2000b）
A, D. Head of worker in full-face view; B, E. Body of worker in lateral view; C, F. Body of worker in dorsal view (D-F. cited from Xu, 2000b)

019 短背短猛蚁
Brachyponera brevidorsa Xu, 1994

Brachyponera brevidorsa Xu, 1994a, Journal of Southwest Forestry College, 14(3): 183, figs. 5, 6 (w.) CHINA (Yunnan).

Holotype worker: TL 3.9, HL 0.93, HW 0.83, CI 89, SL 0.80, S1 97, PW 0.60, AL 1.25, ED 0.15, ML 0.50, NL 0.21, NW 0.48. In full-face view, head longer than broad, narrowing anteriorly, sides weakly convex, posterolateral corners narrowly rounded and relatively prominent, posterior margin straight. Eyes located on the anterior head sides, near the bases of mandibles. Masticatory margin of mandible with 9 teeth. Clypeus with low and broad longitudinal central carina, anterior margin weakly concave in the middle. Antennae 12-segmented, scapes slightly surpassing occipital corners, flagella incrassate towards apex, the apical segment long, the middle segments about as broad as long. In lateral view, promesonotum high, dorsum of pronotum nearly straight. Promesonotal suture notched. Mesonotum weakly convex and lowering down posteriorly. Metanotal groove depressed. Propodeum relatively lower, dorsum weakly convex, slightly lowering down posteriorly, distinctly shorter than declivity; declivity concave in the center, lateral margins extending posterolateraly and prominent; spiracles circular. Subpetiolar process cuneiform; petiolar node high, anteroposteriorly compressed, slightly narrowing dorsally; in dorsal view posterior face flat, anterior face convex; in front view upper margin rounded. Anterior face of gaster vertical, constricted between the 2 basal segments of gaster. Sing extruding. Head and body with exremely weak and fine reticulations, relatively shiny. Sides of alitrunk smooth and shiny. Dorsa of head and body with sparse erect to suberect hairs and abundant decumbent pubescence, mouthparts and gastral apex with abundant erect hairs; sides of alitrunk without pubescence; dorsa of scapes and hind tibiae with abundant decumbent pubescence, without erect hairs. Body color black, mandibles, flagella of antennae and legs yellowish brown. **Paratype workers:** TL 3.6-4.0, HL 0.91-0.98, HW 0.83-0.86, C1 88-92, SL 0.78-0.80, S1 93-97, PW 0.60-0.63, AL 1.25-1.30, ED 0.14-0.15, ML 0.45- 0.50, NL 0.2l-0.25, NW 0.45-0.50 (*n* = 6). As holotype.

Holotype: Worker, China: Yunnan Province, Baoshan County, Chengguan Town, Taibao Park, 1800 m, 1991.Ⅹ.11, collected by Zheng-hui Xu on the ground in conifer-broadleaf mixed forest, No. A91-1022.

Etymology: The species name *brevidorsa* combines Latin *brev-* (short) + word root *dors* (back) + suffix *-a* (feminine form), it refers to the realtively shorter dorsum of propodeum.

正模工蚁： 正面观头长大于宽，前部窄于后部，两侧轻度隆起；后角圆，较突出；后缘较直。复眼位于头前部两侧，接近上颚基部，上颚具9个齿。唇基具低而宽的中央纵脊，前缘中部轻度凹陷。触角12节，柄节稍超过头后角，鞭节向顶端变粗，端节长，中部各节长宽近相等。侧面观前中胸背板高，前胸背面较平。前中胸背板缝凹入，中胸背板轻度隆起，向后降低。后胸沟凹陷。并胸腹节低，背面轻度隆起，向后稍降低，明显短于斜面，斜面中间凹陷，侧缘向侧后方延展，突出，气门圆形。腹柄下突楔形，腹柄结高，前后压扁，向上稍变窄；背面观后面平坦，前面隆起；前面观背缘圆形。腹部前面垂直，基部2节间收缩，螫针伸出。头和体具极细弱的网状刻纹，较光亮。胸部侧板光滑发亮。头和体背面具稀少的直立、亚直立毛和丰富倾斜绒毛被，口器和腹末立毛丰富，胸部侧板缺绒毛被。触角柄节和后足胫节背面具丰富倾斜绒毛被，缺立毛。体黑色，上颚、触角鞭节和足黄棕色。**副模工蚁：** 特征同正模。

正模： 工蚁，中国云南省保山县城关镇太保公园，1800m，1991.Ⅹ.11，徐正会采于针阔混交林地表，No. A91-1022。

词源： 该新种的种名"*brevidorsa*"由拉丁语"*brev-*"（短的）+词根"*dors*"（背）+后缀"*-a*"（阴性形式）组成，指该种的并胸腹节背面较短。

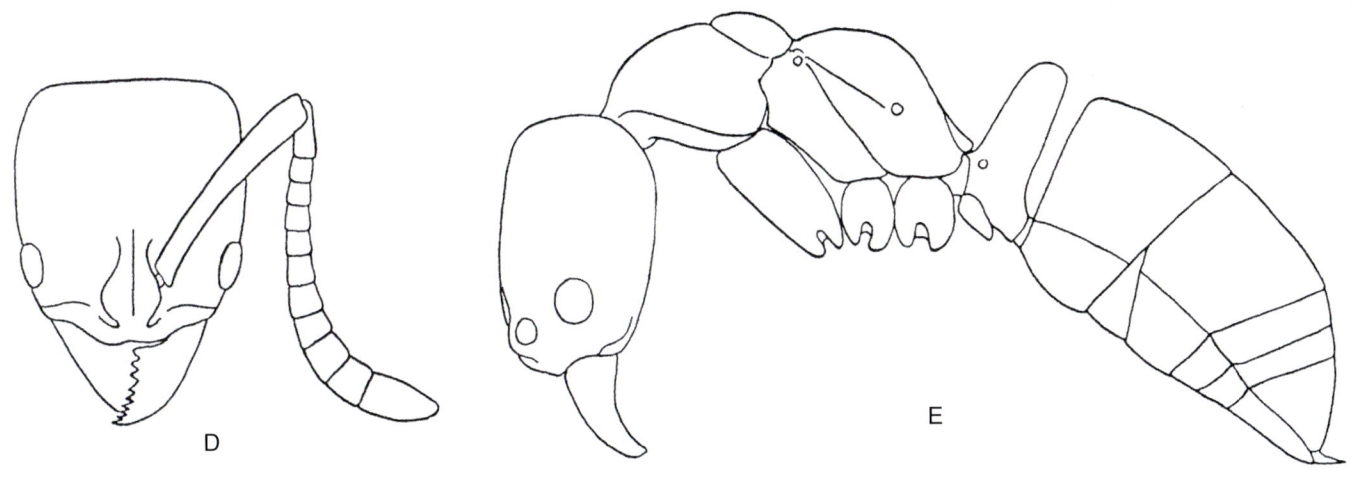

短背短猛蚁 *Brachyponera brevidorsa*

A, D. 工蚁头部正面观；B, E. 工蚁身体侧面观；C. 工蚁身体背面观（D-E. 引自徐正会, 1994a）
A, D. Head of worker in full-face view; B, E. Body of worker in lateral view; C. Body of worker in dorsal view (D-E. cited from Xu, 1994a)

020 直唇隐猛蚁
Cryptopone recticlypea Xu, 1998

Cryptopone recticlypea Xu, 1998a, Zoological Research, 19(2): 162, figs. 4-6 (w.) CHINA (Yunnan).

Holotype worker: TL 3.1, HL 0.59, HW 0.64, CI 109, SL 0.48, SI 75, PW 0.48, AL 0.95, ED 0.00, ML 0.43. In full-face view, head square, slightly broader than long. Lateral margins of head evenly convex, occipital margin shallowly depressed in the middle, occipital corners rounded. Clypeus transverse, narrow; convex in the center, but without longitudinal carina, middle portion of clypeus slightly protruded, its anterior margin slightly convex, nearly straight. Mandibles long triangular, outer face with a shallow elliptic fovea at base; masticatory margin has 4 teeth. Frontal lobes large, close together and covered most of the antennal insertions; a deep central furrow about 2 times as long as the frontal lobes presented between the lobes. Antennae with 12 joints, apex of scape reached to the occipital corner; flagellum incrassate towards apex, the apical 4 joints longer and formed a club. Eyes absent. In lateral view, pronotum and mesonotum a little higher than propodeum, slightly convex. Promesonotal suture complete and shallowly depressed, metanotal groove depressed. Metanotal glands large. Propodeum a little lower, dorsum and declivity slightly convex, nearly straight; dorsum weakly lowering down posteriorly and rounded into declivity; declivity slightly shorter than dorsum. In dorsal view, propodeum nearly triangular, narrowed anteriorly and very narrow near the metanotal groove. In lateral view, petiolar node cuneiform and erect, narrowed upwards; anterior face depressed, posterior face almost straight; dorsal face lowering down posteriorly as a slope; subpetiolar process large, its anteroventral angle roundly extruded. In dorsal view, petiolar node transverse, anterior face depressed, posterior face flat, dorsal face subrectangular; in front view, upper margin of the node rounded. Constriction between the 2 basal segments of gaster distinct. Sing protruded. Mandibles smooth and shining. Head and alitrunk densely and finely punctate, head dull, alitrunk duller. Propodeum, petiole, and gaster with shallow, dense, fine punctures, relatively shining. Head and body with sparse suberect hairs and dense subdecumbent pubescence, but dorsa of head and pronotum without hairs. Dorsa of scapes and hind tibiae with dense decumbent pubescence, without hairs. Outer faces of middle tibiae and tarsi with rich strong setae. Colour brownish yellow; legs yellow; gaster dark yellowish brown.

Holotype: Worker, China: Yunnan Province, Mengla County, Menglun Town, Bakaxiaozhai (21.9°N, 101.2°E), 840 m, 1996Ⅲ.8, collected by Zhi-ping Chen from a soil sample of seasonal rain forest, No. A96-336.

Etymology: The species name *recticlypea* combines Latin *rect-* (straight) + word root *clype* (base of lip) + suffix *-a* (feminine form), it refers to the relatively straight anterior margin of clypeus.

正模工蚁： 正面观，头部方形，宽稍大于长；侧缘均匀隆起；后缘中央浅凹，后角圆。唇基横形，狭窄，中部隆起，但是缺中央纵脊；中部轻度前伸，前缘稍隆起，近平直。上颚长三角形，外侧基部具1个浅的椭圆形凹窝，咀嚼缘具4个齿。额叶大，互相靠近，覆盖触角插入部的大部分；额叶之间具1条深的中央纵沟，长度约为额叶的2倍。触角12节，柄节末端到达头后角；鞭节向端部变粗，端部4节较长，形成触角棒。复眼缺失。侧面观，前胸背板和中胸背板稍高于并胸腹节，轻度隆起；前中胸背板缝完整且浅凹，后胸沟凹陷。后胸侧板腺大。并胸腹节稍低，背面和斜面轻度隆起，近平直；背面向后轻度降低，圆形进入斜面；斜面稍短于背面。背面观，并胸腹节近三角形，向前变窄，近后胸沟处很窄。侧面观，腹柄结楔形直立，向上变窄，前面凹陷，后面近平直，背面向后降低呈坡形；腹柄下突大，前下角圆突。背面观，腹柄结横形，前面凹陷，后面平坦，背面近长方形；前面观，腹柄结背缘圆。后腹部基部2节间缢缩明显。螫针伸出。上颚光滑发亮。头部和胸部具细密刻点，头部暗，胸部较暗；并胸腹节、腹柄和后腹部具浅而密集的细刻点，较光亮。头和身体具稀疏亚直立毛和密集亚倾斜绒毛被，但是头部和前胸背板缺立毛。柄节和后足胫节背面具密集倾斜绒毛被，缺立毛；中足胫节和跗节外侧具丰富的粗刚毛。身体棕黄色，足黄色，后腹部暗黄棕色。

正模： 工蚁，中国云南省勐腊县勐仑镇巴卡小寨（21.9°N，101.2°E），840m，1996.Ⅲ.8，陈志萍采于季节性雨林土壤样中，No. A96-336。

词源： 该新种的种名"*recticlypea*"由拉丁语"*rect-*"（直的）+词根"*clype*"（唇基）+后缀"*-a*"（阴性形式）组成，指该种的唇基前缘较直。

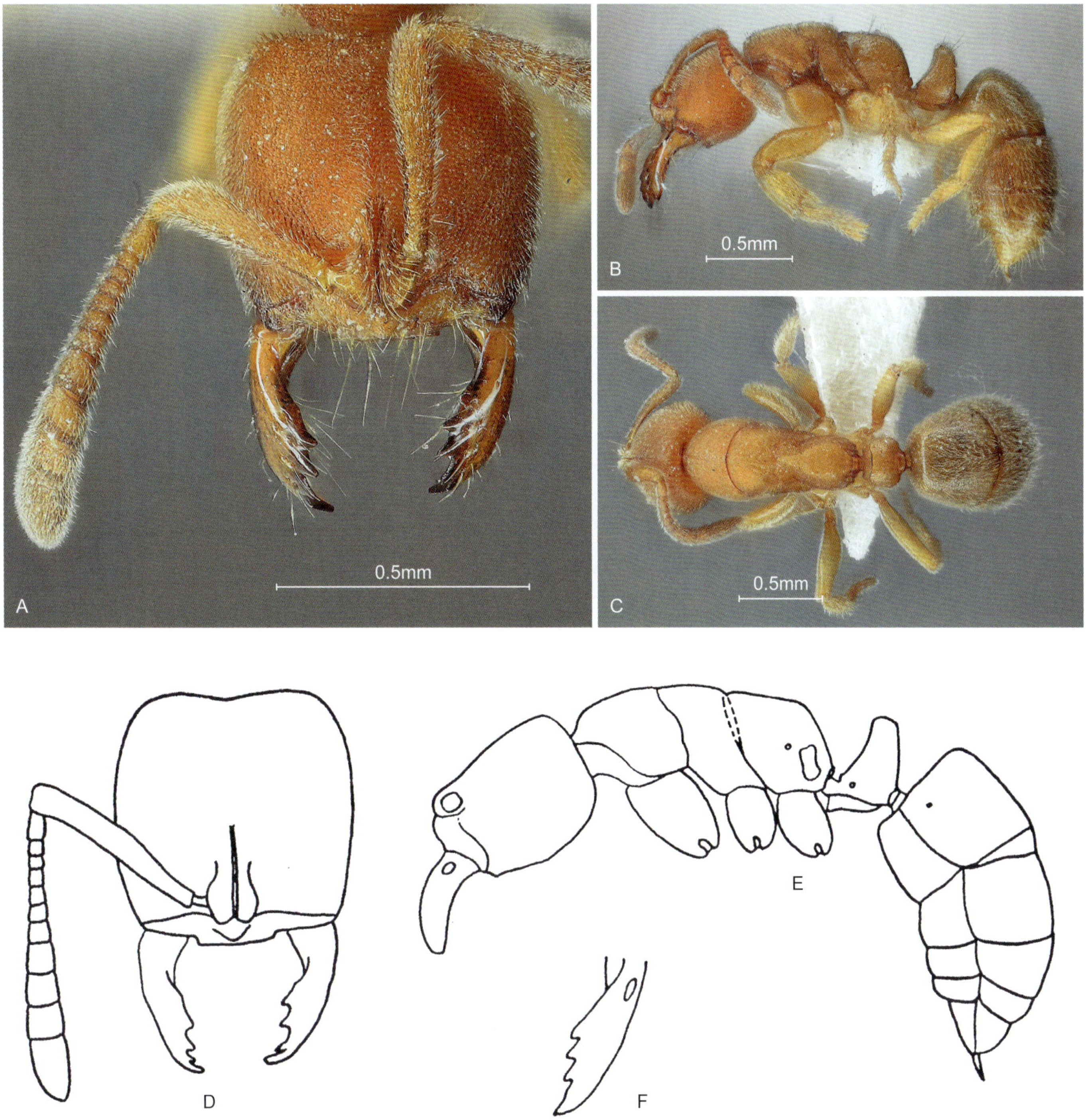

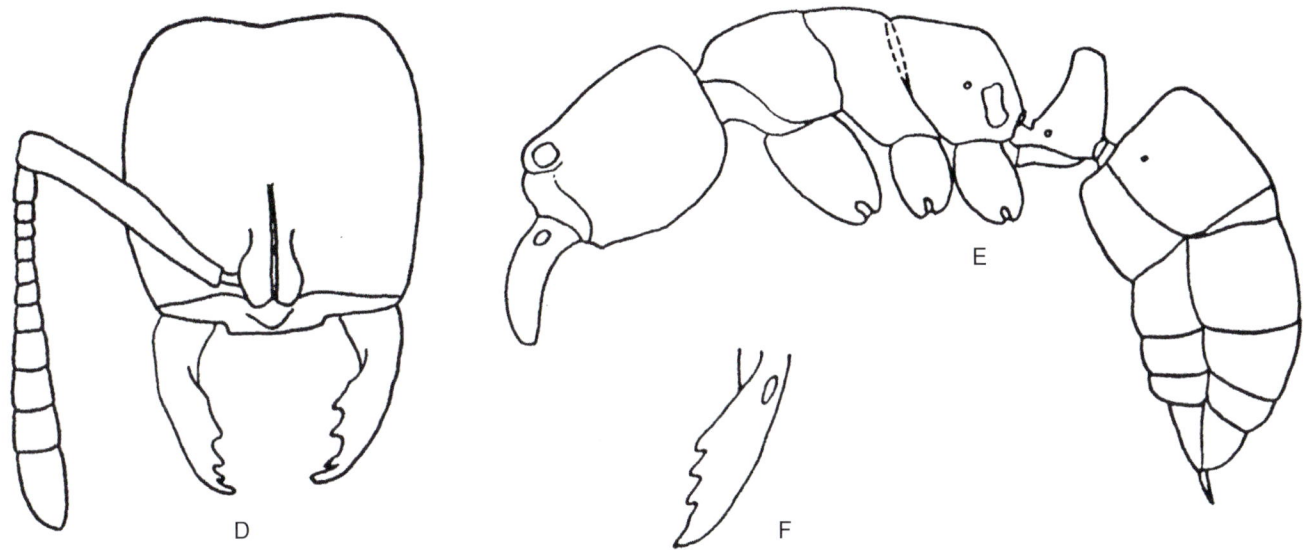

直唇隐猛蚁 Cryptopone recticlypea

A, D. 工蚁头部正面观；B, E. 工蚁身体侧面观；C. 工蚁身体背面观；F. 工蚁上颚背面观（D-F. 引自 Xu, 1998a）
A, D. Head of worker in full-face view; B, E. Body of worker in lateral view; C. Body of worker in dorsal view; F. Mandible of worker in dorsal view (D-F. cited from Xu, 1998a)

黑色埃猛蚁
Emeryopone melaina Xu, 1998

Emeryopone melaina Xu, 1998c, Entomologia Sinica, 5(2): 122, figs. 1, 2 (w.) CHINA (Yunnan).

Holotype worker: TL 4.9, HL 1.10, HW 0.80, CI 80, SL 0.83, SI 104, PW 0.63, AL 1.47, ED 0.10, ML 0.80, MI 80, PNL 0.37, PNW 0.53. In full-face view, head longer than broad, sides slightly convex, occipital margin roundly depressed, occipital corners bluntly protruding. Clypeus with longitudinal central carina, anterior margin extruding into a blunt dent in the middle. Mandibles long and narrow, inner margin with 5 long teeth, apical tooth slender and curved. Eyes small, in full-face view placed on the lateral margins, close to bases of mandibles. Antennal scapes just reach to occipital corners. In lateral view, dorsum of alitrunk weakly convex, lowering down backward. Promesonotal suture distinct, metanotal groove disappear. Dorsum of propodeum very weakly convex, nearly straight; declivity obliquely truncate, shorter than dorsum. Petiolar node thick, in lateral view rectangular, anterior and posterior faces slightly convex, dorsal face roundly convex; subpetiolar process rectangular, with a subtransparent oblique slit, anteroventral angle obliquely truncate, posteroventral angle blunt. Gaster conical, apical half curved downward. Sting long and strong. Mandibles smooth and shining. Head and gaster with moderately dense uniform large punctures, distance between two punctures shorter than diameter of a puncture. Dorsum of alitrunk and petiolar node finely and densely punctate, sides of alitrunk with sparse large punctures and microrugae. Dorsum of head, antennae, and legs with abundant depressed pubescence, without hairs. Dorsa of occiput, alitrunk, petiolar node, and gaster with abundant suberect short hairs and rich depressed pubescence. Body in colour black; mandibles, antennae, legs, and apex of gaster reddish brown. **Paratype dealate female:** TL 5.1, HL 1.03, HW 0.83, CI 81, SL 0.83, SI 100, PW 0.70, AL 1.60, ED 0.17, ML 0.87, MI 84, PNL 0.37, PNW 0.53 (*n*=1). As holotype, but body larger; eyes larger, with 3 ocelli; occipital margin weakly depressed; alitrunk with tegulae, mesopleuron with an oblique furrow; head, pronotum, and gaster dark reddish brown.

Holoytpe: Worker, China: Yunnan Province, Jinghong County, Mengyang Town, Sanchahe, 960 m, 1997.Ⅱ.27, collected by Tian-sheng Li from a soil sample of seasonal rain forest, No. A97-54.

Etymology: The name of the new species is descriptive of the "*melaina*" (Greek, meaning black) color of the body.

正模工蚁：正面观，头长大于宽，侧面轻度隆起；后缘圆形凹陷，后角钝凸。唇基具中央纵脊，前缘中央突出呈1个钝齿。上颚长而窄，内缘具5个长齿，端齿细长而弯曲。复眼小，位于头侧缘上，接近上颚基部。触角柄节刚到达头后角。侧面观，胸部背面轻度隆起，向后降低；前中胸背板缝明显，后胸沟消失。并胸腹节背面轻微隆起，近平直；斜面倾斜平截，短于背面。腹柄结厚，侧面观长方形，前面和后面轻微隆起，背面圆形隆起；腹柄下突长方形，具1条半透明斜缝，前下角斜截，后下角钝。后腹部圆锥形，端半部下弯。螫针长而粗。上颚光滑发亮。头部和后腹部具中等密集的均匀大刻点，刻点间距小于刻点直径；胸部背面和腹柄结具细密刻点，胸部侧面具稀疏大刻点和微皱纹。头部背面、触角和足具丰富平伏绒毛被，无立毛；后头背面、胸部背面、腹柄结和后腹部具丰富亚直立短毛和丰富平伏绒毛被。体黑色，上颚、触角、足和后腹部末端红棕色。**副模脱翅雌蚁**：特征同正模，但身体更大，复眼更大，具3个单眼，头后缘轻度凹陷；胸部具翅基片，中胸侧板具1条斜沟；头部、前胸背板和后腹部暗红棕色。

正模：工蚁，中国云南省景洪县勐养镇三岔河，960m，1997.Ⅱ.27，李天生采于季节性雨林土壤样中，No. A97-54。

词源：该新种的名称是描述其身体颜色为"*melaina*"（希腊语，意为黑色）。

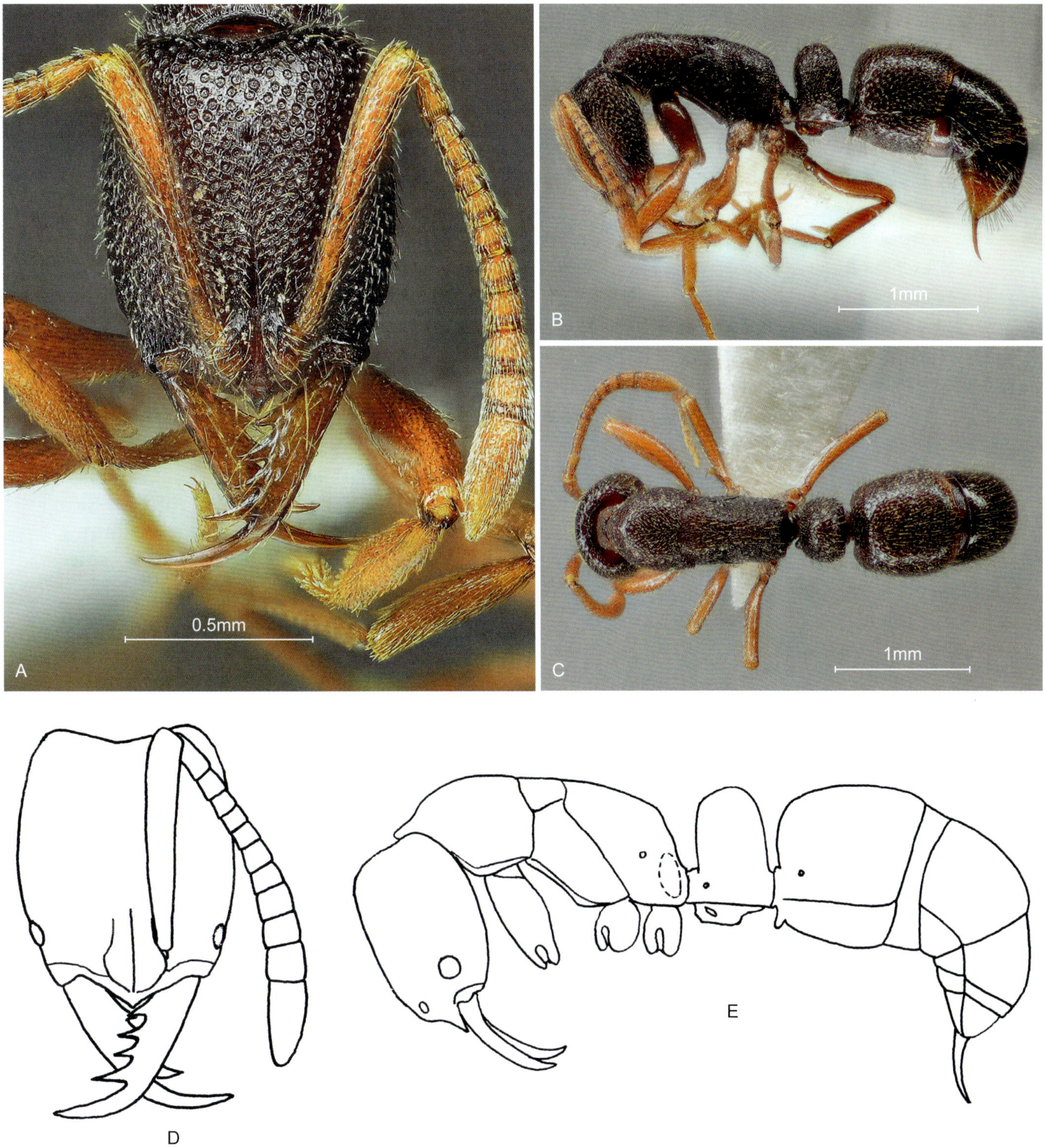

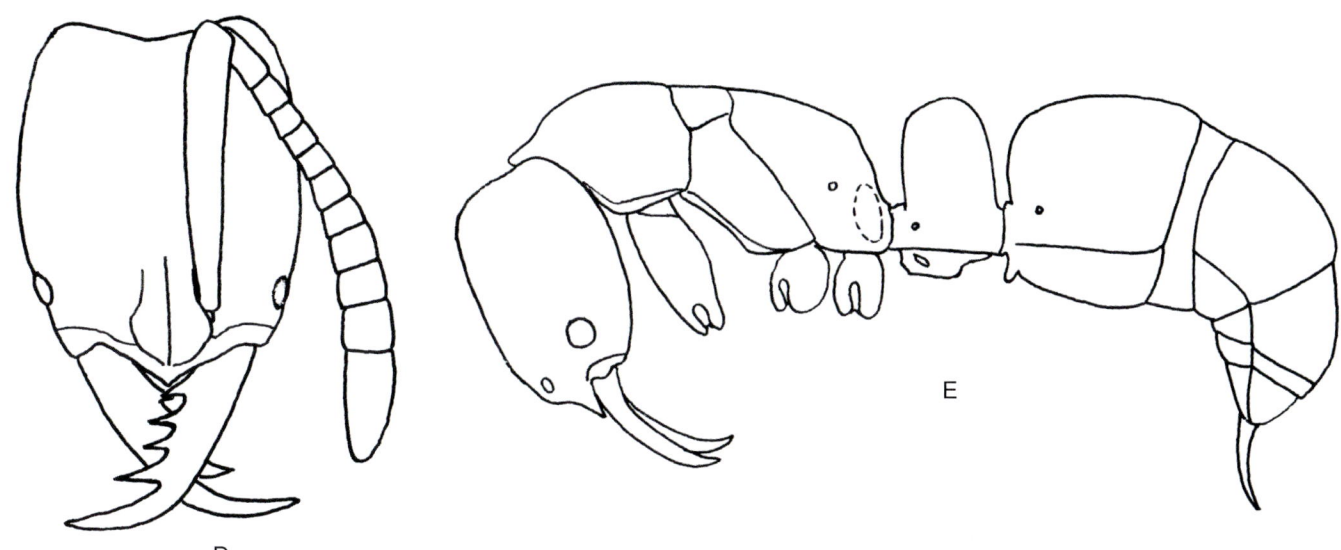

黑色埃猛蚁 *Emeryopone melaina*

A, D. 工蚁头部正面观; B, E. 工蚁身体侧面观; C. 工蚁身体背面观 (D-E. 引自 Xu, 1998c)
A, D. Head of worker in full-face view; B, E. Body of worker in lateral view; C. Body of worker in dorsal view (D-E. cited from Xu, 1998c)

黄帝细颚猛蚁
Leptogenys huangdii Xu, 2000

Leptogenys huangdii Xu, 2000c, Entomologia Sinica, 7(2): 119, figs. 5-8 (w.) CHINA (Yunnan).

Holotype worker: TL 5.2, HL 1.27, HW 1.07, CI 84, SL 1.00, SI 94, PW 0.73, AL 1.93, ED 0.17, ML 0.63, PL 0.40, PH 0.60, DPW 0.43, LPI 160, DPI 108. In full-face view, head nearly square. Occipital margin almost straight, shallowly depressed. Occipital carina distinct. Occipital corners blunt but distinct. Sides of head weakly convex. Inner margin of mandible with 1 small tooth, masticatory margin with 5 large teeth and 3 small teeth. Clypeus has blunt and stout longitudinal central carina, median lobe rounded at apex. Apex of scape just reached to occipital corner, segments 4-9 of flagellum about as broad as long. Eyes smaller. In lateral view dorsum of alitrunk weakly depressed at metanotal groove, promesonotal suture distinct. Dorsum of propodeum straight, about 1.8 times as long as declivity, in lateral view posterodorsal corner rounded, declivity shallowly depressed. Petiolar node anteroposteriorly compressed, in lateral view anterior face evenly convex, dorsal face convex, posterior face straight; in dorsal view the node nearly semicircular, width：length = 5：3. Subpetiolar process large, triangular. Constriction between the two basal gastral segments weak. Mandibles finely and longitudinally striate. Head, alitrunk, petiole and gaster smooth and shining. Anterior 1/3 of head finely and longitudinally striate. Dorsa of head and body with abundant erect or suberect short hairs, without pubescence. Antennal scapes with abundant decumbent hairs and dense decumbent pubescence. Femora and tibiae with dense decumbent hairs. Head and alitrunk black, with blue metallic luster. Petiole and gaster blackish brown. Mandibles, antennae, clypeus, legs and gastral apex reddish brown. Tarsi brownish yellow. **Paratype workers:** TL 4.8-5.2, HL 1.23-1.30, HW 1.00-1.07, CI 79-84, SL 0.97-1.03, SI 94-97, PW 0.70-0.77, AL 1.73-1.90, ED 0.17-0.18, ML 0.60-0.70, PL 0.37-0.40, PH 0.53-0.60, DPW 0.38-0.43, LPI 142-150, DPI 96-109 (*n*=5). As holotype.

Holotype: Worker, China: Yunnan Province, Mengla County, Menglun Town, Menglun, 830 m, 1996.Ⅲ.6, collected by Zheng-hui Xu, No. A96-565.

Etymology: The specific epithet has been named after Huangdi (2717-2599 BC), chief of the ancient Chinese tribes.

正模工蚁： 正面观，头部近方形；后缘几乎平直，浅凹；后头脊明显，后角钝但是明显；侧缘轻度隆起。上颚内缘具1个小齿，咀嚼缘具5个大齿和3个小齿。唇基具粗钝的中央纵脊，中叶顶端圆。触角柄节刚到达头后角，鞭节第4~9节长宽约相等。复眼小。侧面观，胸部背面在后胸沟处轻度凹陷，前中胸背板缝明显。并胸腹节背面直，约为斜面长的1.8倍，后上角圆，斜面浅凹。侧面观，腹柄结前后压扁，前面适度隆起，背面隆起，后面平直；背面观，腹柄结近半圆形，宽：长=5：3；腹柄下突大，三角形。后腹部基部2节间轻度收缩。上颚具细纵条纹。头部、胸部、腹柄和后腹部光滑发亮；头前部1/3具细纵条纹。头部和身体背面具丰富直立或亚直立短毛，缺绒毛被；触角柄节具丰富倾斜立毛和密集倾斜绒毛被；腿节和胫节具密集倾斜立毛。头胸部黑色，具蓝色金属光泽；腹柄和后腹部黑棕色；上颚、触角、唇基、足和后腹部末端红棕色；跗节棕黄色。**副模工蚁：** 特征同正模。

正模： 工蚁，中国云南省勐腊县勐仑镇勐仑，830m，1996.Ⅲ.6，徐正会采集，No. A96-565。

词源： 该新种的名称以中国古代华夏部落首领黄帝（公元前2717—2599年）命名。

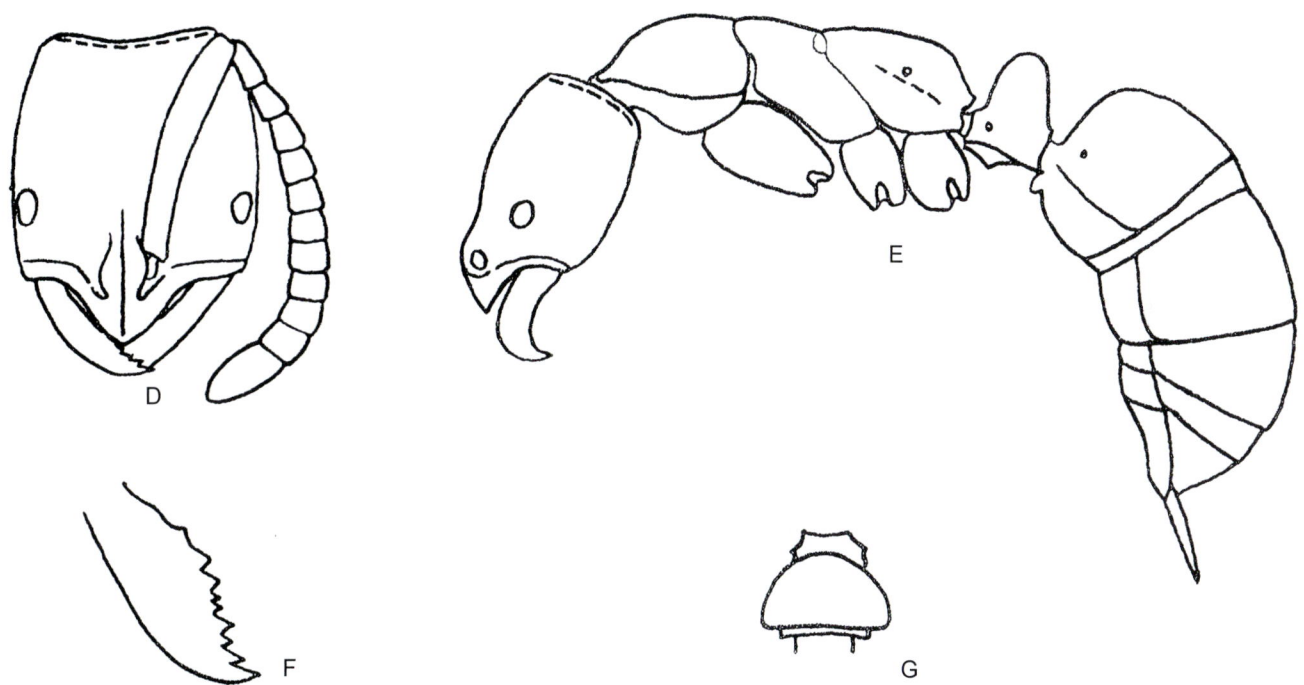

黄帝细颚猛蚁 *Leptogenys huangdii*

A, D. 工蚁头部正面观；B, E. 工蚁身体侧面观；C. 工蚁身体背面观；F. 工蚁上颚背面观；G. 工蚁腹柄背面观（D-G. 引自 Xu, 2000c）

A, D. Head of worker in full-face view; B, E. Body of worker in lateral view; C. Body of worker in dorsal view; F. Mandible of worker in dorsal view; G. Petiole of worker in dorsal view (D-G. cited from Xu, 2000c)

老子细颚猛蚁
Leptogenys laozii Xu, 2000

Leptogenys laozii Xu, 2000c, Entomologia Sinica, 7(2): 123, figs. 41-44 (w.) CHINA (Yunnan).

Holotype worker: TL 4.6, HL 1.00, HW 0.67, CI 67, SL 1.03, SI 155, PW 0.60, AL 1.57, ED 0.17, ML 0.50, PL 0.50, PH 0.60, DPW 0.40, LPI 120, DPI 80. In full-face view, head roughly rectangular, distinctly longer than broad. Occipital margin straight, occipital carina obvious. Occipital corners rounded. Anterior 2/3 of sides of head relatively straight, posterior 1/3 narrowed backward. Mandibles slender, inner margin without tooth, masticatory margin with only 1 apical tooth, basal corner rounded. Clypeus with sharp longitudinal central carina, median lobe extruding at apex. Scape of antenna surpassed occipital corner by about 1/3 of its length. Segments of flagellum longer than broad, the second and third joints about equal. In lateral view dorsum of alitrunk deeply depressed at metanotal groove. Promesonotum evenly convex, promesonotal suture distinct. Dorsum of propodeum slightly convex, about 3 times as long as declivity, declivity weakly convex. In lateral view petiolar node trapezoid, lowering down forward, anterior face nearly straight, about 1/2 as high as posterior face, anterodorsal angle rounded, posterodorsal angle bluntly extruding, dorsal face evenly convex, posterior face straight. In dorsal view, the node as broad as long, narrowed forward. Subpetiolar process small, nearly square, posteroventral corner dentiform. Constriction between the two basal gastral segments distinct. Mandibles and clypeus longitudinally striate. Head, alitrunk and petiole densely punctate. Sides of pronotum smooth. Sides of mesonotum and metanotum longitudinally rugose. Gaster smooth and shining. Dorsum of head and body with abundant erect long hairs, suberect short hairs and decumbent pubescence. Dorsum of head with dense pubescence. Scapes, femora and tibiae with abundant subdecumbent hairs and dense decumbent pubescence. Head, alitrunk and petiole brownish black. Gaster blackish brown. Mandibles, antennae, clypeus and legs reddish brown. **Paratype workers:** TL 4.5-5.0, HL 1.00-1.07, HW 0.63-0.70, CI 63-67, SL 0.97-1.07, SI 152-160, PW 0.58-0.60, AL 1.50-1.57, ED 0.13-0.20, ML 0.47-0.53, PL 0.47 -0.50, PH 0.60-0.63, DPW 0.38-0.42, LPI 120-129, DPI 77-83 (n=4). As holotype, but body reddish brown to brownish black.

Holotype: Worker, China: Yunnan Province, Mengla County, Menglun Town, Menglun, 650 m, 1998.Ⅲ.6, collected by Zheng-hui Xu, No. A98-1010.

Etymology: The specific epithet has been named after Laozi (571 BC-471 BC), founder of the Taoist school, Chinese thinker of the Spring and Autumn period.

正模工蚁： 正面观，头部近长方形，长明显大于宽；后缘平直，后头脊明显，后角圆；侧缘前部2/3较平直，后部1/3向后变窄。上颚细长，内缘缺齿，咀嚼缘仅具1个端齿，基角圆。唇基具锋锐的中央纵脊，中叶顶端突出。触角柄节约1/3超过头后角；鞭节的各节长大于宽，第2节和第3节约等长。侧面观，胸部背面在后胸沟处深凹；前中胸背板均匀隆起，前中胸背板缝明显。并胸腹节背面轻度隆起，约为斜面长度的3倍，斜面轻度隆起。侧面观腹柄结梯形，背面向前降低，前面近平直，约为后面长的1/2，前上角圆，后上角钝凸，背面均匀隆起，后面平直。背面观，腹柄结长宽相等，向前变窄；腹柄下突小，近方形，后下角齿状。后腹部基部2节间明显缢缩。上颚和唇基具纵条纹。头部、胸部和腹柄具密集刻点；前胸背板侧面光滑，中胸和后胸侧面具纵皱纹；后腹部光滑发亮。头部和身体背面具丰富直立长毛、亚直立短毛和倾斜绒毛被，头部背面绒毛被密集；柄节、腿节和胫节具丰富亚倾斜毛和密集倾斜绒毛被。头部、胸和腹柄棕黑色，后腹部黑棕色，上颚、触角、唇基和足红棕色。**副模工蚁：** 特征同正模，但是身体红棕色至棕黑色。

正模： 工蚁，中国云南省勐腊县勐仑镇勐仑，650m，1998.Ⅲ.6，徐正会采集，No. A98-1010。

词源： 该新种的名称以中国春秋时代思想家、道家学派创始人老子（公元前571—471年）命名。

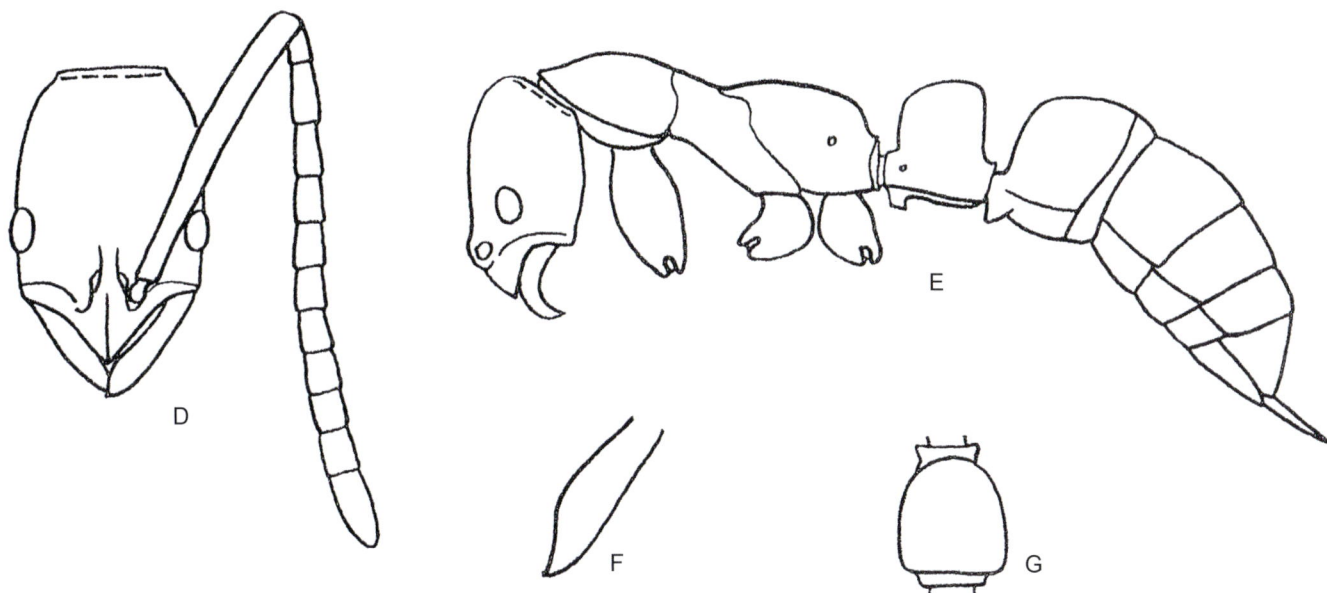

老子细颚猛蚁 *Leptogenys laozii*

A, D. 工蚁头部正面观；B, E. 工蚁身体侧面观；C. 工蚁身体背面观；F. 工蚁上颚背面观；G. 工蚁腹柄背面观（D-G. 引自 Xu, 2000c）

A, D. Head of worker in full-face view; B, E. Body of worker in lateral view; C. Body of worker in dorsal view; F. Mandible of worker in dorsal view; G. Petiole of worker in dorsal view (D-G. cited from Xu, 2000c)

孟子细颚猛蚁
Leptogenys mengzii Xu, 2000

Leptogenys mengzii Xu, 2000c, Entomologia Sinica, 7(2): 124, figs. 45-52 (w.eq.) CHINA (Yunnan).

Holotype worker: TL 5.2, HL 1.10, HW 0.75, CI 68, SL 1.03, SI 138, PW 0.68, AL 1.63, ED 0.22, ML 0.57, PL 0.57, PH 0.63, DPW 0.43, LPI 112, DPI 76. In full-face view, head distinctly longer than broad, nearly rectangular. Occipital margin straight, occipital carina obvious. Occipital corners rounded. Sides of head relatively straight, slightly convex. Mandibles slender, inner margin without tooth, masticatory margin without tooth except the apical one, basal corner bluntly angled. Clypeus with sharp longitudinal central carina, median lobe extruding at apex. Scape of antenna surpassed occipital corner by about 1/3 of its length. Segments of flagellum longer than broad, the second and third joints equal. Eyes moderately large. In lateral view dorsum of alitrunk notched at metanotal groove. Promesonotal suture distinct. Dorsum of propodeum straight for most of its length, about 2.5 times as long as declivity, declivity straight. In lateral view petiolar node roughly trapezoid, lowering down forward, anterior face very short, slightly convex, about 1/3 as high as posterior face, posterior face straight, dorsal face evenly convex, anterodorsal corner rounded, posterodorsal corner roundly extruding. In dorsal view the node about as broad as long, length : width = 6 : 5, narrowed forward. Subpetiolar process nearly triangular, posteroventral corner dentate. Constriction between the two basal gastral segments distinct. Mandibles smooth, sparsely punctate. Clypeus longitudinally striate. Head densely punctate, surface between punctures smooth. Alitrunk longitudinally rugose and sparsely punctate, but pronotum, median portions of sides of mesothorax and metathorax, and dorsum of propodeum smooth and shining. Petiolar node and gaster smooth and shining, but lateroposterior margin of petiolar node with coarse rugae and punctures. Dorsum of head and body with abundant erect long hairs and suberect short hairs, dorsum of head with dense decumbent pubescence. Scapes, femora and tibiae with sparse subdecumbent hairs and dense decumbent pubescence. Body black. Mandibles, clypeus, antennae, legs and gastral apex reddish brown. **Paratype workers:** TL 4.5-5.1, HL 1.03-1.05, HW 0.72-0.73, CI 68-71, SL 0.97-1.02, SI 132-142, PW 0.63-0.68, AL 1.50-1.53, ED 0.18-0.20, ML 0.47-0.50, PL 0.50-0.57, PH 0.60-0.63, DPW 0.40-0.43, LPI 112-120, DPI 71-83 (n=5). As holotype, but body blackish brown to black.

Holotype: Worker, China: Yunnan Province, Menghai County, Meng'a Town, Papo Village, 1600 m, 1997.IX.9, collected by Zheng-hui Xu in deciduous broadleaf forest, No. A97-2249.

Etymology: The specific epithet has been named after Menzi (about 372 BC-289 BC), a leading figure of the Confucian school, a great thinker and educator in the Warring States period of China.

正模工蚁： 正面观，头部长明显大于宽，近长方形；后缘平直，后头脊明显，后角圆；侧缘较直，轻度隆起。上颚细长，内缘缺齿，咀嚼缘除端齿外缺齿，基角钝角状。唇基具锋锐的中央纵脊，中叶顶端突出。触角柄节约1/3超过头后角；鞭节的各节长大于宽，第2节和第3节等长。复眼中等大小。侧面观，胸部背面在后胸沟处切入，前中胸背板缝明显。并胸腹节背面大部平直，约为斜面长的2.5倍长，斜面平直。侧面观，腹柄结近梯形，向前降低；前面很短，轻度隆起，约为后面长度的1/3；后面平直，背面均匀隆起，前上角圆，后上角圆凸。背面观，腹柄结长宽约相等，长：宽=6：5，向前变窄；腹柄下突近三角形，后下角齿状。后腹基部2节间缢缩明显。上颚光滑，具稀疏刻点。唇基具纵条纹。头部具密集刻点，界面光滑。胸部具纵皱纹和稀疏刻点，但是前胸背板、中胸侧面中部、后胸侧面中部和并胸腹节背面光滑发亮。腹柄结和后腹部光滑发亮，但是腹柄结侧后缘具粗糙皱纹和刻点。头部和身体背面具丰富直立长毛和亚直立短毛，头部背面具密集倾斜绒毛被；柄节、腿节和胫节具稀疏亚倾斜毛和密集倾斜绒毛被。身体黑色，上颚、唇基、触角、足和后腹部末端红棕色。**副模工蚁：** 特征同正模，但是身体黑棕色至黑色。

正模： 工蚁，中国云南省勐海县勐阿乡怕迫村，1600m，1997.IX.9，徐正会采于落叶阔叶林中，No. A97-2249。

词源： 该新种的名称以中国战国时期伟大的思想家、教育家，儒家学派的代表人物孟子（公元前372—289年）命名。

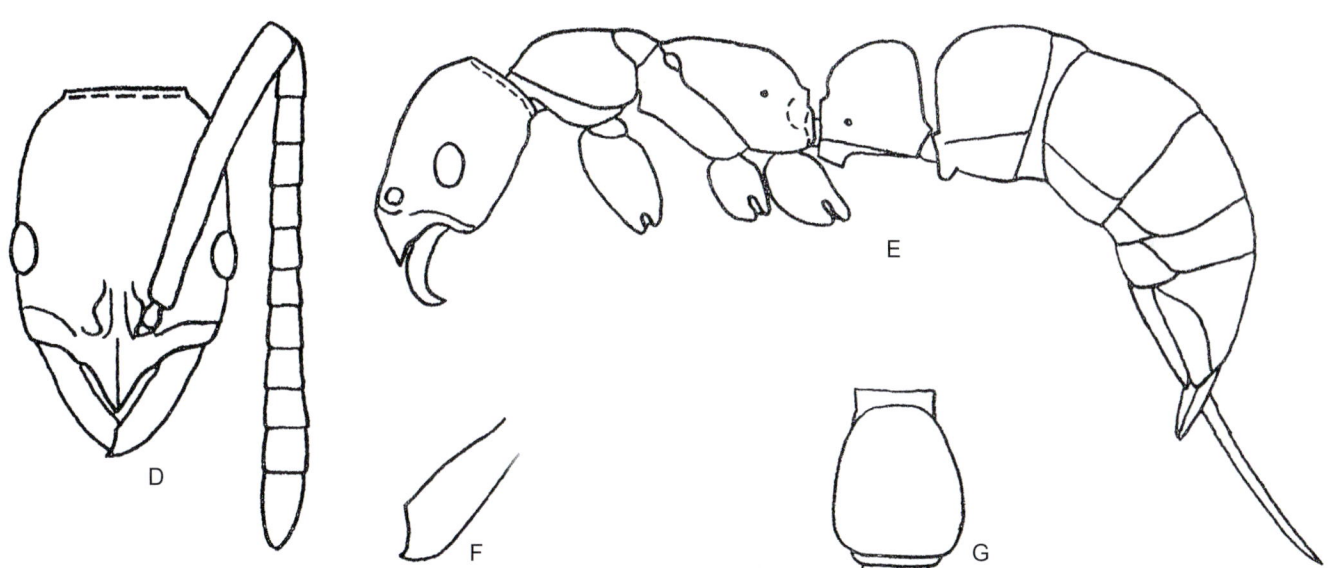

孟子细颚猛蚁 Leptogenys mengzii

A, D. 工蚁头部正面观；B, E. 工蚁身体侧面观；C. 工蚁身体背面观；F. 工蚁上颚背面观；G. 工蚁腹柄背面观（D-G. 引自 Xu, 2000c）
A, D. Head of worker in full-face view; B, E. Body of worker in lateral view; C. Body of worker in dorsal view; F. Mandible of worker in dorsal view; G. Petiole of worker in dorsal view (D-G. cited from Xu, 2000c)

盘古细颚猛蚁
Leptogenys pangui Xu, 2000

Leptogenys pangui Xu, 2000c, Entomologia Sinica, 7(2): 120, figs. 13-16 (w.) CHINA (Yunnan).

Holotype worker: TL 13.5, HL 2.50, HW 1.63, CI 65, SL 3.45, SI 213, PW 1.40, AL 4.50, ED 0.70, ML 1.20, PL 1.57, PH 1.23, DPW 0.67, LPI 79, DPI 43. Body large. In full-face view, head distinctly longer than broad, obviously narrowed backward. Occipital margin nearly straight, shallowly depressed. Occipital carina distinct. Occipital corners bluntly angled. Sides weakly convex. Mandibles slender, inner margin without tooth, masticatory margin with 1 apical tooth and 1 small basal tooth. Clypeus with sharp longitudinal central carina, median lobe truncated at most apex. Scape of antenna surpassed occipital corner by 1/2 of its length. Ratio of length of the basal 3 segments of flagellum expressed as, segment 1 : segment 2 : segment 3 = 4 : 10 : 9. Eyes large. In lateral view dorsum of alitrunk deeply depressed at metanotal groove. Pronotum and mesonotum evenly convex. Promesonotal suture distinct. Dorsum of propodeum long and straight for most of its length, about 3.5 times as long as declivity. Petiolar node strongly and laterally compressed, in lateral view, longer than high, nearly triangular, without distinct anterior face, dorsal face weakly convex, posterior face straight; in dorsal view, about 2 times as long as broad, narrowed forward. Subpetiolar process small, triangular. Constriction between the two basal gastral segments distinct. Mandibles with dense micro-punctures and sparse large punctures. Clypeus finely and longitudinally striate. Head, alitrunk, petiole and gaster smooth and shining, with quite sparse piliferous punctures. Head and body with abundant erect long hairs, suberect short hairs and very sparse decumbent pubescence. Vertex with dense pubescence. Scapes of antennae with sparse suberect long hairs, subdecumbent short hairs and dense decumbent pubescence. Femora and tibiae with decumbent hairs. Body black, with blue metallic luster. Mandibles, flagella of antennae, legs and posterior margins of each gastral segments blackish brown. Pilosity brownish yellow. **Paratype worker:** TL 12.8, HL 2.45, HW 1.60, CI 65, SL 3.30, SI 206, PW 1.45, AL 4.30, ED 0.70, ML 1.15, PL 1.53, PH 1.20, DPW 0.77, LPI 78, DPI 50 (*n*=1). As holotype, but hairs on dorsum of head and body very sparse.

Holotype: Worker, China: Yunnan Province, Mengla County, Menglun Town, Menglun, 650 m, 1998.Ⅸ.15, collected by Zheng-hui Xu, No. A98-995.

Etymology: The specific epithet has been named after Pangu, the god of creation in Chinese mythology.

正模工蚁： 体大。正面观，头部长显著大于宽，向后明显变窄；后缘近平直，浅凹；后头脊明显，后角钝角状；侧缘轻度隆起。上颚细长，内缘缺齿，咀嚼缘具1个端齿和1个小基齿。唇基具锋锐的中央纵脊，前缘中叶顶端平截。触角柄节约1/2超过头后角；鞭节基部第1~3节长度的比值为4：10：9。复眼大。侧面观，胸部背面在后胸沟处深凹，前胸背板和中胸背板均匀隆起，前中胸背板缝明显。并胸腹节背面大部长且平直，约为斜面长度的3.5倍。腹柄结强烈侧扁，侧面观长大于高，近三角形，前面不明显；背面轻度隆起，后面平直。背面观腹柄结长约为宽的2倍，向前变窄。腹柄下突小，三角形。后腹部基2节间缢缩明显。上颚具密集微刻点和稀疏大刻点。唇基具细纵条纹。头部、胸部、腹柄和后腹部光滑发亮，具相当稀疏的具毛刻点。头部和身体具丰富直立长毛、亚直立短毛和非常稀疏的倾斜绒毛被，头顶具密集绒毛被；触角柄节具稀疏亚直立长毛、亚倾斜短毛和密集倾斜绒毛被；腿节和胫节具倾斜毛。身体黑色，具蓝色金属光泽；上颚、触角鞭节、足和后腹部各节后缘黑棕色，毛被棕黄色。**副模工蚁：** 特征同正模，但是头部和身体背面的立毛非常稀疏。

正模： 工蚁，中国云南省勐腊县勐仑镇勐仑，650m，1998.Ⅸ.15，徐正会采集，No. A98-995。

词源： 该新种的名称以中国神话中的创世神盘古命名。

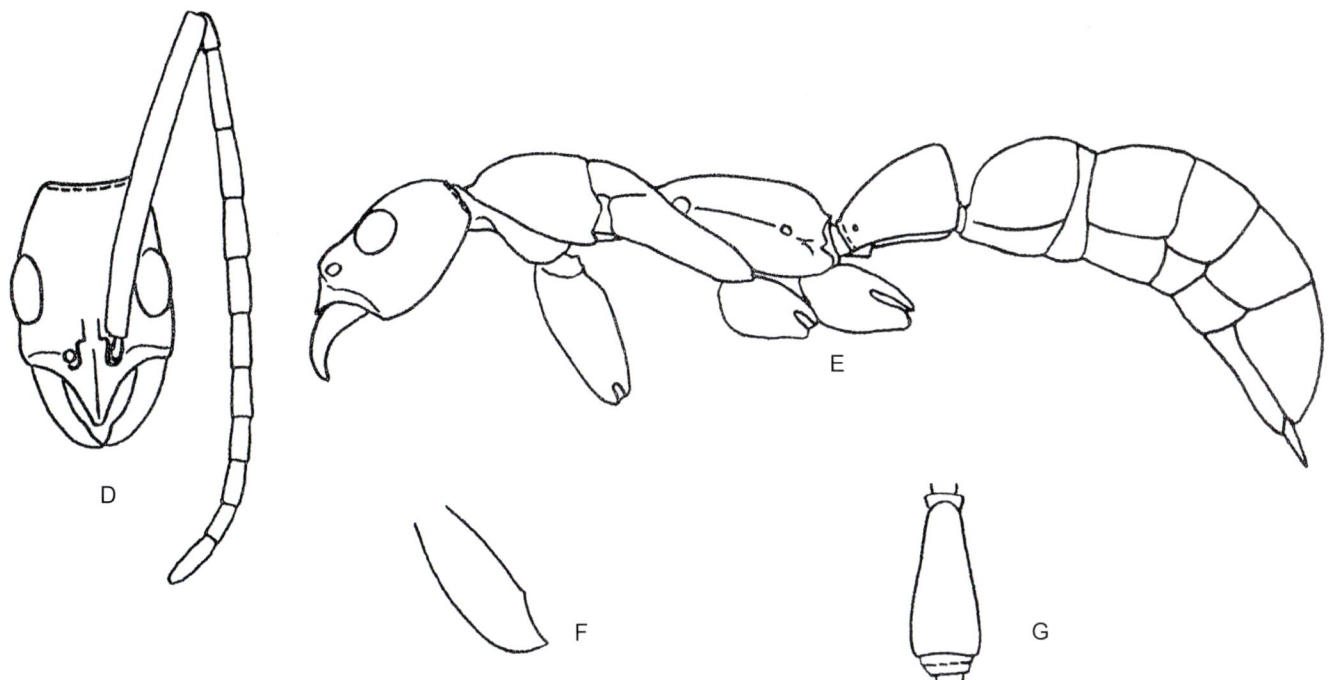

盘古细颚猛蚁 Leptogenys pangui

A, D. 工蚁头部正面观；B, E. 工蚁身体侧面观；C. 工蚁身体背面观；F. 工蚁上颚背面观；G. 工蚁腹柄背面观（D-G. 引自 Xu, 2000c）

A, D. Head of worker in full-face view; B, E. Body of worker in lateral view; C. Body of worker in dorsal view; F. Mandible of worker in dorsal view; G. Petiole of worker in dorsal view (D-G. cited from Xu, 2000c)

孙子细颚猛蚁
Leptogenys sunzii Xu & He, 2015

Leptogenys sunzii Xu & He, 2015, Myrmecological News, 21: 147, figs. 26-43 (w.) CHINA (Yunnan).

Holotype worker: TL 8.2, HL 1.77, HW 1.27, CI 72, SL 1.90, SI 150, ML 1.00, ED 0.37, PW 1.00, MSL 2.93, PL 0.73, PH 0.90, DPW 0.57, LPI 123, DPI 77. In full-face view head trapezoidal, longer than broad and weakly widened anteriorly, posterior margin straight and carinate, posterior corner rounded, lateral margins moderately convex. Mandible relatively broad, masticatory margin edentate, basal corner bluntly angled, inner margin weakly convex. Clypeus acutely longitudinally carinate, bluntly pointed at apex, each side with blunt prominence. Antenna long, two fifths of scape length surpasses posterior head corner, flagellar segments distinctly longer than broad, segment 3 longer than segment 4. Eye moderately large, occupying about one third of lateral cephalic margin, situated slightly in front of midpoint of lateral margin. In lateral view promesonotum moderately convex and distinctly higher than propodeum, promesonotal suture obvious. Metanotal groove deeply angularly impressed. Dorsum of propodeum weakly convex, about 2.5 times as long as declivity, posterodorsal corner rounded. Petiolar node roughly trapezoidal, about 1.4 times higher than long, dorsal and anterior margin weakly convex, posterior margin straight, anterodorsal corner broadly rounded, posterodorsal corner prominent. Subpetiolar process long and narrow, roughly cuneiform. Constriction between abdominal segments III and IV distinct. Sting extruding. In dorsal view mesosoma strongly constricted at mesothorax, distinctly widened posteriorly, sides of pronotum strongly convex. Petiolar node trapezoidal, as broad as long, strongly widened posteriorly, anterior and lateral margins weakly convex, posterior margin almost straight, anterior corner rounded, posterior corner blunt. Mandible smooth and shiny. Clypeus finely longitudinally striate. Head, mesosoma, petiole and gaster smooth and shiny. Dorsal portion of mesopleuron, ventral portions of metapleuron and propodeum striate. Declivity transversely striate. Head dorsum with abundant suberect hairs and decumbent pubescence. Dorsa of mesosoma, petiolar node and gaster with abundant suberect hairs and sparse decumbent pubescence. Scapes and tibiae with sparse subdecumbent hairs and dense decumbent pubescence. Body color black, with a bluish metallic reflection. Mandible, flagellum and leg reddish brown. **Paratype workers:** TL 7.3-8.7, HL 1.50-1.80, HW 1.03-1.30, CI 67-75, SL 1.67-1.93, SI 141-166, ML 0.80-1.07, ED 0.27-0.40, PW 0.87-1.03, MSL 2.47-2.93, PL 0.67-0.73, PH 0.80-0.90, DPW 0.50-0.60, LPI 114-135, DPI 73-86 (*n*=13). As holotype, but paratype workers from Caiyanghe are relatively smaller, dorsum of head with very sparse tiny superficial piliferous punctures, metapleuron with more transverse striations.

Holotype: Worker, China: Yunnan Province, Jingdong County, Wenlong Town, Yichang Village (24.644267°N, 100.731700°E), 1950 m, 2001.XI.11 collected by Zheng-qiang Tong on the ground in conifer-broadleaf mixed forest, No. A4575.

Etymology: The specific epithet refers to "Sunzi" (Wu Sun, 535 B.C. - ?), a famous ancient Chinese strategist.

正模工蚁：正面观，头部梯形，长大于宽，向前轻度变宽；后缘平直，具后缘脊，后角圆；侧缘中度隆起。上颚较宽，咀嚼缘缺齿，基角钝角状，内缘轻度隆起。唇基具锋锐的中央纵脊，前缘中叶顶端钝，每侧具1个钝突。触角长，柄节约2/5超过头后角，鞭节各节长明显大于宽，第3节长于第4节。复眼中等大，约占据头侧缘的1/3，位于头侧缘中点稍前处。侧面观，前中胸背板中度隆起，明显高于并胸腹节，前中胸背板缝明显，后胸沟角状深凹。并胸腹节背面轻度隆起，约为斜面长度的2.5倍，后上角圆。腹柄结近梯形，高约为长的1.4倍，背面和前面轻度隆起，后面平直；前上角宽圆，后上角凸出；腹柄下突长而窄，近楔形。后腹部基部2节间收缩明显。螫针伸出。背面观，胸部在中胸处强烈收缩，向后明显变宽；前胸背板两侧强烈隆起。腹柄结梯形，长宽相等，向后强烈变宽，前缘和侧缘轻度隆起，后缘几乎平直，前侧角圆，后侧角钝。上颚光滑发亮。唇基具细纵条纹。头部、胸部、腹柄和后腹部光滑发亮；中胸侧板上部、后胸侧板下部和并胸腹节具条纹；并胸腹节斜面具横条纹。头部背面具丰富亚直立毛和倾斜绒毛被；胸部、腹柄结和后腹部背面具丰富亚直立毛和稀疏倾斜绒毛被；柄节和胫节具稀疏亚倾斜毛和密集倾斜绒毛被。身体黑色，具蓝色金属光泽；上颚、鞭节和足红棕色。**副模工蚁：**特征同正模，但是采自菜阳河的副模工蚁个体较小，头部背面具非常稀疏的微小肤浅具毛刻点，后胸侧板具更多的横条纹。

正模：工蚁，中国云南省景东县文龙镇义昌村（24.644267°N，100.731700°E），1950m，2001.XI.11，童正强采于针阔混交林地表，No. A4575。

词源：该新种以中国古代著名战略家孙子（孙武，公元前535—？）命名。

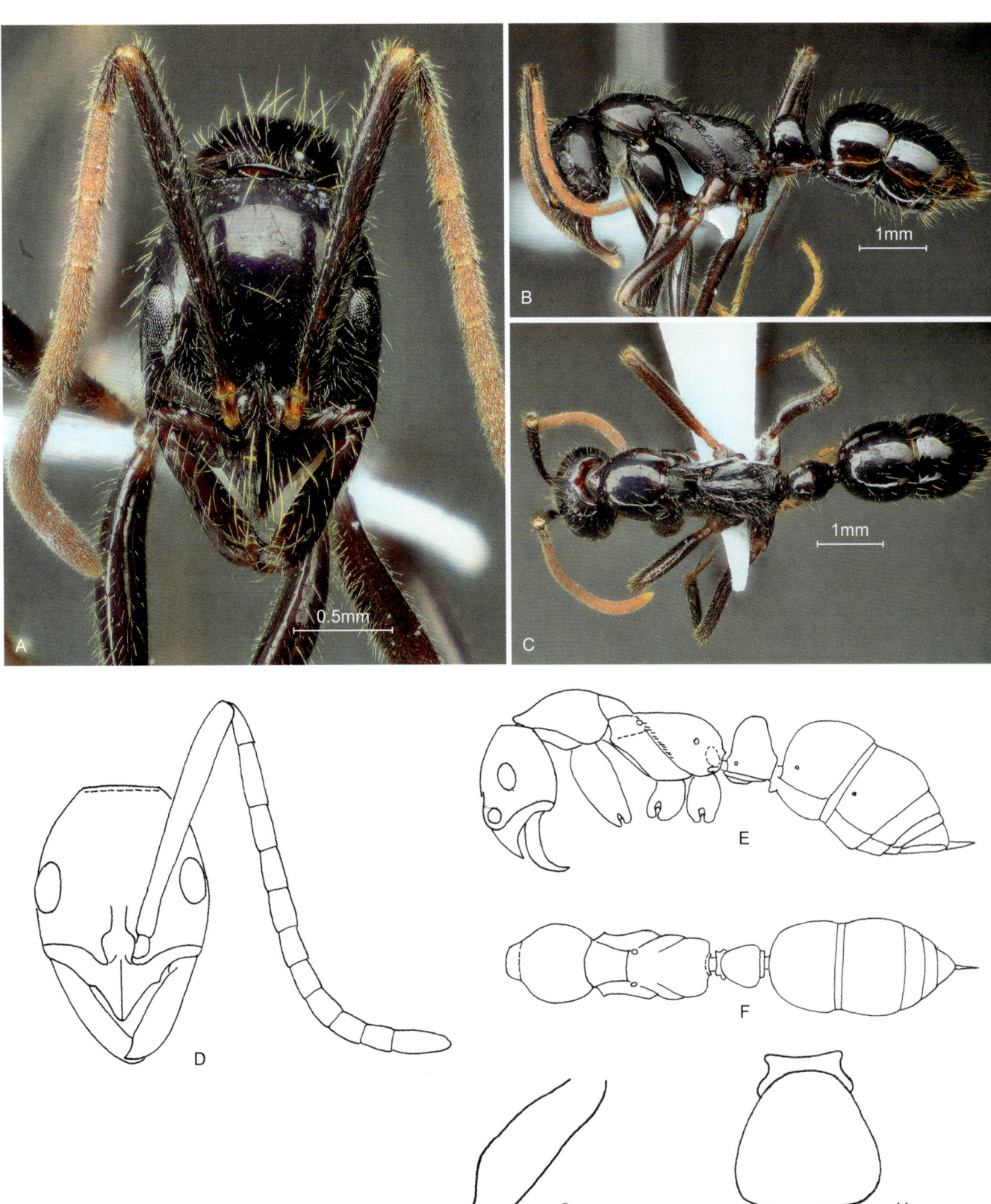

孙子细颚猛蚁 *Leptogenys sunzii*

A, D. 工蚁头部正面观；B, E. 工蚁身体侧面观；C, F. 工蚁身体背面观；G. 工蚁上颚背面观；H. 工蚁腹柄背面观（D-H. 引自 Xu和He, 2015）

A, D. Head of worker in full-face view; B, E. Body of worker in lateral view; C, F. Body of worker in dorsal view; G. Mandible of worker in dorsal view; H. Petiole of worker in dorsal view (D-H. cited from Xu & He, 2015)

炎帝细颚猛蚁
Leptogenys yandii Xu & He, 2015

Leptogenys yandii Xu & He, 2015, Myrmecological News, 21: 145, figs. 28-35 (w.) CHINA (Tibet).

Holotype worker: TL 5.4, HL 1.07, HW 0.73, CI 69, SL 1.07, SI 145, ML 0.53, ED 0.23, PW 0.70, MSL 1.67, PL 0.67, PH 0.70, DPW 0.50, LPI 105, DPI 75. In full-face view, head longer than broad, roughly trapezoidal and widened anteriorly, posterior margin straight and carinate, posterior corner narrowly rounded, lateral margin weakly convex. Mandible narrow and slender, masticatory margin edentate, basal corner bluntly angled. Clypeus acutely longitudinally carinate, weakly convex at apex, each side with blunt tooth. Antenna 12-segmented, scape surpassing posterior head corner by one fourth of its length, flagellar segments longer than broad, segments 3 and 4 about equal. Eye occupying one fourth of lateral cephalic margin, and located in front of midpoint of lateral margin. In lateral view pronotum weakly convex. Promesonotal suture impressed. Dorsum of mesonotum and propodeum almost straight and slightly lower than pronotum, metanotal groove not impressed. Dorsum of propodeum about 1.5 times as long as declivity, posterodorsal corner blunt. Petiolar node trapezoidal, about 1.1 times higher than long, both anterior and posterior margins straight and vertical, dorsal margin weakly convex, anterodorsal corner narrowly rounded, posterodorsal corner blunt. Subpetiolar process short and slender, roughly triangular, posteroventrally pointed. Constriction between abdominal segments III and IV distinct. Sting extruding. In dorsal view, lateral margin of pronotum strongly convex, posterior margin concave. The rest of mesosoma weakly widened posteriorly. Metanotal groove narrow and visible. Petiolar node trapezoidal, weakly widened posteriorly, about 1.1 times as long as broad, anterior and lateral margins slightly convex, posterior margin straight, anterior corners rounded, posterior corners blunt. Mandible finely longitudinally striate. Clypeus longitudinally rugose. Head dorsum largely densely punctate with interspace coarsely retirugose. Mesosoma, petiolar node and first gastral segment fully, largely and deeply punctate, interface smooth and shiny, as broad as or narrower than puncture diameter. Mesopleuron and metapleuron densely punctate with interspace coarsely retirugose. Declivity coarsely transversely striate. Second gastral segment finely sparsely punctate, the rest of gaster smooth and shiny. Head dorsum with abundant suberect hairs and subdecumbent pubescence. Dorsa of mesosoma, petiolar node and gaster with sparse suberect hairs and decumbent pubescence. Scape and tibia with sparse subdecumbent hairs and dense decumbent pubescence. Body color black. Mandible, clypeus, antenna, leg and gastral apex reddish brown. Eye grey. **Paratype workers:** TL 5.2-5.7, HL 1.07-1.10, HW 0.73-0.77, CI 67-72, SL 1.03-1.10, SI 139-145, ML 0.50-0.57, ED 0.20-0.23, PW 0.67-0.73, MSL 1.60-1.80, PL 0.63-0.67, PH 0.70-0.73, DPW 0.47-0.53, LPI 105-119, DPI 70-84 (n=17). As holotype, but in some individuals, apex of clypeus roundly convex, punctures on mesopleuron, metapleuron and side of propodeum are relatively larger with interspace reticulate.

Holotype: Worker, China: Tibet Autonomous Region, Medog County, Medog Town, Yarang Village (29.296000°N, 95.276650°E), 760 m, 2008.V.21, collected by Zheng-hui Xu in a nest inside decayed wood in the valley rainforest, No. A08-1011.

Etymology: The specific epithet refers to "Yandi (Yan Emperor, born about 6000 - 5500 years ago)", one of the two earliest Chinese emperors.

正模工蚁： 正面观，头部长大于宽，近梯形，向前变宽；后缘直，具后头脊，后角窄圆；侧缘轻度隆起。上颚狭窄细长，咀嚼缘缺齿，基角钝角状。唇基具锋锐的中央纵脊，中叶顶端轻度隆起，每侧具1个钝齿。触角12节，柄节约1/4超过头后角，鞭节各节长大于宽，第3节和第4节约等宽。复眼占据头侧缘的1/4，位于头侧缘中点之前。侧面观，前胸背板轻度隆起，前中胸背板缝凹陷；中胸和并胸腹节背面几乎平直，稍低于前胸背板，后胸沟不凹陷。并胸腹节背面约为斜面长的1.5倍，后上角钝。腹柄结梯形，高约为长的1.1倍，前面和后面平直，背面轻度隆起，前上角窄圆，后上角钝；腹柄下突短而细，近三角形，指向后下方。后腹部基部2节间收缩明显。螫针伸出。背面观，前胸背板侧缘强烈突起，后缘凹陷；胸部其余部分向后变宽。后胸沟窄，可见。腹柄结梯形，向后轻度变宽，长约为宽的1.1倍，前缘和侧缘轻度隆起，后缘平直，前侧角圆，后侧角钝。上颚具细纵条纹。唇基具纵皱纹。头部背面具密集大刻点，界面呈粗糙网状皱纹。胸部、腹柄结和后腹部第1节具深凹大刻点，界面光滑发亮，刻点间距等于或小于刻点直径。中胸侧板和后胸侧板具密集刻点，界面呈粗糙网状皱纹。并胸腹节斜面具粗糙横条纹。后腹部第2节具稀疏细刻点，其余腹节光滑发亮。头部背面具丰富亚直立毛和亚倾斜绒毛被；胸部、腹柄结和后腹部背面具稀疏亚直立毛和倾斜绒毛被；柄节和胫节具稀疏亚直立毛和密集倾斜绒毛被。身体黑色，上颚、唇基、触角、足和后腹部末端红棕色，复眼灰色。**副模工蚁：** 特征同正模，但一些个体唇基顶端圆形隆起，胸部侧面刻点较大，

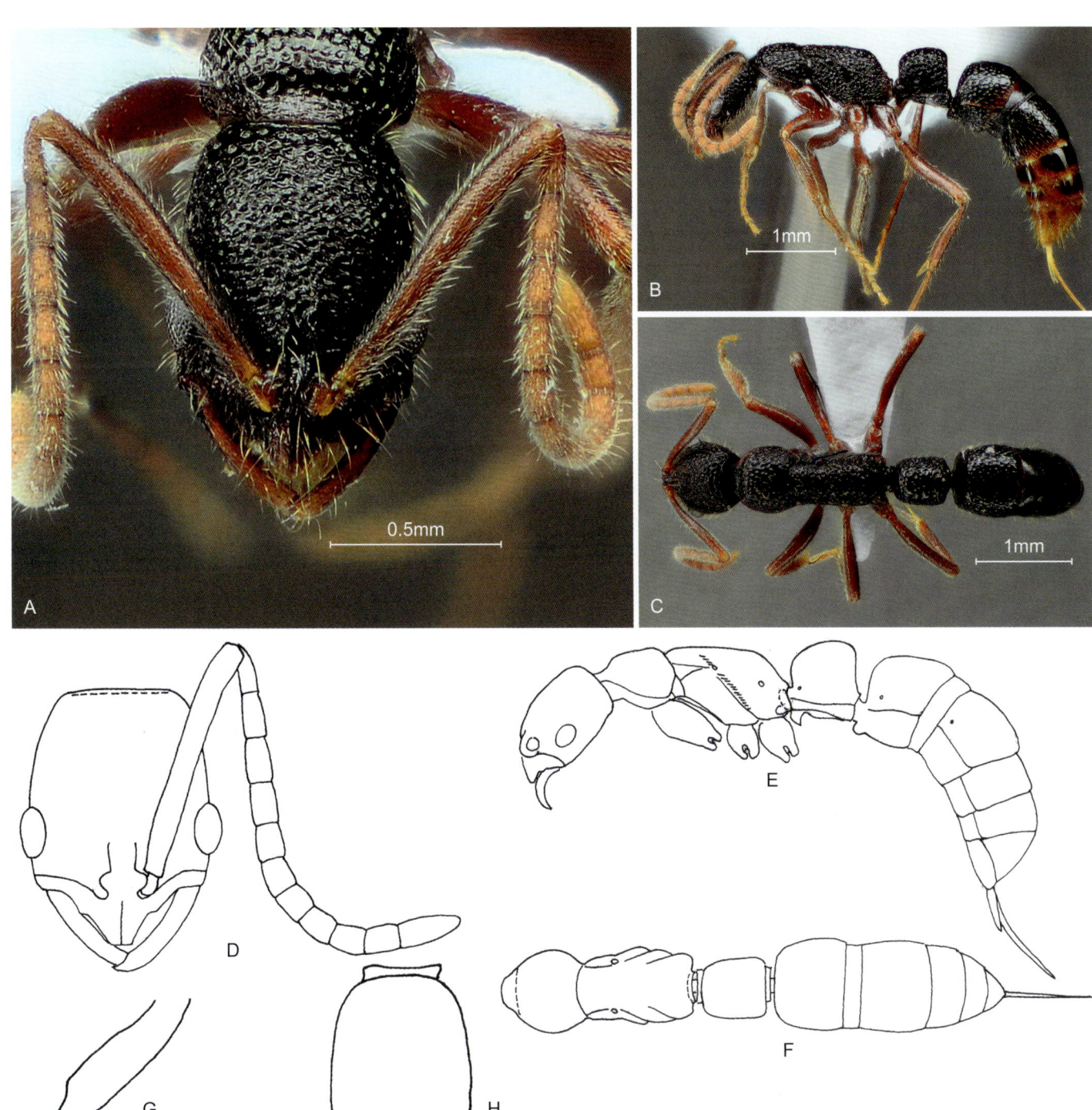

炎帝细颚猛蚁 *Leptogenys yandii*

A, D. 工蚁头部正面观；B, E. 工蚁身体侧面观；C, F. 工蚁身体背面观；G. 工蚁上颚背面观；H. 工蚁腹柄背面观（D-H. 引自 Xu和He, 2015）
A, D. Head of worker in full-face view; B, E. Body of worker in lateral view; C, F. Body of worker in dorsal view; G. Mandible of worker in dorsal view; H. Petiole of worker in dorsal view (D-H. cited from Xu & He, 2015)

界面网状。

正模：工蚁，中国西藏自治区墨脱县墨脱镇亚让村（29.296000°N，95.276650°E），760m，2008.Ⅴ.21，徐正会采于沟谷雨林朽木内巢中，No. A08-1011。

词源：该新种以中国最早的两个皇帝之一"炎帝"（生于6000~5000年前）命名。

庄子细颚猛蚁
Leptogenys zhuangzii Xu, 2000

Leptogenys zhuangzii Xu, 2000c, Entomologia Sinica, 7(2): 122, figs. 37-40 (w.) CHINA (Yunnan).

Holotype worker: TL 7.4, HL 1.70, HW 1.10, CI 65, SL 1.63, SI 148, PW 0.93, AL 2.53, ED 0.25, ML 0.87, PL 0.80, PH 0.93, DPW 0.70, LPI 117, DPI 88. In full-face view, head rectangular, much longer than broad. Occipital margin straight, occipital carina distinct. Occipital corners rounded. Sides of head nearly parallel. Mandibles slender, masticatory margin as long as inner margin, inner margin without tooth, masticatory margin with 1 small tooth in the middle of the blade except the apical tooth, the basal corner rounded. Clypeus with sharp longitudinal central carina, median lobe extruding at apex. Scape of antenna surpassed occipital corner by 1/3 of its length. Joints of flagellum longer than broad, ratio of length of the basal 3 segments expressed as, segment 1 : segment 2 : segment 3 = 4 : 5 : 4. Eyes moderate large. In lateral view dorsum of alitrunk deeply notched at metanotal groove. Promesonotum convex, promesonotal suture distinct. Dorsum of propodeum weakly convex, about 2.5 times as long as declivity, declivity weakly convex. In lateral view, petiolar node roughly trapezoid, anterior face short, slightly convex, about 1/2 as high as posterior face, anterodorsal angle rounded, posterodorsal corner roundly extruding, dorsal face convex, posterior face straight. In dorsal view the node slightly broader than long, width : length = 10 : 9, narrowed forward. Subpetiolar process cuneiform, ventral face with a rounded notch. Constriction between the two basal segments distinct. Mandibles weakly longitudinally rugulose. Clypeus longitudinally striate. Head, alitrunk and petiole with close fine punctures and retirugulae, sides of mesothorax and metathorax coarsely and longitudinally rugose. Gaster smooth and shining, with very sparse piliferous punctures. Dorsum of head and body with abundant erect long hairs, suberect short hairs and sparse decumbent pubescence. Head with dense pubescence. Scapes, femora and tibiae with abundant suberect long hairs, subdecumbent short hairs and dense decumbent pubescence. Body black. Mandibles, flagella, femora, tibiae, tarsi and gastral apex blackish brown to yellowish brown. **Paratype workers:** TL 7.1-7.8, HL 1.63-1.70, HW 1.10-1.13, CI 65-68, SL 1.63-1.67, SI 147-148, PW 0.93-0.97, AL 2.47-2.60, ED 0.23-0.27, ML 0.87-0.90, PL 0.77-0.83, PH 0.87-0.93, DPW 0.67-0.70, LPI 108-122, DPI 83-88 (n=5). As holotype.

Holotype: Worker, China: Yunnan Province, Menghai County, Meng'a Town, Papo Village, 1280 m, 1997.IX.10, collected by Zheng-hui Xu in secondary monsoon evergreen broadleaf forest, No. A97-2331.

Etymology: The specific epithet refers to "Zhuangzi" (369 B.C. – 286 B.C.), a famous thinker, philosopher, litterateur and representative figure of Taoist school in ancient China.

正模工蚁： 正面观，头部长方形，长明显大于宽；后缘平直，后头脊明显，后角圆；侧缘近平行。上颚细长，咀嚼缘与内缘等长，内缘缺齿；咀嚼缘除端齿外，中部具1个小齿；基角圆。唇基具锋锐的中央纵脊，中叶顶端伸出。触角柄节约1/3超过头后角；鞭节各节长大于宽，基部第1～3节的比值为4：5：4。复眼中等大。侧面观，胸部背面在后胸沟处深切，前胸背板隆起，前中胸背板缝明显。并胸腹节背面轻度隆起，约为斜面长的2.5倍；斜面轻度隆起。侧面观腹柄结近梯形，前面短，轻度隆起，约为后面长的1/2；前上角圆，后上角圆凸，背面隆起，后面平直。背面观，腹柄结宽稍大于长，宽：长=10：9，向前变窄；腹柄下突楔形，腹面具1个圆形切口。后腹部基部2节间收缩明显。上颚具弱的纵皱纹。唇基具纵条纹。头部、胸部和腹柄具细密刻点和网状皱纹，中胸侧面和后胸侧面具粗糙纵皱纹；后腹部光滑发亮，具很稀疏的具毛刻点。头部背面和身体具丰富直立长毛、亚直立短立毛和稀疏倾斜绒毛被，头部具密集绒毛被；柄节、腿节和胫节具丰富亚直立长毛，亚倾斜短毛和密集倾斜绒毛被。身体黑色，上颚、鞭节、腿节、胫节、跗节和后腹部末端黑棕色至黄棕色。**副模工蚁：** 特征同正模。

正模： 工蚁，中国云南省勐海县勐阿乡怕迫村，1280m，1997.IX.10，徐正会采于次生季风常绿阔叶林中，No. A97-2331。

词源： 该新种以中国古代著名思想家、哲学家、文学家，道家学派的代表人物庄子（公元前369—286年）命名。

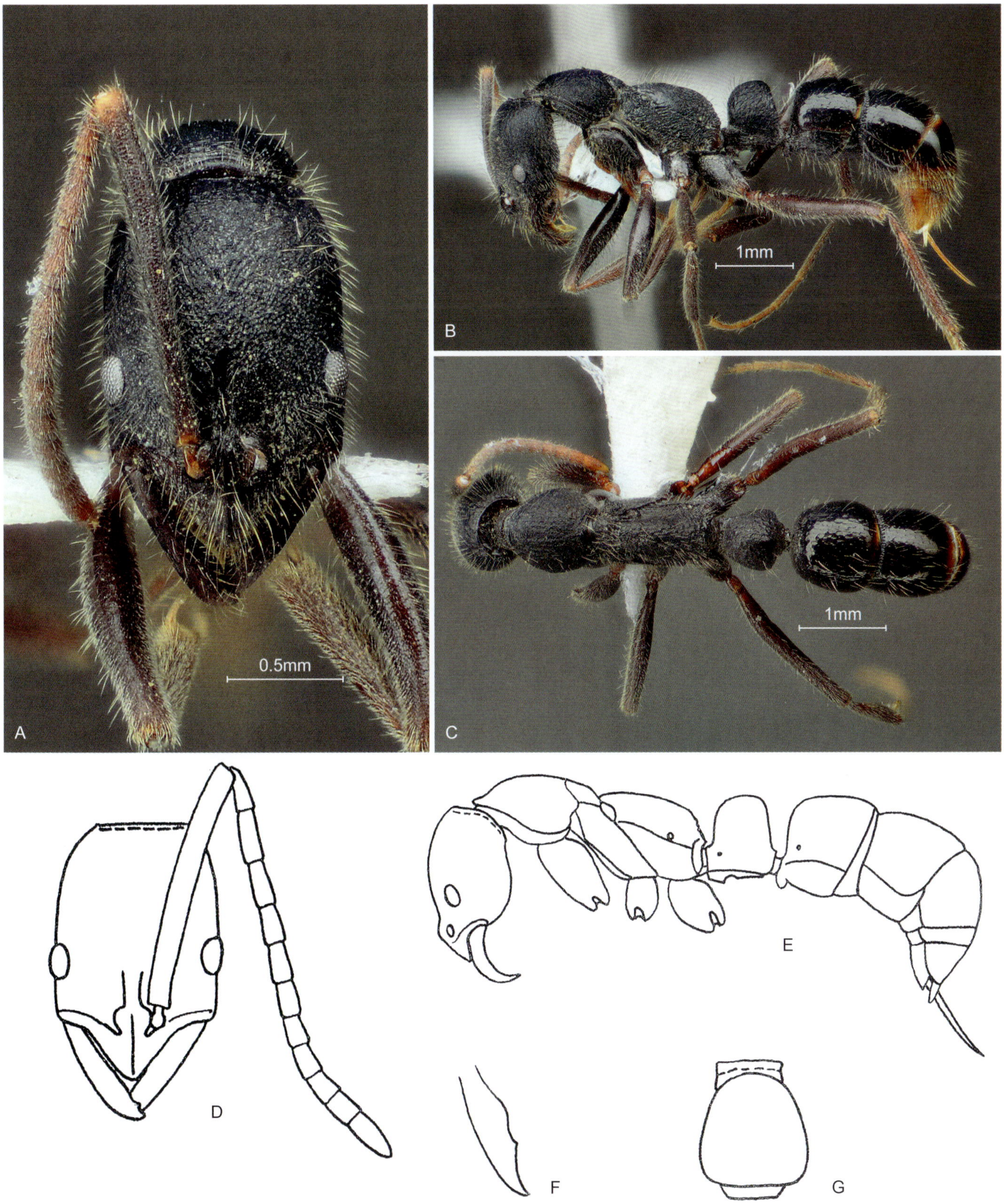

庄子细颚猛蚁 *Leptogenys zhuangzii*

A, D. 工蚁头部正面观；B, E. 工蚁身体侧面观；C. 工蚁身体背面观；F. 工蚁上颚背面观；G. 工蚁腹柄背面观（D-G. 引自 Xu, 2000c）

A, D. Head of worker in full-face view; B, E. Body of worker in lateral view; C. Body of worker in dorsal view; F. Mandible of worker in dorsal view; G. Petiole of worker in dorsal view (D-G. cited from Xu, 2000c)

锥头小眼猛蚁
Myopias conicara Xu, 1998

Myopias conicara Xu, 1998c, Entomologia Sinica, 5(2): 123, figs. 5, 7 (w.) CHINA (Yunnan).

Holotype worker: TL 7.5, HL 1.50, HW 1.40, CI 93, SL 1.33, SI 96, PW 1.00, AL 2.20, ED 0.25, ML 1.35, MI 90. PNL 0.37. In full-face view, head slightly longer than broad, broadest in front, narrowed posteriorly. Occipital margin nearly straight, occipital corners roundly prominent. Eyes placed on sides, close to the bases of mandibles. Mandibles long and slender, slightly curved inward; inner margin with a blunt tooth near the center; masticatory margin narrow, oblique and slightly concave, with a very small denticle at basal 1/3, apical tooth acute, basal tooth triangular. Middle part of clypeus triangular, anterior margin roundly concave, anterolateral corners protruding and dentiform. Frontal lobes large, laterally expanded. Frontal furrow reached to midline of head. Antennal scapes surpass occipital corners by about 1/8 of its length; flagella incrassate towards apex, apical segment slightly longer than the preceding 2 segments combined. Ventral face of head with a conical tubercle in the middle and close to the anterior margin. In lateral view, dorsum of alitrunk at the same level, promesonotal suture distinct and depressed, metanotal groove slightly depressed. Mesonotum short, crescent in dorsal view. Dorsum of propodeum slightly convex, rounded into declivity, the latter nearly truncate, about as long as dorsum. In lateral view, petiolar node quadrate, slightly higher than long; anterior face very weakly concave; posterior face truncate, each side with a vertical narrow depression close to the lateral margin; dorsal face roundly convex, anterodorsal angle and posterodorsal angle rounded, at the same height. In dorsal view, petiolar node about as long as broad, narrowing forwards, anterior margin slightly convex, anterolateral corners rounded, posterior margin straight. Subpetiolar process with a posteroventrally pointed triangular denticle at anteroventral corner, and with a pair of small posteriorly pointed denticles at posteroventral corner. Anterior face of gaster truncate, constriction between the basal two segments distinct. Mandibles sparsely punctate, interspace smooth and shining. Head with abundant moderately large punctures, distance between punctures about equal to diameter of one puncture, interspace smooth and shining; punctures on central dorsum of head sparse. Dorsum of alitrunk, dorsum of petiolar node, and first gastral segment with sparse, large, elliptic punctures, interspace finely longitudinally rugulose, less shining; a longitudinal central strip on dorsa of pronotum and mesonotum without punctures; sides of pronotum, and of petiolar node sparsely punctate; sides of alitrunk finely longitudinally rugulose, opaque; declivity and posterior face of petiolar node superficially transversely rugulose, less shining. Gastral segments 2-6 smooth and shining, but anterior half of 2nd segment sparsely punctate. Head and body with abundant erect or suberect hairs and abundant decumbent pubescence. Antennal scapes with sparse suberect hairs and dense decumbent pubescence; hind tibiae with sparse decumbent hairs and dense depressed pubescence. Body in colour black; mandibles, antennae, legs, and apex of gaster reddish brown; hairs and pubescence light yellow.

Holotype: Worker, China: Yunnan Province, Mengla County, Mengla Town, Longlin Village, 1090 m, 1997.Ⅲ.5, collected by Se-ping Dai, No. A97-186.

Etymology: The species name *conicara* combines Greek *conic-* (conical) + word root *cara* (head) + suffix *-a* (feminine form), it refers to the conical tubercle on the ventral face of head.

正模工蚁： 正面观，头长稍大于头宽，前面最宽，向后变窄；后缘近平直，后角圆凸。复眼位于头侧缘，接近上颚基部。上颚细长，轻度向内弯曲，内缘近中部具1个钝齿；咀嚼缘窄，倾斜，轻度凹陷，在基部1/3处具1个很小的齿，端齿尖锐，基齿三角形。唇基中部三角形，前缘圆形凹陷，前侧角突出呈齿状。额叶大，侧向扩展；额沟到达头中线。触角柄节约1/8超过头后角；鞭节向端部变粗，端节稍长于其前面2节的合长。侧面观，头部腹面中央具1个圆锥形瘤突，接近前缘。胸部背面在同一水平上，前中胸背板缝明显凹陷，后胸沟轻度凹陷。中胸背板短，背面观呈新月形。并胸腹节背面轻度隆起，圆形进入斜面；斜面近平截，约与背面等长。侧面观，腹柄结方形，高稍大于长，前面轻微凹陷；后面平截，两侧接近侧缘处各具1个垂直狭窄的凹陷；背面圆形隆起，前上角和后上角圆且等高。背面观，腹柄结长宽相等，向前变窄；前缘轻度隆起，前侧角圆，后缘平直；腹柄下突在前下角具1个指向后下方的三角形小齿，在后下角具1对指向后方的小齿。后腹部前面平截，基部2节间收缩明显。上颚具稀疏刻点，界面光滑发亮。头部具丰富的中等大小刻点，刻点间距约与刻点直径相等，界面光滑发亮；头部背面中央刻点稀疏。胸部背面、腹柄结背面和后腹部第1节具稀疏的椭圆形大刻点，界面具细纵皱纹，不够光亮；前中胸背板中央纵带无刻点；前胸背板侧面和腹柄结侧面具稀疏刻点；胸部侧面具细纵皱纹，暗；并胸腹节斜面和腹柄结后面具肤浅的横

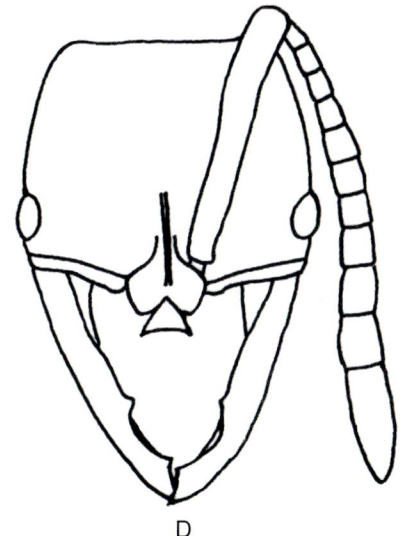
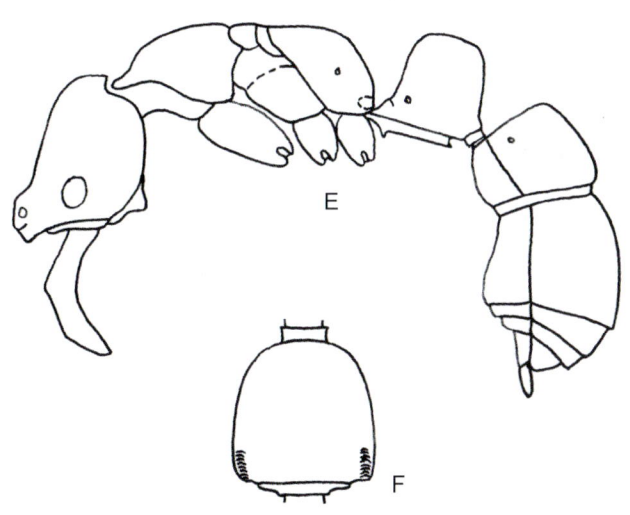

锥头小眼猛蚁 *Myopias conicara*

A, D. 工蚁头部正面观；B, E. 工蚁身体侧面观；C. 工蚁身体背面观；F. 工蚁腹柄背面观（D-F. 引自Xu, 1998c）
A, D. Head of worker in full-face view; B, E. Body of worker in lateral view; C. Body of worker in dorsal view; F. Petiole of worker in dorsal view (D-F. cited from Xu, 1998c)

皱纹，不够光亮。后腹部第2～6节光滑发亮，但是第2节前半部具稀疏刻点。头部和身体具丰富直立、亚直立毛和丰富倾斜绒毛被；触角柄节具稀疏亚直立毛和密集倾斜绒毛被；后足胫节具稀疏倾斜毛和密集平伏绒毛被。身体黑色，上颚、触角、足和后腹部末端红棕色，立毛和绒毛被浅黄色。

正模：工蚁，中国云南省勐腊县勐腊镇龙林村，1090m，1997.Ⅲ.5，代色平采集，No. A97-186。

词源：该新种的种名"*conicara*"由希腊语"*conic-*"（锥形的）+词根"*cara*"（头）+后缀"*-a*"（阴性形式）组成，指该种的头部腹面具锥形瘤突。

傣小眼猛蚁
Myopias daia Xu, Burwell & Nakamura, 2014

Myopias daia Xu et al., 2014b, Sociobiology, 61(2): 166, figs. 28-34 (w.) CHINA (Yunnan).

Holotype worker: TL 4.3, HL 0.93, HW 0.73, CI 79, SL 0.63, SI 86, ED 0.07, ML 0.60, PW 0.59, MSL 1.40, PL 0.40, PH 0.53, DPW 0.40, LPI 133, DPI 100. In full-face view, head roughly rectangular, longer than broad. Posterior margin straight, posterior corners bluntly angled. Sides weakly convex. Mandible elongate triangular, inner margin about 1/3 length of masticatory margin, basal corner bluntly angled; masticatory margin with a large basal tooth, a large middle tooth, a small preapical tooth, and a small apical tooth. Median clypeal lobe trapezoidal and widened forward, broader than long, length : width = 3 : 4, anterior margin weakly convex. Antenna 12-segmented, apex of scape fails to reach posterior head corner by 1/2 apical scape width. Eye small, with 9 facets, located at anterior 1/4 of the head side. In lateral view dorsal outline of mesosoma weakly convex and gently descends posteriorly. Pronotum weakly convex. Promesonotal suture slightly impressed. Mesonotum straight. Metanotal groove weakly narrowly impressed. Dorsum of propodeum straight, rounding into declivity; declivity nearly straight, about 1/2 length of dorsum. Petiolar node nearly rectangular, weakly narrowed dorsally, anterior margin slightly concave, posterior margin slightly convex, dorsal margin weakly convex; anterodorsal corner bluntly angled, posterodorsal corner rounded. Subpetiolar process roughly triangular, ventrally pointed, both anterior and posterior margins oblique, anterior margin short and straight, posterior margin weakly sinuate. Constriction between abdominal segments Ⅲ & Ⅳ distinct. Sting extruding. In dorsal view, pronotum widened posteriorly, sides convex. Sides of propodeum straight and parallel. Petiolar node trapezoidal, narrowed anteriorly, weakly broader than long, length : width = 1 : 1.3; anterior and lateral margins weakly convex, posterior margin weakly concave; anterolateral corners broadly rounded, posterolateral corners narrowly rounded. Whole body surface smooth and shining. Mandibles smooth and shining. Head with sparse erect to suberect hairs and abundant erect to suberect pubescence. Mesosoma, petiole, and gaster with sparse erect to suberect hairs and abundant subdecumbent to decumbent pubescence. Scapes and tibiae with sparse suberect hairs and dense decumbent pubescence. Body color reddish brown. Mandibles, antennae, legs, and gastral apex yellowish brown. Eyes blackish brown.

Holotype: Worker, China: Yunnan Province, Mengla County, Mengla Town, Bubeng Village, QCAS 1000-5 (21.621°N, 101.574°E), 985 m, 2012.Ⅶ, collected by Chris J. Burwell & Aki Nakamura in rainforest using Berlese method, No. A12-1820.

Etymology: The specific epithet has been named after a minority group called "Dai", residing in the Bubeng village locality, Yunnan Province, China.

正模工蚁： 正面观，头部近长方形，长大于宽；后缘平直，后角钝角状；侧缘轻度隆起。上颚长三角形，内缘约为咀嚼缘长的1/3，基角钝角状；咀嚼缘具1个大基齿、1个大中齿、1个小的亚端齿和1个小的端齿。唇基中叶梯形，向前变宽，宽大于长，长：宽=3：4，前缘轻度隆起。触角12节，柄节末端未到达头后角，距头后角的距离约为柄节端部宽的1/2。复眼小，具9个小眼，位于头侧缘前部1/4处。侧面观，胸部背面轻度隆起，向后轻度降低。前胸背板轻度隆起，前中胸背板缝轻微凹陷；中胸背板平直，后胸沟轻度凹陷，狭窄。并胸腹节背面平直，圆形进入斜面；斜面近平直，约为背面长的1/2。腹柄结近长方形，向上轻度变窄，前面轻微凹陷，后面轻微隆起，背面轻度隆起；前上角钝角状，后上角圆；腹柄下突近三角形，指向腹面，前缘和后缘倾斜，前缘短而直，后缘轻度波状。后腹部基部2节间收缩明显。螫针伸出。背面观，前胸背板向后变宽，侧缘隆起。并胸腹节侧缘平直，互相平行。腹柄结梯形，向前变窄，宽稍大于长，长：宽=1：1.3；前面和侧面轻度隆起，后面轻度凹陷；前侧角宽圆，后侧角窄圆。身体表面光滑发亮。上颚光滑发亮。头部具稀疏直立、亚直立毛和丰富直立、亚直立绒毛被。胸部、腹柄和后腹部具稀疏直立、亚直立毛和丰富亚倾斜、倾斜绒毛被。柄节和胫节具稀疏亚直立毛和密集倾斜绒毛被。身体红棕色，上颚、触角、足和后腹部末端黄棕色，复眼黑棕色。

正模： 工蚁，中国云南省勐腊县勐腊镇补蚌村QCAS 1000-5样地（21.621°N，101.574°E），985m，2012.Ⅶ，Chris J. Burwell 和Akihiro Nakamura采用Berlese分离法采于雨林中，No. A12-1820。

词源： 该新种的种名以居住在中国云南省补蚌村的少数民族"傣族"命名。

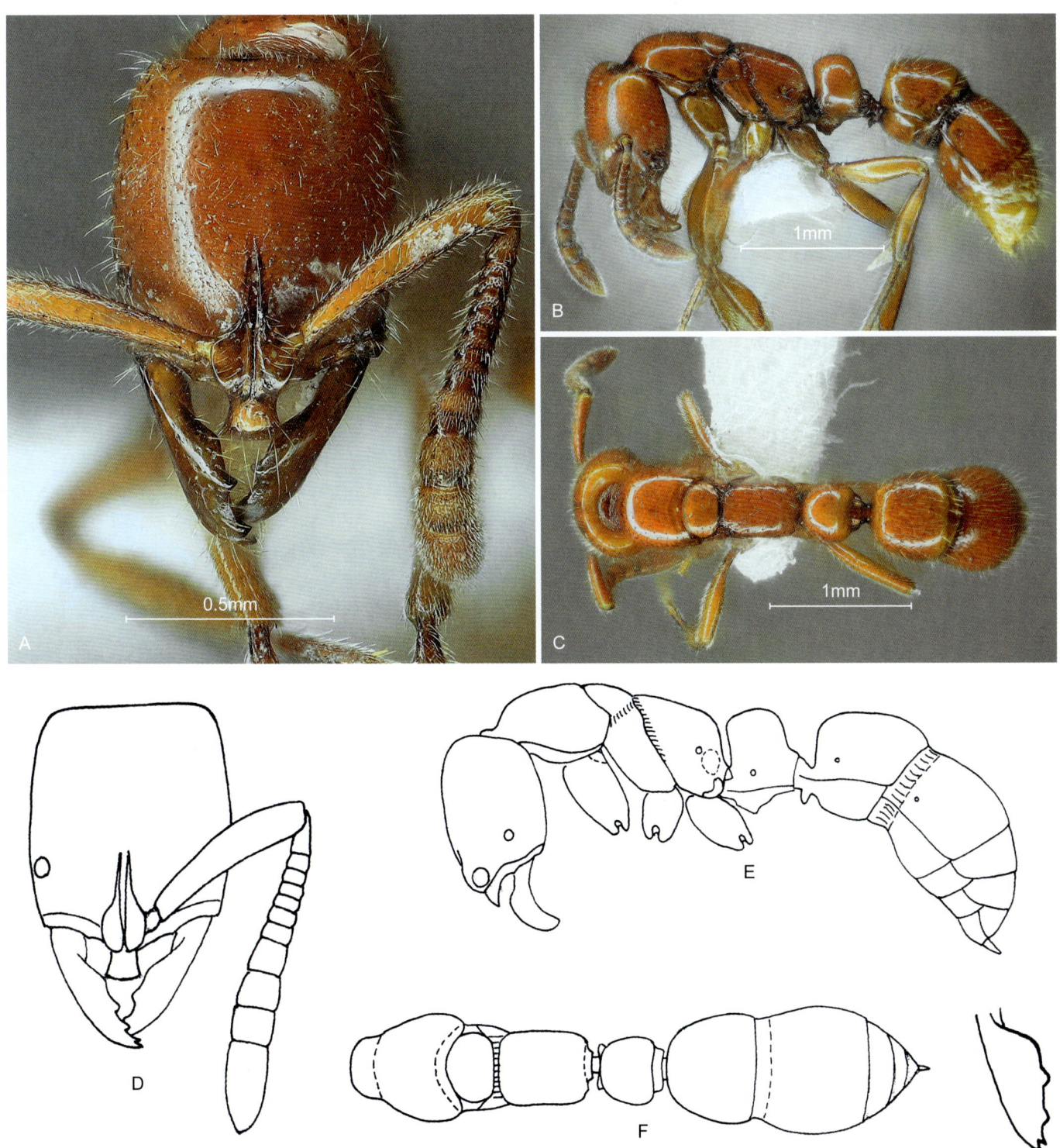

傣小眼猛蚁 *Myopias daia*

A, D. 工蚁头部正面观；B, E. 工蚁身体侧面观；C, F. 工蚁身体背面观；G. 工蚁上颚背面观（D-G. 引自Xu等，2014b）

A, D. Head of worker in full-face view; B, E. Body of worker in lateral view; C, F. Body of worker in dorsal view; G. Mandible of worker in dorsal view (D-G. cited from Xu et al., 2014b)

哈尼小眼猛蚁
Myopias hania Xu & Liu, 2012

Myopias hania Xu & Liu, 2012, Sociobiology, 59(1): 827, figs. 12-15 (w.) CHINA (Yunnan).

Holotype worker: TL 5.9, HL 1.25, HW 1.15, CI 92, SL 0.93, SI 80, ML 1.00, ED 0.18, PW 0.88, AL 1.83, PL 0.60, PH 0.83, DPW 0.68, LPI 138, DPI 113. In full-face view, head nearly square, slightly longer than broad, weakly narrowed forward. Occipital margin slightly concave, occipital corners rounded, sides weakly convex. Mandibles slender and linear, inner margin weakly convex, masticatory margin about 1.3 times as long as inner margin, basal tooth blunt. The basal 1/2 of masticatory margin concave, the apical 1/2 obliquely truncated. Median lobe of clypeus protruding and widened forward, trapezoidal, without longitudinal central carina, length : width = 1 : 1.8, anterior margin weakly concave. Longitudinal furrow between frontal lobes distinct. Antennae 12-segmented, apices of scapes reached to 9/10 of the distance from antennal sockets to occipital corners, antennal clubs 4-segmented. Eyes relatively large, with 11-12 facets in the maximum diameter. In lateral view, anterior 1/2 ventral face of head longitudinally convex, anteroventral corner bluntly angled. In lateral view, dorsum of alitrunk weakly convex, promesonotal suture and metanotal groove weakly depressed. Dorsum of propodeum about as long as declivity, posterodorsal corner rounded, declivity weakly longitudinally depressed. Petiolar node nearly square, anterior face straight, dorsal and posterior faces slightly convex, anterodorsal corner bluntly prominent, posterodorsal corner roundly prominent. Subpetiolar process complex, roughly cuneiform, anteroventral corner acutely toothed, ventral face evenly convex behind the tooth, posteroventral corner toothed. In dorsal view, petiolar node trapezoid and widened backward, length : width = 1 : 1.3, anterior and lateral margins weakly convex, anterolateral corners rounded, posterior margin straight. Constriction between the two basal gastral segments distinct. Sting laterally compressed, apex relatively blunt. Mandibles smooth, with sparse punctures. Head smooth, dorsum with abundant punctures, vertex, occiput, and ventral face with sparse punctures. Alitrunk smooth, with sparse large punctures, sides of mesothorax, metathorax, and propodeum with fine longitudinal striations. Propodeal declivity smooth, the lower portion transversely striate. Petiole smooth, with sparse large punctures, sides finely longitudinally rugose. Gaster smooth, first segment and basal portion of second segment with abundant large punctures. Dorsum of head with sparse suberect hairs and dense decumbent pubescence. Dorsum of body with abundant suberect hairs and abundant decumbent pubescence. Anterior face of petiolar node with dense pubescence. Scapes and tibiae with sparse subdecumbent hairs and dense decumbent pubescence. Color brownish black. Mandibles, antennae, legs, and gastral apex reddish brown. **Paratype worker:** TL 6.1, HL 1.30, HW 1.20, CI 92, SL 0.93, SI 77, ML 1.05, ED 0.20, PW 0.88, AL 1.80, PL 0.63, PH 0.88, DPW 0.73, LPI 140, DPI 116 (*n*=1). As holotype.

Holotype: Worker, China: Yunnan Province, Hekou County, Nanxi Town, Laodoutian, 750 m, 2010.Ⅳ.5, collected by Hui-qin Zhu from a soil sample in the monsoon forest, No. A10-2557.

Etymology: The new species is named after the Chinese minority nationality Hani who commonly lives in southern Yunnan Province.

正模工蚁： 正面观，头部近方形，长稍大于宽，向前轻度变窄；后缘轻微凹陷，后角圆；侧缘轻度隆起。上颚细长，线形，内缘轻度隆起，咀嚼缘约为内缘长的1.3倍，基齿钝；咀嚼缘基部1/2凹陷，端部1/2斜截。唇基中叶伸出，向前变宽，梯形，缺中央纵脊，长：宽=1：1.8，前缘轻度凹陷。额叶之间的纵沟明显。触角12节，柄节末端到达触角窝至头后角间距的9/10处，触角棒4节。复眼较大，最大直径上具11～12个小眼。侧面观，头部腹面前部1/2纵向隆起，前下角钝角状。胸部背面轻度隆起，前中胸背板缝和后胸沟轻度凹陷。并胸腹节背面约与斜面等长，后上角圆，斜面轻度纵向凹陷。腹柄结近方形，前面平直，背面和后面轻度隆起，前上角钝突，后上角圆突；腹柄下突复杂，近楔形，前下角尖齿状，腹面在齿后均匀隆起，后下角具齿。背面观，腹柄结梯形，向后变宽，长：宽=1：1.3，前缘和侧缘轻度隆起，前侧角圆，后缘平直。后腹部基部2节间收缩明显。螯针侧扁，末端较钝。上颚光滑，具稀疏刻点。头部光滑，背面具丰富刻点，头顶、后头和腹面具稀疏刻点。胸部光滑，具稀疏大刻点，中胸侧面、后胸侧面和并胸腹节侧面具细纵条纹；并胸腹节斜面光滑，下部具横条纹。腹柄光滑，具稀疏大刻点，侧面具细纵皱纹。后腹部光滑，第1和第2节基部具丰富大刻点。头部背面具稀疏亚直立毛和密集倾斜绒毛被。身体背面具丰富亚直立毛和丰富倾斜绒毛被，腹柄结前面具密集绒毛被。柄节和胫节具稀疏亚倾斜毛和密集倾斜绒毛被。身体棕黑色，上颚、触角、足和后腹部末端红棕色。**副模工蚁：** 特征同正模。

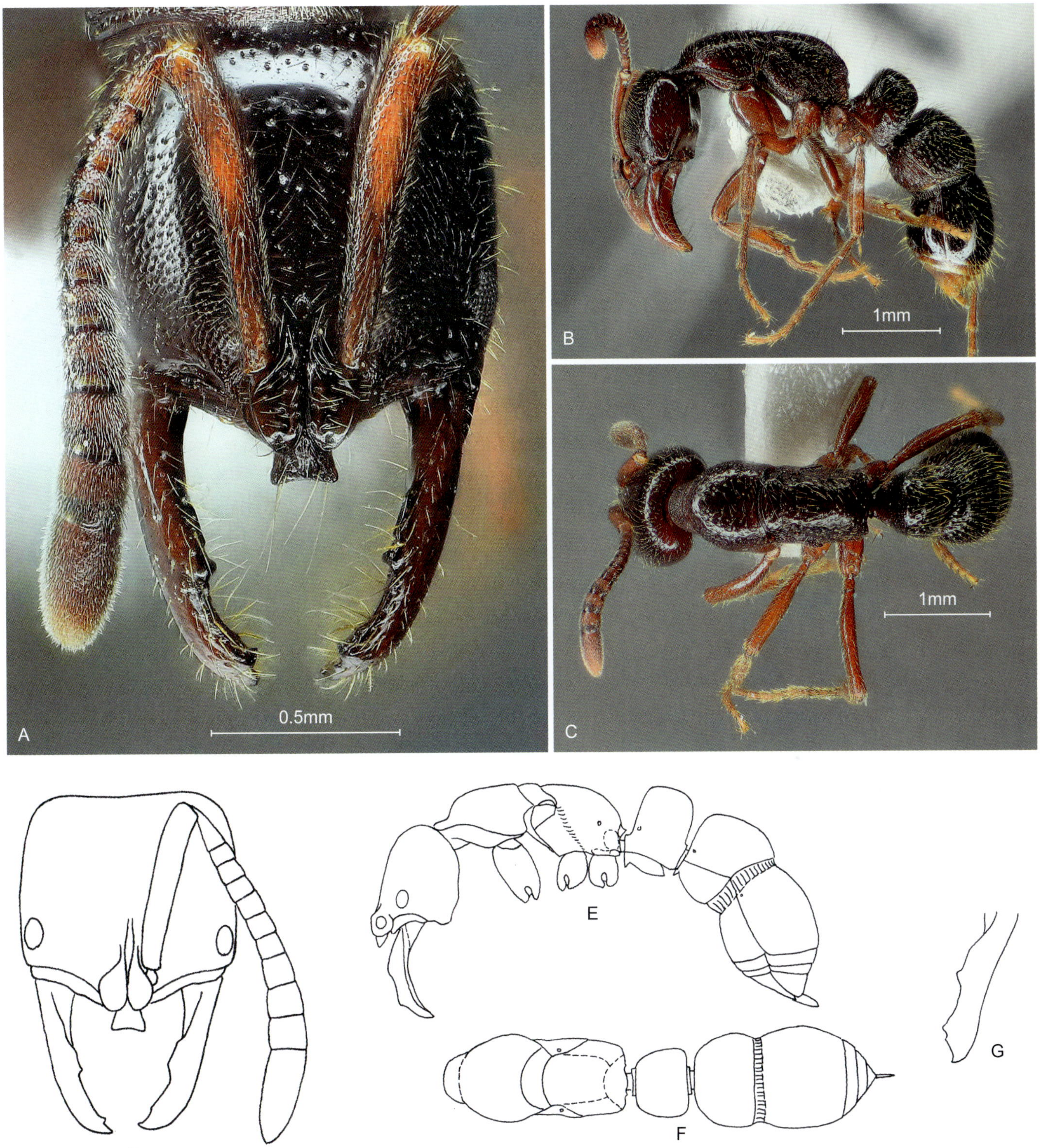

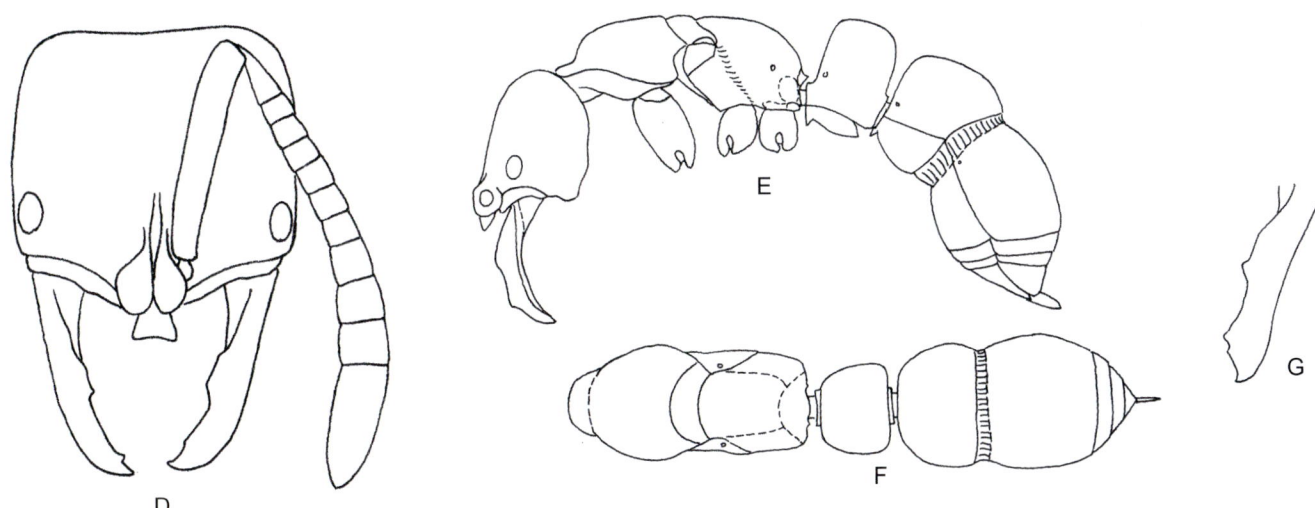

哈尼小眼猛蚁 *Myopias hania*

A, D. 工蚁头部正面观；B, E. 工蚁身体侧面观；C, F. 工蚁身体背面观；G. 工蚁上颚背面观（D-G. 引自Xu和Liu, 2012）
A, D. Head of worker in full-face view; B, E. Body of worker in lateral view; C, F. Body of worker in dorsal view; G. Mandible of worker in dorsal view (D-G. cited from Xu & Liu, 2012)

正模： 工蚁，中国云南省河口县南溪镇老豆田，750m，2010.IV.5，诸慧琴采于季雨林土壤样中，No. A10-2557。

词源： 该新种以普遍居住于中国云南省南部的少数民族"哈尼族"命名。

珞巴小眼猛蚁
Myopias luoba Xu & Liu, 2012

Myopias luoba Xu & Liu, 2012, Sociobiology, 59(1): 822, figs. 1-4 (w.) CHINA (Tibet).

Holotype worker: TL 3.8, HL 0.80, HW 0.65, CI 81, SL 0.58, SI 88, ML 0.53, ED 0.09, PW 0.50, AL 1.08, PL 0.36, PH 0.45, DPW 0.36, LPI 124, DPI 100. In full-face view, head nearly rectangular, longer than broad, weakly widened forward. Occipital margin slightly concave, occipital corners roundly prominent, sides slightly convex. Mandibles elongate triangular, inner margin weakly convex, about as long as masticatory margin. Masticatory margin with 4 teeth, including 1 basal tooth, 1 middle tooth, and 2 minute apical denticles. Median lobe of clypeus protruding forward, nearly triangular, widened forward, length : width = 4 : 5, anterior margin straight. Antennae 12-segmented, apices of scapes almost reach to occipital corners, antennal clubs 4-segmented. Longitudinal furrow between frontal lobes distinct. Eyes small, with 6 facets in the maximum diameter. In lateral view, promesonotum evenly convex, promesonotal suture distinct but not depressed, metanotal groove slightly depressed. Propodeal dorsum slightly convex, about 2 times as long as declivity, posterodorsal corner rounded. Petiolar node nearly square, anterior face straight, dorsal and posterior faces weakly convex, anterodorsal corner blunt, about as high as posterodorsal corner, the latter roundly prominent. Subpetiolar process large, triangular, anteroventral corner blunt. In dorsal view, petiolar node trapezoid, widened backward, length : width = 1 : 1.3, anterior and lateral margins weakly convex, anterolateral corners rounded, posterior margin nearly straight. Constriction between the two basal gastral segments distinct. Sting developed and laterally compressed. Mandibles smooth and shining, with sparse fine punctures. Head densely finely punctured, interfaces smooth. Alitrunk, petiole, and gaster smooth and shining, with sparse fine punctures. Sides of metathorax below propodeal spiracles finely longitudinally striate. Dorsa of head and body with sparse suberect hairs and abundant decumbent pubescence, pubescence on the head dense. Scapes and tibiae with sparse subdecumbent hairs and dense decumbent pubescence. Color blackish brown. Head almost black. Mandibles, clypeus, antennae, legs, and gastral apex brownish yellow.

Holotype: Worker, China: Tibet Autonomous Region, Medog County, Medog Town, Yarang Village, 820 m, 2008.V.20, collected by Zheng-hui Xu on the ground in secondary rain forest, No. A08-875.

Etymology: The new species is named after the Chinese minority nationality Luoba who mainly lives in southeastern Tibet.

正模工蚁： 正面观，头部近长方形，长大于宽，向前轻度变宽；后缘轻微凹陷，后角圆突；侧缘轻微隆起。上颚长三角形，内缘轻度隆起，约与咀嚼缘等长；咀嚼缘具4个齿，包括1个基齿、1个中齿和2个微小的端齿。唇基中叶向前伸出，近三角形，向前变宽，长：宽=4：5，前缘平直。触角12节，柄节末端几乎到达头后角，触角棒4节。额叶间纵沟明显。复眼小，最大直径上具6个小眼。侧面观，前中胸背板均匀隆起，前中胸背板缝明显但不凹陷，后胸沟轻度凹陷。并胸腹节背面轻微隆起，约为斜面长的2倍，后上角圆。腹柄结近方形，前面平直，背面和后面轻度隆起，前上角钝，约与后上角等高，后上角圆突；腹柄下突大，三角形，前下角钝。背面观，腹柄结梯形，向后变宽，长：宽=1：1.3，前缘和侧缘轻度隆起，前侧角圆，后缘近平直。后腹部基部2节间收缩明显。螫针发达，侧扁。上颚光滑发亮，具稀疏细刻点。头部具密集细刻点，界面光滑。胸部、腹柄和后腹部光滑和发亮，具稀疏细刻点；后胸侧面低于并胸腹节气门部分具细纵条纹。头部和身体背面具稀疏亚直毛和丰富倾斜绒毛被，头部绒毛被密集；柄节和胫节具稀疏亚倾斜毛和密集倾斜绒毛被。身体黑棕色，头部几乎呈黑色，上颚、唇基、触角、足和后腹部末端棕黄色。

正模： 工蚁，中国西藏自治区墨脱县墨脱镇亚让村，820m，2008.V.20，徐正会采于次生雨林地表，No. A08-875。

词源： 该新种以主要居住在中国西藏东南部的少数民族"珞巴族"命名。

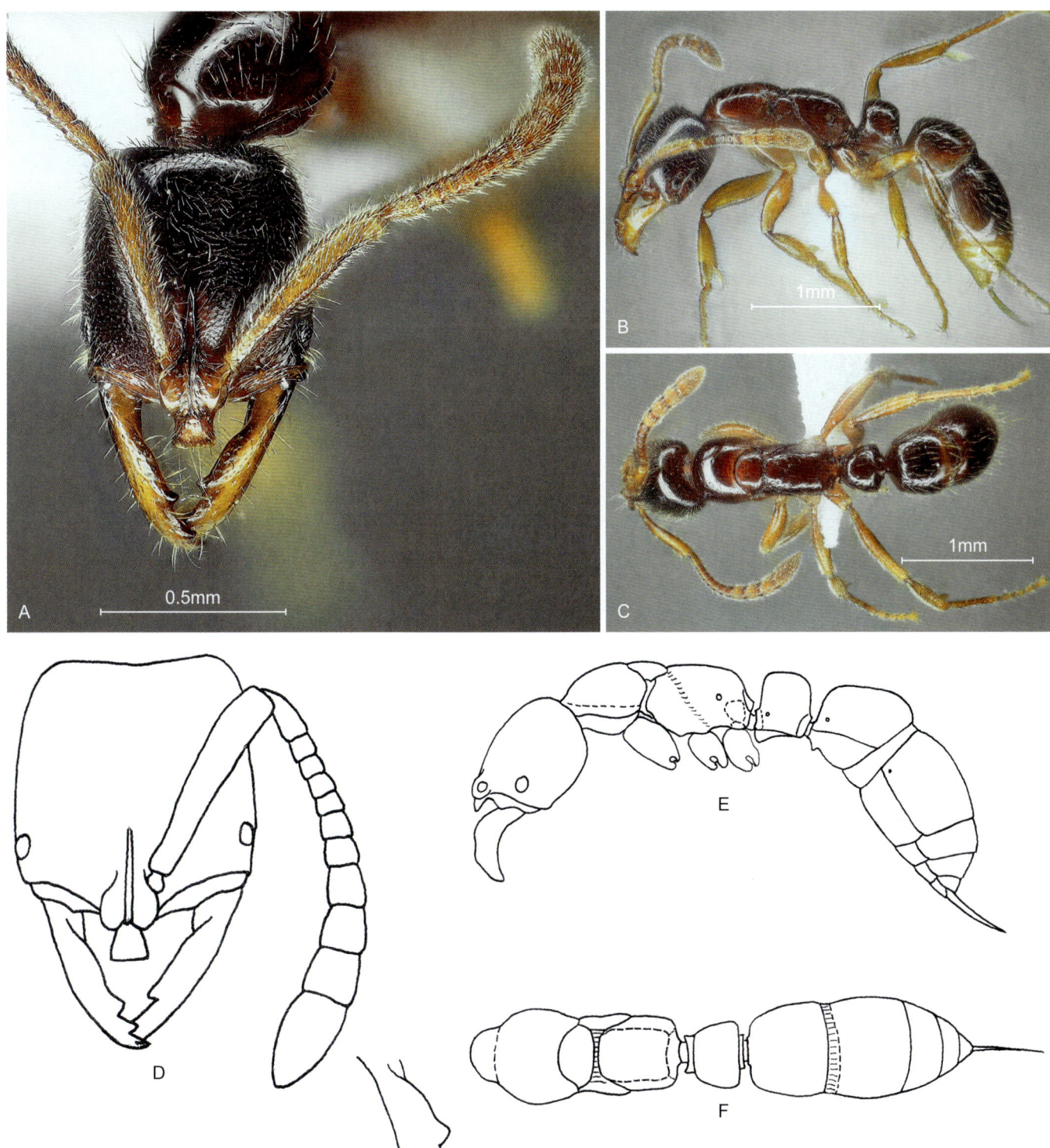

珞巴小眼猛蚁 *Myopias luoba*

A, D. 工蚁头部正面观; B, E. 工蚁身体侧面观; C, F. 工蚁身体背面观; G. 工蚁上颚背面观（D-G. 引自Xu和Liu, 2012）
A, D. Head of worker in full-face view; B, E. Body of worker in lateral view; C, F. Body of worker in dorsal view; G. Mandible of worker in dorsal view (D-G. cited from Xu & Liu, 2012)

门巴小眼猛蚁
Myopias menba Xu & Liu, 2012

Myopias menba Xu & Liu, 2012, Sociobiology, 59(1): 824, figs. 8-11 (w.) CHINA (Tibet).

Holotype worker: TL 3.3, HL 0.73, HW 0.58, CI 79, SL 0.45, SI 78, ML 0.45, ED 0.03, PW 0.41, AL 0.95, PL 0.30, PH 0.40, DPW 0.30, LPI 133, DPI 100. In full-face view, head nearly rectangular, longer than broad, sides nearly parallel. Occipital margin straight, occipital corners bluntly prominent. Mandibles elongate, triangular, inner margin very short, about 1/3 length of masticatory margin. Masticatory margin with 4 teeth, including 1 basal tooth, 1 middle tooth, and 2 minute apical denticles beside the basal corner. Median lobe of clypeus protruding forward, nearly square, length : width = 3 : 4, sides parallel, anterior margin straight. Longitudinal furrow between frontal lobes distinct. Antennae short, 12-segmented, apices of scapes reached to 4/5 of the distance from antennal sockets to occipital corners, antennal clubs 4-segmented. Eyes minute, with only 2 facets. In lateral view, promesonotum weakly convex, promesonotal suture and metanotal groove slightly depressed. Propodeal dorsum straight, about 2 times as long as declivity, posterodorsal corner rounded. Petiolar node very thick, roughly trapezoid, anterior face straight, dorsal face rounded into posterior face, anterodorsal corner bluntly prominent, posterodorsal corner indistinct. Subpetiolar process large, triangular, anteroventral corner bluntly prominent, anterior and posteroventral faces nearly straight. In dorsal view, petiolar node trapezoid, widened backward, length : width = 1 : 1.1, anterior and lateral margins weakly convex, posterior margin weakly concave. Constriction between the two basal gastral segments distinct. Sting laterally compressed, apex relatively blunt. Mandibles smooth and shining, with sparse fine punctures. Head, alitrunk, and petiole densely punctured, sides of alitrunk finely longitudinally rugose, dorsal and posterior faces of petiolar node smooth. Gaster smooth and shining, dorsa of the two basal segments abundantly punctured, the rest segments sparsely punctured. Dorsa of head and body with sparse suberect short hairs and dense decumbent pubescence, gastral apex with abundant longer hairs. Scapes and tibiae with sparse suberect hairs and dense decumbent pubescence. Color brownish yellow, eyes black.

Holotype: Worker, China: Tibet Autonomous Region, Medog County, Medog Town, Yarang Village, 720 m, 2008.Ⅴ.20, collected by Zheng-hui Xu on the ground in secondary rain forest, No. A08-838.

Etymology: The new species is named after the Chinese minority nationality Menba who mainly lives in southeastern Tibet.

正模工蚁： 正面观，头部近长方形，长大于宽，侧缘近平行；后缘平直，后角钝突。上颚伸长，三角形，内缘很短，约为咀嚼缘长的1/3；咀嚼缘除基角外具4个齿，包括1个基齿、1个中齿和2个微小的端齿。唇基中叶向前突出，近方形，长：宽=3：4，侧缘平行，前缘平直。额叶之间纵沟明显。触角短，12节，柄节末端到达触角窝至头后角间距的4/5处，触角棒4节。复眼微小，仅具2个小眼。侧面观，前中胸背板轻度隆起，前中胸背板缝和后胸沟轻微凹陷。并胸腹节背面平直，约为斜面长的2倍，后上角圆。腹柄结很厚，近梯形，前面平直，背面圆形进入后面，前上角钝突，后上角不明显；腹柄下突大，三角形，前下角钝突，前缘和后下缘近平直。背面观，腹柄结梯形，向后变宽，长：宽=1：1.1，前缘和侧缘轻度隆起，后缘轻度凹陷。后腹部基部2节间收缩明显。螯针侧扁，顶端较钝。上颚光滑发亮，具稀疏细刻点。头部、胸部和腹柄节具密集刻点，胸部侧面具细纵皱纹，腹柄结背面和后面光滑。后腹部光滑发亮，基部2节背面具丰富刻点，其余腹节具稀疏刻点。头部和身体背面具稀疏亚直立短毛和密集倾斜绒毛被，后腹部末端具丰富长毛；柄节和胫节具稀疏亚直立毛和密集倾斜绒毛被。身体棕黄色，复眼黑色。

正模： 工蚁，中国西藏自治区墨脱县墨脱镇亚让村，720m，2008.Ⅴ.20，徐正会采于次生雨林地表，No. A08-838。

词源： 该新种以主要居住在中国西藏东南部的少数民族"门巴族"命名。

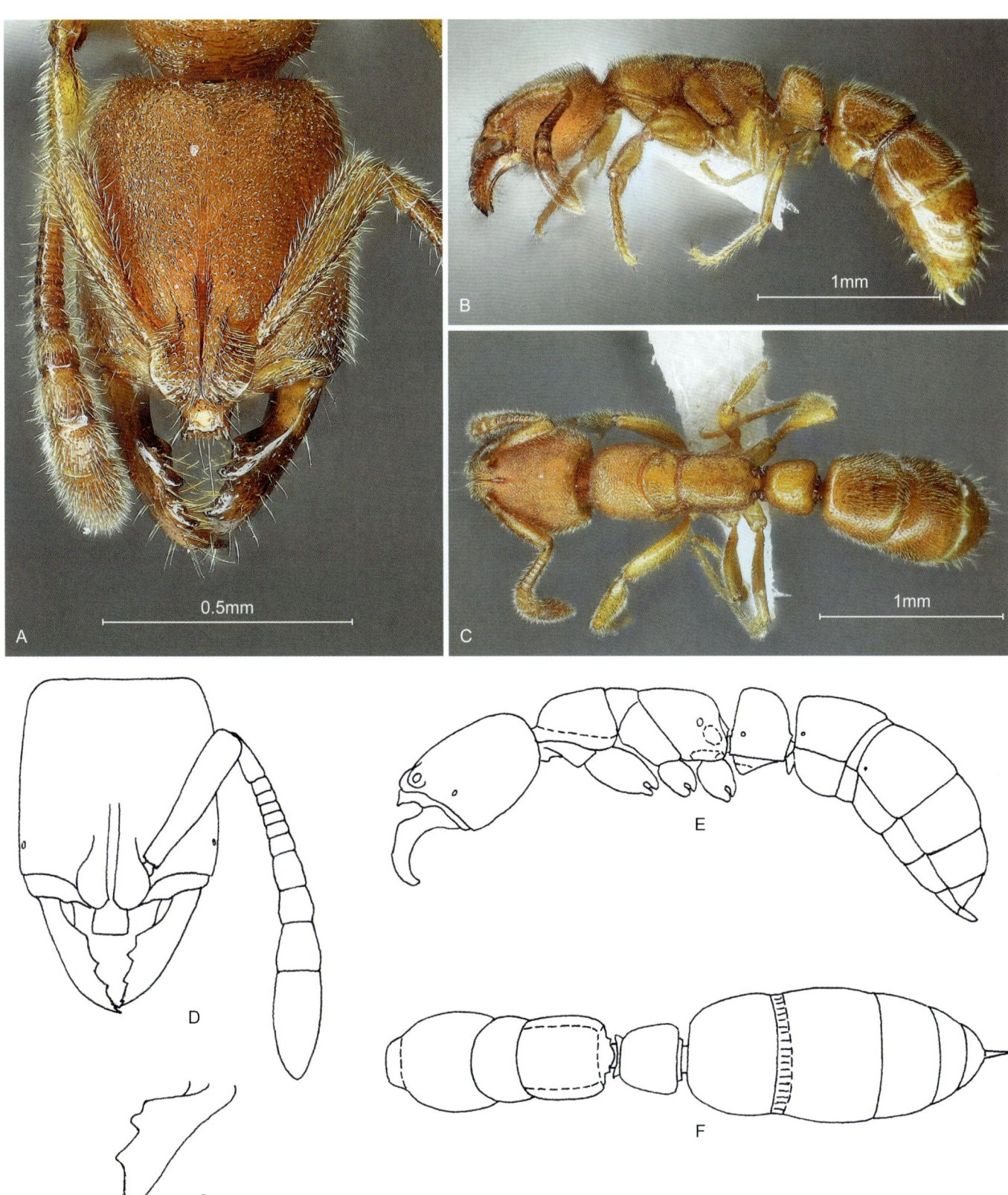

门巴小眼猛蚁 *Myopias menba*

A, D. 工蚁头部正面观；B, E. 工蚁身体侧面观；C, F. 工蚁身体背面观；G. 工蚁上颚背面观（D-G. 引自Xu和Liu, 2012）
A, D. Head of worker in full-face view; B, E. Body of worker in lateral view; C, F. Body of worker in dorsal view; G. Mandible of worker in dorsal view (D-G. cited from Xu & Liu, 2012)

034 片突厚结猛蚁
Pachycondyla lobocarena Xu, 1995

Pachycondyla lobocarena Xu, 1995c, Entomological Research, 1: 109, figs. 15, 16 (w.) CHINA (Yunnan).

Holotype worker: TL 9.4, HL 2.15, HW 2.03, CI 94, SL 1.80, SI 89, PW 1.50, AL 3.00, ED 0.30, ML 1.25, PNL 0.63, PNW 1.10. In full-face view, head slightly longer than broad, sides slightly convex, narrowing anteriorly; occipital margin shallowly emarginate, occipital corners bluntly rounded. Mandibles long triangular, masticatory margin has 9 teeth. Clypeus transverse, anterior margin extruding, concave in the middle. Antenna has 12 segments, apex of scape extending backward just to the occipital corner, funiculi incrassate towards apex. Eyes moderately large. In lateral view, occipital comer produced in a lobe-like process ventrally. Dorsum of alitrunk at the same level, evenly convex; humeral corners of pronotum bluntly angled; promesonotal suture obvious, impressed; metanotal groove indistinct; mesepisternum impressed by a distinct oblique furrow; dorsum of propodeum straight, rounded into declivity, about equal to the latter; declivity concave in the middle. Petiolar node compressed anteroposteriorly, anterior face convex vertically in the middle, upper portion of posterior face bevelled off forwards; subpetiolar process low, cuneiform. Anterior face of gaster truncate, constriction between the basal two segments distinct. Mandibles very finely and longitudinally striate; head and alitrunk densely and longitudinally rugose, rugae on dorsum of head slightly diverging posteriorly from a medial line, rugae on pronotal dorsum concentric, rugae on declivity diverging dorsally; petiolar node with transverse rugae on anterior and posterior faces; gaster finely punctured, not smooth. Dorsum of head and body with abundant erect or suberect hairs and dense subdecumbent pubescence; dorsa of antennal scapes and hind tibiae with sparse suberect hairs and dense subdecumbent pubescence. Colour black, antennal flagella, mandibles, tibiae, tarsi, and apex of gaster dark reddish brown. **Paratype workers:** TL 9.3-10.4, HL 2.03-2.35, HW 1.88-2.23, CI 89-98, SL 1.70-1.90, SI 85-95, PW 1.40-1.60, AL 2.80-3.25, ED 0.25-0.33, ML 1.20-1.40, PNL 0.55-0.70, PNW 0.95-1.15 ($n=7$). As holotype.

Holotype: Worker, China: Yunnan Province, Dali County, Xizhou Town, Hudiequan Spring (25.6°N, 100.1°E), 2000 m, 1991.Ⅹ.10, collected by Zheng-hui Xu on the ground in *Pinus yunnanensis* forest, No. A91-1005.

Etymology: The species name *lobocarena* combines Greek *lob-* (lobe) + word root *caren* (head) + suffix *-a* (feminine form), it refers to the lobe-like process on the posteroventral corner of head.

正模工蚁： 正面观，头长稍大于宽，侧缘轻度隆起，向前变窄；后缘浅凹，后角钝圆。上颚长三角形，咀嚼缘具9个齿。唇基横形，前缘突出，中央凹陷。触角12节，柄节末端刚到达头后角，鞭节向顶端变粗。复眼中等大。侧面观，头后角向腹面延伸呈1个叶状突。胸部背面在同一水平面上，均匀隆起；前胸背板肩角钝角状，前中胸背板缝明显，凹陷；后胸沟不明显，中胸前侧片具1条明显斜沟。并胸腹节背面平直，圆形进入斜面，约与斜面等长，斜面中部凹陷。腹柄结前后压扁，前面中央垂直隆起，后缘上部向前斜切；腹柄下突低，楔形。后腹部前面平截，基部2节间收缩明显。上颚具很细的纵条纹。头胸部具密集纵皱纹，头部背面的皱纹从1条中线向后轻度分歧，前胸背部的皱纹呈同心圆状，并胸腹节斜面的皱纹向上分歧。腹柄结前面和后面具横皱纹；后腹部具细刻点，不光滑。头部和身体背面具丰富直立、亚直立毛和密集倾斜绒毛被；触角柄节和后足胫节背面具稀疏亚直立毛和密集倾斜绒毛被。身体黑色，触角鞭节、上颚、胫节、跗节和后腹部末端暗红棕色。**副模工蚁：** 特征同正模。

正模： 工蚁，中国云南省大理县喜洲镇蝴蝶泉（25.6°N，100.1°E），2000m，1991.Ⅹ.10，徐正会采于云南松林地表，No. A91-1005。

词源： 该新种的种名"*lobocarena*"由希腊语"*lob-*"（叶片）+词根"*caren*"（头）+后缀"*-a*"（阴性形式）组成，指该种的头部后下角具叶状突。

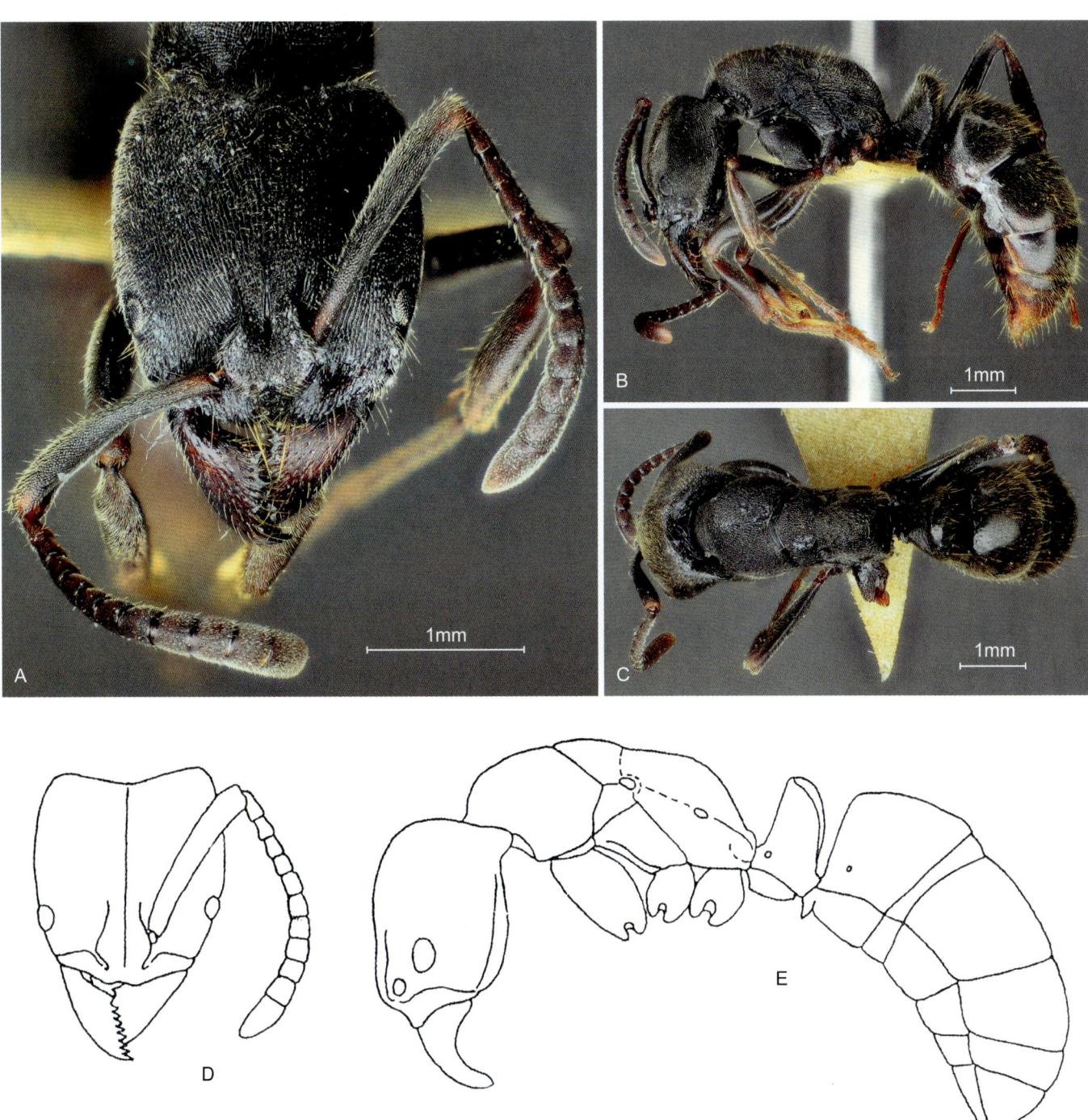

片突厚结猛蚁 Pachycondyla lobocarena

A, D. 工蚁头部正面观；B, E. 工蚁身体侧面观；C. 工蚁身体背面观（D-E. 引自 Xu, 1995c）

A, D. Head of worker in full-face view; B, E. Body of worker in lateral view; C. Body of worker in dorsal view (D-E. cited from Xu, 1995c)

郑氏厚结猛蚁
Pachycondyla zhengi Xu, 1995

Pachycondyla zhengi Xu, 1995c, Entomological Research, 1: 110, figs. 19, 20 (w.) CHINA (Yunnan).

Holotype worker: TL 12.8, HL 2.80, HW 2.60, CI 93, SL 2.40, SI 92, PW 1.80, AL 3.70, ED 0.40, ML 1.80, PNL 0.80, PNW 1.48. In full-face iew, head slightly longer than broad, sides evenly convex, narrowing forwards. Occipital margin angularly excised, occipital comers bluntly angled; dorsum of head raised up as a ridge along the occipital margin. Mandibles long triangular, masticatory margin has 10 teeth. Clypeus transverse, anterior margin extruding, concave in the middle. Antenna has 12 segments, apex of scape feebly extending backward beyond the occipital corner, funiculi incrassate toward the apex. Eyes moderate large. In lateral view, occipital corner extruding in a right angle. Dorsum of alitrunk at the same level; humeral corners of pronotum extruding in right angles; promesonotal suture distinct, metanotal groove indistinct; mesepisternum impressed by a distinct oblique furrow; dorsum of propodeum obviously shorter than declivity, and bluntly rounded into the latter; declivity flat, with sharp lateral margins. Petiolar node compressed anteroposteriorly, anterior face convex vertically in the middle, upper portion of posterior face bevelled off forwards; subpetiolar process low, cuneiform, bluntly angled anteroventrally. Anterior face of gaster truncate, constriction between the basal two segments distinct. Mandibles finely longitudinally striate. Head and alitrunk coarsely longitudinally rugose, rugae on dorsum of head strong, diverging posteriorly from a medial line, rugae on pronotal dorsum concentric; lower portion of mesepisternum with transverse rugae; declivity with longitudinal rugae. Petiolar node with transverse rugae on anterior and posterior faces; gaster densely punctured, opaque. Dorsum of head and body with abundant erect or suberect hairs and dense subdecumbent pubescence; dorsa of antennal scapes and hind tibiae with rich erect long hairs and suberect short ones, pubescence dense and decumbent. Colour black, mandibles, antennal flagella, legs, and gastric apex dark reddish brown. **Paratype workers:** TL 12.0-13.8, HL 2.65-3.15, HW 2.33-3.10, CI 88-98, SL 2.10-2.55, SI 81-94, PW 1.65-2.00, AL 3.55-4.40, ED 0.35-0.44, ML 1.65-1.95, PNL 0.65-0.80, PNW 1.30-1.55 (n=7). As holotype.

Holotype: Worker, China: Yunnan Province, Baoshan County, Chengguan Town, Taibao Park (25.1°N, 99.1°E), 1800 m, 1991.Ⅹ.11, collected by Zheng-hui Xu on the ground in conifer-broadleaf mixed forest, No. A91-1052.

Etymology: This new species is named in honor of Professor Zhe-min Zheng for his outstanding contribution to systematic entomology.

正模工蚁： 正面观，头长稍大于宽，侧面均匀隆起，向前变窄；后缘角状凹陷，后角钝角状；头部背面沿后缘隆起呈脊状。上颚长三角形，咀嚼缘具10个齿。唇基横形，前缘突出，中央凹陷。触角12节，柄节末端稍超过头后角，鞭节向顶端变粗。复眼中等大。侧面观，头部后下角突出呈直角形。胸部背面在同一水平上，前胸背板肩角突出呈直角形，前中胸背板缝明显，后胸沟不明显。中胸前侧片具1条明显斜沟。并胸腹节背面明显短于斜面，钝圆进入斜面；斜面平坦，具锋利的侧边。腹柄结前后压扁，前面中央垂直隆起，后缘上部向前斜切；腹柄下突低，楔形，前下角钝角状。后腹部前面平截，基部2节间收缩明显。上颚具细纵条纹。头胸部具粗糙纵皱纹，头部背面皱纹粗，从1条中线向后分歧；前胸背面皱纹同心圆状，中胸前侧片下部具横皱纹，并胸腹节斜面具纵皱纹。腹柄结前面和后面具横皱纹，后腹部具密集刻点，暗。头部和身体背面具丰富直立、亚直毛和密集倾斜绒毛被；触角柄节和后足胫节背面具丰富直立长毛、亚直立短毛和密集倾斜绒毛被。身体黑色，上颚、触角鞭节、足和后腹部末端暗红棕色。**副模工蚁：** 特征同正模。

正模： 工蚁，中国云南省保山县城关镇太保公园（25.1°N，99.1°E），1800m，1991.Ⅹ.11，徐正会采于针阔混交林地表，No. A91-1052。

词源： 该新种以郑哲民教授姓氏命名，以纪念他在昆虫分类学领域作出的杰出贡献。

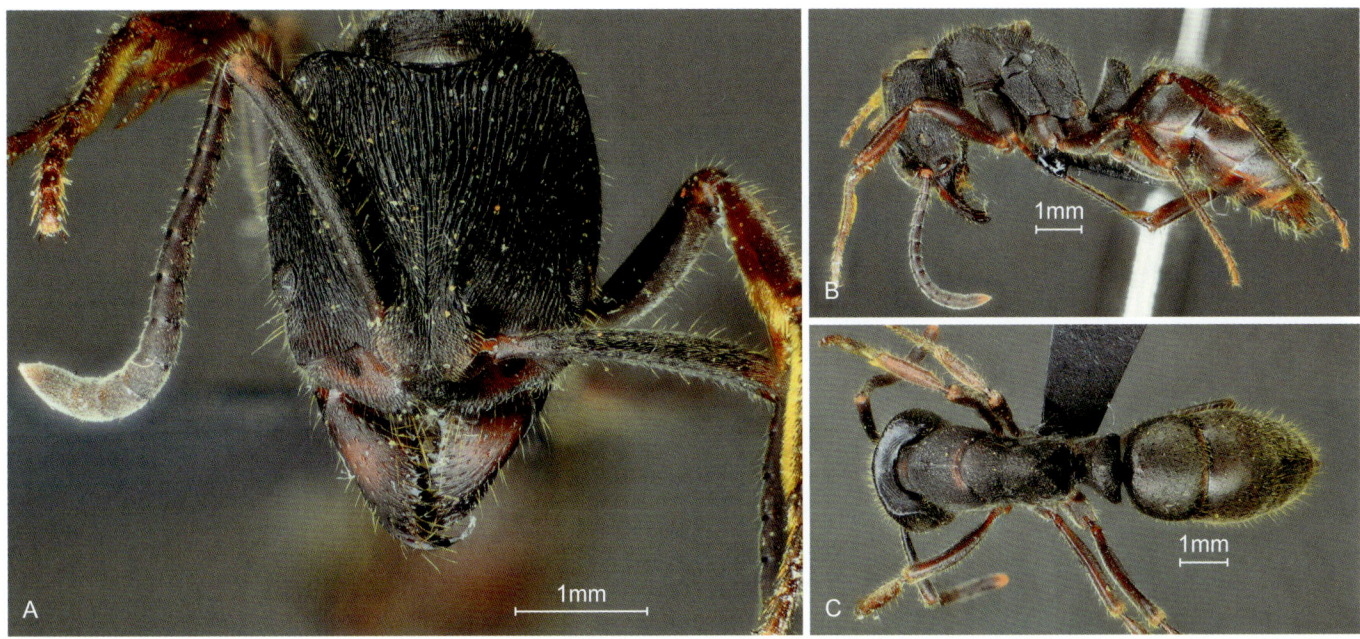
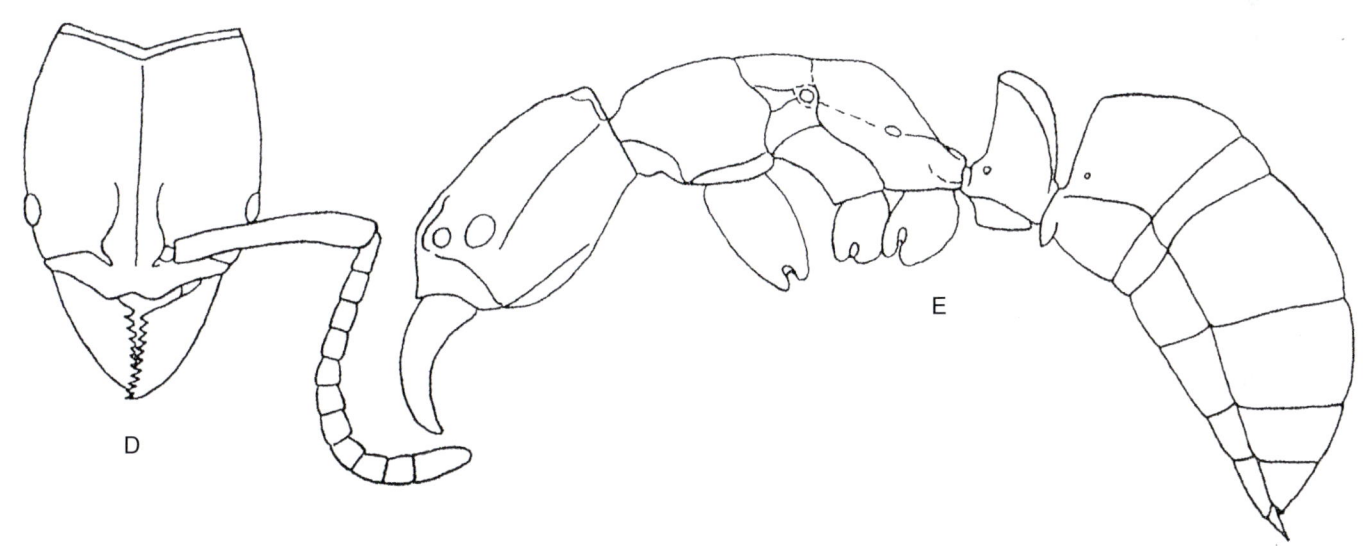

郑氏厚结猛蚁 *Pachycondyla zhengi*

A, D. 工蚁头部正面观；B, E. 工蚁身体侧面观；C. 工蚁身体背面观（D-E. 引自Xu, 1995c）

A, D. Head of worker in full-face view; B, E. Body of worker in lateral view; C. Body of worker in dorsal view (D-E. cited from Xu, 1995c)

036 巴卡猛蚁
Ponera baka Xu, 2001

Ponera baka Xu, 2001a, Entomotaxonomia, 23(1): 57, figs. 22-24 (w.) CHINA (Yunnan).

Holotype worker: TL 1.9, HL 0.43, HW 0.33, CI 76, SL 0.25, SI 77, PW 0.25, AL 0.53, ED 0.03, ML 0.18, PNL 0.15, DPW 0.18, PH 0.25, PNI 70, LPI 60. In full-face view, head rectangular, distinctly longer than broad. Occipital margin weakly concave, occipital corners bluntly prominent, sides weakly convex. Mandible only with 3 apical teeth, the basal portion of masticatory margin without teeth or denticles. Anterior margin of clypeus convex and very bluntly angled in the middle. Antennae short, apex of scape reached to 4/5 of the distance from antennal socket to occipital corner, antennal club consisted of the 4 apical segments. Eye with one facet. In lateral view dorsum of alitrunk weakly convex, promesonotal suture distinct, metanotal groove only with trace. Dorsum of propodeum longer than declivity, posterodorsal corner of propodeum very bluntly angled. In lateral view, petiolar node thick, tapering upward, anterior, posterior and dorsal faces straight, anterior face vertical and formed a right angle with dorsal face, posterior face steeply sloped and formed a more blunter angle with dorsal face. Subpetiolar process with small circular fenestra, anteroventral corner blunt, posteroventral corner with a small tooth. In dorsal view, the node trapezoid, narrowed forwards, anterior border straight, lateral borders weakly convex, posterior border slightly concave. Gaster slightly constricted between the two basal segments. Mandibles smooth and shining, very sparsely punctured. Head and gaster densely and finely punctured, relatively dim. Alitrunk and petiole abundantly and superficially punctured, relatively shining. Surface of whole body and appendages with dense decumbent pubescence, erect hairs only present on anterior portion of head and apex of gaster. Body in color yellowish brown.

Holotype: Worker, China: Yunnan Province, Mengla County, Menglun Town, Bakaxiaozhai, 840 m, 1997.XII.8, collected by Bi-lun Yang in a soil sample of seasonal rain forest, No. A97-2990.

Etymology: The new species is named after the type locality "Bakaxiaozhai".

正模工蚁： 正面观，头部长方形，长明显大于宽；后缘轻度凹陷，后角钝突；侧缘轻度隆起。上颚咀嚼缘仅具3个端齿，基部缺齿和小齿。唇基前缘隆起，中部呈很钝的角状。触角短，柄节末端到达触角窝至头后角间距的4/5处，触角棒4节。复眼仅具1个小眼。侧面观，胸部背面轻度隆起，前中胸背板缝明显，后胸沟仅具痕迹。并胸腹节背面长于斜面，后上角成极钝的角状。侧面观，腹柄结厚，向上变窄，前面、后面和背面平直；前面垂直，与背面形成直角；后面陡坡状，与背面形成钝角；腹柄下突具1个小的圆形窗斑，前下角钝，后下角具1个小齿。背面观，腹柄结梯形，向前变窄，前缘平直，侧缘轻度隆起，后缘轻微凹陷。后腹部基部2节间轻度缢缩。上颚光滑光亮，具很稀疏的刻点。头部和后腹部具细密刻点，较暗。胸部和腹柄具丰富的肤浅刻点，较光亮。全身表面和附肢具密集倾斜绒毛被，仅头前部和后腹部末端具立毛。身体黄棕色。

正模： 工蚁，中国云南省勐腊县勐仑镇巴卡小寨，840m，1997.XII.8，杨比伦采于季节性雨林土壤样中，No. A97-2990。

词源： 该新种以模式产地"巴卡小寨"命名。

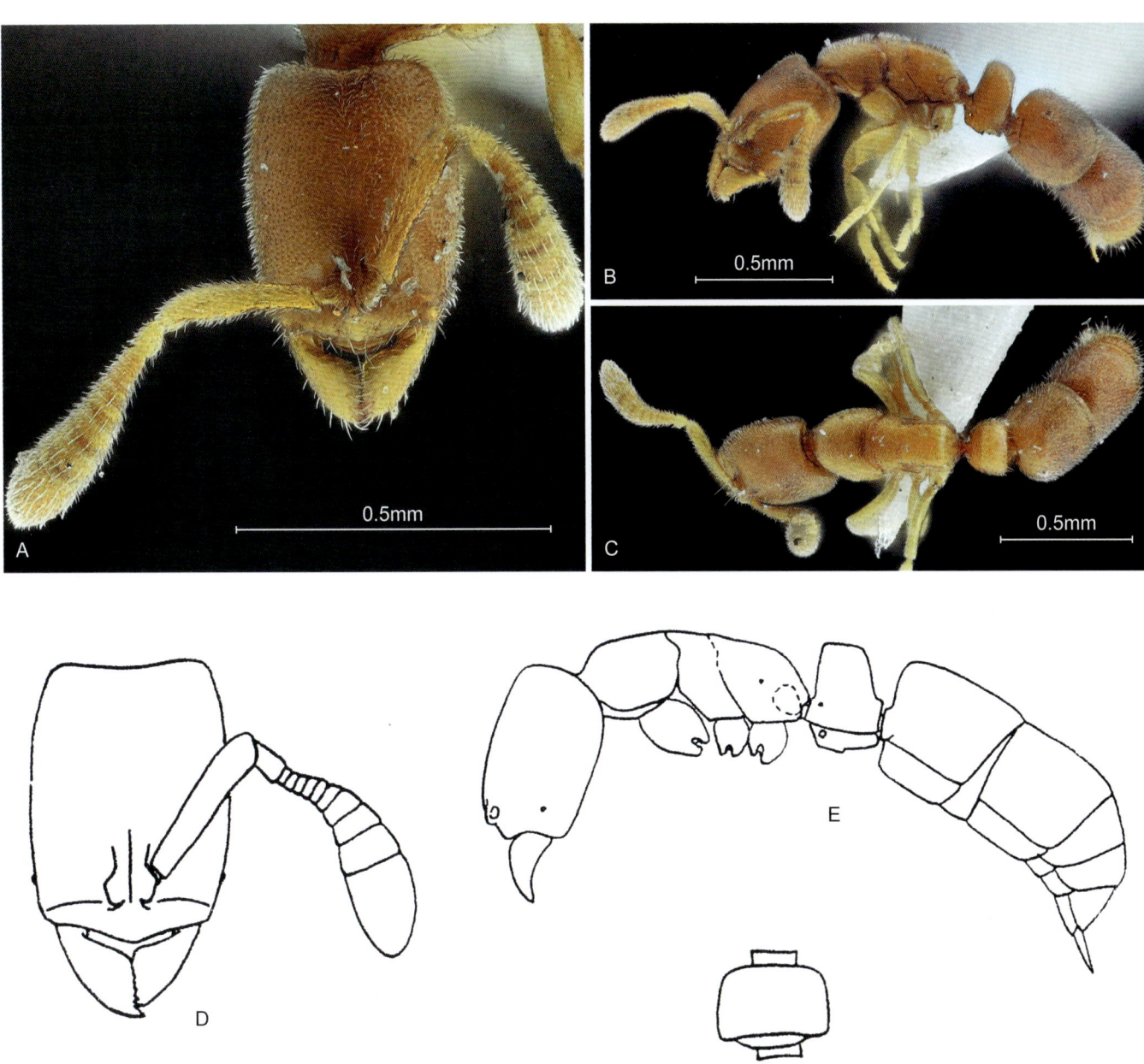

巴卡猛蚁 *Ponera baka*

A, D. 工蚁头部正面观; B, E. 工蚁身体侧面观; C. 工蚁身体背面观; F. 工蚁腹柄背面观（D-F. 引自 Xu, 2001a）
A, D. Head of worker in full-face view; B, E. Body of worker in lateral view; C. Body of worker in dorsal view; F. Petiole of worker in dorsal view (D-F. cited from Xu, 2001a)

坝湾猛蚁
Ponera bawana Xu, 2001

Ponera bawana Xu, 2001c, Entomotaxonomia, 23(3): 220, figs. 19-21 (w.q.) CHINA (Yunnan).

Holotype worker: TL 1.8, HL 0.67, HW 0.53, CI 80, SL 0.43, SI 87, PW 0.42, AL 0.80, ED 0.03, ML 0.30, DPW 0.30, PNI 72, PH 0.40, PNL 0.20, LPI 50. In full-face view, head roughly rectangular, distinctly longer than broad, weakly narrowed forward. Occipital margin nearly straight, occipital corners roundly prominent, sides weakly convex. Masticatory margin of mandible with 3 apical teeth and followed by a row of indistinct minute denticles. Anterior margin of clypeus bluntly angled. Antennae short, apex of scape failed to reach occipital corner by 1/7 of its length, antennal club with 5 segments. Eye with 4 facets. In lateral view, dorsum of thorax evenly convex, promesonotal suture depressed, metanotal groove distinct. Dorsum of propodeum slightly longer than declivity, posterodorsal corner rounded. In lateral view, petiolar node trapezoid, narrowed upward, anterior and posterior faces straight, dorsal face evenly convex, anterodorsal and posterodorsal corners rounded and about at same level. Subpetiolar process cuneiform, fenestra small and circular, anteroventral corner obliquely truncate, posteroventral border with a minute blunt denticle. In dorsal view, petiolar node semicircular, anterior and lateral borders roundly convex, posterior border straight, length : width = 1 : 2. Mandibles smooth and shining. Head closely and finely punctured, dull. Thorax densely and finely punctured, relatively shining. Petiole and gaster superficially, densely and finely punctured, shining. Dorsum of head and thorax with dense decumbent pubescence, without erect hairs. Petiolar node and gaster with sparse suberect hairs and dense decumbent pubescence. Scapes with sparse suberect hairs and dense decumbent pubescence. Dorsa of tibiae with dense decumbent pubescence, without erect hairs. Head, propodeum and upper portion of petiole blackish brown; pronotum, mesonotum, lower portion of petiole and gaster reddish brown. Mandibles, antennae and legs brownish yellow. **Paratype females:** TL 3.5-3.6, HL 0.75-0.77, HW 0.60, CI 78-80, SL 0.50-0.52, SI 83-86, PW 0.57, AL 1.10, ED 0.20-0.22, ML 0.37-0.40, DPW 0.38, PNI 68, PH 0.50, PNL 0.23-0.25, LPI 47-50 (*n*=2). As holotype, but with body much larger, eyes normal and large, with 3 ocelli. Mesothorax and metathorax complete and winged, the wings shedable, mesopleuron with a transverse furrow. In lateral view petiolar node thinner. Color of body similar to holotype but head, propodeum and petiolar node black.

Holotype: Worker, China: Yunnan Province, Baoshan City, Bawan Town, Bawan Village, 1500 m, 1998.Ⅷ.11, collected by Qi-zhen Long in *Pinus yunnanensis* forest, No. A98-2097.

Etymology: The new species is named after the type locality "Bawan".

正模工蚁： 正面观，头部近长方形，长明显大于宽，向前轻度变窄；后缘近平直，后角圆凸；侧缘轻度隆起。上颚咀嚼缘具3个端齿和1列不明显的细齿。唇基前缘钝角状。触角短，柄节末端至头后角的距离为柄节长的1/7，触角棒5节。复眼具4个小眼。侧面观，胸部背面均匀隆起，前中胸背板缝凹陷，后胸沟明显。并胸腹节背面稍长于斜面，后上角圆。侧面观，腹柄结梯形，向上变窄，前面和后面平直，背面均匀隆起；前上角和后上角圆，约在同一水平上；腹柄下突楔形，窗斑小，圆形，前下角斜截，后下缘具1个小钝齿。背面观腹柄结半圆形，前缘和侧缘圆形隆起，后缘平直，长：宽=1：2。上颚光滑发亮。头部具细密刻点，暗。胸部具细密刻点，较光亮。腹柄和后腹部具细密肤浅刻点，光亮。头部背面和胸部具密集倾斜绒毛被，缺立毛。腹柄结和后腹部具稀疏亚直立毛和密集倾斜绒毛被。柄节具稀疏亚直立毛和密集倾斜绒毛被。胫节背面具密集倾斜绒毛被，缺立毛。头部、并胸腹节和腹柄上部黑棕色；前胸背板、中胸背板、腹柄下部和后腹部红棕色；上颚、触角和足棕黄色。**副模雌蚁：** 与正模相似，但是身体大得多；复眼正常且大，有3个单眼；中胸和后胸完整，具翅，翅可脱落，中胸侧板具横沟，侧面观，腹柄结较薄；体色与正模相似，但是头部、并胸腹节和腹柄结黑色。

正模： 工蚁，中国云南省保山市坝湾镇坝湾村，1500m，1998.Ⅷ.11，龙启珍采于云南松林中，No. A98-2097。

词源： 该新种以模式产地坝湾命名。

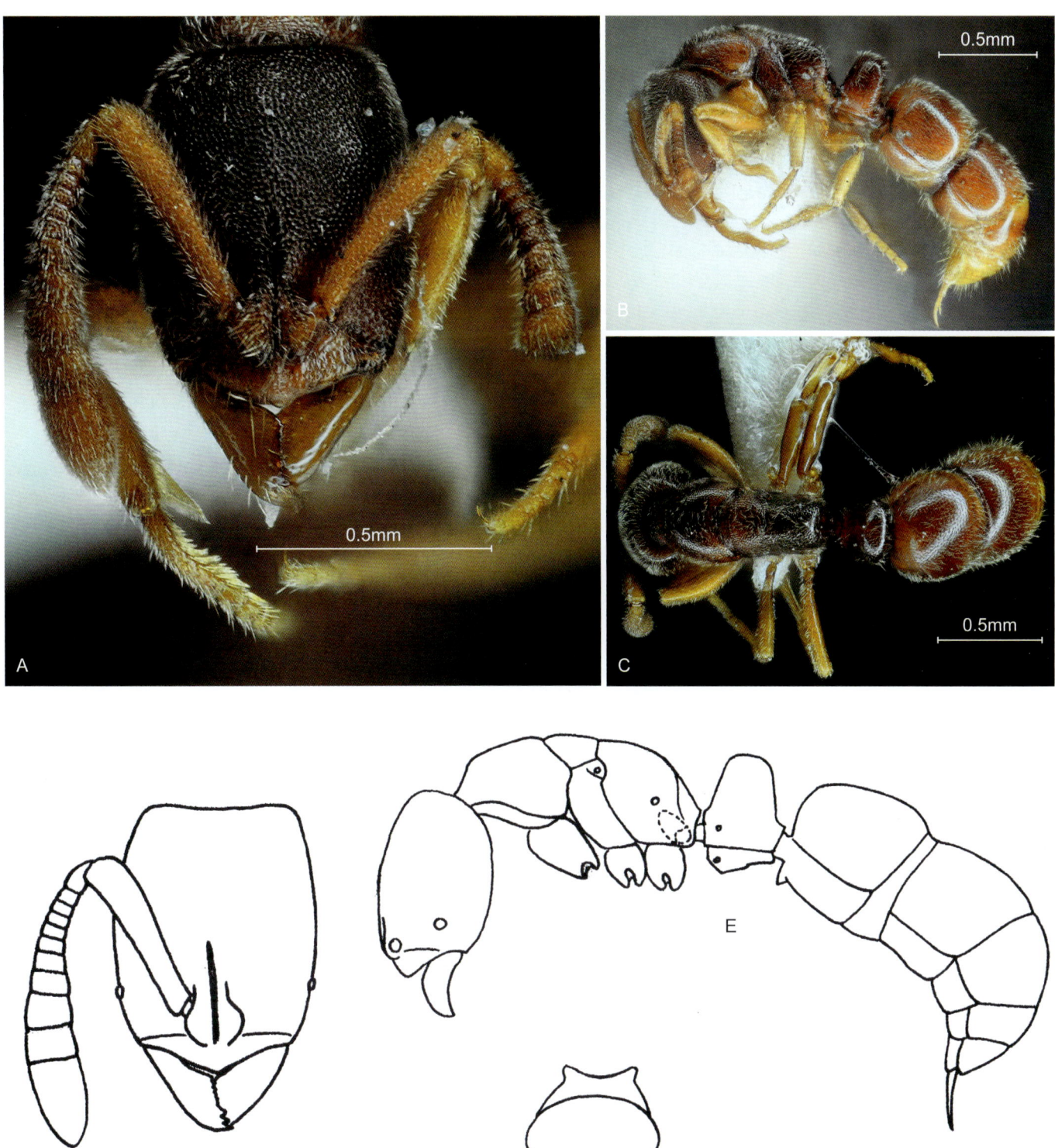

坝湾猛蚁 *Ponera bawana*

A, D. 工蚁头部正面观；B, E. 工蚁身体侧面观；C. 工蚁身体背面观；F. 工蚁腹柄背面观（D-F. 引自 Xu, 2001c）
A, D. Head of worker in full-face view; B, E. Body of worker in lateral view; C. Body of worker in dorsal view; F. Petiole of worker in dorsal view (D-F. cited from Xu, 2001c)

二齿猛蚁
Ponera diodonta Xu, 2001

Ponera diodonta Xu, 2001c, Entomotaxonomia, 23(3): 221, figs. 25-27 (w.) CHINA (Yunnan).

Holotype worker: TL 2.5, HL 0.60, HW 0.50, CI 83, SL 0.43, SI 87, PW 0.38, AL 0.77, ED 0.02, ML 0.27, DPW 0.32, PNI 83, PH 0.43, PNL 0.20, LPI 46. In full-face view, head roughly rectangular, longer than broad, weakly narrowed forward. Occipital margin weakly emarginate, occipital corners roundly prominent, sides weakly convex. Masticatory margin of mandible with 3 apical teeth and followed by a row of indistinct minute denticles. Anterior margin of clypeus evenly convex. Antennae short, apex of scape failed to reach occipital corner by 1/7 of its length, antennal club with 5 segments. Eye with only one facet. In lateral view, dorsum of thorax weakly convex, promesonotal suture depressed, metanotal groove fine and distinct. Dorsum of propodeum as long as declivity, posterodorsal corner rounded. In lateral view, petiolar node roughly rectangular, slightly narrowed upward, anterior and posterior faces straight, dorsal face evenly convex, anterodorsal corner prominent and higher than posterodorsal corner, the latter rounded. Subpetiolar process cuneiform, fenestra small and circular, anteroventral corner rounded, posteroventral border with 2 teeth. In dorsal view, petiolar node semicircular, anterior and lateral borders roundly convex, posterior border slightly concave, length : width = 5 : 9. Mandibles smooth and shining. Head closely and finely punctured, dull. Pronotum, mesonotum and gaster densely and finely punctured, relatively shining. Propodeum and petiolar node superficially and finely punctured, shining. Dorsum of head and thorax with sparse short erect hairs and dense decumbent pubescence. Petiolar node and gaster with abundant erect hairs and dense decumbent pubescence. Scapes with sparse erect hairs and dense decumbent pubescence. Dorsa of tibiae with dense decumbent pubescence, without erect hairs. Body in color reddish brown. Head and middle portion of gaster blackish brown. Mandibles, antennae and legs brownish yellow.

Holotype: Worker, China: Yunnan Province, Lushui County, Shangjiang Town, Shangjiang Village, 1600 m, 2000.Ⅲ.23, collected by Zheng-hui Xu in monsoon evergreen broadleaf forest, No. A008.

Etymology: The species name *diodonta* combines Greek *di-* (two) + word root *odont* (teeth) + suffix *-a* (feminine form), it refers to the two small teeth on the posteroventral boder of subpetiolar process.

正模工蚁： 正面观，头部近长方形，长大于宽，向前轻度变窄；后缘轻度凹陷，后角圆凸；侧缘轻度隆起。上颚咀嚼缘具3个端齿和1列不明显的细齿。唇基前缘均匀隆起。触角短，柄节未到达头后角，柄节末端至头后角距离为柄节长的1/7，触角棒5节。复眼仅具1个小眼。侧面观，胸部背面轻度隆起，前中胸背板缝凹陷，后胸沟细而明显。并胸腹节背面与斜面等长，后上角圆。侧面观，腹柄结近长方形，向上轻微变窄，前面和后面平直，背面均匀隆起；前上角突出，高于后上角，后上角圆；腹柄下突楔形，窗斑小而圆，前下角圆，后下缘具2个齿。背面观，腹柄结半圆形，前缘和侧缘圆形隆起，后缘轻微凹陷，长：宽=5：9。上颚光滑发亮。头部具密集细刻点，暗。前胸背板、中胸背板和后腹部具密集细刻点，较光亮。并胸腹节和腹柄结具肤浅细刻点，光亮。头部背面和胸部具稀疏直立短毛和密集倾斜绒毛被；腹柄结和后腹部具丰富立毛和密集倾斜绒毛被；柄节具稀疏立毛和密集倾斜绒毛被；胫节背面具密集倾斜绒毛被，缺立毛。身体红棕色，头部和后腹部中部黑棕色，上颚、触角和足棕黄色。

正模： 工蚁，中国云南省泸水县上江乡上江村，1600m，2000.Ⅲ.23，徐正会采于季风常绿阔叶林中，No. A008。

词源： 该新种的种名"*diodonta*"由希腊语"*di-*"（二）+词根"*odont*"（齿）+后缀"*-a*"（阴性形式）组成，指腹柄下突后下缘具2个小齿。

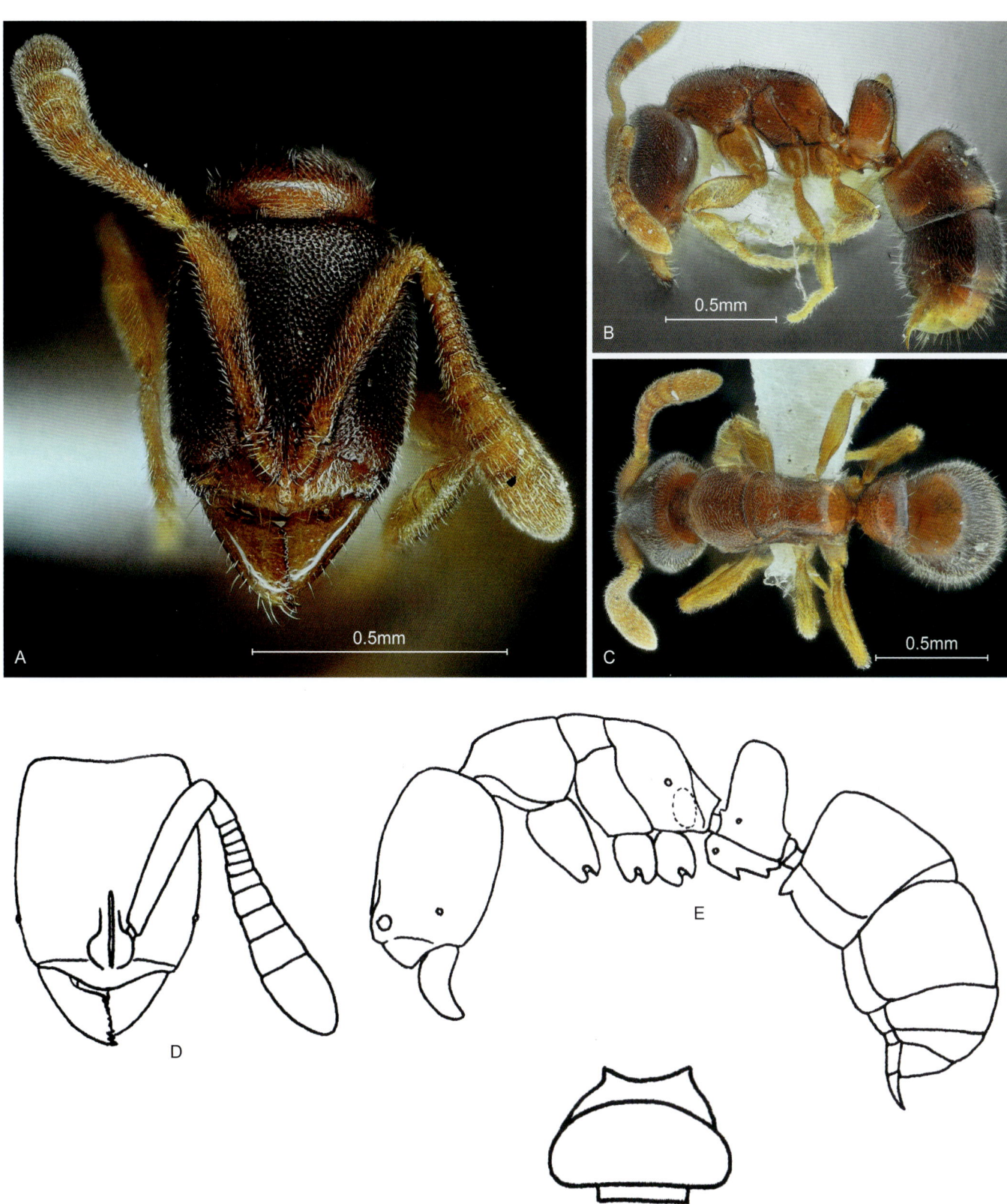

二齿猛蚁 *Ponera diodonta*

A, D. 工蚁头部正面观；B, E. 工蚁身体侧面观；C. 工蚁身体背面观；F. 工蚁腹柄背面观（D-F. 引自Xu, 2001c）
A, D. Head of worker in full-face view; B, E. Body of worker in lateral view; C. Body of worker in dorsal view; F. Petiole of worker in dorsal view (D-F. cited from Xu, 2001c)

龙林猛蚁
Ponera longlina Xu, 2001

Ponera longlina Xu, 2001a, Entomotaxonomia, 23(1): 56, figs. 16-18 (w.) CHINA (Yunnan).

Holotype worker: TL 2.2, HL 0.50, HW 0.45, CI 90, SL 0.35, SI 78, PW 0.35, AL 0.63, ED 0.03, ML 0.25, PNL 0.18, DPW 0.25, PH 0.33, PNI 71, LPI 54. In full-face view, head roughly square, slightly longer than broad, narrowed forward. Occipital margin weakly concave, occipital corners roundly prominent, sides evenly convex. Mandible with 3 enlarged apical teeth followed by a series of minute denticles. Anterior margin of clypeus weakly convex. Apex of scape reached to 9/10 of the distance from antennal socket to occipital corner, antennal club consisted of the 4 apical segments. Eye with one facet. In lateral view dorsum of alitrunk weakly convex, promesonotal suture distinct, metanotal groove absent. Dorsum of propodeum about as long as declivity, declivity flat, posterodorsal corner quite blunt. In lateral view, petiolar node weakly tapering upward, anterior and posterior faces straight, dorsal face evenly convex, anterodorsal corner blunt, posterodorsal corner more blunter. Subpetiolar process with small circular fenestra, anteroventral corner blunt, posteroventral corner with an enlarged tooth. In dorsal view, the node roughly semicircular, anterior and lateral borders formed a single arch, posterior border weakly concave. Gaster weakly constricted between the two basal segments. Mandibles smooth and shining, sparsely punctured. Head densely and finely punctured, dim. Prothorax, mesothorax and gaster densely and weakly punctured, less shining. Propodeum and petiole smooth and shining, sparsely punctured. Surface of whole body and appendages with dense decumbent pubescence, erect hairs only present on anterior portion of head and posterior half of gaster. Body in color reddish brown, head black, mandibles, antennae and legs yellowish brown.

Holotype: Worker, China: Yunnan Province, Mengla County, Mengla Town, Longlin Village, 1050 m, 1997.Ⅷ.11, collected by Yun-feng He in a ground sample of mountain rain forest, No. A97-1315.

Etymology: The new species is named after the type locality "Longlin".

正模工蚁： 正面观，头部近方形，长稍大于宽，向前变窄；后缘轻度凹陷，后角圆凸；侧缘均匀隆起。上颚具3个大的端齿和1列细齿。唇基前缘轻度隆起。柄节末端到达触角窝至头后角间距的9/10处，触角棒4节。复眼具1个小眼。侧面观，胸部背面轻度隆起，前中胸背板缝明显，缺后胸沟。并胸腹节背面约与斜面等长，斜面平坦，后上角很钝。侧面观，腹柄结向上轻度变窄，前面和后面平直，背面均匀隆起，前上角钝，后上角钝圆；腹柄下突具小的圆形窗斑，前下角钝，后下角具1个大齿。背面观，腹柄结近半圆形，前缘和侧缘形成弓形，后缘轻度凹陷。后腹部基部2节间轻度缢缩。上颚光滑发亮，具稀疏刻点。头部具细密刻点，暗。前胸、中胸和后腹部具密集弱刻点，不太光亮。并胸腹节和腹柄光滑发亮，具稀疏刻点。全身表面和附肢具密集倾斜绒毛被，仅头前部和后腹部的后半部具立毛。身体红棕色，头部黑色，上颚、触角和足黄棕色。

正模： 工蚁，中国云南省勐腊县勐腊镇龙林村，1050m，1997.Ⅷ.11，何云峰采于山地雨林地表样中，No. A97-1315。

词源： 该新种以模式产地"龙林"命名。

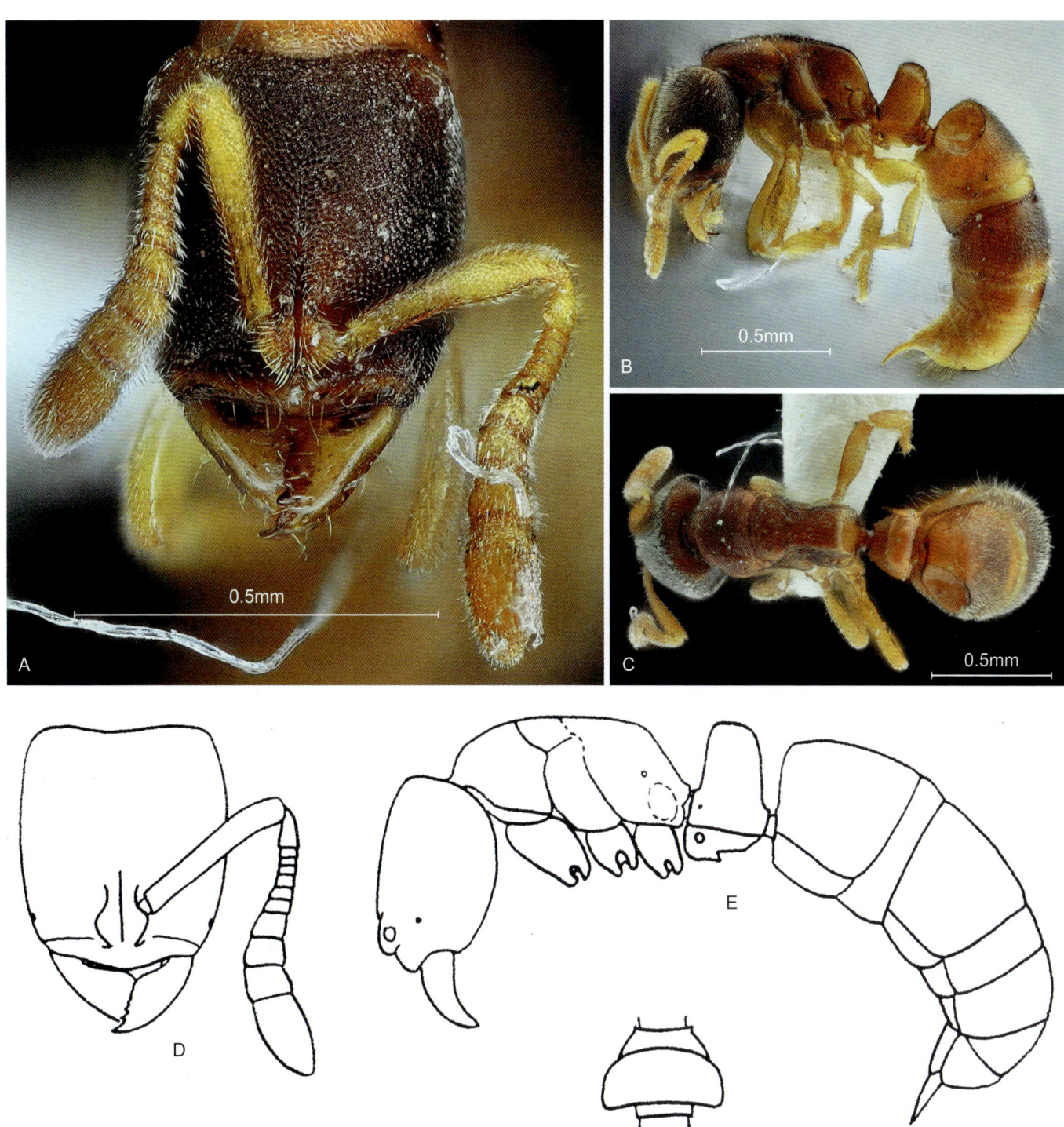

龙林猛蚁 *Ponera longlina*

A, D. 工蚁头部正面观；B, E. 工蚁身体侧面观；C. 工蚁身体背面观；F. 工蚁腹柄背面观（D-F. 引自 Xu, 2001a）

A, D. Head of worker in full-face view; B, E. Body of worker in lateral view; C. Body of worker in dorsal view; F. Petiole of worker in dorsal view (D-F. cited from Xu, 2001a)

勐腊猛蚁
Ponera menglana Xu, 2001

Ponera menglana Xu, 2001a, Entomotaxonomia, 23(1): 54, figs. 7-9 (w.) CHINA (Yunnan).

Holotype worker: TL 2.9, HL 0.68, HW 0.60, CI 89, SL 0.50, SI 83, PW 0.50, AL 0.90, ED 0.03, ML 0.40, PNL 0.23, DPW 0.43, PH 0.50, PNI 85, LPI 45. In full-face view, head roughly square, lightly longer than broad. Occipital margin slightly concave, occipital corners blunt, sides weakly convex. Mandible with 3 enlarged apical teeth followed by a series of minute denticles. Anterior margin of clypeus evenly convex. Apex of scape reached to 9/10 of the distance from antennal socket to occipital corner, antennal club consisted of the apical 5 segments. Eye with one facet. In lateral view, dorsum of alitrunk slightly convex, promesonotal suture distinct, metanotal groove very weak with fine visible trace. Dorsum of propodeum about as long as declivity, posterodorsal corner of propodeum rounded, sides of propodeum weakly depressed, declivity obviously depressed, sides of declivity distinctly marginate. In lateral view petiolar node higher than long, anterior face straight and vertical, dorsal and posterior faces formed a single arched surface, anterodorsal corner blunt. Subpetiolar process with small circular fenestra, anteroventral corner obliquely truncate, posteroventral corner with a minute denticle. In dorsal view, petiolar node roughly semicircular, anterior and lateral borders formed a single arch, posterior face weakly concave. Gaster distinctly constricted between the two basal segments. Mandibles smooth and shining. Head, alitrunk and the two basal segments of gaster densely and finely punctured. Petiole with anterior and lateral faces weakly finely punctured, posterior face smooth. Segments 3-6 of gaster smooth and shining. Whole body surface with sparse erect or suberect hairs and dense decumbent pubescence. Scapes and tibiae with dense decumbent pubescence, but without erect hairs. Body in color black. Mandibles, antennae, legs, subpetiolar process and apex of gaster yellowish brown. **Paratype workers:** TL 2.7-3.1, HL 0.65-0.68, HW 0.58-0.60, CI 85-89, SL 0.48-0.53, SI 83-88, PW 0.45-0.48, AL 0.85-0.90, ED 0.03, ML 0.33-0.38, PNL 0.20-0.23, DPW 0.40-0.43, PH 0.45-0.50, PNI 84-94, LPI 43-50 (n=5). As holotype.

Holotype: Worker, China: Yunnan Province, Mengla County, Mengla Town, Bubang Village, 730 m, 1997.Ⅷ.17, collected by Guang Zeng in a soil sample of seasonal rain forest, No. A97-2046.

Etymology: The new species is named after the type locality "Mengla County".

正模工蚁： 正面观，头部近方形，长稍大于宽；后缘轻微凹陷，后角钝；侧缘轻度隆起。上颚具3个大的端齿和1列细齿。唇基前缘均匀隆起。柄节末端到达触角窝至头后角间距的9/10处，触角棒5节。复眼具1个小眼。侧面观，胸部背面轻微隆起，前中胸背板缝明显，后胸沟可见细弱的痕迹。并胸腹节背面约与斜面等长，后上角圆，侧面轻度凹陷；斜面明显凹陷，斜面两侧具明显边缘。侧面观腹柄结高大于长，前面平直且垂直，背面和后面形成1个弓形面，前上角钝；腹柄下突具小的圆形窗斑，前下角斜截，后下角具1个小齿。背面观腹柄结近半圆形，前缘和侧缘形成1个弓形面，后缘轻度凹陷。后腹部基部2节间明显缢缩。上颚光滑发亮。头部、胸部和后腹部基部2节具细密刻点。腹柄前面和侧面具弱的细刻点，后面光滑；后腹部第3~6节光滑发亮。全身表面具稀疏直立、亚直立毛和密集倾斜绒毛被；柄节和胫节具密集倾斜绒毛被，缺立毛。身体黑色，上颚、触角、足、腹柄下突和后腹部末端黄棕色。**副模工蚁：** 特征同正模。

正模： 工蚁，中国云南省勐腊县勐腊镇补蚌村，730m，1997.Ⅷ.17，曾光采于季节性雨林土壤样中，No. A97-2046。

词源： 该新种以模式产地"勐腊县"命名。

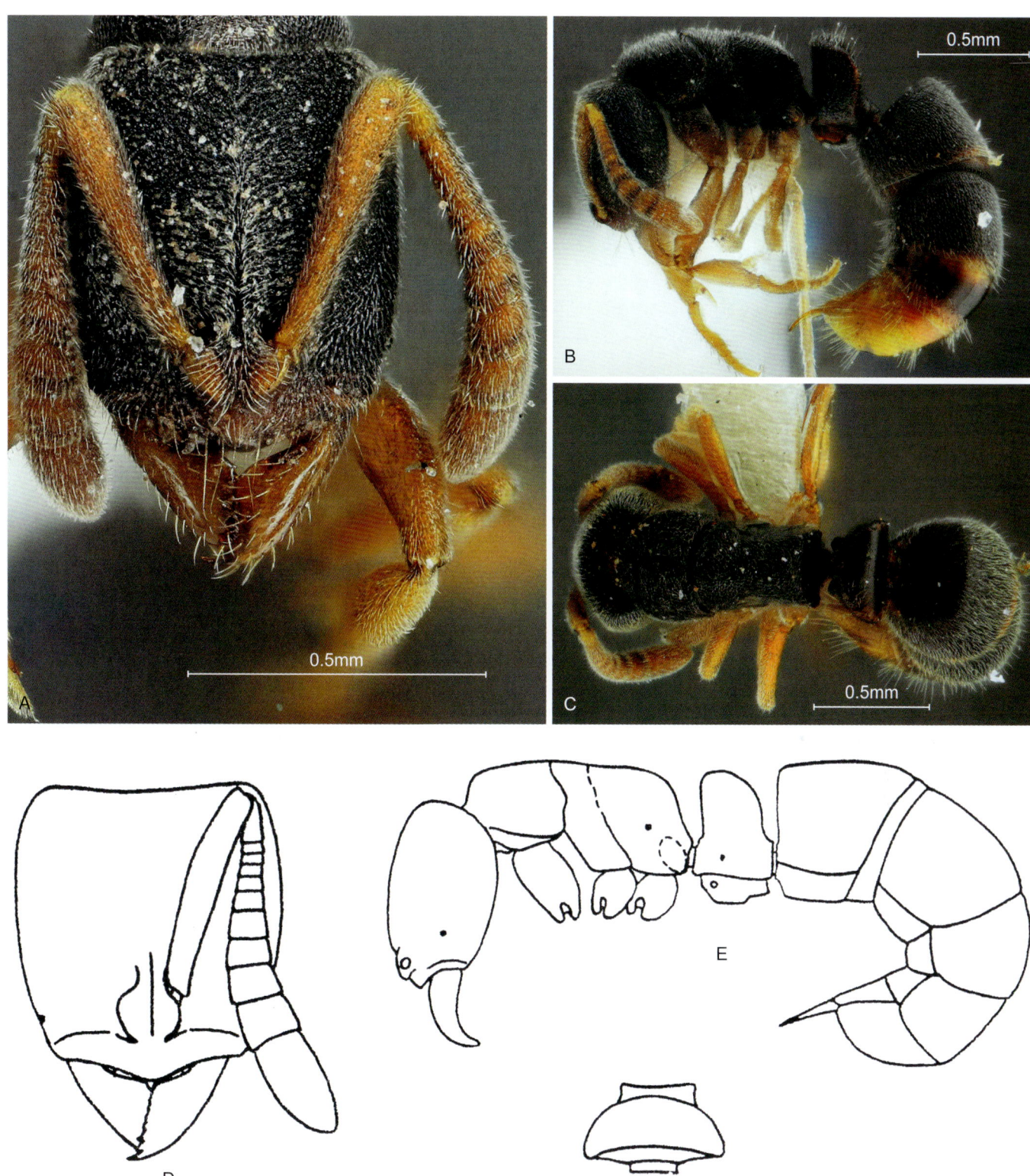

勐腊猛蚁 *Ponera menglana*

A, D. 工蚁头部正面观；B, E. 工蚁身体侧面观；C. 工蚁身体背面观；F. 工蚁腹柄背面观（D-F. 引自Xu, 2001a）
A, D. Head of worker in full-face view; B, E. Body of worker in lateral view; C. Body of worker in dorsal view; F. Petiole of worker in dorsal view (D-F. cited from Xu, 2001a)

南贡山猛蚁
Ponera nangongshana Xu, 2001

Ponera nangongshana Xu, 2001a, Entomotaxonomia, 23(1): 55, figs. 13-15 (w.) CHINA (Yunnan).

Holotype worker: TL 2.6, HL 0.58, HW 0.48, Cl 83, SL 0.40, SI 84, PW 0.38, AL 0.75, ED 0.03, ML 0.25, PNL 0.20, DPW 0.28, PH 0.38, PNI 73, LPI 53. In full-face view, head roughly rectangular, longer than broad, weakly narrowed anteriorly. Occipital margin weakly concave, occipital corners bluntly prominent, sides weakly convex. Mandible with 3 enlarged apical teeth followed by a series of minute denticles. Anterior margin of clypeus convex. Apex of scape reached to 9/10 of the distance from antennal socket to occipital corner, antennal club consisted of the apical 5 segments. Eye with one facet. In lateral view, dorsum of alitrunk weakly convex, promesonotal suture and metanotal groove distinct. Dorsum of propodeum longer than declivity, declivity flat, posterodorsal corner of propodeum bluntly angled. In lateral view, petiolar node thick, roughly square, anterior and posterior faces nearly straight and parallel, dorsal face weakly convex, anterodorsal and posterodorsal corners blunt. Subpetiolar process with small circular fenestra, anteroventral corner obliquely truncate, posteroventral corner without tooth. In dorsal view petiolar node roughly rectangular, anterior and lateral borders convex, posterior border nearly straight. Gaster distinctly constricted between the two basal segments. Mandibles smooth and shining, sparsely punctured. Head closely and finely punctured and dim. Alitrunk and gaster densely and finely punctured, less shining. Petiole shining, with very weak punctures. Surface of whole body with very sparse erect short hairs and dense decumbent pubescence. Appendages with dense decumbent pubescence, but without erect hairs. Body in color reddish brown. **Paratype workers:** TL 2.4-2.8 , HL 0.55-0.60, HW 0.48, CI 79-83, SL 0.40-0.43, SI 84-89, PW 0.38-0.40, AL 0.73-0.78, ED 0.03-0.04, ML 0.28, PNL 0.20, DPW 0.28-0.30, PH 0.35-0.38, PNI 69-80, LPI 53-57 (*n*=5). As holotype, but body in color yellowish brown to reddish brown.

Holotype: Worker, China: Yunnan Province, Mengla County, Mengla Town, Nangongshan Mountain, 1620 m, 1998.Ⅲ.15, collected by Yun-feng He in a soil sample of monsoon evergreen broadleaf forest, No. A98-824.

Etymology: The new species is named after the type locality "Nangongshan Mountain".

正模工蚁：正面观，头部近长方形，长大于宽，向前轻度变窄；后缘轻度凹陷，后角钝凸；侧缘轻度隆起。上颚具3个大的端齿和1列细齿。唇基前缘隆起。触角柄节末端到达触角窝至头后角间距的9/10处，触角棒5节。复眼具1个小眼。侧面观，胸部背面轻度隆起，前中胸背板缝和后胸沟明显。并胸腹节背面长于斜面，斜面平坦，后上角钝角状。侧面观，腹柄结厚，近方形；前面和后面近平直，互相平行；背面轻度隆起，前上角和后上角钝；腹柄下突具小的圆形窗斑，前下角斜截，后下角缺齿。背面观，腹柄结近长方形，前缘和侧缘隆起，后缘近平直。后腹部基部2节间明显收缩。上颚光滑发亮，具稀疏刻点。头部具细密刻点，暗。胸部和后腹部具细密刻点，不太光亮。腹柄光亮，具很弱的刻点。全身表面具很稀疏的直立短毛和密集倾斜绒毛被；附肢具密集倾斜绒毛被，缺立毛。身体红棕色。**副模工蚁：**特征同正模，但是身体黄棕色至红棕色。

正模：工蚁，中国云南省勐腊县勐腊镇南贡山，1620m，1998.Ⅲ.15，何云峰采于季风常绿阔叶林土壤样中，No. A98-824。

词源：该新种以模式产地"南贡山"命名。

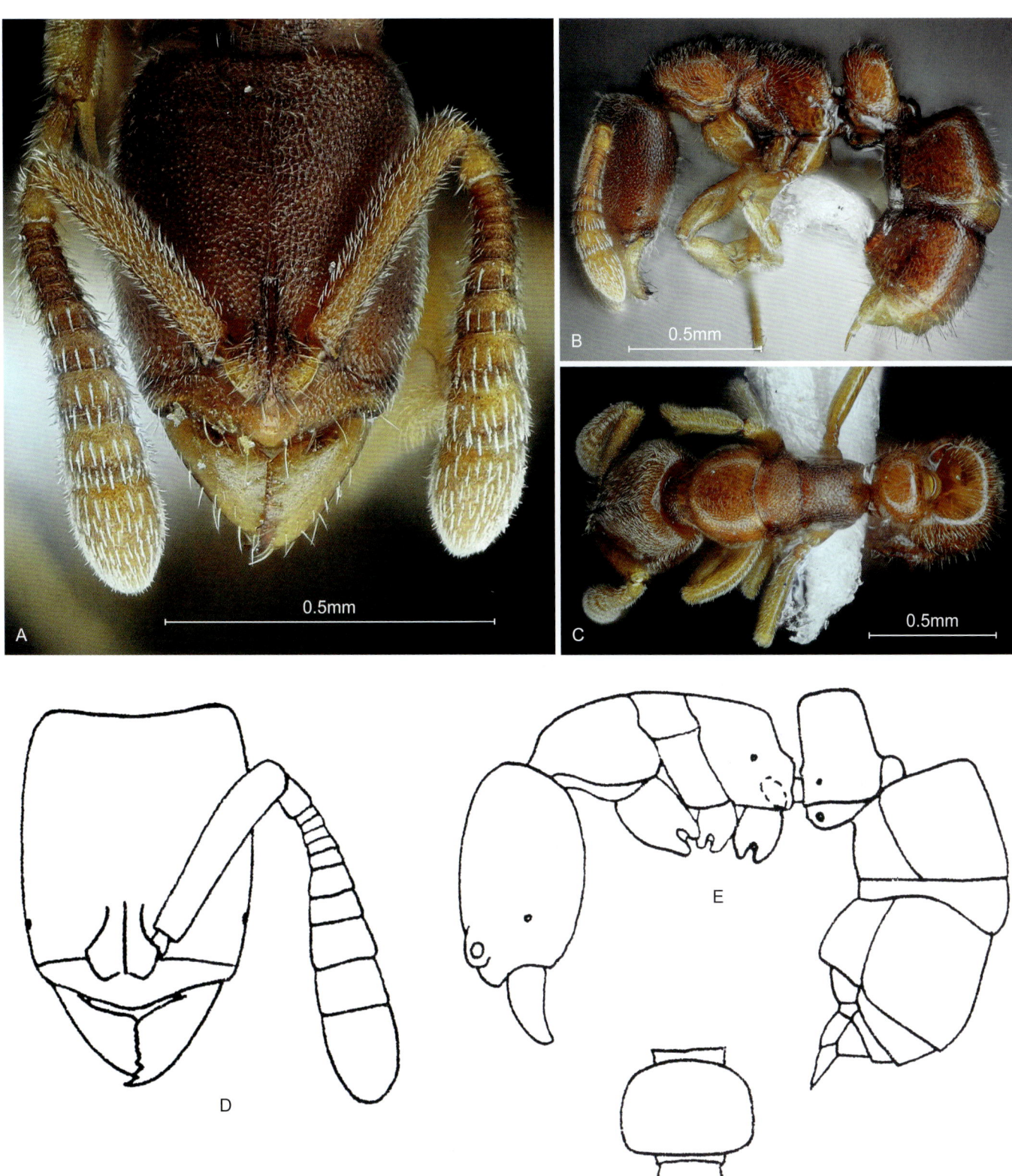

南贡山猛蚁 *Ponera nangongshana*

A, D. 工蚁头部正面观；B, E. 工蚁身体侧面观；C. 工蚁身体背面观；F. 工蚁腹柄背面观（D-F. 引自Xu, 2001a）
A, D. Head of worker in full-face view; B, E. Body of worker in lateral view; C. Body of worker in dorsal view; F. Petiole of worker in dorsal view (D-F. cited from Xu, 2001a)

五齿猛蚁
Ponera pentodontos Xu, 2001

Ponera pentodontos Xu, 2001a, Entomotaxonomia, 23(1): 53, figs. 1-3 (w.) CHINA (Yunnan).

Holotype worker: TL 2.7, HL 0.55, HW 0.53, CI 95, SL 0.45, SI 86, PW 0.40, AL 0.80, ED 0.03, ML 0.35, PNL 0.20, DPW 0.33, PH 0.40, PNI 81, LPI 50. In full-face view, head nearly square, occipital margin weakly concave, occipital corners blunt, sides weakly convex. Mandible with 5 distinct subequal large teeth. Anterior margin of clypeus convex and very bluntly angled in the middle. Apex of scape reached to 19/20 of the distance from antennal socket to occipital corner, antennal club consisted of the apical 5 segments. Eye with one facet. In lateral view, alitrunk weakly convex, promesonotal suture and metanotal groove distinct. Dorsum of propodeum and declivity subequal and flat, posterodorsal corner blunt. In lateral view, petiolar node distinctly tapering upward, anterior face straight and vertical, dorsal and posterior faces formed a single arched surface, anterodorsal corner blunt. Subpetiolar process cuneiform, fenestra circular and small, anteroventral corner rounded, posteroventral corner with a large tooth. In dorsal view, petiolar node nearly semicircular, anterior and lateral borders formed a single arch, posterior border straight. Gaster weakly constricted between the two basal segments. Mandibles smooth and shining, with very sparse punctures. Head finely and closely punctured. Alitrunk, petiole and gaster finely and densely punctured. Surface of whole body and appendages with dense decumbent pubescence, anterior portion of head, petiolar node and gaster with sparse erect short hairs. Body in color black. Mandibles, antennae and legs yellowish brown. **Paratype workers:** TL 2.6-2.8, HL 0.55-0.58, HW 0.53-0.55, CI 95-98, SL 0.45-0.48, SI 82-86, PW 0.40-0.43, AL 0.80-0.83, ED 0.03-0.04, ML 0.35, PNL 0.20, DPW 0.33, PH 0.38-0.40, PNI 76-81, LPI 50-53 (*n*=5). As holotype.

Holotype: Worker, China: Yunnan Province, Mengla County, Mengla Town, Bubang Village, 730 m, 1997.Ⅷ.17, collected by Guang Zeng in a soil sample of seasonal rain forest, No. A97-2046.

Etymology: The species name *pentodontos* combines Greek *pent-* (five) + word root *odontos* (teeth), it refers to the five distinct teeth on the masticatory margin of mandible.

正模工蚁： 正面观，头部近方形；后缘轻度凹陷，后角钝；侧缘轻度隆起。上颚具5个明显近等长的大齿。唇基前缘隆起，中央呈极钝的角状。触角柄节末端到达触角窝至头后角间距的19/20处，触角棒5节。复眼具1个小眼。侧面观，胸部背面轻度隆起，前中胸背板缝和后胸沟明显。并胸腹节背面和斜面近等长，均平直，后上角钝。侧面观，腹柄结向上明显变窄，前面平直且垂直，背面和后面形成1个弓形面，前上角钝；腹柄下突楔形，具小的圆形窗斑，前下角圆，后下角具1个大齿。背面观，腹柄结近半圆形，前缘和侧缘形成1个弓形面，后缘平直。后腹部基部2节间轻度收缩。上颚光滑发亮，具很稀疏的刻点。头部具稠密细刻点。胸部、腹柄和后腹部具细密刻点。全身表面和附肢具密集倾斜绒毛被，头前部、腹柄结和后腹部具稀疏直立短毛。身体黑色，上颚、触角和足黄棕色。**副模工蚁：** 特征同正模。

正模： 工蚁，中国云南省勐腊县勐腊镇补蚌村，730m，1997.Ⅷ.17，曾光采于季节性雨林土壤样中，No. A97-2046。

词源： 该新种的种名"*pentodontos*"由希腊语"*pent-*"（五）+词根"*odontos*"（齿）组成，指上颚咀嚼缘具5个明显的齿。

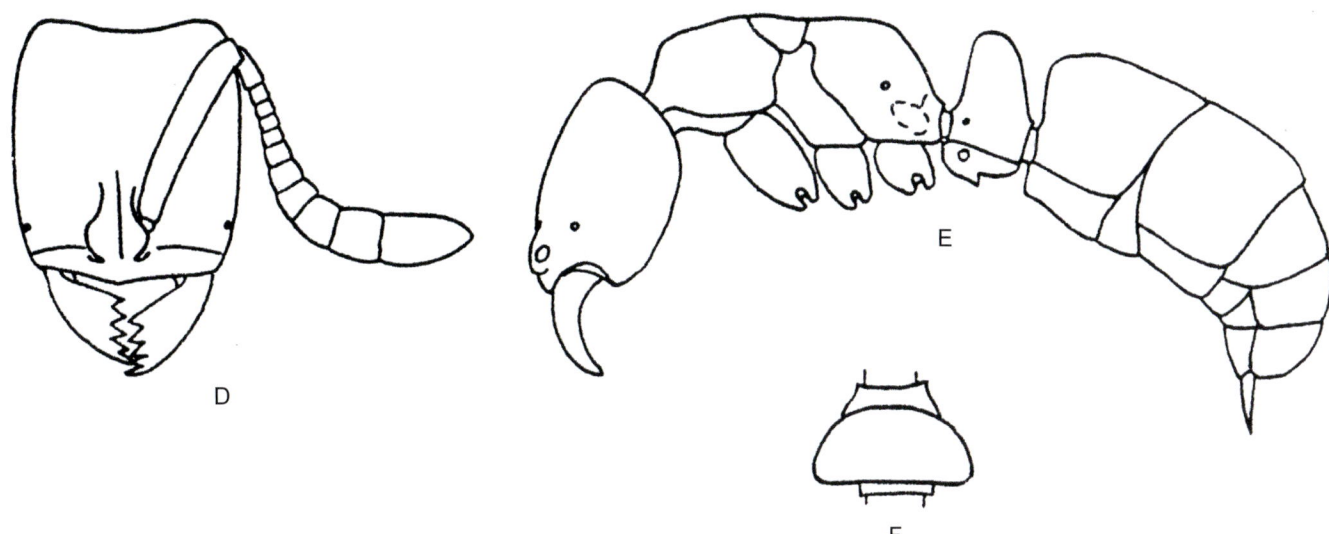

五齿猛蚁 *Ponera pentodontos*
A, D. 工蚁头部正面观；B, E. 工蚁身体侧面观；C. 工蚁身体背面观；F. 工蚁腹柄背面观（D-F. 引自Xu, 2001a）
A, D. Head of worker in full-face view; B, E. Body of worker in lateral view; C. Body of worker in dorsal view; F. Petiole of worker in dorsal view (D-F. cited from Xu, 2001a)

片马猛蚁
Ponera pianmana Xu, 2001

Ponera pianmana Xu, 2001c, Entomotaxonomia, 23(3): 223, figs. 31-33 (w.) CHINA (Yunnan).

Holotype worker: TL 2.2, HL 0.52, HW 0.40, CI 77, SL 0.33, SI 33, PW 0.32, AL 0.67, ED 0.02, ML 0.23, DPW 0.25, PNI 79, PH 0.30, PNL 0.17, LPI 56. In full-face view, head roughly rectangular, distinctly longer than broad, feebly narrowed forward. Occipital margin weakly emarginate, occipital corners roundly prominent, sides weakly convex. Masticatory margin of mandible with 3 apical teeth and followed by a row of indistinct minute denticles. Anterior margin of clypeus strongly convex. Antennae short, apex of scape failed to reach occipital corner by 1/4 of its length, antennal club with 5 segments. Eye with only one facet. In lateral view, dorsum of thorax weakly convex, promesonotal suture depressed, metanotal groove fine and distinct. Dorsum of propodeum as long as declivity, posterodorsal corner blunt. In lateral view, petiolar node trapezoid, distinctly narrowed upward, anterior face nearly vertical, posterior face steeply slope, dorsal face convex, anterodorsal corner higher than posterodorsal corner. Subpetiolar process cuneiform, fenestra medium size, anteroventral corner obliquely truncate, posteroventral border with an acute tooth. In dorsal view, petiolar node crescent, anterior and lateral borders roundly convex, posterior border weakly concave, length∶width = 1∶2. Mandibles smooth and shining, with very sparse fine punctures. Head densely and finely punctured, relatively dull. Thorax, petiole and gaster weakly, densely and finely punctured, relatively shining. Dorsum of head and thorax with dense decumbent pubescence, without erect hairs. Petiole and gaster with sparse suberect hairs and dense decumbent pubescence, apex of gaster with abundant hairs. Scapes with abundant suberect short hairs and dense decumbent pubescence. Dorsa of tibiae with dense decumbent pubescence, without erect hairs. Body in color brown, antennae and legs brownish yellow.

Holotype: Worker, China: Yunnan Province, Lushui County, Pianma Town, Pianma Village, 1650 m, 1999.Ⅳ.26, collected by Ji-guai Li in monsoon evergreen broadleaf forest, No. A99-47.

Etymology: The new species is named after the type locality "Pianma".

正模工蚁： 正面观，头部近长方形，长明显大于宽，向前轻度变窄；后缘轻度凹陷，后角圆凸；侧缘轻度隆起。上颚咀嚼缘具3个端齿和1列不明显的细齿。唇基前缘强烈隆起。触角短，柄节未到达头后角，柄节末端至头后角的距离约为柄节长的1/4，触角棒5节。复眼仅具1个小眼。侧面观，胸部背面轻度隆起，前中胸背板缝凹陷，后胸沟细而明显。并胸腹节背面与斜面等长，后上角钝。侧面观，腹柄结梯形，向上明显变窄；前面近垂直，后面陡坡状，背面隆起，前上角高于后上角；腹柄下突楔形，窗斑中等大小，前下角斜截，后下缘具1个锐齿。背面观，腹柄结新月形，前缘和侧缘圆形隆起，后缘轻度凹陷，长∶宽=1∶2。上颚光滑发亮，具很稀疏的细刻点。头部具细密刻点，较暗。胸部、腹柄和后腹部具弱的细密刻点，较光亮。头胸部背面具密集倾斜绒毛被，缺立毛；腹柄和后腹部具稀疏亚直立毛和密集倾斜绒毛被，后腹部末端具丰富立毛；柄节具丰富亚直立短毛和密集倾斜绒毛被；胫节背面具密集倾斜绒毛被，缺立毛。身体棕色，触角和足棕黄色。

正模： 工蚁，中国云南省泸水县片马镇片马村，1650m，1999.Ⅳ.26，李继乖采于季风常绿阔叶林中，No. A99-47。

词源： 该新种以模式产地"片马"命名。

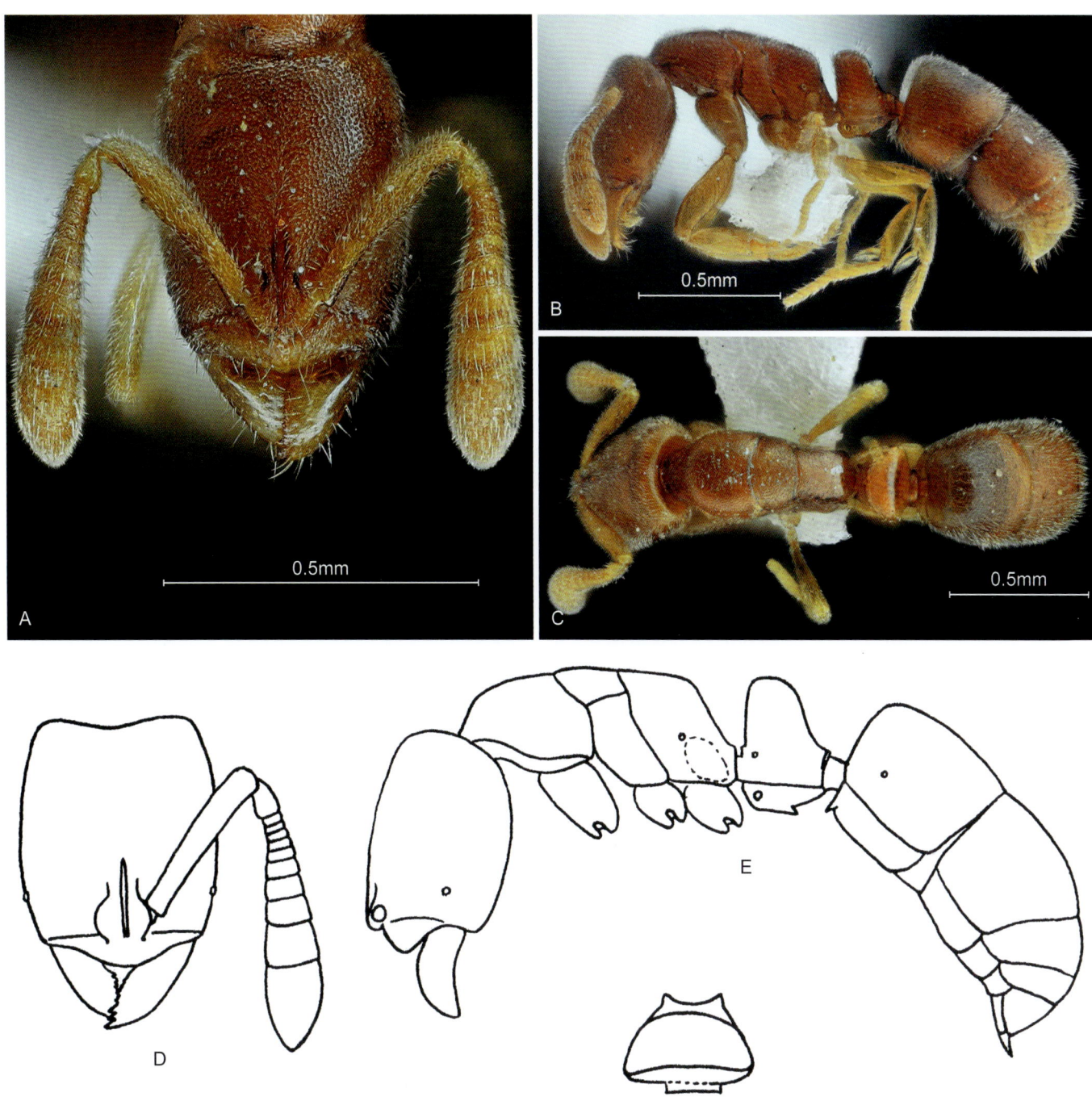

片马猛蚁 *Ponera pianmana*
A, D. 工蚁头部正面观；B, E. 工蚁身体侧面观；C. 工蚁身体背面观；F. 工蚁腹柄背面观（D-F. 引自Xu, 2001c）
A, D. Head of worker in full-face view; B, E. Body of worker in lateral view; C. Body of worker in dorsal view; F. Petiole of worker in dorsal view (D-F. cited from Xu, 2001c)

044 黄色猛蚁
Ponera xantha Xu, 2001

Ponera xantha Xu, 2001c, Entomotaxonomia, 23(3): 219, figs. 13-15 (w.) CHINA (Yunnan).

Holotype worker: TL 1.9, HL 0.45, HW 0.35, CI 78, SL 0.27, SI 76, PW 0.25, AL 0.57, ED 0.02, ML 0.23, DPW 0.18, PNI 73, PH 0.27, PNL 0.17, LPI 63. In full-face view, head roughly rectangular, distinctly longer than broad, weakly narrowed forward. Occipital margin weakly emarginate, occipital corners roundly prominent, sides weakly convex. Masticatory margin of mandible with 3 apical teeth and followed by a row of indistinct minute denticles. Anterior margin of clypeus evenly convex. Antennae short, apex of scape failed to reach occipital corner by 1/4 of its length, antennal club with 4 segments. Eye with only one facet. In lateral view, dorsum of thorax weakly convex, promesonotal suture depressed, metanotal groove fine and distinct. Dorsum of propodeum as long as declivity, posterodorsal corner blunt. In lateral view, petiolar node thick and nearly trapezoid, narrowed upward, anterior face steeper than posterior face, anterodorsal corner higher and more prominent, posterodorsal corner lower and more blunt. Subpetiolar process cuneiform, fenestra medium size, ventral face convex, anteroventral corner roundly blunt, posteroventral border without tooth. In dorsal view, anterior and lateral borders of petiolar node roundly convex, posterior border straight, length : width = 8 : 9, almost as broad as long. Mandibles smooth and shining, with very sparse fine punctures. Head densely and finely punctured, relatively dull. Thorax, petiole and gaster weakly, densely and finely punctured, relatively shining. Dorsum of head and body with sparse subdecumbent hairs and dense decumbent pubescence, hairs on gaster abundant. Scapes with sparse subdecumbent hairs and dense decumbent pubescence. Dorsa of tibiae with dense decumbent pubescence, without erect hairs. Whole body yellow, gaster brownish yellow, eyes black.

Holotype: Worker, China: Yunnan Province, Tengchong County, Jietou Town, Daying Village, 2000 m, 1999.Ⅴ.2, collected by Lei Fu in subalpine moist evergreen broadleaf forest, No. A99-270.

Etymology: The species name *xantha* combines Greek *xanth-* (yellow) + suffix *-a* (feminine form), it refers to the yellow color of the body.

正模工蚁： 正面观，头部近长方形，长明显大于宽，向前轻度变窄；后缘轻度凹陷，后角圆凸；侧缘轻度隆起。上颚咀嚼缘具3个端齿和1列不明显的细齿。唇基前缘均匀隆起。触角短，柄节未到达头后角，柄节末端至头后角的距离约为柄节长度的1/4，触角棒4节。复眼仅具1个小眼。侧面观，胸部背面轻度隆起，前中胸背板缝凹陷，后胸沟细而明显。并胸腹节背面与斜面等长，后上角钝。侧面观，腹柄结厚，近梯形，向上变窄；前面比后面陡，前上角较高、较突；后上角较低、较钝；腹柄下突楔形，窗斑中等大小，腹面隆起，前下角圆钝，后下缘缺齿。背面观，腹柄结前缘和侧缘圆形隆起，后缘平直，长：宽=8：9，长宽几乎相等。上颚光滑发亮，具很稀疏的细刻点。头部具细密刻点，较暗。胸部、腹柄和后腹部具弱的细密刻点，较光亮。头部和身体背面具稀疏亚倾斜毛和密集倾斜绒毛被，后腹部立毛丰富；柄节具稀疏亚倾斜毛和密集倾斜绒毛被；胫节背面具密集倾斜绒毛被，缺立毛。全身黄色，后腹部棕黄色，复眼黑色。

正模： 工蚁，中国云南省腾冲县界头乡大营村，2000m，1999.Ⅴ.2，付磊采于中山湿性常绿阔叶林中，No. A99-270。

词源： 该新种的种名"*xantha*"由希腊语"*xanth-*"（黄色）+后缀"*-a*"（阴性形式）组成，指身体颜色为黄色。

黄色猛蚁 Ponera xantha

A, D. 工蚁头部正面观；B, E. 工蚁身体侧面观；C. 工蚁身体背面观；F. 工蚁腹柄背面观（D-F. 引自 Xu, 2001c）
A, D. Head of worker in full-face view; B, E. Body of worker in lateral view; C. Body of worker in dorsal view; F. Petiole of worker in dorsal view (D-F. cited from Xu, 2001c)

长柄小盲猛蚁
Probolomyrmex longiscapus Xu & Zeng, 2000

Probolomyrmex longiscapus Xu & Zeng, 2000, Entomologia Sinica, 7(3): 216, figs. 9-11 (w.) CHINA (Yunnan).

Holotype worker: TL 3.0, HL 0.70, HW 0.45, CI 64, SL 0.60, SI 133, PW 0.36, AL 0.96, PNL 0.38, PNW 0.20, PH 0.26. In full-face view, head distinctly longer than broad, anterior 1/4 narrowed obviously. Occipital margin nearly straight, very shallowly emarginate, occipital corner blunt. Sides of head evenly convex. Anterior margin of clypeus convex. Frontal carinae erect, close to each other and parallel for most of their length, anterior apices reached to anterior margin of clypeus. Apex of scape almost reached to occipital margin. Antennal segments 3-6 about as broad as long, segments 7-11 broader than long. In full-face view, mandibles invisible. In lateral view, dorsum of alitrunk straight, promesonotal suture and metanotal groove absent. Propodeum with a pair of triangular teeth, declivity depressed and marginate laterally. In lateral view, petiolar node longer than high, anterior and dorsal faces formed an even arch, posterodorsal corner formed an acute angle, posterior face depressed and marginate laterally. Ventral face of petiole convex, anteroventral corner with a slender tooth. In dorsal view, petiolar node rectangular, sides nearly parallel, posterior 1/3 narrowed backward, posterior border straight. Constriction between the two basal segments of gaster distinct. Surfaces of head and whole body shagreened. Head and alitrunk weakly and sparsely punctured. Petiole and gaster distinctly punctured. Dorsum of head and body, and appendages without erect hairs, but with dense decumbent short pubescence. Body in color reddish brown. **Paratype worker:** TL 2.8, HL 0.70, HW 0.44, CI 63, SL 0.60, SI 136, PW 0.34, AL 0.94, PNL 0.38, PNW 0.19, NH 0.24 (*n*=1). As holotype.

Holotype: Worker, China: Yunnan Province, Mengla County, Shangyong Town, Nanqian Village, 820 m, 1998.Ⅲ.11, collected by Guang Zeng from a soil sample of hilly land shrub, No. A98-537.

Etymology: The species name *longiscapus* combines Latin *long-* (long) + word root *scap* (scape) + suffix *-us* (masculine form), it refers to the relatively longer scapes of the antennae.

正模工蚁： 正面观，头长明显大于宽，前部1/4明显变窄；后缘近平直，轻微凹陷，后角钝；侧缘均匀隆起。唇基前缘隆起。额脊直立，互相接近，大部互相平行，前端到达唇基前缘。触角柄节末端几乎到达头后缘，第3~6节长宽约相等，第7~11节宽大于长。正面观看不见上颚。侧面观，胸部背面平直，前中胸背板缝和后胸沟消失。并胸腹节具1对三角形齿，斜面凹陷，两侧具边缘。侧面观，腹柄结长大于高，前面和背面形成1个均匀的弓形面，后上角呈锐角；后面凹陷，两侧具边缘；腹柄腹面隆起，前下角具1个细齿。背面观，腹柄结长方形，侧缘近平行，后部1/3向后变窄，后缘平直。后腹部基部2节间收缩明显。头部和全身表面具细粒纹，头胸部具弱的稀疏刻点，腹柄和后腹部具明显刻点。头部和身体背面及附肢具密集倾斜短绒毛被，缺立毛。身体红棕色。**副模工蚁：** 特征同正模。

正模： 工蚁，中国云南省勐腊县尚勇乡南欠村，820m，1998.Ⅲ.11，曾光采于山地灌丛土壤样中，No. A98-537。

词源： 该新种的种名"*longiscapus*"由拉丁语"*long-*"（长的）+词根"*scap*"（柄节）+后缀-"*us*"（阳性形式）组成，指触角的柄节相对较长。

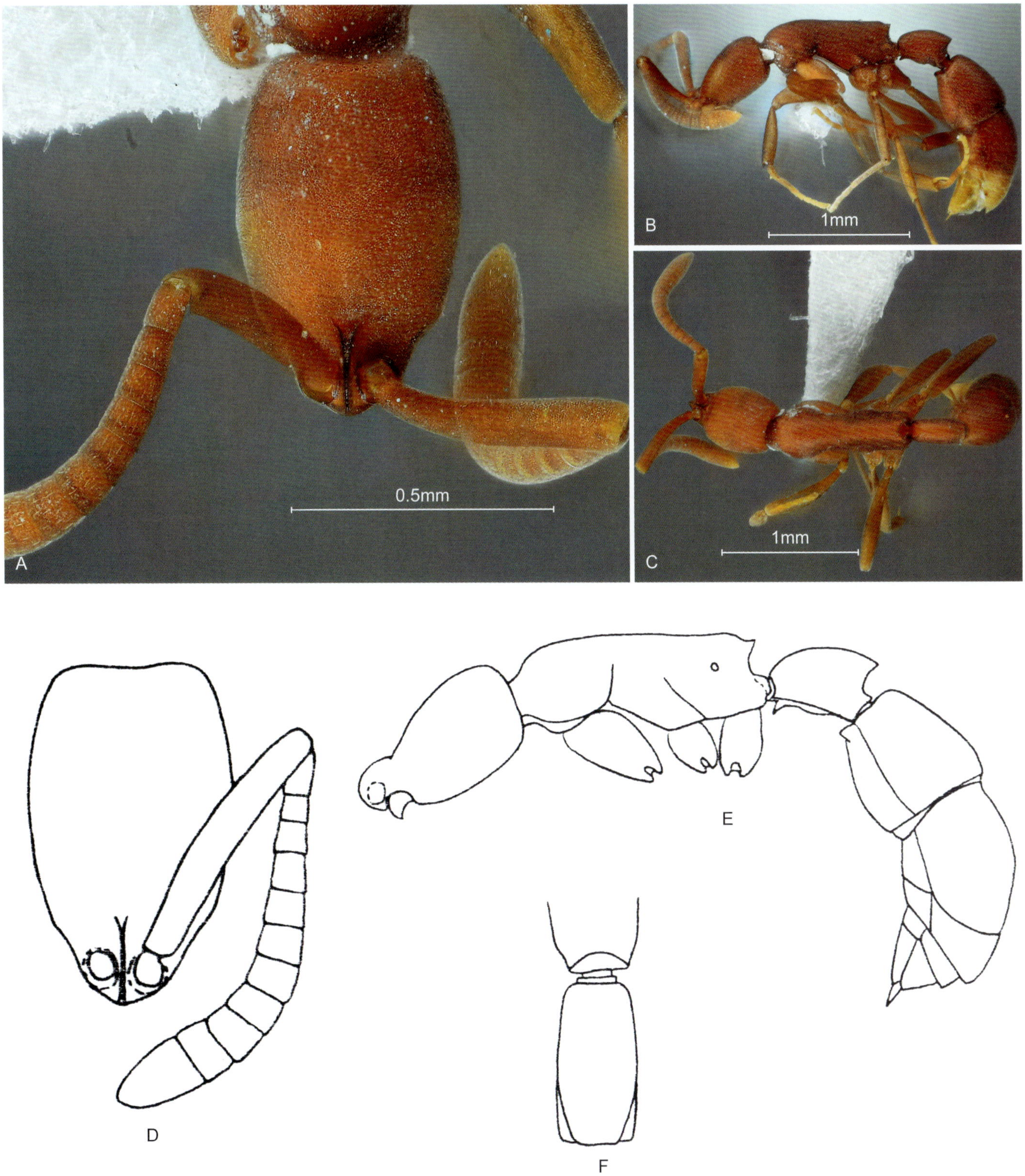

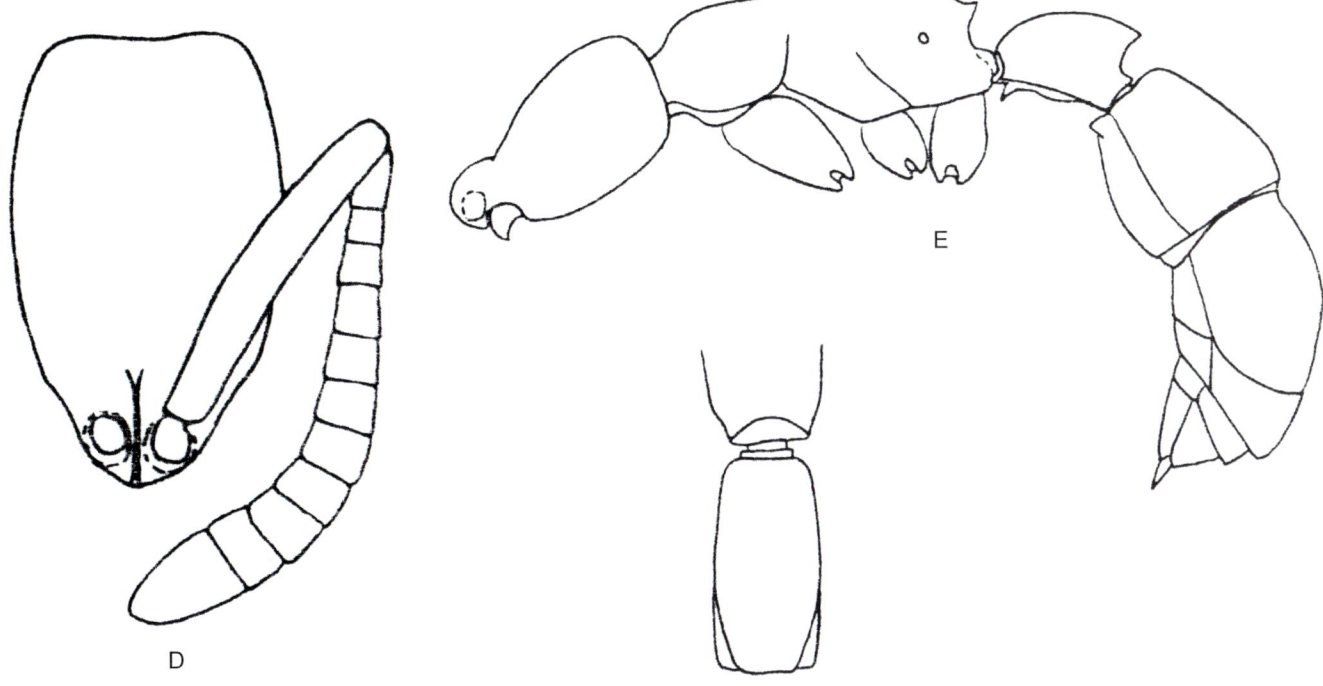

长柄小盲猛蚁 *Probolomyrmex longiscapus*

A, D. 工蚁头部正面观；B, E. 工蚁身体侧面观；C. 工蚁身体背面观；F. 工蚁并胸腹节和腹柄背面观（D-F. 引自Xu和Zeng, 2000）

A, D. Head of worker in full-face view; B, E. Body of worker in lateral view; C. Body of worker in dorsal view; F. Propodeum and petiole of worker in dorsal view (D-F. cited from Xu & Zeng, 2000)

046 六刺云行军蚁
Yunodorylus sexspinus Xu, 2000

Yunodorylus sexspinus Xu, 2000a, Zoological Research, 21(4): 298, figs. 1-6 (w.) CHINA (Yunnan).

Holotype worker: TL 3.4, HL 0.77, HW 0.70, CI 91, SL 0.37, SI 52, PW 0.50, AL 0.93, ML 0.40, PL 0.27, PH 0.40, DPW 0.37, LPI 150, DPI 138. In full-face view, head nearly square, slightly longer than broad, narrowed anteriorly. Occiput shallowly depressed, occipital corners roundly prominent, sides evenly convex. Anterior portion of gena convex. Mandibles elongate triangular, the large apical tooth followed by a smaller preapical tooth and 4 minute denticles. Clypeus reduced and only visible on sides. Frontal lobes suberect and reached to anterior margin of head. Antennae with 12 segments, apex of scape reached to 7/15 of the distance from antennal socket to occipital corner, flagellum distinctly incrassate towards apex. Eyes and ocelli absent. In lateral view, dorsum of alitrunk straight and flat, very slightly convex, posterodorsal corner of propodeum rounded. In dorsal view, alitrunk roughly rectangular, sides weakly impressed at mid-length. In lateral view, petiolar node roughly rectangular, very thick, anterior face straight, dorsal and posterior faces roundly convex, anterodorsal angle blunt, posterodorsal angle indistinct. Subpetiolar process large, roughly rectangular, posteroventrally pointed. In dorsal view, petiolar node broader than long, width : length = 11 : 8, anterior face straight, posterior face and sides roundly convex. Dorsum of pygidium slightly depressed, lateroposterior margins with 6 minute peg-like spines on each side. Sting extruding. Mandibles, head and alitrunk with sparse large punctures, interspace smooth and shining, distance between punctures larger than or equal to diameter of a puncture. Sides of alitrunk and petiolar node with sparse fine punctures, interspace granulate, less shining. Gaster with punctures from which setae arising, interspace smooth. Dorsum of whole body with a few suberect hairs and dense decumbent pubescence. Antennal scapes and tibiae of legs with a few suberect hairs and dense decumbent pubescence. Body in color yellowish brown, mandibles and antennae dark reddish brown, legs brownish yellow.

Paratype workers: TL 2.3-3.6, HL 0.53-0.77, HW 0.43-0.73, CI 79-91, SL 0.27-0.40, SI 50-62, PW 0.30-0.53, AL 0.67-1.00, ML 0.27-0.43, PL 0.20-0.30, PH 0.23-0.43, DPW 0.22-0.40, LPI 117-150, DPI 108-138 (n=10). As holotype, but body varying in size, color brownish yellow to dark reddish brown.

Holotype: Worker, China: Yunnan Province, Mengla County, Mengla Town, Bubang Village, 730 m, 1997.VIII.17, collected by Guang Zeng from a soil nest in seasonal rain forest, No. A97-2064.

Etymology: The species name *sexspinus* combines Latin *sex-* (six) + word root *spin* (spine) + suffix *-us* (masculine form), it refers to the six minute peg-like spines on the lateroposterior margin of pygidium.

正模工蚁：正面观，头部近方形，长稍大于宽，向前变窄；后缘浅凹，后角圆凸；侧缘均匀隆起；颊区前部隆起。上颚长三角形，咀嚼缘具1个大端齿、1个较小的亚端齿和4个小齿。唇基退化，仅在两侧可见。额叶亚直立，到达头部前缘。触角12节，柄节末端到达触角窝至头后角间距的7/15处，鞭节向顶端明显变粗。缺复眼和单眼。侧面观，胸部背面直且平坦，轻微隆起，并胸腹节后上角圆。背面观，胸部近长方形，两侧在中部轻度凹陷。侧面观，腹柄结近长方形，较厚，前面平直，背面和后面圆形隆起，前上角钝，后上角不明显；腹柄下突大，近长方形，指向后下方。背面观腹柄结宽大于长，宽：长=11：8，前缘平直，后缘和侧缘圆形隆起。臀板背面轻度凹陷，后侧缘每侧具6个木钉状小刺。螫针伸出。上颚、头部和胸部具稀疏大刻点，界面光滑发亮，刻点间距大于或等于刻点直径；胸部和腹柄结侧面具稀疏细刻点，界面具细粒纹，不太光亮。后腹部有具毛刻点，界面光滑。全身背面具少数亚直立毛和密集倾斜绒毛被；触角柄节和足的胫节具少数亚直立毛和密集倾斜绒毛被。身体黄棕色，上颚和触角暗红棕色，足棕黄色。**副模工蚁：**特征同正模，但是身体大小有变化，体色棕黄色至暗红棕色。

正模：工蚁，中国云南省勐腊县勐腊镇补蚌村，730m，1997.VIII.17，曾光采于季节性雨林的1个土壤巢中，No. A97-2064。

词源：该新种的种名"*sexspinus*"由拉丁语"*sex-*"（六）+词根"*spin*"（刺）+后缀"*-us*"（阳性形式）组成，指臀板的后侧缘具6个木钉状小刺。

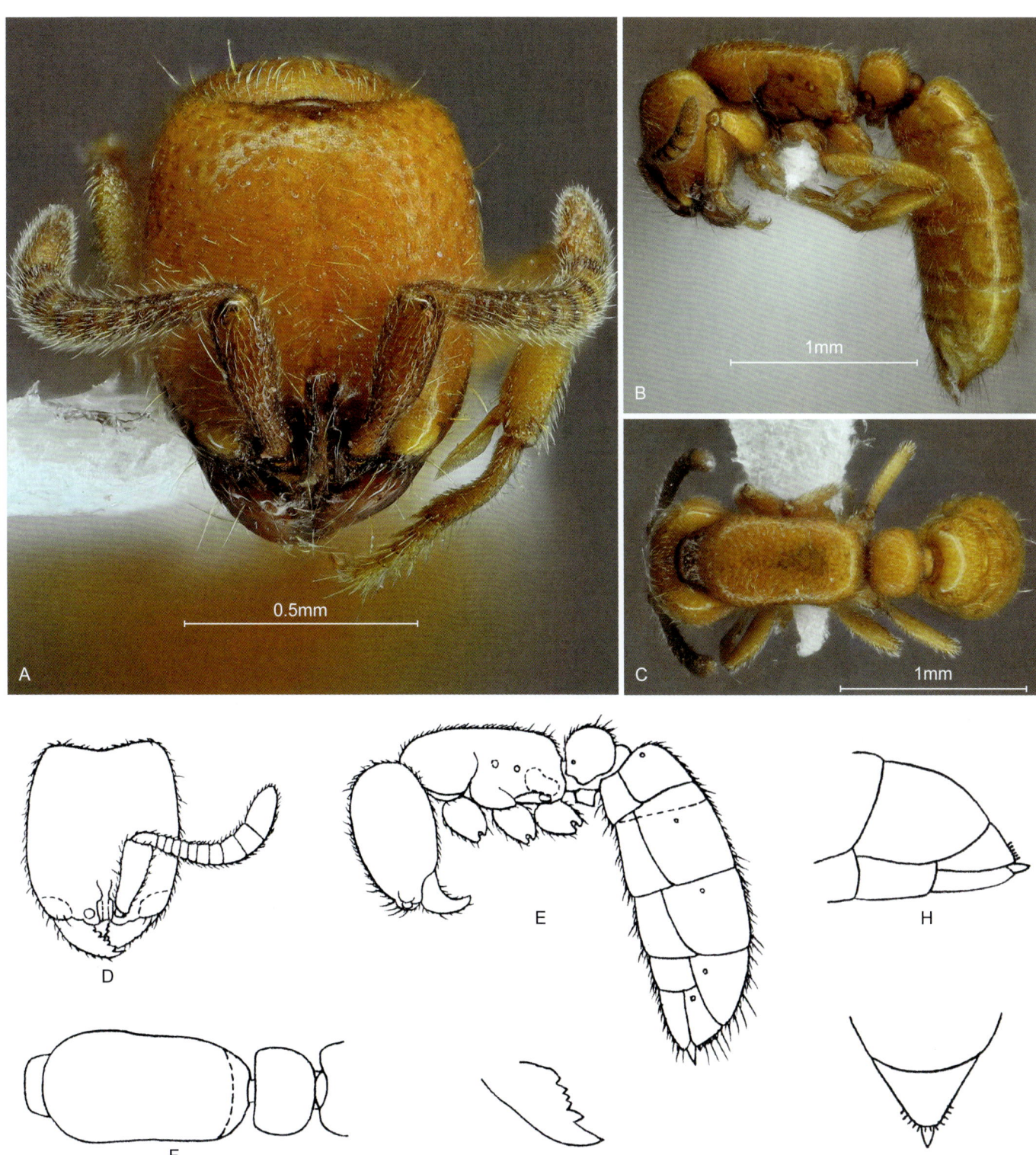

六刺云行军蚁 *Yunodorylus sexspinus*

A, D. 工蚁头部正面观；B, E. 工蚁身体侧面观；C, F. 工蚁身体背面观；G. 工蚁上颚背面观；H. 工蚁后腹末侧面观；I. 工蚁后腹末背面观（D-I. 引自Xu, 2000a）

A, D. Head of worker in full-face view; B, E. Body of worker in lateral view; C, F. Body of worker in dorsal view; G. Mandible of worker in dorsal view; H. Gastral apex of worker in lateral view; I. Gastral apex of worker in dorsal view (D-I. cited from Xu, 2000a)

北京细蚁
Leptanilla beijingensis Qian, Xu, Man & Liu, 2024

Leptanilla beijingensis Qian et al., 2024, Myrmecological News, 34: 23, fig. 1 (w.) CHINA (Beijing).

Holotype worker: TL 1.6, HL 0.29, HW 0.25, CI 86, SL 0.16, SI 64, PW 0.18, WL 0.40, PL 0.12, PH 0.12, DPW 0.10, PPL 0.12, PPH 0.14, PPW 0.10. In full-face view, head longer than broad, roughly rectangular, posterior margin weakly concave, posterolateral corners narrowly rounded, lateral margins moderately convex. Mandibles elongate triangular, masticatory margin about 2/3 length of basal margin, with three teeth, the apical one long and sharp; the basal one shorter, nearly triangular; the middle one minute and shorter than the basal one, about locating at middle of the apical and basal ones. Median lobe of clypeus weakly broadly protruding anteriorly and surpassing antennal sockets, anterior margin moderately convex with a narrow deep notch in the center. Antennae 12-segmented, apex of scape reaching 2/5 of the distance from antennal socket to posterolateral corner of head, flagella obviously incrassate toward apex. In lateral view dorsal and ventral margins of head slightly convex. Mesosoma moderately constricted at promesonotal suture, the suture weakly impressed on dorsum. Pronotum moderately convex. Dorsa of mesonotum and propodeum slightly convex and weakly sloping down posteriorly, metanotal groove absent, posterodorsal corner of propodeum very broadly rounded; declivity steeply sloping and very short, nearly straight, about 1/4 length of dorsum; spiracle circular and locating at about midpoint of propodeal side. Metapleural bulla large and distinct, roughly elliptical. Petiolar node widening posteriorly, dorsum moderately convex, anterodorsal corner broadly rounded, posterodorsal corner indistinct; ventral margin strongly convex. Postpetiole weakly higher than petiole, dorsum moderately convex, posterodorsal corner broadly rounded and higher than anterodorsal corner, the latter narrowly rounded; sternite strongly convex and rounded apically, strongly inclining anteriorly. Gaster strongly elongate and roughly elliptical, first segment occupying about 2/3 length of the gaster, apex with sting. In dorsal view, pronotum broadest, lateral margins moderately convex, humeral corners indistinct. Promesonotal suture moderately constricting and weakly impressing. Mesonotum nearly square, lateral margins weakly convex. Metanotal groove absent. Propodeum longer than broad, roughly trapezoidal and narrowing posteriorly, lateral margins weakly convex, posterolateral corners broadly rounded. Petiolar node longer than broad and roughly elliptical, about 1.3 time as long as broad, slightly narrowing posteriorly, lateral margins moderately convex. Postpetiolar node as broad as long, about as broad as petiolar node, roughly trapezoidal and widening posteriorly, anterior and posterior margins nearly straight, lateral margins moderately convex. Gaster elongate and roughly elliptical. Body surface smooth and shiny. Body dorsum with dense decumbent pubescence, without subdecumbent hairs. Scapes with sparse subdecumbent hairs and dense decumbent pubescence. Tibiae with dense decumbent pubescence. Body color yellow, antennae and legs light yellow.

Paratype workers: TL 1.4-1.8, HL 0.30-0.37, HW 0.25-0.32, CI 83-86, SL 0.15-0.16, SI 50-64, PW 0.15-0.19, WL 0.40-0.54, PL 0.10-0.15, PH 0.10-0.12, DPW 0.08-0.12, PPL 0.10-0.11, PPH 0.12-0.13, PPW 0.08-0.13 (n=20). As holotype, but body size slightly variable, body color light yellow to yellow.

Holotype: Worker, China: Beijing City, Yanqing District, Badaling Town, Xibozi Village (40.373889°N, 115.971667°E), 550 m, 2015.Ⅹ.15, collected by Pei Man from an underground trap in deciduous broadleaf forest, No. A15-2355.

Etymology: The specific epithet refers to the type locality, Beijing City.

正模工蚁： 正面观，头部长大于宽，近长方形；后缘轻度凹陷，后角窄圆；侧缘中度隆起。上颚长三角形，咀嚼缘约为基缘长度的2/3，具3个齿，端齿长而锋利；基齿短，近三角形；中齿微小，短于基齿，大约位于端齿和基齿的中间。唇基中叶轻度宽形前伸，超过触角窝，前缘中度隆起，中央具1个深切口。触角12节，柄节末端到达触角窝至头后角间距的2/5处，鞭节向顶端明显变粗。侧面观，头部背面和腹面轻微隆起。胸部在前中胸背板缝处中度收缩，前中胸背板缝在背面轻度凹陷。前胸背板中度隆起。中胸背板和并胸腹节背面轻微隆起，向后轻度降低，缺后胸沟，并胸腹节后上角极宽圆；斜面陡坡状，很短，近平直，约为背面长度的1/4；气门圆形，大约位于并胸腹节侧面中部。后胸腺泡大而明显，近椭圆形。腹柄结向后变高，背面中度隆起，前上角宽圆，后上角不明显；腹面强烈隆起。后腹柄轻度高于腹柄，背面中度隆起，后上角宽圆，高于前上角，前上角窄圆；腹板强烈隆起，顶端圆，强烈前倾。后腹部强烈伸长，近椭圆形，第1节占据后腹部长度的2/3，末端具螯针。背面观，前胸背板最宽，侧缘中度隆起，肩角不明显。前中胸背板缝中度收缩，轻度凹陷。中胸背板近方形，侧缘轻度隆起。缺后胸沟。并胸腹

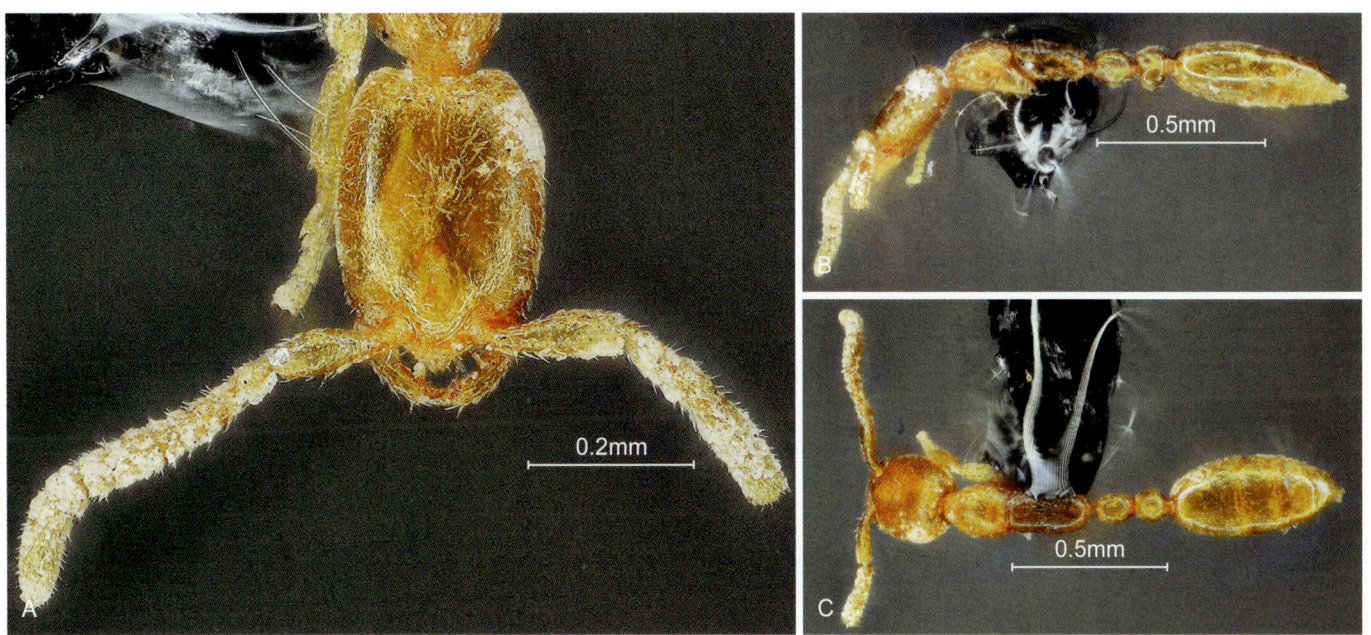

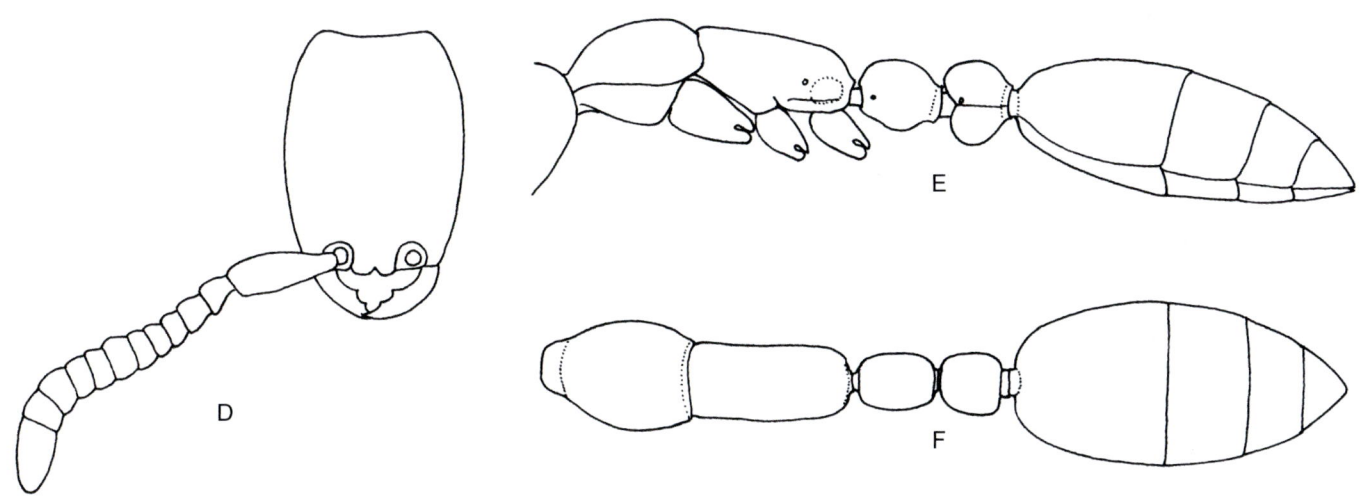

北京细蚁 *Leptanilla beijingensis*

A, D. 工蚁头部正面观；B, E. 工蚁身体侧面观；C, F. 工蚁身体背面观（D-F. 引自Qian等，2024）

A, D. Head of worker in full-face view; B, E. Body of worker in lateral view; C, F. Body of worker in dorsal view (D-F. cited from Qian et al., 2024)

节长大于宽，近梯形，向后变窄，侧面轻度隆起，后侧角宽圆。腹柄结长大于宽，近椭圆形，长约为宽的1.3倍，向后轻微变窄，侧缘中度隆起。后腹柄结长宽相等，约与腹柄结等宽，近梯形，向后变宽，前缘和后缘近平直，侧缘中度隆起。后腹部伸长，近椭圆形。身体表面光滑发亮。身体背面具密集倾斜绒毛被，缺亚倾斜毛。柄节具稀疏亚倾斜毛和密集倾斜绒毛被。胫节具密集倾斜绒毛被。身体黄色，触角和足浅黄色。**副模工蚁：**特征同正模，但身体大小稍有差异，身体浅黄色至黄色。

正模：工蚁，中国北京市延庆区八达岭镇西拨子村（40.373889°N，115.971667°E），550m，2015.X.15，满沛采用地下陷阱法采于落叶阔叶林中，No. A15-2355。

词源：该种的种名以模式产地"北京"命名。

048 德宏细蚁
Leptanilla dehongensis Qian, Xu, Man & Liu, 2024

Leptanilla dehongensis Qian et al., 2024, Myrmecological News, 34: 26, fig. 3 (w.) CHINA (Yunnan).

Holotype worker: TL 1.7, HL 0.50, HW 0.38, CI 76, SL 0.18, SI 47, PW 0.23, WL 0.57, PL 0.15, PH 0.15, DPW 0.12, PPL 0.11, PPH 0.18, PPW 0.13. In full-face view, head longer than broad, roughly trapezoidal and weakly narrowing anteriorly, posterior margin almost straight, posterolateral corners narrowly rounded, lateral margins moderately convex. Mandibles elongate triangular, basal margin about 4/5 length of masticatory margin; masticatory margin with three teeth, the apical one long and sharp, the middle one relatively shorter and acute, located in the middle of the apical and basal ones; the basal one shortest and nearly triangular. Median lobe of clypeus strongly protruding anteriorly and nearly square, significantly surpassing antennal sockets, dorsum longitudinally convex, anterior margin straight, anterolateral corners rightly angled. Antennae 12-segmented, apex of scape reaching 1/2 of the distance from antennal socket to posterolateral corner of head, flagella weakly incrassate toward apex. In lateral view, dorsal and ventral margins of head slightly convex. Mesosoma strongly constricted at promesonotal suture, the suture moderately impressed on dorsum. Pronotum moderately convex. Dorsa of mesonotum and propodeum straight and weakly sloping down posteriorly, metanotal groove absent. Posterodorsal corner of propodeum broadly rounded; declivity steeply sloping and short, about 1/4 length of dorsum; spiracles circular and locating at about midpoint of side. Metapleural bulla large and distinct, roughly elliptical. Dorsum of petiolar node strongly convex and nearly semicircular, weakly inclining anteriorly, anterodorsal corner broadly rounded, posterodorsal corner indistinct; ventral margin strongly convex, anteroventral corner bluntly angled. Postpetiole distinctly higher than petiole, dorsum strongly convex and strongly inclining anteriorly, anterodorsal corner narrowly rounded, posterodorsal corner indistinct; sternite large and nearly trapezoidal, strongly inclining anteriorly, ventral margin strongly convex and rounded. Gaster strongly elongate and roughly elliptical, first segment occupying about 2/3 length of the gaster, apex with strong sting. In dorsal view, pronotum broadest, lateral margins moderately convex, humeral corners broadly rounded. Promesonotal suture strongly constricted and moderately impressed. Mesonotum short and widening posteriorly, lateral margins weakly convex. Metanotal groove absent. Propodeum longer than broad, nearly rectangular, lateral margins weakly convex, posterolateral corners broadly rounded. Petiolar node longer than broad, about 1.2 times as long as broad, nearly trapezoidal and weakly widening posteriorly, anterior margin strongly convex, lateral margins moderately convex, posterior margin weakly convex. Postpetiolar node weakly broader than petiolar node, about as broad as long, roughly trapezoidal and widening posteriorly, anterior margin strongly convex, lateral margins moderately convex, posterior margin straight. Gaster strongly elongate and roughly elliptical. Body surface smooth and shiny. Body dorsum with sparse subdecumbent hairs and abundant decumbent pubescence. Scapes with abundant subdecumbent hairs and dense decumbent pubescence. Tibiae with dense decumbent pubescence. Body color brownish yellow, antennae and legs yellow. **Paratype workers:** TL 1.6-2.3, HL 0.40-0.54, HW 0.30-0.42, CI 75-78, SL 0.15-0.23, SI 47-67, PW 0.20-0.29, WL 0.45-0.67, PL 0.10-0.16, PH 0.10-0.15, DPW 0.10-0.14, PPL 0.10-0.12, PPH 0.15-0.22, PPW 0.10-0.14 (n=21). As holotype, but body size weakly variable, body color yellow to yellowish brown.

Holotype: Worker, China: Yunnan Province, Dehong Dai and Jingpo Autonomous Prefecture, Longchuan County, Husa Town, Baoping Village (24.382222°N, 97.833611°E), 1537 m, 2015.Ⅲ.23, collected by Ying Zheng from a soil sample in monsoon evergreen broadleaf forest, No. A15-548.

Etymology: The specific epithet refers to the type locality, Dehong Dai and Jingpo Autonomous Prefecture in Yunnan Province.

正模工蚁： 正面观，头部长大于宽，近梯形，向前轻度变窄；后缘几乎平直，后角窄圆；侧缘中度隆起。上颚长三角形，基缘约为咀嚼缘长度的4/5；咀嚼缘具3个齿，端齿长而锋利；中齿较短较尖，位于端齿和基齿中间；基齿最短，近三角形。唇基中叶强烈前伸，近方形，显著超过触角窝，背面纵向隆起，前缘平直，前侧角直角形。触角12节，柄节末端到达触角窝至头后角间距的1/2处，鞭节向顶端轻度变粗。侧面观，头部背面和腹面轻度隆起。胸部在前中胸背板缝处强烈收缩，前中胸背板缝在背面中度凹陷。前胸背板中度隆起。中胸背板和并胸腹节背面平直，向后轻度降低，缺后胸沟。并胸腹节后上角宽圆，斜面陡而短，约为背面长度的1/4；气门圆形，位于并胸腹节侧面中部。后胸腺泡大而明显，近椭圆形。腹柄结背面强烈隆起，近半圆形，轻度向前倾斜，前上角宽圆，后上角不明显；腹面强烈隆起，前下角钝角状。后腹柄明显高于腹柄，背面强烈隆起，强烈前倾，前上角窄圆，后上角不

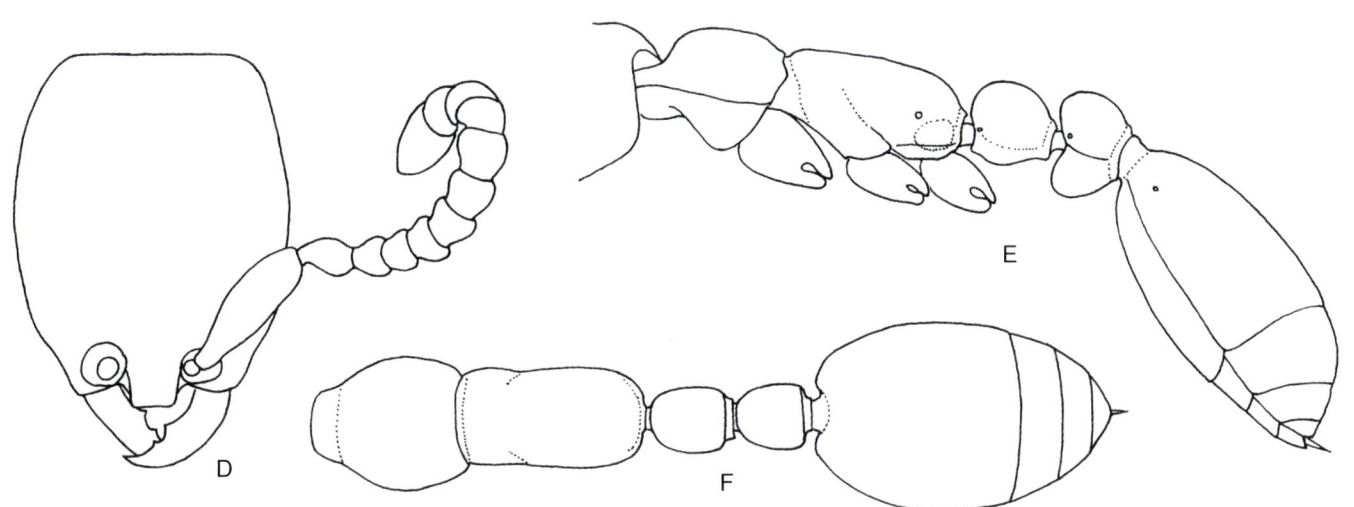

德宏细蚁 Leptanilla dehongensis

A, D. 工蚁头部正面观；B, E. 工蚁身体侧面观；C, F. 工蚁身体背面观（D-F. 引自Qian等, 2024）
A, D. Head of worker in full-face view; B, E. Body of worker in lateral view; C, F. Body of worker in dorsal view (D-F. cited from Qian et al., 2024)

明显；腹板大，近梯形，强烈前倾，腹面强烈隆起，钝圆。后腹部强烈伸长，近椭圆形，第1节约占据后腹部长度的2/3，末端具发达的螫针。背面观，前胸背板最宽，侧缘中度隆起，肩角宽圆。前中胸背板缝强烈收缩，中度凹陷。中胸背板短，向后变宽，侧缘轻度隆起。缺后胸沟。并胸腹节长大于宽，近长方形，侧缘轻度隆起，后侧角宽圆。腹柄结长大于宽，长约为宽的1.2倍，近梯形，向后轻度变宽，前缘强烈隆起，侧缘中度隆起，后缘轻度隆起。后腹柄结轻度宽于腹柄结，长宽约相等，近梯形，向后变宽，前面强烈隆起，侧面中度隆起，后面平直。后腹部强烈伸长，近椭圆形。身体表面光滑发亮。身体背面具稀疏亚直立毛和丰富倾斜绒毛被。柄节具丰富亚直立毛和密集倾斜绒毛被。胫节具密集倾斜绒毛被。身体棕黄色，触角和足黄色。**副模工蚁**：特征同正模，但是身体大小有轻度差异，身体黄色至黄棕色。

正模：工蚁，中国云南省德宏傣族景颇族自治州陇川县户撒乡保平村（24.382222°N，97.833611°），1537m，2015.Ⅲ.23，郑莹采于季风常绿阔叶林土壤样中，No. A15-548。

词源：该种的种名以模式产地"德宏"命名。

昆明细蚁

(049) *Leptanilla kunmingensis* Xu & Zhang, 2002

Leptanilla kunmingensis Xu & Zhang, 2002a, Acta Zootaxonomica Sinica, 27(1): 142, figs. 19-21 (w.) CHINA (Yunnan).

Holotype worker: TL 2.2, HL 0.47, HW 0.37, Cl 79, SL 0.27, SI 73, PW 0.25, AL 0.63, ML 0.22, PNL 0.17, PNH 0.15, PNW 0.13, PPNL 0.13, PPNH 0.20, PPNW 0.13. In full-face view, head longer than broad, distinctly narrowed forward. Occipital margin weakly concave, occipital corners bluntly prominent, sides roundly convex. Mandible narrow and slender, masticatory margin with 3 teeth. Clypeus protruding forward and bicarinated, anterior margin evenly concave. Antenna short, with 12 segments, apex of scape reached to 2/3 of the distance from antennal socket to occipital corner, segments 4-10 about as broad as long, antennal club indistinct. In lateral view, pronotum evenly convex, promesonotal suture distinct. Mesothorax constricted at its anterior portion, mesonotum convex. Metanotal groove depressed. Propodeum weakly convex, about 2 times as long as declivity, posterodorsal corner rounded. In lateral view, dorsum of petiolar node roundly convex, anterior face short and sloped, dorsal and posterior faces formed a single arch. Subpetiolar process very narrow, weakly convex, anteroventral corner acutely angled. Postpetiolar node shorter than petiolar node, with dorsum roundly convex, the sternite formed a large broad subpostpetiolar process with blunt anteroventrally pointed corner. In dorsal view, petiolar node longer than broad, sides weakly convex. Postpetiolar node nearly square, sides evenly convex, as broad as petiolar node. Mandibles, head and whole body smooth and shining. Head and body with dense decumbent pubescence. Petiole, postpetiole and gaster with sparse subdecumbent hairs. Scapes and tibiae with dense decumbent pubescence. Body in color brownish yellow. **Paratype workers:** TL 2.1-2.2, HL 0.47-0.50, HW 0.37-0.40, Cl 76-80, SL 0.23-0.30, SI 64-77, PW 0.25-0.27, AL 0.57-0.63, ML 0.20-0.23, PNL 0.15-0.17, PNH 0.13-0.15, PNW 0.12-0.13, PPNL 0.13, PPNH 0.18-0.22, PPNW 0.12-0.13 ($n=8$). As holotype.

Holotype: Worker, China: Yunnan Province, Kunming City, Xishan Forest Park, Longmen, 2150 m, 2001.Ⅴ.4, collected by Yu-xiang Zhao from a soil sample in evergreen broadleaf forest, No. A00506.

Etymology: The new species is named after the type locality "Kunming".

正模工蚁： 正面观，头长大于宽，向前明显变窄；后缘轻度凹陷，后角钝突；侧缘圆形隆起。上颚窄而细长，咀嚼缘具3个齿。唇基前伸，具双脊，前缘均匀凹陷。触角短，12节，柄节末端到达触角窝至头后角间距的2/3处，第4～10节长宽近相等，触角棒不明显。侧面观，前胸背板均匀隆起，前中胸背板缝明显。中胸在前部收缩，中胸背板隆起。后胸沟凹陷。并胸腹节轻度隆起，背面为斜面长度的2倍，后上角圆。侧面观，腹柄结背面圆形隆起；前面短，坡形；背面和后面形成1个弓形面；腹柄下突很窄，轻度隆起，前下角锐角状。后腹柄结短于腹柄结，背面圆形隆起；腹板形成宽大的后腹柄下突，指向前下方，顶端钝。背面观，腹柄结长大于宽，侧缘轻度隆起。后腹柄结近方形，侧面均匀隆起，与腹柄结等宽。上颚、头部和身体光滑发亮。头部和身体具密集倾斜绒毛被，腹柄、后腹柄和后腹部具稀疏亚倾斜毛；柄节和胫节具密集倾斜绒毛被。身体棕黄色。**副模工蚁：** 特征同正模。

正模： 工蚁，中国云南省昆明市西山森林公园龙门，2150m，2001.Ⅴ.4，赵宇翔采于常绿阔叶林土壤样中，No. A00506。

词源： 该新种以模式产地"昆明"命名。

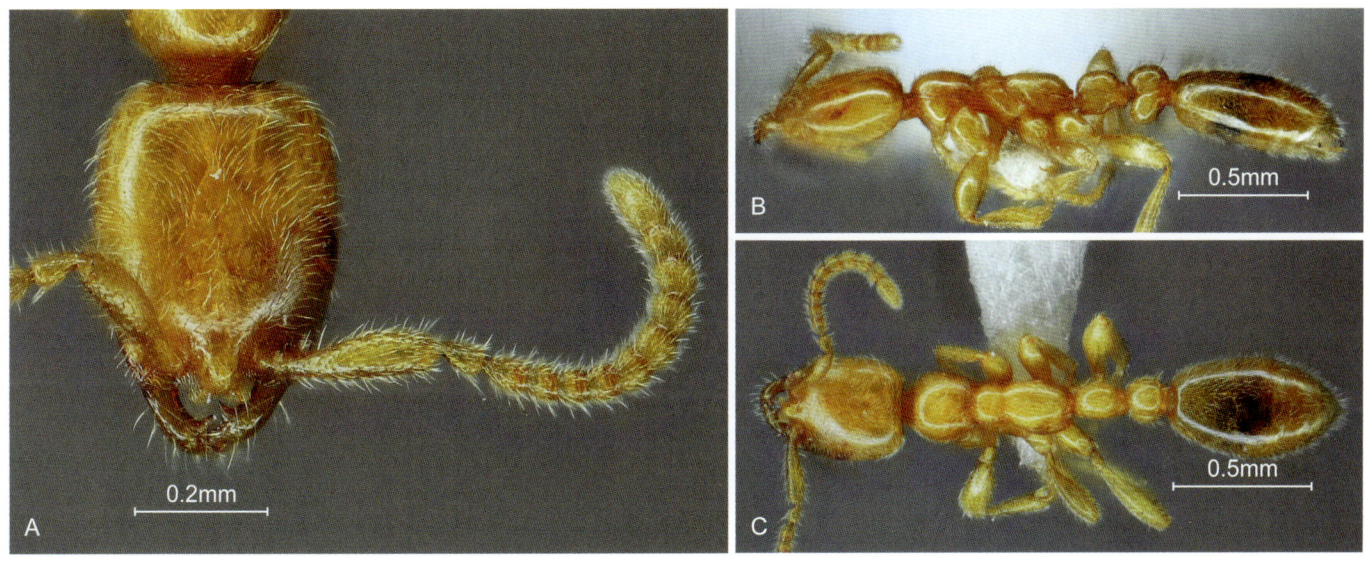
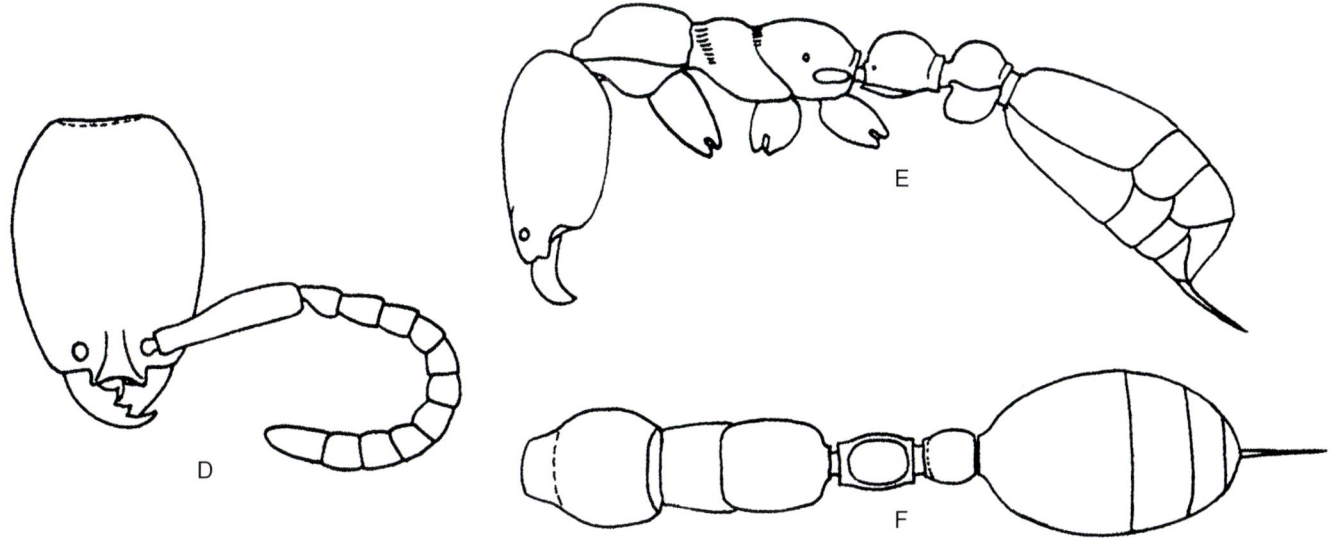

昆明细蚁 *Leptanilla kunmingensis*

A, D. 工蚁头部正面观；B, E. 工蚁身体侧面观；C, F. 工蚁身体背面观（D-F. 引自Xu和Zhang, 2002a）
A, D. Head of worker in full-face view; B, E. Body of worker in lateral view; C, F. Body of worker in dorsal view (D-F. cited from Xu & Zhang, 2002a)

秦岭细蚁
Leptanilla qinlingensis Qian, Xu, Man & Liu, 2024

Leptanilla qinlingensis Qian et al., 2024, Myrmecological News, 34: 28, fig. 5 (w.) CHINA (Shaanxi).

Holotype worker: TL 1.5, HL 0.33, HW 0.25, CI 77, SL 0.20, SI 80, PW 0.17, WL 0.45, PL 0.12, PH 0.10, DPW 0.10, PPL 0.10, PPH 0.10, PPW 0.11. In full-face view, head longer than broad and nearly rectangular, posterior margin almost straight, posterolateral corners narrowly rounded, lateral margins moderately convex. Mandibles elongate triangular, basal margin about as long as masticatory margin; masticatory margin with three teeth, the apical one long and sharp; the basal one shorter and acute, nearly perpendicular to the basal margin; the middle one shortest and nearly triangular, located at basal 2/5 of masticatory margin and closer to the basal one. Median lobe of clypeus broadly protruding anteriorly and surpassing antennal sockets, anterior margin strongly convex with a narrow deep notch in the center. Antennae 12-segmented, apex of scape reaching about 1/2 of the distance from antennal socket to posterolateral corner of head, flagella obviously incrassate toward apex. In lateral view, dorsal and ventral margins of head slightly convex. Mesosoma moderately constricted at promesonotal suture, the suture moderately impressed on dorsum. Pronotum weakly convex. Mesonotum straight. Metanotal groove absent. Propodeal dorsum long and weakly convex, weakly sloping down posteriorly, posterodorsal corner very broadly rounded; declivity steeply sloping and short, weakly convex, about 1/4 length of dorsum; spiracle circular and locating at about midpoint of propodeal side. Metapleural bulla large and distinct, roughly elliptical. Petiolar node roughly trapezoidal and widening posteriorly, dorsum moderately convex, anterodorsal corner broadly rounded, posterodorsal corner indistinct; posteroventral corner strongly convex and broadly rounded, anteroventral margin nearly straight. Postpetiole as high as petiole, dorsum slightly convex, posterodorsal corner broadly rounded, weakly higher than anterodorsal corner; sternite narrow and inclining anteriorly, ventral margin moderately convex, anteroventral corner narrowly rounded. Gaster strongly elongate and roughly elliptical, first segment occupying about 2/3 length of the gaster, apex with sting. In dorsal view, pronotum broadest, lateral margins moderately convex, humeral corners broadly rounded. Promesonotal suture strongly constricting and moderately impressed. Mesonotum nearly trapezoidal and widening posteriorly, lateral margins weakly convex. Metanotal groove absent. Propodeum about as broad as long, narrowing posteriorly, lateral margins moderately convex, posterolateral corners broadly rounded. Petiolar node longer than broad, about 1.3 times as long as broad, roughly trapezoidal and widening posteriorly, anterior margin strongly convex, lateral margins moderately convex, posterior margin weakly convex. Postpetiolar node shorter and broader than petiolar node, about as broad as long, roughly trapezoidal and widening posteriorly, anterior and lateral margins moderately convex, posterior margin nearly straight. Gaster strongly elongate and roughly elliptical. Body surface smooth and shiny. Body dorsum with abundant decumbent pubescence, without subdecumbent hairs, pubescence on head and gaster denser. Scapes with abundant subdecumbent hairs and dense decumbent pubescence. Tibiae with dense decumbent pubescence. Body color brownish yellow, antennae and legs yellow, mandibular teeth brown. **Paratype workers:** TL 1.4-2.0, HL 0.30-0.42, HW 0.23-0.31, CI 74-77, SL 0.15-0.22, SI 67-80, PW 0.15-0.21, WL 0.40-0.54, PL 0.08-0.14, PH 0.08-0.11, DPW 0.08-0.11, PPL 0.08-0.10, PPH 0.10-0.13, PPW 0.08-0.12 ($n=5$). As holotype, but body size weakly variable, body color yellow to brownish yellow.

Holotype: Worker, China: Shaanxi Province, Xi'an City, Chang'an District, Xiangyu Forest Park (33.962085°N, 108.805718°E), 1200 m, 2020.Ⅴ.18, collected by Guan-Lin Liu from a soil nest in conifer-broadleaf mixed forest, No. A20-3892.

Etymology: The specific epithet refers to the type locality, Mt. Qinling in Shaanxi Province.

正模工蚁： 正面观，头部长大于宽，近长方形；后缘几乎平直，后角窄圆，侧缘中度隆起。上颚长三角形，基缘约与咀嚼缘等长；咀嚼缘具3个齿，端齿长而锋利；基齿较短，尖锐，几乎与基缘垂直；中齿最短，近三角形，位于咀嚼缘基部2/5处，接近基齿。唇基中叶宽形前伸，超过触角窝，前缘强烈隆起，中间具1个窄而深的缺口。触角12节，柄节末端到达触角窝至头后角间距的1/2处，鞭节向顶端明显变粗。侧面观，头部背面和腹面轻微隆起。中胸在前中胸背板缝处中度收缩，前中胸背板缝在背面中度凹陷。前胸背板轻度隆起，中胸背板平直，缺后胸沟。并胸腹节背面长，轻度隆起，向后轻度降低，后上角极宽圆；斜面短且陡，轻度隆起，约为背面长度的1/4；气门圆形，位于并胸腹节侧面近中部。后胸腺泡大而明显，近椭圆形。腹柄结近梯形，向后变高，背面中度隆起，前上角宽圆，后上角不明显；后下角强烈隆起，宽圆，前下缘近平直。后腹柄与腹柄等高，背面轻微隆起，后上角宽圆，

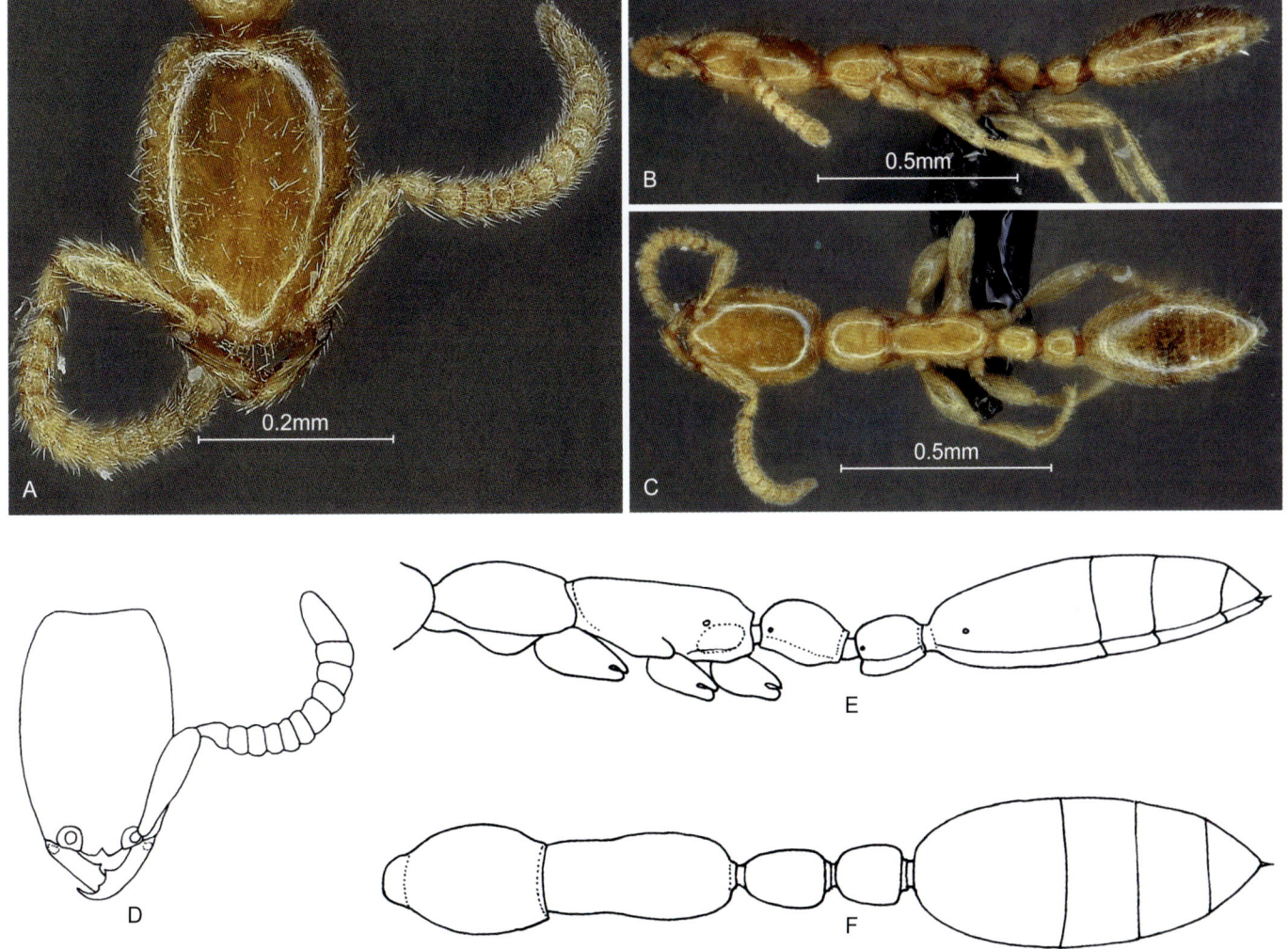

秦岭细蚁 *Leptanilla qinlingensis*

A, D. 工蚁头部正面观；B, E. 工蚁身体侧面观；C, F. 工蚁身体背面观（D-F. 引自Qian等, 2024）
A, D. Head of worker in full-face view; B, E. Body of worker in lateral view; C, F. Body of worker in dorsal view (D-F. cited from Qian et al., 2024)

稍高于前上角；腹板窄，前倾，腹面中度隆起，前下角窄圆。后腹部强烈伸长，近椭圆形，第1节约占后腹部长度的2/3，末端具螫针。背面观，前胸背板最宽，侧缘中度隆起，肩角宽圆。前中胸背板缝强烈收缩，中度凹陷。中胸背板近梯形，向后变宽，侧缘轻度隆起。缺后胸沟。并胸腹节长宽约相等，向后变窄，侧缘中度隆起，后侧角宽圆。腹柄结长大于宽，长约为宽的1.3倍，近梯形，向后变宽；前缘强烈隆起，侧缘中度隆起，后缘轻度隆起。后腹柄结短于腹柄结但是宽于腹柄结，长宽约相等，近梯形，向后变宽；前缘和侧缘轻度隆起，后缘近平直。后腹部强烈伸长，近椭圆形。身体表面光滑发亮。身体背面具丰富倾斜绒毛被，缺亚倾斜毛，头部和后腹部绒毛被较密集；柄节具丰富亚倾斜毛和密集倾斜绒毛被；胫节具密集倾斜绒毛被。身体棕黄色，触角和足黄色，上颚齿棕色。**副模工蚁**：特征同正模，但是身体大小有轻度变化，身体黄色至棕黄色。

正模：工蚁，中国陕西省西安市长安区祥峪森林公园（33.962085°N，108.805718°E），1200m，2020.V.18，刘冠临采于针阔混交林土壤巢中，No. A20-3892。

词源：该种的种名以模式产地陕西省的"秦岭"命名。

云南细蚁
Leptanilla yunnanensis Xu, 2002

Leptanilla yunnanensis Xu, 2002b, Acta Entomologica Sinica, 45(1): 116, figs. 10-15 (w.q.) CHINA (Yunnan).

Holotype worker: TL 1.4, HL 0.30, HW 0.24, CI 80, SL 0.14, SI 58, PW 0.16, AL 0.36, PNL 0.08, PNW 0.11, PNH 0.12, PPNL 0.08, PPNW 0.13, PPNH 0.14. In full-face view, head longer than broad, roughly rectangular, narrowed forward. Occipital margin shallowly emarginated, occipital comers roundly prominent. Sides weakly convex, anterolateral corners rounded. Anterior margin of clypeus weakly convex and complete. Anterior portion of head convex between antennal sockets. Mandible with 3 teeth, the large apical tooth followed by 2 small teeth. Antenna 12-segmented, apex of scape reached to 1/2 of the distance from antennal socket to occipital comer, segments 3-11 broader than long. In lateral view, dorsum of alitrunk relatively straight, promesonotal suture distinct, metanotal groove absent. In lateral view, petiolar node roughly rectangular, anterodorsal and posterodorsal comers blunt and distinct, dorsum slightly convex. Sternite of petiole roughly triangular, bluntly angled in ventral direction. Postpetiolar node also rectangular and similar to petiolar node in lateral view, sternite of postpetiole large and anteroventrally pointed, rounded at apex. In dorsal view, petiolar node roughly rectangular and broader than long, postpetiolar node broader than long and narrowed forward. Head, alitrunk, petiole, postpetiole and gaster smooth and shining. Head and body with sparse subdecumbent hairs and dense decumbent pubescence. Appendages with dense decumbent pubescence. Body in color orange yellow. **Paratype workers:** TL 1.2-1.4, HL 0.28-0.30, HW 0.23-0.24, CI 79-86, SL 0.13-0.14, SI 57-58, PW 0.16, AL 0.34-0.36, PNL 0.07-0.08, PNW 0.11, PNH 0.11-0.13, PPNL 0.07-0.08, PPNW 0.12-0.13, PPNH 0.14-0.15 ($n = 5$). As holotype.

Holotype: Worker, China: Yunnan Province, Menghai County, Meng'a Town, Papo Village, 1600 m, 1997.Ⅸ.9, collected by Zheng-hui Xu from a soil nest in deciduous broadleaf forest, No. A97-2306.

Etymology: The new species is named after the type locality "Yunnan".

正模工蚁： 正面观，头部长大于宽，近长方形，向前变窄；后缘浅凹，后角圆凸；侧缘轻度隆起，前侧角圆。唇基前缘轻度隆起，完整。头前部在触角窝之间隆起。上颚具3个齿，大端齿后跟着2个小齿。触角12节，柄节末端到达触角窝至头后角间距的1/2处，第3～11节宽大于长。侧面观，胸部背面较平直，前中胸背板缝明显，缺后胸沟。侧面观，腹柄结近长方形，前上角和后上角钝而明显，背面轻微隆起；腹板近三角形，腹面钝角状。后腹柄结也近长方形，与腹柄结相似；腹板大，指向前下方，顶端圆。背面观，腹柄结近长方形，宽大于长；后腹柄结宽大于长，向前变窄。头部、胸部、腹柄、后腹柄和后腹部光滑发亮。头部和身体具稀疏亚倾斜毛和密集倾斜绒毛被；附肢具密集倾斜绒毛被。身体橙黄色。**副模工蚁：** 特征同正模。

正模： 工蚁，中国云南省勐海县勐阿乡怕迫村，1600m，1997.Ⅸ.9，徐正会采于落叶阔叶林土壤巢中，No. A97-2306。

词源： 该新种以模式产地"云南"命名。

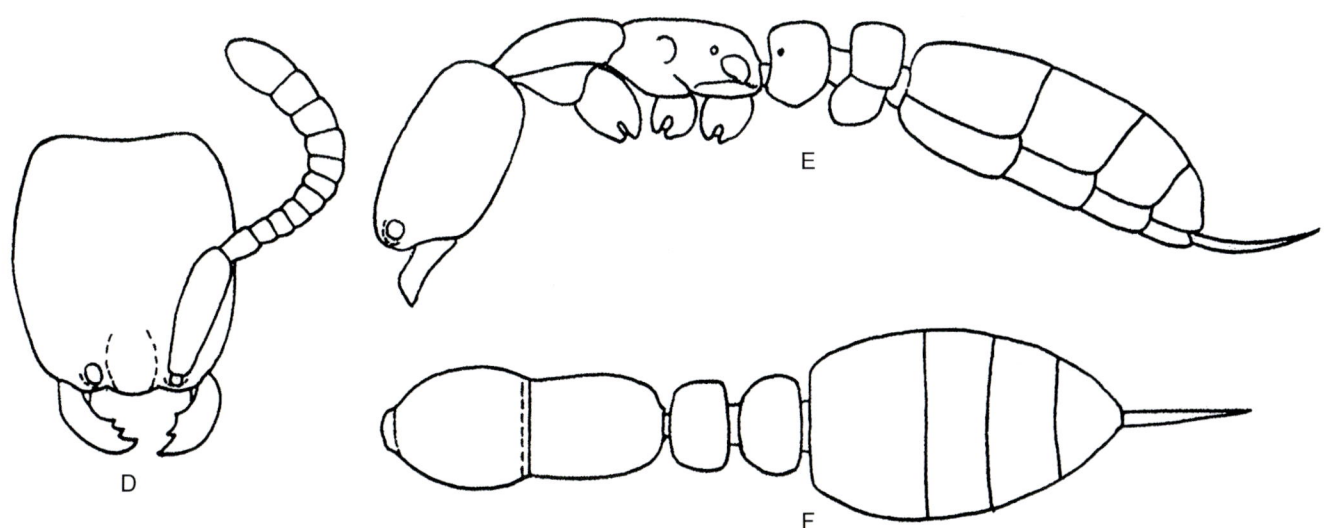

云南细蚁 *Leptanilla yunnanensis*

A, D. 工蚁头部正面观；B, E. 工蚁身体侧面观；C, F. 工蚁身体背面观（D-F. 引自Xu, 2002b）
A, D. Head of worker in full-face view; B, E. Body of worker in lateral view; C, F. Body of worker in dorsal view (D-F. cited from Xu, 2002b)

北京原细蚁
Protanilla beijingensis Man, Ran, Chen & Xu, 2017

Protanilla beijingensis Man et al., 2017, Asian Myrmecology, 9: 6, figs. 4-9 (w.q.) CHINA (Beijing).

Holotype worker: TL 4.0, HL 0.70, HW 0.68, CI 97, SL 0.61, SI 90, ML 0.51, PW 0.44, AL 1.21, PNL 0.25, PNH 0.44, PNW 0.29, PI 116, PPNL 0.28, PPNH 0.40, PPNW 0.29, PPI 104. In full-face view, head roughly trapezoidal and slightly longer than broad, anterior 1/3 of the head distinctly narrowed anteriorly and strongly constricted at antennal socket position, lateral margins evenly convex. Posterior margin weakly concave, posterior corners rounded. Mandibles elongate and curving downwards apically, lateral surface with a longitudinal groove, basal corners prominently round, masticatory margin with 19 peg-like teeth. Clypeus nearly trapezoidal, with a depressed longitudinal central furrow, anterior margin weakly concave. Apex of labrum moderately convex, with a peg-like tooth and a pair of stout long hairs. Antennae 12-segmented, apex of scape surpassed posterior head corner by about 1/6 of its length, flagella segments 4-9 about as broad as long. In lateral view, dorsum of mandible strongly convex. Mesosoma strongly constricted at middle position. Dorsum of pronotum weakly convex. Promesonotal suture complete and weakly depressed. Dorsum of mesonotum straight, weakly sloping down posteriorly. Metanotal groove moderately impressed. Dorsum of propodeum weakly convex, posterodorsal corner rounded; declivity slightly convex, about 1/2 length of the dorsum. Petiolar node nearly trapezoidal and narrowed dorsally, anterior face weakly convex, posterior face nearly straight, dorsal face roundly convex; anterodorsal corner rounded, posterodorsal corner relatively prominent. Subpetiolar process large and triangular, with an elliptical semitransparent fenestra, anteroventral corner blunt, anterior and posteroventral margins weakly convex. Postpetiolar node roughly rectangular and weakly widened dorsally, dorsal face weakly convex, anterior face strongly convex, anterodorsal corner rounded, posterodorsal corner blunt. Subpostpetiolar process large and lobe-like, anteroventrally pointed and rounded at apex. Gaster roughly elliptical, first gastral segment occupies about 1/2 length of gaster. Sting well-developed and extruding. In dorsal view, pronotum wide with strongly convex sides. Mesonotum strongly constricted and nearly square. Propodeum relatively narrow and rectangular, with weakly convex sides. Petiolar node nearly rectangular, slightly broader than long, sides evenly convex, anterior face almost straight, posterior face slightly concave. Postpetiolar node trapezoidal and widened posteriorly, as broad as long; anterior face, sides and posterior face weakly convex. Anterior margin of gaster weakly concave. Mandibles finely retirugose. Head and body smooth and shining. Body dorsum with sparse subdecumbent hairs and abundant decumbent pubescence. Scapes with sparse subdecumbent hairs and abundant decumbent pubescence. Tibiae with abundant decumbent pubescence, mandibles and clypeus with relatively abundant stouter and longer hairs, apex of each mandible with a very long stout hair on ventral portion. Body color reddish brown with the exception of the black parts of the posterior half of mesothorax and anterior half of the metathorax, mandibles, antennae, pronotum, legs, and posterior 2/3 of gaster brownish yellow. **Paratype workers:** TL 3.9-4.0, HL 0.65-0.70, HW 0.63-0.68, CI 96-97, SL 0.61-0.63, SI 90-98, ML 0.47-0.51, PW 0.43-0.44, AL 1.21-1.23, PNL 0.31-0.35, PNH 0.42-0.44, PNW 0.28-0.30, PI 86-90, PPNL 0.25-0.28, PPNH 0.38-0.40, PPNW 0.28-0.30, PPI 107-112 (*n*=2). As holotype, but slightly vary in total length and body color darker.

Holotype: Worker, China: Beijing City, Mentougou District, Xiaolongmen National Forest Park (39.973611°N, 115.425000°E), 1247 m, 2015.Ⅹ.15, collected by Pei Man in monsoon deciduous forest using subterranean pitfall trap, No. IOZ(E) 227911.

Etymology: The new species is named after the type locality "Beijing".

正模工蚁： 正面观，头部近梯形，长稍大于宽，前1/3向前明显变窄，在触角窝处强烈收缩；侧缘均匀隆起；后缘轻度凹陷，后角圆。上颚伸长，顶端下弯，侧面具1条纵沟，基角窄圆，咀嚼缘具19个木钉状齿。唇基近梯形，具1条凹陷的中央纵沟，前缘轻度凹陷。上唇端部中度隆起，具1个木钉状齿和1对粗长刚毛。触角12节，柄节约1/6超过头后角，鞭节第4～9节长宽约相等。侧面观，上颚背面强烈隆起。胸部在中部强烈收缩。前胸背板轻度隆起。前中胸背板缝完整，轻度凹陷。中胸背板平直，向后轻度降低。后胸沟中度凹陷。并胸腹节背面轻度隆起，后上角圆；斜面轻度隆起，约为背面长的1/2。腹柄结近梯形，向上变窄；前面轻度隆起，后面近平直，背面圆形隆起，前上角圆，后上角较突出；腹柄下突大，三角形，具1个椭圆形半透明窗斑，前下角钝，前缘和后下缘轻度隆起。后腹柄结近长方形，向上轻度变宽；背面轻度隆起，前面强烈隆起，后上角钝；腹柄下突大，叶状，指向前下方，末端圆。后腹部近椭圆形，第1节约占据后腹部长度的1/2。螫针发达，伸出。背面观，前胸背板宽，侧缘强

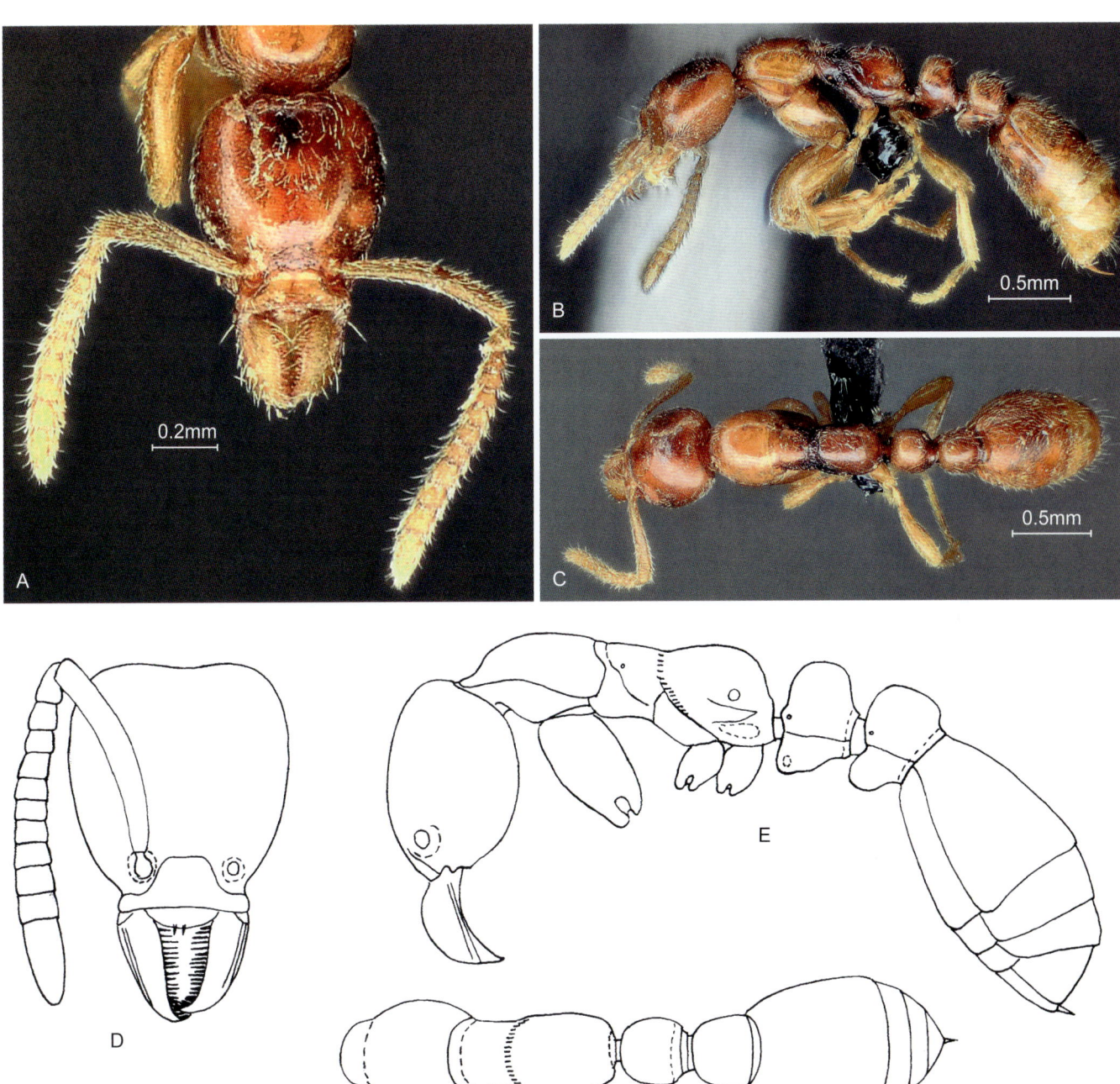

北京原细蚁 *Protanilla beijingensis*

A, D. 工蚁头部正面观；B, E. 工蚁身体侧面观；C, F. 工蚁身体背面观（D-F. 引自Man等，2017）

A, D. Head of worker in full-face view; B, E. Body of worker in lateral view; C, F. Body of worker in dorsal view (D-F. cited from Man et al., 2017)

烈隆起。中胸背板强烈收缩，近方形。并胸腹节较窄，长方形，侧缘轻度隆起。腹柄结近长方形，宽稍大于长，侧缘均匀隆起，前缘几乎平直，后缘轻度凹陷。后腹柄结梯形，向后变宽，长宽相等，前缘、侧缘和后缘轻度隆起。后腹部前缘轻度凹陷。上颚具细网纹。头部和身体光滑发亮。身体背面具稀疏亚倾斜毛和丰富倾斜毛被；柄节具稀疏亚倾斜毛和丰富倾斜绒毛被；胫节具丰富倾斜绒毛被。上颚和唇基具相对丰富的粗长立毛，每个上颚顶端腹面具1根很长的粗毛。身体红棕色，中胸后半部和后胸前半部黑色，上颚、触角、前胸背板、足和后腹部后部2/3棕黄色。**副模工蚁：** 特征同正模，但是体长稍有变化，体色较暗。

正模： 工蚁，中国北京市门头沟区小龙门国家森林公园（39.973611°N，115.425000°），1247m，2015.Ⅹ.15，满沛采用地下陷阱法采于季风落叶林中，No. IOZ(E)227911。

词源： 该新种以模式产地"北京"命名。

双色原细蚁
Protanilla bicolor Xu, 2002

Protanilla bicolor Xu, 2002b, Acta Entomologica Sinica, 45(1): 119, figs. 21-23 (w.) CHINA (Yunnan).

Holotype worker: TL 3.0, HL 0.53, HW 0.43, CI 81, SL 0.47, SI 108, PW 0.37, AL 0.87, PNL 0.24, PNW 0.19, PNH 0.30, PPNL 0.24, PPNW 0.21, PPNH 0.32. In full-face view, head longer than broad, narrowed forward. Occipital margin shallowly emarginate, occipital comers roundly prominent. Sides evenly convex, each side with a prominence near the antennal socket position. Clypeus longitudinally depressed, without a depressed longitudinal central line, anterior margin emarginate. Mandibles long triangular, curved down at apex, masticatory margin with 11 peg-like teeth. Antennae stout, apex of antennal scape just reached occipital comer, segments 4-10 broader than long. In lateral view, promesonotum higher than propodeum, mesothorax weakly constricted. Promesonotal suture distinct, metanotal groove shallowly depressed. Dorsum of propodeum straight, about 2 times as long as declivity, posterodorsal comer rounded. In lateral view, petiolar node narrowed upward, both anterior and posterior faces slope-like, dorsum convex, anterodorsal and posterodorsal comers rounded. Anteroventral comer of petiolar sternite bluntly extruded, with a circular subtransparent fovea. Postpetiolar node strongly inclined forward, anterodorsal comer roundly prominent, sternite longer than high. In dorsal view, both petiolar node and postpetiolar node elliptic and longer than broad. First gastral segment large and about 3/5 as long as the gaster. Mandibles, head, alitrunk, petiole, postpetiole and gaster smooth and shining. Head and body with sparse suberect hairs and dense subdecumbent pubescence. Scapes and tibiae with sparse suberect hairs and dense decumbent pubescence. Body in color brownish yellow, posterior portion of mesothorax, metathorax, propodeum, petiole, postpetiole and first gastral segment black. **Paratype workers:** TL 2.7-3.0, HL 0.50-0.53, HW 0.42-0.45, CI 78-87, SL 0.43-0.47, SI 96-108, PW 0.33-0.38, AL 0.83-0.87, PNL 0.22-0.24, PNW 0.19-0.21, PNH 0.30-0.33, PPNL 0.24-0.26, PPNW 0.21-0.24, PPNH 0.32-0.35 ($n = 8$). As holotype.

Holotype: Worker, China: Yunnan Province, Menghai County, Meng'a Town, Papo Village, 1600 m, 1997.IX.9, collected by Zheng-hui Xu from a soil nest in deciduous broadleaf forest, No. A97-2240.

Etymology: The species name *bicolor* combines Latin *bi-* (two) + word root *color* (color) (feminine form), it refers to the obviously bicolored body.

正模工蚁： 正面观，头部长大于宽，向前变窄；后缘浅凹，后角圆凸；侧缘均匀隆起，每侧在触角窝处具1个突起。唇基纵向凹陷，缺中央纵沟，前缘凹陷。上颚长三角形，顶端下弯，咀嚼缘具11个木钉状齿。触角粗，柄节末端刚到达头后角，第4～10节宽大于长。侧面观，前中胸背板高于并胸腹节，中胸轻度收缩。前中胸背板缝明显，后胸沟浅凹。并胸腹节背面平直，约为斜面长的2倍，后上角圆。侧面观，腹柄结向上变窄，前面和后面均为坡形，背面隆起，前上角和后上角圆；腹板前下角钝凸，具1个圆形半透明窗斑。后腹柄结强烈前倾，前上角圆凸，腹板长大于高。背面观，腹柄结和后腹柄结均为椭圆形，且长大于宽。后腹部第1节大，约为后腹部长的3/5。上颚、头部、胸部、腹柄、后腹柄和后腹部光滑发亮。头部和身体具稀疏亚直立毛和密集倾斜绒毛被；柄节和胫节具稀疏亚直立毛和密集倾斜绒毛被。身体棕黄色，中胸后部、后胸、并胸腹节、腹柄、后腹柄和后腹部第1节黑色。**副模工蚁：** 特征同正模。

正模： 工蚁，中国云南省勐海县勐阿乡怕迫村，1600m，1997.IX.9，徐正会采于落叶阔叶林土壤巢中，No. A97-2240。

词源： 该新种的种名"*bicolor*"由拉丁语"*bi-*"（二，双）+词根"*color*"（颜色）（阴性形式）组成，指身体明显呈双色。

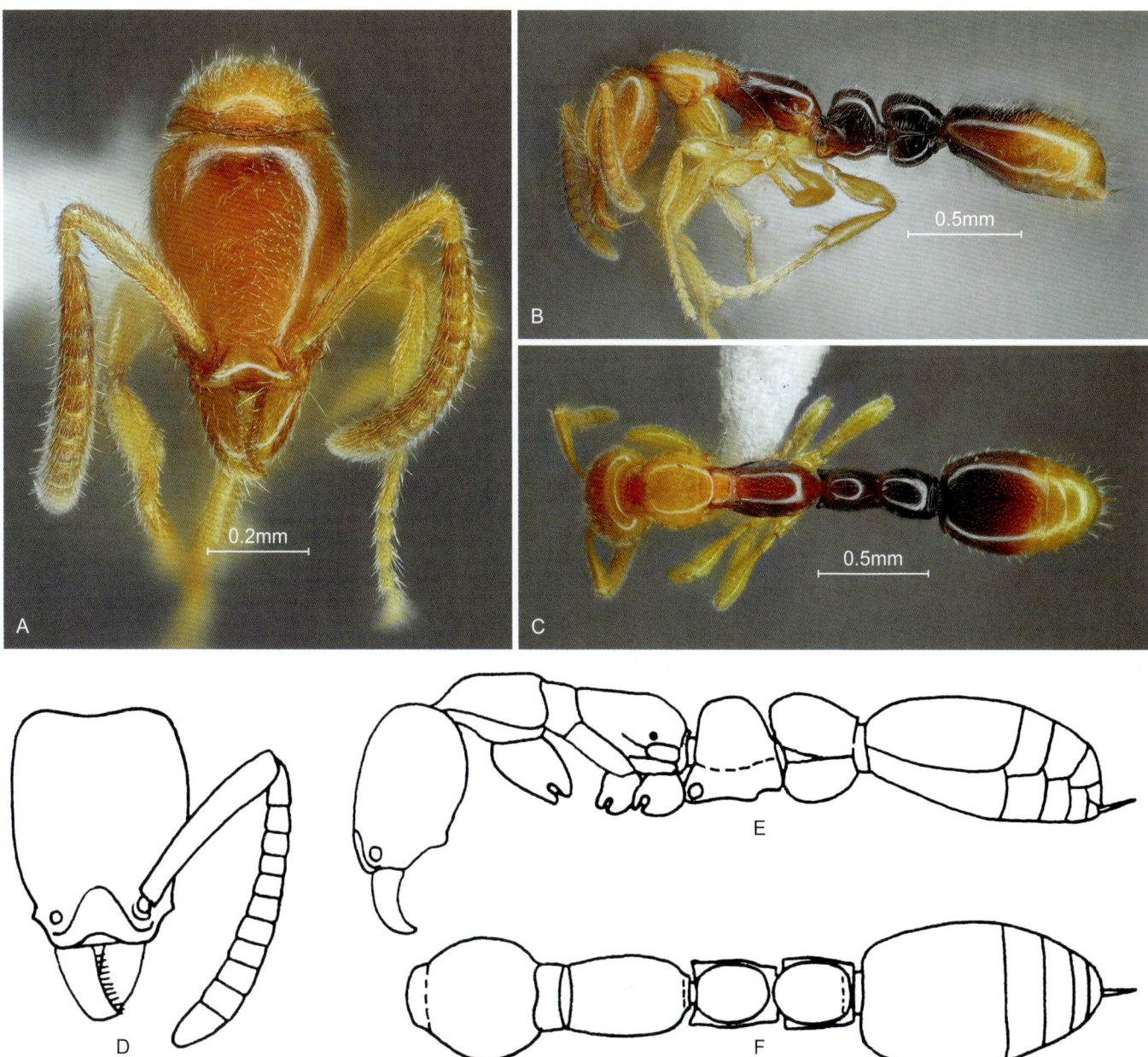

双色原细蚁 *Protanilla bicolor*

A, D. 工蚁头部正面观；B, E. 工蚁身体侧面观；C, F. 工蚁身体背面观（D-F. 引自Xu, 2002b）
A, D. Head of worker in full-face view; B, E. Body of worker in lateral view; C, F. Body of worker in dorsal view (D-F. cited from Xu, 2002b)

054 单色原细蚁
Protanilla concolor Xu, 2002

Protanilla concolor Xu, 2002b, Acta Entomologica Sinica, 45(1): 118, figs. 18-20 (w.) CHINA (Yunnan).

Holotype worker: TL 2.9, HL 0.57, HW 0.47, CI 82, SL 0.47, SI 100, PW 0.38, AL 0.80, PNL 0.22, PNW 0.23, PNH 0.33, PPNL 0.20, PPNW 0.26, PPNH 0.34. In full-face view, head longer than broad, narrowed forward. Occipital margin weakly emarginate. Occipital comers rounded. Sides of head roundly convex and constricted at the antennal socket position. Clypeus with a depressed longitudinal central line, anterior margin straight. Mandibles long triangular and curved down at apex, masticatory margin with 13 peg-like teeth. Scape of antenna surpassed occipital comer by about 1/10 of its length, segments 4-10 about as broad as long. In lateral view, promesonotum higher than propodeum, mesothorax weakly constricted. Promesonotal suture distinct, metanotal groove shallowly depressed. Dorsum of propodeum straight and longer than declivity, posterodorsal comer rounded. In lateral view, petiolar node narrowed upward, anterior face straight, dorsal and posterior faces convex, anterodorsal comer roundly prominent and higher than posterodorsal comer, the latter rounded. Anteroventral comer of petiolar sternite bluntly extruded, with a circular subtransparent fovea. Postpetiolar node inclined forward, anterodorsal comer roundly prominent, sternite higher than long and inclined forward. In dorsal view, both petiolar node and postpetiolar node transverse and broader than long, narrowed forward. First segment of gaster large, about 3/5 as long as gaster. Mandibles sparsely and finely punctured. Head, alitrunk, petiole, postpetiole and gaster smooth and shining. Head and body with sparse suberect hairs and abundant decumbent pubescence. Scapes and tibiae with sparse erect hairs and dense decumbent pubescence. Body in color reddish brown. Mandibles and clypeus yellow. Antennae and legs light yellowish brown.

Holotype: Worker, China: Yunnan Province, Mengla County, Mengla Town, Peak of Nangongshan Mountain, 1980 m, 1998.Ⅲ.16, collected by Zheng-hui Xu from a soil sample of the mossy evergreen broadleaf forest, No. A98-993.

Etymology: The species name *concolor* combines Latin *con-* (same) + word root *color* (color) (feminine form), it refers to the unicolored body.

正模工蚁： 正面观，头部长大于宽，向前变窄；后缘轻度凹陷，后角圆；侧缘圆形隆起，在触角窝处收缩。唇基具1条凹陷的中央纵沟，前缘平直。上颚长三角形，顶端下弯，咀嚼缘具13个木钉状齿。触角柄节约1/10超过头后角，第4~10节长宽约相等。侧面观，前中胸背板高于并胸腹节，中胸轻度收缩。前中胸背板缝明显，后胸沟浅凹。并胸腹节背面平直，长于斜面，后上角圆。侧面观，腹柄结向上变窄，前面平直，背面和后面隆起；前上角圆凸，高于后上角，后上角圆；腹板前下角钝突，具1个圆形半透明窗斑。后腹柄结前倾，前上角圆凸；腹板高大于长，前倾。背面观，腹柄结和后腹柄结均横向，且宽大于长，向前变窄。后腹部第1节大，约为后腹部长的3/5。上颚具稀疏细刻点。头部、胸部、腹柄、后腹柄和后腹部光滑发亮。头部和身体具稀疏亚直立毛和丰富倾斜绒毛被；柄节和胫节具稀疏直立毛和密集倾斜绒毛被。身体红棕色，上颚和唇基黄色，触角和足浅黄棕色。

正模： 工蚁，中国云南省勐腊县勐腊镇南贡山顶，1980m，1998.Ⅲ.16，徐正会采于苔藓常绿阔叶林土壤样中，No. A98-993。

词源： 该新种的种名"*concolor*"由拉丁语"*con-*"（相同的）+词根"*color*"（颜色）（阴性形式）组成，指身体颜色为单色。

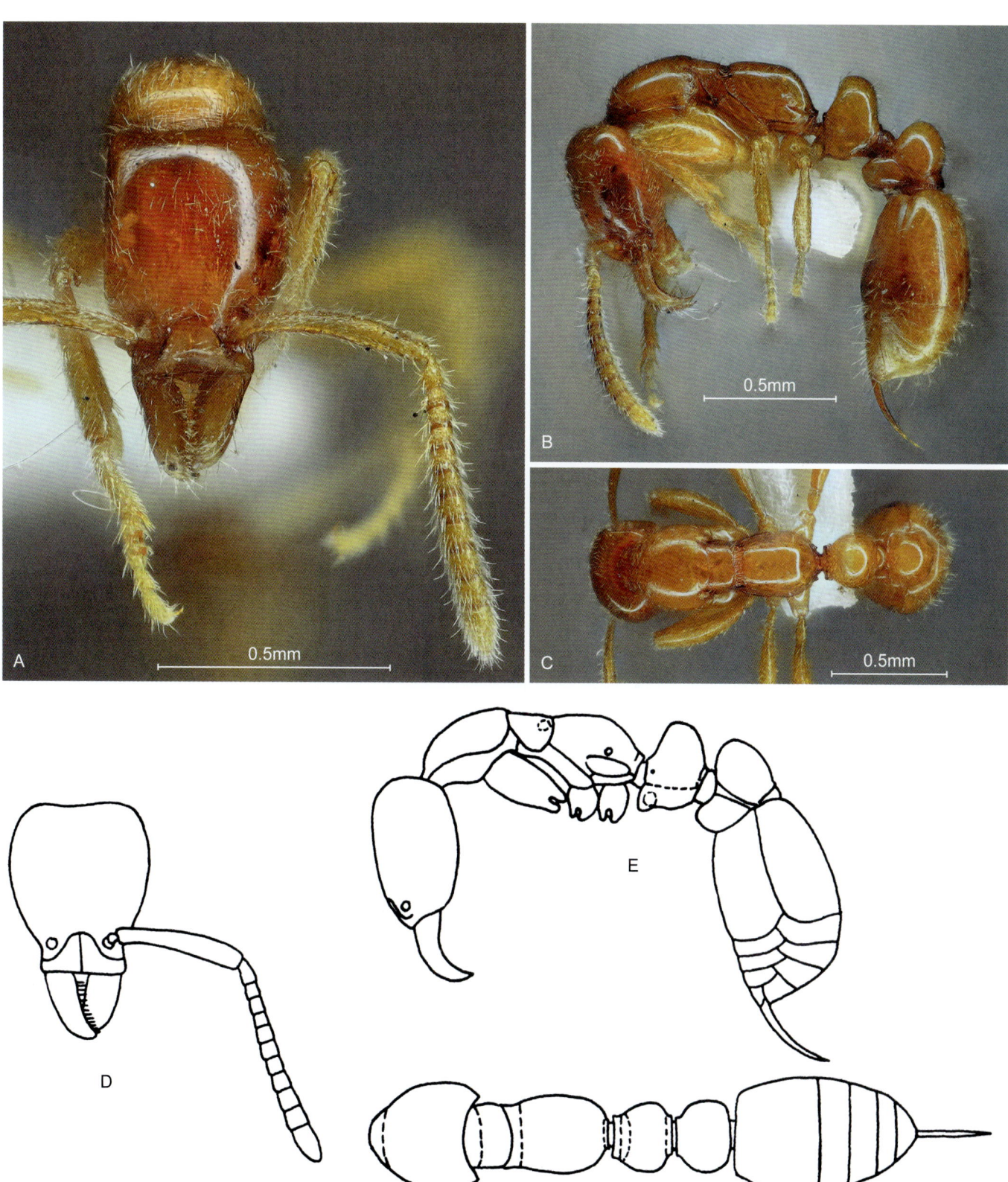

单色原细蚁 *Protanilla concolor*

A, D. 工蚁头部正面观；B, E. 工蚁身体侧面观；C, F. 工蚁身体背面观（D-F. 引自Xu, 2002b）
A, D. Head of worker in full-face view; B, E. Body of worker in lateral view; C, F. Body of worker in dorsal view (D-F. cited from Xu, 2002b)

叉颚原细蚁
Protanilla furcomandibula Xu & Zhang, 2002

Protanilla furcomandibula Xu & Zhang, 2002a, Acta Zootaxonomica Sinica, 27(1): 140, figs. 1-3 (w.) CHINA (Yunnan).

Holotype worker: TL 4.1, HL 0.77, HW 0.60, CI 78, SL 0.67, SI 111, PW 0.47, AL 1.17, ML 0.50, PNL 0.33, PNH 0.43, PNW 0.32, PPNL 0.40, PPNH 0.43, PPNW 0.30. In full-face view, head distinctly longer than broad, narrowed forward. Occipital margin straight, occipital corners rounded. Sides of head evenly convex and with a tooth-like prominence at the antennal socket position, below the prominence with a deep notch. In dorsal view mandible long triangular, masticatory margin with 15 spine-like teeth. In lateral view mandible thick, lateroventral margin with 2 teeth, the basal one short and oblique, the apical one long and erect. Clypeus roughly triangular, anterior margin obviously concave in the center. Antenna with 12 segments, scape surpassed occipital corner by 1/5 of its length, segments 5-11 longer than broad, antennal club indistinct. In lateral view, thorax distinctly constricted at mesothorax, pronotum roundly convex, mesonotum straight. Promesonotal suture distinct, metanotal groove depressed. Dorsum of propodeum evenly convex, longer than declivity, posterodorsal corner rounded. In lateral view, petiolar node nearly rectangular, anterodorsal corner bluntly prominent, higher than posterodorsal corner, the latter rounded, anterior face vertical, posterior face sloped, dorsal face weakly convex. Subpetiolar process long and anteroventrally pointed, with a circular subtransparent fovea. Postpetiolar node weakly inclined forward, anterior face convex, dorsal face straight, anterodorsal corner rounded. Sternite of postpetiole deeply concave on the ventral face. In dorsal view, petiolar node roughly square, width : length = 9 : 8, anterior and posterior borders weakly convex, lateral borders evenly convex. Postpetiolar node trapezoid, narrowed forward and longer than broad, anterior and posterior faces roundly convex, sides straight. In lateral view, anterior margin of gaster with a narrow deep notch between tergite and sternite of the first segment. In dorsal view, anterior margin of gaster deeply concave, with anterolateral corners protruding and surrounded the postpetiole. Mandibles, head and whole body smooth and shining. Head and body with sparse suberect hairs and dense decumbent pubescence. Scapes and tibiae with sparse subdecumbent hairs and dense decumbent pubescence. Body in color yellowish brown, legs brownish yellow. **Paratype worker:** TL 4.0, HL 0.73, HW 0.57, CI 77, SL 0.67, SI 118, PW 0.43, AL 1.13, ML 0.47, PNL 0.33, PNH 0.43, PNW 0.30, PPNL 0.43, PPNH 0.43, PPNW 0.28 (*n*=1). As holotype.

Holotype: Worker, China: Yunnan Province, Kunming City, Xishan Forest Park, Huatingsi Temple, 2250 m, 2001.Ⅲ.31, collected by Zheng-hui Xu from a soil sample in conifer-broadleaf mixed forest, No. A00250.

Etymology: The species name *furcomandibula* combines Latin *furc-* (bifurcated) + word root *mandibul* (mandible) + suffix *-a* (feminine form), it refers to the mandible with two bifurcated teeth on ventral margin in lateral view.

正模工蚁： 正面观，头部长明显大于宽，向前变窄；后缘平直，后角圆；侧缘均匀隆起，在触角窝处具1个齿状突起，突起之前具1个深切口。背面观，上颚长三角形，咀嚼缘具15个刺状齿；侧面观，上颚厚，侧下缘具2个齿，基部的齿短而倾斜，端部的齿长而直立。唇基近三角形，前缘中央明显凹陷。触角12节，柄节约1/5超过头后角，第5～11节长大于宽，触角棒不明显。侧面观，胸部在中胸处明显收缩，前胸背板圆形隆起，中胸背板平直。前中胸背板缝明显，后胸沟凹陷。并胸腹节背面均匀隆起，长于斜面，后上角圆。侧面观，腹柄结近长方形，前上角钝凸，高于后上角，后上角圆；前面垂直，后面坡形，背面轻度隆起；腹柄下突长，指向前下方，具1个圆形半透明窗斑。后腹柄结轻度前倾，前面隆起，背面平直，前上角圆；腹板腹面深凹。背面观，腹柄结近方形，宽：长=9：8，前缘和后缘轻度隆起，侧缘均匀隆起。后腹柄结梯形，向前变窄，长大于宽，前缘和后缘圆形隆起，侧缘平直。侧面观，后腹部前缘在第1节背板和腹板之间具1个窄而深缺口。背面观，后腹部前缘深凹，前侧角突出并包围后腹柄。上颚、头部和整个身体光滑发亮。头部和身体具稀疏亚直立毛和密集倾斜绒毛被；柄节和胫节具稀疏亚倾斜毛和密集倾斜绒毛被。身体黄棕色，足棕黄色。**副模工蚁：** 特征同正模。

正模： 工蚁，中国云南省昆明市西山森林公园华亭寺，2250m，2001.Ⅲ.31，徐正会采于针阔混交林土壤样中，No. A00250。

词源： 该新种的种名"*furcomandibula*"由拉丁语"*furc-*"（分叉的）+词根"*mandibul*"（上颚）+后缀"*-a*"（阴性形式）组成，指侧面观上颚下缘具2个叉状的齿。

Individual Theory 分 论

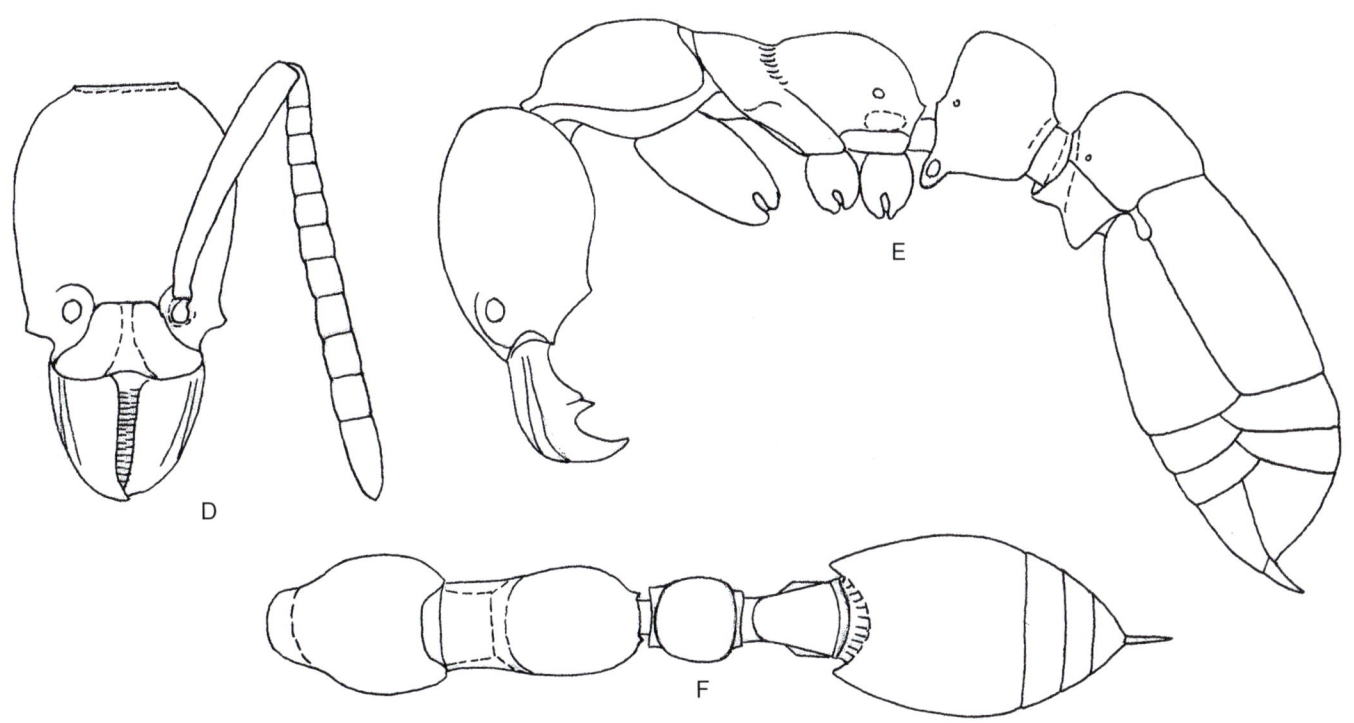

叉颚原细蚁 *Protanilla furcomandibula*

A, D. 工蚁头部正面观；B, E. 工蚁身体侧面观；C, F. 工蚁身体背面观（D-F. 引自Xu和Zhang, 2002a）
A, D. Head of worker in full-face view; B, E. Body of worker in lateral view; C, F. Body of worker in dorsal view (D-F. cited from Xu & Zhang, 2002a)

056 耿马原细蚁
Protanilla gengma Xu, 2012

Protanilla gengma Xu, 2012c, Sociobiology, 59(2): 485, figs. 13-16 (w.) CHINA (Yunnan).

Holotype worker: TL 4.1, HL 0.73, HW 0.60, CI 83, SL 0.68, SI 113, ML 0.43, PW 0.48, AL 1.20, PNL 0.38, PNH 0.48, PNW 0.26, PPNL 0.35, PPNH 0.48, PPNW 0.28. In full-face view, head longer than broad, anterior 1/2 of the head distinctly narrowed forward. Sides evenly convex, anterolateral corners of head prominent and tooth-like. Occipital margin weakly concave, occipital corners rounded. Mandibles elongate and down-curved apically, dorsolateral surface without a longitudinal groove, basal corners roundly prominent, masticatory margin with about 21 peg-like teeth. Clypeus nearly trapezoidal, anterior margin evenly concave. Apex of labrum roundly convex, with 4 peg-like teeth. Antennae 12-segmented, apices of scapes surpassed occipital corners by about 1/6 of its length, flagella segments 4-9 about as broad as long. In lateral view, dorsum of pronotum nearly straight. Promesonotal suture complete but not depressed. Dorsum of mesonotum straight, slope down backward. Metanotal groove moderately notched. Dorsum of propodeum slightly convex, weakly slope down backward, posterodorsal corner evenly convex; declivity weakly convex, about 1/2 length of the dorsum. Petiolar node nearly trapezoidal, anterior and posterior faces nearly straight, dorsal face roundly convex; anterodorsal corner slightly higher the posterodorsal corner, both are rounded. Ventral face of petiole roundly convex, anteroventral corner roundly extruding, with a circular semitransparent fenestra; posteroventral corner acutely angled. Postpetiolar node strongly inclined forward, with dorsum roundly convex, anterodorsal corner bluntly prominent, anterior face straight; ventral face roundly convex, anteroventral corner bluntly angled. Sting strong and extruding. In dorsal view, mesonotum strongly constricted. Petiolar node nearly rectangular, longer than broad, length∶width = 1.4∶1, sides evenly convex, anterior face nearly straight, posterior face roundly convex. Postpetiolar node trapezoidal and widened backward, longer than broad, length∶width = 1.3∶1; sides weakly convex, anterior face roundly convex, posterior face nearly straight. Anterior margin of gaster straight. Mandibles sparsely finely punctured. Head and body smooth and shining. Dorsa of head and gaster with abundant erect to suberect hairs and dense decumbent pubescence. Alitrunk, petiole, and postpetiole with sparse erect to suberect hairs and abundant decumbent pubescence. Scapes and tibiae with abundant suberect to subdecumbent hairs and dense decumbent pubescence. Mouthparts with abundant longer hairs, apex of mandible with a stout long hair. Head light black. Mandibles, antennae, prothorax, legs, and gastral segments 2-4 yellowish brown. Mesothorax, metathorax, propodeum, petiole, postpetiole, and first gastral segment black. **Paratype workers:** TL 4.1-4.5, HL 0.70-0.78, HW 0.60-0.65, CI 83-87, SL 0.65-0.73, SI 104-113, ML 0.43-0.48, PW 0.48-0.51, AL 1.18-1.30, PNL 0.35-0.38, PNH 0.48-0.50, PNW 0.26-0.29, PPNL 0.33-0.35, PPNH 0.48-0.51, PPNW 0.28-0.30 (*n*=6). As holotype, but head light black to blackish brown.

Holotype: Worker, China: Yunnan Province, Gengma County, Mengding Town, Nantianmen, 1760 m, 2011.Ⅲ.10, collected by Hai-bin Li from a soil sample in monsoon evergreen broadleaf forest, No. A11-29.

Etymology: The new species is named after the type locality "Gengma"

正模工蚁： 正面观，头部长大于宽，前部1/2向前明显变窄，侧面均匀隆起，前侧角突起呈齿状；后缘轻度凹陷，后角圆。上颚伸长，顶端下弯，背侧面缺纵沟，基角圆突，咀嚼缘约具21个木钉状齿。唇基近梯形，前缘均匀凹陷。上唇顶端圆形隆起，具4个木钉状齿。触角12节，柄节约1/6超过头后角，鞭节第4～9节长宽约相等。侧面观前胸背板背面近平直，前中胸背板缝完整但不凹陷。中胸背板平直，向后降低，后胸沟中度切入。并胸腹节背面轻微隆起，向后轻度降低，后上角均匀隆起；斜面轻度隆起，约为背面长的1/2。腹柄结近梯形，前面和后面近平直，背面圆形隆起，前上角稍高于后上角，均为圆形；腹柄腹面圆形隆起，前下角圆形突出，具1个圆形半透明窗斑，后下角锐角状。后腹柄结强烈前倾，背面圆形隆起，前上角钝突，前面平直；腹面圆形隆起，前下角钝角状。螫针发达，伸出。背面观，中胸强烈收缩。腹柄结近长方形，长大于宽，长∶宽=1.4∶1，侧面均匀隆起，前面近平直，后面圆形隆起。后腹柄结梯形，向后变宽，长大于宽，长∶宽=1.3∶1；侧面轻度隆起，前面圆形隆起，后面近平直。后腹部前缘平直。上颚具稀疏细刻点。头部和身体光滑发亮。头部和后腹部背面具丰富直立、亚直立毛和密集倾斜绒毛被；胸部、腹柄和后腹柄具稀疏直立、亚直立毛和丰富倾斜绒毛被；柄节和胫节具丰富亚直立、亚倾斜毛和密集倾斜绒毛被；口器具丰富长毛，上颚顶端具1根粗长毛。头部浅黑色；上颚、触角、前胸、足和后腹部第2～4节黄棕色；中胸、后胸、并胸腹节、腹柄、后腹柄和后腹部第1节黑色。**副模工蚁：** 特征同正模，但是头部浅

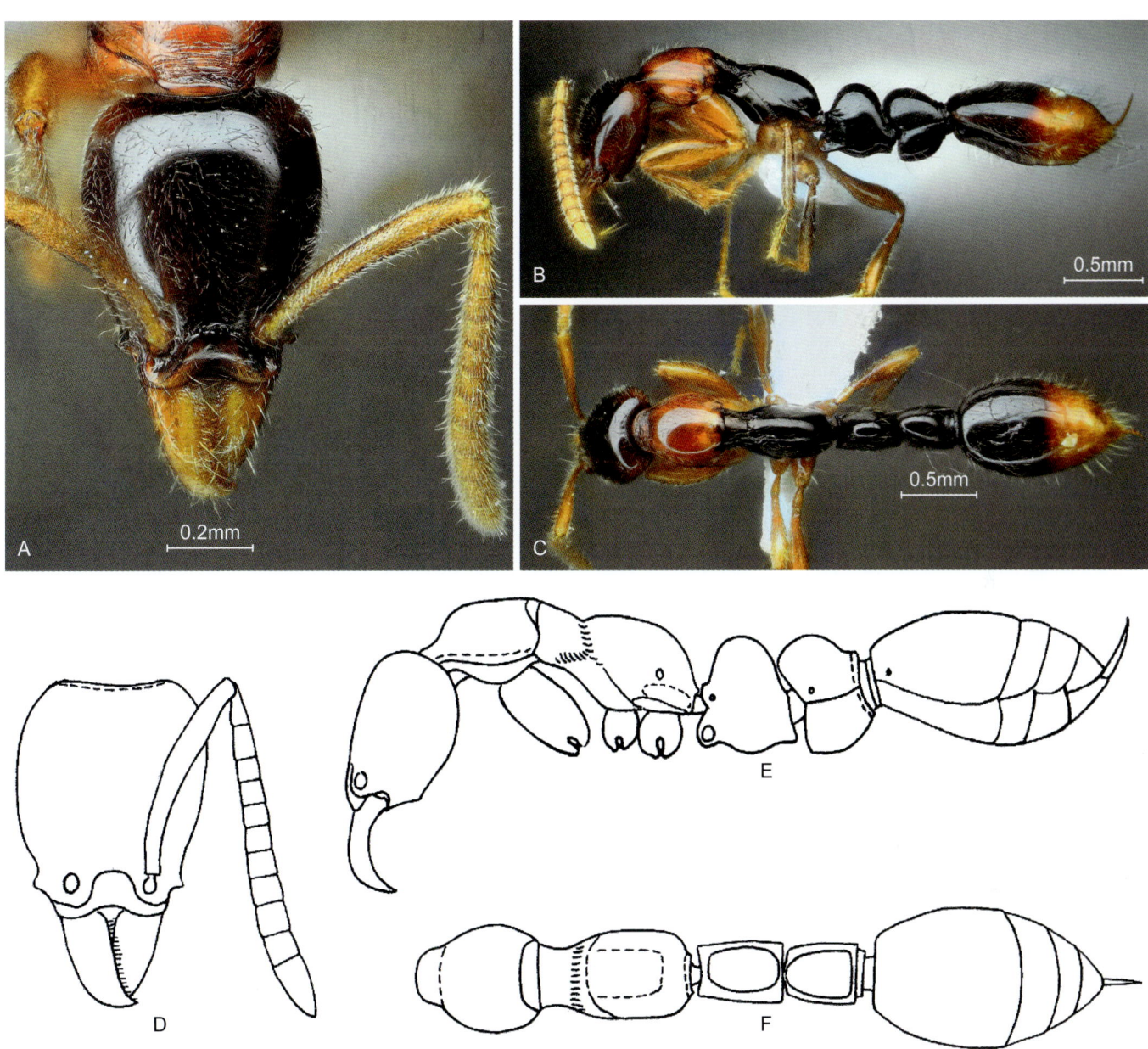

耿马原细蚁 *Protanilla gengma*

A, D. 工蚁头部正面观；B, E. 工蚁身体侧面观；C, F. 工蚁身体背面观（D-F.引自Xu, 2012c）
A, D. Head of worker in full-face view; B, E. Body of worker in lateral view; C, F. Body of worker in dorsal view (D-F. cited from Xu, 2012c)

黑色至黑棕色。

正模： 工蚁，中国云南省耿马县孟定镇南天门，1760m，2011.Ⅲ.10，李海斌采于季风常绿阔叶林土壤样中，No. A11-29。

词源： 该新种以模式产地"耿马"命名。

西藏原细蚁
Protanilla tibeta Xu, 2012

Protanilla tibeta Xu, 2012c, Sociobiology, 59(2): 487, figs. 17-20 (w.) CHINA (Tibet).

Holotype worker: TL 2.6, HL 0.53, HW 0.40, CI 76, SL 0.43, SI 106, ML 0.28, PW 0.33, AL 0.75, PNL 0.18, PNH 0.30, PNW 0.23, PPNL 0.20, PPNH 0.30, PPNW 0.23. In full-face view, head longer than broad, anterior 1/3 of the head distinctly narrowed forward. Sides weakly convex, anterolateral corners prominent and tooth-like, strongly concave before the teeth. Occipital margin weakly concave. Occipital corners roundly prominent. Mandibles elongate and down-curved apically, laterodorsal surface with a longitudinal groove, basal corners roundly prominent, masticatory margin with about 12 peg-like teeth. Clypeus roughly trapezoidal, anterior margin nearly straight. Antennae 12-segmented, apices of scapes just reached to occipital corners, flagella segments 2-10 about as broad as long. In lateral view, pronotum weakly convex. Promesonotal suture complete but not depressed. Dorsum of mesonotum straight and weakly slope down backward. Metanotal groove shallowly depressed. Dorsum of propodeum weakly convex, posterodorsal corner evenly convex; declivity weakly convex, about 1/2 length of the dorsum. Propodeal spiracle circular and small, lower down on the side. Metapleural bulla elongate and roughly elliptical, close to the spiracle. Petiolar node nearly trapezoidal, anterior face nearly straight, about as long as dorsal face, dorsal and posterior faces weakly convex; anterodorsal corner blunt, higher than posterodorsal corner, the latter rounded; ventral face oblique and straight, anteroventral process roughly square, with circular semitransparent fenestra. Postpetiolar node strongly inclined forward, dorsum roundly convex, anterodorsal corner bluntly prominent, anterior face straight; ventral face roundly convex, anteroventral corner roundly protruding. Sting strong and extruding. In dorsal view, petiolar node broader than long, width : length = 1.4 : 1, slightly widened backward; sides weakly convex, anterior and posterior faces nearly straight. Postpetiolar node about as broad as long, width : length = 1.1 : 1, weakly widened backward, sides and anterior face roundly convex, posterior face nearly straight. Anterior margin of gaster nearly straight. Mandibles, head, and body smooth and shining. Dorsa of head and body with sparse erect to suberect hairs and dense decumbent pubescence. Scapes and tibiae with abundant subdecumbent short hairs and dense decumbent pubescence. Mouthparts with abundant longer hairs, apex of mandible with a long stout hair on the ventral face. Color reddish brown. Mandibles, antennae, and legs brownish yellow. **Paratype worker:** TL 2.7, HL 0.51, HW 0.40, CI 78, SL 0.40, SI 100, ML 0.25, PW 0.33, AL 0.83, PNL 0.18, PNH 0.31, PNW 0.23, PPNL 0.20, PPNH 0.30, PPNW 0.23 (n=1). As holotype.

Holotype: Worker, China: Tibet Autonomous Region, Medog County, Damu Town, Damu Village, 1200 m, 2011.Ⅶ.20, collected by Xia Liu from a soil sample in valley tropical rain forest, No. A11-3925.

Etymology: The new species is named after the type locality "Tibet".

正模工蚁： 正面观，头部长大于宽，前部1/3向前明显变窄；侧缘轻度隆起，前侧角突起呈齿状，齿之前深凹；后缘轻度凹陷，后角圆突。上颚伸长，顶端下弯，背侧面具1条纵沟，基角圆突，咀嚼缘约具12个木钉状齿。唇基近梯形，前缘近平直。触角12节，柄节末端刚到达头后角，鞭节第2~10节长宽约相等。侧面观，前胸背板轻度隆起，前中胸背板缝完整，但不凹陷。中胸背板平直，向后轻度降低，后胸沟浅凹。并胸腹节背面轻度隆起，后上角均匀隆起；斜面轻度隆起，约为背面长的1/2；气门小而圆，位于侧面中下部。后胸侧板腺泡伸长，近椭圆形，接近并胸腹节气门。腹柄结近梯形，前面近平直，约与背面等长，背面和后面轻度隆起；前上角钝，高于后上角，后上角圆；腹面倾斜，平直，前下突近方形，具圆形半透明窗斑。后腹柄结强烈前倾，背面圆形隆起，前上角钝突，前面平直；腹面圆形隆起，前下角圆凸。螫针发达，伸出。背面观，腹柄结宽大于长，宽：长=1.4：1，向后轻微变宽；侧面轻度隆起，前面和后面近平直。后腹柄结长宽约相等，宽：长=1.1：1，向后轻度变宽，侧面和前面圆形隆起，后面近平直。后腹部前缘近平直。上颚、头部和身体光滑发亮。头部和身体背面具稀疏直立、亚直立毛和密集倾斜绒毛被；柄节和胫节具丰富亚倾斜短毛和密集倾斜绒毛被；口器具丰富长毛，上颚顶端腹面具1根粗长毛。身体红棕色，上颚、触角和足棕黄色。**副模工蚁：** 特征同正模。

正模： 工蚁，中国西藏自治区墨脱县达木乡达木村，1200m，2011.Ⅶ.20，刘霞采于沟谷热带雨林土壤样中，No. A11-3925。

词源： 该新种以模式产地"西藏"命名。

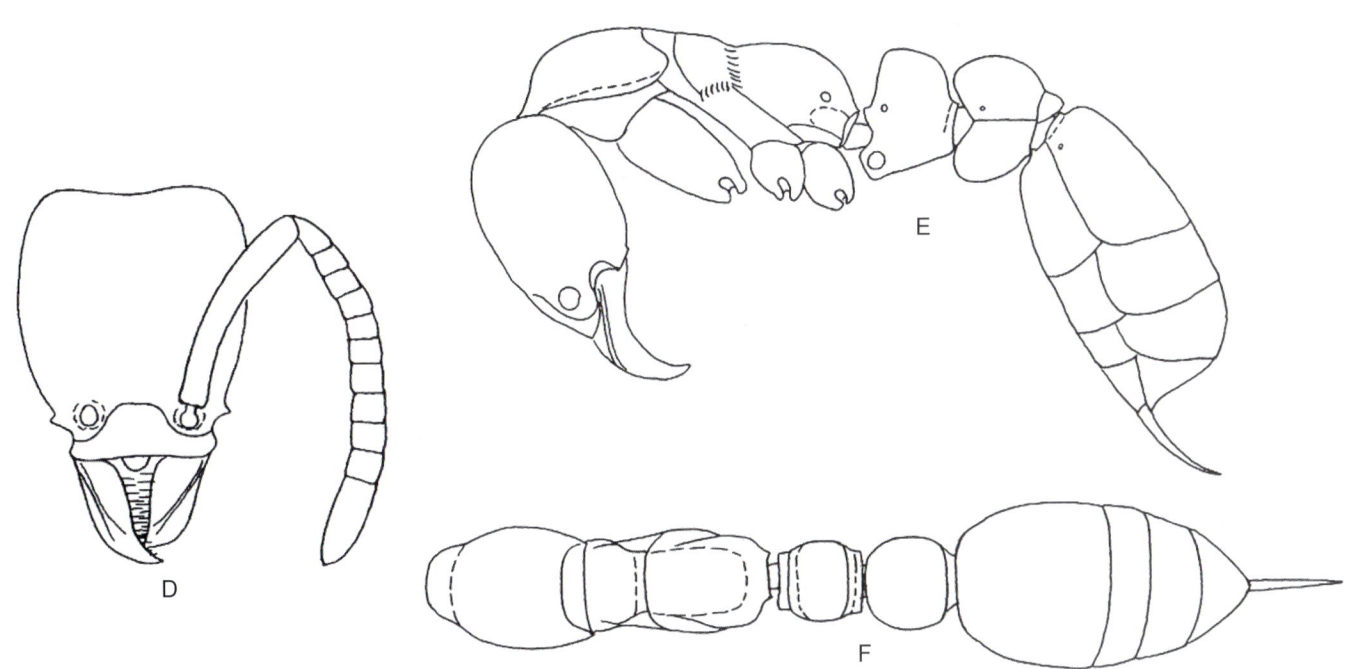

西藏原细蚁 *Protanilla tibeta*

A, D. 工蚁头部正面观; B, E. 工蚁身体侧面观; C, F. 工蚁身体背面观 (D-F. 引自Xu, 2012c)
A, D. Head of worker in full-face view; B, E. Body of worker in lateral view; C, F. Body of worker in dorsal view (D-F. cited from Xu, 2012c)

无缘细长蚁
Tetraponera amargina Xu & Chai, 2004

Tetraponera amargina Xu & Chai, 2004, Acta Zootaxonomica Sinica, 29(1): 73, figs. 58-62 (w.q.) CHINA (Yunnan).

Holotype worker: TL 3.5, HL 0.70, HW 0.55, CI 79, SL 0.30, SI 55, PW 0.37, AL 0.97, EL 0.25, REL 36, PDH 0.28, MTW 0.27, PDI 106, PL 0.43, PH 0.22, DPW 0.17, PLI 50, PWI 38. In full-face view, head rectangular, distinctly longer than broad and slightly narrowed forward. Occipital margin nearly straight, occipital corners rounded. Anterior 1/2 of lateral side slightly concave, posterior 1/2 weakly convex. Mandible narrow and slender, with 3 teeth on the masticatory margin and 1 denticle on the basal margin. Median lobe of clypeus narrow and protruding, anterior margin with 3 small denticles arrange in a line. Apex of scape reached to 1/2 of the distance from antennal socket to occipital corner. Eyes relatively larger, just reached to sides of head. Ocelli absent. In lateral view, pronotum weakly convex, lateral borders bluntly rounded and lack distinct margins. Mesonotum weakly convex and slope down backward. Metanotal groove widely and deeply impressed. Anterior face of propodeum short and sloped, dorsum weakly convex and slope down backward, posterodorsal corner rounded and lacks a distinct boundary between dorsum and declivity, declivity relatively straight and seems longer than dorsum. In dorsal view, pronotum narrowed backward, humeral corners rounded. Metanotum raised up as a small conical prominence between the spiracles. In posterior view, dorsum of pronotum nearly straight, laterodorsal corners roundly prominent and lack distinct margins. In lateral view, anterior peduncle of petiole moderately long and weakly curved down, ventral face moderately concave, petiolar node roundly convex and symmetrical. Postpetiolar node about as high as petiolar node. Basal half of mandible smooth, apical half longitudinally striate. Head, alitrunk, petiole, postpetiole and gaster smooth and shining. Whole body with sparse decumbent pubescence, standing hairs very sparse: 2 pairs on frontal carinae, 1 pair on vertex, 2 pairs on pronotum and 1 pair on postpetiole. Mesonotum, propodeum and petiole lack standing hairs. Gaster with sparse erect hairs. Scapes and tibiae with abundant depressed pubescence, but lack erect hairs. Body color yellow, gaster black.
Paratype workers: TL 3.2-3.6, HL 0.68-0.75, HW 0.48-0.53, CI 69-76, SL 0.30-0.33, SI 58-65, PW 0.33-0.37, AL 0.93-1.00, EL 0.23-0.27, REL 33-39, PDH 0.26-0.32, MTW 0.25-0.28, PDI 94-113, PL 0.43-0.50, PH 0.20-0.23, DPW 0.16-0.18, PLI 46-50, PWI 37-42 (*n*=10). As holotype, but 7 workers with black gaster, 11 workers with yellow gaster.

Holotype: Worker, China: Yunnan Province, Jinghong County, Dadugang Town, Guanping Village, 1120 m, 1997.Ⅷ.7, collected by Guang Zeng from a canopy sample in mountain rain forest of Xishuangbanna National Nature Reserve, No. A97-884.

Etymology: The species name *amargina* combines Latin *a-* (without) + word root *margin* (margin) + suffix *-a* (feminine form), it refers to the pronotum without marginated sides on the dorsum.

正模工蚁： 正面观，头部长方形，长明显大于宽，向前轻微变窄；后缘近平直，后角圆；侧缘前部1/2轻微凹陷，后部1/2轻度隆起。上颚狭窄细长，咀嚼缘具3个齿，基缘具1个小齿。唇基中叶狭窄，突出，其前缘具3个小齿，排成1行。触角柄节末端到达触角窝至头后角间距的1/2处。复眼较大，刚到达头侧缘。缺单眼。侧面观，前胸背板轻度隆起，两侧钝圆，缺明显的隆起边缘。中胸背板轻度隆起，向后坡形降低，后胸沟宽形深凹。并胸腹节前面短，坡形；背部轻度隆起，向后降低；后上角圆，背面和斜面之间缺明显的分界；斜面较平直，长于背面。背面观，前胸背板向后变窄，肩角圆。后胸背板在气门之间升高呈1个小的锥形突。后面观，前胸背板背面近平直，侧上角圆凸，缺明显的边缘。侧面观，腹柄前面小柄中等长，轻度下弯，腹面中度凹陷；腹柄结圆形隆起，对称。后腹柄结约与腹柄结等高。上颚基半部光滑发亮，端半部具纵条纹。头部、胸部、腹柄、后腹柄和后腹部光滑发亮。全身具稀疏倾斜绒毛被，立毛很稀疏：额脊具2对，头顶具1对，前胸背板具2对，后腹柄具1对；中胸背板、并胸腹节和腹柄缺立毛，后腹部具稀疏立毛；柄节和胫节具丰富平伏绒毛被，缺立毛。身体黄色，后腹部黑色。**副模工蚁：** 特征同正模，但是7头副模工蚁后腹部黑色，11头副模工蚁后腹部黄色。

正模： 工蚁，中国云南省景洪县大渡岗乡关坪村，1120m，1997.Ⅷ.7，曾光采于西双版纳国家级自然保护区山地雨林中，No. A97-884。

词源： 该新种的种名"*amargina*"由拉丁语"*a-*"（无）+词根"*margin*"（边缘）+后缀"*-a*"（阴性形式）组成，指前胸背板背面缺隆起的边缘。

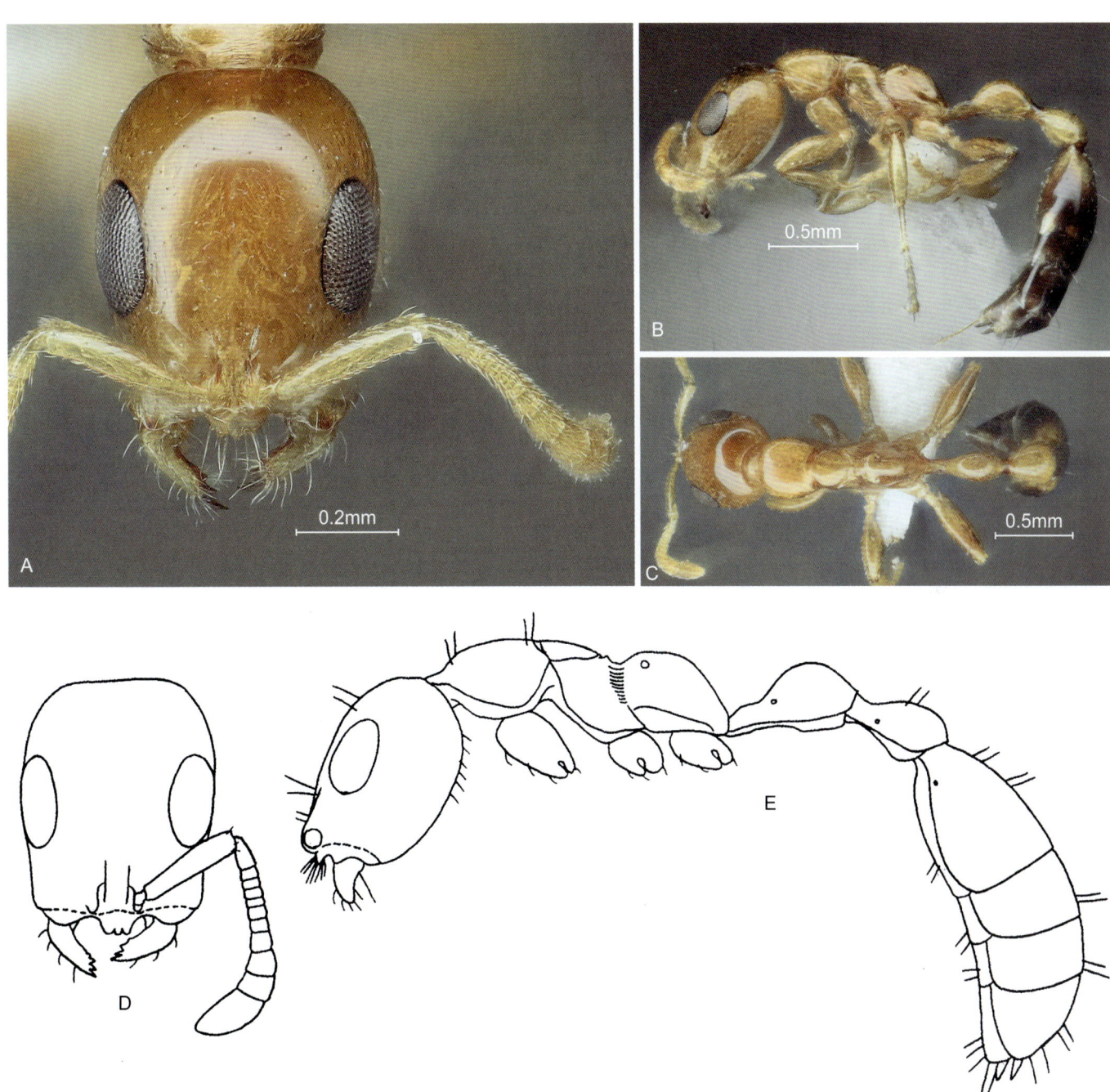

无缘细长蚁 Tetraponera amargina

A, D. 工蚁头部正面观；B, E. 工蚁身体侧面观；C. 工蚁身体背面观（D-E. 引自Xu和Chai, 2004）
A, D. Head of worker in full-face view; B, E. Body of worker in lateral view; C. Body of worker in dorsal view (D-E. cited from Xu & Chai, 2004)

059 凹唇细长蚁
Tetraponera concava Xu & Chai, 2004

Tetraponera concava Xu & Chai, 2004, Acta Zootaxonomica Sinica, 29(1): 65, figs. 6-10 (w.) CHINA (Yunnan).

Holotype worker: TL 9.7, HL 1.90, HW 1.73, CI 91, SL 0.97, SI 56, PW 1.17, AL 2.80, EL 0.60, REL 32, PDH 0.87, MTW 0.97, PDI 90, PL 1.23, PH 0.53, DPW 0.57, PLI 43, PWI 46. In full-face view, head nearly square and weakly narrowed forwards, slightly longer than broad. Occipital margin straight, occipital corners rounded. Anterior half of lateral side slightly concave and posterior half weakly convex. Mandible broad and robust, with 5 teeth on the masticatory margin and 1 denticle on the basal margin. Median lobe of clypeus distinct, broad and shorter, anterior margin concave in the middle, lateral lobelets roundly convex. Apex of scape reached to 3/5 of the distance from antennal socket to occipital corner. Eyes relatively smaller and not reached to sides of head. Vertex with 3 distinct ocelli. In lateral view, pronotum moderately convex, lateral margins blunt. Promesonotal suture impressed. Mesonotum flat and slope down backward. Metanotal groove deeply impressed. Anterior end of propodeum raised up, dorsum relatively straight and weakly slope down backward. Posterodorsal corner blunt. Declivity straight and shorter than the dorsum. In dorsal view, humeral corners of pronotum bluntly angled and much more than 90°. Dorsum of propodeum relatively flat. In posterior view, pronotum moderately convex, laterodorsal corners bluntly angled. In lateral view, petiole with short down-curved peduncle, anteroventral corner with a blunt tooth. Anterior and posterior faces of petiolar node sloped, dorsum long and moderately convex. Postpetiolar node as high as petiolar node. Mandible with longitudinal rugae. Head and pronotum densely and coarsely punctate, distance between punctures about equal to puncture diameter. Punctures on occipital area sparse, distance between punctures about 2-3 times of the puncture diameter. Interspace appears as microreticulations. Sides of pronotum, mesonotum and dorsum of propodeum densely and finely punctured. Sides of propodeal dorsum with dense large punctures. Metanotal groove with short longitudinal rugae. Mesopleuron, metapleuron, and sides of propodeum densely and longitudinally striate. Declivity and dorsum of petiolar node sparsely and finely punctate. Sides of petiole with sparse large punctures. Postpetiole and gaster relatively smooth. Head, prothorax and mesothorax with sparse erect hairs and dense decumbent pubescence. Propodeum, petiole, postpetiole and gaster with abundant erect hairs and dense decumbent pubescence. Scapes and tibiae with sparse suberect hairs and dense decumbent pubescence. Head, pronotum, procoxae, postpetiole and gaster black. Mesothorax, metathorax, propodeum and petiole orange. Mandibles, scapes, coxae, femora and tibiae blackish brown. Flagella and tarsi yellowish brown. **Paratype worker:** TL 10.2, HL 2.10, HW 1.87, CI 89, SL 1.07, SI 57, PW 1.30, AL 3.07, EL 0.63, REL 30, PDH 1.07, MTW 1.13, PDI 94, PL 1.37, PH 0.67, DPW 0.70, PLI 49, PWI 51 (*n*=1). As holotype, but mandibles reddish brown, scapes and lower half of postpetiole yellowish brown, upper half of postpetiole blackish brown.

Holotype: Worker, China: Yunnan Province, Yingjiang County, Tongbiguan Town, Tongbiguan Provincial Nature Reserve, 1100 m, 1995.Ⅳ.23, collected by Bi-lun Yang in mountain rain forest, No. A95-53.

Etymology: The specific epithet refers to the concave anterior margin of median lobe of clypeus.

正模工蚁： 正面观头部近方形，向前轻度变窄，长稍大于宽；后缘平直，后角圆；侧缘前半部轻微凹陷，后半部轻度隆起。上颚宽而粗壮，咀嚼缘具5个齿，基缘具1个小齿。唇基中叶明显，宽而短，前缘中央凹陷，侧叶圆形隆起。柄节顶端到达触角窝至头后角间距的3/5处。复眼较小，未到达头侧缘。头顶具3个明显的单眼。侧面观，前胸背板中度隆起，侧缘钝。前中胸背板缝凹陷。中胸背板平坦，向后坡形降低。后胸沟深凹。并胸腹节前端升高，背面较平直，向后轻度降低，后上角钝；斜面平直，短于背面。背面观，前胸背板肩角钝角状，远大于90°。并胸腹节背面较平坦。后面观，前胸背板中度隆起，侧背角钝角状。侧面观，腹柄前面具短而下弯的小柄，前下角具1个钝齿；腹柄结前面和后面坡形，背面长，中度隆起。后腹柄结与腹柄结等高。上颚具纵皱纹。头部和前胸背板具密集粗糙刻点，刻点间距约等于刻点直径。后头区域刻点稀疏，刻点间距为刻点直径的2～3倍，界面为微网纹。前胸背板侧面、中胸背板和并胸腹节背面具密集细刻点，并胸腹节背面两侧具密集大刻点。后胸沟具短的纵皱纹。中胸侧板、后胸侧板和并胸腹节侧面具密集纵条纹。并胸腹节斜面和腹柄结背面具稀疏细刻点，腹柄侧面具稀疏大刻点。后腹柄和后腹部相对光滑。头部、前胸和中胸具稀疏直立毛和密集倾斜绒毛被；并胸腹节、腹柄、后腹柄和后腹部具丰富直立毛和密集倾斜绒毛被；柄节和胫节具稀疏亚直立毛和密集倾斜绒毛被。头部、前胸背板、前足基节、后腹柄和后腹部黑色；中胸、后胸、并胸腹节和腹柄橙黄色；上颚、柄节、基节、腿节和胫节黑棕色；鞭节和

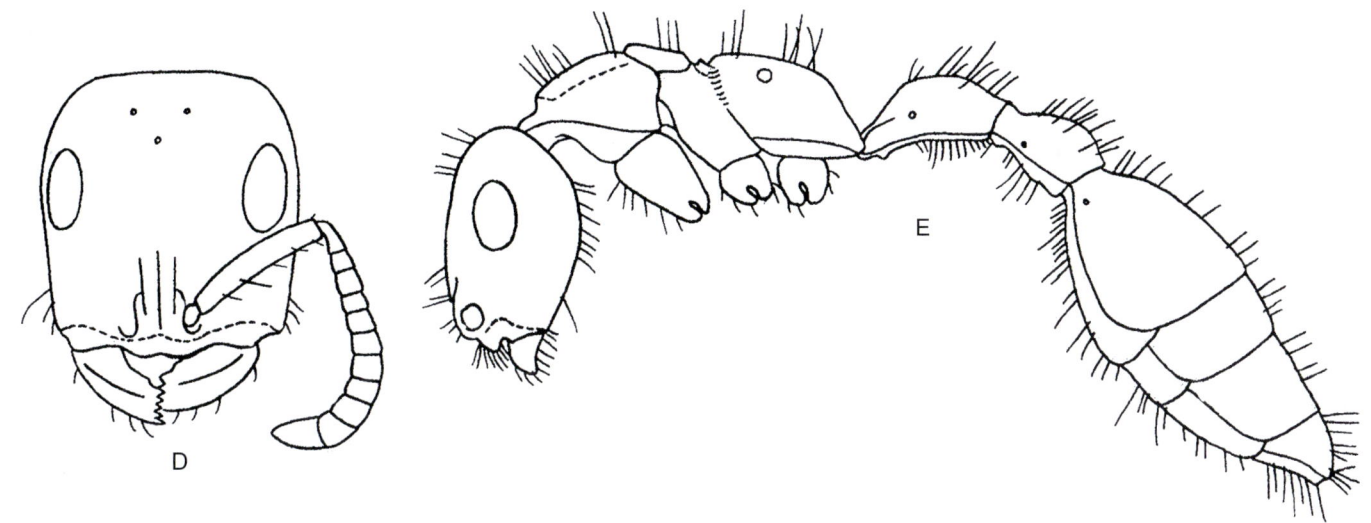

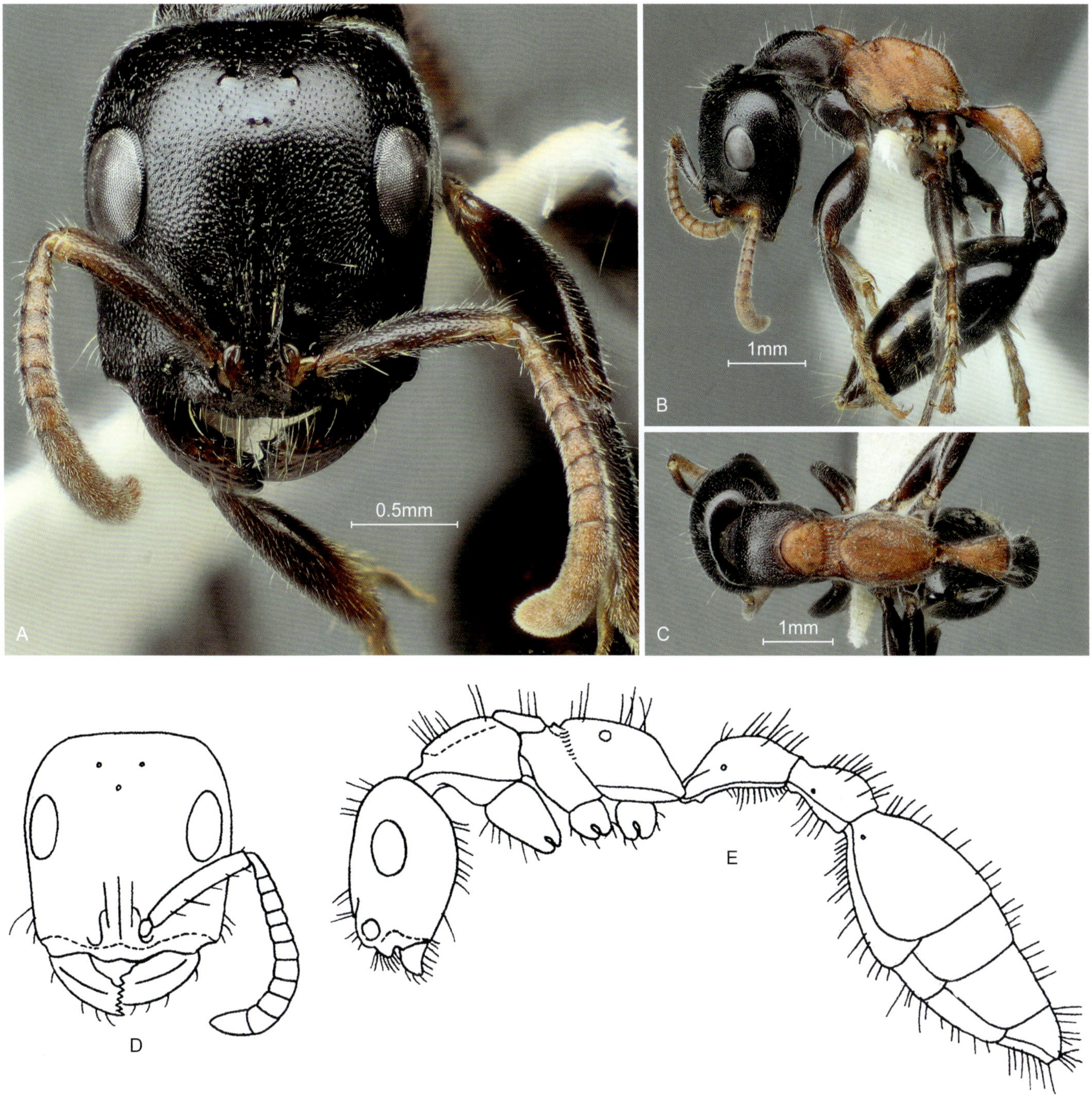

凹唇细长蚁 *Tetraponera concava*

A, D. 工蚁头部正面观；B, E. 工蚁身体侧面观；C. 工蚁身体背面观（D-E. 引自Xu和Chai, 2004）
A, D. Head of worker in full-face view; B, E. Body of worker in lateral view; C. Body of worker in dorsal view (D-E. cited from Xu & Chai, 2004)

跗节黄棕色。**副模工蚁**：特征同正模，但是上颚红棕色，柄节和后腹柄下半部黄棕色，后腹柄上半部黑棕色。

　　正模：工蚁，中国云南省盈江县铜壁关镇铜壁关省级自然保护区，1100m，1995.Ⅳ.23，杨比伦采于山地雨林中，No. A95-53。

　　词源：该新种的种名是指唇基中叶前缘凹陷。

060 隆背细长蚁
Tetraponera convexa Xu & Chai, 2004

Tetraponera convexa Xu & Chai, 2004, Acta Zootaxonomica Sinica, 29(1): 69, figs. 33-37 (w.) CHINA (Yunnan).

Holotype worker: TL 5.7, HL 1.13, HW 0.87, CI 76, SL 0.57, SI 65, PW 0.63, AL 1.67, EL 0.37, REL 32, PDH 0.52, MTW 0.50, PDI 103, PL 0.87, PH 0.35, DPW 0.28, PLI 40, PWI 33. In full-face view, head rectangular, distinctly longer than broad. occipital margin straight, occipital corners rounded. Anterior half of lateral side weakly concave, posterior half moderately convex. Head broadest at the position of posterior margins of eyes, narrowed forward and backward. Mandible narrow and slender, with 3 teeth on the masticatory margin and 2 denticles on the basal margin. Median lobe of clypeus narrow and very short, indistinct, anterior margin straight, without teeth. Apex of scape reached to 4/7 of the distance from antennal socket to occipital corner. Eyes larger, slightly surpassed sides of head. Ocelli absent. In lateral view, pronotum strongly convex, lateral margins distinct. Mesonotum nearly straight and slope down backward, posterior end down curved. Metanotal groove widely and deeply impressed. Anterior face of propodeum short and sloped, dorsum weakly convex and slope down backward, posterodorsal corner rounded, declivity relatively straight and about equal to dorsum. In dorsal view, pronotum narrowed backward, humeral corners roundly prominent. Metanotum raised up as a longitudinal central ridge between the spiracles. In posterior view, pronotum roundly convex, lateral margins distinct and bluntly angled. In lateral view, anterior peduncle of petiole slender and weakly down curved, ventral face weakly concave, petiolar node roundly convex and symmetrical. Postpetiolar node lower than petiolar node. Mandibles finely and longitudinally striate on basal half, coarsely and longitudinally striate on apical half. Head, pronotum and mesonotum very sparsely and finely punctate, interspace smooth. Sides of pronotum smooth. Mesopleuron, metapleuron and sides of propodeum weakly, finely and longitudinally rugulose. Metanotal groove with short longitudinal ridges. Petiole, postpetiole and gaster smooth. Dorsum of anterior peduncle of petiole densely and finely punctate. Whole body with abundant decumbent pubescence, standing hairs very sparse: 2 pairs on frontal carinae, 1 pair on vertex, 2 pair on pronotum, 1 pair on petiole and 1 pair on postpetiole. Mesonotum and propodeum lack standing hairs. Gaster with sparse erect hairs. Scapes and tibiae with sparse suberect hairs and dense subdecumbent pubescence. Body color black. Mandibles, antennae, tibiae and tarsi brownish yellow. Clypeus and femora brownish black. **Paratype workers:** TL 5.2-6.0, HL 1.03-1.10, HW 0.80-0.87, CI 76-79, SL 0.53-0.57, SI 64-67, PW 0.57-0.60, AL 1.47-1.63, EL 0.37-0.40, REL 34-36, PDH 0.47-0.50, MTW 0.47-0.50, PDI 93-100, PL 0.80-0.90, PH 0.30-0.35, DPW 0.25-0.30, PLI 35-40, PWI 31-35 (*n*=6). As holotype.

Holotype: Worker, China: Yunnan Province, Jinghong County, Dadugang Town, Guanping Village, 1120 m, 1997.Ⅷ.7, collected by Tai-yong Liu from a canopy sample in mountain rain forest of Xishuangbanna National Nature Reserve, No. A97-975.

Etymology: The specific epithet refers to the strongly convex pronotum.

正模工蚁： 正面观，头部长方形，长明显大于宽；后缘平直，后角圆；侧缘前半部轻度凹陷，后半部中度隆起。头部在复眼后缘位置最宽，向前与向后均变窄。上颚窄，细长，咀嚼缘具3个齿，基缘具2个小齿。唇基中叶窄且很短，不明显，前缘平直，缺齿。柄节末端到达触角窝至头后角间距的4/7处。复眼较大，稍超出头侧缘。缺单眼。侧面观，前胸背板强烈隆起，侧边缘明显。中胸背板近平直，向后坡形降低，后端下弯。后胸沟宽形深凹。并胸腹节前面短，坡形；背面轻度隆起，向后坡形降低，后上角圆；斜面较平直，约与背面等长。背面观，前胸背板向后变窄，肩角圆形突出。后胸背板在气门之间升高成1条纵脊。后面观，前胸背板圆形隆起，侧边缘明显，钝角状。侧面观，腹柄前面小柄细长，轻度下弯，腹面轻度凹陷；腹柄结圆形隆起，对称。后腹柄结低于腹柄结。上颚基半部具细纵条纹，端半部具粗糙纵条纹。头部、前胸背板和中胸背板具很稀疏的细刻点，界面光滑，前胸背板侧面光滑。中胸侧板、后胸侧板和并胸腹节侧面具弱的细纵皱纹。后胸沟具短的纵脊。腹柄、后腹柄和后腹部光滑。腹柄小柄背面具细密刻点。全身具丰富倾斜绒毛被，立毛很稀疏：额脊具2对，头顶具1对，前胸背板具2对，腹柄具1对，后腹柄具1对。中胸背板和并胸腹节缺立毛，后腹部具稀疏立毛。柄节和胫节具稀疏亚直立毛和密集倾斜绒毛被。身体黑色，上颚、触角、胫节和跗节棕黄色，唇基和腿节棕黑色。**副模工蚁：** 特征同正模。

正模： 工蚁，中国云南省景洪县大渡岗乡关坪村，1120m，1997.Ⅷ.7，柳太勇采于西双版纳国家级自然保护区山地雨林树冠样中，No. A97-975。

词源： 该新种的种名是指前胸背板强烈隆起。

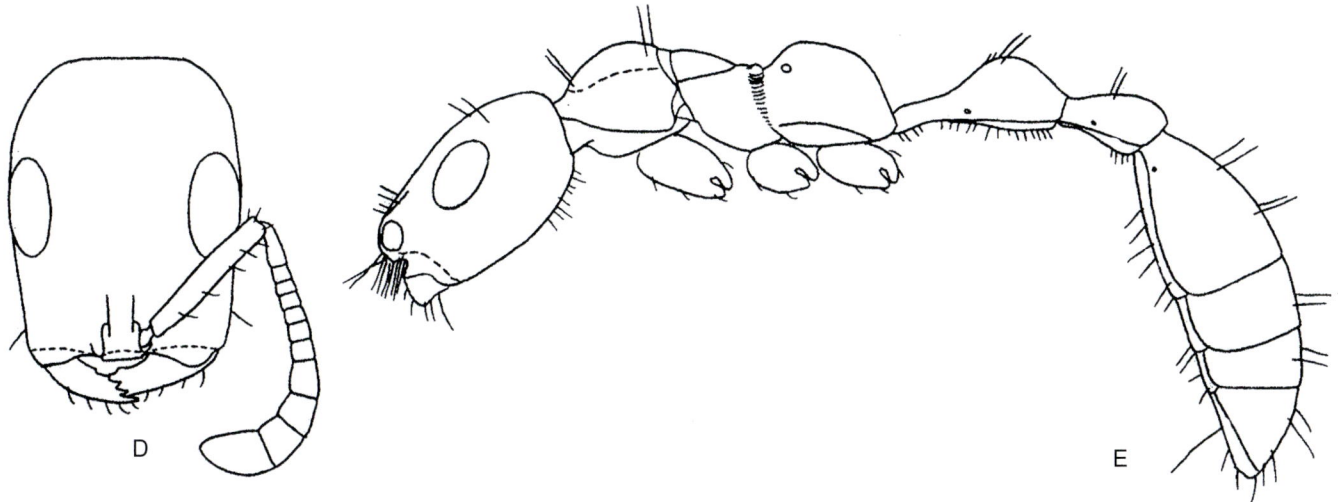

隆背细长蚁 Tetraponera convexa

A, D. 工蚁头部正面观；B, E. 工蚁身体侧面观；C. 工蚁身体背面观（D-E. 引自Xu和Chai, 2004）
A, D. Head of worker in full-face view; B, E. Body of worker in lateral view; C. Body of worker in dorsal view (D-E. cited from Xu & Chai, 2004)

061 叉唇细长蚁
Tetraponera furcata Xu & Chai, 2004

Tetraponera furcata Xu & Chai, 2004, Acta Zootaxonomica Sinica, 29(1): 70, figs. 43-47 (w.) CHINA (Yunnan).

Holotype worker: TL 4.4, HL 0.90, HW 0.62, Cl 69, SL 0.40, SI 65, PW 0.43, AL 1.17, EL 0.30, REL 33, PDH 0.40, MTW 0.35, PDI 114, PL 0.63, PH 0.26, DPW 0.23, PLI 42, PWI 37. In full-face view, head roughly rectangular, distinctly longer than broad. Occipital margin short and relatively straight, occipital corners rounded. Anterior 2/3 of lateral side weakly concave. Mandible narrow and slender, with 3 teeth on the masticatory margin and 1 denticle on basal margin. Median lobe of clypeus narrow and protruding, anterolateral corners furnished with a pair of elongate teeth appears fork-like, anterior margin between the teeth straight. Apex of scape reached to 3/5 of the distance from antennal socket to occipital corner. Eyes large and surpassed sides of head. Ocelli absent. In lateral view, pronotum slightly convex, lateral borders distinctly marginate. Mesonotum straight and slope down backward. Metanotal groove widely and deeply impressed. Anterior face of propodeum short and sloped, dorsum slightly convex and slope down backward, posterodorsal corner rounded, declivity straight and slightly longer than dorsum. In dorsal view, pronotum narrowed backward, humeral corners roundly prominent. Metanotum raised up as a small conical prominence between the spiracles. In posterior view, pronotum weakly convex, laterodorsal corners sharp and rightly angled. In lateral view, anterior peduncle of petiole slender and weakly curved down, ventral face weakly concave, petiolar node roundly convex with anterodorsal corner a little higher. Postpetiolar node slightly lower than petiolar node. Mandibles sparsely, finely and longitudinally striate. Head, pronotum and mesonotum very sparsely and finely punctate, interspace smooth. Mesopleuron, metapleuron and sides of propodeum weakly, finely and longitudinally rugulose. Metanotal groove with short longitudinal ridges. Petiole, postpetiole and gaster smooth. Dorsum of anterior peduncle of petiole densely and finely punctate. Whole body with sparse decumbent pubescence, standing hairs very sparse: 2 pairs on frontal carinae, 1 pair on vertex, 2 pairs on pronotum, 1 pair on petiole and 1 pair on postpetiole. Mesonotum and propodeum lack standing hairs. Gaster with sparse erect hairs. Scapes and tibiae with abundant decumbent pubescence, but lack standing hairs. Body color black. Mandibles and clypeus reddish brown. Mesofemora, metafemora and petiole blackish brown. Antennae, tibiae and tarsi yellow. Profemora, mesocoxae, metacoxae and postpetiole yellowish brown. **Paratype workers:** TL 4.1-4.7, HL 0.85-0.97, HW 0.62-0.70, Cl 71-77, SL 0.40-0.43, SI 60-65, PW 0.43-0.50, AL 1.17-1.30, EL 0.30-0.33, REL 34-39, PDH 0.37-0.40, MTW 0.33-0.40, PDI 100-110, PL 0.57-0.70, PH 0.25-0.30, DPW 0.22-0.25, PLT 43-45, PWI 35-38 (*n*=6). As holotype.

Holotype: Worker, China: Yunnan Province, Jinghong County, Dadugang Town, Guanping Village, 1120 m, 1998.Ⅲ.5, collected by Tai-yong Liu from a canopy sample in mountain rain forest of Xishuangbanna National Nature Reserve, No. A98-130.

Etymology: The species name *furcata* combines Latin *furc-* (fork) + suffix *-ata* (feminine form, with), it refers to the paired fork-like teeth on the median lobe of clypeus.

正模工蚁： 正面观，头部近长方形，长明显大于宽；后缘短，较平直，后角圆；侧缘前部2/3轻度凹陷。上颚窄而细长，咀嚼缘具3个齿，基缘具1个小齿。唇基中叶窄而突出，前侧角具1对伸长的齿，呈叉状，前缘在两齿之间平直。柄节末端到达触角窝至头后角间距的3/5处。复眼大，超出头侧缘。缺单眼。侧面观，前胸背板轻度隆起，侧缘明显具边缘。中胸背板平直，向后坡形降低。后胸沟宽形深凹。并胸腹节前面短坡形，背部轻度隆起，向后坡形降低，后上角圆；斜面平直，稍长于背部。背面观，前胸背板向后变窄，肩角圆突。后胸背板在气门之间升高形成1个小的圆锥形隆起。后面观，前胸背板轻度隆起，背侧角锋利，直角形。侧面观，腹柄前面小柄细长，轻度下弯，腹面轻度凹陷；腹柄结圆形隆起，前上角稍高。后腹柄结稍低于腹柄结。上颚具稀疏细纵条纹。头部、前胸背板和中胸背板具很稀疏的细刻点，界面光滑。中胸侧板、后胸侧板和并胸腹节侧面具弱的细纵皱纹。后胸沟具短纵脊。腹柄、后腹柄和后腹部光滑。腹柄的小柄背面具细密刻点。全身具稀疏倾斜绒毛被，立毛很稀疏：额脊具2对，头顶具1对，前胸背板具2对，腹柄具1对，后腹柄具1对。中胸背板和并胸腹节缺立毛，后腹部具稀疏直立。柄节和胫节具丰富倾斜绒毛被，缺立毛。身体黑色，上颚和唇基红棕色，中足腿节、后足腿节和腹柄黑棕色，触角、胫节和跗节黄色，前足腿节、中足基节、后足基节和后腹柄黄棕色。**副模工蚁：** 特征同正模。

正模： 工蚁，中国云南省景洪县大渡岗乡关坪村，1120m，1998.Ⅲ.5，柳太勇采于西双版纳国家级自然保护区山地雨林树冠样中，No. A98-130。

叉唇细长蚁 *Tetraponera furcata*

A, D. 工蚁头部正面观；B, E. 工蚁身体侧面观；C. 工蚁身体背面观（D-E. 引自Xu和Chai, 2004）
A, D. Head of worker in full-face view; B, E. Body of worker in lateral view; C. Body of worker in dorsal view (D-E. cited from Xu & Chai, 2004)

词源： 该新种的种名"*furcata*"由拉丁语"*furc-*"（叉）+ 后缀"*-ata*"（阴性形式，具有）组成，指唇基中叶具成对的叉状齿。

062 尖唇细长蚁
Tetraponera protensa Xu & Chai, 2004

Tetraponera protensa Xu & Chai, 2004, Acta Zootaxonomica Sinica, 29(1): 71, figs. 48-52 (w.) CHINA (Yunnan).

Holotype worker: TL 4.9, HL 0.97, HW 0.77, CI 79, SL 0.40, SI 52, PW 0.50, AL 1.30, EL 0.30, REL 31, PDH 0.43, MTW 0.40, PDI 108, PL 0.63, PH 0.30, DPW 0.27, PLI 47, PWI 42. In full-face view, head rectangular and distinctly longer than broad. Occipital margin weakly concave, occipital corners roundly prominent. Anterior half of lateral side slightly concave, posterior half weakly convex. Mandible narrow and slender, with 3 teeth on the masticatory margin and 2 denticles on basal margin. Median lobe of clypeus narrow and protruding, anterior margin with 3 teeth, the middle tooth exceptionally developed and much longer than the lateral ones. Apex of scape reached to 1/2 of the distance from antennal socket to occipital corner. Eyes moderately large and just reached to sides of head. Ocelli absent. In lateral view, pronotum moderately convex, lateral borders distinctly marginate. Mesonotum weakly convex and slope down backward. Metanotal groove widely and deeply impressed. Anterior face of propodeum sloped, dorsum moderately convex and slope down backward, anterodorsal and posterodorsal corners rounded, declivity relatively straight. In dorsal view, pronotum narrowed backward, humeral corners rounded. Metanotum raised up as a small conical prominence between the spiracles. In posterior view, pronotum moderately convex, laterodorsal corners bluntly angled. In lateral view, anterior peduncle of petiole moderately long and slightly curved down, ventral face weakly concave, petiolar node roundly convex and symmetrical. Postpetiolar node slightly lower than petiolar node. Mandibles finely and longitudinally striate. Head, prothorax and mesothorax very sparsely and finely punctate, interspace smooth. Metathorax and sides of propodeum weakly, finely and longitudinally rugulose. Dorsum of propodeum, petiole, postpetiole and gaster smooth and shining. Metanotal groove with short longitudinal ridges. Dorsum of peduncle of petiole densely and finely punctate. Whole body with sparse decumbent pubescence, standing hairs very sparse: 2 pairs on frontal carinae, 1 pair on vertex, 2 pairs on pronotum, 1 pair on petiole and 2 pairs on postpetiole. Mesonotum and propodeum lack standing hairs. Gaster with sparse erect hairs. Scapes and tibiae with abundant decumbent pubescence and sparse suberect hairs. Body color black. Clypeus and femora blackish brown. Mandibles and tibiae brown. Antennae and tarsi brownish yellow. **Paratype worker:** TL 4.8, HL 1.00, HW 0.78, CI 78, SL 0.42, SI 53, PW 0.50, AL 1.33, EL 0.30, REL 30, PDH 0.43, MTW 0.42, PDI 104, PL 0.63, PH 0.32, DPW 0.28, PLI 50, PWI 45 (*n*=1). As holotype.

Holotype: Worker, China: Yunnan Province, Jinghong County, Dadugang Town, Guanping Village, 1120 m, 1998.Ⅲ.5, collected by Tai-yong Liu from a canopy sample of mountain rain forest of Xishuangbanna National Nature Reserve, No. A98-129.

Etymology: The species name *protensa* combines Latin *protens-* (extend) + suffix *-a* (feminine form), it refers to the exceptionally extended middle tooth of the median lobe of clypeus which is much longer than the lateral teeth.

正模工蚁： 正面观，头部长方形，长明显大于宽；后缘轻度凹陷，后角圆凸；侧缘前半部轻度凹陷，后半部轻度隆起。上颚窄而细长，咀嚼缘具3个齿，基缘具2个小齿。唇基中叶窄而突出，前缘具3个齿，中齿非常发达，比侧齿长得多。柄节顶端到达触角窝至头后角间距的1/2处。复眼中等大小，刚到达头侧缘。缺单眼。侧面观，前胸背板中度隆起，侧缘明显具边缘。中胸背板轻度隆起，向后坡形降低。后胸沟宽形深凹。并胸腹节前面坡形，背面中度隆起，向后坡形降低，前上角和后上角圆，斜面较平直。背面观，前胸背板向后变窄，肩角圆。后胸背板在气门之间升高形成1个小的圆锥形隆起。后面观，前胸背板中度隆起，背侧角钝角状。侧面观，腹柄前面小柄中等长，轻微下弯，腹面轻度凹陷；腹柄结圆形隆起，对称。后腹柄结稍低于腹柄结。上颚具细纵条纹。头部、前胸和中胸具很稀疏的细刻点，界面光滑。后胸和并胸腹节侧面具弱的细纵皱纹。并胸腹节背面、腹柄、后腹柄和后腹部光滑发亮。后胸沟具短纵脊。腹柄的小柄背面具细密刻点。全身具稀疏倾斜绒毛被，立毛很稀疏：额脊具2对，头顶具1对，前胸背板具2对，腹柄具1对，后腹柄具2对。中胸背板和并胸腹节缺立毛。后腹部具稀疏立毛。柄节和胫节具丰富倾斜绒毛被和稀疏亚直立毛。身体黑色，唇基和腿节黑棕色，上颚和胫节棕色，触角和跗节棕黄色。**副模工蚁：** 特征同正模。

正模： 工蚁，中国云南省景洪县大渡岗乡关坪村，1120m，1998.Ⅲ.5，柳太勇采于西双版纳国家级自然保护区山地雨林树冠样中，No. A98-129。

词源： 该新种的种名"*protensa*"由拉丁语"*protens-*"（伸长）+后缀"*-a*"（阴性形式）组成，指唇基中叶的中齿特别伸长，比侧齿长得多。

Individual Theory 分 论

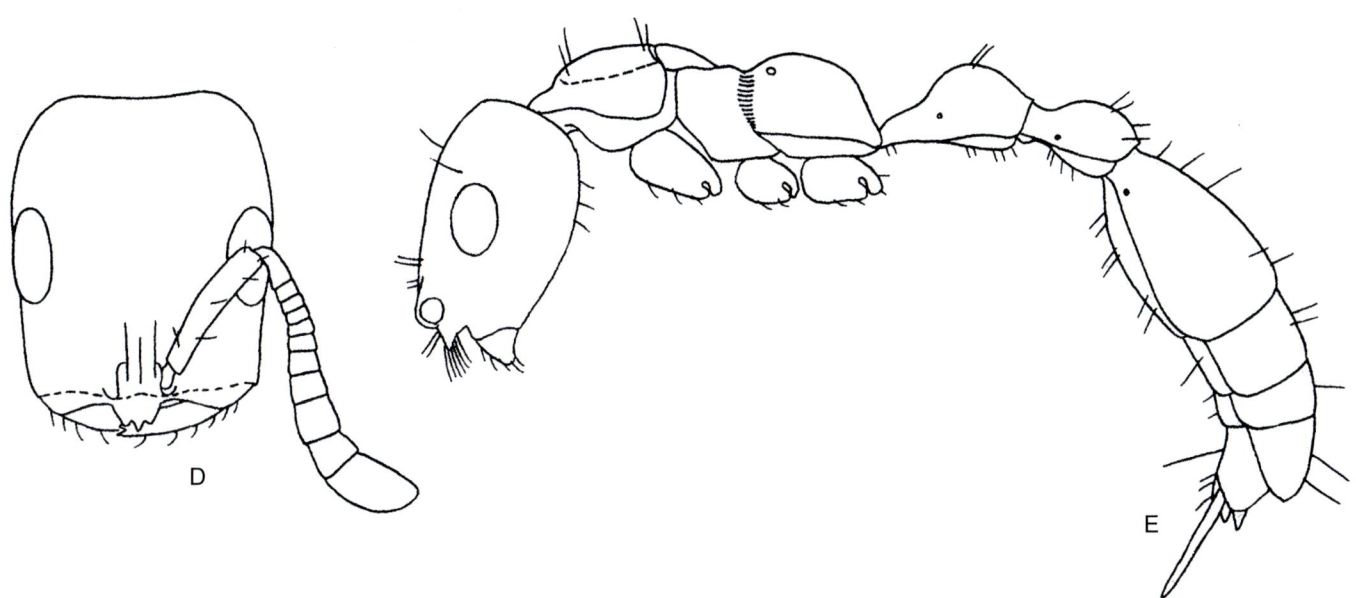

尖唇细长蚁 *Tetraponera protensa*

A, D. 工蚁头部正面观；B, E. 工蚁身体侧面观；C. 工蚁身体背面观（D-E. 引自Xu和Chai, 2004）
A, D. Head of worker in full-face view; B, E. Body of worker in lateral view; C. Body of worker in dorsal view (D-E. cited from Xu & Chai, 2004)

063 大禹圆鳞蚁
Epitritus dayui Xu, 2000

Epitritus dayui Xu, 2000d, Entomotaxonomia, 22(4): 299, figs. 9-12 (w.) CHINA (Yunnan).

Holotype worker: TL 2.6, HL 0.63, HW 0.60, CI 96, SL 0.28, SI 46, PW 0.33, AL 0.70, ED 0.04. In full-face view, head about as broad as long, anterior half distinctly narrower than the posterior half. Occipital margin roundly incised in the middle. Occipital corners rounded. Mandibles slender and roundly curved, masticatory margin with 7 teeth, the basal and apical teeth elongate and spine-like, the basal tooth about as long as the width of the masticatory margin, the apical tooth about as long as half length of the basal one. The other 5 teeth between the basal and apical ones minute. Labrum visible, anterior margin with a pair of cone-like projections. Anterior margin of clypeus nearly straight, slightly convex in the middle. Antenna with 6 segments. Scape depressed, subbasal lobe nearly formed a right angle. The 3rd and 4th segments small and about equal, length of the two segments combined about equal to length of the 2nd and 5th ones. The apical segment very long, about 1.5 times as long as the rest of the flagellum. Eye minute, with 4 facets. In lateral view, promesonotum relatively flat, mesonotum rounded and lowered down posteriorly. Promesonotal suture fine and distinct. Metanotal groove shallowly depressed. Propodeum slope-like and weakly convex. Propodeal lobes blunt at apex. Dorsum of petiolar node roundly convex. Dorsum of postpetiolar node evenly convex. In dorsal view, petiolar node nearly square, postpetiolar node broader than long and nearly semicircular, petiolar node width : postpetiolar node width = 2 : 3. Sides of declivity of propodeum with spongiform longitudinal ridges. Ventral face of petiole, ventral and posterior faces of postpetiole with spongiform appendage. Head, alitrunk and petiolar node densely and finely puncatured, dim. Sides of mesothorax and metathorax, postpetiolar node and gaster smooth and shining. Head, mandibles, scapes, alitrunk, petiolar node, dosum of femora and tibiae with circular or subcircular flattened hairs. Sides of pronotum, postpetiolar node and gaster with clavate hairs. Ventral face of head, mandibles, flagella, legs and ventral face of gaster with dense decumbent pubescence. Body in color yellowish brown, gaster dark yellowish brown.

Holotype: Worker, China: Yunnan Province, Mengla County, Mengla Town, Nangongshan Mountain, 1380 m, 1997.IX.16, collected by Zheng-hui Xu in monsoon evergreen broadleaf forest, No. A97-2677.

Etymology: The new species is named after Dayu, the founder of the Xia Dynasty (2070 B.C. -1600 B.C.), who was famous for controlling flooding of the Yellow River.

正模工蚁： 正面观，头部长宽约相等，前半部明显窄于后半部；后缘中部圆形凹陷，后角圆。上颚细长，圆形弯曲，咀嚼缘具7个齿，基齿和端齿伸长呈刺状，基齿长度约等于咀嚼缘宽度，端齿长度约为基齿长度的一半；基齿和端齿之间的其他5个齿微小。唇基可见，前缘具1对锥状突起。唇基前缘近平直，中部轻微隆起。触角6节，柄节压扁，亚基叶近直角形；第3节和第4节小，近等长，这两节的合长约与第2节和第5节等长；端节很长，约为鞭节其余各节合长的1.5倍。复眼微小，具4个小眼。前中胸背板较平坦，中胸背板圆，向后降低。前中胸背板缝细而明显，后胸沟浅凹。并胸腹节坡形，轻度隆起。并胸腹节侧叶顶端钝。腹柄结背面圆形隆起。后腹柄结背面均匀隆起。背面观，腹柄结近方形；后腹柄结宽大于长，近半圆形；腹柄结宽：后腹柄结宽=2：3。并胸腹节斜面两侧具海绵状纵脊；腹柄腹面、后腹柄的腹面和后面具海绵状附属物。头部、胸部和腹柄结具细密刻点，暗。中后胸侧面、后腹柄结和后腹部光滑发亮。头部、上颚、柄节、胸部、腹柄结、腿节背面和胫节具圆形或近圆形扁平毛；前胸背板侧面、后腹柄结和后腹部具棒状毛；头部腹面、上颚、鞭节、足和后腹部腹面具密集倾斜绒毛被。身体黄棕色，后腹部暗黄棕色。

正模： 工蚁，中国云南省勐腊县勐腊镇南贡山，1380m，1997.IX.16，徐正会采于季风常绿阔叶林中，No. A97-2677。

词源： 该新种以夏朝（公元前2070—1600年）开国君主大禹命名，他以治理黄河水患而闻名。

Individual Theory 分 论

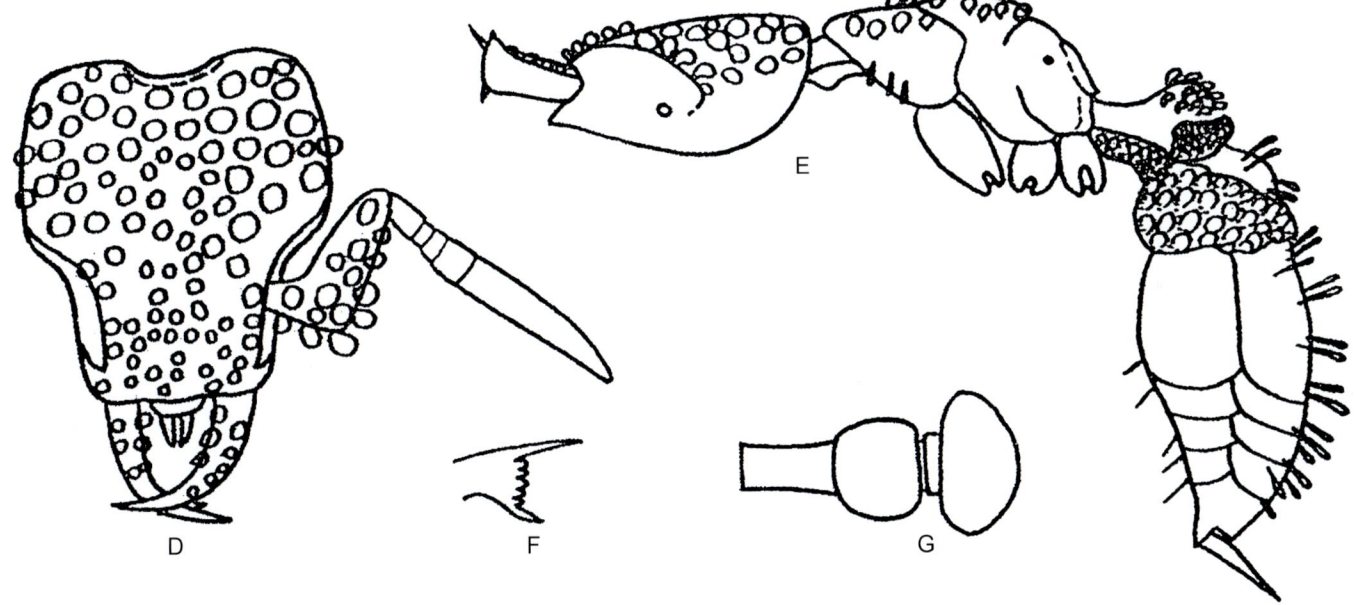

大禹圆鳞蚁 *Epitritus dayui*

A, D. 工蚁头部正面观；B, E. 工蚁身体侧面观；C. 工蚁身体背面观；F. 工蚁上颚端部前面观；G. 工蚁腹柄和后腹柄背面观（D-G.引自Xu, 2000d）

A, D. Head of worker in full-face view; B, E. Body of worker in lateral view; C. Body of worker in dorsal view; F. Mandible apex of worker in front view; G. Petiole and postpetiole of worker in dorsal view (D-G. cited from Xu, 2000d)

平背高黎贡蚁
Gaoligongidris planodorsa Xu, 2012

Gaoligongidris planodorsa Xu, 2012b, Sociobiology, 59(2): 336, figs. 1-7 (w.) CHINA (Yunnan).

Holotype worker: TL 2.6, HL 0.67, HW 0.65, CI 98, SL 0.43, SI 67, ED 0.10, PW 0.42, AL 0.73. In full-face view, head square, as broad as long. Occipital margin nearly straight, occipital corners roundly prominent. Sides weakly convex. Mandibles subtriangular, masticatory margin with 6 teeth, which decrease in size from apex to base. Median portion of clypeus bicarinate and extruding forward, anterior margin nearly straight. Posterior clypeal extension about as broad as the frontal lobes. Frontal carinae and antennal scrobes absent. Antennae short, 11-segmented, apex of scape reached to 3/4 of the distance from antennal socket to occipital corner. The apical 3 segments form the antennal club, apical segment 2.5 times as long as the preceding one; the third segment counting from apex weakly enlarged, about 1/2 length of the second one. Eyes moderately large, situated in front of the midpoints of the sides of the head, with about 12 ommatidia in the maximum diameter. In lateral view, pronotum and mesonotum form a high plateau which gently slopes down backward, posterodorsal corner of mesonotum bluntly angled and steeply slopes down to the metanotal groove. Promesonotal suture absent. Metanotal groove deeply impressed. Propodeum low, dorsum short and straight. Propodeal spines long, sharp, and straight, laterally compressed, about as long as propodeal dorsum. Declivity weakly concave, sides marginate. Propodeal spiracles large and circular, well before the declivity margins, and high up on the sides. Propodeal lobes small and bluntly angled at apex. Metapleural gland bullae large and roughly triangular. Petiolar node triangular, anterior face straight, posterior face weakly convex, anterior peduncle longer than the node, ventral face weakly concave under the node, and weakly convex before the concavity, subpetiolar process absent. Postpetiolar dorsum roundly convex, slightly lower than petiolar node, ventral face with 2 small convexities. In dorsal view, promesonotum nearly triangular, narrowed backward. Sides of mesonotum marginate, lateroposterior corners rightly angled. Postpetiolar node about 1.4 times as broad as petiolar node, sides of postpetiole roundly convex. Mandibles densely longitudinally striate. Head densely reticulate, but the vertex longitudinally striate, interfaces finely punctured. Alitrunk densely coarsely punctured, but pronotal dorsum densely reticulate. Petiole and postpetiole densely finely punctured, but dorsum of postpetiolar node finely longitudinally striate. Gaster smooth and shining, but the first tergite with short basal costulae, which distinctly shorter than the postpetiolar node. Dorsum of head with dense erect short hairs. Dorsum of alitrunk with sparse erect to suberect longer hairs and abundant decumbent pubescence. Petiole, postpetiole, and gaster with abundant suberect hairs and decumbent pubescence. Scapes and tibiae with dense subdecumbent to decumbent short hairs. Color yellowish brown, but dorsum of head and middle portion of gaster blackish brown. **Paratype workers:** TL 2.3-2.6, HL 0.60-0.67, HW 0.57-0.65, CI 89-100, SL 0.37-0.43, SI 63-67, ED 0.10-0.12, PW 0.37-0.42, AL 0.70-0.80 (*n*=10). As holotype, but color brownish yellow to yellowish brown.

Holotype: Worker, China: Yunnan Province, Tengchong County, Jietou Town, Datang Village, 2000 m, 1999.V.1, collected by Ji-guai Li from a soil sample in subalpine moist evergreen broadleaf forest on the west slope of Gaoligong Mountain, No. A99-195.

Etymology: The name of the new species is descriptive of the "*plane*" promesonotal "*dorsum*" of alitrunk in lateral view.

正模工蚁： 正面观，头部方形，长宽相等；后缘近平直，后角圆凸；侧缘轻度隆起。上颚亚三角形，咀嚼缘具6个齿，从端部向基部依次变小。唇基中部具双脊，向前突出，前缘近平直；唇基后延至额叶之间部分约与额叶等宽。缺额脊和触角沟。触角短，11节，柄节顶端到达触角窝至头后角间距的3/4处；触角棒3节，端节长为亚端节的2.5倍；从端部数第3节轻度膨大，约为第2节长度的1/2。复眼中等大，位于头侧缘中点之前，最大直径上约具12个小眼。侧面观，前胸背板和中胸背板形成1个高台，向后缓坡形降低。中胸背板后上角钝角状，陡坡状进入后胸沟。前中胸背板缝缺失，后胸沟深凹。并胸腹节低，背面短而直；并胸腹节刺长而直，尖锐，左右压扁，约与背面等长；斜面轻度凹陷，两侧具边缘；并胸腹节气门大而圆，远在斜面边缘之前，在侧面中上部。并胸腹节侧叶小，末端钝角状。后胸侧板腺泡大，近三角形。腹柄结三角形，前面平直，后面轻度隆起；前面小柄长于腹柄结，腹柄结下方腹面轻度凹陷，凹陷之前轻度隆起，缺腹柄下突。后腹柄背面圆形隆起，稍低于腹柄结，腹面具2个小的隆起。背面观，前中胸背板近三角形，向后变窄。中胸背板两侧具边缘，侧后角直角形。后腹柄结约为腹柄结宽的1.4倍，后腹柄侧面圆形隆起。上颚具密集纵条纹。头部具密集网纹，但头顶具纵条纹，界面具细刻点。胸部具密集粗糙刻点，但前胸背板背面具密集网状。腹柄和后腹柄具细密刻点，但后腹柄结背面具细纵条纹。后腹部光滑发亮，但第1节背板具短的基纵脊，基纵脊明显短于后腹柄结。头部背面具密集直立短毛。胸部背面具稀疏直立、亚直立长毛和

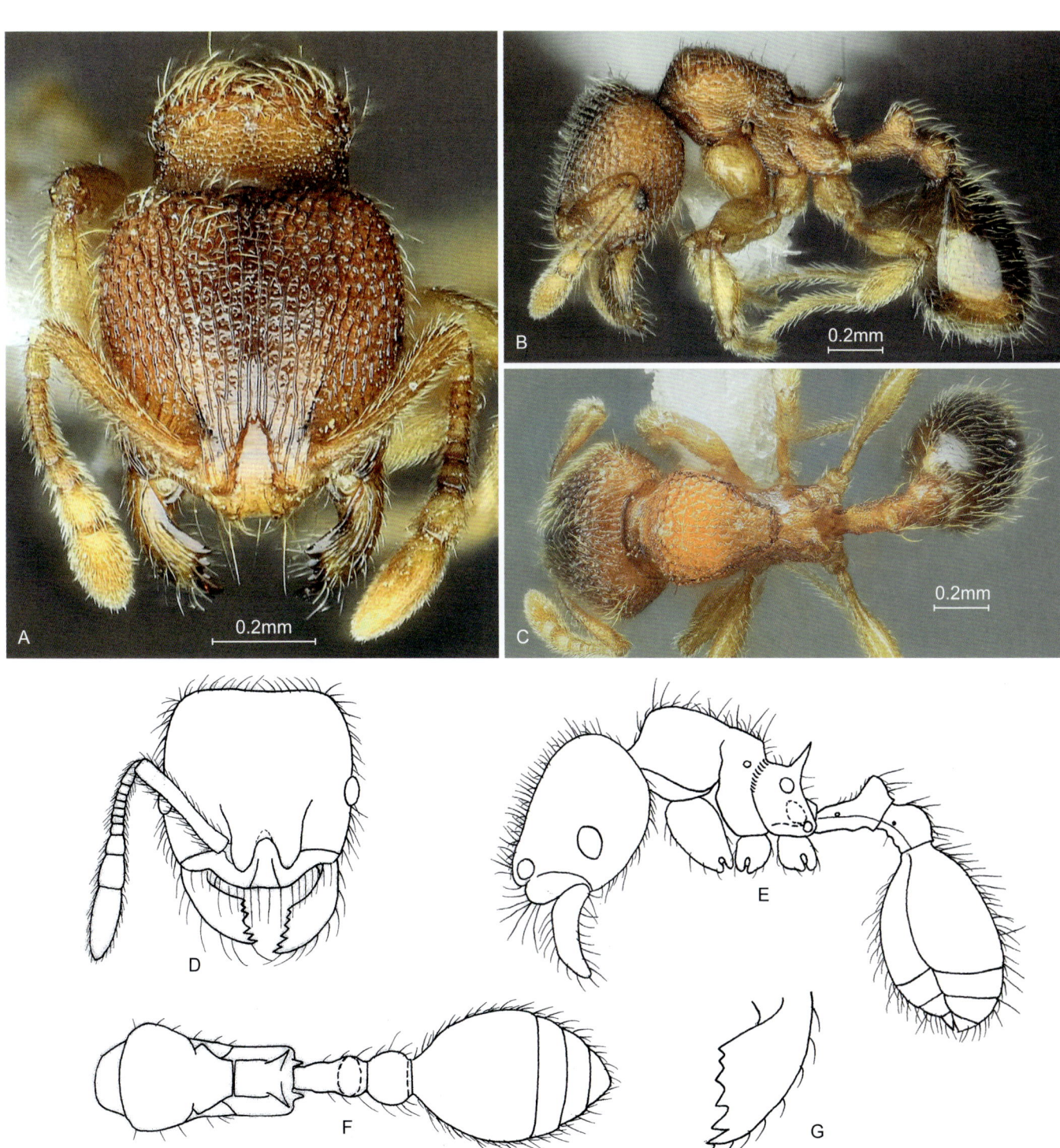

平背高黎贡蚁 *Gaoligongidris planodorsa*

A, D. 工蚁头部正面观；B, E. 工蚁身体侧面观；C, F. 工蚁身体背面观；G. 工蚁上颚背面观（D-G. 引自Xu, 2012b）

A, D. Head of worker in full-face view; B, E. Body of worker in lateral view; C, F. Body of worker in dorsal view; G. Mandible of worker in dorsal view (D-G. cited from Xu, 2012b)

丰富倾斜绒毛被。腹柄、后腹柄和后腹部具丰富亚直立毛和倾斜绒毛被。柄节和胫节具密集亚倾斜、倾斜短毛。身体黄棕色，但头部背面和后腹部中部黑棕色。**副模工蚁**：特征同正模，但身体颜色棕黄色至黄棕色。

正模：工蚁，中国云南省腾冲县界头乡大塘村，2000m，1999.Ⅴ.1，李继乖采于高黎贡山西坡中山湿性常绿阔叶林土壤样中，No. A99-195。

词源：该新种的名称描述侧面观胸部具有"平坦"的前中胸背板"背面"。

阿诗玛无刺蚁
Kartidris ashima Xu & Zheng, 1995

Kartidris ashima Xu & Zheng, 1995, Entomotaxonomia, 17(2): 144, figs. 6, 8, 9 (w.) CHINA (Yunnan).

Holotype worker: TL 3.5, HL 0.85, HW 0.83, CI 97, SL 0.78, SI 94, PW 0.53, AL 1.10, ED 0.18. In full-face view, head longer than broad, occipital margin roundly convex. Mandibles have fine longitudinal striae, and 5 teeth on the masticatory margin. Median portion of clypeus strongly convex, anterior margin roundly extruding. Antennae have 12 segments, the last 3 forming the clubs, the apical one the longest, about 2 times the length of the segment next to it. Maximum diameter of eye with 8-9 ommatidia. Vertex with a depressed pit in the center, in lateral view, the vertex slightly impressed. Pronotum high, dorsum roundly convex. Mesonotum lowering down as a slope posteriorly. Metanotal groove deeply depressed. Propodeum low, dorsum flat in lateral view. Propodeal spines absent, the posterolateral angles of propodeum more extruding, median portion between them longitudinally depressed. Petiole with long peduncle anteriorly, and a low triangular process anteroventrally; the node high, rounded above. Postpetiolar node high and convex, inclined posteriorly, with the dorsum slightly longitudinally depressed in the middle. Head and body smooth and shining. The portions between the antennal fossae and the eyes have weak reticulate-rugulae; mesopleura with fine dense reticulate-rugulae; metapleura with longitudinal striae. Sternites of pedicel segments with fine dense punctulations. Head and body with abundant erect or suberect hairs, hairs on the head dense. Eyes with abundant anteriorly curved short hairs arising between the facets. Dorsal surfaces of scapes and hind tibiae with suberect long and short hairs. Body yellowish brown, head and gaster brown. **Paratype workers:** TL 3.4-3.8, HL 0.83-0.88, HW 0.80-0.85, CI 93-100, SL 0.76-0.80, SI 94-97, PW 0.50-0.54, AL 1.05-1.15, ED 0.18-0.20 (*n*=7). As holotype, but some specimens with lower ventral process of petiole.

Holotype: Worker, China: Yunnan Province, Anning County, Wenquan Town, Yangjiaocun Village (24.9°N, 102.4°E), 1820 m, 1991.Ⅶ.19, collected by Zheng-hui Xu on the ground in conifer-broadleaf mixed forest, No. A91-298.

Etymology: The new species is named after "Ashima", the name of a beautiful and kind girl in the classic legend of the Sani people of the Yi nationality in Yunnan Province, China.

正模工蚁： 正面观，头长大于宽，后缘圆形隆起。上颚具细纵条纹，咀嚼缘具5个齿。唇基中部强烈隆起，前缘圆形突出。触角12节，触角棒3节，端节最长，大约是其前面1节长度的2倍。复眼最大直径上具8～9个小眼。头顶中央具凹坑，侧面观头顶轻度凹陷。前胸背板高，背面圆形隆起。中胸背板向后坡状降低，后胸沟深凹。并胸腹节低，侧面观背面平坦；缺并胸腹节刺，后上角较突出，中间纵向凹陷。腹柄前面具长的小柄，前下角具小三角形的腹柄下突；腹柄结高，背面圆。后腹柄结高而隆起，后倾，背面中央轻度纵向凹陷。头部和身体光滑发亮。触角窝和复眼之间具弱的网纹，中胸侧板具细密网纹，后胸侧板具纵条纹。腹柄和后腹柄的腹板具细密刻点。头部和身体具丰富直立、亚直立毛，头部立毛密集。复眼的小眼间具丰富的向前弯曲短毛。柄节和后足胫节背面具直立、亚直立长毛和短毛。身体黄棕色，头部和后腹部棕色。**副模工蚁：** 特征同正模，但一些标本的腹柄具短的腹柄下突。

正模： 工蚁，中国云南省安宁县温泉镇羊角村（24°9′N，102°4′E），1820m，1991.Ⅶ.19，徐正会采于针阔混交林地表，No. A91-298。

词源： 该新种以中国云南彝族撒尼人经典传说中一个美丽善良的姑娘"阿诗玛"的名字命名。

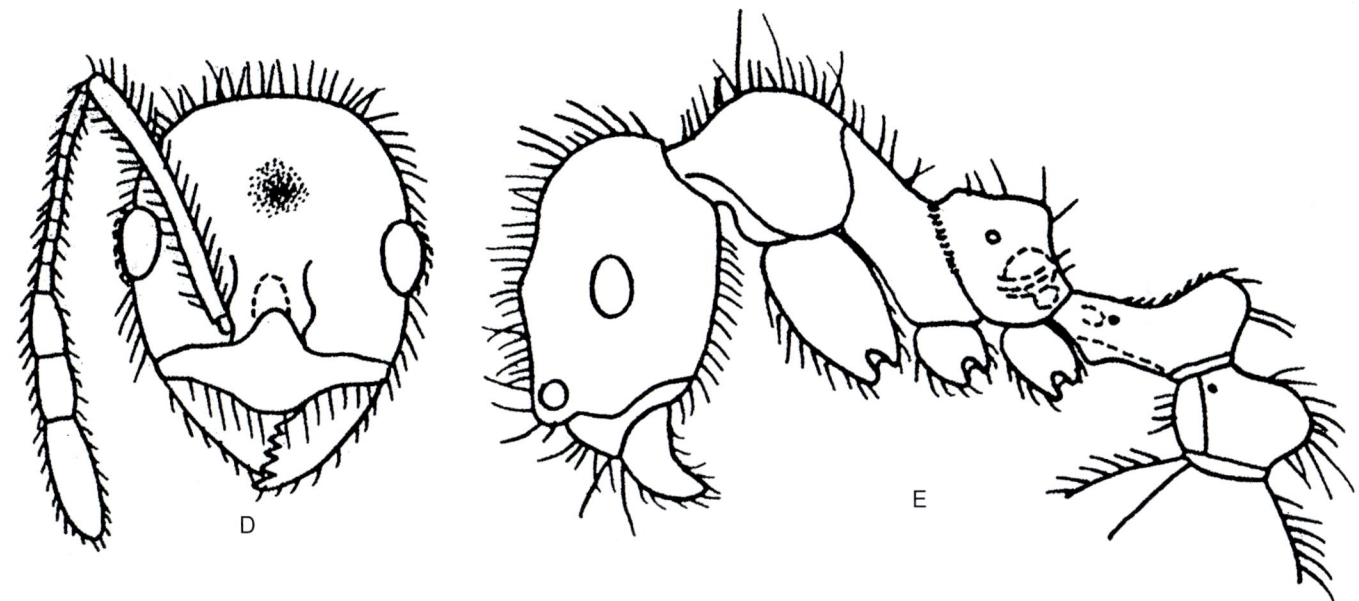

阿诗玛无刺蚁 *Kartidris ashima*

A, D. 工蚁头部正面观；B, E. 工蚁身体侧面观；C. 工蚁身体背面观（D-E. 引自Xu和Zheng, 1995）
A, D. Head of worker in full-face view; B, E. Body of worker in lateral view; C. Body of worker in dorsal view (D-E. cited from Xu & Zheng, 1995)

疏毛无刺蚁
Kartidris sparsipila Xu, 1999

Kartidris sparsipila Xu, 1999, Acta Biologica Plateau Sinica, 14: 134, figs. 2, 15, 16 (w.) CHINA (Yunnan).

Holotype worker: TL 4.1, HL 0.88, HW 0.78, CI 89, SL 0.85, SI 110, PW 0.55, AL 1.18, ED 0.23. In full-face view, head nearly rectanguiar, longer than broad. Occipital margin and sides feebly convex, occipital corners rounded. Mandibles with 5 teeth. Clypeus convex in center, anterior margin complete, roundly convex. Antennal scapes surpassing occipital corners by about 1/4 of its length, the apical 3 segments forming the antennal club. Eyes situated behind midline of head. Vertex of head distinctly depressed in lateral view. Pronotum high and roundly convex. Mesonotum lowering down as a slope. Promesonotal suture only visible on sides, metanotal groove deeply depressed. Dorsum of propodeum straight in lateral view, lowering down posteriorly, posterodorsal corner forming a distinct obtuse angle. Metapleural lobes small, rounded apically. In lateral view, anterodorsal angle of petiolar node higher than posterodorsal one, subpetiolar process low. Postpetiolar node inclined posteriorly. Mandibles and clypeus finely longitudinally striate. Cephalic dorsum, pronotum, mesonotum, petiole and sides of postpetiole with weak fine reticulation; distinct fine reticulation visible on propodeum and between eyes and frontal carinae, sides of mesothorax with rough reticulation, petiolar node, postpetiolar node and gaster with very weak superficial reticulation, shining. Head and body with sparse erect or suberect hairs and abundant decumbent pubescence, hairs on alitrunk very sparse, propodeum and petiolar node without erect hairs. Antennal scapes and tibiae with dense decumbent pubescence, but without erect hairs. Eyes without distinct projecting short hairs. Head and gaster dark brown, antennae, alitrunk, legs, petiole and postpetiole reddish brown. **Paratype workers:** TL 3.6-4.1, HL 0.80-0.91, HW 0.73-0.80, CI 88-91, SL 0.78-0.90, SI 103-113, PW 0.50-0.63, AL 1.08-1.25, ED 0.20-0.24 (*n*=6). As holotype, but in some individuals petiolar node with a pair of erect hairs or postpetiolar node without erect hairs.

Holotype: Worker, China: Yunnan Province, Jinghong County, Mengyang Town, Sanchahe River, 980 m, 1997.Ⅲ.1, collected by Yu-chu Lai from a ground sample of seasonal rain forest, No. A97-90.

Etymology: The species name *sparsipila* combines Latin *spars-* (sparse) + word root *pil* (hair) + suffix *-a* (feminine form), it refers to the sparse standing hairs on the dorsum of body.

正模工蚁： 正面观，头部近长方形，长大于宽；后缘和侧缘轻微隆起，后角圆。上颚具5个齿。唇基中央隆起，前缘完整，圆形隆起。触角柄节约1/4超过头后角，触角棒3节。复眼位于头中线之后。侧面观，头顶明显凹陷。前胸背板高，圆形隆起；中胸背板向后坡形降低，前中胸背板缝仅在侧面可见，后胸沟深凹。并胸腹节背面平直，向后降低，后上角形成明显的钝角；后侧叶小，末端圆。侧面观，腹柄结前上角高于后上角，腹柄下突低。后腹柄结后倾。上颚和唇基具细纵条纹。头部背面、前胸背板、中胸背板、腹柄和后腹柄侧面具弱的细网纹；头部复眼和额脊之间和并胸腹节具明显的细网纹；中胸侧面具粗网纹；腹柄结、后腹柄结和后腹部具很弱的肤浅网纹，光亮。头部和身体具稀疏直立、亚直立毛和密集倾斜绒毛被，胸部立毛很稀疏，并胸腹节和腹柄结缺立毛；触角柄节和胫节具密集倾斜绒毛被，缺立毛；复眼缺明显伸出的短毛。头部和后腹部黑棕色，触角、胸部、足、腹柄和后腹柄红棕色。**副模工蚁：** 特征同正模，但一些个体腹柄结具1对立毛，或后腹柄结缺立毛。

正模： 工蚁，中国云南省景洪县勐养镇三岔河，980m，1997.Ⅲ.1，赖玉初采于季节性雨林地表样中，No. A97-90。

词源： 该新种的种名"*sparsipila*"由拉丁语"*spars-*"（稀疏的）+ 词根"*pil*"（毛发）+ 后缀"*-a*"（阴性形式）组成，指身体背面立毛稀疏。

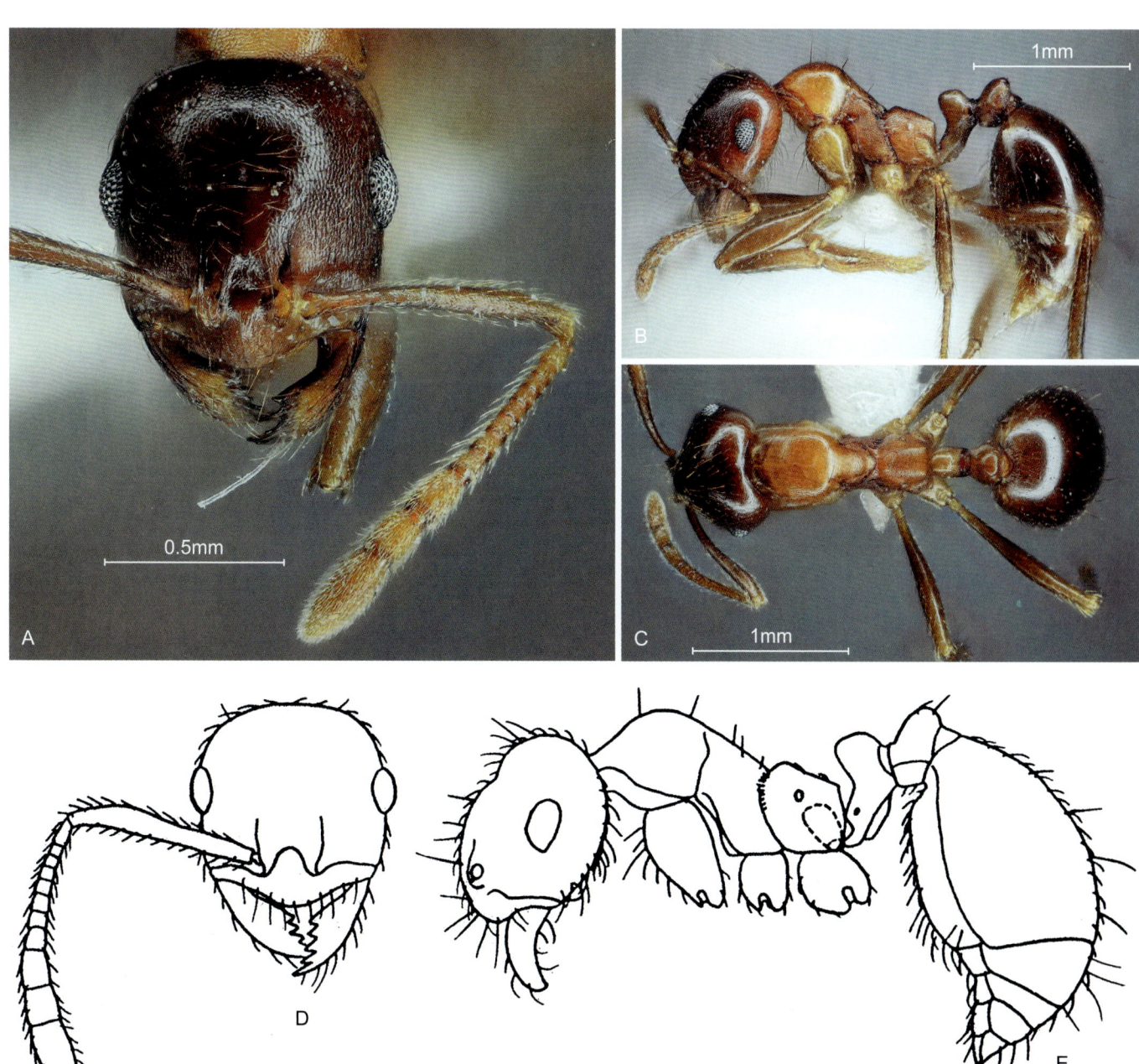

疏毛无刺蚁 *Kartidris sparsipila*

A, D. 工蚁头部正面观；B, E. 工蚁身体侧面观；C. 工蚁身体背面观（D-E. 引自Xu, 1999）

A, D. Head of worker in full-face view; B, E. Body of worker in lateral view; C. Body of worker in dorsal view (D-E. cited from Xu, 1999)

景颇弯蚁

067 *Lordomyrma jingpo* Liu, Xu & Hita Garcia, 2021

Lordomyrma jingpo Liu et al., 2021, Asian Myrmecology, 14: 3, figs. 1-7 (w.q.aq.) CHINA (Yunnan).

Holotype worker: TL 4.1, HL 0.90, HW 0.83, CI 92, SL 0.63, SI 76, ED 0.16, PW 0.70, WL 1.25, PL 0.40, PH 0.35, DPW 0.29, PI 88, PPL 0.25, PPH 0.33, PPW 0.35, PPI 130. In full-face view, head roughly rectangular, longer than broad, posterior margin weakly convex, posterior corners narrowly rounded, lateral margins convex. Mandibles elongate triangular, masticatory margin with about 10 indistinct crenate denticles. Clypeus with a pair of anteriorly divergent carinae which are located close to each other, anterior margin roundly convex. Frontal lobes well developed with concealed antennal sockets. Antennae 12-segmented, apices of scapes just reaching to posterior head corners, antennal clubs consisted of the apical three segments; antennal scrobes well developed, deeply concave. Eyes small, located before midpoints of head sides. In lateral view, promesonotum strongly convex, roundly arched and sloping posteriorly. Promesonotal suture absent. Metanotal groove moderately notched. Propodeal dorsum slightly convex and sloping posteriorly, slightly longer than declivity; propodeal spines long and sharp, weakly curving down posteriorly; declivity moderately concave. Propodeal lobes triangular, acutely toothed apically, about 1/3 length of propodeal spines. Petiolar node triangular, anterior margin weakly concave, posterior margin weakly convex, apex angled to right, anterior peduncle indistinct; subpetiolar process narrow and cuneiform, anteroventrally pointed, with concavity after the process. Postpetiolar node weakly inclined anteriorly, dorsum roundly convex, anterodorsal corner narrowly rounded; ventral margin with two notches, anteroventral corner acutely toothed. Gaster ovate, first segment very large and occupying 4/5 of the gaster, sting extruding. In dorsal view, promesonotum roughly trapezoidal and narrowing posteriorly, lateral margins moderately convex, humeral corners rightly angled, promesonotal suture absent. Metanotal groove impressed. Propodeum short and broad, lateral margins almost straight, propodeal spines straight and lateroposteriorly pointed. Petiole longer than broad, weakly widening posteriorly, sides of the node weakly convex. Postpetiole broader than long, and broader than petiole, narrowing posteriorly, anterior margin straight, sides weakly convex. Anterior margin of gaster deeply concave. Mandibles smooth and shining, with very sparse elongate punctures. Head, mesosoma, petiole and postpetiole coarsely uniformly reticulate, clypeus finely reticulate. Gaster with sparse piliferous punctures, interface microreticulate and relatively shining. Body dorsum with abundant erect to suberect long hairs and abundant subdecumbent short pubescence, pubescence on gaster relatively denser; scapes and tibiae with sparse suberect hairs and dense decumbent pubescence. Body color blackish brown; mandibles, antennae, legs and gaster reddish brown; eyes grey. **Paratype workers:** TL 4.1-4.3, HL 0.85-0.90, HW 0.80-0.83, CI 92-94, SL 0.65-0.68, SI 79-82, ED 0.15-0.18, PW 0.68-0.70, WL 1.20-1.28, PL 0.35-0.40, PH 0.33-0.35, DPW 0.28-0.30, PI 88-93, PPL 0.23-0.25, PPH 0.33-0.34, PPW 0.33-0.35, PPI 130-144 (*n*=10). As holotype, but sometimes body color reddish brown with antennae and legs yellowish brown.

Holotype: Worker, China: Yunnan Province, Yingjiang County, Nabang Town, Palan Village (24.6831°N, 97.5858°E), 590 m, 2015.Ⅲ.22, collected by Ying Zheng from a soil nest in valley lowland rainforest, No. A15-381.

Etymology: The species is named after a minority nationality called "Jingpo", residing in the type specimen locality, Yingjiang County, Yunnan Province, China. The species epithet is a noun in apposition and thus invariant.

正模工蚁： 正面观，头部近长方形，长大于宽；后缘轻度隆起，后角窄圆；侧缘隆起。上颚长三角形，咀嚼缘约具10个不明显的圆钝细齿。唇基具1对向前分歧彼此靠近的纵脊，前缘圆形隆起。额叶发育良好，遮盖触角窝。触角12节，柄节末端刚到达头后角，触角棒3节；触角沟发育良好，深凹。复眼小，位于头侧缘中点之前。侧面观，前中胸背板强烈隆起，圆弓形，向后坡形降低。缺前中胸背板缝，后胸沟中度切入。并胸腹节背面轻微隆起，向后坡形降低，稍长于斜面；并胸腹节刺长而尖锐，向后轻度下弯；斜面中度凹陷。并胸腹节侧叶三角形，顶端尖齿状，约为并胸腹节刺长度的1/3。腹柄结三角形，前缘轻度凹陷，后缘轻度隆起，顶端直角形；前面小柄不明显；腹柄下突窄，楔形，指向前下角，腹柄下突之后凹陷。后腹柄结轻度前倾，背面圆形隆起，前上角窄圆；腹缘具2个切口，前下角尖齿状。后腹部卵圆形，第1节很大，占据后腹部的4/5。螫针伸出。背面观，前中胸背板近梯形，向后变窄，侧缘中度隆起，肩角直角形。缺前中胸背板缝，后胸沟凹陷。并胸腹节短而宽，侧缘几乎平直；并胸腹节刺直，指向侧后方。腹柄长大于宽，向后轻度变宽，腹柄结两侧轻度隆起。后腹柄宽大于长，宽于腹柄，向后变

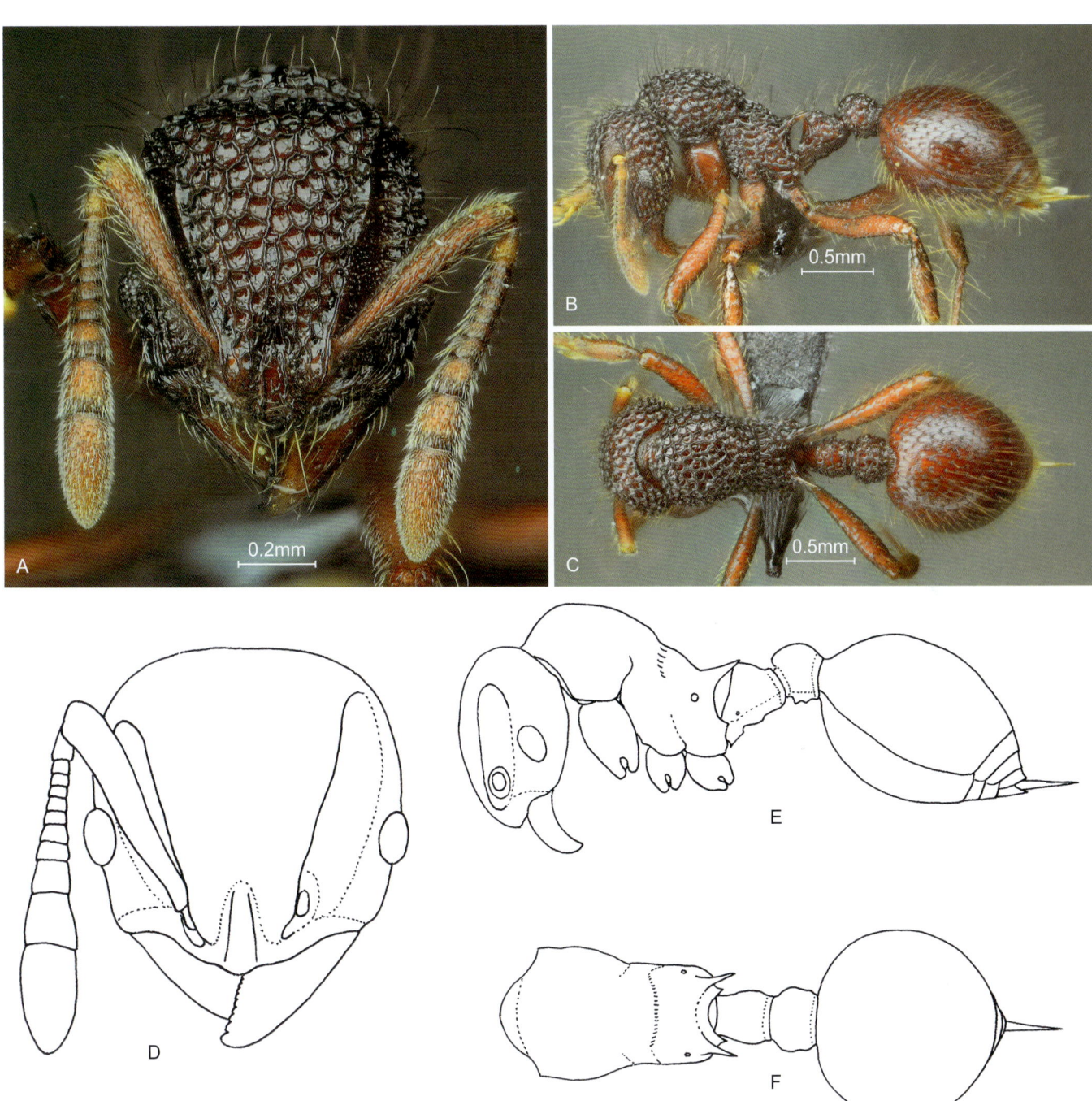

景颇弯蚁 *Lordomyrma jingpo*

A, D. 工蚁头部正面观；B, E. 工蚁身体侧面观；C, F. 工蚁身体背面观（A-C. 引自Liu等，2021）
A, D. Head of worker in full-face view; B, E. Body of worker in lateral view; C, F. Body of worker in dorsal view (A-C. cited from Liu et al., 2021)

窄，前缘平直，侧缘轻度隆起。后腹部前缘深凹。上颚光滑发亮，具很稀疏的伸长刻点。头部、胸部、腹柄和后腹柄具粗糙一致的网纹，唇基具细网纹。后腹部具稀疏具毛刻点，界面具微网纹，较光亮。身体背面具丰富直立、亚直立长毛和丰富倾斜短绒毛被，后腹部绒毛被较密集；柄节和胫节具稀疏亚直立毛和密集倾斜绒毛被。身体黑棕色，上颚、触角、足和后腹部红棕色，复眼灰色。**副模工蚁：** 特征同正模，但有时身体红棕色，触角和足黄棕色。

正模： 工蚁，中国云南省盈江县那邦镇帕烂村（24.6831°N，97.5858°E），590m，2015.Ⅲ.22，郑莹采于沟谷低地雨林土壤巢中，No. A15-381。

词源： 该新种以中国云南省盈江县模式标本产地聚居的少数民族"景颇族"命名，物种的修饰词是同位语中的名词，因而是不变的。

尼玛弯蚁
068 *Lordomyrma nima* Liu, Xu & Hita Garcia, 2021

Lordomyrma nima Liu et al., 2021, Asian Myrmecology, 14: 6, figs. 8-14 (w.q.) CHINA (Tibet).

Holotype worker: TL 3.8, HL 0.83, HW 0.73, CI 88, SL 0.60, SI 83, ED 0.13, PW 0.50, WL 1.13, PL 0.45, PH 0.33, DPW 0.20, PI 72, PPL 0.25, PPH 0.25, PPW 0.25, PPI 100. In full-face view, head roughly rectangular, longer than broad, slightly narrowed anteriorly, posterior margin slightly convex laterally and slightly concave medially, posterior corners rounded, lateral margins weakly convex. Mandibles elongate triangular, masticatory margin with 8 teeth, basal margin with 5 minute teeth. Clypeus with pair of anteriorly diverging carinae, anterior margin roundly convex with tiny tooth medially. Antennae 12-segmented, apices of scapes reaching to 9/10 of distance from antennal socket to posterior head corner, antennal clubs consisting of apical 3 antennomeres. Eyes small, located before midpoint of head, with 6 ommatidia in maximum diameter. In lateral view, promesonotum moderately convex and sloping posteriorly. Promesonotal suture absent. Mesopleuron with an oblique furrow. Metanotal groove deeply notched. Propodeal dorsum straight and sloping posteriorly, propodeal spines acutely toothed, declivity weakly concave, propodeal lobes triangular and shorter than propodeal spines. Petiolar node thick, roughly trapezoidal and narrowed dorsally, with roundly convex dorsum; anterior peduncle relatively longer, about as long as the node, anteroventral corner protruding as a cuneiform ridge with bluntly angled corner. Postpetiolar node weakly inclined posteriorly, with roundly convex dorsum; ventral margin weakly concave, anteroventral corner acutely toothed. Gaster elongate oval, sting extruding. In dorsal view, pronotum broadest, lateral margins moderately convex, humeral corners narrowly rounded. Promesonotal suture absent. Mesonotum narrowest, lateral margins slightly convex. Metanotal groove deeply impressed. Propodeum widened posteriorly, lateral margins almost straight. Anterior peduncle of petiole widened posteriorly, petiolar node roughly elliptical, slightly broader than long. Postpetiole widened posteriorly, lateral margins moderately convex, broader than petiolar node. Mandibles longitudinally striate. Head dorsum with dense posteriorly divergent rugae, the rugae become reticulate-rugose on the sides. Clypeus relatively smooth and shining. Mesosoma longitudinally rugose; sides of pronotum obliquely rugose; mesopleura reticulate-rugose; propodeal dorsum and declivity transversely rugose, propodeal sides obliquely rugose. Petiolar node and postpetiolar node finely transversely rugose, sides of petiolar node obliquely rugose, lower portions of petiole and postpetiole densely punctured. Gaster smooth and shining. Body dorsum with abundant erect to suberect hairs and abundant decumbent pubescence, hairs on head dorsum relatively shorter and denser. Scapes with dense subdecumbent hairs and decumbent pubescence, tibiae with dense decumbent pubescence. Body color blackish brown; petiole, postpetiole and gaster reddish brown; mandibles, antennae and legs yellowish brown. **Paratype workers:** TL 3.4-4.3, HL 0.75-0.93, HW 0.68-0.80, CI 85-91, SL 0.53-0.65, SI 78-86, ED 0.11-0.18, PW 0.48-0.56, WL 0.78-1.25, PL 0.40-0.48, PH 0.28-0.33, DPW 0.19-0.20, PI 58-75, PPL 0.20-0.28, PPH 0.20-0.26, PPW 0.24-0.28, PPI 90-105 (*n*=10). As holotype, but anteroventral corner of petiole variable, either cuneiform convexity, blunt angle or ventrally pointed acute tooth. Body color yellowish brown, reddish brown or blackish brown. Head length varies between 0.75 mm to 0.93 mm.

Holotype: Worker, China: Tibet Autonomous Region, Medog County, Dagmo Town, 62K (29.7096°N, 95.5823°E), 2620 m, 2008.Ⅴ.13, collected by Zheng-hui Xu from a nest under stone in *Alnus nepalensis* forest, No. A08-325.

Etymology: The specific epithet refers to "Nima", a common male name widely used in Tibet and Qinghai. The species epithet is a noun in apposition and thus invariant.

正模工蚁： 正面观，头部近长方形，长大于宽，向前轻微变窄；后缘两侧轻微隆起，中央轻微凹陷；后角圆，侧缘轻度隆起。上颚长三角形，咀嚼缘具8个齿，基缘具5个小齿，唇基具1对向前分歧的纵脊，前缘圆形隆起，具微小的中齿。触角12节，柄节到达触角窝至头后角间距的9/10处，触角棒3节。复眼小，位于头部中点之前，最大直径上具6个小眼。侧面观，前中胸背板中度隆起，向后坡形降低。缺前中胸背板缝，中胸侧板具1条斜沟，后胸沟深切。并胸腹节背面平直，向后坡形降低；并胸腹节刺尖齿状，斜面轻度凹陷。并胸腹节侧叶三角形，短于并胸腹节刺。腹柄结厚，近梯形，向上变窄，背面圆形隆起；前面小柄较长，约与腹柄结等长，前下角突出呈楔形脊，钝角状。后腹柄结轻度后倾，背面圆形隆起；腹面轻度凹陷，前下角尖齿状。后腹部长卵形。螫针伸出。背面观，前胸背板最宽，侧缘中度隆起，肩角窄圆。缺前中胸背板缝。中胸背板最窄，侧缘轻度隆起。后胸沟深凹。并

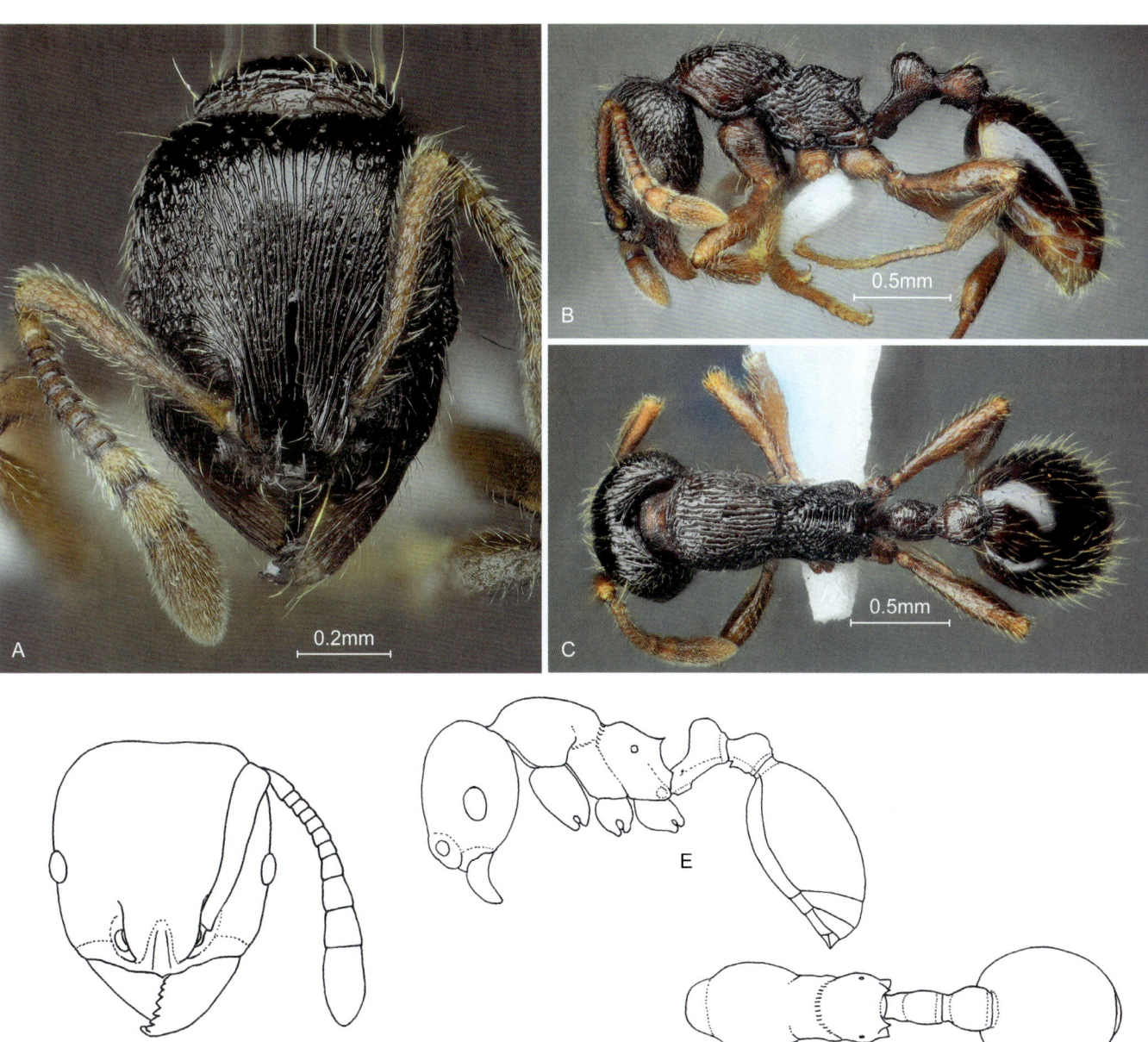

尼玛弯蚁 *Lordomyrma nima*

A, D. 工蚁头部正面观；B, E. 工蚁身体侧面观；C, F. 工蚁身体背面观（A-C. 引自Liu等, 2021）
A, D. Head of worker in full-face view; B, E. Body of worker in lateral view; C, F. Body of worker in dorsal view (A-C. cited from Liu et al., 2021)

胸腹节向后变宽，侧缘几乎平直。腹柄前面小柄向后变宽，腹柄结近椭圆形，宽稍大于长。后腹柄向后变宽，侧缘中度隆起，宽于腹柄结。上颚具纵条纹。头部背面具向后分歧的密集皱纹，在侧面变成网纹。唇基较光滑，发亮。胸部具纵皱纹，前胸背板侧面具倾斜皱纹，中胸侧面具网纹；并胸腹节背面和斜面具横皱纹，侧面具倾斜皱纹。腹柄结和后腹柄结具细的横皱纹，腹柄结侧面具倾斜皱纹；腹柄和后腹柄下部具密集刻点。后腹部光滑光亮。身体背面具丰富直立、亚直立毛和丰富倾斜绒毛被，头部背面立毛较短较密。柄节具密集亚倾斜毛和密集倾斜绒毛被，胫节具密集倾斜绒毛被。身体黑棕色，腹柄、后腹柄和后腹部红棕色，上颚、触角和足黄棕色。**副模工蚁**：特征同正模，但是腹柄的前下角可变，可以呈楔形、钝角或指向腹面的锐齿。体色黄棕色、红棕色或黑棕色。头长为0.75~0.93mm。

正模：工蚁，中国西藏自治区墨脱县达木乡62K（29.7096°N，95.5823°E），2620m，2008.V.13，徐正会采于尼泊尔桤木林石下巢中，No. A08-325。

词源：该新种以中国西藏和青海广泛使用的男性名字"尼玛"命名，物种修饰词是同位语中的名词，因而是不变的。

069 尖刺稀切叶蚁
Oligomyrmex acutispinus Xu, 2003

Oligomyrmex acutispinus Xu, 2003, Acta Zootaxonomica Sinica, 28(2): 315, figs. 16-19 (s.w.) CHINA (Yunnan).

Holotype soldier: TL 1.4, HL 0.54, HW 0.40, CI 80, SL 0.23, SI 56, PW 0.23, AL 0.40, PL 0.15, PH 0.11, DPW 0.10. Body small. In full-face view, head rectangular, longer than broad. Occipital margin moderately concave, occipital corners roundly prominent. Sides nearly straight. Mandible with 5 teeth. Median portion of clypeus longitudinally depressed, bicarinate and divergent forward, anterior margin weakly concave. Antenna 9 segments with a 2-segmented club, apex of scape reaching to 1/2 of the distance from socket to occipital corner. Eye with 1 facet. In lateral view, occiput with a pair of large acute horns. Dorsum of head weakly convex. Promesonotum high and roundly convex. Promesonotal suture obsolete on the dorsum. Metanotum absent. Metanotal groove shallowly impressed. Propodeum with a pair of acute teeth, dorsum convex at anterior portion and sloping down rearwards, declivity concave with narrow thin lateral laminae. Petiole pedunculate anteriorly, ventral face straight. Petiolar node thick, anterior face gently sloping, posterior face steeply sloping, dorsal face roundly prominent. Postpetiolar node roundly convex and lower than petiolar node. In dorsal view, petiolar node as broad as postpetiolar node. Mandibles and median portion of clypeus smooth and shiny. Head finely reticulate, anterior 3/4 of head dorsum and anterior 1/5 of genae finely longitudinally striate. Reticulations present between occipital horns but without transverse ridge or striations. Alitrunk, petiole and postpetiole densely and finely punctate, interspace appears as fine microreticulations. Gaster smooth and shiny. Head, alitrunk, petiole and postpetiole with sparse erect or suberect hairs and dense decumbent pubescence. Gaster with sparse suberect hairs and abundant decumbent pubescence. Frontal carina with 4 erect long hairs. Scapes and tibiae with dense decumbent pubescence. Head, alitrunk, petiole and postpetiole yellow. Antennae, legs and gaster light yellow. **Paratype soldier:** TL 1.4, HL 0.53, HW 0.40, CI 76, SL 0.23, SI 56, PW 0.23, AL 0.38, PL 0.13, PH 0.11, DPW 0.10 (*n*=1). As holotype. **Paratype workers:** TL 0.8-0.9, HL 0.30-0.33, HW 0.28, CI 85-92, SL 0.18-0.19, SI 64-68, PW 0.18-0.20, AL 0.28-0.30, PL 0.09-0.10, PH 0.08-0.09, DPW 0.08 (*n*=5). As holotype, but body much smaller. Head nearly square, occiput without horns. Occipital margin shallowly concave in the middle, sides weakly convex. Apex of scape reaching 2/3 of the distance from socket to occipital corner. In lateral view, promesonotum weakly convex. Body in color light yellow.

Holotype: Soldier, China: Yunnan Province, Mengla County, Mengla Town, Nangongshan Mountain, 1380 m, 1997.Ⅸ.15, collected by Zheng-hui Xu in monsoon evergreen broadleaf forest, No. A97-2591.

Etymology: The species name *acutispinus* combines Latin *acut-* (sharp) + word root *spin* (spine) + suffix *-us* (masculine form), it refers to the acute spines of propodeum of soldier.

正模兵蚁： 体小。正面观，头部长方形，长大于宽；后缘中度凹陷，后角圆突；侧缘近平直。上颚具5个齿。唇基中部纵向凹陷，具向前分歧的双脊，前缘轻度凹陷。触角9节，触角棒2节，柄节末端到达触角窝至头后角间距的1/2处。复眼具1个小眼。侧面观，头顶具1对大的尖角突，头部背部轻度隆起。前中胸背板高，圆形隆起。前中胸背板缝在背部退化，缺后胸背板，后胸沟浅凹。并胸腹节具1对尖齿，背面前部隆起，向后坡形降低；斜面凹陷，具窄而薄的侧脊。腹柄前面具小柄，腹面平直；腹柄结厚，前面缓坡形，后面陡坡形，背面圆突。后腹柄结圆形隆起，低于腹柄结。背面观，腹柄结与后腹柄结等宽。上颚和唇基中部光滑发亮。头部具细网纹，背面前部3/4和颊区前部1/5具细纵条纹；头后缘角突之间具网纹，但是缺横脊或横条纹。胸部、腹柄和后腹柄具细密刻点，界面呈细微网状。后腹部光滑发亮。头部、胸部、腹柄和后腹柄具稀疏直立、亚直立毛和密集倾斜绒毛被；后腹部具稀疏亚直立毛和丰富倾斜绒毛被；额脊具4根直立长毛。柄节和胫节具密集倾斜绒毛被。头部、胸部、腹柄和后腹柄黄色；触角、足和后腹部浅黄色。**副模兵蚁：** 特征同正模。**副模工蚁：** 特征同正模，但是身体小很多；头部近方形，头顶缺角突，头后缘中央浅凹，侧缘轻度隆起；柄节末端到达触角窝至头后角间距的2/3处；侧面观，前中胸背板轻度隆起；身体浅黄色。

正模： 兵蚁，中国云南省勐腊县勐腊镇南贡山，1380m，1997.Ⅸ.15，徐正会采于季风常绿阔叶林中，No. A97-2591。

词源： 该新种的种名"*acutispinus*"由拉丁语"*acut-*"（尖锐的）+词根"*spin*"（刺）+后缀"*-us*"（阳性形式）组成，指兵蚁并胸腹节具尖锐的刺。

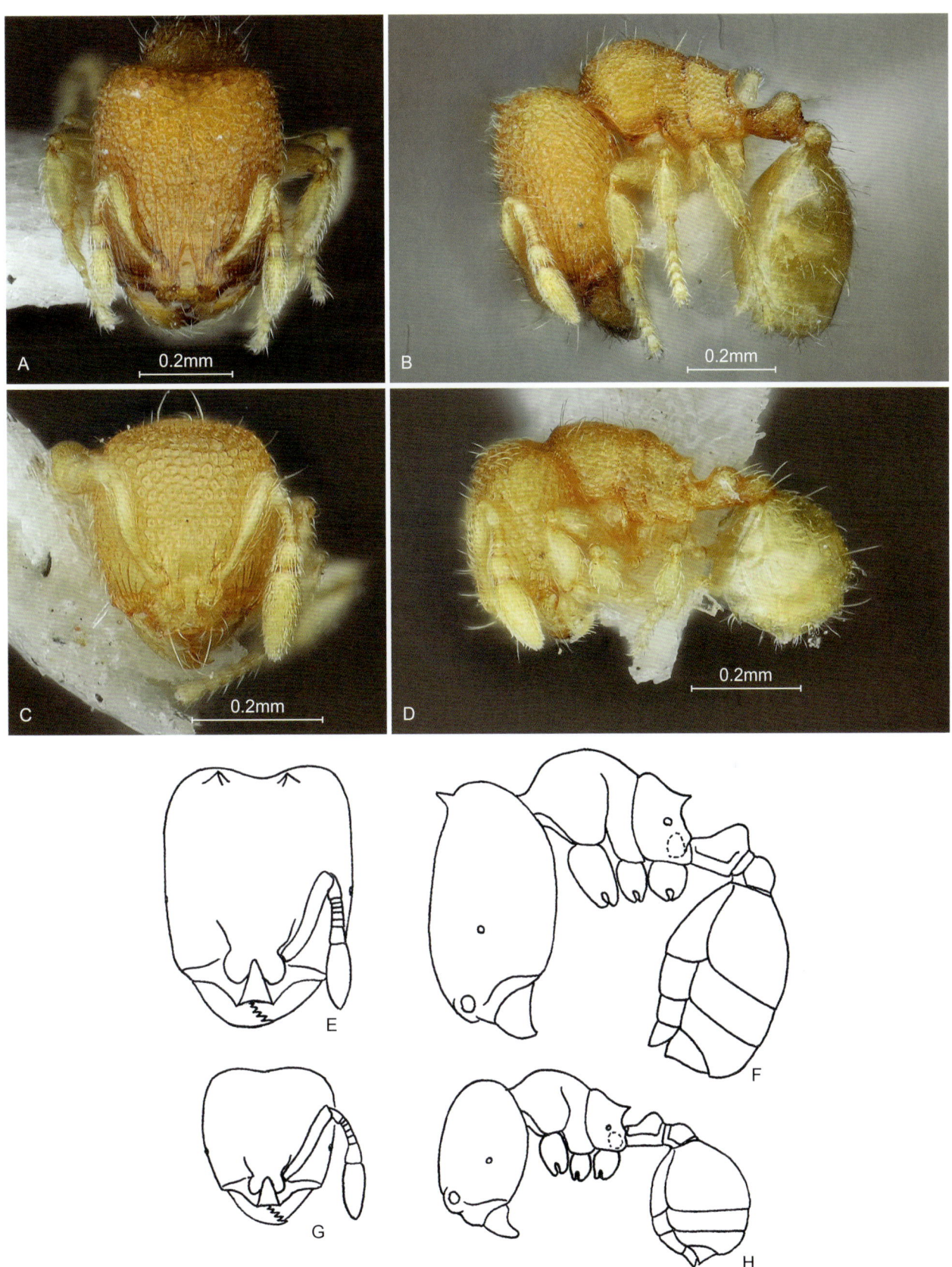

尖刺稀切叶蚁 *Oligomyrmex acutispinus*

A, E. 兵蚁头部正面观；B, F. 兵蚁身体侧面观；C, G. 工蚁头部正面观；D, H. 工蚁身体侧面观（E-H. 引自 Xu, 2003）
A, E. Head of soldier in full-face view; B, F. Body of soldier in lateral view; C, G. Head of worker in full-face view; D, H. Body of worker in lateral view (E-H. cited from Xu, 2003)

070 高结稀切叶蚁
Oligomyrmex altinodus Xu, 2003

Oligomyrmex altinodus Xu, 2003, Acta Zootaxonomica Sinica, 28(2): 312, figs. 5-8 (s.w.) CHINA (Yunnan).

Holotype soldier: TL 3.7, HL 1.05, HW 0.95, CI 90, SL 0.50, SI 53, PW 0.58, AL 0.90, PL 0.45, PH 0.35, DPW 0.28. In full-face view, head slightly longer than broad, roughly trapezoid and moderately narrowed forward. Occipital margin moderately concave. Occipital corners roundly prominent. Sides slightly convex. Mandible with 5 teeth. Median portion of clypeus depressed longitudinally, bicarinate and divergent forward, anterior margin nearly straight. Antenna 11 segments with a 2-segmented club, apex of scape reached to 4/7 of the distance from antennal socket to occipital corner. Eye with 7 facets. In lateral view, occiput with a pair of minute horns. Dorsum of head evenly convex. Promesonotum roundly convex. Promesonotal suture distinct. Mesonotum with a transverse furrow, posterior portion prominent and forming a transverse ridge behind the furrow. Metanotum present and tongue-like. Metanotal groove impressed. Propodeum with posterodorsal corner bluntly angled, dorsum straight and depressed longitudinally in the middle, declivity concave. Petiole pedunculate anteriorly, ventral face straight, anteroventral corner with a rightly angled tooth. Petiolar node narrow and high, anterior and posterior faces steeply sloped, dorsal face narrow and convex. Postpetiolar node roundly and anterodorsally convex, lower than petiolar node. In dorsal view, petiolar node width : postpetiolar node width = 1.0 : 1.2, petiolar node transverse, postpetiolar node semicircular. Anterior border of gaster deeply concave in the middle in order to accept postpetiole. Mandibles and median portion of clypeus smooth and shiny. Head densely and largely punctate, distance between punctures about equal to diameter of a puncture. Punctures on dorsum of head much denser. Anterior 1/3 of head longitudinally striate. Dorsum of promesonotum sparsely punctate, sides of pronotum smooth and shining. Sides of mesothorax, metathorax, propodeum, petiole and postpetiole densely and finely punctate. Dorsal face of petiolar node smooth and shining. Gaster sparsely and largely punctate, distance between punctures about 2-3 times of the diameter of a puncture. Head and body with dense decumbent pubescence. Frontal carina with 4 long erect hairs. Mesonotum, petiolar node, postpetiolar node and apex of gaster with sparse erect hairs. Scapes and tibiae with dense decumbent pubescence, without erect hairs. Head and gaster blackish brown, alitrunk, petiole and postpetiole reddish brown. Mandibles, antennae and legs yellow. **Paratype soldiers:** TL 3.6-4.5, HL 1.00-1.08, HW 0.90-1.00, CI 90-93, SL 0.50, SI 50-56, PW 0.53-0.63, AL 0.90-1.03, PL 0.40-0.45, PH 0.33-0.38, DPW 0.25-0.30 ($n=5$). As holotype, but occipital horns minute to absent, posterodorsal corner of propodeum bluntly angled to rightly angled. **Paratype workers:** TL 1.3-1.5, HL 0.38-0.43, HW 0.35-0.38, CI 88-100, SL 0.25-0.28, SI 67-79, PW 0.23-0.24, AL 0.40-0.43, PL 0.18-0.20, PH 0.15, DPW 0.09-0.10 ($n=5$). Similar to the holotype, but much smaller, head normal. Occipital margin weakly concave. Apex of scape reached to 5/6 of the distance from antennal socket to occipital corner. Eye with 2 facets. Occiput without horns. Promesonotum moderately convex. Promesonotal suture obsolete on the dorsum. Mesonotum without transverse furrow and ridge. Metanotum absent. Metanotal groove deeply impressed. Posterodorsal corner of propodeum rounded. Head, pronotum, petiolar node, postpetiolar node and gaster smooth and shining. Mesonotum, petiole and postpetiole without erect hairs. Head and body with short, sparse and depressed pubescence. Body yellow in color, head and gaster yellowish brown.

Holotype: Soldier, China: Yunnan Province, Jingdong County, Huashan Town, Wencha Village, 1500 m, 2002.Ⅳ.13, collected by Zheng-qun Chai in conifer-broadleaf mixed forest on the west slope of Ailao Mountain, No. A1194.

Etymology: The species name *altinodus* combines Latin *alt-* (tall) + word root *nod* (node) + suffix *-us* (masculine form), it refers to the tall node of petiole of soldier.

正模兵蚁： 正面观，头部长稍大于宽，近梯形，向前中度变窄；后缘中度凹陷，后角圆突；侧缘轻微隆起。上颚具5个齿。唇基中部纵向凹陷，具向前分歧的双脊，前缘近乎直。触角11节，触角棒2节，柄节末端到达触角窝至头后角间距的4/7处。复眼具7个小眼。侧面观，头顶具1对微小的角突，头背面均匀隆起。前中胸背板圆形隆起，前中胸背板缝明显。中胸背板具1横沟，后部突起，在沟后形成1条横脊。后胸背板存在，呈舌尖状。后胸沟凹陷。并胸腹节后上角钝角状，背面平直，中间纵向凹陷；斜面凹陷。腹柄前面具小柄，腹面平直，前下角具1个直角形齿；腹柄结窄而高，前面和后面陡坡状，背面窄而隆起。后腹柄结前上角圆形隆起，低于腹柄结。背面观，腹柄结宽：后腹柄结宽=1.0：1.2，腹柄结横形，后腹柄结半圆形。后腹部前缘中央深凹，以便接纳后腹柄。上颚和唇基中部光滑发亮。头部具密集大刻点，刻点间距约等于刻点直径，背面刻点更加密集；前部1/3具纵条纹。前中胸背

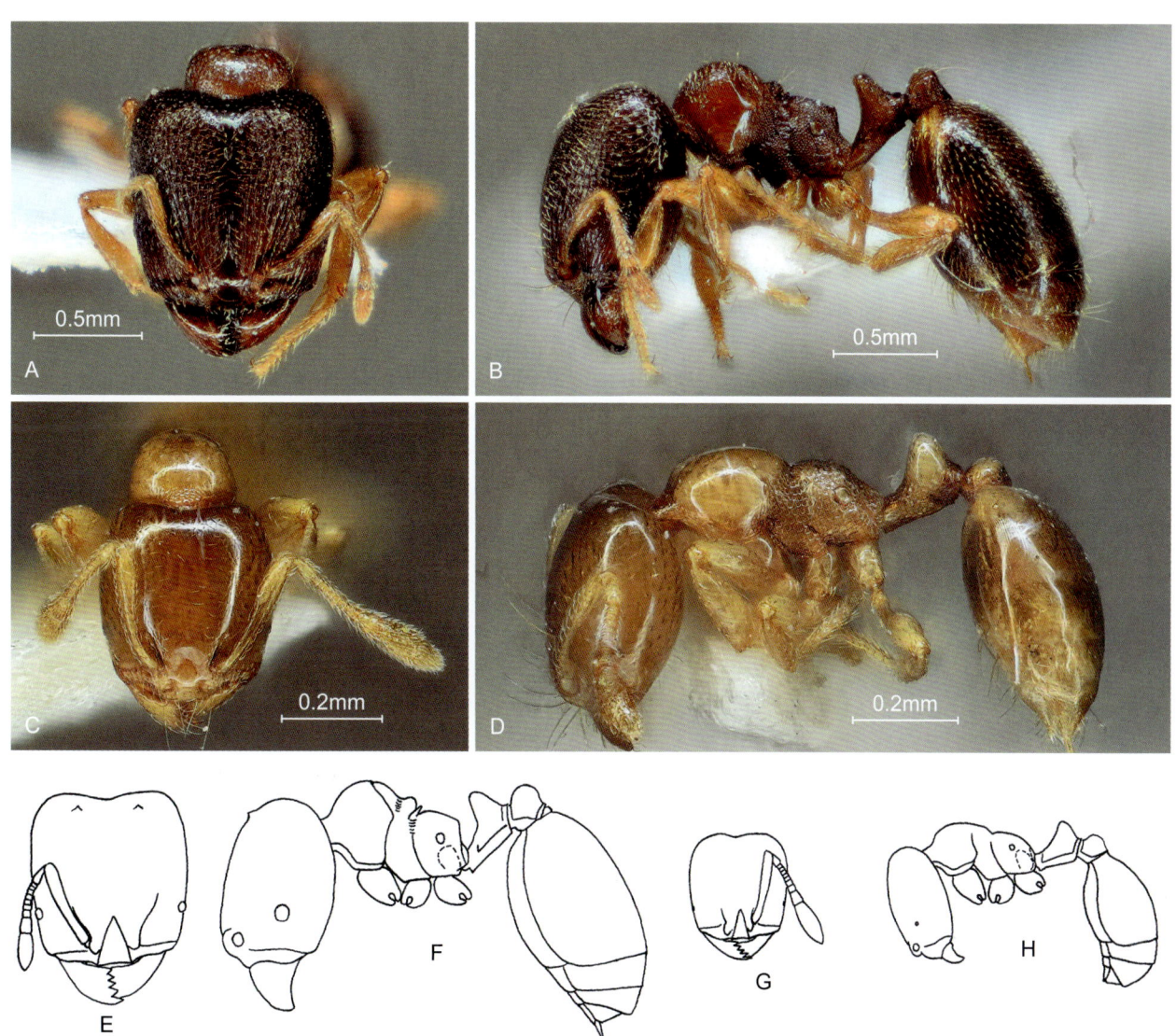

高结稀切叶蚁 *Oligomyrmex altinodus*

A, E. 兵蚁头部正面观；B, F. 兵蚁身体侧面观；C, G. 工蚁头部正面观；D, H. 工蚁身体侧面观（E-H. 引自Xu, 2003）
A, E. Head of soldier in full-face view; B, F. Body of soldier in lateral view; C, G. Head of worker in full-face view; D, H. Body of worker in lateral view (E-H. cited from Xu, 2003)

板背面具稀疏刻点，前胸背板侧面光滑发亮；中胸侧面、后胸、并胸腹节、腹柄和后腹柄具细密刻点；腹柄结背面光滑发亮。后腹部具稀疏大刻点，刻点间距为刻点直径的2~3倍。头部和身体具密集倾斜绒毛被。额脊具4根直立长毛。中胸背板、腹柄结、后腹柄结和后腹部末端具稀疏立毛。柄节和胫节具密集倾斜绒毛被，缺立毛。头部和后腹部黑棕色；胸部、腹柄和后腹柄红棕色；上颚、触角和足黄色。**副模兵蚁**：特征同正模，但是头顶角突微小至缺失，并胸胸节后上角钝角状至直角形。**副模工蚁**：与正模相似，但身体小很多，头部正常；头后缘轻度凹陷，柄节末端到达触角窝至头后角间距的5/6处，复眼具2个小眼，头顶缺角突；前中胸背板中度隆起，前中胸背板缝在背面消失，中胸背板缺横沟和横脊，缺后胸背板，后胸沟深凹，并胸腹节后上角圆；头部、前胸背板、腹柄结、后腹柄结和后腹部光滑发亮；中胸背板、腹柄和后腹柄缺直立毛；头部和身体具短而稀疏的平伏绒毛被；身体黄色，头部和后腹部黄棕色。

正模：兵蚁，中国云南省景东县花山乡文岔村，1500m，2002.Ⅳ.13，柴正群采于哀牢山西坡针阔混交林中，No. A1194。

词源：该新种的种名"*altinodus*"由拉丁语"*alt-*"（高的）+词根"*nod*"（结）+后缀"-*us*"（阳性形式）组成，指兵蚁的腹柄具高的结。

双角稀切叶蚁

Oligomyrmex bihornatus Xu, 2003

Oligomyrmex bihornatus Xu, 2003, Acta Zootaxonomica Sinica, 28(2): 317, figs. 24-27 (s.w.) CHINA (Yunnan).

Holotype soldier: TL 2.4, HL 0.70, HW 0.50, CI 71, SL 0.28, SI 55, PW 0.35, AL 0.63, PL 0.23, PH 0.18, DPW 0.16. In full-face view, head longer than broad, roughly rectangular. Occipital margin deeply and roundly concave, occipital corners protruding into a pair of developed acute horns, a transverse ridge present on the occiput and connecting the 2 horns. Sides moderately convex. Mandible with 5 teeth. Median portion of clypeus longitudinally depressed, bicarinate and divergent forward, anterior margin straight. Antenna 9 segments with a 2-segmented club, apex of scape reaching to 5/11 of the distance from antennal socket to occipital corner. Eye with 4 facets. In lateral view, occiput with a pair of well developed large acute horns. Dorsum of head straight. Promesonotum high and roundly convex. Promesonotal suture present. Mesonotum suddenly down curved at posterior end. Metanotum present but narrow, dorsally pointed. Mesometanotal suture and metapropodeal suture depressed. Propodeum with posterodorsal corner bluntly angled, dorsal face straight and sloping down backward, longitudinally depressed in the middle. Declivity straight and vertically depressed in the middle. Petiole pedunculate anteriorly, ventral face straight. Petiolar node thick, anterior and posterior faces sloping, dorsal face roundly prominent. Postpetiolar node roundly convex and lower than petiolar node. In dorsal view, petiolar node about as broad as postpetiolar node. Mandibles and median portion of clypeus smooth and shiny. Head smooth and shiny, weakly and very sparsely punctate. Alitrunk, petiole, postpetiole and gaster smooth and shiny. Head, alitrunk, petiole and postpetiole with abundant erect or suberect hairs. Gaster with dense decumbent hairs. Scapes and tibiae with dense decumbent pubescence. Body color yellow. Masticatory margins of mandibles and eyes black. **Paratype soldiers:** TL 2.1-3.1, HL 0.70-0.75, HW 0.50-0.53, CI 69-71, SL 0.28-0.30, SI 55-57, PW 0.33-0.35, AL 0.60-0.63, PL 0.23-0.25, PH 0.18-0.20, DPW 0.15-0.18 (*n*=5). As holotype. **Paratype workers:** TL 1.1-1.2, HL 0.33-0.35, HW 0.26-0.28, CI 79-81, SL 0.20, SI 73-76, PW 0.18-0.20, AL 0.33-0.35, PL 0.09-0.10, PH 0.08-0.09, DPW 0.08 (*n*=4). As holotype, but much smaller. Head normal and rectangular. Occiput without horns, occipital margin slightly concave, occipital corners prominent. Sides slightly convex. Apex of scape reaching 2/3 of the distance from antennal socket to occipital corner. Eyes absent. Promesonotum weakly convex. Promesonotal suture obsolete on the dorsum. Metanotum absent. Head sparsely and finely punctured. Body color light yellow.

Holotype: Soldier, China: Yunnan Province, Baoshan City, Mangkuan Town, Mangkuan Village, 1000 m, 1998.Ⅷ.7, collected by Yuan-chao Zhao in dry-hot valley shrub on the east slope of Gaoligong Mountain, No. A98-1537.

Etymology: The species name *bihornatus* combines Latin *bi-* (pair) + word root *horn* (horn) + suffix *-atus* (masculine form, with), it refers to the paired horns on the posterior head corners of soldier.

正模兵蚁： 正面观，头部长大于宽，近长方形；后缘圆形深凹，后角突出成1对发达的尖角突，后缘具1条连接2个角突的横脊；侧缘中度隆起。上颚具5个齿。唇基中部纵向凹陷，具向前分歧的双脊，前缘平直。触角9节，触角棒2节，柄节顶端到达触角窝至头后角间距的5/11处。复眼具4个小眼。侧面观，头后缘具1对发达的大形尖锐角突，头部背面平直。前中胸背板高，圆形隆起。前中胸背板缝存在。中胸背板后端急剧下弯。后胸背板存在，但是狭窄，指向背面。中后胸背板缝和后胸沟凹陷。并胸腹节后上角钝角状，背面平直，向后坡形降低，中部纵向凹陷；斜面平直，中部垂直凹陷。腹柄前面具小柄，腹面平直；腹柄结厚，前面和后面坡形，背面圆凸。后腹柄结圆形隆起，低于腹柄结。背面观，腹柄结约与后腹柄结等宽。上颚和唇基中部光滑发亮。头部光滑发亮，具很稀疏的弱刻点。胸部、腹柄、后腹柄和后腹部光滑发亮。头部、胸部、腹柄和后腹柄具丰富直立、亚直立毛；后腹部具密集倾斜绒毛被；柄节和胫节具密集倾斜绒毛被。身体黄色，上颚咀嚼缘和复眼黑色。**副模兵蚁：** 特征同正模。**副模工蚁：** 特征同正模，但是身体小很多；头部正常，长方形；头后缘轻度凹陷，缺角突，头后角突出，侧缘轻微隆起；柄节末端到达触角窝至头后角间距的2/3处；缺复眼；前中胸背板轻度隆起，前中胸背板缝在背面消失，缺后胸背板；头部具稀疏细刻点，身体浅黄色。

正模： 兵蚁，中国云南省保山市芒宽乡芒宽村，1000m，1998.Ⅷ.7，赵远潮采于高黎贡山东坡干热河谷稀树灌丛中，No. A98-1537。

词源： 该新种的种名"*bihornatus*"由拉丁语"*bi-*"（双）+词根"*horn*"（角突）+后缀"*-atus*"（阳性形式，具有）组成，指兵蚁的头后角具成对的角突。

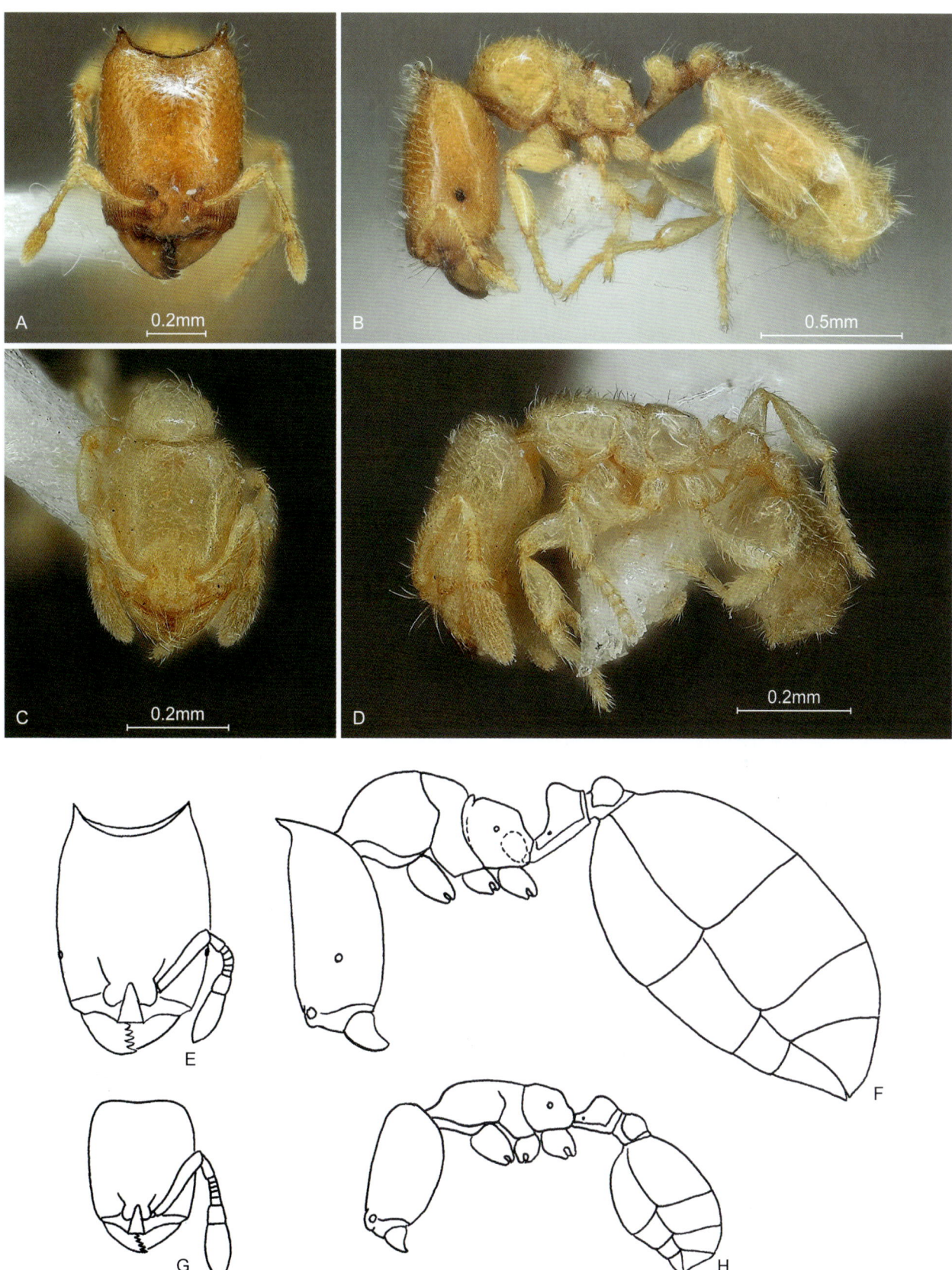

双角稀切叶蚁 *Oligomyrmex bihornatus*

A, E. 兵蚁头部正面观; B, F. 兵蚁身体侧面观; C, G. 工蚁头部正面观; D, H. 工蚁身体侧面观（E-H. 引自Xu, 2003）
A, E. Head of soldier in full-face view; B, F. Body of soldier in lateral view; C, G. Head of worker in full-face view; D, H. Body of worker in lateral view (E-H. cited from Xu, 2003)

072 弯刺稀切叶蚁
Oligomyrmex curvispinus Xu, 2003

Oligomyrmex curvispinus Xu, 2003, Acta Zootaxonomica Sinica, 28(2): 313, figs. 9-12 (s.w.) CHINA (Yunnan).

Holotype soldier: TL 3.0, HL 0.93, HW 0.73, CI 78, SL 0.38, SI 52, PW 0.48, AL 0.85, PL 0.23, PH 0.23, DPW 0.23. In full-face view, head distinctly longer than broad, nearly rectangular, slightly narrowed forward. Occipital margin moderately concave. Occipital corners rounded. Sides nearly straight. Mandible with 6 teeth. Median portion of clypeus depressed longitudinally, bicarinate and divergent forward, anterior margin weakly concave. Antenna 11 segments with a 2-segmented club, apex of scape reached to 3/7 of the distance from antennal socket to occipital corner. Eyes absent. In lateral view, occiput with a pair of small dent-like horns. Dorsum of head straight. Promesonotum roundly convex. Promesonotal suture obsolete on the dorsum. Dorsum of mesonotum straight. Metanotum present, triangular and convex. Mesometanotal suture and metapropodeal suture present. Propodeum with a pair of laterally compressed stout dents which curve down at apex, dorsum weakly concave and down-sloping backward, declivity straight. Petiole pedunculate anteriorly, ventral face straight, anteroventral corner with an acute tooth. Petiolar node triangular, anterior face weakly concave, posterior face weakly convex, the top rightly angled. Postpetiolar node roundly convex and slightly lower than petiolar node. In dorsal view, petiolar node as broad as postpetiolar node. Mandibles and median portion of clypeus smooth and shiny. Head with anterior portion longitudinally striate, posterior portion finely reticulate. Alitrunk, petiole, postpetiole and gaster densely and finely punctate. Head and body with abundant suberect short hairs and dense decumbent pubescence. Scapes and tibiae with dense decumbent pubescence. Body color yellow, masticatory margin of mandible black. **Paratype soldiers:** TL 3.3-3.6, HL 0.90-0.98, HW 0.70-0.75, CI 76-78, SL 0.35-0.38, SI 50-52, PW 0.48-0.50, AL 0.85-0.90, PL 0.23-0.25, PH 0.23, DPW 0.23-0.25 (*n*=5). As holotype. **Paratype workers:** TL 1.3-1.4, HL 0.43, HW 0.30-0.33, CI 71-76, SL 0.25-0.28, SI 77-85, PW 0.20, AL 0.40-0.43, PL 0.13, PH 0.10, DPW 0.08 (*n*=5). As holotype, but much smaller, head normal. Occipital margin nearly straight, slightly concave in the middle. Mandible with 5 teeth. Apex of scape reaching 2/3 of the distance from socket to occipital corner. Promesonotum slightly convex, nearly straight. Metanotum absent. Metanotal groove shallowly impressed. Propodeum with a pair of acute straight teeth, dorsum weakly convex. Petiolar node thicker, with the top bluntly prominent. Anteroventral corner of petiole with a rightly angled tooth.

Holotype: Soldier, China: Yunnan Province, Jinghong County, Puwen Town, Songshanling, 1270 m, 1998.Ⅲ.4, collected by Yun-feng He in warm conifer forest, No. A98-39.

Etymology: The species name *curvispinus* combines Latin *curv-* (curved) + word root *spin* (spine) + suffix *-us* (masculine form), it refers to the down-curved spines of propodeum of soldier.

正模兵蚁： 正面观，头部长明显大于宽，近长方形，向前轻微变窄；后缘中度凹陷，后角圆；侧缘近平直。上颚具6个齿。唇基中部纵向凹陷，具向前分歧的双脊，前缘轻度凹陷。触角11节，触角棒2节，柄节末端到达触角窝至头后角间距的3/7处。缺复眼。侧面观，头后缘具1对小齿状角突，头部背面平直。前中胸背板圆形隆起，前中胸背板缝在背面消失。中胸背板背面平直。后胸背板存在，三角形，隆起。中后胸背板缝和后胸沟存在。并胸腹节具1对侧扁粗齿，顶端下弯；背面轻度凹陷，向后坡形降低；斜面平直。腹柄前面具小柄，腹面平直，前下角具1个尖齿；腹柄结三角形，前面轻度凹陷，后面轻度隆起，顶端直角形。后腹柄结圆形隆起，稍低于腹柄结。背面观，腹柄结与后腹柄结等宽。上颚和唇基中部光滑发亮。头前部具纵条纹，后部具细网纹。胸部、腹柄、后腹柄和后腹部具细密刻点。头部和身体具丰富亚直立短毛和密集倾斜绒毛被；柄节和胫节具密集倾斜绒毛被。身体黄色，上颚咀嚼缘黑色。**副模兵蚁：** 特征同正模。**副模工蚁：** 特征同正模，但是身体小很多，头部正常；头后缘近平直，中央轻微凹陷；上颚具5个齿，触角柄节末端到达触角窝至头后角间距的2/3处；前中胸背板轻微隆起，近平直；缺后胸背板，后胸沟浅凹；并胸腹节具1对直的尖齿，背面轻度隆起；腹柄结较厚，顶端钝凸；腹柄前下角具1个直角形齿。

正模： 兵蚁，中国云南省景洪县普文镇松山岭，1270m，1998.Ⅲ.4，何云峰采于暖性针叶林中，No. A98-39。

词源： 该新种的种名"*curvispinus*"由拉丁语"*curv-*"（弯曲的）+词根"*spin*"（刺）+后缀"*-us*"（阳性形式）组成，指兵蚁的并胸腹节具下弯的刺。

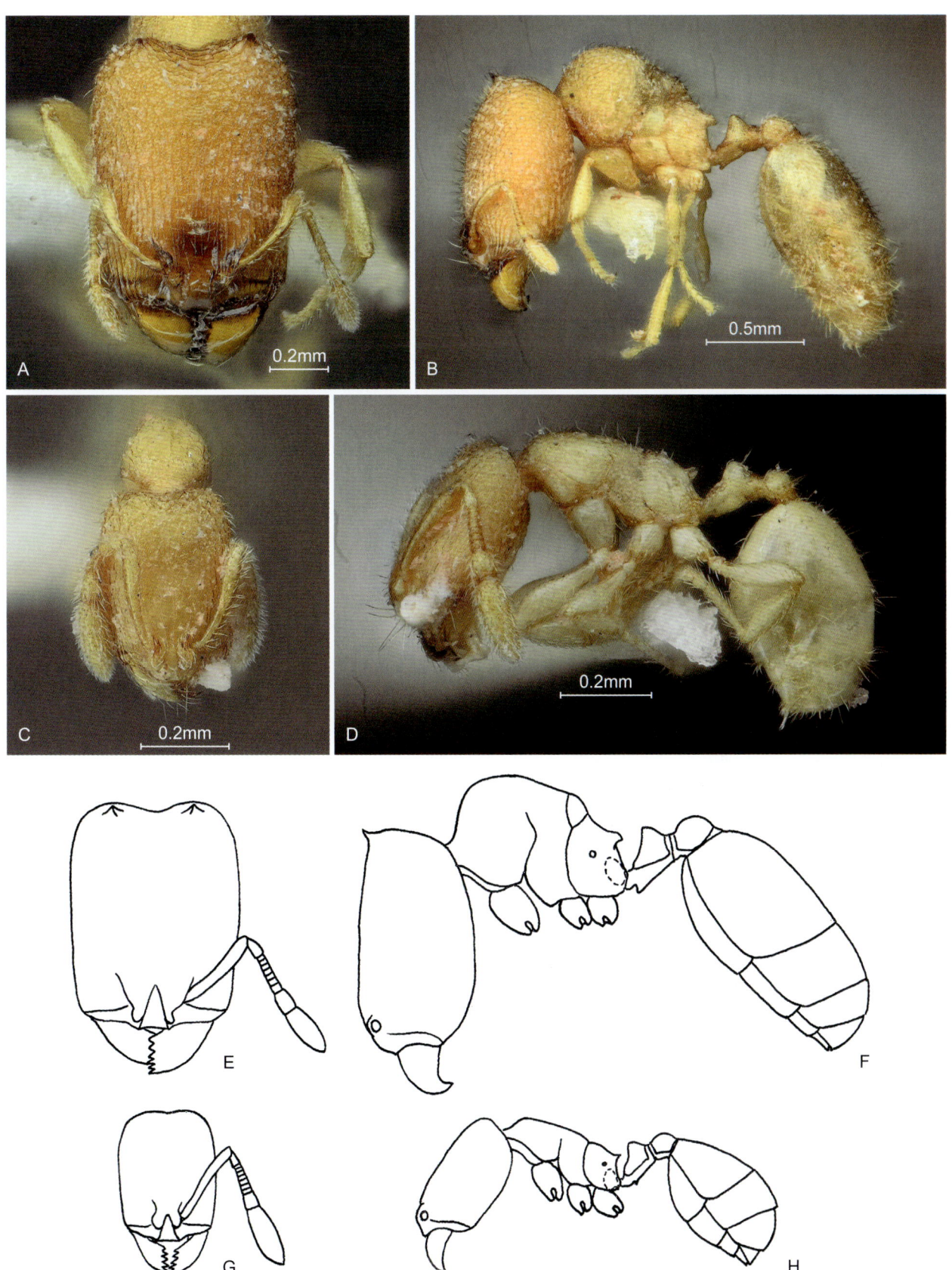

弯刺稀切叶蚁 *Oligomyrmex curvispinus*

A, E. 兵蚁头部正面观；B, F. 兵蚁身体侧面观；C, G. 工蚁头部正面观；D, H. 工蚁身体侧面观（E-H. 引自Xu, 2003）
A, E. Head of soldier in full-face view; B, F. Body of soldier in lateral view; C, G. Head of worker in full-face view; D, H. Body of worker in lateral view (E-H. cited from Xu, 2003)

钝齿稀切叶蚁
Oligomyrmex obtusidentus Xu, 2003

Oligomyrmex obtusidentus Xu, 2003, Acta Zootaxonomica Sinica, 28(2): 316, figs. 20-23 (s.w.) CHINA (Yunnan).

Holotype soldier: TL 2.6, HL 0.73, HW 0.58, CI 79, SL 0.30, SI 52, PW 0.33, AL 0.58, PL 0.23, PH 0.16, DPW 0.15. In full-face view, head longer than broad, rectangular. Occipital margin moderately concave, occipital corners rounded. Sides straight and parallel. Mandible with 5 teeth. Median portion of clypeus longitudinally depressed, bicarinate and divergent forward, anterior margin concave. Antenna 9 segments with a 2-segmented club, apex of scape extending to 1/2 of the distance from socket to occipital corner. Eye with 1 facet. In lateral view, occiput with a pair of acute small horns, dorsum of head lightly convex. Promesonotum high and roundly convex. Promesonotal suture obsolete on the dorsum. Metanotum absent. Metanotal groove deeply impressed. Propodeum with a pair of rightly angled teeth, dorsum straight and sloping down rearwards, declivity concave with thin lateral laminae. Petiole pedunculate anteriorly, ventral face weakly concave, anteroventral corner acutely dentate. Petiolar node thick, anterior and posterior faces sloping, dorsal face roundly convex. Postpetiolar node roundly convex and lightly lower than petiolar node. In dorsal view, petiolar node about as broad as postpetiolar node. Mandibles and median portion of clypeus smooth and shiny. Anterior 2/3 of head dorsum densely and finely striate, posterior 1/3 of head dorsum and sides reticulate. Transverse striations present between occipital horns, but without a developed transverse ridge. Alitrunk and petiole densely punctate, interspace appearing as microreticulations. Postpetiole and gaster smooth and shiny. Head with dense suberect hairs. Alitrunk, petiole and postpetiole with sparse suberect hairs and dense decumbent pubescence. First segment of gaster with dense decumbent hairs, the other segments with decumbent hairs posteriorly. Scapes and tibiae with dense decumbent pubescence. Head yellowish brown. Alitrunk, petiole and postpetiole yellow. Antennae, tibiae and gaster light yellow. **Paratype soldiers:** TL 2.1-2.6, HL 0.65-0.73, HW 0.53-0.58, CI 76-81, SL 0.28-0.33, SI 50-57, PW 0.28-0.33, AL 0.53-0.58, PL 0.21-0.25, PH 0.15-0.16, DPW 0.13-0.15 (*n*=5). As holotype, but occipital horns vary in size from minute to small. **Paratype workers:** TL 1.2-1.3, HL 0.38-0.40, HW 0.33-0.34, CI 87-93, SL 0.23-0.25, SI 68-71, PW 0.23-0.24, AL 0.38, PL 0.13-0.15, PH 0.10, DPW 0.08-0.09 (*n*=5). As holotype, but much smaller. Head normal, nearly square, without occipital horns, occipital margin shallowly concave in the middle, sides lightly convex. Apex of scape reached to 3/4 of the distance from socket to occipital corner. Posterior 1/2 of head dorsum reticulate. Promesonotum moderately convex. Propodeum with a pair of acute teeth, dorsum convex.

Holotype: Soldier, China: Yunnan Province, Tengchong County, Jietou Town, Daying Village, 2000 m, 1999.Ⅴ.2, collected by Lei Fu in subalpine moist evergreen broadleaf forest on the west slope of Gaoligong Mountain, No. A99-259.

Etymology: The species name *obtusidentus* combines Latin *obtus-* (blunt) + word root *dent* (tooth) + suffix *-us* (masculine form), it refers to the blunt teeth of propodeum of soldier.

正模兵蚁： 正面观，头部长大于宽，长方形；后缘中度凹陷，后角圆；侧缘平直，互相平行。上颚具5个齿。唇基中部纵向凹陷，具向前分歧的双脊，前缘凹陷。触角9节，触角棒2节，柄节末端到达触角窝至头后角间距的1/2处。复眼具1个小眼。侧面观，头顶后部具1对尖锐的小角突，头部背面轻微隆起。前中胸背板高，圆形隆起。前中胸背板缝在背面消失，缺后胸背板，后胸沟深凹。并胸腹节具1对直角形齿，背面平直，向后坡形降低；斜面凹陷，具薄的侧脊。腹柄前面具小柄，腹面轻度凹陷，前下角具尖齿；腹柄结厚，前面和后面坡形，背面圆形隆起。后腹柄结圆形隆起，稍低于腹柄结。背面观，腹柄结约与后腹柄结等宽。上颚和唇基中部光滑发亮。头部背面前部2/3具细密条纹，后部1/3和侧面具网纹；头顶后部角突之间具横条纹，但是缺1条发达的横脊。胸部和腹柄具密集刻点，界面呈微网纹。后腹柄和后腹部光滑发亮。头部具密集亚直立毛；胸部、腹柄和后腹柄具稀疏亚直立毛和密集倾斜绒毛被；后腹部第1节具密集倾斜毛，其余各节后部具倾斜毛；柄节和胫节具密集倾斜绒毛被。头部黄棕色，胸部、腹柄和后腹柄黄色，触角、胫节和后腹部浅黄色。**副模兵蚁：** 特征同正模，但是头顶后部的角突有变化，微小至小形。**副模工蚁：** 特征同正模，但是身体小很多。头部正常，近方形，头顶后部缺角突，头后缘中央浅凹，侧缘轻微隆起。柄节末端到达触角窝至头后角间距的3/4处。头部背面后部1/2具网纹。前中胸背板中度隆起。并胸腹节具1对尖齿，背面隆起。

正模： 兵蚁，中国云南省腾冲县界头乡大营村，2000m，1999.Ⅴ.2，付磊采于高黎贡山西坡中山湿性常绿阔叶林中，No. A99-259。

词源： 该新种的种名"*obtusidentus*"由拉丁语"*obtus-*"（钝的）+词根"*dent*"（齿）+后缀"*-us*"（阳性形式）组成，指兵蚁的并胸腹节具钝的齿。

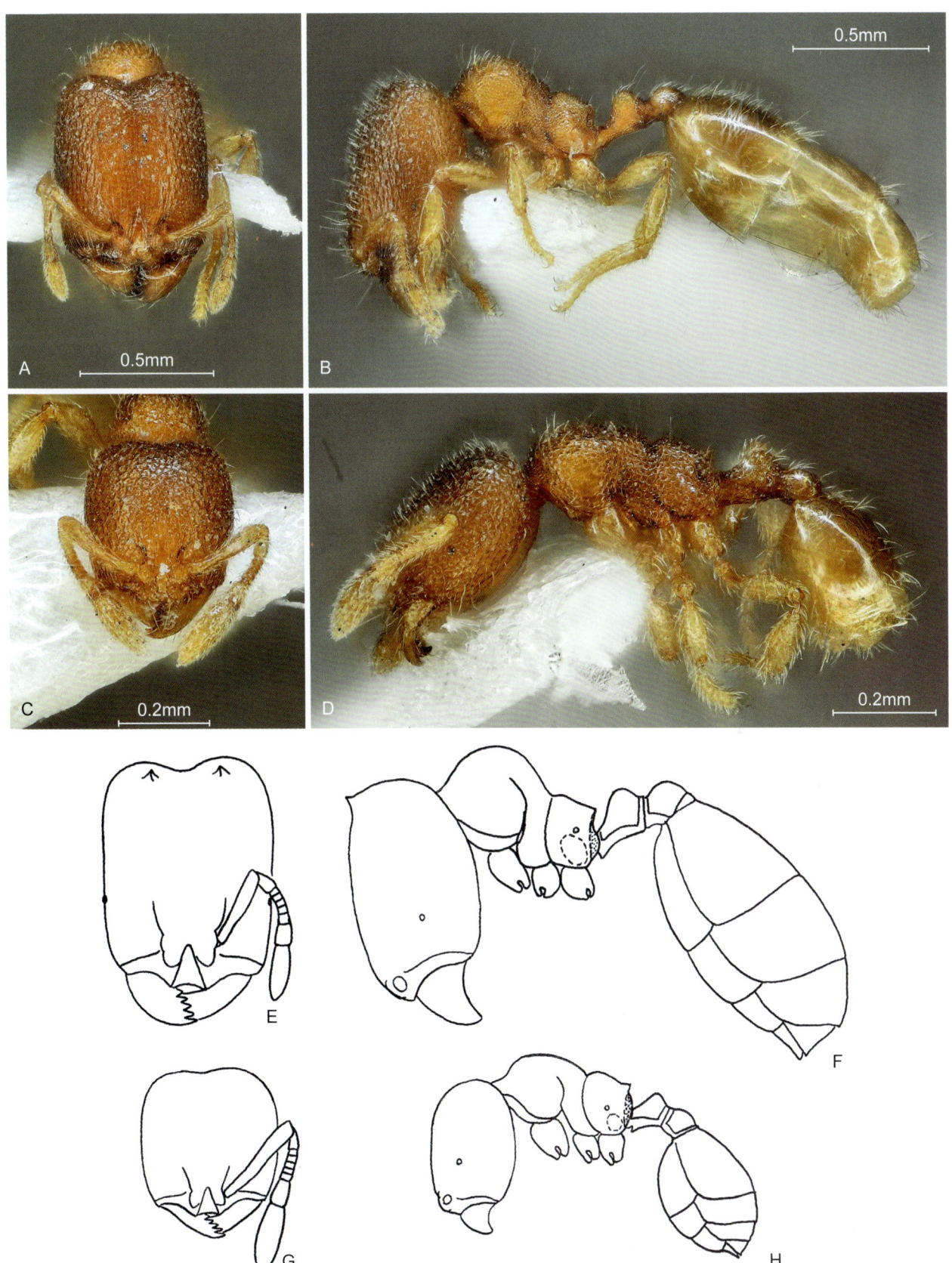

钝齿稀切叶蚁 *Oligomyrmex obtusidentus*

A, E. 兵蚁头部正面观；B, F. 兵蚁身体侧面观；C, G. 工蚁头部正面观；D, H. 工蚁身体侧面观（E-H. 引自Xu, 2003）
A, E. Head of soldier in full-face view; B, F. Body of soldier in lateral view; C, G. Head of worker in full-face view; D, H. Body of worker in lateral view (E-H. cited from Xu, 2003)

直背稀切叶蚁
Oligomyrmex rectidorsus Xu, 2003

Oligomyrmex rectidorsus Xu, 2003, Acta Zootaxonomica Sinica, 28(2): 319, figs. 32-35 (s.w.) CHINA (Yunnan).

Holotype soldier: TL 1.7, HL 0.63, HW 0.48, CI 76, SL 0.25, SI 53, PW 0.28, AL 0.45, PL 0.18, PH 0.13, DPW 0.10. In full-face view, head longer than broad, rectangular, slightly narrowed forward. Occipital margin moderately and angularly concave in the middle, occipital corners roundly prominent. Sides weakly convex. Mandible with 5 teeth. Median portion of clypeus longitudinally depressed, bicarinate and divergent forward, anterior margin weakly concave. Antenna 9 segments with a 2-segmented club, apex of scape reaching 1/2 of the distance from socket to occipital corner. Eye with 2 facets. In lateral view, occiput with a pair of small acute horns, dorsum of head weakly convex. Promesonotum high and roundly convex. Promesonotal suture obsolete on the dorsum. Metanotum absent. Metanotal groove deeply impressed. Propodeum with posterodorsal corner roundly prominent, dorsum straight and sloping down rearwards, declivity concave with thin lateral laminae. Petiole pedunculate anteriorly, ventral face straight, anteroventral corner weakly and bluntly angled. Petiolar node thick, anterior and posterior faces sloping, dorsal face roundly prominent. Postpetiolar node roundly convex and lower than petiolar node. In dorsal view, petiolar node as broad as postpetiolar node. Mandibles and median portion of clypeus smooth and shiny. Anterior 2/5 of head finely and longitudinally striate, middle portion smooth and shiny, occiput with sparse transverse striations. Transverse striations present between occipital horns, but without a developed ridge. Pronotum and mesonotum smooth and shiny. Sides of mesothorax, metathorax, sides of propodeum, petiole and postpetiole finely punctured. Dorsum of propodeum, dorsum of petiolar node, dorsum of postpetiolar node and gaster smooth and shiny. Head and body with abundant erect or suberect hairs and abundant decumbent pubescence. Scapes and tibiae with dense decumbent pubescence. Body color yellow. Head brownish yellow. Masticatory margins of mandibles, eyes and occipital horns black. **Paratype soldiers:** TL 1.4-1.7, HL 0.55-0.60, HW 0.43-0.46, CI 75-78, SL 0.23-0.25, SI 49-56, PW 0.24-0.25, AL 0.38-0.45, PL 0.15-0.18, PH 0.11-0.13, DPW 0.10 (*n*=5). As holotype. **Paratype workers:** TL 1.1-1.2, HL 0.35-0.38, HW 0.30, CI 80-86, SL 0.20-0.23, SI 67-75, PW 0.20, AL 0.33-0.38, PL 0.13, PH 0.09-0.10, DPW 0.08 (*n*=5). As holotype, but body much smaller. Head normal and nearly square, slightly longer than broad, without occipital horns. Occipital margin shallowly concave in the middle, sides of head moderately convex. Apex of scape reaching to 3/4 of the distance from socket to occipital corner. Eye with 1 facet. In lateral view, promesonotum weakly convex. Head smooth and shining. Body color light yellow.

Holotype: Soldier, China: Yunnan Province, Mengla County, Shangyong Town, Nanqian Village, 820 m, 1997.Ⅷ.15, collected by Zheng-hui Xu in secondary shrub, No. A97-1956.

Etymology: The species name *rectidorsus* combines Latin *rect-* (straight) + word root *dors* (back) + suffix *-us* (masculine form), it refers to the straight dorsum of propodeum of soldier.

正模兵蚁：正面观，头部长大于宽，长方形，向前轻微变窄；后缘中央角状中度凹陷，后角圆凸；侧缘轻度隆起。上颚具5个齿。唇基中部纵向凹陷，具向前分歧的双脊，前缘轻度凹陷。触角9节，触角棒2节，柄节末端到达触角窝至头后角间距的1/2处。复眼具2个小眼。侧面观，头顶后部具1对小的尖角突，头部背面轻度隆起。前中胸背板高，圆形隆起。前中胸背板缝在背面消失，缺后胸背板，后胸沟深凹。并胸腹节后上角圆，背面平直，向后坡形降低；斜面凹陷，具薄的侧脊。腹柄前面具小柄，腹面平直，前下角具低的钝角；腹柄结厚，前面和后面坡形，背面圆突。后腹柄结圆形隆起，低于腹柄结。背面观，腹柄结与后腹柄结等宽。上颚和唇基中部光滑发亮。头前部2/5具细纵条纹，中部光滑发亮，后部具稀疏横条纹；头顶后部角突之间具横条纹，但是缺1条发达的横脊。前胸背板和中胸背板光滑发亮；中胸侧面、后胸、并胸腹节侧面、腹柄和后腹柄具细刻点；并胸腹节背面、腹柄结背面、后腹柄结背面和后腹部光滑发亮。头部和身体具丰富直立、亚直立毛和丰富倾斜绒毛被；柄节和胫节具密集倾斜绒毛被。身体黄色，头部棕黄色，上颚咀嚼缘、复眼和头顶角突黑色。**副模兵蚁：**特征同正模。**副模工蚁：**特征同正模，但是身体小很多；头部正常，近方形，长稍大于宽；头顶缺角突，后缘中央浅凹，侧缘中度隆起；柄节末端到达触角窝至头后角间距的3/4处；复眼具1个小眼；侧面观，前中胸背板轻度隆起；头部光滑发亮，身体浅黄色。

正模：兵蚁，中国云南省勐腊县尚勇乡南欠村，820m，1997.Ⅷ.15，徐正会采于次生灌丛中，No. A97-1956。

词源：该新种的种名"*rectidorsus*"由拉丁语"*rect-*"（直的）+词根"*dors*"（背）+后缀"*-us*"（阳性形式）组成，指兵蚁的并胸腹节背面平直。

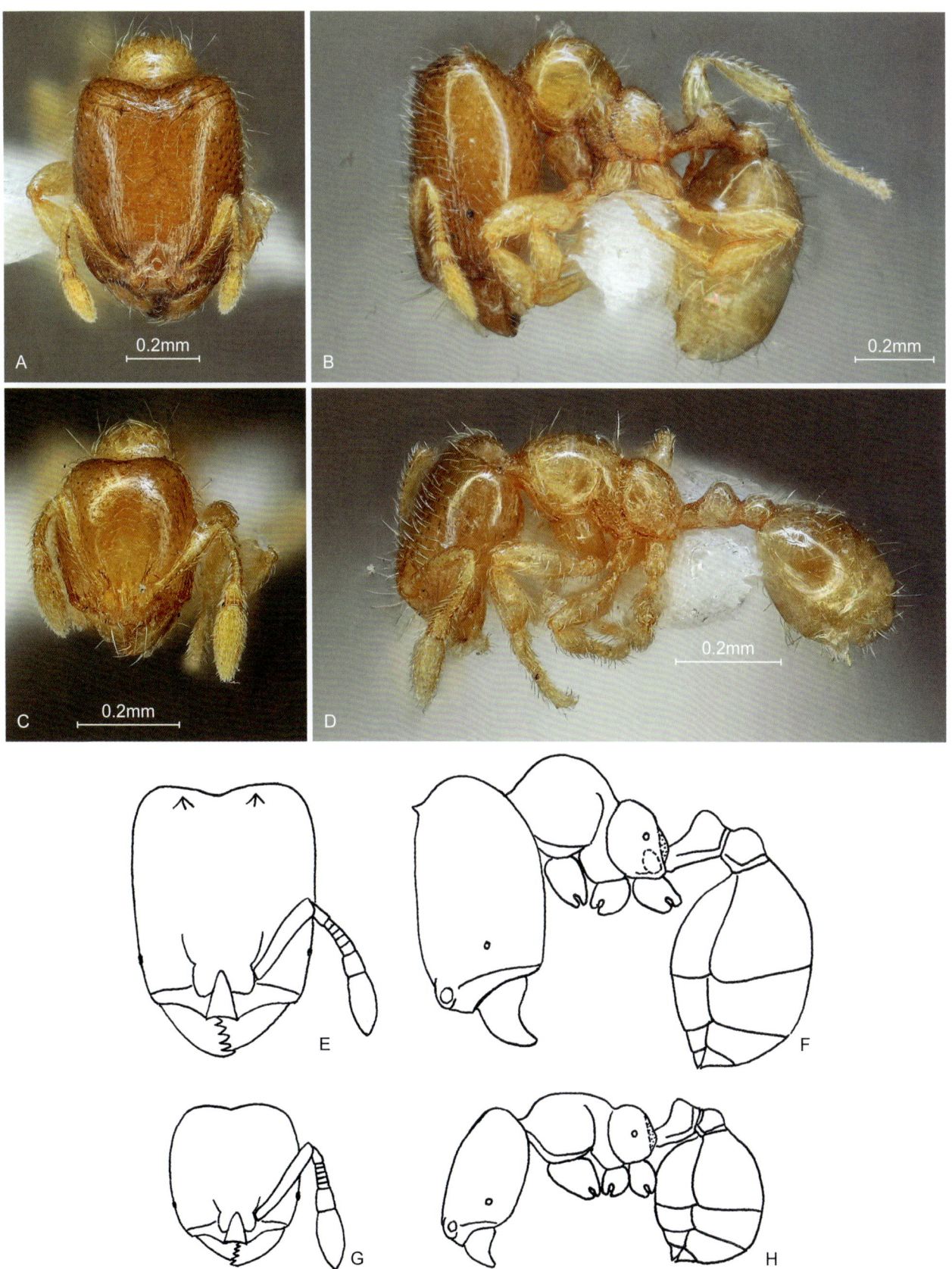

直背稀切叶蚁 *Oligomyrmex rectidorsus*

A, E. 兵蚁头部正面观；B, F. 兵蚁身体侧面观；C, G. 工蚁头部正面观；D, H. 工蚁身体侧面观（E-H. 引自Xu, 2003）
A, E. Head of soldier in full-face view; B, F. Body of soldier in lateral view; C, G. Head of worker in full-face view; D, H. Body of worker in lateral view (E-H. cited from Xu, 2003)

纹头稀切叶蚁
Oligomyrmex reticapitus Xu, 2003

Oligomyrmex reticapitus Xu, 2003, Acta Zootaxonomica Sinica, 28(2): 319, figs. 38-41 (s.w.) CHINA (Yunnan).

Holotype soldier: TL 1.7, HL 0.58, HW 0.45, CI 78, SL 0.25, SI 56, PW 0.28, AL 0.45, PL 0.18, PH 0.13, DPW 0.13. In full-face view, head longer than broad, rectangular. Occipital margin moderately and angularly concave, occipital corners roundly prominent. Sides straight and parallel. Mandible with 5 teeth. Median portion of clypeus longitudinally depressed, bicarinate and divergent forward, anterior margin nearly straight. Antenna 9 segments with a 2-segmented club, apex of scape reaching 5/9 of the distance from socket to occipital corner. Eye with 3 facets. In lateral view, occiput without horns, dorsum of head slightly convex. Promesonotum high and roundly convex. Promesonotal suture obsolete on the dorsum. Metanotum absent. Metanotal groove deeply impressed. Propodeum with a pair of acute teeth, dorsum straight and sloping down rearward, declivity concave with thin lateral laminae. Petiole pedunculate anteriorly, ventral face straight, anteroventral corner minutely toothed. Petiolar node thick, anterior and posterior faces sloping, dorsal face roundly prominent. Postpetiolar node roundly convex and lower than petiolar node. In dorsal view, petiolar node about as broad as postpetiolar node. Mandibles and median portion of clypeus smooth and shiny. Head finely reticulate, occiput finely and transversely striate. Alitrunk and petiole with microreticulations. Postpetiole and gaster smooth and shiny. Head and body with sparse suberect hairs and dense decumbent pubescence. Scapes and tibiae with dense decumbent pubescence. Body in color yellow. Head brownish yellow. **Paratype soldiers:** TL 1.5-2.0, HL 0.55-0.60, HW 0.40-0.48, CI 73-79, SL 0.20-0.28, SI 50-58, PW 0.25-0.30, AL 0.43-0.48, PL 0.18-0.20, PH 0.13-0.15, DPW 0.11-0.13 (*n*=4). As holotype, but eye with 1-3 facets. **Paratype workers:** TL 1.1-1.2, HL 0.30-0.38, HW 0.28-0.30, CI 80-92, SL 0.20-0.23, SI 73-75, PW 0.18-0.21, AL 0.33-0.38, PL 0.11-0.13, PH 0.10, DPW 0.08-0.09 (*n*=5). As holotype, but body much smaller. Head normal, slightly longer than broad and weakly narrowed forward, sides weakly convex, occipital margin shallowly concave in the middle. Apex of scape reaching 2/3 of the distance from socket to occipital corner. Eye with 1 facet. Promesonotum weakly convex. Propodeum with a pair of small acute teeth, dorsum moderately convex. Body color yellow.

Holotype: Soldier, China: Yunnan Province, Menghai County, Meng'a Town, Papo Village, 1600 m, 1997.IX.9, collected by Zheng-hui Xu in warm deciduous broadleaf forest, No. A97-2282.

Etymology: The species name *reticapitus* combines Latin *ret-* (net) + word root *capit* (head) + suffix *-us* (masculine form), it refers to the reticulate head of soldier.

正模兵蚁： 正面观，头部长大于宽，长方形；后缘中度角状凹陷，后角圆凸；侧缘平直，互相平行。上颚具5个齿。唇基中部纵向凹陷，具向前分歧的双脊，前缘近平直。触角9节，触角棒2节，柄节末端到达触角窝至头后角间距的5/9处。复眼具3个小眼。侧面观，头顶后部缺角突，头部背面轻微隆起。前中胸背板高，圆形隆起。前中胸背板缝在背面消失，缺后胸背板，后胸沟深凹。并胸腹节具1对尖齿，背面平直，向后坡形降低；斜面凹陷，具薄的侧脊。腹柄前面具小柄，腹面平直，前下角具小齿；腹柄结厚，前面和后面坡形，背面圆突。后腹柄结圆形隆起，低于腹柄结。背面观，腹柄结约与后腹柄结等宽。上颚和唇基中部光滑发亮。头部具细网纹，头后部具细横条纹。胸部和腹柄具微网纹。后腹柄和后腹部光滑发亮。头部和身体具稀疏亚直立毛和密集倾斜绒毛被；柄节和胫节具密集倾斜绒毛被。身体黄色，头部棕黄色。**副模兵蚁：** 特征同正模，但是复眼具1～3个小眼。**副模工蚁：** 特征同正模，但是身体小很多；头部正常，长稍大于宽，向前轻度变窄，侧缘轻度隆起，后缘中央浅凹；触角柄节末端到达触角窝至头后角间距的2/3处；复眼具1个小眼；前中胸背板轻度隆起；并胸腹节具1对小的尖齿，背面中度隆起；身体黄色。

正模： 兵蚁，中国云南省勐海县勐阿乡怕迫村，1600m，1997.IX.9，徐正会采于暖性落叶阔叶林中，No. A97-2282。

词源： 该新种的种名"*reticapitus*"由拉丁语"*ret-*"（网）+词根"*capit*"（头）+后缀"*-us*"（阳性形式）组成，指兵蚁的头部具网纹。

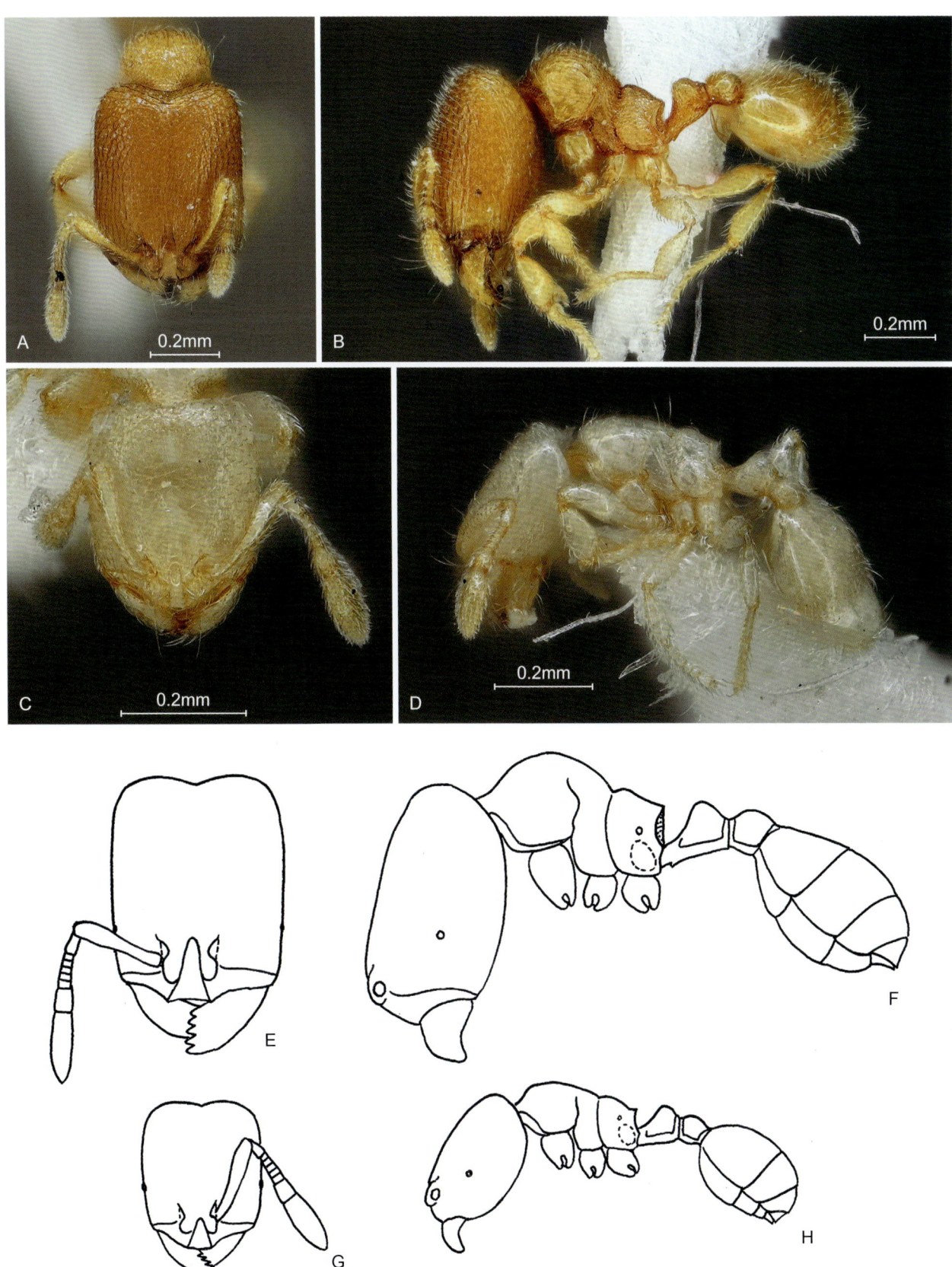

纹头稀切叶蚁 *Oligomyrmex reticapitus*

A, E. 兵蚁头部正面观; B, F. 兵蚁身体侧面观; C, G. 工蚁头部正面观; D, H. 工蚁身体侧面观（E-H. 引自Xu, 2003）
A, E. Head of soldier in full-face view; B, F. Body of soldier in lateral view; C, G. Head of worker in full-face view; D, H. Body of worker in lateral view (E-H. cited from Xu, 2003)

条纹稀切叶蚁
Oligomyrmex striatus Xu, 2003

Oligomyrmex striatus Xu, 2003, Acta Zootaxonomica Sinica, 28(2): 314, figs. 13-15 (s.) CHINA (Yunnan).

Holotype soldier: TL 2.5, HL 0.78, HW 0.60, CI 77, SL 0.30, SI 50, PW 0.43, AL 0.75, PL 0.25, PH 0.23, DPW 0.28. In full-face view, head longer than broad, roughly rectangular and moderately narrowed forward. Occipital margin moderately concave, occipital corners rounded. Sides straight. Mandible with 6 teeth. Median portion of clypeus depressed longitudinally, bicarinate and divergent forward, anterior margin concave. Antenna 11 segments with a 2-segmented club, apex of scape reaching 3/7 of the distance from socket to occipital corner. Eye minute, with 1 facet. In lateral view, occiput with a pair of short triangular horns. Dorsum of head straight. Frontal lobes protruding. Promesonotum high and roundly convex. Promesonotal suture obsolete on the dorsum. Dorsum of mesonotum straight, suddenly down-curved at posterior end. Metanotum present, narrow and acutely pointed backward. Mesometanotal groove impressed, metapropodeal groove present. Propodeum with a pair of strong teeth, dorsum concave and down sloping, declivity concave. Petiole pedunculate anteriorly, ventral face weakly depressed, anteroventral corner with an acute tooth, sides with sharp horizontal ridge. Petiolar node triangular, anterior face slope, posterior face nearly vertical, dorsal face narrow and rounded. Postpetiolar node roundly convex and lightly lower than petiolar node, sides also with sharp horizontal ridge. In dorsal view, petiolar node as broad as postpetiolar node, both with quite developed roundly convex lateral ridges. Mandibles and median portion of clypeus smooth and shiny. Head longitudinally striate, occiput with reticulations. Dorsum of pronotum and postpetiole with reticulations. Mesonotum finely longitudinally striate. Sides of alitrunk, declivity of propodeum and petiole finely densely punctate. Dorsum of gaster densely and longitudinally striate, ventral face smooth. Head and body with dense erect or suberect hairs and dense decumbent pubescence. Scapes and tibiae with dense decumbent pubescence. Body color yellow. Antennae and legs light yellow. Gaster yellowish brown. Apical segments of gaster behind the first one lost. **Workers:** unknown.

Holotype: Soldier, China: Yunnan Province, Mengla County, Shangyong Town, Nanqian Village, 820 m, 1997.Ⅷ.15, collected by Zheng-hui Xu in secondary shrub, No. A97-1925.

Etymology: The species name *striatu* combines Latin *stri-* (stripe) + suffix *-atus* (masculine form, with), it refers to the striate gaster of soldier.

正模兵蚁： 正面观，头部长大于宽，近长方形，向前中度变窄；后缘中度凹陷，后角圆，侧缘平直。上颚具6个齿。唇基中部纵向凹陷，具向前分歧的双脊，前缘凹陷。触角11节，触角棒2节，柄节末端到达触角窝至头后角间距的3/7处。复眼微小，具1个小眼。侧面观，头顶后部具1对短的三角形角突，头部背面平直，额叶突出。前中胸背板高，圆形隆起。前中胸背板缝在背面消失。中胸背板背面平直，在后端急剧下弯。后胸背板存在，窄而尖锐，指向后方。中后胸沟凹陷，后胸沟存在。并胸腹节具1对发达的齿，背面凹陷，向后坡形降低；斜面凹陷。腹柄前面具小柄，腹面轻度凹陷，前下角具尖齿，两侧具水平的锐脊；腹柄结三角形，前面坡形，后面近垂直，背面窄而圆。后腹柄结圆形隆起，稍低于腹柄结，两侧同样具水平锐脊。背面观，腹柄结与后腹柄结等宽，均具相当发达圆形隆起的侧脊。上颚和唇基中部光滑发亮。头部具纵条纹，头后部具网纹。前胸背板背面和后腹柄具网纹；中胸背板具细纵条纹；胸部侧面、并胸腹节斜面和腹柄具细密刻点；后腹部背面具密集纵条纹，腹面光滑。头部和身体具密集直立、亚直立毛和密集倾斜绒毛被；柄节和胫节具密集倾斜绒毛被。身体黄色，触角和足浅黄色，后腹部黄棕色。后腹部第1节之后各节缺失。**工蚁：** 未知。

正模： 兵蚁，中国云南省勐腊县尚勇乡南欠村，820m，1997.Ⅷ.15，徐正会采于次生灌丛中，No. A97-1925。

词源： 该新种的种名"*striatus*"由拉丁语"*stri-*"（条纹）+后缀"*-atus*"（阳性形式，具有）组成，指兵蚁的后腹部具条纹。

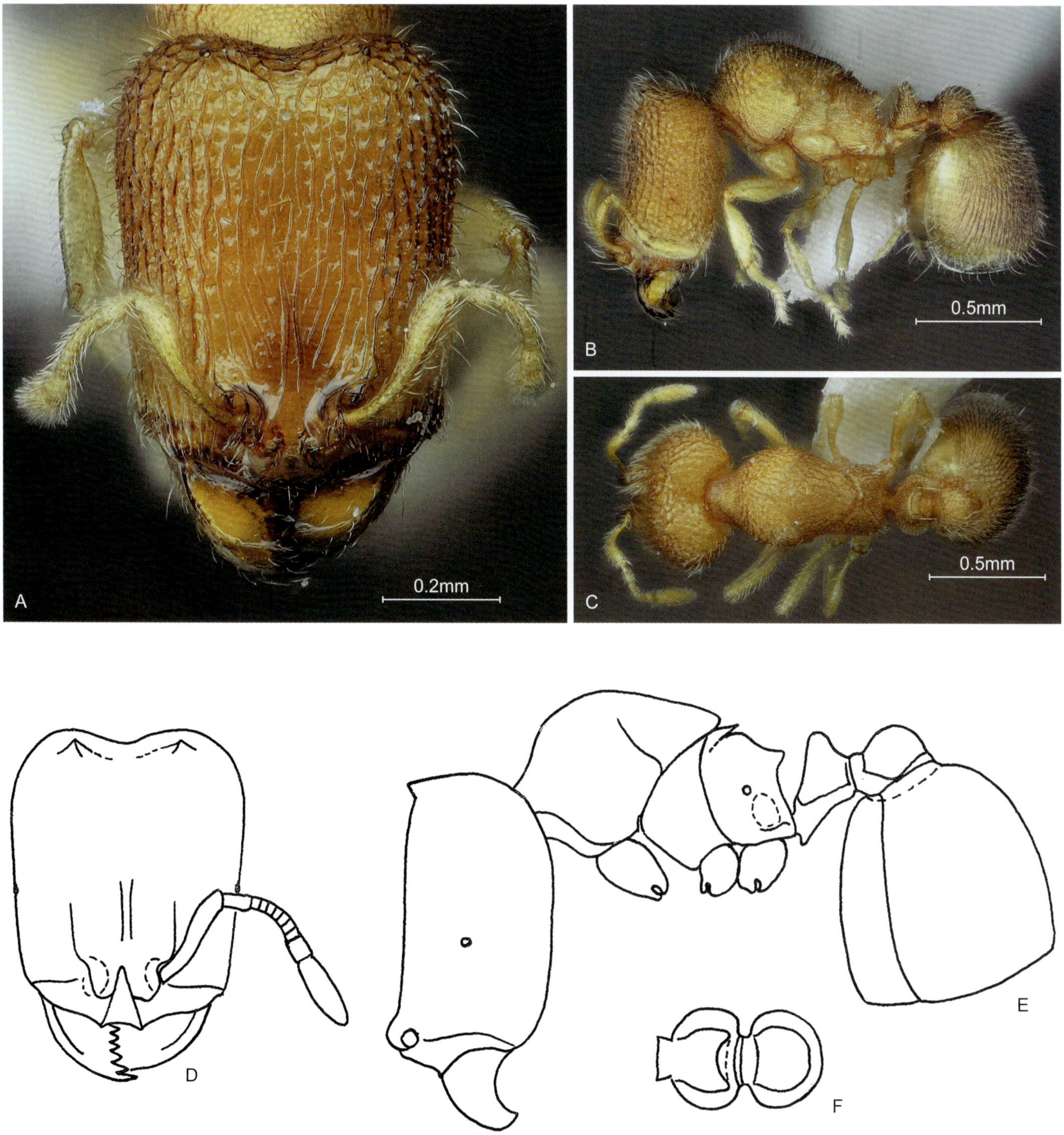

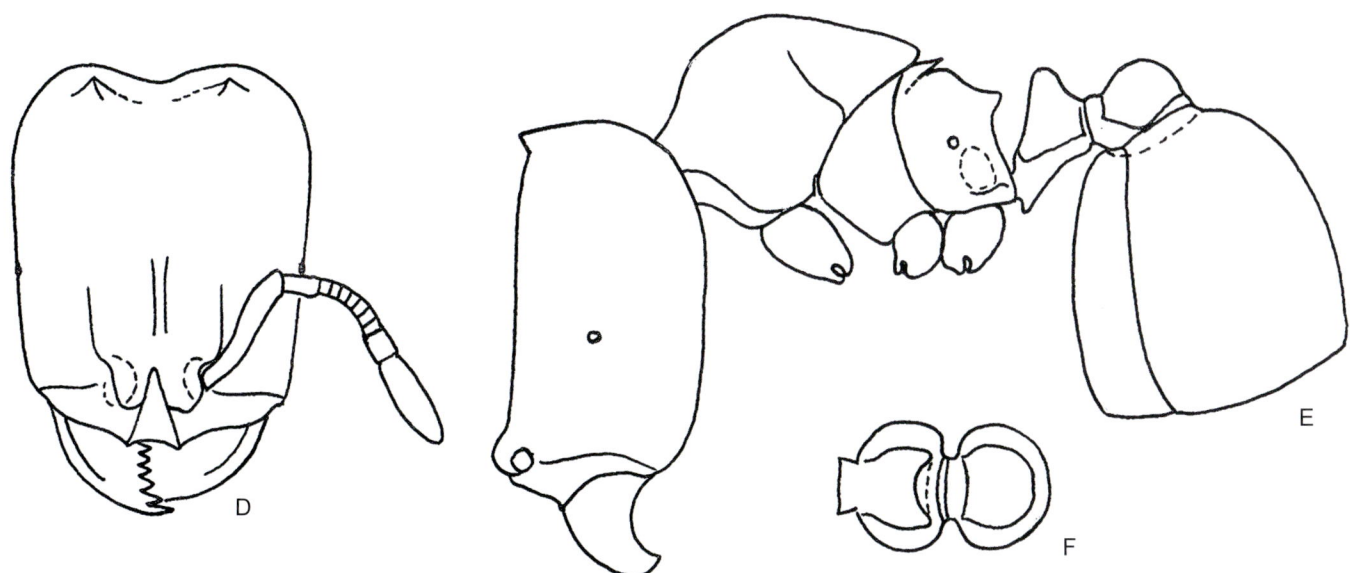

条纹稀切叶蚁 *Oligomyrmex striatus*

A, D. 兵蚁头部正面观；B, E. 兵蚁身体侧面观；C. 兵蚁身体背面观；F. 兵蚁腹柄和后腹柄背面观（D-F. 引自Xu, 2003）

A, D. Head of soldier in full-face view; B, E. Body of soldier in lateral view; C. Body of soldier in dorsal view; F. Petiole and postpetiole of soldier in dorsal view (D-F. cited from Xu, 2003)

郑氏华丽蚁
Paratopula zhengi Xu & Xu, 2011

Paratopula zhengi Xu & Xu, 2011, Acta Zootaxonomica Sinica, 36(3): 595, figs. 1-7 (w.) CHINA (Tibet).

Holotype worker: TL 6.6, HL 1.55, HW 1.25, CI 81, SL 0.98, SI 78, PW 0.90, AL 2.08, ED 0.30. In full-face view, head rectangular, distinctly longer than broad. Occipital margin evenly concave, occipital corners roundly prominent. Sides nearly parallel, slightly narrowed forward. Mandibles triangular, masticatory margin with 9 teeth which decreased in size from apex to base. Anterior margin of clypeus distinctly notched in the middle, posterior median portion between frontal lobes distinctly wider than the latter. Antennae 12-segmented, antennal club 3-segmented. Apexes of scapes reached to 4/5 of the distance from antennal sockets to occipital corners. Frontal carinae weakly developed and parallel, about as long as antennal scapes. Eyes developed, located slightly before the midpoints of sides of head, with about 13 facets across the maximum diameter. In lateral view, promesonotum weakly convex and formed a weak arch. Promesonotal suture indistinct, metanotal groove distinctly notched. Propodeal dorsum feebly convex, slope down backwards, slightly longer than declivity, the latter straight. Propodeal spines slender and acute, slightly curved upwards apically, about 1/2 as long as propodeal dorsum. Propodeal lobes short and truncated apically. Femora of legs obviously swelled in the middle. In lateral view, ventral face of petiole weakly concave, anteroventral corner with a minute prominent. Petiolar node nearly trapezoid, dorsal face weakly convex, anterodorsal and posterodorsal corners roundly prominent. Anterior peduncle shorter than petiolar node. Dorsum of postpetiole roundly convex, anterior 2/3 of sternite roundly strongly convex ventrally. In dorsal view, both petiolar and postperiolar nodes roughly trapezoid, distinctly narrowed forwards. First gastral tergite large, occupied about 3/5 of the length of gaster. Sting extruding. Mandibles relatively smooth, with sparse fine punctures. Head, alitrunk, petiole, and postpetiole with similar relatively coarse reticulations. Clypeus with sparse fine longitudinal striations. Dorsum of head with relatively coarse longitudinal striations between frontal carinae. Dorsal surfaces of middle and hind tibiae with longitudinal striations. Gaster smooth, basigastral costulae present on basal 1/3 of the first tergite. Dorsum of whole body with abundant similar short blunt erect to suberect hairs, sparse depressed pubescence visible on the gaster. Antennal scapes with sparse short blunt erect hairs and dense depressed pubescence. Tibiae with abundant decumbent pubescence, but without erect hairs. Color orange yellow; mandibles, tarsi, and middle gaster brown; eyes black; coxae, femora, and tibiae yellow.

Holotype: Worker, China: Tibet Autonomous Region, Medog County, Medog Town, Medog (29.316667°N, 95.316667°E), 1080 m, 2008.V.18, collected by Zheng-hui Xu on the ground in monsoon forest area, No. A08-682.

Etymology: The new species is named in honor of Professor Zhe-min Zheng for his outstanding contribution to the systematic entomology.

正模工蚁： 正面观，头部长方形，长明显大于宽，后缘均匀凹陷，后角圆凸，侧缘近平行，向前轻微变窄。上颚三角形，咀嚼缘具9个齿，从端部向基部变小。唇基前缘中央明显切入，后延至额叶之间的中部明显宽于额叶。触角12节，触角棒3节，柄节末端到达触角窝至头后角间距的4/5处。额脊不太发达，互相平行，约与触角柄节等长。复眼发达，位于头侧缘中点稍前处，最大直径上具13个小眼。侧面观，前中胸背板轻度隆起，形成弱弓形。前中胸背板缝不明显，后胸沟明显切入。并胸腹节背面轻度隆起，向后坡形降低，稍长于斜面，斜面平直；并胸腹节刺细长尖锐，顶端轻微上弯，约为并胸腹节背面长的1/2。并胸腹节侧叶短，顶端平截。足的腿节中部明显膨大。侧面观，腹柄腹面轻度凹陷，前下角具1个微小突起；腹柄结近梯形，背面轻度隆起，前上角和后上角圆凸；前面小柄短于腹柄结。后腹柄背面圆形隆起，腹板前部2/3向腹面强烈圆形隆起。背面观，腹柄结和后腹柄结均近梯形，向前明显变窄。后腹部第1节背板大，约占据后腹部长度的3/5。螫针伸出。上颚较光滑，具稀疏细刻点。头部、胸部、腹柄和后腹柄具较粗糙的相似网纹；唇基具稀疏细纵条纹，头部背面额脊之间具较粗糙的纵条纹。中后足胫节背面具纵条纹。后腹部光滑，第1节背板基部1/3具基纵脊。整个身体背面具丰富的相似短钝直立、亚直立毛，稀疏的平伏绒毛被仅在后腹部可见；触角柄节具稀疏短钝直立毛和密集平伏绒毛被；胫节具丰富倾斜绒毛被，缺立毛。身体橙黄色，上颚、跗节和后腹部中部棕色，复眼黑色，基节、腿节和胫节黄色。

正模： 工蚁，中国西藏自治区墨脱县墨脱镇墨脱（29.316667°N，95.316667°E），1080m，2008.V.18，徐正会采于季雨林地表，No. A08-682。

词源： 该新种以郑哲民教授姓氏命名，以纪念他在昆虫分类学领域作出的杰出贡献。

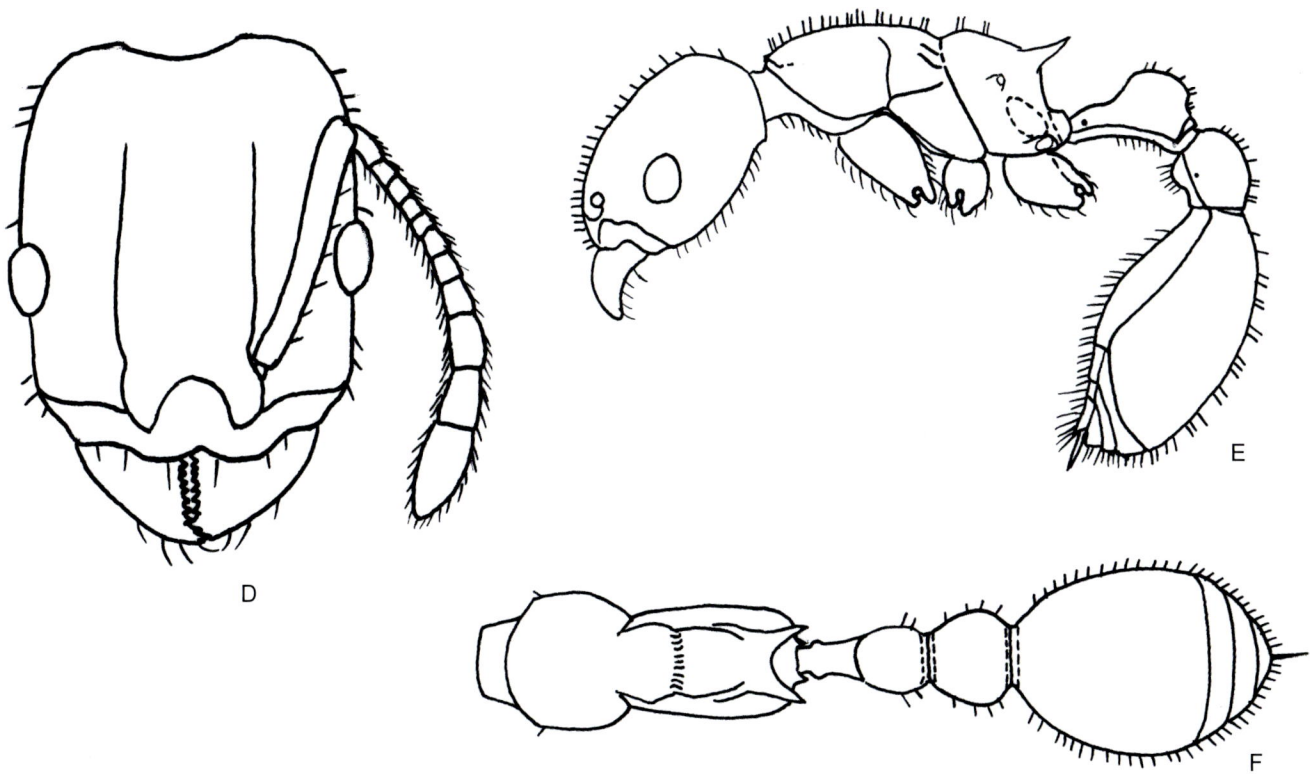

郑氏华丽蚁 *Paratopula zhengi*

A, D. 工蚁头部正面观；B, E. 工蚁身体侧面观；C, F. 工蚁身体背面观（D-F. 引自Xu和Xu, 2011）
A, D. Head of worker in full-face view; B, E. Body of worker in lateral view; C, F. Body of worker in dorsal view (D-F. cited from Xu & Xu, 2011)

裂唇奇蚁
Perissomyrmex fissus Xu & Wang, 2004

Perissomyrmex fissus Xu & Wang, 2004, Entomotaxonomia, 26(3): 218, figs. 7-10 (w.) CHINA (Yunnan).

Holotype worker: TL 3.1, HL 0.83, HW 0.80, CI 97, SL 0.73, SI 91, EL 0.10, PW 0.50, AL 0.88, PL 0.33, PH 0.26, DPW 0.15, PPL 0.23, PPH 0.28, PPW 0.21, PI 46, PPI 94. In full-face view, head nearly square, about as broad as long, widened forward. Occipital margin slightly concave, occipital corners rounded, sides weakly convex. Mandible rectangular, inner margin with one tooth in the middle, masticatory margin with 3 teeth and a diastema, arranged as apical tooth, preapical tooth, diastema and basal tooth. Central notch of anterior margin of clypeus deep and U-shaped, its depth about 2/5 of clypeus length. Sides of the central notch formed a pair of acute teeth. Anterolateral border of clypeus with two denticles. Frontal lobes absent, antennal sockets completely exposed. Antenna 9-segmented, scape surpassed occipital corner by 1/5 of its length, antennal club consisted of 3 apical segments. Eyes relatively smaller and prominent outside the sides of head, with 6 ommatidia along the maximum diameter. In lateral view, pronotum weakly convex. Promesonotal suture absent. Mesonotum sloped backward and angularly prominent at posterior 1/3. Metanotal groove depressed. Propodeal spines strong and acute, dorsum of propodeum straight. Propodeal lobes small and angularly prominent at apices. In lateral view, petiole without subpetiolar process. Petiolar node relatively high and thin, weakly narrowed upward, dorsum moderately convex, anterodorsal corner higher than posterodorsal corner. Anteroventral corner of postpetiole rightly angled, postpetiolar node inclined posteriorly, as high as petiolar node. Apical 1/4 of mandible smooth and shining, basal 3/4 finely longitudinally striate. Clypeus smooth and shining. Head and alitrunk sparsely coarsely and longitudinally striate. Sides of mesothorax, metathorax and propodeum smooth and shining, with very sparse longitudinal striations. Dorsum of propodeum smooth. Lower portions of petiole and postpetiole finely reticulate, petiolar node and postpetiolar node smooth and shining. Gaster smooth and shining. Dorsal surfaces of head and body with abundant erect or suberect long hairs and subdecumbent short hairs. Scapes and tibiae with abundant subdecumbent long hairs and decumbent short hairs. Head, mandibles, antennae and alitrunk brown. Legs, petiole and postpetiole yellowish brown. Gaster black, anterior face and apex blackish brown.

Holotype: Worker, China: Yunnan Province, Xinping County, Gasa Town, Jinshan Pass, 2500 m, 2003.Ⅲ.15, collected by Zheng-hui Xu from a ground sample in the primary subalpine moist evergreen broadleaf forest of Ailao Mountain National Nature Reserve, No. A2826.

Etymology: The species name *fissus* from the Latin *fissus* (masculine form, split open), it refers to the deep central notch on the anterior margin of clypeus.

正模工蚁： 正面观，头部近方形，长宽约相等，向前变宽；后缘轻微凹陷，后角圆，侧缘轻度隆起。上颚长方形，内缘中央具1个齿，咀嚼缘具3个齿和1个齿间隙，排列顺序为端齿、亚端齿、齿间隙和基齿。唇基前缘中央的缺口深，"U"形，其深度约为唇基长度的2/5；中央缺口两侧形成1对尖齿；唇基前侧缘各具2个小齿。缺额叶，触角窝完全外露。触角9节，柄节约1/5超过头后角，触角棒3节。复眼较小，突出于头侧缘之外，最大直径上具6个小眼。侧面观，前胸背板轻度隆起，缺前中胸背板缝。中胸背板向后坡形降低，在后部1/3处角状突出。后胸沟凹陷。并胸腹节刺发达而尖锐，并胸腹节背面平直。并胸腹节侧叶小，顶端角状突出。侧面观，腹柄缺腹柄下突；腹柄结较高较薄，向上轻度变窄，背面中度隆起，前上角高于后上角。后腹柄前下角直角形，后腹柄结后倾，与腹柄结等高。上颚端部1/4光滑发亮，基部3/4具细纵条纹。唇基光滑发亮。头部和胸部具稀疏粗糙纵条纹；中胸侧面、后胸侧面和并胸腹节侧面光滑发亮，具很稀疏的纵条纹；并胸腹节背面光滑。腹柄和后腹柄下部具细网纹，腹柄结和后腹柄结光滑发亮。后腹部光滑发亮。头部和身体背面具丰富直立、亚直立长毛和亚倾斜短毛；柄节和胫节具丰富亚倾斜长毛和倾斜短毛。头部、上颚、触角和胸部棕色；足、腹柄和后腹柄黄棕色；后腹部黑色，前面和末端黑棕色。

正模： 工蚁，中国云南省新平县戛洒镇金山垭口，2500m，2003.Ⅲ.15，徐正会采于哀牢山国家级自然保护区原始中山湿性常绿阔叶林地表样内，No. A2826。

词源： 该新种的种名"*fissus*"来自拉丁语"*fissus*"（阳性形式，裂开），指唇基前缘有深的中央裂口。

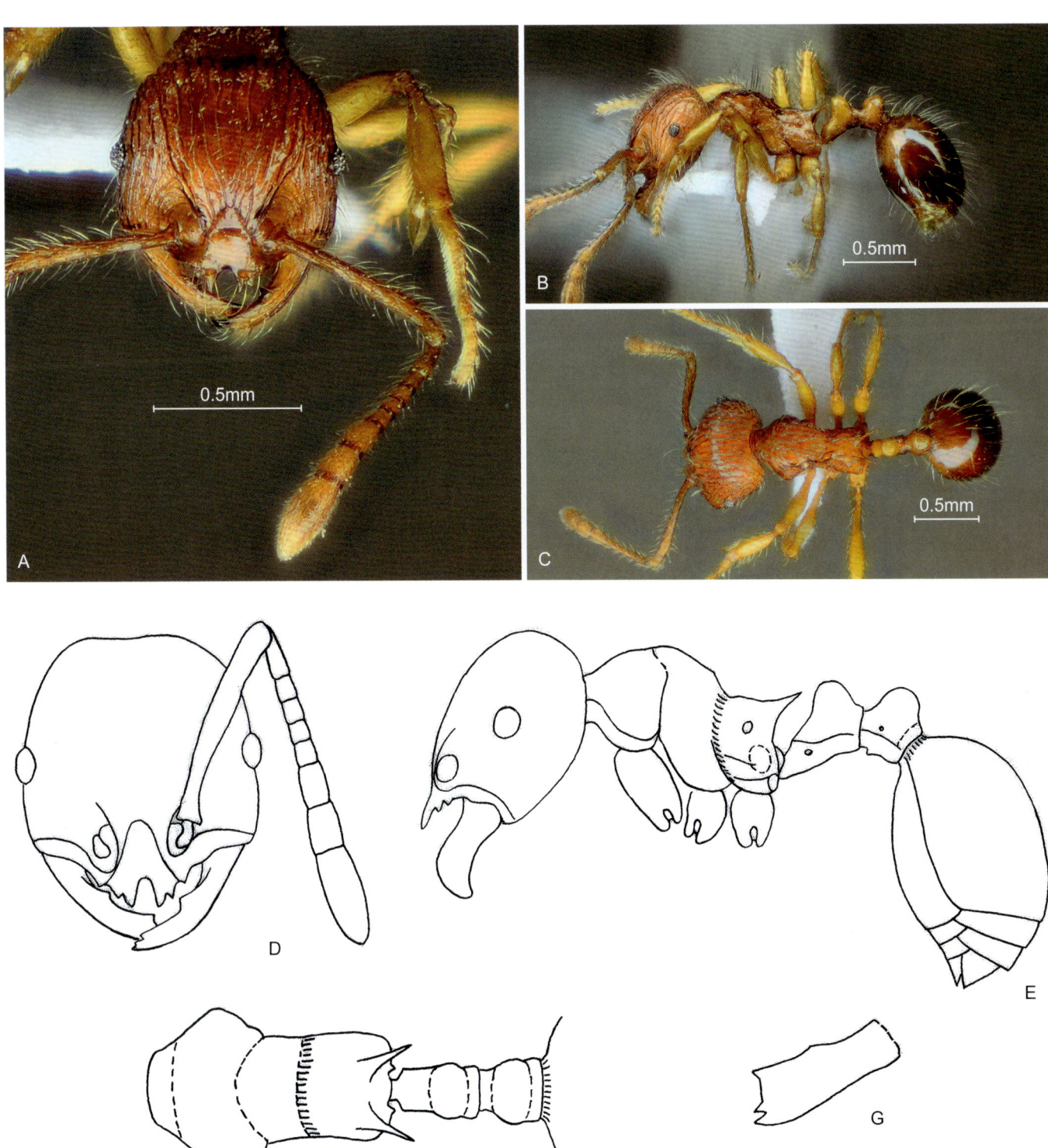

裂唇奇蚁 *Perissomyrmex fissus*

A, D. 工蚁头部正面观；B, E. 工蚁身体侧面观；C, F. 工蚁身体背面观；G. 工蚁上颚背面观（D-G. 引自Xu和Wang, 2004）
A, D. Head of worker in full-face view; B, E. Body of worker in lateral view; C, F. Body of worker in dorsal view; G. Mandible of worker in dorsal view (D-G. cited from Xu & Wang, 2004)

墨脱奇蚁
Perissomyrmex medogensis Xu & Zhang, 2012

Perissomyrmex medogensis Xu & Zhang, 2012, Myrmecological News, 17: 149, figs. 6-21 (s.w.q.) CHINA (Tibet).

Holotype worker: TL 4.0, HL 1.00, HW 1.00, CI 100, SL 0.88, SI 88, ED 0.16, PW 0.58, AL 1.10. In full-face view, head nearly square, as broad as long. Occipital margin weakly convex, occipital corners rounded. Sides evenly convex, widened forward. Mandibles rectangular, inner margin with one tooth, masticatory margin with three teeth; with diastema between subapical and basal tooth. Backward extended median portion of clypeus triangular, posterior corner acutely angled, and followed by short longitudinal furrow. Anterior margin of clypeus with one pair of large triangular median teeth, apices curved outward, with deep notch between teeth. Antennae nine segmented, scape surpassing occipital corners by about 2/7 of its length, antennal clubs three-segmented. Eyes circular and convex, located slightly before midpoints of sides of head. In lateral view, promesonotum roundly convex, sides of pronotum weakly marginate. Promesonotal suture distinct but not depressed, consisting of row of large punctures on the dorsum. Metanotal groove deeply depressed. Mesopleuron with oblique furrow. Dorsum of propodeum straight, weakly sloped down backward. Propodeal spines short and stout, apices slightly curved down, about 1/3 length of propodeal dorsum. Declivity weakly concave, about half length of propodeal dorsum. Propodeal lobes very short, rounded apically. Propodeal spiracle circular, high up on the lateral side, ventral and posterior sides of spiracle depressed. Petiolar node roughly triangular, anterodorsal corner prominent, posterodorsal corner weakly convex, anterior peduncle about as long as node. Ventral face of petiole weakly concave, anteroventral corner weakly convex. Postpetiolar node inclined backward, dorsum roundly convex, posterior face nearly vertical, posterodorsal corner prominent; ventral face of postpetiole nearly straight, anteroventral corner toothed. In dorsal view, sides of pronotum roundly prominent. Propodeal spines slightly curved inward. Petiole rectangular, petiolar node about as broad as long. Postpetiole widened backward, broader than petiole, postpetiolar node broader than long. Mandibles sparsely coarsely longitudinally striate and with sparse large punctures. Head smooth and shiny, lateral areas between genae and antennal sockets sparsely coarsely longitudinally striate. Mesosoma smooth and shiny, metanotal groove with short costulae. Petiole and postpetiole smooth and shiny. Gaster smooth and shiny, anterior margin with short basal costulae. Head and body with abundant erect to suberect hairs, almost without pubescence except for cervicum. Scapes with abundant subdecumbent hairs and abundant decumbent pubescence. Tibiae with abundant subdecumbent to decumbent hairs, but without pubescence. Color black. Mandibles, antennae, and legs brown. **Paratype workers:** TL 3.9-4.1, HL 0.93-1.05, HW 0.95-1.05, CI 98-103, SL 0.85-0.93, SI 83-92, ED 0.15-0.16, PW 0.55-0.60, AL 1.08-1.18 (*n*=10). As holotype, but in some individuals, apices of median anterior clypeal teeth truncated or weakly bifid. **Paratype soldiers:** TL 4.4-4.8, HL 1.28-1.45, HW 1.33-1.50, CI 103-104, SL 0.95-1.00, SI 67-72, ED 0.18, PW 0.65-0.73, AL 1.25-1.30 (*n*=2). As holotype, but body relatively larger, head proportionally larger and broader. In full-face view, occipital margin weakly concave in the middle. Teeth on masticatory margins of mandibles very blunt or whole margin edentate. Median teeth of anterior clypeal margin truncated, or teeth very short and closely approximated, with very shallow notch between them. Antennae relatively short, apices of scapes just reaching to occipital corners. In lateral view, promesonotum relatively flat, promesonotal suture weakly depressed. Sides of mesonotum with pair of prominences. Frons of head sparsely longitudinally striate. Mandibles, antennae, and legs blackish brown.

Holotype: Worker, China: Tibet Autonomous Region, Medog County, Damu Town, 70K (29.703733°N, 95.522433°E), 2750 m, 2011.Ⅶ.22, collected by Cheng-lin Zhang from a nest inside decayed wood in a ground sample in *Abies* forest on the south slope of Himalaya Mountain, No. A11-4178.

Etymology: The new species is named after the type locality, Medog County.

正模工蚁： 正面观，头部近方形，长宽相等；后缘轻度隆起，后角圆；侧缘均匀隆起，向前变宽。上颚长方形，内缘具1个齿，咀嚼缘具3个齿，亚端齿和基齿之间具1个齿间隙。唇基后延的中部三角形，后角锐角状，之后具短纵沟。唇基前缘具1对大的三角形中齿，顶端外弯，中齿之间具深缺口。触角9节，柄节约2/7超过头后角，触角棒3节。复眼圆形，隆起，位于头侧缘中点稍前处。侧面观，前中胸背板圆形隆起，前胸背板两侧具弱边缘。前中胸背板缝明显但不凹陷，在背面由1列大刻点组成。后胸沟深凹，中胸侧板具斜沟。并胸腹节背面平直，向后轻度降低；并胸腹节刺短而粗，顶端轻微下弯，约为背面长的1/3；斜面轻度凹陷，约为背面长的一半。并胸腹节侧叶很短，顶端圆。并胸腹节气门圆形，位于侧面上部，气门的下面和后面凹陷。腹柄结近三角形，前上角突出，后上角

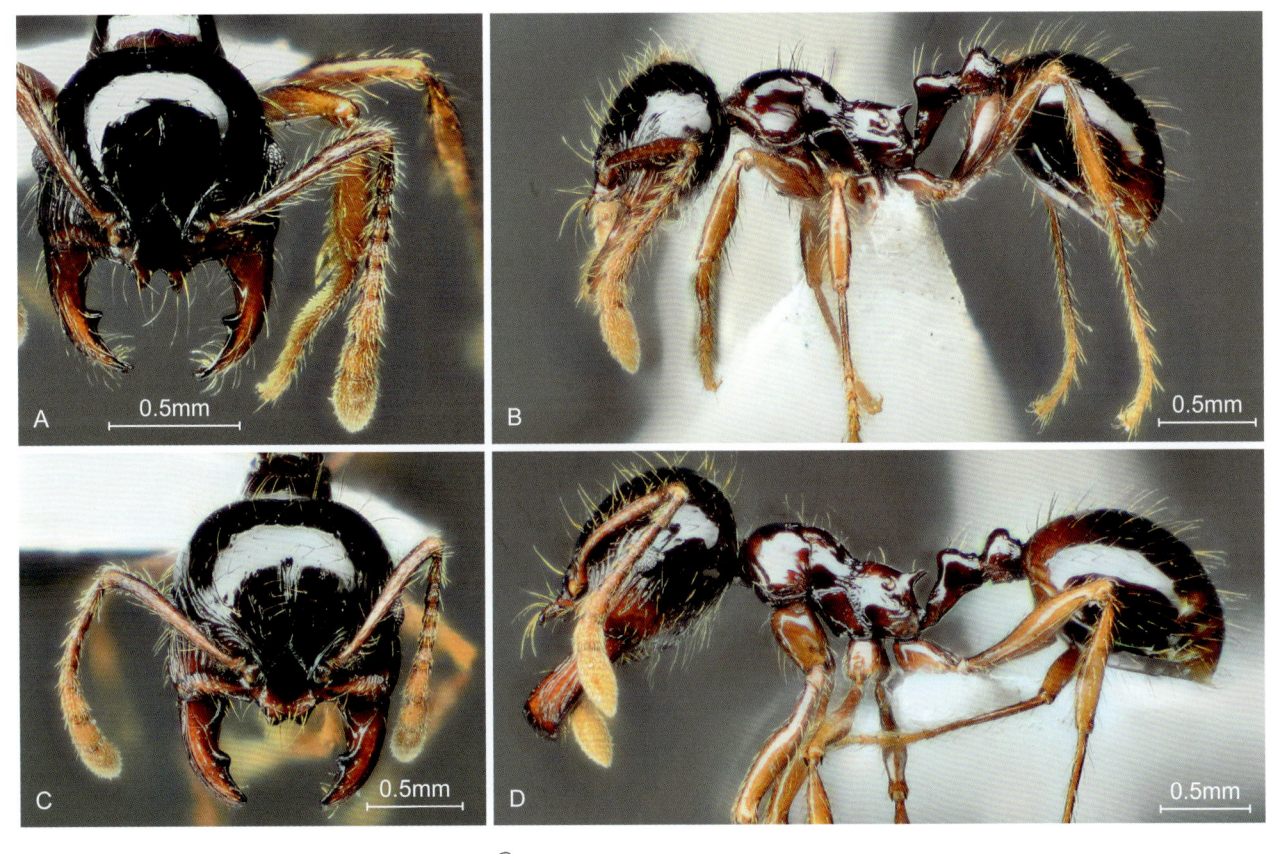

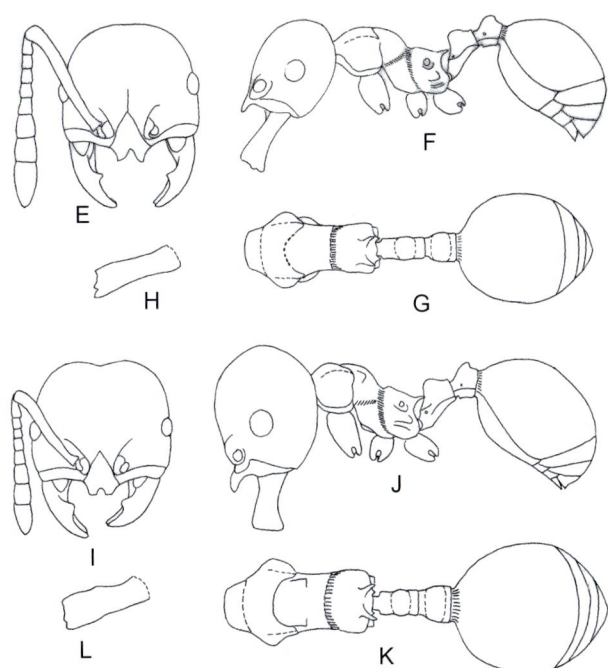

墨脱奇蚁 *Perissomyrmex medogensis*

A, E. 工蚁头部正面观；B, F. 工蚁身体侧面观；C, I. 兵蚁头部正面观；D, J. 兵蚁身体侧面观；G. 工蚁身体背面观；H. 工蚁上颚背面观；K. 兵蚁身体背面观；L. 兵蚁上颚背面观（E-L. 引自Xu和Zhang, 2012）
A, E. Head of worker in full-face view; B, F. Body of worker in lateral view; C, I. Head of soldier in full-face view; D, J. Body of soldier in lateral view; G. Body of worker in dorsal view; H. Mandible of worker in dorsal view; K. Body of soldier in dorsal view; L. Mandible of soldier in dorsal view (E-L. cited from Xu & Zhang, 2012)

轻度隆起，前面小柄约与腹柄结等长；腹面轻度凹陷，前下角轻度隆起。后腹柄结后倾，背面圆形隆起，后面近垂直，后上角突出；腹面近平直，前下角齿状。背面观，前胸背板侧面圆凸。并胸腹节刺轻微内弯。腹柄长方形，腹柄结长宽约相等。后腹柄向后变宽，宽于腹柄，后腹柄结宽大于长。上颚具稀疏粗纵条纹和稀疏大刻点。头部光滑发亮，颊区和触角窝之间具稀疏粗纵条纹。胸部光滑发亮，后胸沟具短纵脊。腹柄和后腹柄光滑发亮。后腹部光滑发亮，前缘具短的基纵脊。头部和身体具丰富直立、亚直立毛，除颈部外几乎无绒毛。柄节具丰富亚倾斜毛和丰富倾斜绒毛被。胫节具丰富亚倾斜、倾斜毛，缺绒毛被。身体黑色，上颚、触角和足棕色。**副模工蚁**：特征同正模，但是在一些个体中，唇基前缘的1对中齿顶端平截或轻度二裂。**副模兵蚁**：与正模相似，但是身体较大。头部更大更宽，正面观后缘中央轻度凹陷；上颚咀嚼缘的齿很钝或完全缺齿；唇基前缘中央的1对中齿顶端平截，或齿很短且互相接近，齿间缺口很浅；触角较短，柄节末端刚到达头后角；侧面观，前中胸背板较平坦，前中胸背板缝轻度凹陷；中胸背板侧面具1对突起；头部额区具稀疏纵条纹；上颚、触角和足黑棕色。

正模：工蚁，中国西藏自治区墨脱县达木乡70K（29.703733°N，95.522433°E），2750m，2011.Ⅶ.22，张成林采于喜马拉雅山南坡冷杉林地表样朽木内巢中，No. A11-4178。

词源：该新种以模式产地"墨脱"命名。

弯钩棱胸蚁
Pristomyrmex hamatus Xu & Zhang, 2002

Pristomyrmex hamatus Xu & Zhang, 2002b, Entomologia Sinica, 9(4): 70, figs. 9-12 (w.) CHINA (Yunnan).

Holotype worker: TL 3.2, HL 0.95, HW 0.90, CI 95, SL 0.88, SI 97, PW 0.63, AL 0.90, ED 0.16. In full-face view, head nearly circular, slightly longer than broad. Occipital margin nearly straight, weakly emarginated in the middle, occipital corners rounded, sides roundly convex. Inner margin of mandible with a tooth in the center, masticatory margin with 4 teeth and a short diastema between the 2nd and 3rd teeth. Clypeus with a longitudinal central carina, anterior margin roundly convex and armed with 8 blunt teeth. Frontal carinae long, extending backward beyond the posterior margin level of the eyes. Antennal scrobes shallow and distinct. Antenna with 11 segments, scape surpassing occipital corner by about 1/4 of its length, antennal club consisted of the 3 apical segments. Eyes situated on the midline of the head and extruding outwards laterally. In lateral view, dorsum of alitrunk roundly convex, sloping downwards at rear. Anterior margin of pronotum submarginate. Promesonotal suture and metanotal groove absent. Propodeal spines long, curved inward at apices and hook-like. Propodeal lobes elongate and spine-like, about 1/3 times as long as propodeal spine. In lateral view, petiole with short and stout peduncle anteriorly, petiolar node inclined backward, anterior corner higher than posterior one, dorsal face slightly convex, anterior face slope-like. Postpetiolar node short and high, strongly inclined backward. In dorsal view, petiolar node nearly square, postpetiolar node transverse and rectangular. Mandibles sparsely, longitudinally and coarsely striate. Head and alitrunk largely foveolate, interface formed a coarse reticulation system. Petiole and postpetiole longitudinally and coarsely striate. Gaster smooth and shining. Head and alitrunk with abundant erect long hairs and suberect short hairs. Petiolar node with a pair of erect hairs, postpetiolar node with 2 pairs of suberect long hairs. Gaster with sparse depressed short pubescence, without erect hairs. Scapes, femorae and tibiae with abundant suberect hairs. Body reddish brown in colour, appendages yellowish brown. **Paratype workers:** TL 3.2-3.4, HL 0.93-0.95, HW 0.88-0.90, CI 95-97, SL 0.88-0.90, SI 97-100, PW 0.60-0.63, AL 0.88-0.90, ED 0.15-0.16 (*n*=5). As holotype, but body yellow to reddish brown in colour.

Holotype: Worker, China: Yunnan Province, Mengla County, Menglun Town, Cuipingfeng, 660 m, 1997.Ⅷ.10, collected by Tai-yong Liu in karst monsoon forest, No. A97-1165.

Etymology: The species name *hamatus* combines Latin *ham*- (hook) + suffix *-atus* (masculine form, with), it refers to the long, curved inward and hook-like propodeal spines.

正模工蚁： 正面观，头部近圆形，长稍大于宽；后缘近平直，中央浅凹，后角圆；侧缘圆形隆起。上颚内缘中央具1个齿；咀嚼缘具4个齿，第2齿和第3齿之间具1个短的齿间隙。唇基具中央纵脊，前缘圆形隆起，具8个钝齿。额脊长，向后延伸超过复眼后缘水平。触角沟浅而明显。触角11节，柄节约1/4超过头后角，触角棒3节。复眼位于头中线上，突出侧缘之外。侧面观，胸部背面圆形隆起，向后坡形降低。前胸背板前缘具棱边，前中胸背板缝和后胸沟缺失。并胸腹节刺长，顶端内弯呈钩状。并胸腹节侧叶伸长，刺状，约为并胸腹节刺长的1/3。侧面观，腹柄前面具短粗的小柄，腹柄结后倾，前上角高于后上角，背面轻微隆起，前面坡形。后腹柄结短而高，强烈后倾。背面观，腹柄结近方形；后腹柄结横形，长方形。上颚具稀疏粗纵条纹。头部和胸部具大凹坑，界面形成粗糙网纹。腹柄和后腹柄具粗糙纵条纹。后腹部光滑发亮。头部和胸部具丰富直立长毛和亚直立短毛；腹柄结具1对直立毛，后腹柄结具2对亚直立长毛；后腹部具稀疏平伏短绒毛被，缺立毛；柄节、腿节和胫节具丰富亚直立毛。身体红棕色，附肢黄棕色。**副模工蚁：** 特征同正模，但身体黄色至红棕色。

正模： 工蚁，中国云南省勐腊县勐仑镇翠屏峰，660m，1997.Ⅷ.10，柳太勇采于石灰岩季雨林中，No. A97-1165。

词源： 该新种的种名"*hamatus*"由拉丁语"*ham-*"（钩）+后缀"*-atus*"（阳性形式，具有）组成，指并胸腹节刺内弯呈钩状。

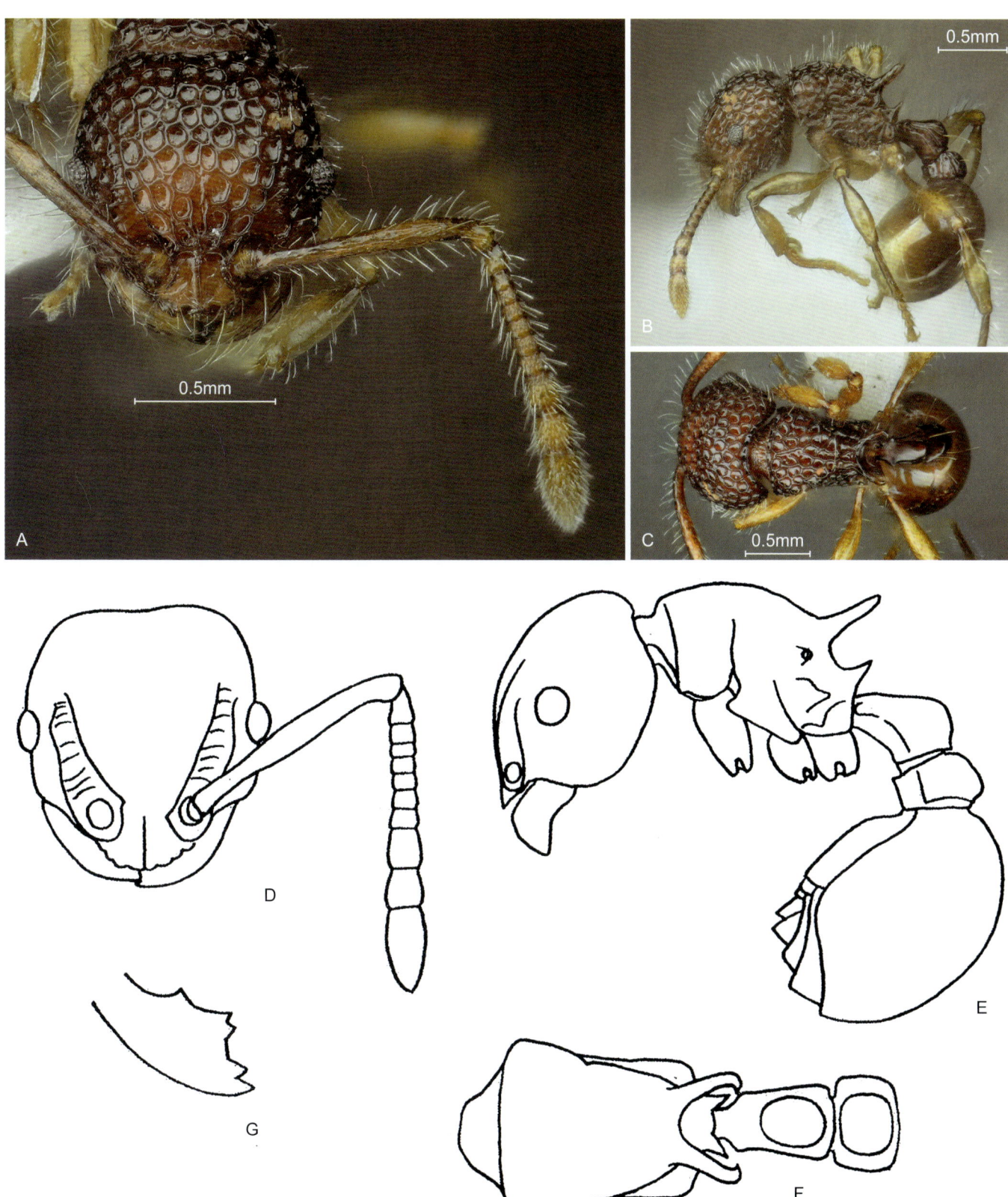

弯钩棱胸蚁 *Pristomyrmex hamatus*

A, D. 工蚁头部正面观；B, E. 工蚁身体侧面观；C, F. 工蚁身体背面观；G. 工蚁上颚背面观（D-G. 引自Xu和Zhang, 2002b）
A, D. Head of worker in full-face view; B, E. Body of worker in lateral view; C, F. Body of worker in dorsal view; G. Mandible of worker in dorsal view (D-G. cited from Xu & Zhang, 2002b)

哀牢山塔蚁
Pyramica ailaoshana Xu & Zhou, 2004

Pyramica ailaoshana Xu & Zhou, 2004, Acta Zootaxonomica Sinica, 29(3): 445, figs. 19, 20 (w.q.) CHINA (Yunnan).

Holotype worker: TL 2.6, HL 0.70, HW 0.50, CI 71, ML 0.10, MI 14, SL 0.33, SI 65, PW 0.29, AL 0.68. In full-face view, head elongate triangular, distinctly longer than broad and narrowed forward. Occipital margin widely and deeply concave. Occipital corners prominent and triangular. Sides of head prominent and bluntly angled at posterior 1/4. Mandible triangular and down curved at apex, masticatory margin with about 10 spine-like slender teeth. Clypeus rhombic, anterior margin roundly prominent in the middle. Antennal scrobe distinct. Antenna with 6 segments, antennal club 2-segmented, apex of scape reached to 5/8 of the distance from antennal socket to occipital corner. Eye with 4-5 ommatidia along the maximum diameter. Sides of alitrunk dorsum distinctly marginate. In lateral view, ventral face of head deeply concave. Pronotum flat, promesonotal suture indistinct. Mesonotum slightly convex. Metanotal groove absent. Dorsum of propodeum weakly convex and slope down backward. Propodeal spines long and acute, with apex slightly curved upward. Sides of declivity with developed curtain-like spongiform lamellae, the upper margin connecting the propodeal spine, with posterior margin deeply concave. In lateral view, petiole with large longitudinal curtain-like subpetiolar spongiform lobe, petiolar node long and low, with dorsum roundly convex, sides with wing-like spongiform lobes. Postpetiole with large semicircular subpostpetioiar spongiform lobe, dorsum of postpetiolar node weakly convex, sides with wing-like spongiform lobes. In dorsal view, petiolar node rectangular, distinctly longer than broad. Postpetiolar node nearly square, slightly broader than long. Mandibles punctuate. Head finely reticulate, occiput finely longitudinally striate. Antennal scrobes densely finely punctuate, interface appears as microreticulations. Dorsum of alitrunk finely longitudinally striate and finely reticulate. Sides of pronotum and propodeum smooth, sparsely striate. Sides of mesothorax and metathorax sparsely striate and densely finely punctuate, interface appears as reticulations. Petiole finely reticulate, dorsum of petiolar node sparsely transversely striate. Postpetiolar node smooth and shining. Gaster smooth and shining, with longitudinal basal costulae. Head with dense depressed pubescence, occipital margin with a pair of erect hairs in the middle, ventral face with a pair of erect hair at the concaved position. In full-face view, lateral side of occipital lobe with 4 decumbent hairs. Alitrunk, petiole and postpetiole with sparse depressed pubescence. Dorsum of pronotum without erect hairs, but humeral corners each with a laterally pointed long hair. Mesonotum, propodeum, petiole and postpetiole with sparse erect to decumbent hairs, 2 pairs on mesonotum, 3 pairs on propodeum, 3 pairs on petiole, and 2 pairs on postpetiole. Gaster with abundant erect hairs, hairs on the basal dorsum anterodorsally pointed, pubescence almost absent. Scapes with dense depressed pubescence. Femora and tibiae with abundant decumbent longer hairs. Body color brown, eyes and lateral margins of alitrunk black.

Paratype workers: TL 2.5-2.7, HL 0.68-0.73, HW 0.48-0.53, CI 68-74, ML 0.09-0.10, MI 13-15, SL 0.30-0.33, SI 60-68, PW 0.25-0.30, AL 0.63-0.70 (*n*=7). As holotype.

Holotype: Worker, China: Yunnan Province, Jingdong County, Jinping Town, Ailao Mountain, 1250 m, 2002.Ⅳ.14, collected by Zheng-qun Chai in *Pinus kesiya* var. *langbianensis* forest, No. A1037.

Etymology: The new species is named after the Ailao Mountain where the type specimens collected.

正模工蚁： 正面观，头部长三角形，长明显大于宽，向前变窄；后缘宽形深凹，后角突出呈三角形；侧缘在后部1/4处突出成钝角。上颚三角形，顶端下弯，咀嚼缘约具10个细长的刺状齿。唇基菱形，前缘中央圆形突起。触角沟明显，触角6节，触角棒2节，柄节末端到达触角窝至头后角间距的5/8处。复眼最大直径上具4~5个小眼。胸部背面两侧明显具边缘。侧面观，头部腹面深凹。前胸背板平坦，前中胸背板缝不明显。中胸背板轻微隆起，缺后胸沟。并胸腹节背面轻度隆起，向后坡形降低；并胸腹节刺长而尖锐，末端轻微上弯；斜面侧缘具发达的帘状海绵体纵脊，上缘连接并胸腹节刺，后缘深凹。侧面观，腹柄具大形纵向帘状海绵体腹柄下叶，腹柄结长而低，背面圆形隆起，侧面具翼状海绵体叶片。后腹柄具大的半圆形海绵体后腹柄下叶，后腹柄结背面轻度隆起，侧面具翼状海绵叶。背面观，腹柄结长方形，长明显大于宽。后腹柄结近方形，宽稍大于长。上颚具刻点。头部具细网纹，头后部具细纵条纹。触角沟具细密刻点，界面呈微网状。胸部背面具细纵条纹和细网纹。前胸背板侧面和并胸腹节侧面光滑，具稀疏条纹。中胸和后胸侧面具稀疏条纹和细密刻点，界面成网纹。腹柄具细网纹，腹柄结背面具稀疏横条纹。后腹柄结光滑发亮。后腹部光滑发亮，基部具短纵脊。头部具密集平伏绒毛被，后缘中部具1对直立毛，腹面在凹陷处具1对直立毛。正面观，头后叶侧缘各具4根倾斜毛。胸部、腹柄和后腹柄

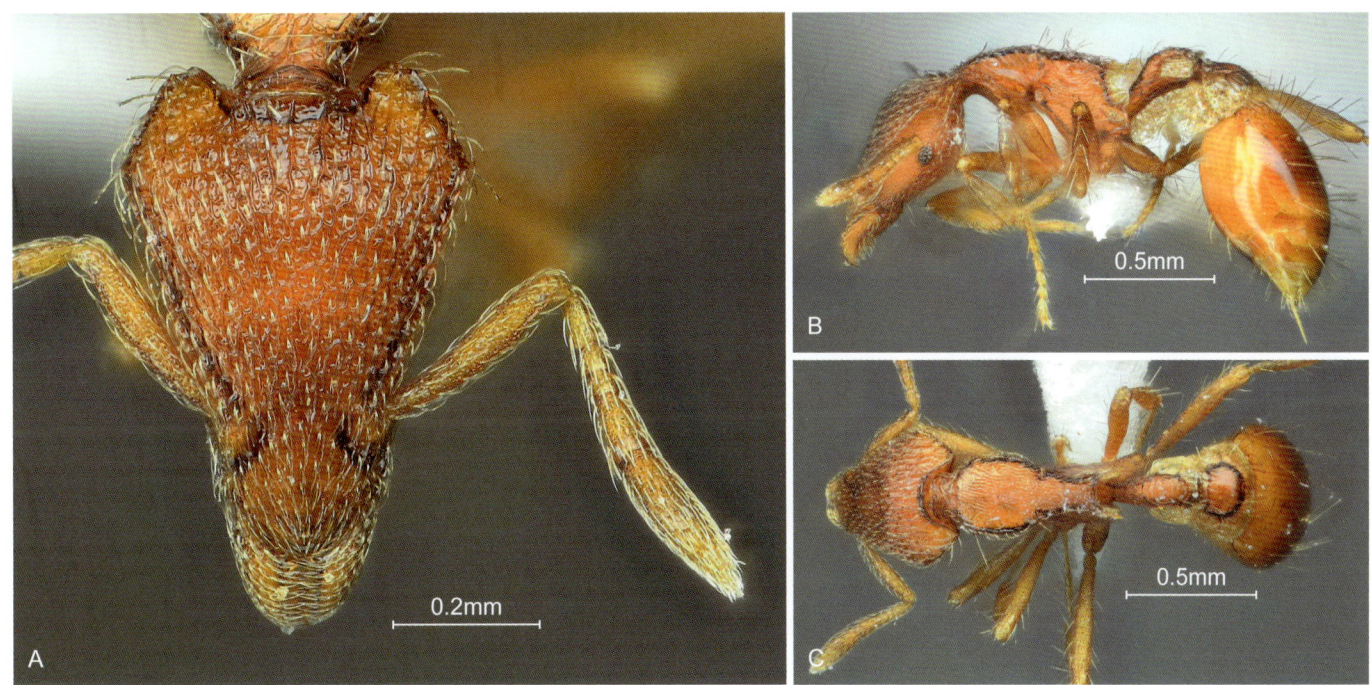

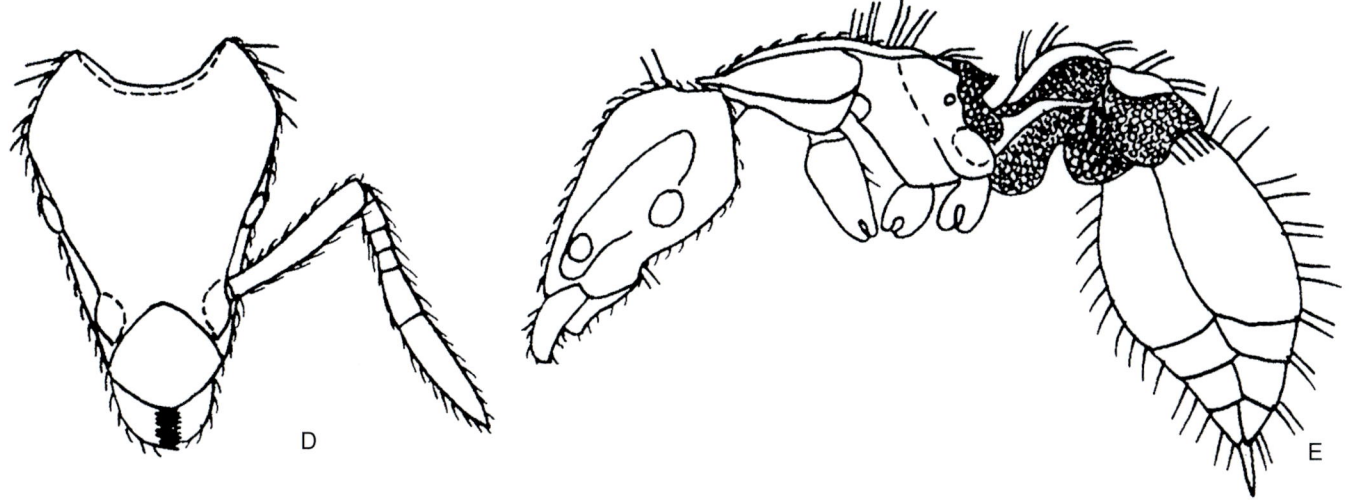

哀牢山塔蚁 *Pyramica ailaoshana*

A, D. 工蚁头部正面观；B, E. 工蚁身体侧面观；C. 工蚁身体背面观（D-E. 引自Xu和Zhou, 2004）
A, D. Head of worker in full-face view; B, E. Body of worker in lateral view; C. Body of worker in dorsal view (D-E. cited from Xu & Zhou, 2004)

具稀疏平伏绒毛被，前胸背板背面缺立毛，每个肩角具1根指向侧面的长毛。中胸背板、并胸腹节、腹柄和后腹柄具稀疏直立、倾斜毛，中胸背板具2对，并胸腹节具3对，腹柄具3对，后腹柄具2对。后腹部具丰富直立毛，基部背面立毛指向前上方，几乎缺绒毛被。柄节具密集平伏绒毛被。腿节和胫节具丰富倾斜长毛。身体棕色，复眼和胸部侧缘黑色。**副模工蚁：** 特征同正模。

正模： 工蚁，中国云南省景东县锦屏镇哀牢山，1250m，2002.Ⅳ.14，柴正群采于思茅松林中，No. A1037。

词源： 该新种以模式标本采集地"哀牢山"命名。

082 弄巴塔蚁
Pyramica nongba Xu & Zhou, 2004

Pyramica nongba Xu & Zhou, 2004, Acta Zootaxonomica Sinica, 29(3): 443, figs. 7, 8 (w.) CHINA (Yunnan).

Holotype worker: TL 1.8, HL 0.48, HW 0.40, CI 84, ML 0.11, MI 24, SL 0.28, SI 69, PW 0.28, AL 0.48. In full-face view, head triangular, longer than broad and narrowed forward. Occipital margin weakly roundly concave in the middle, occipital corners rounded. Sides weakly concave. Mandible elongate triangular, basal corner angled, basal half of masticatory margin without teeth, apical half with about 10 slender spine-like teeth. Clypeus nearly triangular, anterior margin roundly convex. Antennal scrobe distinctly depressed. Antenna with 6 segments, antennal club 2-segmented, apex of scape reached to 3/4 of the distance from antennal socket to occipital corner. Eye small, with 6 ommatidia. In lateral view, vertex roundly prominent. Pronotum roundly convex, promesonotal suture indistinct. Dorsum of mesonotum straight, metanotal groove shallowly depressed. Dorsum of propodeum convex anteriorly, slope down backward and rounded into declivity. Declivity weakly concave, sides with narrow vertical ridge-like spongiform appendages, posterodorsal corner of the spongiform appendage bluntly angled. Petiole without spongiform appendages, ventral face with a narrow longitudinal ridge, dorsum of petiolar node roundly convex. Postpetiolar node roundly convex, posterior border with a narrow spongiform appendage, sides each with a triangular spongiform appendage. Mandibles smooth and shining. Head, petiole and postpetiole densely finely punctuate and opaque. Alitrunk densely micro-reticulate and less shining. Gaster smooth and shining. Head, antennae and legs with abundant depressed pubescence. Alitrunk, petiole, postpetiole and gaster with sparse depressed pubescence. Body surface with sparse pilosity. Leading edge of antennal scape with a row of spatulate hairs, five of them curved apically, three of them curved basally. Vertex, mesonotum, petiole and postpetiole each with a pair of clavate hairs. Gaster with 2 pairs of clavate hairs on first tergite, 1 pair similar hairs on tergites 2-3 separately. Body color yellow, mandibles brownish yellow, eyes black. **Paratype workers:** TL 1.9-2.0, HL 0.50, HW 0.41-0.43, CI 83-85, ML 0.13, MI 25, SL 0.28, SI 65-67, PW 0.29-0.30, AL 0.49-0.51 (*n*=3). As holotype.

Holotype: Worker, China: Yunnan Province, Longchuan County, Nongba Town, Leiliang Village, 1200 m, 1995.XII.29, collected by Zheng-hui Xu in Tongbiguan Provincial Nature Reserve, No. A95-236.

Etymology: The new species is named after the type locality Nongba Town.

正模工蚁： 正面观，头部三角形，长大于宽，向前变窄；后缘在中央轻度圆形凹陷，后角圆；侧缘轻度凹陷。上颚长三角形，基角呈角状，咀嚼缘基半部缺齿，端半部约具10个细长的刺状齿。唇基近三角形，前缘圆形隆起。触角沟明显凹陷。触角6节，触角棒2节，柄节末端到达触角窝至头后角间距的3/4处。复眼小，具6个小眼。侧面观，头顶圆形突起。前胸背板圆形隆起，前中胸背板缝不明显。中胸背板背面平直，后胸沟浅凹。并胸腹节背面前部隆起，向后坡形降低，圆形进入斜面；斜面轻度凹陷，两侧具狭窄垂直的脊状海绵体，海绵体的后上角钝角状。腹柄缺海绵状附属物，腹面具1个狭窄纵脊，腹柄结背面圆形隆起。后腹柄结圆形隆起，后缘有1个狭窄的海绵状体，两侧各有1个三角形海绵体。上颚光滑发亮。头部、腹柄和后腹柄具细密刻点，暗。胸部具密集微网纹，不太光亮。后腹部光滑发亮。头部、触角和足具丰富平伏绒毛被；胸部、腹柄、后腹柄和后腹部具稀疏平伏绒毛被。身体表面具稀疏立毛。触角柄节前缘有1列匙形毛，其中，5根弯向柄节端部，3根弯向基部。头顶、中胸背板、腹柄和后腹柄各具1对棒状毛。后腹部第1节背板具2对棒状毛，第2~3节背板各具1对相似的毛。身体黄色，上颚棕黄色，复眼黑色。**副模工蚁：** 特征同正模。

正模： 工蚁，中国云南省陇川县弄巴镇垒良村，1200m，1995.XII.29，徐正会采于铜壁关省级自然保护区。No. A95-236。

词源： 该新种以模式产地"弄巴镇"命名。

弄巴塔蚁 *Pyramica nongba*

A, D. 工蚁头部正面观；B, E. 工蚁身体侧面观；C. 工蚁身体背面观（D-E. 引自Xu和Zhou, 2004）
A, D. Head of worker in full-face view; B, E. Body of worker in lateral view; C. Body of worker in dorsal view (D-E. cited from Xu & Zhou, 2004)

杨氏塔蚁
Pyramica yangi Xu & Zhou, 2004

Pyramica yangi Xu & Zhou, 2004, Acta Zootaxonomica Sinica, 29(3): 447, figs. 25, 26 (w.) CHINA (Yunnan).

Holotype worker: TL 1.8, HL 0.45, HW 0.40, CI 89, ML 0.13, MI 28, SL 0.23, SI 56, PW 0.24, AL 0.50. In full-face view, head violin-like, longer than broad, posterior portion distinctly broader than anterior portion. Occipital margin deeply and roundly concave, occipital corners roundly prominent. Sides of posterior 2/3 roundly convex, sides of anterior 1/3 straight and slightly widened forward, and angularly notched just behind the antennal sockets. Mandible triangular, dorsum depressed, with a transverse basal rim and an oblique dorsolateral ridge along the depressed dorsal surface, masticatory margin with about 10 small acute teeth. Dorsum of clypeus depressed, anterior margin weakly convex. Antennal scrobe distinctly depressed. Antenna with 6 segments, antennal club 2-segmented, basal corner of scape prominent and bluntly angled, apex of scape reached to 2/3 of the distance from antennal socket to occipital corner. Eye indistinct and with only 1 ommatidium. In lateral view, vertex strongly prominent and formed nearly a right angle, dorsal face and posterior face straight. Dorsum of pronotum nearly straight, dorsum of mesonotum roundly convex, promesonotal suture and metanotal groove weakly impressed. Dorsum of propodeum nearly straight and slope down backward, posterodorsal corner rounded. Declivity roundly and deeply concave, sides with narrow vertical spongiform lobes, posterodorsal corner of the lobe bluntly angled. In dorsal view, pronotum weakly depressed in central area and with blunt lateral margins. In lateral view, petiole with rectangular ventral spongiform appendage, petiolar node roundly prominent, posterior border with narrow spongiform appendage. Postpetiolar node roundly convex, sides and posterior border with large triangular spongiform appendages. Mandibles and clypeus with micro-punctures and relatively shining. Head densely coarsely punctured and opaque. Dorsum of alitrunk with micro-punctures and relatively shining. Sides of alitrunk smooth and shining. Petiolar node with micro-punctures and relatively shining. Postpetiolar node and gaster smooth and shining, first tergite of gaster with longitudinal basal costulae. Body surface with sparse pubescence, pilosity rare. Antennae and legs with abundant pubescence. Leading edge of scape with a row of rightly curved and apically directed spatulate hairs. Head, alitrunk, petiole and postpetiole without standing hairs. First tergite of gaster with 2 pairs of clavate erect hairs, tergites 2-4 each with 1 pair of apically blunt hairs. Body color yellow, the spongiform appendages light yellow.

Holotype: Worker, China: Yunnan Province, Mengla County, Menglun Town, Manyangguang Village, 700 m, 1998.Ⅲ.15, collected by Xiao-dong Yang, No. A98-1001.

Etymology: The new species is named after the type specimen collector Xiao-dong Yang.

正模工蚁： 正面观，头部似小提琴状，长大于宽，后部明显宽于前部；后缘圆形深凹，后角圆突；后部2/3的侧面圆形隆起，前部1/3的侧缘平直，向前轻微变宽，触角窝后具角状缺口。上颚三角形，背面凹陷，基部具1条横脊，沿着凹陷的背表面具1条倾斜的背侧脊；咀嚼缘约具10个小尖齿。唇基背面凹陷，前缘轻度隆起。触角沟明显凹陷。触角6节，触角棒2节；柄节的基角突出，钝角状；柄节末端到达触角窝至头后角间距的2/3处。复眼不明显，仅具1个小眼。侧面观，头顶强烈突起，近直角形，背面和后面平直。前胸背板背面近平直，中胸背板背面圆形隆起，前中胸背板缝和后胸沟轻度凹陷。并胸腹节背面近平直，向后坡形降低，后上角圆；斜面圆形深凹，两侧具狭窄垂直的海绵叶，海绵叶后上角钝角状。背面观，前胸背板中部轻度凹陷，具钝的背侧缘。侧面观，腹柄腹面具长方形海绵体，腹柄结圆形突起，后缘具狭窄的海绵体。后腹柄结圆形隆起，侧缘和后缘具大的三角形海绵体。上颚和唇基具微刻点，较光亮。头部具密集粗刻点，暗。胸部背面具微刻点，较光亮。胸部侧面光滑发亮。腹柄结具微刻点，较光亮。后腹柄结和后腹部光滑发亮，后腹部第1节背板具基纵脊。身体表面具稀疏绒毛被，立毛稀少。触角和足具丰富绒毛被。柄节前缘具1列直角形弯曲、指向柄节顶端的匙状毛。头部、胸部、腹柄和后腹柄缺立毛。后腹部第1节背板具2对棒状直立毛，第2~4节背板各具1对钝毛。身体黄色，海绵体浅黄色。

正模： 工蚁，中国云南省勐腊县勐仑镇曼养广村，700m，1998.Ⅲ.15，杨效东采集，No. A98-1001。

词源： 该新种以模式标本采集人杨效东的姓氏命名。

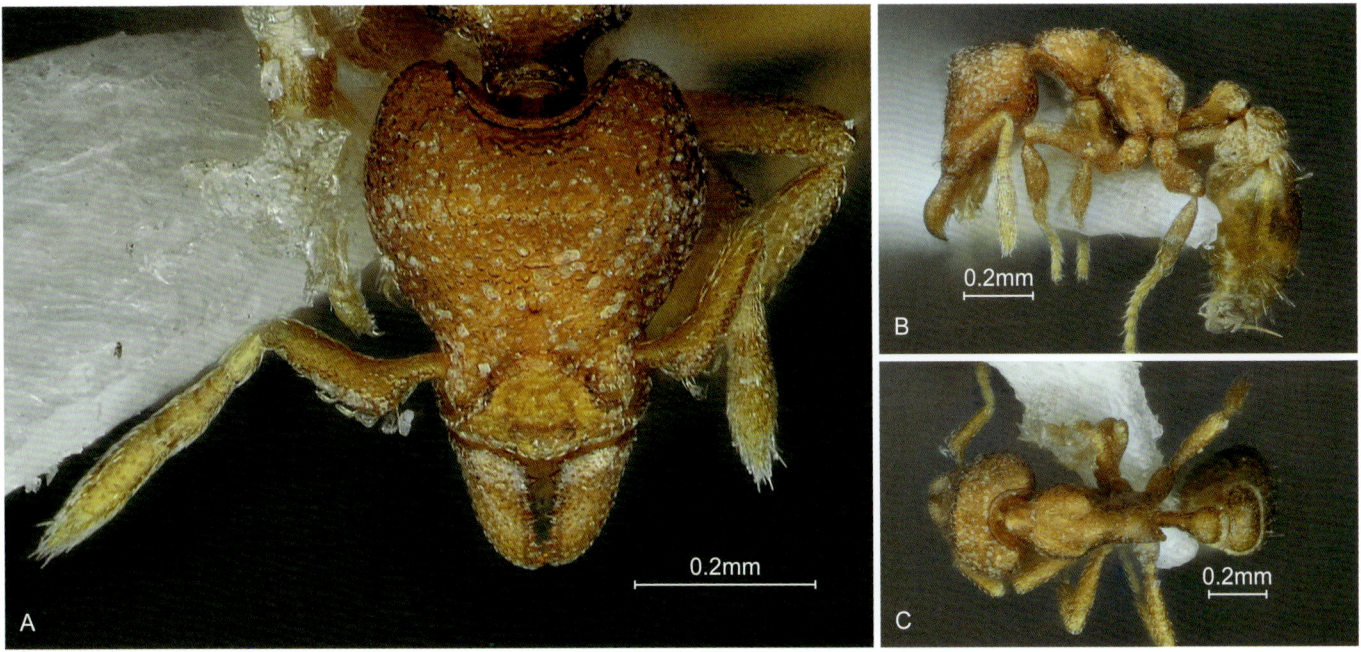
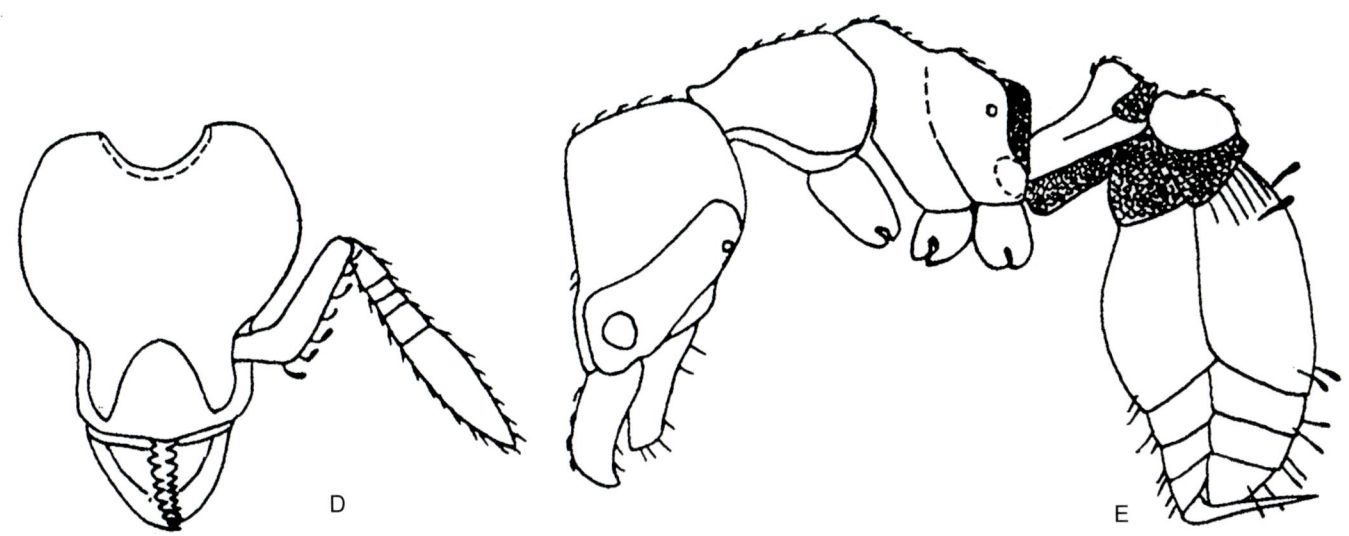

杨氏塔蚁 Pyramica yangi

A, D. 工蚁头部正面观；B, E. 工蚁身体侧面观；C. 工蚁身体背面观（D-E. 引自Xu和Zhou, 2004）
A, D. Head of worker in full-face view; B, E. Body of worker in lateral view; C. Body of worker in dorsal view (D-E. cited from Xu & Zhou, 2004)

084 女娲角腹蚁
Recurvidris nuwa Xu & Zheng, 1995

Recurvidris nuwa Xu & Zheng, 1995, Entomotaxonomia, 17(2): 143, figs. 2-4 (w.) CHINA (Guizhou).

Holotype worker: TL 1.9, HL 0.48, HW 0.44, CI 92, SL 0.36, SI 82, PW 0.25, AL 0.65, ED 0.13. Body slender. In full-face view, head roughly rectangular, longer than broad. Occipital margin straight. The anterior portion of the head strongly convex, so that the clypeus nearly vertical. Median portion of clypeus convex, anterior margin straight, with a narrow edge. Mandibles with fine longitudinal rugulae, and 4 teeth which nearly equal in size on the masticatary margin. Frontal carinae short, divergent posteriorly. Scrobes absent. Antennae have 11 segments, the scapes not reaching the occipital angles; the last 3 segments forming the clubs, the apical one the longest, about 2.3 times as long as the segment next to it. The eyes extending as an angle anteroventrally, the maximum diameter of eye with 7-8 ommatidia. In lateral view, pronotum high, dorsum comparatively flat. Mesonotum lowering down as a slope posteriorly. Metanotal groove depressed. Propodeum low, with the dorsum convex. Propodeal spines long and strong, compressed laterally, and curved anterodorsally. Petiole with peduncle anteriorly, a long spine-like process are present underneath, pointing posteroventrally; the node triangular, with dorsum slightly longitudinally depressed. Postpetiole with strong peduncle anteriorly, the node very low, trapezoid in dorsal view, wider than long, without constriction posteriorly, and widely attacking to the gaster. Gaster in lateral view triangular, flat above and convex as a round angle below. Head, alitrunk and pedicel have fine dense reticulate-rugulae. Dorsal median portion of the head, clypeus, dorsum and lateral portions of pronotum, dorsum of mesonotum, and gaster smooth and shining. Head and body with abundant erect or suberect short blunt hairs. Dorsal surfaces of scapes and hind tibiae with short curved pubescence. Body yellow, the posterior half of gaster brownish yellow. **Paratype workers:** TL 1.8-2.0, HL 0.48-0.50, HW 0.43-0.44, CI 88-89, SL 0.36-0.38, SI 83-88, PW 0.25, AL 0.63-0.64, ED 0.13 (*n*=2). As holotype.

Holotype: Worker, China: Guizhou Province, Duyun City, Duyun (26.266667°N, 107.533333°E), 780 m, 1991.Ⅸ.10, collected by Zheng-hui Xu on the ground in conifer-broadleaf mixed forest, No. A91-693.

Etymology: The new species is named after Nüwa, the goddess of creation in ancient Chinese mythology, who created man, patched the sky and redeemed the world.

正模工蚁： 身体细长。正面观，头部近长方形，长大于宽，后缘平直。头前部强烈隆起，致使唇基近垂直。唇基中部隆起，前缘平直，具窄的边缘。上颚具细纵皱纹，咀嚼缘具4个近等大的齿。额脊短，向后分歧。缺触角沟，触角11节，柄节未到达头后角，触角棒3节，端节最长，约为亚端节长度的2.3倍。复眼前下角延伸呈角状，最大直径上具7~8个小眼。侧面观，前胸背板高，背面较平。中胸背板向后降低，后胸沟凹陷。并胸腹节低，背面隆起；并胸腹节刺长而发达，侧扁，向前上方弯曲。腹柄前面具小柄，腹面具长刺状腹柄下突，指向后下方；腹柄结三角形，背面轻微纵向凹陷。后腹柄前面具粗的小柄，后腹柄结很低，背面观梯形，宽大于长，后部不收缩，与后腹部宽阔连接。侧面观，后腹部三角形，背面平坦，腹面隆起呈圆角。头部、胸部、腹柄和后腹柄具细密网纹。头部背面中部、唇基、前胸背板背面和侧面、中胸背板背面和后腹部光滑发亮。头部和身体具丰富直立、亚直立短钝毛；柄节背面和后足胫节具平伏短绒毛被。身体黄色，后腹部的后半部棕黄色。**副模工蚁：** 特征同正模。

正模： 工蚁，中国贵州省都匀市都匀（26.266667°N，107.533333°E），780m，1991.Ⅸ.10，徐正会采于针阔混交林地表，No. A91-693。

词源： 该新种的名称以中国上古神话中的创世女神女娲命名，她造物造人，补天救世。

Individual Theory 分 论

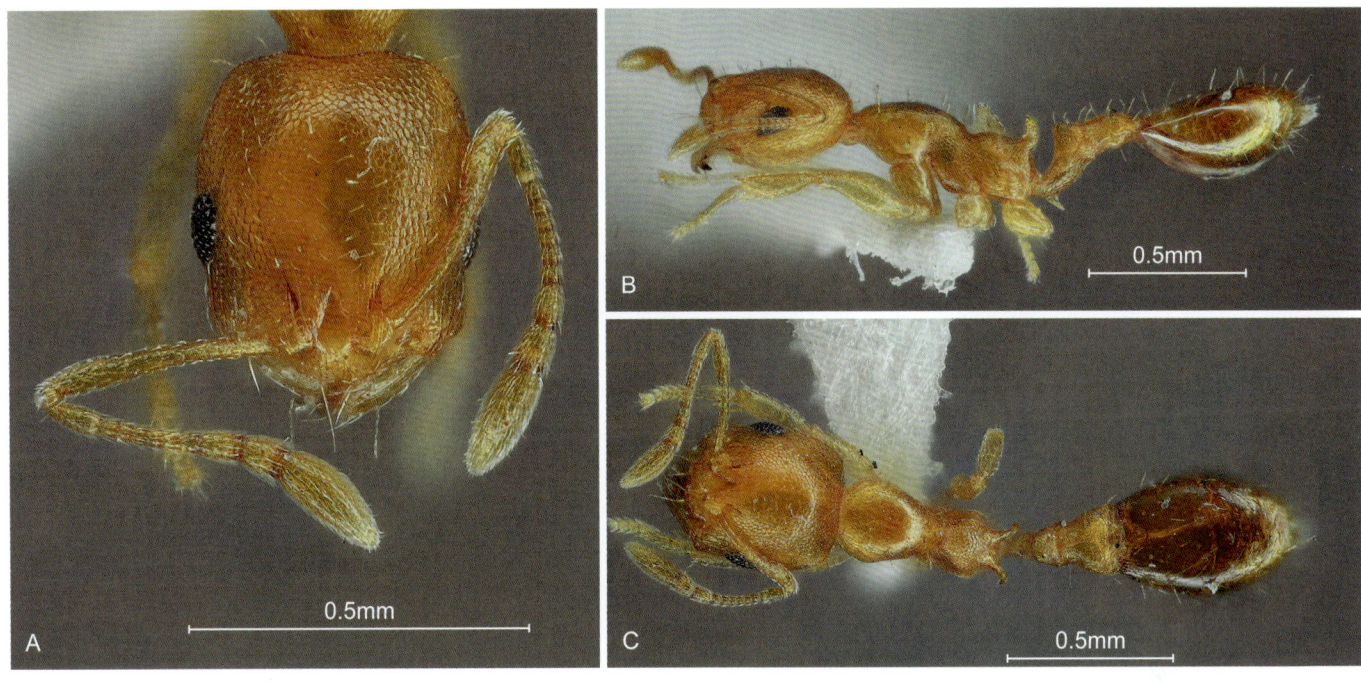

女娲角腹蚁 *Recurvidris nuwa*

A, D. 工蚁头部正面观；B, E. 工蚁身体侧面观；C. 工蚁身体背面观；F. 工蚁腹柄和后腹柄背面观（D-F. 引自Xu和Zheng, 1995）

A, D. Head of worker in full-face view; B, E. Body of worker in lateral view; C. Body of worker in dorsal view; F. Petiole and postpetiole of worker in dorsal view (D-F. cited from Xu & Zheng, 1995)

暗首棒角蚁
Rhopalomastix umbracapita Xu, 1999

Rhopalomastix umbracapita Xu, 1999, Acta Biologica Plateau Sinica, 14: 131, figs. 1, 4-6 (w.) CHINA (Yunnan).

Holotype worker: TL 2.2, HL 0.53, HW 0.50, CI 95, SL 0.23, SI 45, PW 0.35, AL 0.65, ED 0.09. In full-face view, head nearly square, narrowered anteriorly, sides slightly convex. Occipital margin slightly concave in the center, occipital corners rounded. Mandibles with 4 teeth, the basal 2 small and blunt. Clypeus without longitudinal central carina, anterior margin straight in the center, without a central seta, but with a pair of subcentral setae. Eye with 15 facets, placed at anterior 1/4 of the sides of head. Frontal furrow short, feebly longer than frontal carinae. Antennae 10 segmented, apex of scape reached to 5/9 of the distance from antennal fossa to occipital corner, segments 3-8 very short, apical 2 segments forming the antennal club. Dorsum of alitrunk relatively flat, submarginated laterally, without sutures. In lateral view, outline of alitrunk complete, slighdy convex. Promesonotal suture and metanotal groove only visible on sides below spiracles. Declivity of propodeum steep. Petiolar node subtriangular in lateral view, inclined posteriorly, anterior face sloping, posterior face nearly vertical, dorsal face roundly convex; subpetiolar process small, bluntly angled, subtransparent. Postpetiolar node rounded dorsally, ventral process indistinct. In dorsal view, both petiolar node and postpetiolar node are transverse. Gaster elongate ovate. Mandibles smooth and shining. Dorsa of head and alitrunk with dense fine longitudinal striae, dorsum of head with sparse small punctures in addition, striae of sides of head and occipital margin weakened, sides of alitrunk weakly finely reticulate. Petiolar node, postpetiolar node and gaster relatively smooth, shining. Dosa of head and body with sparse suberect hairs and abundant decumbent pubescence, pubescence on gaster dense. Antennal scapes with abundant decumbent pubescence and several suberect long hairs. Tibiae only with rich pubescence. Head dark reddish brown, dorsum of alitrunk and gaster reddish brown, antennae, sides of alitrunk, legs, petiole and postpetiole brownish yellow.

Holotype: Worker, China: Yunnan Province, Jinghong County, Mengyang Town, Sanchahe River, 960 m, 1997.Ⅱ.27, collected by Zheng-hui Xu from a canopy sample of seasonal rain forest, No. A97-58.

Etymology: The species name *umbracapita* combines Latin *umbra-* (dull) + word root *capit* (head) + suffix *-a* (feminine form), it refers to the relatively darker color of head.

正模工蚁：正面观，头部近方形，向前变窄，侧缘轻微隆起；后缘中央轻微凹陷，后角圆。上颚具4个齿，基部2个齿小而钝。唇基缺中央纵脊，前缘中部平直，缺中央刚毛，但是具1对近中央刚毛。复眼具15个小眼，位于头侧缘前部1/4处。额沟短，稍长于额脊。触角10节，柄节末端到达触角窝至头后角间距的5/9处，第3～8节很短，触角棒2节。胸部背面较平坦，两侧具钝边缘，缺沟缝。侧面观，胸部轮廓完整，轻度隆起，前中胸背板缝和后胸沟仅在侧面气门之下可见。并胸腹节斜面陡峭。侧面观腹柄结亚三角形，后倾，前面坡形，后面近垂直，背面圆形隆起；腹柄下突小，钝角状，半透明。后腹柄结背面圆，腹面突起不明显。背面观，腹柄结和后腹柄结均为横形。后腹部长卵形。上颚光滑发亮。头部和胸部背面具细密纵条纹，此外头部背面还具稀疏细刻点；头后缘和侧面的条纹变弱，胸部侧面具弱的细网纹。腹柄结、后腹柄结和后腹部较光滑，发亮。头部和身体背面具稀疏亚直立毛和丰富倾斜绒毛被，后腹部绒毛被密集；触角柄节具丰富倾斜绒毛被和几根亚直立长毛；胫节具丰富绒毛被。头部暗红棕色，胸部背面和后腹部红棕色，触角、胸部侧面、足、腹柄和后腹柄棕黄色。

正模：工蚁，中国云南省景洪县勐养镇三岔河，960m，1997.Ⅱ.27，徐正会采于季节性雨林树冠样中，No. A97-58。

词源：该新种的种名"*umbracapita*"由拉丁语"*umbra-*"（暗的）+词根"*capit*"（头）+后缀"*-a*"（阴性形式）组成，指头部颜色较暗。

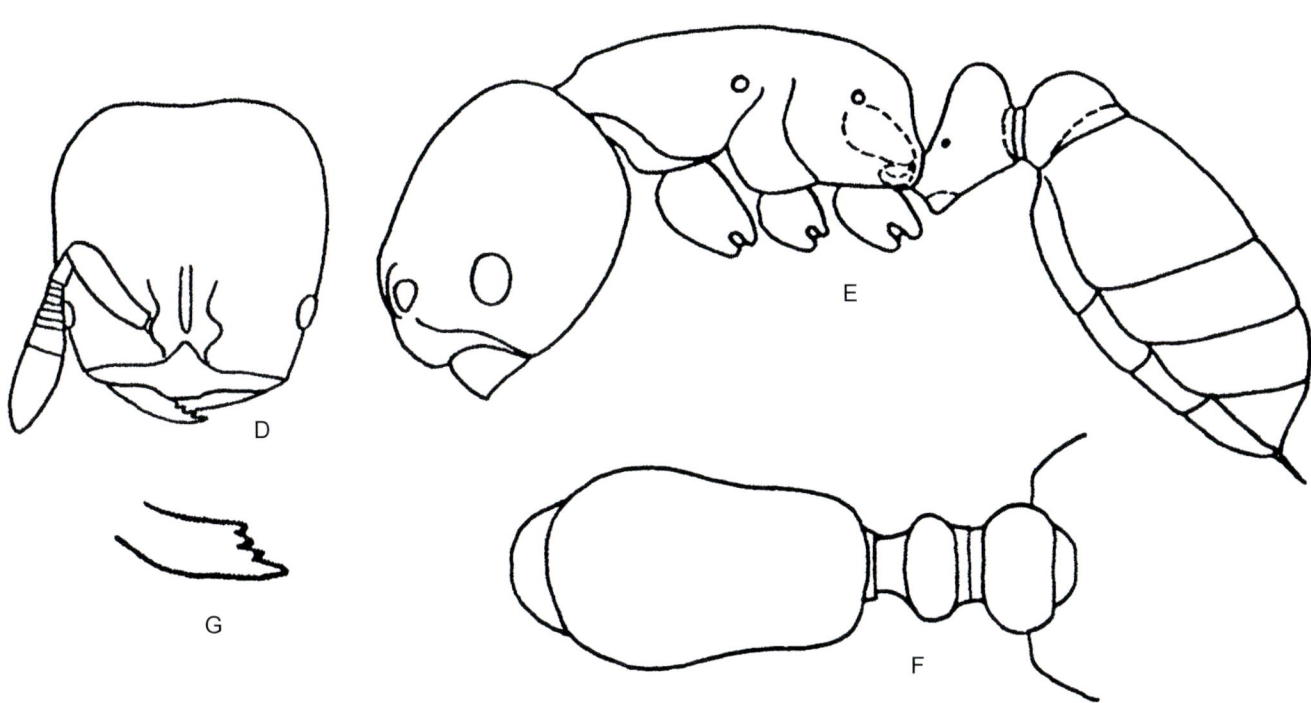

暗首棒角蚁 *Rhopalomastix umbracapita*

A, D. 工蚁头部正面观；B, E. 工蚁身体侧面观；C, F. 工蚁身体背面观；G. 工蚁上颚背面观（D-G. 引自Xu, 1999）
A, D. Head of worker in full-face view; B, E. Body of worker in lateral view; C, F. Body of worker in dorsal view; G. Mandible of worker in dorsal view (D-G. cited from Xu, 1999)

086 哀牢窄结蚁
Stenamma ailaoense Liu & Xu, 2011

Stenamma ailaoense Liu & Xu, 2011, Sociobiology, 58(3): 740, figs. 1-3 (w.) CHINA (Yunnan).

Holotype worker: TL unknown, HL 1.03, HW 0.83, CI 80, SL 0.90, SI 109, PW 0.60, AL 1.38, ED 0.11, PL 0.60, PH 0.28, PI 46, DPW 0.21. In full-face view, head roughly rectangular, longer than broad. Occipital margin straight, occipital corners roundly prominent, sides weakly convex. Anterior margin of clypeus convex, and concave in the middle. Mandibles with 3 distinct apical teeth and followed by 8 indistinct denticles. Antennae long, 12-segmented, scapes surpassing occipital corners by 1/6 of its length, antennal clubs 5-segmented. Eyes located before the midpoints of sides of head, with 5 ommatidia in the maximum diameter. In lateral view, promesonotum high and convex, nearly arched, the middle portion relatively straight, with trace of promesonotal suture. Metanotal groove wide, but shallowly depressed. Propodeum distinctly lower than promesonotum, dorsum straight and formed a gentle slope. Propodeal spines slender, about 1/2 length of declivity. Declivity straight, about 1/2 length of dorsum. Propodeal plates broad, nearly trapezoidal, slightly shorter than propodeal spines, posterodorsal corner bluntly angled, posteroventral corner rounded. Petiole long, length: height: width = 3: 1.4: 1. Petiolar node low, shorter than anterior peduncle, and roundly prominent at top. Anteroventral corner of petiole weakly convex, anterior 2/5 of ventral face straight, posterior 3/5 depressed. Postpetiole and gaster lost. Mandibles finely longitudinally striate. Head retirugose, but longitudinally rugose before eyes and between frontal carinae. Median portion of clypeus smooth and shining. Alitrunk retirugose. Posterior 2/3 of pronotum with posteriorly divergent longitudinal rugae. Sides of mesothorax and propodeum finely retirugose. Metapleuron with 3 coarse longitudinal rugae, interspaces smooth. Petiole with interweaved fine longitudinal rugae, and densely finely punctuate behind petiolar node. Fore coxae transversely rugose, middle and hind coxae finely reticulate. Dorsa of head and body with sparse erect to suberect hairs and abundant decumbent pubescence. Dorsum of petiolar peduncle without erect hairs. Antennal scapes and hind tibiae with dense decumbent pubescence, but without erect hairs. Head and body reddish brown, appendages yellowish brown, eyes and masticatory margins of mandibles black.

Holotype: Worker, China: Yunnan Province, Jingdong County, Taizhong Town, Xujiaba, 2500 m, 2002.Ⅳ.7, collected by Zheng-qun Chai from a soil sample in primitive subalpine moist evergreen broadleaf forest of Ailao Mountain, No. A00831.

Etymology: The new species is named after the type locality Mt. Ailao in central Yunnan Province.

正模工蚁： 正面观，头部近长方形，长大于宽；后缘平直，后角圆凸；侧缘轻度隆起。唇基前缘隆起，中部凹陷。上颚具3个明显的端齿和8个不明显的小齿。触角长，12节，柄节1/6超过头后角，触角棒5节。复眼位于头两侧中点之前，最大直径上具5个小眼。侧面观，前中胸背板高，隆起近弓形，中部较平直，有前中胸背板缝的痕迹。后胸沟宽，浅凹。并胸腹节明显低于前中胸背板，背部平直，缓坡形；并胸腹节刺细长，约为斜面长度的1/2；斜面平直，约为背面长度的1/2。并胸腹节侧叶宽，近梯形，稍短于并胸腹节刺，后上角钝角状，后下角圆。腹柄长，长：高：宽=3：1.4：1；腹柄结低，短于前面小柄，顶端圆形突起；腹柄前下角轻度隆起，腹面前部2/5平直，后部3/5凹陷。后腹柄和后腹部丢失。上颚具细纵条纹。头部具网纹，但是在复眼之前和额脊之间具纵皱纹。唇基中部光滑发亮。胸部具网纹；前胸背板后部2/3具向后分歧的纵皱纹；中胸和并胸腹节侧面具细网纹；后胸侧板具3条粗纵皱纹，界面光滑。腹柄具互相交织的细纵皱纹，腹柄结之后具细密刻点。前足基节具横皱纹，中足和后足基节具细网纹。头部和身体背面具稀疏直立、亚直立毛和丰富倾斜绒毛被；腹柄前面小柄的背面缺立毛。触角柄节和后足胫节具密集倾斜绒毛被，缺立毛。头部和身体红棕色，附肢黄棕色，复眼和上颚咀嚼缘黑色。

正模： 工蚁，中国云南省景东县太忠乡徐家坝，2500m，2002. Ⅳ.7，柴正群采于哀牢山原始中山湿性常绿阔叶林土壤样中，No. A00831。

词源： 该新种以云南省中部的模式产地"哀牢山"命名。

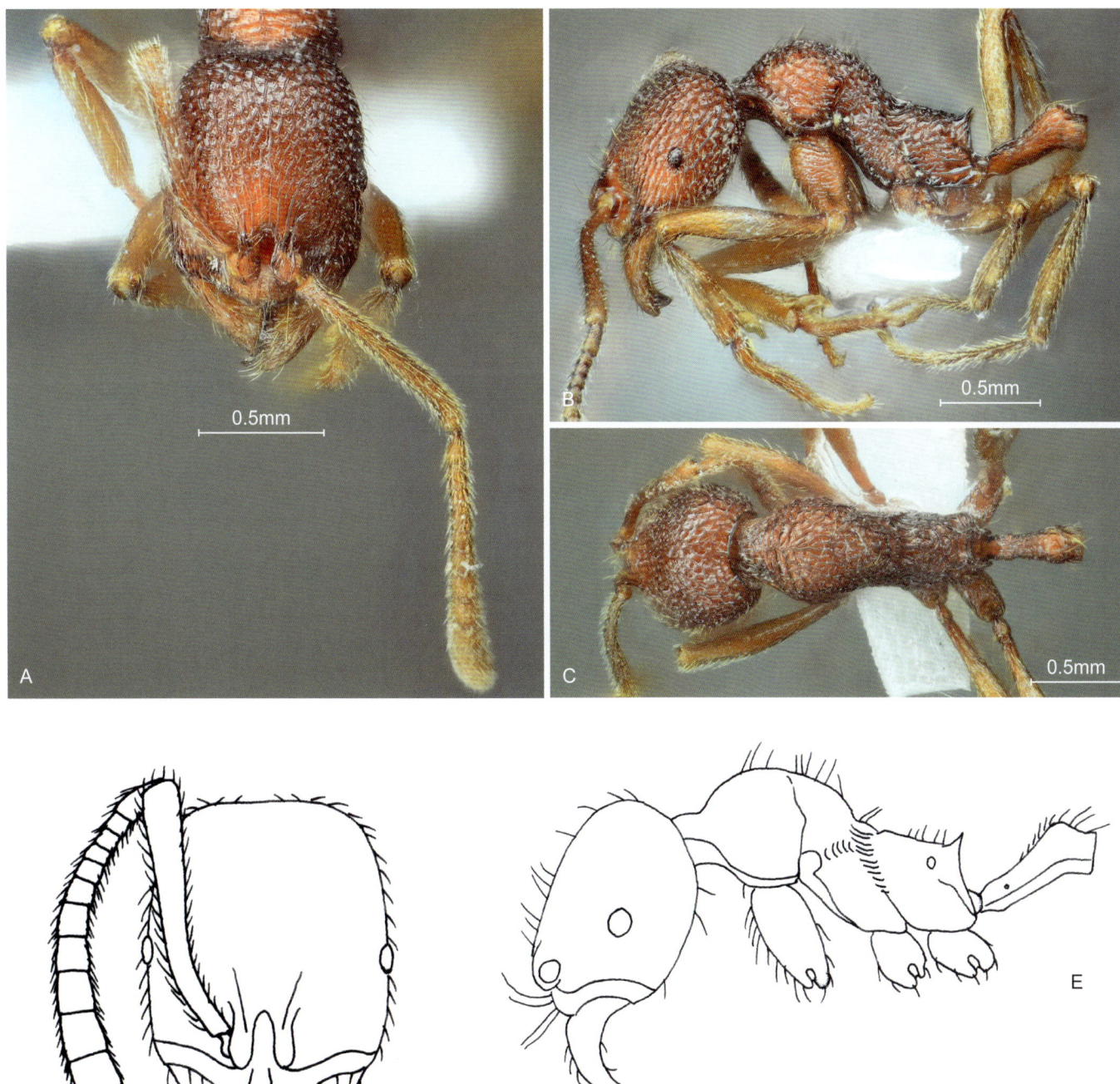

哀牢窄结蚁 *Stenamma ailaoense*

A, D. 工蚁头部正面观；B, E. 工蚁身体侧面观；C. 工蚁身体背面观；F. 工蚁上颚背面观（D-F. 引自Liu和Xu, 2011）

A, D. Head of worker in full-face view; B, E. Body of worker in lateral view; C. Body of worker in dorsal view; F. Mandible of worker in dorsal view (D-F. cited from Liu & Xu, 2011)

乌蒙窄结蚁
Stenamma wumengense Liu & Xu, 2011

Stenamma wumengense Liu & Xu, 2011, Sociobiology, 58(3): 742, figs. 4-6 (w.) CHINA (Yunnan).

Holotype worker: TL 3.9, HL 0.95, HW 0.75, CI 79, SL 0.78, SI 103, PW 0.55, AL 1.20, ED 0.08, PL 0.50, PH 0.25, PI 50, DPW 0.19, PPL 0.33, PPH 0.24, PPI 73, PPW 0.25. In full-face view, head roughly rectangular, longer than broad. Occipital margin nearly straight, occipital corners roundly prominent, sides weakly convex. Anterior margin of clypeus convex, and concave in the middle. Mandibles with 2 distinct apical teeth and followed by 6 indistinct denticles. Antennae 12-segmented, scapes surpassed occipital corners by 1/15 of its length, antennal clubs 4-segmented. Eyes located before midpoints of sides of head, with 4 ommatidia in the maximum diameter. In lateral view, promesonotum high, anterior portion roundly convex, middle portion relatively straight, posterior portion steeply sloped. Promesonotal suture absent. Metanotal groove narrow, shallowly depressed. Propodeum distinctly lower than promesonotum, dorsum straight and formed a gentle slope. Propodeal spines very short, acutely toothed, about 1/6 length of declivity. Declivity straight, about 3/5 length of dorsum. Propodeal plates short and broad, nearly trapezoid, slightly longer than propodeal spines. Petiole long, length∶height∶width = 4∶2∶1.5, gradually thickened backwards. Petiolar node bluntly prominent at top, shorter than anterior peduncle, slightly depressed between them, but without distinct boundary line. Ventral face of petiole nearly straight, anteroventral corner bluntly angled. Postpetiolar dorsum roundly convex, length∶height∶width = 2∶1.5∶1.1, ventral face weakly concave, anteroventral corner extruding and tooth-like. Mandibles finely longitudinally striate. Head retirugose, but longitudinally rugose before eyes and between frontal carinae. Median portion of clypeus smooth and shining. Dorsum of alitrunk retirugose. Pronotum with a longitudinal central carina. Sides of alitrunk longitudinally rugose, but sides of pronotum finely retirugose. Propodeal declivity smooth, upper portion sparsely transversely rugose. Fore coxae transversely rugose. Petiole and postpetiole with interweaved fine longitudinal rugae and fine punctures. Dorsum of petiolar peduncle with a fine longitudinal central carina. Gaster smooth and shining, basal carinae about 1/3 length of postpetiole. Head with sparse erect to suberect hairs and dense decumbent pubescence. Alitrunk and petiole with sparse erect to suberect hairs and sparse decumbent pubescence. Postpetiole and gaster with abundant erect to suberect hairs and abundant decumbent pubescence. Dorsa of propodeum and petiolar peduncle without erect hairs. Antennal scapes and hind tibiae with dense decumbent pubescence, but without erect hairs. Head and body reddish brown, appendages and gaster yellowish brown, eyes and masticatory margins of mandibles black.

Holotype: Worker, China: Yunnan Province, Yongshan County, Xisha Town, Xiaoyanfang, 2070 m, 2006.Ⅶ.19, collected by Zheng-hui Xu on the ground in broadleaf forest, No. A06-643.

Etymology: The new species is named after the type locality Mt. Wumeng in northeastern Yunnan Province, China.

正模工蚁： 正面观，头部近长方形，长大于宽；后缘近平直，后角圆凸；侧缘轻度隆起。唇基前缘隆起，中央凹陷。上颚具2个明显端齿和6个不明显的小齿。触角12节，柄节1/15超过头后角，触角棒4节。复眼位于头侧缘中点之前，最大直径上具4个小眼。侧面观，前中胸背板高，前部圆形隆起，中部较平直，后部陡坡状。缺前中胸背板缝。后胸沟窄，浅凹。并胸腹节明显低于前中胸背板，背部平直，缓坡形；并胸腹节刺很短，尖齿状，约为斜面长度的1/6；斜面平直，约为背面长度的3/5。并胸腹节侧叶短而宽，近梯形，稍长于并胸腹节刺。腹柄长，长∶高∶宽=4∶2∶1.5，向后逐渐变厚；腹柄结顶端钝突，短于前面小柄，二者之间轻微凹陷，缺明显的界线；腹柄腹面近平直，前下角钝角状。后腹柄背面圆形隆起，长∶高∶宽=2∶1.5∶1.1，腹面轻度凹陷，前下角突出呈齿状。上颚具细纵条纹。头部具网纹，但是在复眼之前、额脊之间具纵皱纹。唇基中部光滑发亮。胸部背面具网纹；前胸背板具1条中央纵脊；胸部侧面具纵皱纹，但是前胸背板侧面具细网纹；并胸腹节斜面光滑，上部具稀疏横皱纹。前足基节具横皱纹。腹柄和后腹柄具互相交织的细纵皱纹和细刻点，腹柄前面小柄的背面具1条细的中央纵脊。后腹部光滑发亮，基纵脊约为后腹柄长度的1/3。头部具稀疏直立、亚直立毛和密集倾斜绒毛被。胸部和腹柄具稀疏直立、亚直立毛和稀疏倾斜绒毛被。后腹柄和后腹部具丰富直立、亚直立毛和丰富倾斜绒毛被。并胸腹节背面和腹柄前面小柄缺立毛。触角柄节和后足胫节具密集倾斜绒毛被，缺立毛。头部和身体红棕色，附肢和后腹部黄棕色，复眼和上颚咀嚼缘黑色。

正模： 工蚁，中国云南省永善县细沙乡小岩方，2070m，2006.Ⅶ.19，徐正会采于阔叶林地表，No. A06-643。

词源： 该新种以模式产地中国云南省东北部的"乌蒙山"命名。

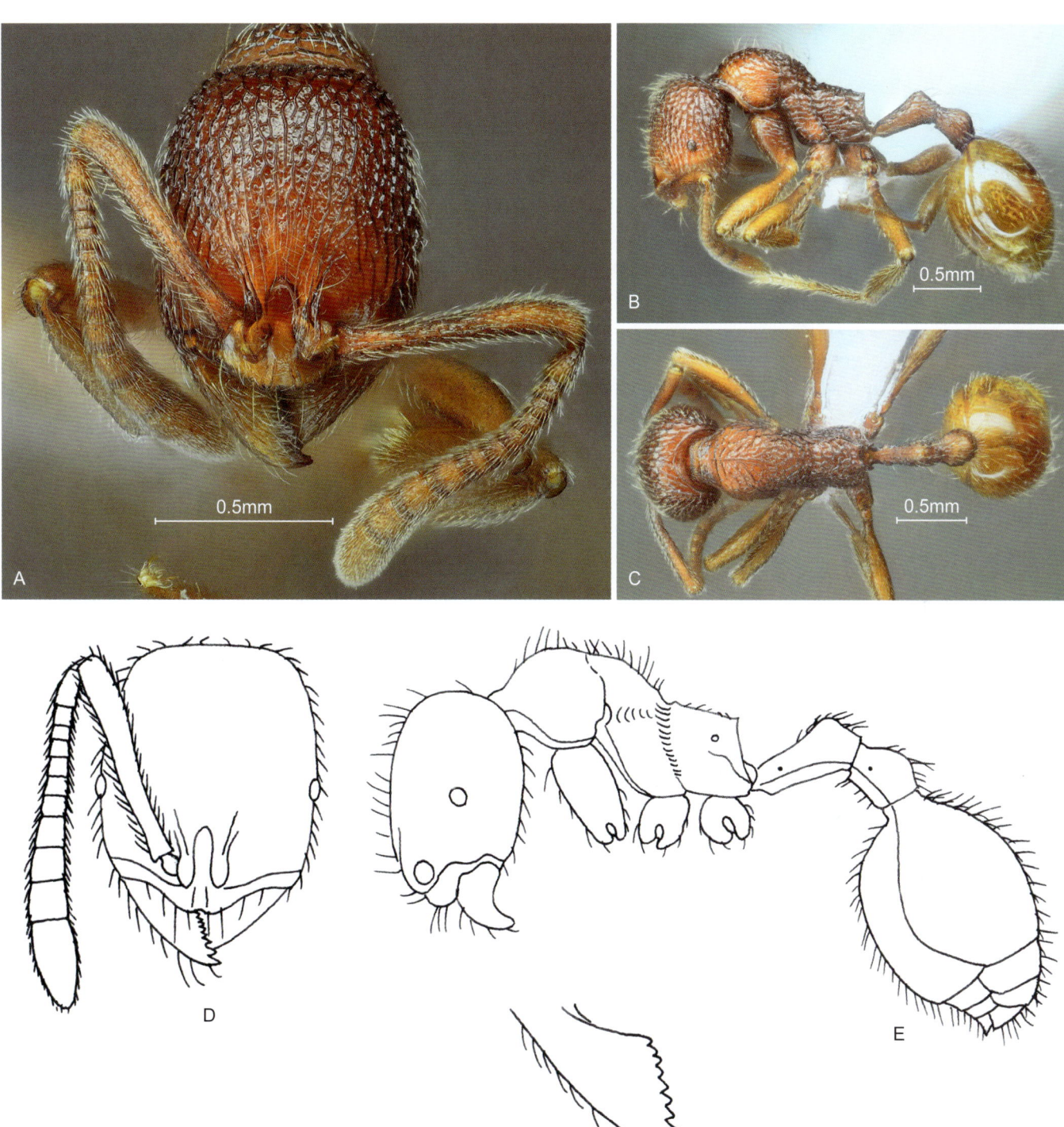

乌蒙窄结蚁 Stenamma wumengense

A, D. 工蚁头部正面观; B, E. 工蚁身体侧面观; C. 工蚁身体背面观; F. 工蚁上颚背面观（D-F. 引自Liu和Xu, 2011）
A, D. Head of worker in full-face view; B, E. Body of worker in lateral view; C. Body of worker in dorsal view; F. Mandible of worker in dorsal view (D-F. cited from Liu & Xu, 2011)

雅鲁藏布窄结蚁
Stenamma yaluzangbum Liu & Xu, 2011

Stenamma yaluzangbum Liu & Xu, 2011, Sociobiology, 58(3): 744, figs. 7-9 (w.) CHINA (Tibet).

Holotype worker: TL 3.8, HL 0.88, HW 0.78, CI 89, SL 0.65, SI 84, PW 0.55, AL 1.13, ED 0.10, PL 0.48, PH 0.25, PI 53, DPW 0.18, PPL 0.28, PPH 0.24, PPI 86, PPW 0.24. In full-face view, head roughly rectangular, longer than broad. Occipital margin straight, occipital corners rounded, sides weakly convex. Anterior margin of clypeus convex, and concave in the middle. Mandibles with 2 distinct apical teeth and followed by 7 indistinct denticles. Antennae short, 12-segmented; apices of scapes reached to 9/10 of the distance from antennal sockets to occipital corners; antennal clubs 4-segmented. Eyes located before the midpoints of sides of head, with 4 ommatidia in the maximum diameter. In lateral view, promesonotum high, roundly convex and arched. Promesonotal suture absent. Metanotal groove wide, deeply depressed, and with a convex tubercle in the groove. Propodeum distinctly lower than promesonotum, dorsum straight and steeply sloped. Propodeal spines slender and acute, about 1/3 length of declivity. Declivity straight, about 3/4 length of dorsum. Propodeal plates broad, rounded at apices, about as long as propodeal spines. Petiole relatively long, length: height: width = 2: 1: 0.7. Petiolar node cone-shaped, about as long as anterior peduncle, and distinctly depressed between the node and peduncle. Ventral face of petiole nearly straight, but weakly concave at posterior 1/3, anteroventral corner weakly prominent. Postpetiolar dorsum roundly convex, ventral face weakly concave, anteroventral corner extruding and toothed. Mandibles finely longitudinally striate. Head retirugose, but longitudinally rugose on central dorsum and before eyes, median portion of clypeus smooth and shining. Anterior 1/3 of pronotum with interweaved transverse rugae. Posterior 2/3 of pronotum and sides with posteriorly divergent longitudinal rugae. Mesonotum, metapleura, and sides of propodeum with interweaved longitudinal rugae. Mesopleura densely and coarsely punctuate. Metanotal groove with short longitudinal rugae. Propodeal dorsum finely retirugose. Petiole and postpetiole densely finely punctuate, dorsal faces with sparse fine longitudinal rugae, dorsum of postpetiolar node relatively smooth. Gaster smooth and shining, basal carinae about 1/2 length of postpetiole. Dorsa of head and body with abundant erect to suberect hairs and dense decumbent pubescence, but alitrunk dorsum with sparse pubescence. Dorsa of propodeum and petiolar peduncle without erect hairs. Antennal scapes and hind tibiae with dense decumbent pubescence, but without erect hairs. Head and body blackish brown, appendages and gaster yellowish brown, eyes and masticatory margins of mandibles black.

Paratype workers: TL 2.9-3.8, HL 0.73-0.88, HW 0.60-0.73, CI 82-84, SL 0.53-0.65, SI 88-93, PW 0.41-0.53, AL 0.88-1.08, ED 0.06-0.09, PL 0.35-0.48, PH 0.20-0.23, PI 47-57, DPW 0.15-0.16, PPL 0.23-0.28, PPH 0.20-0.21, PPI 73-89, PPW 0.19-0.21 ($n=4$). As holotype, but in some specimens with anteroventral corner of petiole bluntly angled, some specimens with smaller body size and reddish brown in body color, some specimens with pronotum retirugose, upper portions of mesopleura sparsely longitudinally rugose.

Holotype: Worker, China: Tibet Autonomous Region, Milin County, Lilong Town, Lilongou, 3085 m, 2008.Ⅷ.10, collected by Zheng-hui Xu under stone in *Pinus densata* forest, No. A08-3122.

Etymology: The new species is named after the big river Yarlungzangbo which flows over the main type localities in southeastern Tibet, China.

正模工蚁：正面观，头部近长方形，长大于宽；后缘平直，后角圆；侧缘轻度隆起。唇基前缘隆起，中央凹陷。上颚具2个明显的端齿和7个不明显的小齿。触角短，12节，柄节末端到达触角窝至头后角间距的9/10处，触角棒4节。复眼位于头侧缘中点之前，最大直径上具4个小眼。侧面观，前中胸背板高，圆形隆起呈弓形。缺前中胸背板缝。后胸沟宽，深凹，沟内具1个隆起的小瘤突。并胸腹节明显低于前中胸背板，背部平直，陡坡状；并胸腹节刺细长且尖锐，约为斜面长度的1/3；斜面平直，约为背面长度的3/4。并胸腹节侧叶宽，顶端圆，约与并胸腹节刺等长。腹柄较长，长：高：宽=2：1：0.7；腹柄结圆锥形，约与前面小柄等长，腹柄结与小柄之间明显凹陷；腹柄腹面近平直，但是在后部1/3轻度凹陷，前下角轻度突出。后腹柄背面圆形隆起，腹面轻度凹陷，前下角突出呈齿状。上颚具细纵条纹。头部具网纹，但是在复眼之前的背面中央具纵皱纹。唇基中部光滑发亮。前胸背板前部1/3具互相交织的横皱纹，后部2/3和侧面具向后分歧的纵皱纹。中胸背板、后胸侧板和并胸腹节侧面具互相交织的纵皱纹，中胸侧板具密集粗刻点。后胸沟具短纵脊，并胸腹节背面具细网纹。腹柄和后腹柄具细密刻点，背面具稀疏的细纵皱纹，后腹柄结背面较光滑。后腹部光滑发亮，基纵脊约为后腹柄长度的1/2。头部和身体背面具丰富直立、亚直立

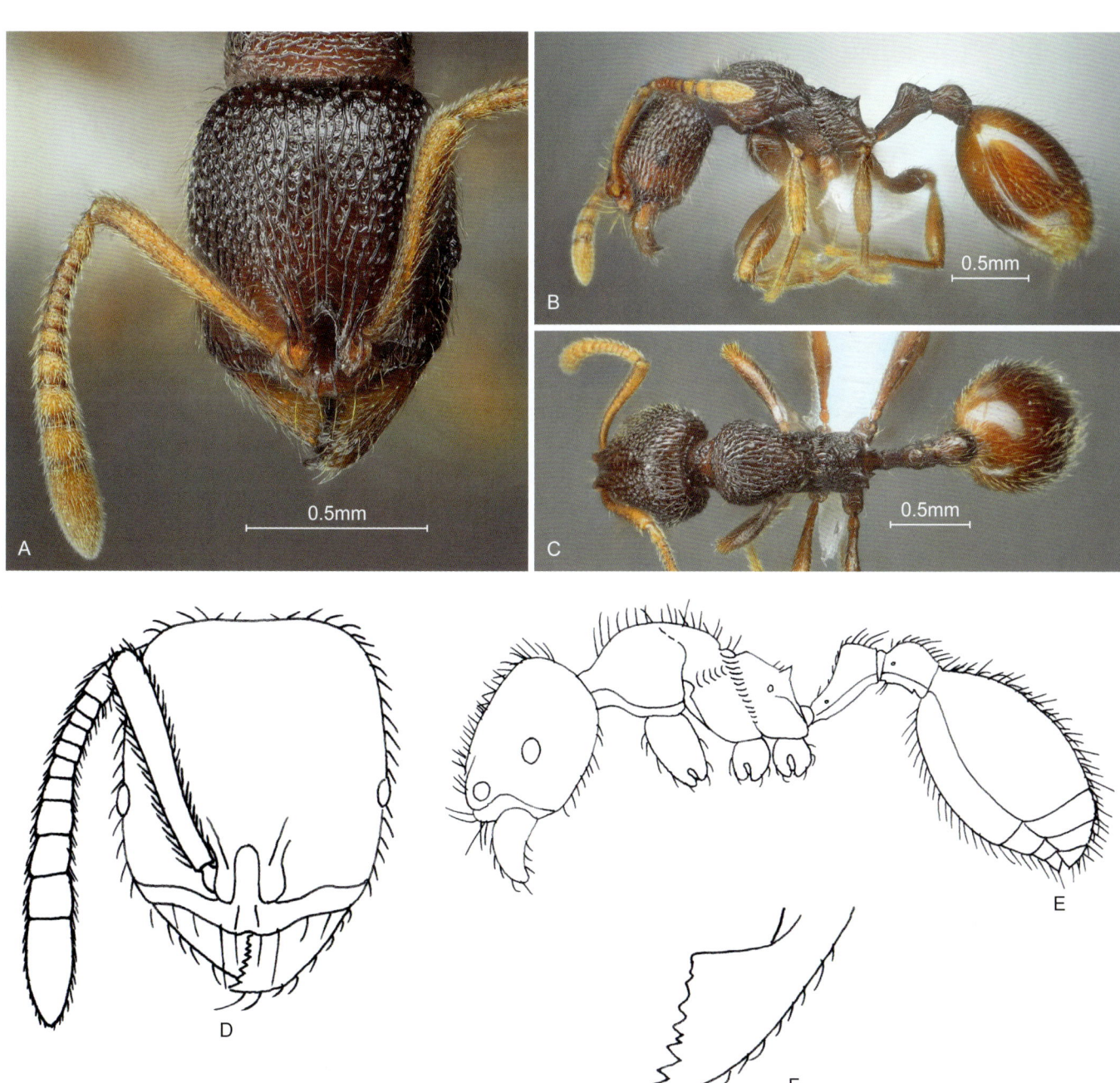

雅鲁藏布窄结蚁*Stenamma yaluzangbum*

A, D. 工蚁头部正面观; B, E. 工蚁身体侧面观; C. 工蚁身体背面观; F. 工蚁上颚背面观(D-F. 引自Liu和Xu, 2011)

A, D. Head of worker in full-face view; B, E. Body of worker in lateral view; C. Body of worker in dorsal view; F. Mandible of worker in dorsal view (D-F. cited from Liu & Xu, 2011)

毛和密集倾斜绒毛被，但是胸部背面绒毛被稀疏，并胸腹节背面和腹柄前面小柄缺立毛。触角柄节和后足胫节具密集倾斜绒毛被，缺立毛。头部和身体黑棕色，附肢和后腹部黄棕色，复眼和上颚咀嚼缘黑色。**副模工蚁**：特征同正模，但是有的标本腹柄前下角钝角状；有的标本体形较小，身体红棕色；有的标本前胸背板具网纹，中胸侧板上部具稀疏纵皱纹。

正模：工蚁，中国西藏自治区米林县里龙乡里龙沟，3085m，2008.Ⅷ.10，徐正会采于高山松林石下，No. A08-3122。

词源：该新种以流经中国西藏东南部主要模式产地的"雅鲁藏布江"命名。

白露切胸蚁
Temnothorax bailu Qian & Xu, 2024

Temnothorax bailu Qian & Xu, 2024, European Journal of Taxonomy, 936: 7, fig. 3 (w.) CHINA (Tibet).

Holotype worker: TL 3.6, HL 0.76, HW 0.66, CI 88, SL 0.64, SI 97, ED 0.14, PW 0.49, WL 1.00, PL 0.42, PH 0.25, DPW 0.20. In full-face view, head subrectangular, longer than broad, posterior margin almost straight, posterior corners narrowly rounded, lateral margins weakly convex. Mandibles triangular, masticatory margin with 5 teeth. Clypeal dorsum weakly convex, with median carina, anterior margin moderately convex. Frontal lobes broad, concealing antennal sockets. Frontal carinae short, reaching to the level of midpoint of eyes. Antennae 12-segmented, scapes almost reaching to posterior head margin, club 3-segmented. Eyes located at midpoint of lateral head margin, occupying 1/4 of lateral margin. In lateral view, pronotum weakly convex, promesonotal suture absent. Mesonotum almost straight and gently sloping down posteriorly, metanotal groove shallowly impressed. Propodeal dorsum slightly concave; propodeal spines moderately long and acute, about 1/3 length of dorsum, back-curved and pointed posterodorsally; declivity weakly concave, about 1/2 length of dorsum; propodeal lobes small and rounded apically. Petiole with long anterior peduncle, about as long as petiolar node; petiolar node roughly trapezoidal, dorsal and anterior margins almost straight, posterior margin weakly convex, anterodorsal corner rightly angled, posterodorsal corner lower and narrowly rounded; ventral margin almost straight, anteroventral corner largely acutely toothed. Postpetiole as high as petiolar node, anterior margin weakly convex, dorsum strongly convex, posterior margin weakly concave; ventral margin short and weakly concave. In dorsal view, pronotum broadest, anterior margin moderately convex, humeral corners bluntly angled, lateral margin weakly convex, promesonotal suture absent. Mesothorax moderately constricted, lateral margins weakly concave, metanotal groove shallowly impressed. Propodeum roughly rectangular, lateral margins weakly convex; spines moderately long and pointed posterolaterally, weakly in-curved. Petiole roughly rectangular, slightly longer than broad, lateral margins slightly convex. Postpetiole roughly trapezoidal and narrowing posteriorly, about 1.4 times as broad as petiole, anterolateral corners narrowly rounded, lateral margins slightly convex. Gaster elongate oval. Mandibles longitudinally striate. Head dorsum loosely longitudinally rugose and reticulate laterally, the longitudinal central strip smooth and shiny, posteroventral part of lateral head margin relatively smooth. Clypeus smooth, each side with two short rugae. Mesosoma finely reticulate; pronotal sides longitudinally rugose; mesonotal dorsum smooth; propodeal dorsum coarsely reticulate. Petiole densely punctate, the node micro-reticulate; postpetiole densely punctate, dorsum smooth and shiny. Gaster smooth and shiny. Body dorsum with abundant erect to suberect apically blunt short hairs and abundant decumbent pubescence, hairs on mesosoma sparse. Scapes and tibiae with dense decumbent pubescence. Body color blackish brown; head dorsum and apical half of gaster black; mandibles, basal segments of flagella, trochanters and tarsi brownish yellow. **Paratype workers:** TL 2.6-3.6, HL 0.70-0.75, HW 0.55-0.63, CI 79-83, SL 0.49-0.50, SI 80-91, ED 0.13-0.15, PW 0.40-0.50, WL 0.80-1.00, PL 0.30-0.40, PH 0.20-0.25, DPW 0.13-0.18 (*n*=12). As holotype, in some individuals mesonotum weakly punctate, metanotal groove shallowly or indistinctly impressed, lateral head margins weakly rugose.

Holotype: Worker, China: Tibet Autonomous Region, Chayu County, Zhuwagen Town, Jiangtuo Village (28.612850°N, 97.306850°E), 2050 m, 2010.Ⅷ.30, collected by Xia Liu from a canopy sample of *Pinus yunnanensis* forest, No. A10-3265.

Etymology: The specific epithet refers to "bailu", one of the Twenty-four Solar Terms of China.

正模工蚁： 正面观，头部近长方形，长大于宽；后缘几乎平直，后角窄圆；侧缘轻度隆起。上颚三角形，咀嚼缘具5个齿。唇基背面轻度隆起，具中央纵脊，前缘中度隆起。额叶宽，遮盖触角窝。额脊短，到达复眼中点水平。触角12节，柄节几乎到达头后缘，触角棒3节。复眼位于头侧缘中点处，占据侧缘的1/4。侧面观，前胸背板轻度隆起，缺前中胸背板缝。中胸背板几乎平直，向后坡形降低，后胸沟浅凹。并胸腹节背面轻度凹陷；并胸腹节刺中等长，尖锐，约为背面长度的1/3，后弯，指向后上方；斜面轻度凹陷，约为背面长度的1/2。并胸腹节侧叶小，末端圆。腹柄前面具长的小柄，约与腹柄结等长；腹柄结近梯形，背面和前面几乎平直，后面轻度隆起；前上角直角形，后上角较低，窄圆；腹柄腹面近平直，前下角具大而尖的齿。后腹柄与腹柄结等高，前面轻度隆起，背面强烈隆起，后面轻度凹陷；腹面短，轻度凹陷。背面观，前胸背板最宽，前缘中度隆起，肩角钝角状，侧面轻度隆起，缺前中胸背板缝。中胸中度收缩，侧缘轻度凹陷，后胸沟浅凹。并胸腹节近长方形，侧面轻度隆起；并胸腹节刺中等长，指向侧后方，轻度内弯。腹柄近长方形，长稍大于宽，侧缘轻度隆起。后腹柄近梯形，向后变窄，约为腹柄

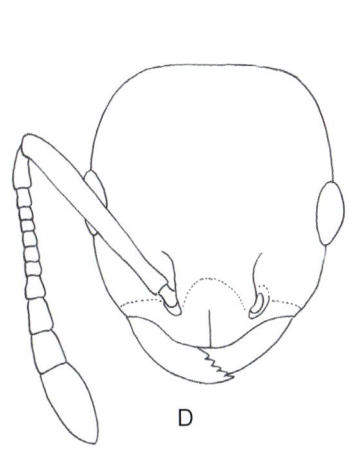
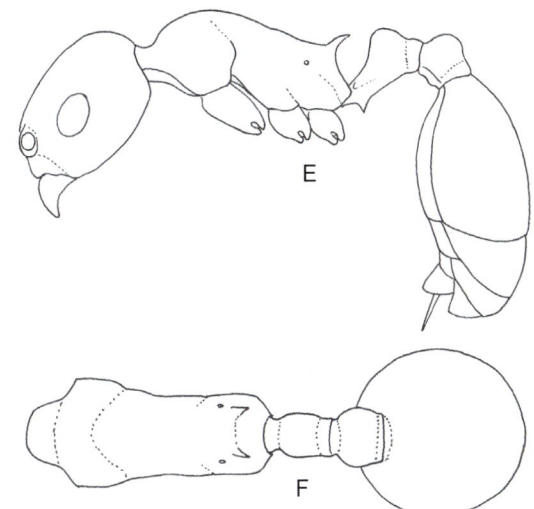

白露切胸蚁 Temnothorax bailu

A, D. 工蚁头部正面观；B, E. 工蚁身体侧面观；C, F. 工蚁身体背面观（A-C. 引自Qian和Xu, 2024）
A, D. Head of worker in full-face view; B, E. Body of worker in lateral view; C, F. Body of worker in dorsal view (A-C. cited from Qian & Xu, 2024)

宽的1.4倍，前侧角窄圆，侧面轻度隆起。腹部长卵圆形。上颚具纵条纹。头部背面具松散纵皱纹，侧面具网纹，中央纵带光滑发亮，头侧缘后下部较光滑。唇基光滑，每侧具2条短皱纹。胸部具细网纹；前胸侧面具纵皱纹，中胸背面光滑，并胸腹节背面具粗糙网纹。腹柄具密集刻点，腹柄结具微网纹；后腹柄具密集刻点，背面光滑发亮。后腹部光滑发亮。身体背面具丰富直立、亚直立短钝毛和丰富倾斜绒毛被，胸部立毛稀疏；柄节和胫节具密集倾斜绒毛被。身体黑棕色，头部背面和后腹部后半部黑色，上颚、鞭节基部各节、转节和跗节棕黄色。**副模工蚁**：特征同正模，一些个体中胸背板具弱刻点，后胸沟浅或凹陷不明显，头侧缘具弱纵纹。

正模：工蚁，中国西藏自治区察隅县竹瓦根镇江拖村（28.612850°N，97.306850°E），2050m，2010.Ⅷ.30，刘霞采于云南松林树冠样中，No. A10-3265。

词源：该新种以中国二十四节气中的"白露"命名。

春切胸蚁
Temnothorax chun Qian & Xu, 2024

Temnothorax chun Qian & Xu, 2024, European Journal of Taxonomy, 936: 10, fig. 5 (w.) CHINA (Yunnan).

Holotype worker: TL 3.1, HL 0.74, HW 0.62, CI 84, SL 0.60, SI 97, ED 0.16, PW 0.43, WL 0.91, PL 0.33, PH 0.21, DPW 0.16. In full-face view, head roughly rectangular, longer than broad, posterior margin almost straight, posterior corners narrowly rounded, lateral margins weakly convex. Mandibles triangular, masticatory margin with 5 teeth. Clypeal dorsum weakly convex, with median carina, anterior margin moderately convex. Frontal lobes broad, concealing antennal sockets. Frontal carinae short, reaching to the level of anterior eye margins. Antennae 12-segmented, scapes just reaching to posterior head margin. Eyes located at midpoint of lateral head margin, nearly occupying 1/3 of lateral margin. In lateral view, pronotum weakly convex, promesonotal suture present but not impressed. Mesonotum almost straight and gently sloping down posteriorly, metanotal groove absent. Propodeal dorsum slightly convex and gently sloping down posteriorly; propodeal spines short and stout, about 1/3 length of dorsum and longer than their basal width, weakly down-curved; declivity almost straight, shorter than dorsum; propodeal lobes short and rounded apically. Petiole with long anterior peduncle, about 3/4 length of petiolar node; petiolar node roughly semicircular, anterior and posterior margins weakly convex, dorsal margin narrowly rounded; ventral margin straight anteriorly, moderately concave posteriorly, anteroventral corner triangularly toothed. Postpetiole as high as petiolr node, anterodorsal margin rounded, posterior margin weakly concave, ventral margin angularly concave. In dorsal view, pronotum broadest, anterior and lateral margins moderately convex, humeral corners broadly rounded, promesonotal suture present. Mesothorax weakly constricted, lateral margins weakly concave, metanotal groove absent. Propodeum roughly rectangular, lateral margins almost straight, spines short and pointed posterolaterally. Petiole roughly trapezoidal and widening posteriorly, anterior peduncle narrower than the node. Postpetiole roughly trapezoidal and narrowing posteriorly, about 1.6 times as broad as petiole, anterolateral corners narrowly rounded, lateral margins weakly convex. Gaster elongate oval. Mandibles longitudinally striate. Head dorsum loosely longitudinally rugose and reticulate laterally, with the longitudinal central strip relatively smooth; genae longitudinally rugose. Clypeus smooth, each side with 3-4 short rugae. Mesosoma reticulate, pronotal sides longitudinally rugose, mesonotal dorsum smooth. Petiole and postpetiole densely finely punctate, dorsa of petiolar node and postpetiole smooth. Gaster smooth and shiny. Body dorsum with abundant erect to suberect apically blunt short hairs and abundant decumbent pubescence. Scapes and tibiae with dense decumbent pubescence. Body color blackish brown; head dorsum, eyes and gaster black; mandibles and tarsi brownish yellow.

Holotype: Worker, China: Yunnan Province, Yuanmou County, Laocheng Town, Naneng Village (25.638800°N, 101.958307°E), 2000 m, 2013.Ⅲ.12, collected by Wen-tao Yang from a ground sample of conifer-broadleaf mixed forest, No. A13-38.

Etymology: The specific epithet refers to "chun", one of the four seasons.

正模工蚁： 正面观，头部近长方形，长大于宽；后缘几乎平直，后角窄圆；侧面轻度隆起。上颚三角形，咀嚼缘具5个齿。唇基背面轻度隆起，具中央纵脊，前缘中度隆起。额叶宽，遮盖触角窝。额脊短，到达复眼前缘水平。触角12节，柄节刚到达头后缘。复眼位于头侧缘中点，约占据侧缘的1/3。侧面观，前胸背板轻度隆起，前中胸背板缝存在但不凹陷。中胸背板几乎平直，缓坡形，缺后胸沟。并胸腹节背面轻微隆起，缓坡形；并胸腹节刺短而粗，约为背面长度的1/3，长于其基部宽，轻度下弯；斜面近平直，短于背面。并胸腹节侧叶短，顶端圆。腹柄前面具长的小柄，约为腹柄结长的3/4；腹柄结近半圆形，前面和后面轻度隆起，背面窄圆；腹面前部平直，后部中度凹陷，前下角具三角形齿。后腹柄与腹柄结等高，前上缘圆，后缘轻度凹陷，腹缘角状凹陷。背面观，前胸背板最宽，前面和侧面轻度隆起，肩角宽圆，前中胸背板缝存在。中胸轻度收缩，侧面轻度凹陷，缺后胸沟。并胸腹节近长方形，侧缘近平直；并胸腹节刺短，指向侧后方。腹柄近梯形，向后变宽，前面小柄窄于腹柄结。后腹柄近梯形，向后变窄，约为腹柄宽的1.6倍，前侧角窄圆，侧面轻度隆起。后腹部长卵形。上颚具纵条纹。头部背面具松散纵皱纹，侧面具网纹，中央纵带较光滑，颊区具纵皱纹。唇基光滑，每侧具3~4条短皱纹。胸部具网纹，前胸背板侧面具纵皱纹，中胸背板背面光滑。腹柄和后腹柄具细密刻点，腹柄结背面和后腹柄背面光滑。后腹部光滑发亮。身体背面具丰富直立、亚直立短钝毛和丰富倾斜绒毛被。柄节和胫节具密集倾斜绒毛被。身体黑棕色，头部背面、复眼和后腹部黑色，上颚和跗节棕黄色。

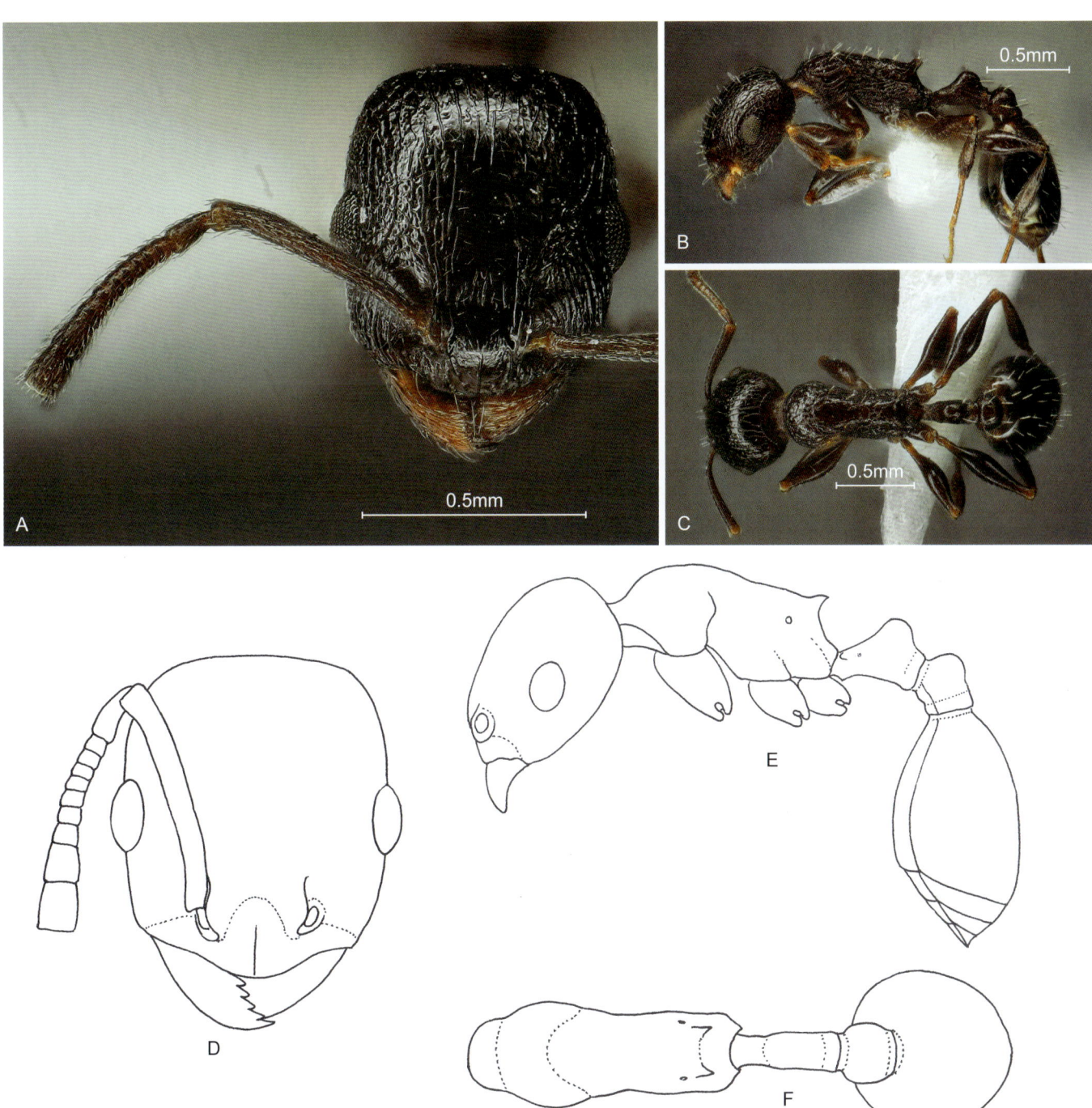

春切胸蚁 *Temnothorax chun*

A, D. 工蚁头部正面观；B, E. 工蚁身体侧面观；C, F. 工蚁身体背面观（A-C. 引自Qian和Xu, 2024）
A, D. Head of worker in full-face view; B, E. Body of worker in lateral view; C, F. Body of worker in dorsal view (A-C. cited from Qian & Xu, 2024)

正模： 工蚁，中国云南省元谋县老城乡那能村（25.638800°N，101.958307°E），2000m，2013.Ⅲ.12，杨文涛采于针阔混交林地表样中，No. A13-38。

词源： 该新种以四个季节中的"春"命名。

春分切胸蚁
Temnothorax chunfen Qian & Xu, 2024

Temnothorax chunfen Qian & Xu, 2024, European Journal of Taxonomy, 936: 11, fig. 6 (w.) CHINA (Yunnan).

Holotype worker: TL 2.4, HL 0.58, HW 0.48, CI 82, SL 0.46, SI 96, ED 0.13, PW 0.33, WL 0.74, PL 0.25, PH 0.20, DPW 0.15. In full-face view, head roughly rectangular, longer than broad, posterior margin slightly convex, posterior corners broadly rounded, lateral margins weakly convex. Mandibles triangular, masticatory margin with 5 teeth. Clypeal dorsum weakly convex, with median carina, anterior margin moderately convex. Frontal lobes broad, concealing antennal sockets. Frontal carinae short, about as long as frontal lobes. Antennae 12-segmented, scapes just reaching to posterior head margin, antennal clubs 3-segmented. Eyes located at midpoint of lateral head margin, occupying 1/4 of lateral margin. In lateral view, pronotum weakly convex, anterodorsal corner narrowly rounded. Promesonotal suture and metanotal groove absent. Mesonotum almost straight and gently sloping down posteriorly. Propodeal dorsum slightly convex and gently sloping down posteriorly; propodeal spines short and triangular, blunt apically, shorter than their basal width; declivity slightly concave; propodeal lobes broad, truncated apically. Petiole with relatively short anterior peduncle, about 2/3 length of petiolar node; petiolar node roughly conical, anterior margin slightly convex, posterior margin weakly convex, dorsum narrowly rounded; ventral margin slightly concave, anteroventral corner bluntly toothed. Postpetiole as high as petiolar node, anterodorsal margin rounded, posterior margin slightly concave, ventral margin almost straight. In dorsal view, pronotum broadest, anterior margin weakly convex, humeral corners broadly rounded, lateral margin moderately convex. Promesonotal suture and metanotal groove absent. Mesothorax weakly constricted, lateral margins weakly concave. Propodeum roughly rectangular, slightly widening posteriorly, lateral margins almost straight; spines short and triangular, pointed posterolaterally. Petiole roughly trapezoidal and widening posteriorly, longer than broad, lateral margins weakly convex. Postpetiole roughly trapezoidal and narrowing posteriorly, about 1.5 times as broad as petiole, anterolateral corners narrowly rounded, lateral margins weakly convex. Gaster elongate oval. Mandibles longitudinally striate. Head dorsum smooth and shiny, each side with 3 oblique rugae above frontal carinae. Clypeus smooth, each side with 1 short longitudinal ruga. Pronotum smooth and shiny, the rest of mesosoma finely reticulate, mesonotal dorsum smooth. Petiole and postpetiole densely finely punctate, postpetiolar dorsum smooth. Gaster smooth and shiny. Body dorsum with abundant erect to suberect apically blunt short hairs and abundant decumbent pubescence. Scapes with dense subdecumbent hairs and dense decumbent pubescence, tibiae with dense decumbent pubescence. Body brownish black; gaster black; mandibles, antennae, trochanters, tibiae and tarsi brownish yellow.

Holotype: Worker, China: Yunnan Province, Deqin County, Benzilan Town, Benzilan Village (28.242147°N, 99.281039°E), 2260 m, 2004.Ⅹ.12, collected by Jun-wu Yang from a canopy sample of dry-warm valley shrub, No. A04-713.

Etymology: The specific epithet refers to "chunfen", one of the Twenty-four Solar Terms of China.

正模工蚁： 正面观，头部近长方形，长大于宽；后缘轻微隆起，后角宽圆；侧缘轻度隆起。上颚三角形，咀嚼缘具5个齿。唇基背面轻度隆起，具中央纵脊，前缘中度隆起。额叶宽，遮盖触角窝。额脊短，约与额叶等长。触角12节，柄节刚到达头后缘，触角棒3节。复眼位于头侧缘中点，占据侧缘的1/4。侧面观，前胸背板轻度隆起，前上角窄圆。缺前中胸背板缝和后胸沟。中胸背板近平直，向后轻度降低。并胸腹节背面轻微隆起，向后轻度坡形降低；并胸腹节刺短，三角形，末端钝，短于基部宽；斜面轻微凹陷。并胸腹节侧叶宽，末端平截。腹柄前面小柄较短，约为腹柄结长的2/3；腹柄结近锥形，前面轻微隆起，后面轻度隆起，背面窄圆；腹面轻微凹陷，前下角钝齿状。后腹柄与腹柄结等高，前上缘圆，后缘轻微凹陷，腹缘近平直。背面观，前胸背板最宽，前缘轻度隆起，肩角宽圆，侧缘中度隆起。缺前中胸背板缝和后胸沟。中胸轻度收缩，侧面轻度凹陷。并胸腹节近长方形，向后轻微变宽，侧缘近平直；并胸腹节刺短，三角形，指向后侧方。腹柄近梯形，向后变宽，长大于宽，侧缘轻度隆起。后腹柄近梯形，向后变窄，约为腹柄宽的1.5倍，前侧角窄圆，侧缘轻度隆起。后腹部长卵形。上颚具纵条纹。头部背面光滑发亮，额脊上方每侧具3条倾斜皱纹。唇基光滑，每侧具1条短纵皱纹。前胸光滑发亮，胸部其余部分具细网纹，中胸背板背面光滑。腹柄和后腹柄具细密刻点，后腹柄背面光滑。后腹部光滑发亮。身体背面具丰富直立、亚直立短钝毛和丰富倾斜绒毛被；柄节具密集亚倾斜毛和密集倾斜绒毛被，胫节具密集倾斜绒毛被。身体棕黑色，后腹部黑色，上颚、触角、转节、胫节和跗节棕黄色。

正模： 工蚁，中国云南省德钦县奔子栏乡奔子栏村（28.242147°N，99.281039°E），2260m，

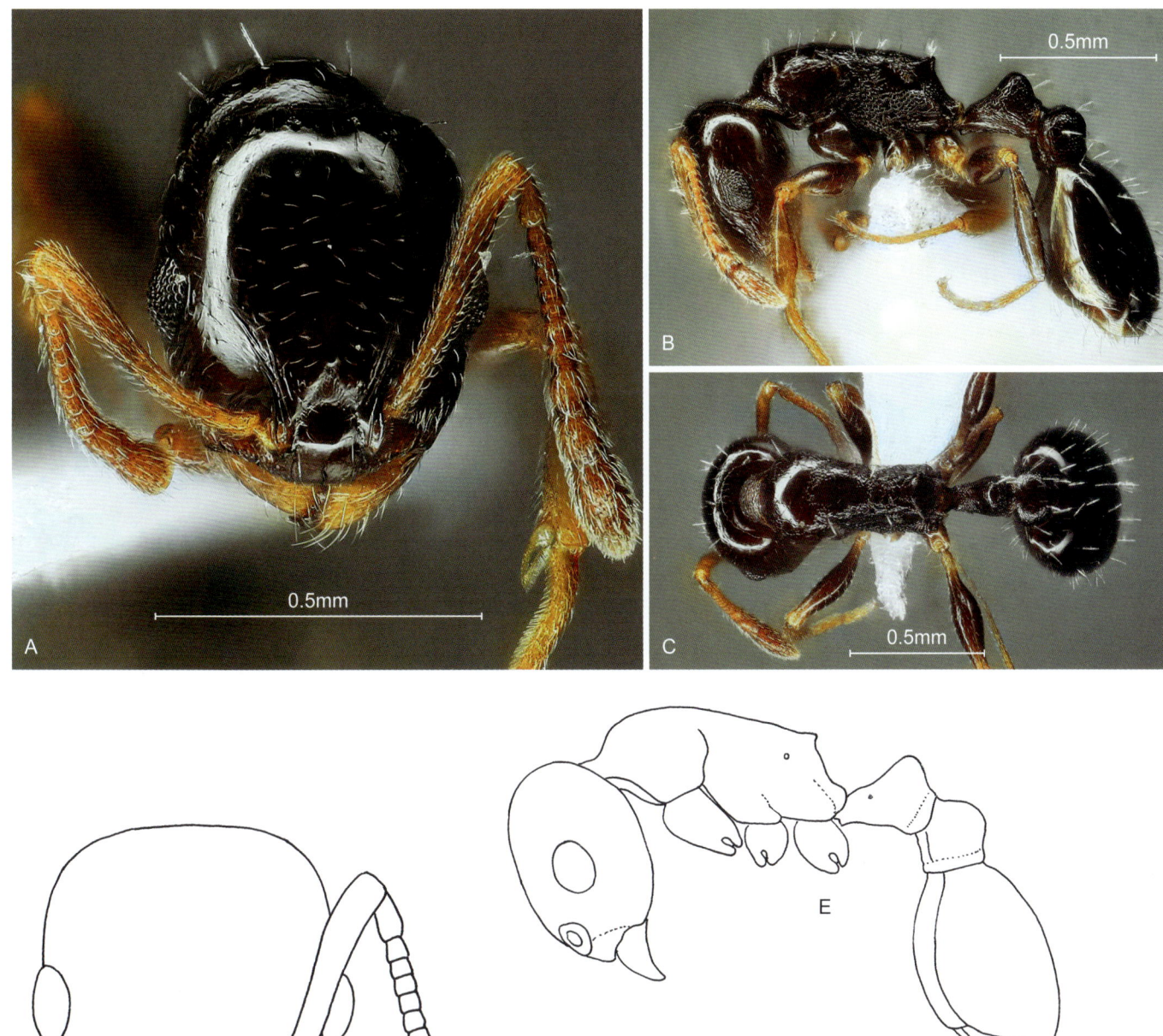

春分切胸蚁 *Temnothorax chunfen*

A, D. 工蚁头部正面观；B, E. 工蚁身体侧面观；C, F. 工蚁身体背面观（A-C. 引自Qian和Xu, 2024）
A, D. Head of worker in full-face view; B, E. Body of worker in lateral view; C, F. Body of worker in dorsal view (A-C. cited from Qian & Xu, 2024)

2004.Ⅹ.12，杨俊伍采于干暖河谷灌丛中，No. A04-713。

词源： 该新种以中国二十四节气中的"春分"命名。

处暑切胸蚁
Temnothorax chushu Qian & Xu, 2024

Temnothorax chushu Qian & Xu, 2024, European Journal of Taxonomy, 936: 14, fig. 8 (w.) CHINA (Yunnan).

Holotype worker: TL 2.9, HL 0.67, HW 0.55, CI 82, SL 0.63, SI 115, ED 0.15, PW 0.41, WL 0.83, PL 0.32, PH 0.18, DPW 0.16. In full-face view, head roughly rectangular, longer than broad, posterior margin moderately convex and slightly concave in the middle, posterior corners broadly rounded, lateral margins weakly convex. Mandibles triangular, masticatory margin with 5 teeth. Clypeal dorsum weakly convex, with median carina, anterior margin moderately convex. Frontal lobes narrow, concealed half of antennal socket. Frontal carinae short, reaching to midpoint level of eyes. Antennae 12-segmented, scapes surpassing posterior margin by 1/12 of its length, club 3-segmented. Eyes slightly before midpoint of lateral head margin, occupying 1/4 of lateral margin. In lateral view, pronotum moderately convex, promesonotal suture slightly impressed. Mesonotum slightly convex and gently sloping down posteriorly. Metanotal groove widely moderately impressed and distinct. Propodeal dorsum nearly straight and gently sloping down posteriorly; propodeal spines very short and acutely toothed, shorter than their basal width; declivity moderately concave, about 1/2 length of dorsum; propodeal lobes large, narrowly rounded apically. Petiole with anterior peduncle, about 2/3 length of petiolar node; petiolar node roughly trapezoidal, anterior margin straight, dorsal margin roundly convex, posterior margin weakly convex; ventral margin weakly convex in the middle, moderately concave posteriorly, anteroventral corner sharply toothed. Postpetiole about as high as petiole, anterodorsal margin roundly convex, posterior margin slightly concave, ventral margin weakly concave. In dorsal view, pronotum broadest, anterior margin moderately convex and weakly marginated, humeral corners broadly rounded, lateral margins strongly convex. Promesonotal suture very shallowly impressed. Mesothorax moderately constricted and narrow, lateral margins strongly concave. Metanotal groove widely impressed. Propodeum roughly rectangular, lateral margins weakly convex; propodeal spines very short and bluntly toothed. Petiole longer than broad, widening posteriorly; petiolar node roughly trapezoidal and widening posteriorly, lateral margins moderately convex. Postpetiole roughly trapezoidal and narrowing posteriorly, about 1.3 times as broad as petiole, anterolateral corners narrowly rounded, lateral margins slightly convex. Gaster elongate oval. Mandibles longitudinally striate. Head dorsum longitudinally rugose before midline of eyes, gradually reticulate posteriorly and laterally with interface densely punctate, genae longitudinally rugose. Clypeus with 3-4 rugae on each side with interface smooth. Mesosoma densely punctate with interface appearing finely reticulate; pronotal sides longitudinally rugose with rugae interweaved; lower part of metapleuron with two strong costulae. Petiole and postpetiole densely weakly punctate. Gaster smooth and shiny. Body dorsum with abundant erect to suberect apically blunt hairs and abundant decumbent pubescence, hairs and pubescence on the head relatively denser. Scapes and tibiae with dense subdecumbent hairs and dense decumbent pubescence. Body color brownish yellow, gaster blackish brown, legs light yellow. **Paratype workers:** TL 2.4-3.1, HL 0.55-0.75, HW 0.45-0.60, CI 80-82, SL 0.50-0.60, SI 100-111, ED 0.12-0.13, PW 0.30-0.35, WL 0.55-0.83, PL 0.29-0.30, PH 0.15-0.20, DPW 0.10-0.15 ($n=14$). As holotype, in some individuals propodeal spines slightly shorter or longer than in holotype, body color brown to blackish brown.

Holotype: Worker, China: Yunnan Province, Lushui City, Pianma Town, Pianma Village (25.993789°N, 98.660772°E), 2500 m, 1999.Ⅳ.25, collected by Zheng-hui Xu from a nest inside dead branch in subalpine moist evergreen broadleaf forest, No. A99-29.

Etymology: The specific epithet refers to "chushu", one of the Twenty-four Solar Terms of China.

正模工蚁： 正面观，头部近长方形，长大于宽；后缘中度隆起，中央轻微凹陷，后角宽圆；侧缘轻度隆起。上颚三角形，咀嚼缘具5个齿。唇基背面轻度隆起，具中央纵脊，前缘中度隆起。额叶窄，遮盖触角窝的一半。额脊短，到达复眼中点水平。触角12节，柄节1/12超过头后缘，触角棒3节。复眼位于头侧缘中点稍前处，占据侧缘的1/4。侧面观，前胸背板中度隆起，前中胸背板缝轻微凹陷。中胸背板轻度隆起，向后缓坡形降低。后胸沟中度凹陷，宽而明显。并胸腹节背面近平直，向后缓坡形降低；并胸腹节刺很短，尖齿状，短于基部宽；斜面中度凹陷，约为背面长的1/2。并胸腹节侧叶大，顶端窄圆。腹柄前面具小柄，约为腹柄结长的2/3；腹柄结近梯形，前面平直，背面圆形隆起，后面轻度隆起；腹面中部轻度隆起，后部中度凹陷，前下角具锐齿。后腹柄约与腹柄等高，前上缘圆形隆起，后缘轻微凹陷，腹面轻度凹陷。背面观，前胸背板最宽，前缘中度隆起，具弱边缘，肩角宽圆，侧缘强烈隆起。前中胸背板缝浅凹。中胸中度收缩，狭窄，侧缘强烈凹陷。后胸沟宽形凹陷。并胸腹节近长方形，侧缘轻度隆起；并胸腹节刺很短，钝齿状。腹柄长大于宽，向后变宽；腹柄结近梯形，向后变宽，侧缘中度隆起。后

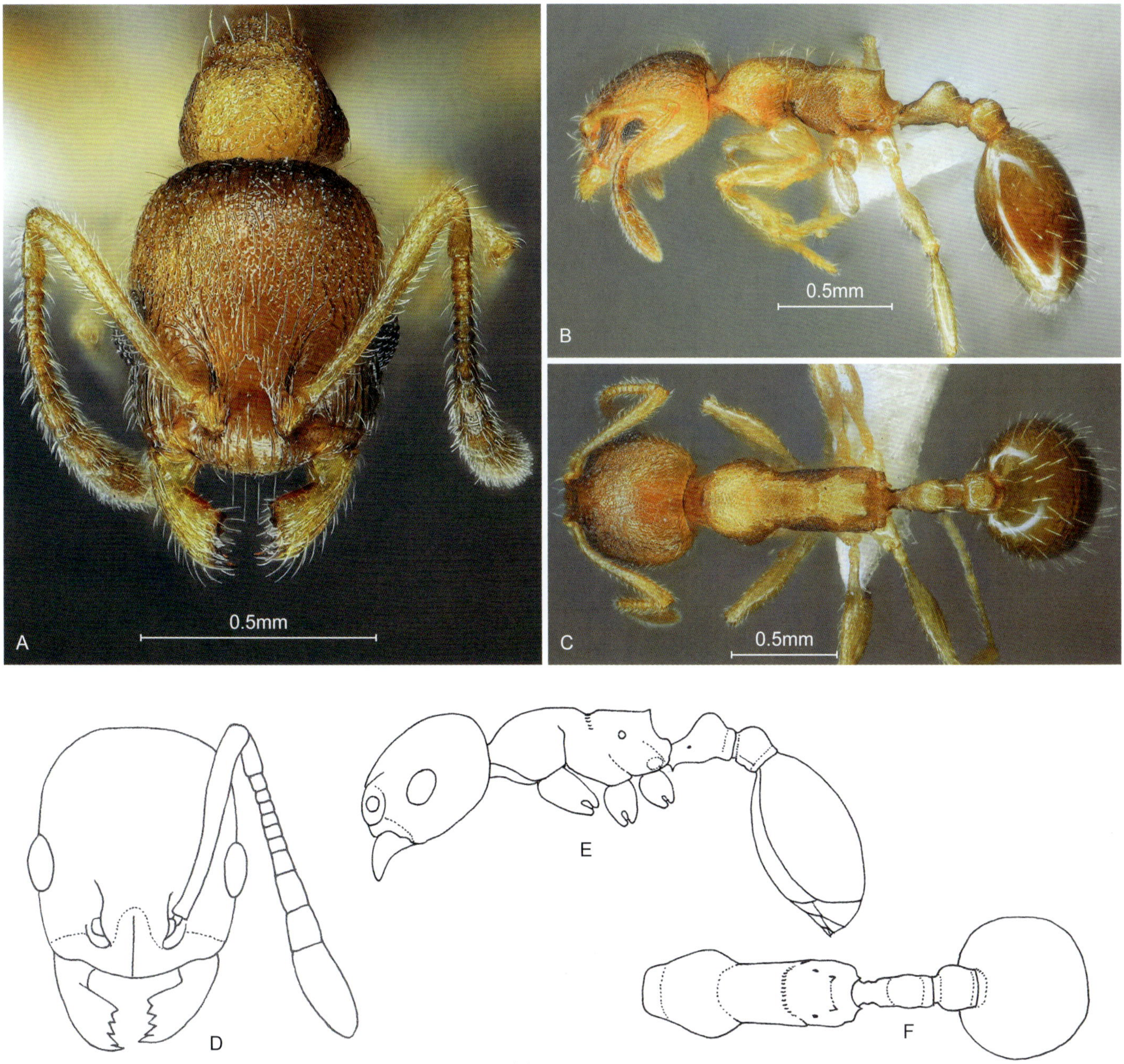

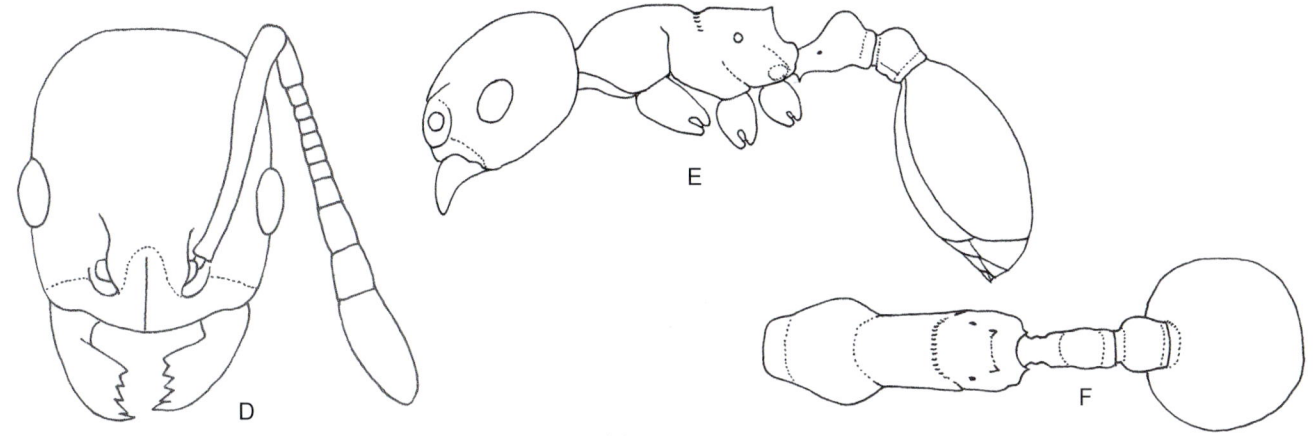

处暑切胸蚁 Temnothorax chushu

A, D. 工蚁头部正面观；B, E. 工蚁身体侧面观；C, F. 工蚁身体背面观（A-C. 引自Qian和Xu, 2024）

A, D. Head of worker in full-face view; B, E. Body of worker in lateral view; C, F. Body of worker in dorsal view (A-C. cited from Qian & Xu, 2024)

腹柄近梯形，向后变窄，约为腹柄宽的1.3倍，前侧角窄圆，侧缘轻微隆起。后腹部长卵形。上颚具纵条纹。头背面在复眼中线之前具纵皱纹，后部和两侧逐渐变成网纹，界面具密集刻点，颊区具纵皱纹。唇基光滑，每侧具3~4条短纵皱纹。胸部具密集刻点，界面呈细网纹；前胸背板侧面具互相交织的纵皱纹；后胸侧板下部具2条发达的隆脊。腹柄和后腹柄具密集弱刻点。后腹部光滑发亮。身体背面具丰富直立、亚直立钝毛和丰富倾斜绒毛被，头部的立毛和绒毛被较密集；柄节和胫节具密集亚倾斜毛和密集倾斜绒毛被。身体棕黄色，后腹部黑棕色，足浅黄色。**副模工蚁**：特征同正模，一些个体的并胸腹节刺比正模稍短或稍长，身体棕色至黑棕色。

正模：工蚁，中国云南省泸水市片马镇片马（25.993789°N, 98.660772°E），2500m，1999.Ⅳ.25，徐正会采于中山湿性常绿阔叶林枯枝内巢中，No. A99-29。

词源：该新种以中国二十四节气中的"处暑"命名。

大寒切胸蚁
Temnothorax dahan Qian & Xu, 2024

Temnothorax dahan Qian & Xu, 2024, European Journal of Taxonomy, 936: 16, fig. 10 (w.) CHINA (Yunnan).

Holotype worker: TL 2.8, HL 0.62, HW 0.53, CI 85, SL 0.53, SI 100, ED 0.14, PW 0.38, WL 0.77, PL 0.29, PH 0.20, DPW 0.16. In full-face view, head roughly rectangular, longer than broad, posterior margin nearly straight, posterior corners broad rounded, lateral margins weakly convex. Mandibles triangular, masticatory margin with 5 teeth. Clypeal dorsum weakly convex, with median carina, anterior margin weakly convex. Frontal lobes broad, concealing antennal sockets. Frontal carinae short, reaching to the level of midpoint of eyes. Antennae 12-segmented, scapes just reaching to posterior head margin. Eyes located at midpoint of lateral head margin, occupying 1/4 of lateral margin. In lateral view, pronotum moderately convex, promesonotal suture absent. Dorsa of mesonotum and propodeum slightly convex and gently sloping down posteriorly. Metanotal groove absent. Propodeal spines short, about as long as their basal width; declivity straight. Propodeal lobes rounded apically. Petiole with short anterior peduncle, about 1/3 length of petiolar node; petiolar node roughly triangular, anterior margin straight, posterior margin weakly convex, top corner blunt angled; ventral margin weakly concave posteriorly, anteroventral corner toothed. Postpetiole as high as petiole, anterodorsal margin roundly convex, posterior margin almost straight; ventral margin short and almost straight, anteroventral corner rightly angled. In dorsal view, pronotum broadest, anterior margin moderately convex, humeral corners broadly rounded, lateral margin weakly convex. Mesothorax moderately constricted and narrowed posteriorly, lateral margins strongly concave. Propodeum roughly square, lateral margins weakly concave, spines short, pointed posterolaterally. Petiole elongate trapezoidal and widening posteriorly, longer than broad; lateral margins weakly convex posteriorly. Postpetiole roughly trapezoidal, about 1.5 times as broad as petiole, narrowing posteriorly, anterior margin moderately convex, lateral margins strongly convex. Gaster elongate oval. Mandible longitudinally striate. Head dorsum longitudinally rugose with interweaved fine rugulae between the main rugae, and gradually reticulate laterally, interface abundantly punctate. Clypeus with three longitudinal rugae on each side, interface smooth. Dorsum of mesosoma coarsely reticulate, dorsum of mesonotum relatively smooth in the center. Pronotal sides coarsely reticulate anteriorly and longitudinally rugose and micro-reticulate posteriorly. Mesopleura coarsely reticulate anteriorly and micro-reticulate posteriorly; metapleura and propodeal sides densely reticulate. Petiole and postpetiole densely punctate. Gaster smooth and shiny. Head dorsum with sparse erect to suberect apically blunt short hairs and dense decumbent pubescence; mesosoma, petiole and postpetiole with sparse suberect apically blunt short hairs and sparse decumbent pubescence; gaster with abundant suberect apically blunt short hairs and abundant decumbent pubescence. Scapes with dense decumbent pubescence; tibiae with abundant decumbent pubescence. Body color brownish black; mandibles, antennae and legs brown to brownish yellow. **Paratype workers:** TL 2.5-2.7, HL 0.60-0.65, HW 0.45-0.55, CI 75-85, SL 0.55-0.60, SI 109-122, ED 0.10-0.15, PW 0.35-0.40, WL 0.70-0.75, PL 0.25-0.30, PH 0.15-0.2, DPW 0.13-0.15 (n=15). As holotype, but in some individuals mesonotum smooth, petiolar node and postpetiolar node smooth, mesosomal dorsum weakly reticulate.

Holotype: Worker, China: Yunnan Province, Wenshan City, Bozhu Town, Paomatangpo (23.389739°N, 105.933033°E), 2530 m, 2019.Ⅹ.12, collected by Hong Du from a ground sample of subalpine moist evergreen broadleaf forest, No. A19-6318.

Etymology: The specific epithet refers to "dahan", one of the Twenty-four Solar Terms of China.

正模工蚁：正面观，头部近长方形，长大于宽；后缘近平直，后角宽圆；侧缘轻度隆起。上颚三角形，咀嚼缘具5个齿。唇基背面轻度隆起，具中央纵脊，前缘轻度隆起。额叶宽，遮盖触角窝。额脊短，到达复眼中点水平。触角12节，柄节刚到达头后缘。复眼位于头侧缘中点上，占据侧缘的1/4。侧面观，前胸背板中度隆起，缺前中胸背板缝。中胸背板和并胸腹节背面轻微隆起，向后缓坡形降低，缺后胸沟；并胸腹节刺短，长度约与基部宽相等；斜面平直。并胸腹节侧叶顶端圆。腹柄前面小柄短，约为腹柄结长的1/3；腹柄结近三角形，前面平直，后面轻度隆起，顶端钝角状；腹面后部轻度凹陷，前下角齿状。后腹柄与腹柄等高，前上缘圆形隆起，后面近平直；腹面短，近平直，前下角直角形。背面观，前胸背板最宽，前缘中度隆起，肩角宽圆，侧缘轻度隆起。中胸中度收缩，向后变窄，侧缘强烈凹陷。并胸腹节近方形，侧缘轻度凹陷；并胸腹节刺短，指向后侧方。腹柄长梯形，向后变宽，长大于宽，侧缘后部轻度隆起。后腹柄近梯形，约为腹柄宽的1.5倍，向后变窄，前面中度隆起，侧面强烈隆起。后腹部长卵形。上颚具纵条纹。头部背面具纵皱纹，纵皱纹之间具交织的细皱纹，侧面逐渐变为网纹，界面具丰富刻

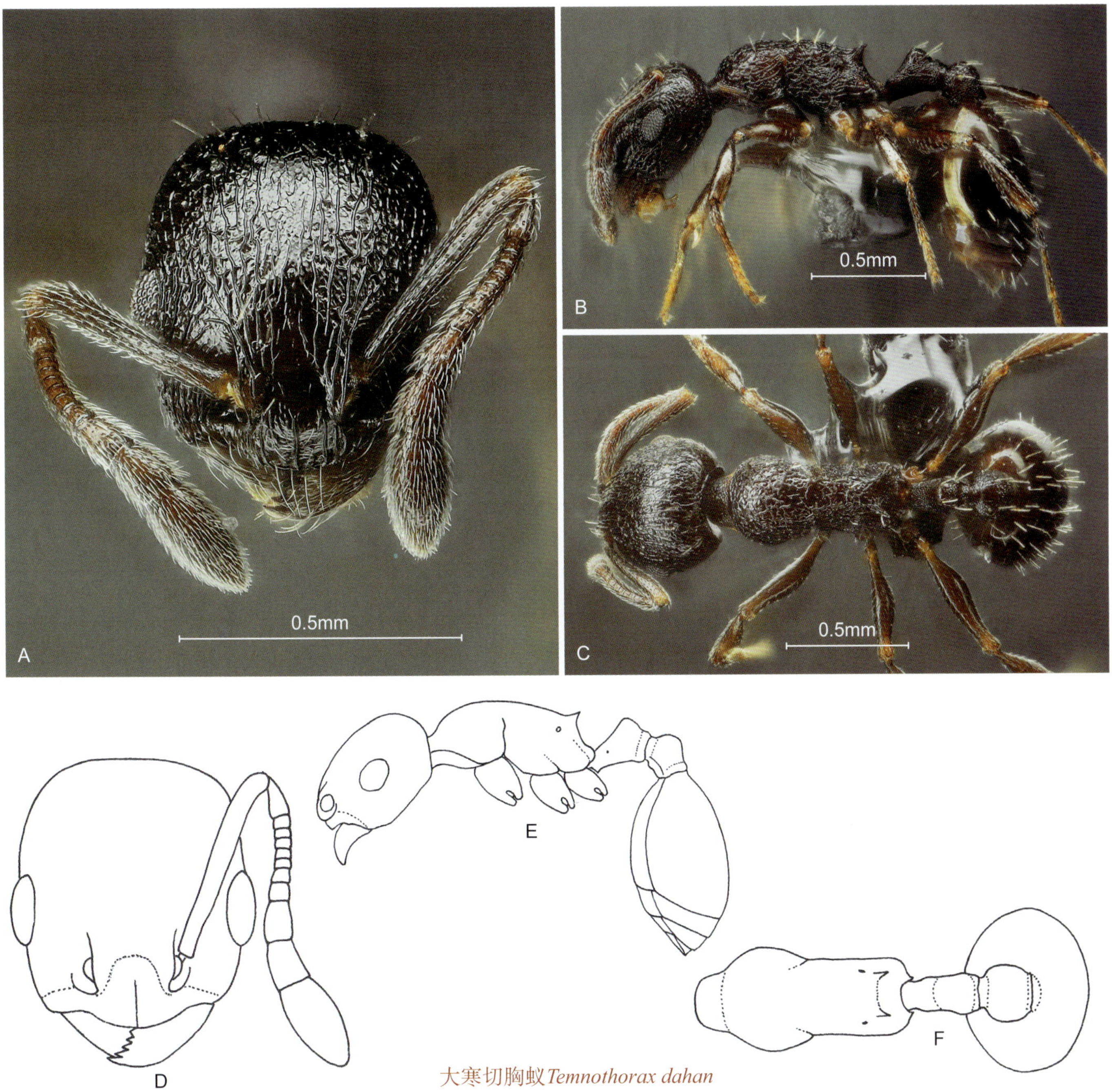

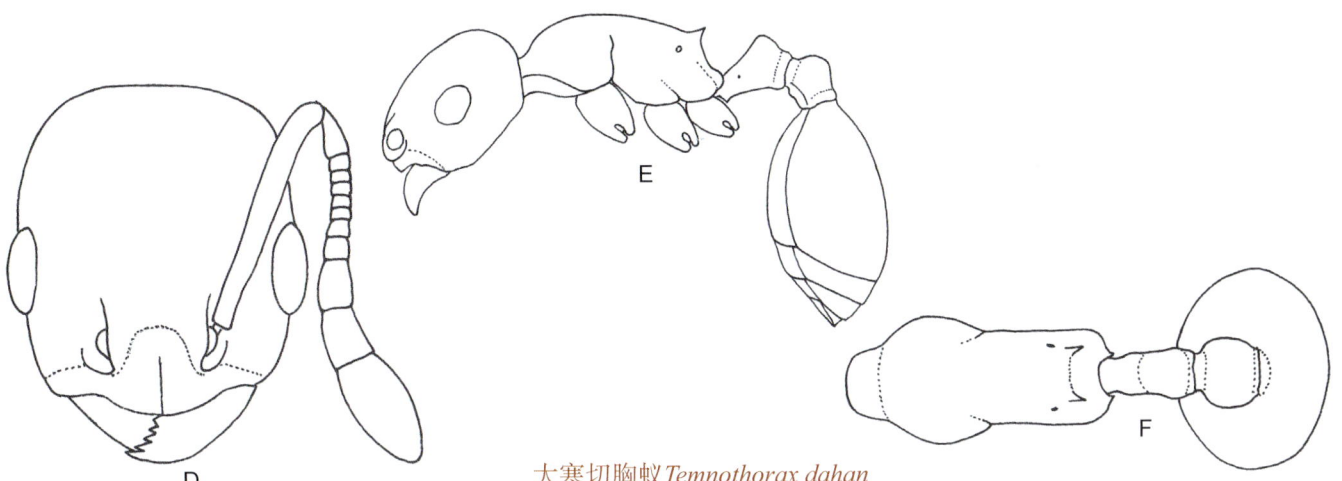

大寒切胸蚁 *Temnothorax dahan*

A, D. 工蚁头部正面观；B, E. 工蚁身体侧面观；C, F. 工蚁身体背面观（A-C. 引自Qian和Xu, 2024）
A, D. Head of worker in full-face view; B, E. Body of worker in lateral view; C, F. Body of worker in dorsal view (A-C. cited from Qian & Xu, 2024)

点。唇基每侧具3条短纵皱纹，界面光滑。胸部背面具粗糙网纹，中胸背板背面中央较光滑；前胸背板侧面前部具粗网纹，后部具纵皱纹和微网纹；中胸侧板前部具粗网纹，后部具微网纹；后胸侧板和并胸腹节侧面具密集网纹。腹柄和后腹柄具密集刻点。后腹部光滑发亮。头部背面具稀疏直立、亚直立短钝毛和密集倾斜绒毛被；胸部、腹柄和后腹柄具稀疏亚直立短钝毛和稀疏倾斜绒毛被；后腹部具丰富亚直立短钝毛和丰富倾斜绒毛被；柄节具密集倾斜绒毛被；胫节具丰富倾斜绒毛被。身体棕黑色，上颚、触角和足棕色至棕黄色。**副模工蚁**：特征同正模，但是一些个体中胸背板光滑，腹柄结和后腹柄结光滑，胸部背面具弱网纹。

正模：工蚁，中国云南省文山市薄竹镇跑马塘坡（23.389739°N，105.933033°E），2530m，2019.X.12，都红采于中山湿性常绿阔叶林地表样中，No. A19-6318。

词源：该新种以中国二十四节气中的"大寒"命名。

094 大暑切胸蚁
Temnothorax dashu Qian & Xu, 2024

Temnothorax dashu Qian & Xu, 2024, European Journal of Taxonomy, 936: 19, fig. 12 (w.) CHINA (Yunnan).

Holotype worker: TL 2.8, HL 0.68, HW 0.57, CI 84, SL 0.62, SI 109, ED 0.15, PW 0.40, WL 0.84, PL 0.28, PH 0.21, DPW 0.16. In full-face view, head roughly rectangular, longer than broad, posterior margin slightly convex, posterior corners narrowly rounded, lateral margins weakly convex. Mandibles triangular, masticatory margin with 5 teeth. Clypeal dorsum weakly convex, with median carina, anterior margin moderately convex. Frontal lobes narrow, concealing half of antennal sockets. Frontal carinae short, reaching to the level of anterior eye margin. Antennae 12-segmented, scapes reaching to posterior head margin, club 3-segmented. Eyes located at midpoint of lateral head margin, occupying 1/5 of lateral margin. In lateral view, pronotum moderately convex, promesonotal suture absent. Mesonotum slightly convex and gently sloping down posteriorly. Metanotal groove widely shallowly impressed and distinct, with a blunt prominence in the groove. Propodeal dorsum straight and gently sloping down posteriorly, anterodorsal corner lowly bluntly toothed; propodeal spines short and stout, roughly triangular, blunt apically; declivity weakly concave, about 2/3 length of dorsum; propodeal lobes small, rounded apically. Petiole with anterior peduncle, about 2/3 length of petiolar node; petiolar node roughly triangular, anterior margin straight, posterior margin moderately convex, anterodorsal corner narrowly rounded; ventral margin almost straight, slightly concave posteriorly, anteroventral corner acutely toothed. Postpetiole slightly lower than petiole, anterodorsal margin rounded, posterior margin weakly convex, ventral margin almost straight. In dorsal view, pronotum broadest, anterior margin almost straight, humeral corners slightly convex, lateral margins strongly convex. Promesonotal suture absent. Mesothorax weakly constricted and narrow, lateral margins almost straight. Metanotal groove shallowly impressed. Propodeum roughly trapezoidal and widening posteriorly, lateral margins almost straight; spines short and stout, slightly in-curved. Petiole longer than broad, the node roughly circular, lateral margins weakly convex. Postpetiole roughly trapezoidal and narrowing posteriorly, about 1.4 times as broad as petiole, anterolateral corners bluntly angled, lateral margins nearly straight. Gaster elongate oval. Mandibles longitudinally striate. Head dorsum longitudinally rugose anteriorly before midline of eyes with rugae divergent posteriorly, gradually reticulate posteriorly and laterally, interface densely punctate; genae longitudinally rougose. Clypeus with 3-4 rugae on each side, interface smooth. Promesonotal dorsum and pronotal sides reticulate, with three transverse rugae near metanotal groove; mesopleura and metapleura finely reticulate; propodeum, petiole and postpetiole densely punctate, postpetiolar dorsum smooth. Gaster smooth and shiny. Body dorsum with abundant suberect to subdecumbent apically blunt short hairs and abundant decumbent pubescence, hairs and pubescence on the head relatively denser. Scapes and tibiae with dense subdecumbent hairs and dense decumbent pubescence. Body color yellowish brown, head dorsum and median portion of gaster blackish brown; mandibles, antennae and legs brownish yellow. **Paratype workers:** TL 2.4-2.9, HL 0.55-0.65, HW 0.45-0.60, CI 82-92, SL 0.50-0.55, SI 91-111, ED 0.13-0.15, PW 0.30-0.31, WL 0.60-0.84, PL 0.20-0.30, PH 0.15-0.18, DPW 0.10-0.11 (*n*=9). As holotype, in some individuals propodeal spines slightly shorter or longer than in holotype, body color lighter than in holotype.

Holotype: Worker, China: Yunnan Province, Luquan County, Wumeng Town, Jiaozi Snow Mountain (26.059833°N, 102.824917°E), 3250 m, 2013.Ⅲ.22, collected by Xin Chen from a ground sample of alpine *Rhododendron* forest, No. A13-479.

Etymology: The specific epithet refers to "dashu", one of the Twenty-four Solar Terms of China.

正模工蚁：正面观，头部近长方形，长大于宽；后缘轻度隆起，后角窄圆；侧缘轻度隆起。上颚三角形，咀嚼缘具5个齿。唇基背面轻度隆起，具中央纵脊，前缘中度隆起。额叶窄，遮盖触角窝的一半。额脊短，到达复眼前缘水平。触角12节，柄节到达头后缘，触角棒3节。复眼位于头侧缘中点上，占据侧缘的1/5。侧面观，前胸背板中度隆起，缺前中胸背板缝。中胸背板轻度隆起，向后缓坡形降低。后胸沟宽形浅凹，明显，沟内具1个钝突。并胸腹节背面平直，向后缓坡形降低，前上角呈低的钝齿状；并胸腹节刺短粗，近三角形，末端钝；斜面浅凹，约为背面长的2/3。并胸腹节侧叶小，顶端圆。腹柄前面具小柄，约为腹柄结长的2/3；腹柄结近三角形，前面平直，后面中度隆起，前上角窄圆；腹面近平直，后部轻微凹陷，前下角尖齿状。后腹柄稍低于腹柄，前上缘圆，后缘轻度隆起，腹面近平直。背面观，前胸背板最宽，前缘近平直，肩角轻微隆起，侧缘强烈隆起。缺前中胸背板缝。中胸轻度收缩，较窄，侧缘近平直。后胸沟浅凹。并胸腹节近梯形，向后变宽，侧缘近平直；并胸腹节刺短粗，轻微内弯。腹柄长大于宽，腹柄结近圆形，侧缘轻度隆起。后腹柄近梯形，向后变窄，约为腹柄宽的1.4倍，前侧角钝角状，侧缘

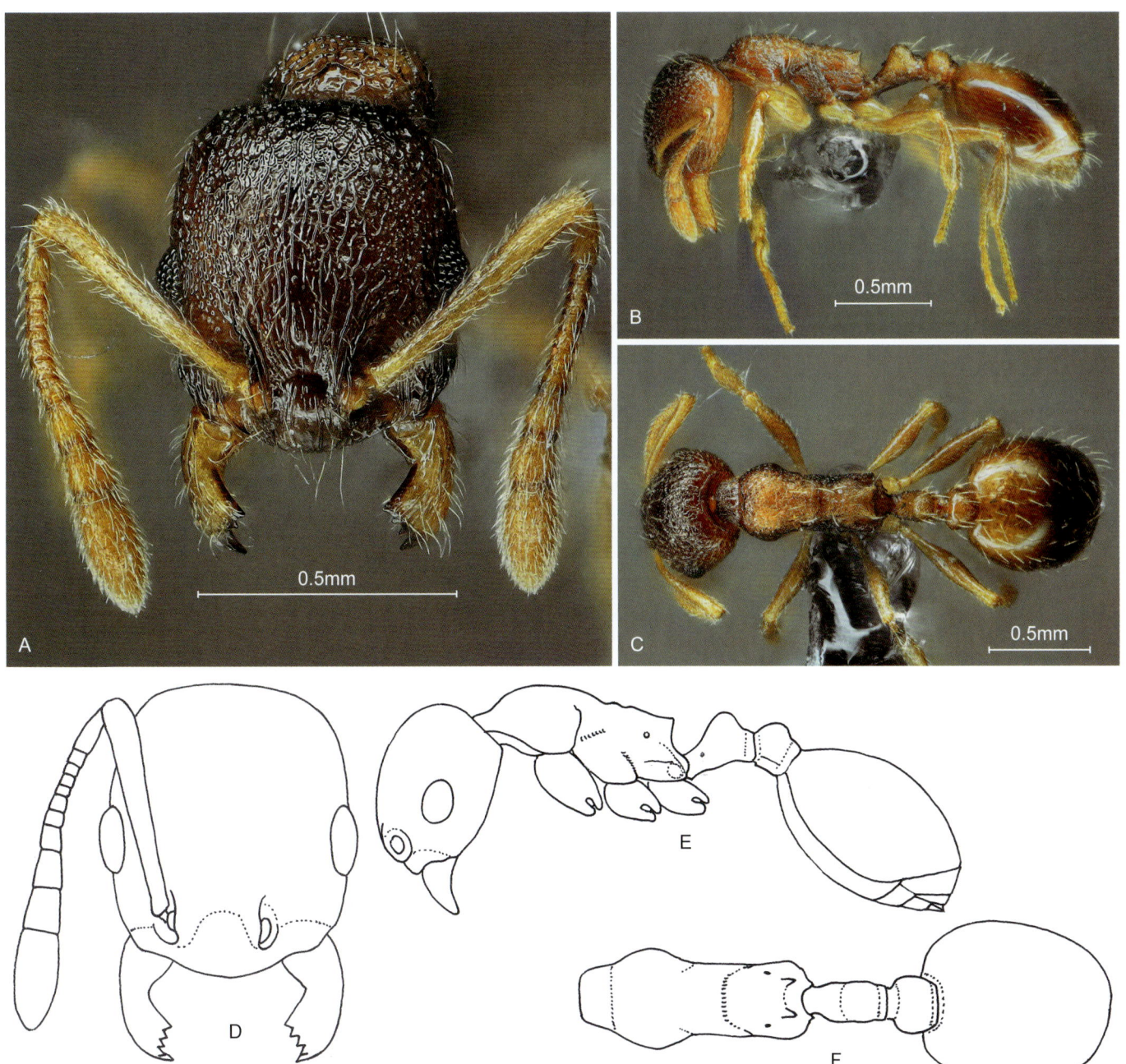

大暑切胸蚁 *Temnothorax dashu*

A, D. 工蚁头部正面观；B, E. 工蚁身体侧面观；C, F. 工蚁身体背面观（A-C. 引自Qian和Xu, 2024）
A, D. Head of worker in full-face view; B, E. Body of worker in lateral view; C, F. Body of worker in dorsal view (A-C. cited from Qian & Xu, 2024)

近平直。后腹部长卵形。上颚具纵条纹。头部背面复眼中线之前具向后发散的纵皱纹，后部和侧面逐渐变成网纹，界面具密集刻点，颊区具纵皱纹。唇基每侧具3~4条纵皱纹，界面光滑。前中胸背板背面和前胸背板侧面具网纹，后胸沟附近具3条横皱纹；中胸侧板和后胸侧板具细网纹；并胸腹节、腹柄和后腹柄具密集刻点，后腹柄背面光滑。后腹部光滑发亮。身体背面具丰富亚直立、亚倾斜短钝毛和丰富倾斜绒毛被，头部立毛和绒毛被较密集；柄节和胫节具密集亚倾斜毛和密集倾斜绒毛被。身体黄棕色，头背面和后腹部中部黑棕色，上颚、触角和足棕黄色。**副模工蚁**：特征同正模，一些个体并胸腹节刺稍短或稍长，身体颜色较浅。

正模：工蚁，中国云南省禄劝县乌蒙乡轿子雪山（26.059833°N，102.824917°E），3250m，2013.Ⅲ.22，陈鑫采于高山杜鹃林地表样中，No. A13-479。

词源：该新种以中国二十四节气中的"大暑"命名。

大雪切胸蚁
Temnothorax daxue Qian & Xu, 2024

Temnothorax daxue Qian & Xu, 2024, European Journal of Taxonomy, 936: 21, fig. 13 (w.) CHINA (Yunnan).

Holotype worker: TL 3.5, HL 0.83, HW 0.69, CI 84, SL 0.65, SI 94, ED 0.16, PW 0.47, WL 1.03, PL 0.37, PH 0.28, DPW 0.18. In full-face view, head roughly rectangular, longer than broad, posterior margin almost straight, posterior corners narrowly rounded, lateral margins weakly convex. Mandibles triangular, masticatory margin with 5 teeth. Clypeal dorsum weakly convex, without median carina, anterior margin almost straight in the center. Frontal lobes narrow, concealing half of antennal socket. Frontal carinae short, reaching to the level of anterior eye margin. Antennae 12-segmented, scapes failing to reach posterior head margin, club 3-segmented. Eyes located at midpoint of lateral head margin, occupying 1/5 of lateral margin. In lateral view, promesonotum moderately convex and weakly arched, gently sloping down posteriorly, promesonotal suture and metanotal groove absent. Propodeal dorsum almost straight, gently sloping down posteriorly; propodeal spines long and slender, about as long as declivity, weakly back-curved; declivity weakly concave, propodeal lobes short and rounded apically. Petiole with short anterior peduncle, about 1/2 length of petiolar node; petiolar node roughly conical, anterior margin almost straight, posterior margin weakly convex, apex narrowly rounded; ventral margin weakly concave, anteroventral corner acutely toothed. Postpetiole as high as petiolar node, anterior margin moderately convex, posterior margin almost straight, apex broadly rounded; ventral margin deeply notched, anteroventral corner acutely toothed, posteroventral corner rightly angled. In dorsal view, pronotum broadest, anterior margin moderately convex, humeral corners broadly rounded, lateral margin moderately convex. Promesonotal suture absent. Mesothorax weakly constricted, lateral margins weakly concave. Propodeum roughly rectangular, lateral margins almost straight, spines acute and pointed posterolaterally. Petiole roughly rectangular, slightly longer than broad, lateral margins slightly convex. Postpetiole roughly trapezoidal and weakly narrowing posteriorly, about 1.4 times as broad as petiole, anterolateral corners broadly rounded, lateral margins slightly concave. Gaster elongate oval. Mandibles longitudinally striate. Head dorsum nearly smooth in the central strip, loosely longitudinally rugose beside the central smooth strip, reticulate laterally, longitudinally rugose below eyes, genae with arched rugae. Clypeus smooth, each side with 4-5 short rugae. Mesosomal dorsum coarsely reticulate; pronotal sides longitudinally rugose; mesopleura and metapleura coarsely reticulate; propodeal sides and petiole finely reticulate; postpetiole densely finely punctate, dorsum smooth and shiny. Gaster smooth and shiny. Body dorsum with sparse erect to suberect short apically blunt hairs and sparse decumbent pubescence, hairs on the head abundant. Scapes with abundant subdecumbent hairs and dense decumbent pubescence; tibiae with abundant decumbent pubescence. Body color black; mandible, antennae and legs blackish brown.

Holotype: Worker, China: Yunnan Province, Lanping County, Lajing Town, Lajing Village (26.497406°N, 99.280276°E), 2600 m, 2003.Ⅹ.9, collected by You Chen from a canopy sample of conifer-broadleaf mixed forest, No. A3245.

Etymology: The specific epithet refers to "daxue", one of the Twenty-four Solar Terms of China.

正模工蚁： 正面观，头部近长方形，长大于宽；后缘近平直，后角窄圆；侧缘轻度隆起。上颚三角形，咀嚼缘具5个齿。唇基背面轻度隆起，缺中央纵脊，前缘中部近平直。额叶窄，遮盖触角窝的一半。额脊短，到达复眼前缘水平。触角12节，柄节未到达头后缘，触角棒3节。复眼位于头侧缘中点，占据侧缘的1/5。侧面观，前中胸背板中度隆起呈弱弓形，向后缓坡形降低，缺前中胸背板缝和后胸沟。并胸腹节背面近平直，向后缓坡形降低；并胸腹节刺细长，约与斜面等长，轻度后弯；斜面浅凹；并胸腹节侧叶短，顶端圆。腹柄前面具短的小柄，约为腹柄结长的1/2；腹柄结近锥形，前面近平直，后面轻度隆起，顶端窄圆；腹面浅凹，前下角尖齿状。后腹柄与腹柄结等高，前面中度隆起，后面近平直，顶部宽圆，腹面深切，前下角尖齿状，后下角直角形。背面观，前胸背板最宽，前缘中度隆起，肩角宽圆，侧缘中度隆起。缺前中胸背板缝。中胸轻度收缩，侧缘浅凹。并胸腹节近长方形，侧缘近平直；并胸腹节刺尖锐，指向后侧方。腹柄近长方形，长稍大于宽，侧缘轻微隆起。后腹柄近梯形，向后轻度变窄，约为腹柄宽的1.4倍，前侧角宽圆，侧缘轻微凹陷。后腹部长卵形。上颚具纵条纹。头部背面中央纵带近光滑，纵带两侧具松散纵皱纹，侧面具网纹，复眼下方具纵皱纹，颊区具弧形皱纹。唇基光滑，每侧具4~5条短皱纹。胸部背面具粗糙网纹；前胸背板两侧具纵皱纹，中胸侧板和后胸侧板具粗糙网纹；并胸腹节侧面和腹柄具细网纹；后腹柄具细密刻点，背面光滑发亮。后腹部光滑发亮。身体背面具稀疏直立、亚直立短钝毛和稀疏倾斜绒毛被，头部立毛丰富；柄节具丰富亚倾斜毛和密集倾斜绒毛被；胫节具丰富倾斜绒毛被。身体黑色，上颚、触角和足黑棕色。

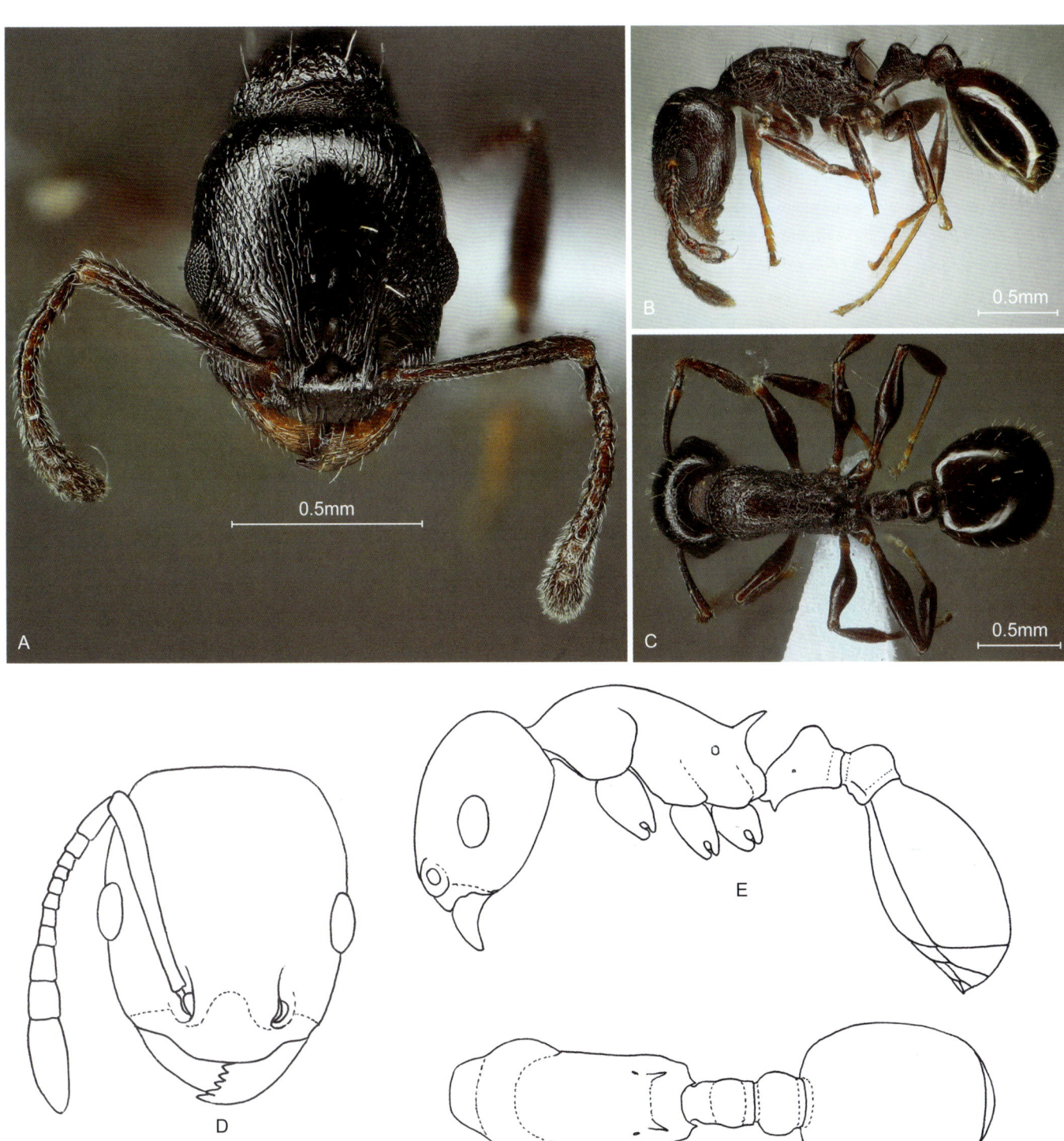

大雪切胸蚁 Temnothorax daxue

A, D. 工蚁头部正面观；B, E. 工蚁身体侧面观；C, F. 工蚁身体背面观（A-C. 引自Qian和Xu, 2024）
A, D. Head of worker in full-face view; B, E. Body of worker in lateral view; C, F. Body of worker in dorsal view (A-C. cited from Qian & Xu, 2024)

正模： 工蚁，中国云南省兰坪县啦井乡啦井村（26.497406°N，99.280276°E），2600m，2003.Ⅹ.9，陈友采于针阔混交林树冠样中，No. A3245。

词源： 该新种以中国二十四节气中的"大雪"命名。

冬切胸蚁
Temnothorax dong Qian & Xu, 2024

Temnothorax dong Qian & Xu, 2024, European Journal of Taxonomy, 936: 23, fig. 15 (w.) CHINA (Yunnan).

Holotype worker: TL 2.8, HL 0.68, HW 0.57, CI 84, SL 0.60, SI 105, ED 0.13, PW 0.41, WL 0.78, PL 0.30, PH 0.21, DPW 0.16. In full-face view, head roughly rectangular, longer than broad, posterior and lateral margins weakly convex, posterior corners broadly rounded. Mandibles triangular, masticatory margin with 5 teeth. Clypeal dorsum weakly convex, with median carina, anterior margin moderately convex. Frontal lobes broad, concealing most of antennal sockets. Frontal carinae short, reaching to the level of anterior eye margins. Antennae 12-segmented, scapes slightly surpassing posterior head margin, club 3-segmented. Eyes located at midpoint of lateral head margin, occupying 1/4 of lateral margin. In lateral view, mesosomal dorsum weakly convex and gently sloping down posteriorly, anterodorsal corner of pronotum bluntly angled, promesonotal suture and metanotal groove absent. Propodeal spines moderately long, about 2/3 length of declivity and about 2 times as long as their basal width; declivity weakly concave; propodeal lobes short and rounded apically. Petiole with long anterior peduncle, about as long as petiolar node; petiolar node roughly triangular, anterior margin almost straight, posterior margin weakly convex, top corner rightly angled; ventral margin straight, weakly concave posteriorly, anteroventral corner acutely toothed. Postpetiole about as high as petiolar node, anterior margin almost straight, anterodorsal corner bluntly angled, dorsal margin weakly convex, ventral margin angularly concave. In dorsal view, pronotum broadest, anterior margin moderately convex, humeral corners broadly rounded, lateral margins strongly convex, promesonotal suture absent. Mesothorax moderately constricted, lateral margins weakly concave, metanotal groove absent. Propodeum roughly rectangular and slightly widening posteriorly, lateral margins almost straight; spines moderately long and straight, pointed posterolaterally. Petiole roughly trapezoidal and widening posteriorly, longer than broad, lateral margins weakly convex. Postpetiole roughly rectangular, about 1.4 times as broad as petiole, anterolateral corners narrowly rounded, lateral margins weakly convex. Gaster elongate oval. Mandibles longitudinally striate. Head dorsum densely finely reticulate, interface densely punctate; frontal area longitudinally rugose. Clypeus longitudinally rugose. Mesosoma densely reticulate; pronotal sides and lower portion of metapleura longitudinally rugose. Propodeum, petiole and postpetiole densely finely punctate. Gaster smooth and shiny. Body dorsum with abundant erect to suberect apically blunt short hairs and abundant decumbent pubescence. Scapes and tibiae with dense decumbent pubescence. Body color black; antennae and legs blackish brown; mandibles and tarsi brownish yellow. **Paratype worker:** TL 2.4, HL 0.65, HW 0.50, CI 77, SL 0.50, SI 100, ED 0.10, PW 0.35, WL 0.60, PL 0.25, PH 0.18, DPW 0.15 (*n*=1). As holotype.

Holotype: Worker, China: Yunnan Province, Tengchong County, Jietou Town, Daying Village (25.504974°N, 98.687047°E), 1900 m, 1999.Ⅴ.2, collected by Lei Fu from a ground sample of conifer-broadleaf mixed forest, No. A99-428.

Etymology: The specific epithet refers to "dong", one of the four seasons.

正模工蚁： 正面观，头部近长方形，长大于宽；后缘和侧缘轻度隆起，后角宽圆。上颚三角形，咀嚼缘具5个齿。唇基背面轻度隆起，具中央纵脊，前缘中度隆起。额叶宽，遮盖触角窝大部。额脊短，到达复眼前缘水平。触角12节，柄节稍超过头后缘，触角棒3节。复眼位于头侧缘中点，占据侧缘的1/4。侧面观，胸部背面轻度隆起，向后缓坡形降低；前胸背板前上角钝角状，缺前中胸背板缝和后胸沟。并胸腹节刺中等长，约为斜面长的2/3，为基部宽的2倍；斜面浅凹。并胸腹节侧叶短，顶端圆。腹柄前面具长的小柄，约与腹柄结等长；腹柄结近三角形，前面近平直，后面轻度隆起，顶端直角形；腹面平直，后部浅凹，前下角尖齿状。后腹柄约与腹柄结等高，前面近平直，前上角钝角状，背面轻度隆起，腹面角状凹陷。背面观，前胸背板最宽，前缘中度隆起，肩角宽圆，侧缘强烈隆起。缺前中胸背板缝。中胸中度收缩，侧缘浅凹，缺后胸沟。并胸腹节近长方形，向后稍变宽，侧缘近平直；并胸腹节刺中等长且直，指向后侧方。腹柄近梯形，向后变宽，长大于宽，侧缘轻度隆起。后腹柄近长方形，约为腹柄宽的1.4倍，前侧角窄圆，侧缘轻度隆起。后腹部长卵形。上颚具纵条纹。头部背面具细密网纹，界面具密集刻点，额区具纵皱纹。唇基具纵皱纹。胸部具密集网纹；前胸背板两侧和后胸侧板下部具纵皱纹。并胸腹节、腹柄和后腹柄具细密刻点。后腹部光滑发亮。身体背面具丰富直立、亚直立短钝毛和丰富倾斜绒毛被；柄节和胫节具密集倾斜绒毛被。身体黑色，触角和足黑棕色，上颚和跗节棕黄色。**副模工蚁：** 特征同正模。

正模： 工蚁，中国云南省腾冲县界头乡大营村（25.504974°N，98.687047°E），1900m，1999.Ⅴ.2，付

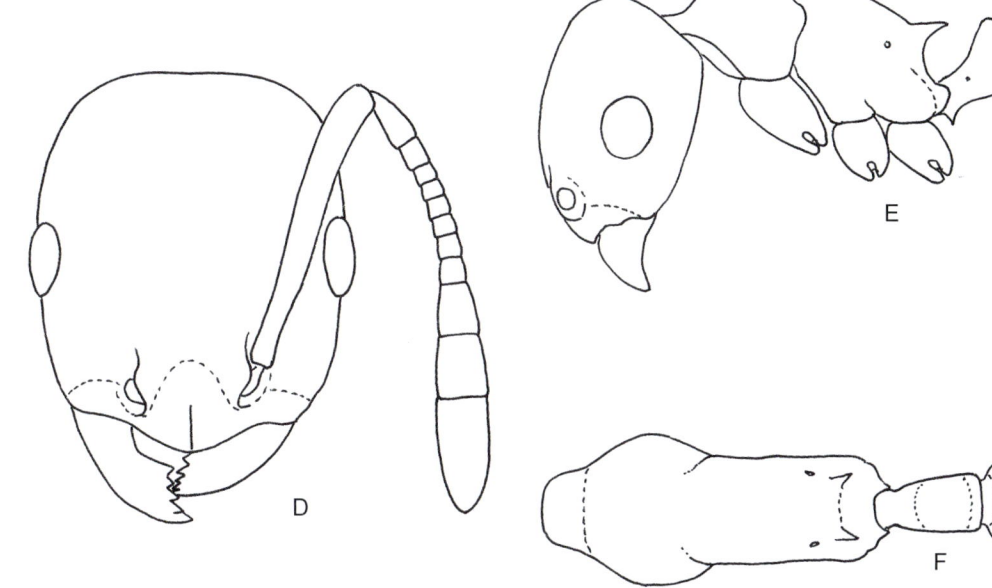

冬切胸蚁 *Temnothorax dong*

A, D. 工蚁头部正面观；B, E. 工蚁身体侧面观；C, F. 工蚁身体背面观（A-C. 引自Qian和Xu, 2024）
A, D. Head of worker in full-face view; B, E. Body of worker in lateral view; C, F. Body of worker in dorsal view (A-C. cited from Qian & Xu, 2024)

磊采于针阔混交林地表样中，No. A99-428。

词源： 该新种以四季中的"冬"命名。

冬至切胸蚁
Temnothorax dongzhi Qian & Xu, 2024

Temnothorax dongzhi Qian & Xu, 2024, European Journal of Taxonomy, 936: 26, fig. 17 (w.) CHINA (Sichuan)

Holotype worker: TL 4.0, HL 0.88, HW 0.80, CI 91, SL 0.63, SI 79, ED 0.17, PW 0.57, WL 1.18, PL 0.47, PH 0.30, DPW 0.21. In full-face view, head roughly rectangular, slightly longer than broad, posterior margin slightly concave, posterior corners narrowly rounded, lateral margins weakly convex. Mandibles triangular, masticatory margin with 5 teeth. Clypeal dorsum weakly convex, with median carina, anterior margin moderately convex, narrowly notched in the middle. Frontal lobes broad, concealing antennal sockets. Frontal carinae short, reaching to the level of anterior eye margin. Antennae 12-segmented, scapes failing to reach posterior head margin by 1/5 of its length, antennal clubs 3-segmented. Eyes located slightly before midpoint of lateral head margin, occupying 1/4 of lateral margin. In lateral view, pronotum almost straight and sloping down anteriorly, anterodorsal corner rightly angled. Promesonotal suture distinct on the sides. Mesonotum straight and gently sloping down posteriorly, weakly convex anteriorly. Metanotal groove absent. Propodeal dorsum straight and sloping down posteriorly, forming a very blunt angle with mesonotum; propodeal spines relatively short, suberect and slightly back-curved, about 1/2 length of declivity; declivity almost straight; propodeal lobes broad, truncated apically. Petiole with relatively long anterior peduncle, about 3/4 length of petiolar node; petiolar node roughly triangular, anterior margin almost straight, posterior margin slightly convex, dorsal margin weakly convex, anterodorsal corner bluntly angled, posterodorsal corner narrowly rounded; ventral margin weakly concave, anteroventral corner bluntly toothed. Postpetiole slightly lower than petiolar node, anterior margin weakly convex, anterodorsal corner rounded, dorsal margin strongly convex, posterior margin weakly concave, ventral margin weakly concave. In dorsal view, pronotum broadest, anterior margin strongly convex, humeral corners very bluntly angled, lateral margins moderately convex. Promesonotal suture and metanotal groove absent. Mesothorax weakly constricted, lateral margins weakly concave. Propodeum roughly rectangular, lateral margins slightly convex, posterolateral corners bluntly angled; spines relatively short, pointed posterolaterally. Petiole roughly trapezoidal and widening posteriorly, longer than broad, lateral margins slightly convex. Postpetiole roughly trapezoidal and narrowing posteriorly, about 1.5 times as broad as petiole, anterolateral corners narrowly rounded, lateral margins weakly convex. Gaster elongate oval. Mandibles longitudinally striate. Head dorsum loosely longitudinally rugose, gradually reticulate laterally with interface densely punctate. Clypeus smooth, each side with 4-5 oblique rugae. Mesosomal dorsum reticulate, sides coarsely longitudinally rugose. Propodeal sides obliquely rugose. Petiolar node finely reticulate, ventral face of petiole and postpetiole densely punctate, anterior face and dorsum of postpetiole smooth. Gaster smooth and shiny. Body dorsum with sparse erect to suberect apically blunt short hairs and sparse decumbent pubescence. Scapes and tibiae with dense decumbent pubescence. Body color blackish brown, head dorsum and gaster black. **Paratype worker:** TL 3.3, HL 0.75, HW 0.68, CI 92, SL 0.60, SI 88, ED 0.16, PW 0.46, WL 0.94, PL 0.33, PH 0.25, DPW 0.18 (n=1). As holotype.

Holotype: Worker, China: Sichuan Province, Muli County, Xiamaidi Town, Xiaochanggou Village (27.776050°N, 101.216714°E), 2513 m, 2018.Ⅶ.28, collected by Xin-min Zhang from a canopy sample of *Pinus yunnanensis* forest, No. C18-962.

Etymology: The specific epithet refers to "dongzhi", one of the Twenty-four Solar Terms of China.

正模工蚁： 正面观，头部近长方形，长稍大于宽；后缘轻微凹陷，后角窄圆；侧缘轻度隆起。上颚三角形，咀嚼缘具5个齿。唇基背面轻度隆起，具中央纵脊，前缘中度隆起，中央狭窄切入。额叶宽，遮盖触角窝。额脊短，到达复眼前缘水平。触角12节，柄节末端至头后缘距离为柄节长的1/5，触角棒3节。复眼位于头侧缘中点稍前处，占据侧缘的1/4。侧面观，前胸背板近平直，向前坡形降低，前上角直角形。前中胸背板缝在侧面明显。中胸背板平直，向后缓坡形降低，前部轻度隆起。缺后胸沟。并胸腹节背面平直，向后坡形降低，与中胸背板形成1个极钝的角；并胸腹节刺较短，亚直立，轻微后弯，约为斜面长的1/2；斜面近平直。并胸腹节侧叶宽，顶端平截。腹柄前面具较长小柄，约为腹柄结长的3/4；腹柄结近三角形，前面近平直，后面轻微隆起，背面轻度隆起，前上角钝角状，后上角窄圆；腹面轻度凹陷，前下角钝齿状。后腹柄稍低于腹柄结，前面轻度隆起，前上角圆，背面强烈隆起，后面轻度凹陷，腹面轻度凹陷。背面观，前胸背板最宽，前缘强烈隆起，肩角极钝的角状，侧缘中度隆起。缺前中胸背板缝和后胸沟。中胸轻度收缩，侧缘轻度凹陷。并胸腹节近长方形，侧缘轻微隆起，后侧角钝角状；并胸腹节刺较短，指向后侧方。腹柄近梯形，向后变宽，长大于宽，侧缘轻微隆起。后腹柄近梯形，向后变窄，约为腹柄宽的

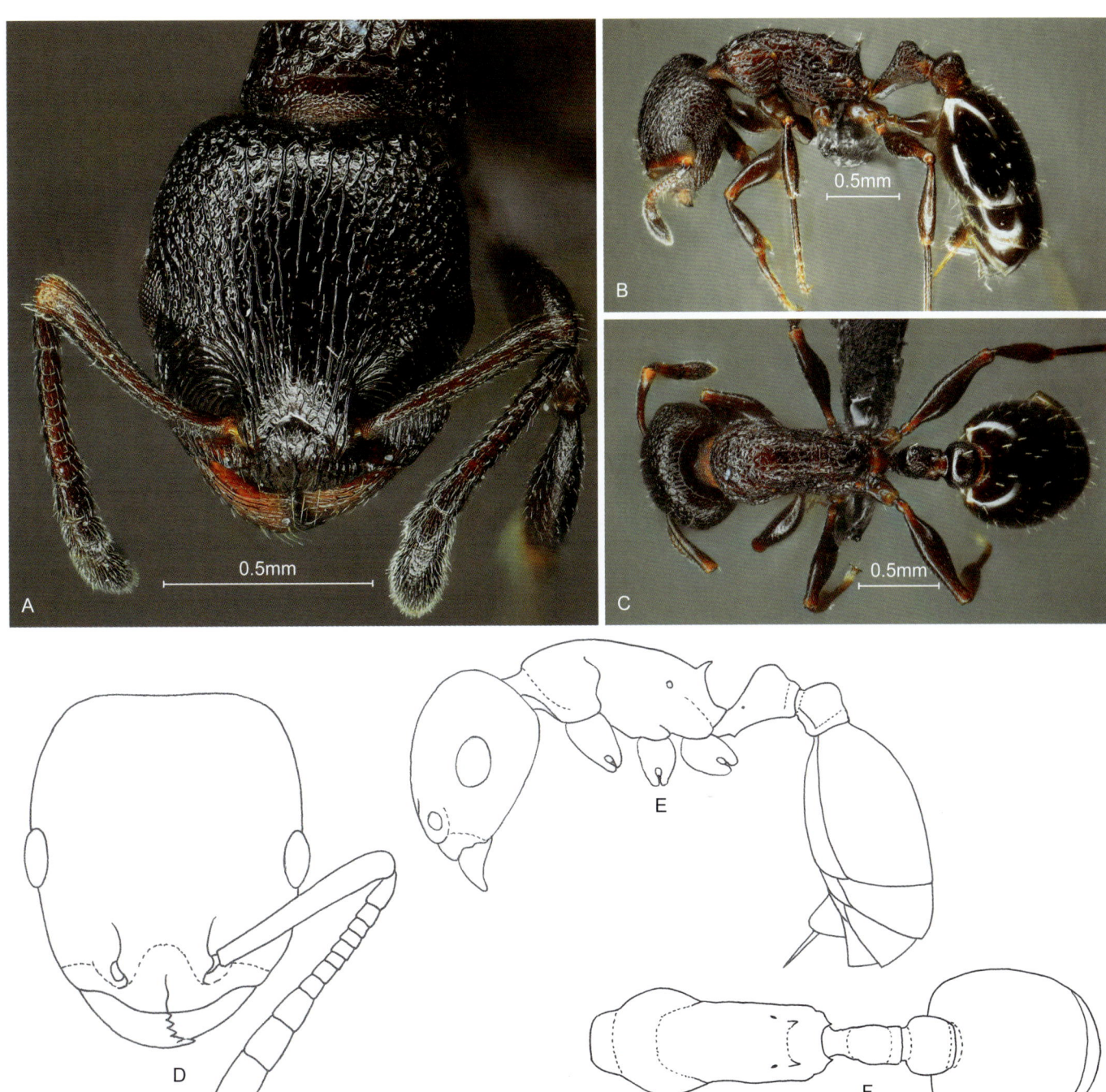

冬至切胸蚁 *Temnothorax dongzhi*

A, D. 工蚁头部正面观；B, E. 工蚁身体侧面观；C, F. 工蚁身体背面观（A-C. 引自Qian和Xu, 2024）
A, D. Head of worker in full-face view; B, E. Body of worker in lateral view; C, F. Body of worker in dorsal view (A-C. cited from Qian & Xu, 2024)

1.5倍，前侧角窄圆，侧缘轻度隆起。后腹部长卵形。上颚具纵条纹。头部背面具松散的纵皱纹，向两侧逐渐变成网纹，界面具密集刻点。唇基光滑，每侧具4~5条倾斜皱纹。胸部背面具网纹，侧面具粗糙纵皱纹；并胸腹节侧面具倾斜皱纹。腹柄结具细网纹，腹柄腹面和后腹柄具密集刻点，后腹柄前面和背面光滑。后腹部光滑发亮。身体背面具稀疏直立、亚直立短钝毛和稀疏倾斜绒毛被；柄节和胫节具密集倾斜绒毛被。身体黑棕色，头部背面和后腹部黑色。**副模工蚁：** 特征同正模。

正模： 工蚁，中国四川省木里县下麦地乡硝厂沟（27.776050°N，101.216714°E），2513m, 2018.Ⅶ.28，张新民采于云南松林树冠样中，No. C18-962。

词源： 该种以中国二十四节气中的"冬至"命名。

谷雨切胸蚁
Temnothorax guyu Qian & Xu, 2024

Temnothorax guyu Qian & Xu, 2024, European Journal of Taxonomy, 936: 28, fig. 19 (w.) CHINA (Yunnan).

Holotype worker: TL 2.5, HL 0.56, HW 0.48, CI 85, SL 0.50, SI 104, ED 0.12, PW 0.36, WL 0.71, PL 0.26, PH 0.21, DPW 0.16. In full-face view, head roughly rectangular, longer than broad, posterior margin moderately convex, posterior corners broadly rounded, lateral margins weakly convex. Mandibles triangular, masticatory margin with 5 teeth. Clypeal dorsum weakly convex, without median carina, anterior margin moderately convex. Frontal lobes broad, concealing antennal sockets. Frontal carinae relatively short, reaching to the level of midpoint of eyes. Antennae 12-segmented, scapes slightly surpassing posterior head margin, antennal clubs 3-segmented. Eyes located slightly before midpoint of lateral head margin, occupying 1/4 of lateral margin. In lateral view, pronotum moderately convex and sloping down anteriorly, promesonotal suture absent. Mesonotum slightly convex and gently sloping down posteriorly, metanotal groove absent. Propodeal dorsum almost straight and gently sloping down posteriorly; propodeal spines very short and dent-like, about as long as propodeal lobes; declivity slightly concave; propodeal lobes short, bluntly angled apically. Petiole with anterior peduncle, about 3/4 length of petiolar node; petiolar node roughly conical, anterior margin almost straight, posterior margin weakly convex, dorsum rounded; ventral margin almost straight, anteroventral corner shortly bluntly angled. Postpetiole as high as petiolar node, anterodorsal margin rounded, posterior margin almost straight, ventral margin weakly concave. In dorsal view, pronotum broadest, anterior margin moderately convex, humeral corners very broadly rounded, lateral margins strongly convex, promesonotal suture absent. Mesothorax slightly constricted, lateral margins weakly concave. Propodeum roughly rectangular, lateral margins almost straight; propodeal spines very short and dent-like. Petiole roughly trapezoidal and widening posteriorly, longer than broad, lateral margins weakly convex. Postpetiole roughly rectangular, slightly narrowing posteriorly, about 1.5 times as broad as petiole, anterolateral corners narrowly rounded, lateral margins almost straight. Gaster elongate oval. Mandibles longitudinally striate. Head dorsum smooth medially, longitudinally rugose above frontal lobes and on the genae, reticulate laterally, and relatively smooth below eyes. Clypeus smooth medially, longitudinally rugose laterally. Mesosoma, petiole and postpetiole densely coarsely punctate, interface finely reticulate. Gaster smooth and shiny. Body dorsum with abundant erect to suberect apically blunt short hairs and abundant decumbent pubescence. Scapes and tibiae with dense decumbent pubescence. Body color blackish brown; mandibles, antennae, trochanters and tibiae yellowish brown. **Paratype workers:** TL 2.4-2.7, HL 0.55-0.65, HW 0.40-0.55, CI 73-85, SL 0.45-0.55, SI 100-113, ED 0.10-0.15, PW 0.30-0.40, WL 0.60-0.70, PL 0.20-0.25, PH 0.15-0.23, DPW 0.10-0.13 (*n*=8). As holotype, in some individuals body size slightly different, pronotum and mesonotum reticulate rugose weakly.

Holotype: Worker, China: Yunnan Province, Lijiang City, Gucheng District, Heilongtan Park (26.888242°N, 100.166922°E), 2400 m, 1991.Ⅹ.8, collected by Zheng-hui Xu on the ground in *Pinus yunnanensis* forest, No. A91-945.

Etymology: The specific epithet refers to "guyu", one of the Twenty-four Solar Terms of China.

正模工蚁： 正面观，头部近长方形，长大于宽；后缘中度隆起，后角宽圆，侧缘轻度隆起。上颚三角形，咀嚼缘具5个齿。唇基背面轻度隆起，缺中央纵脊，前缘中度隆起。额叶宽，遮盖触角窝。额脊较短，到达复眼中点水平。触角12节，柄节稍超过头后缘，触角棒3节。复眼位于头侧缘中点稍前处，占据侧缘的1/4。侧面观，前胸背板中度隆起，向前坡形降低，缺前中胸背板缝。中胸背板轻度隆起，向后缓坡形降低，缺后胸沟。并胸腹节背面近平直，向后缓坡形降低；并胸腹节刺很短，齿状，约与并胸腹节侧叶等长；斜面轻度凹陷。并胸腹节侧叶短，顶端钝角状。腹柄前面具小柄，约为腹柄结长的3/4；腹柄结近锥形，前面近平直，后面轻度隆起，背面圆；腹面近平直，前下角短，钝角状。后腹柄与腹柄结等高，前上缘圆，后缘近平直，腹面浅凹。背面观，前胸背板最宽，前缘中度隆起，肩角极宽圆，侧缘强烈隆起。缺前中胸背板缝。中胸轻微收缩，侧缘轻度凹陷。并胸腹节近长方形，侧缘近平直；并胸腹节刺很短，齿状。腹柄近梯形，后部变宽，长大于宽，侧缘轻度隆起。后腹柄近长方形，向后稍变窄，约为腹柄宽的1.5倍，前侧角窄圆，侧缘近平直。后腹部长卵形。上颚具纵条纹。头部背面中部光滑，额叶上方和颊区具纵皱纹，侧面具网纹，复眼下方较光滑。唇基中部光滑，两侧具纵皱纹。胸部、腹柄和后腹柄具密集粗刻点，界面具细网纹。后腹部光滑发亮。身体背面具丰富直立、亚直立短钝毛和密集倾斜绒毛被；柄节和胫节具密集倾斜绒毛被。身体黑棕色，上颚、触角、转节和胫节黄棕色。**副模工蚁：** 特征同正模，一些个体身体大小稍有差

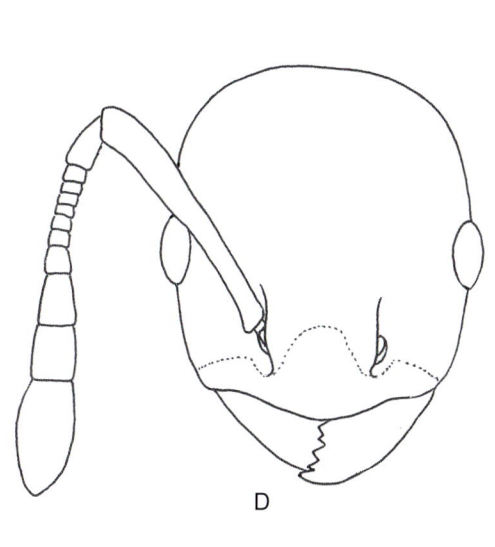
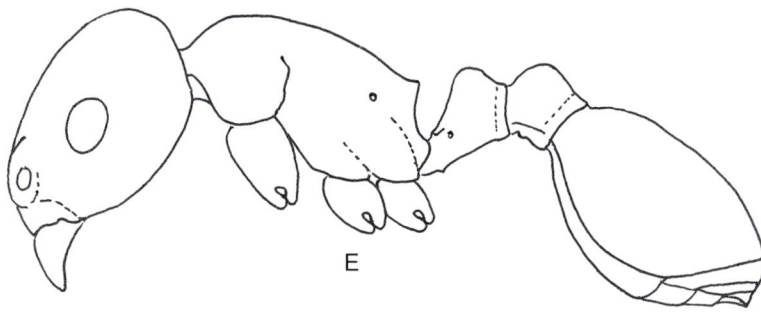
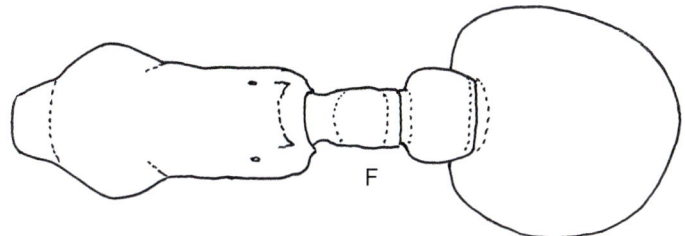

谷雨切胸蚁 *Temnothorax guyu*

A, D. 工蚁头部正面观；B, E. 工蚁身体侧面观；C, F. 工蚁身体背面观（A-C. 引自Qian和Xu, 2024）
A, D. Head of worker in full-face view; B, E. Body of worker in lateral view; C, F. Body of worker in dorsal view (A-C. cited from Qian & Xu, 2024)

异，前中胸背板具弱的网状皱纹。

正模： 工蚁，中国云南省丽江市古城区黑龙潭公园（26.888242°N，100.166922°E），2400m，1991.X.8，徐正会采于云南松林地表，No. A91-945。

词源： 该新种以中国二十四节气中的"谷雨"命名。

099 寒露切胸蚁
Temnothorax hanlu Qian & Xu, 2024

Temnothorax hanlu Qian & Xu, 2024, European Journal of Taxonomy, 936: 31, fig. 21 (w.) CHINA (Yunnan).

Holotype worker: TL 2.4, HL 0.59, HW 0.47, CI 79, SL 0.52, SI 111, ED 0.12, PW 0.37, WL 0.74, PL 0.28, PH 0.2, DPW 0.16. In full-face view, head roughly rectangular, longer than broad, posterior margin weakly convex, posterior corners narrowly rounded, lateral margins weakly convex. Mandibles triangular, masticatory margin with 5 teeth. Clypeal dorsum weakly convex, with median carina, anterior margin weakly convex. Antennae 12-segmented, scapes just reaching to posterior head margin. Eyes located at midpoint of lateral head margin, occupying 1/4 of lateral margin. In lateral view, promesonotum weakly convex and gently sloping down posteriorly, promesonotal suture and metanotal groove absent. Dorsum of propodeum slightly convex and sloping down posteriorly; propodeal spines long, weakly curved and pointed posteriorly, longer than propodeal dorsum; declivity weakly concave. Petiole with short anterior peduncle, about 1/2 length of petiolar node; petiolar node roughly trapezoidal, anterior margin and posterior margin almost straight, dorsal margin weakly convex; ventral margin weakly concave, anteroventral corner toothed. Postpetiole weakly lower than petiolar node, anterodorsal corner rounded, posterior margin straight, ventral margin short and almost straight. In dorsal view, pronotum broadest, lateral margin strongly convex, humeral corners rounded. Mesothorax moderately constricted and narrow, lateral margins weakly concave. Propodeum roughly rectangular, lateral margins almost straight, spines long and pointed posterolaterally. Petiole roughly trapezoidal and widening posteriorly, longer than broad, lateral margins slightly convex. Postpetiole roughly trapezoidal, about 1.4 times as broad as petiole, narrowing posteriorly, anterior margin moderately convex, anterolateral corners narrowly rounded, lateral margins nearly straight. Gaster elongate oval. Mandibles longitudinally striate. Head dorsum coarsely reticulate, longitudinally rugose between frontal carinae. Pronotum, mesonotum and propodeal dorsum coarsely reticulate; mesopleura, metapleura, sides of propodeum, petiole and postpetiole finely reticulate. Gaster smooth and shiny. Body dorsum with sparse erect to suberect apically blunt short hairs and abundant decumbent pubescence, hairs on head relatively abundant, pubescence on gaster sparse; scapes and tibiae with dense decumbent pubescence. Body color brownish yellow, mandibels, scapes and legs yellow; gaster black, anterior face brownish yellow. **Paratype worker:** TL 2.3, HL 0.55, HW 0.45, CI 82, SL 0.50, SI 111, ED 0.11, PW 0.30, WL 0.60, PL 0.25, PH 0.20, DPW 0.10 (*n*=1). As holotype.

Holotype: Worker, China: Yunnan Province, Yuanyang County, Nansha Town, Shicaichang (23.213064°N, 102.822789°E), 518 m, 2019.Ⅹ.17, collected by Zheng-hui Xu from a ground sample of dry evergreen broadleaf forest, No. A19-7128.

Etymology: The specific epithet refers to "hanlu", one of the Twenty-four Solar Terms of China.

正模工蚁： 正面观，头部近长方形，长大于宽；后缘轻度隆起，后角窄圆；侧缘轻度隆起。上颚三角形，咀嚼缘具5个齿。唇基背面轻度隆起，具中央纵脊，前缘轻度隆起。触角12节，柄节刚到达头后缘。复眼位于头侧缘中点处，占据侧缘的1/4。侧面观，前中胸背板轻度隆起，向后缓坡形降低，缺前中胸背板缝和后胸沟。并胸腹节背面轻微隆起，向后缓坡形降低；并胸腹节刺长，轻度弯曲，指向后方，长于并胸腹节背面；斜面浅凹。腹柄前面具短的小柄，约为腹柄结长的1/2；腹柄结近梯形，前面和后面近平直，背面轻度隆起；腹面浅凹，前下角齿状。后腹柄轻度低于腹柄结，前上角圆，后面平直；腹面短，近平直。背面观，前胸背板最宽，侧缘强烈隆起，肩角圆。中胸中度收缩，狭窄，侧缘浅凹。并胸腹节近长方形，侧缘近平直；并胸腹节刺长，指向后侧方。腹柄近梯形，向后变宽，长大于宽，侧缘轻微隆起。后腹柄近梯形，约为腹柄宽的1.4倍，向后变窄，前面中度隆起，前侧角窄圆，侧缘近平直。后腹部长卵形。上颚具纵条纹。头部背面具粗网纹，额脊之间具纵皱纹。前胸背板、中胸背板和并胸腹节背面具粗网纹，中胸侧板、后胸侧板、并胸腹节侧面、腹柄和后腹柄具细网纹。后腹部光滑发亮。身体背面具稀疏直立、亚直立短钝毛和丰富倾斜绒毛被，头部直立较丰富，后腹部绒毛被稀疏；柄节和胫节具密集倾斜绒毛被。身体棕黄色，上颚、柄节和足黄色；后腹部黑色，前面棕黄色。**副模工蚁：** 特征同正模。

正模： 工蚁，中国云南省元阳县南沙镇石材厂（23.213064°N，102.822789°E），518m，2019.Ⅹ.17，徐正会采于干性常绿阔叶林地表样中，No. A19-7128。

词源： 该新种以中国二十四节气中的"寒露"命名。

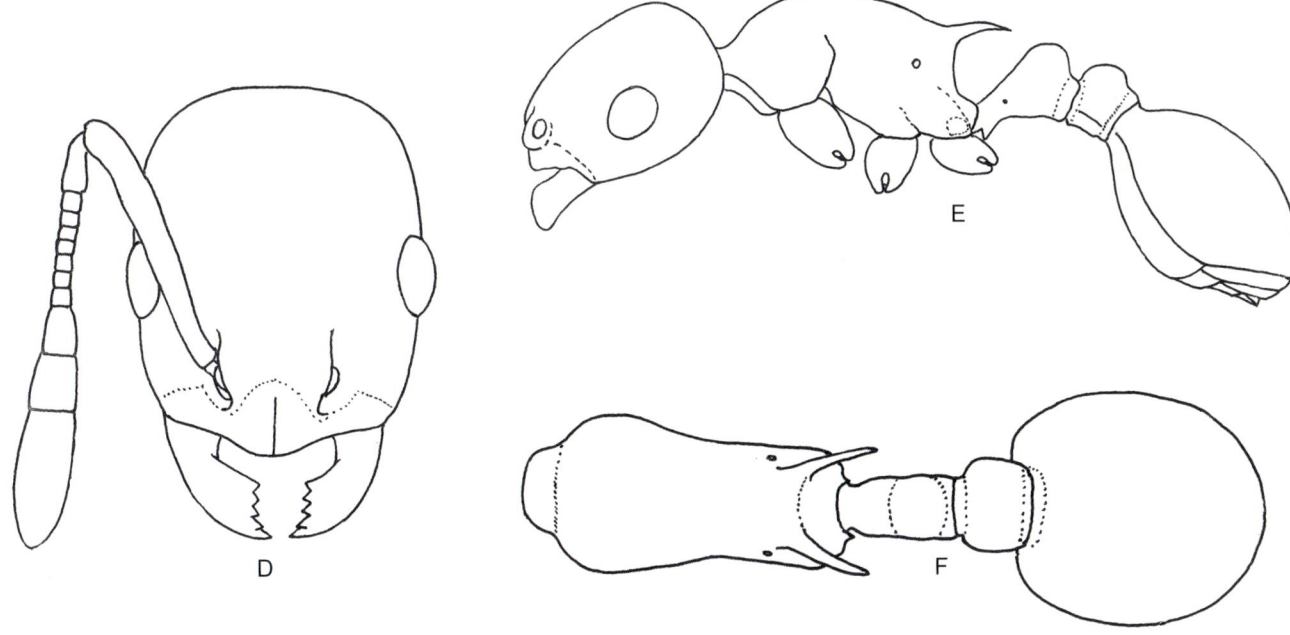

寒露切胸蚁 *Temnothorax hanlu*

A, D. 工蚁头部正面观；B, E. 工蚁身体侧面观；C, F. 工蚁身体背面观（A-C. 引自Qian和Xu, 2024）

A, D. Head of worker in full-face view; B, E. Body of worker in lateral view; C, F. Body of worker in dorsal view (A-C. cited from Qian & Xu, 2024)

100 惊蛰切胸蚁
Temnothorax jingzhe Qian & Xu, 2024

Temnothorax jingzhe Qian & Xu, 2024, European Journal of Taxonomy, 936: 33, fig. 22 (w.) CHINA (Yunnan).

Holotype worker: TL 3.1, HL 0.72, HW 0.63, CI 87, SL 0.60, SI 95, ED 0.15, PW 0.42, WL 0.85, PL 0.32, PH 0.20, DPW 0.15. In full-face view, head roughly rectangular, longer than broad, posterior and lateral margins weakly convex, posterior corners broadly rounded. Mandibles triangular, masticatory margin with 5 teeth. Clypeal dorsum weakly convex, median carina indistinct, anterior margin moderately convex. Frontal lobes narrow, concealing half of antennal sockets. Frontal carinae short, reaching to the level of anterior eye margin. Antennae 12-segmented, scapes almost reaching to posterior margin, club 3-segmented. Eyes located slightly before midpoint of lateral head margin, occupying 1/5 of lateral margin. In lateral view, pronotum weakly convex, dorsa of mesonotum and propodeum straight and gently sloping down posteriorly, promesonotal suture and metanotal groove absent. Propodeal spines very short and stout, roughly triangular, about as long as their basal width; declivity almost straight; propodeal lobes short and rounded apically. Petiole with long anterior peduncle, about as long as petiolar node; petiolar node roughly triangular, anterior and posterior margins weakly convex, top corner narrowly rounded; ventral margin almost straight, anteroventral corner minutely toothed. Postpetiole as high as petiolar node, anterodorsal corner rounded, posterior margin almost straight, ventral margin weakly concave. In dorsal view, pronotum broadest, anterior margin moderately convex, humeral corners broadly rounded, lateral margin strongly convex, promesonotal suture absent. Mesothorax moderately constricted, lateral margins weakly concave. Propodeum roughly rectangular, lateral margins almost straight; spines short and stout, roughly triangular, pointed posterolaterally. Petiole roughly trapezoidal and widening posteriorly, longer than broad, lateral margins weakly convex. Postpetiole roughly trapezoidal and narrowing posteriorly, about 1.4 times as broad as petiole, anterolateral corners broadly rounded. Gaster elongate oval. Mandibles longitudinally striate. Head dorsum densely longitudinally rugose, with interface densely punctate posteriorly and laterally, posteroventral part of lateral head margins smooth and shiny. Clypeus densely longitudinally rugose. Mesosoma reticulate; propodeal dorsum coarsely reticulate, metapleura longitudinally rugose. Propodeal sides, petiole and postpetiole finely reticulate, postpetiolar dorsum relatively smooth. Gaster smooth and shiny. Body dorsum with abundant suberect to subdecumbent short hairs and abundant decumbent pubescence, hairs on mesosoma subdecumbent. Scapes and tibiae with dense decumbent pubescence. Body color blackish brown, head and gaster black; mandibles, apices and bases of scapes, basal segments of flagella, trochanters, bases and apices of femora, apices of tibiae, and tarsi brownish yellow.

Holotype: Worker, China: Yunnan Province, Cangyuan County, Menglai Town, Wokan Mountain (23.332309°N, 99.19831°E), 2450 m, 2012.Ⅳ.19, collected by Chun-liang Li on the tree in moss evergreen broadleaf forest, No. A12-1144.

Etymology: The specific epithet refers to "jingzhe", one of the Twenty-four Solar Terms of China.

正模工蚁： 正面观，头部近长方形，长大于宽；后缘和侧缘轻度隆起，后角宽圆。上颚三角形，咀嚼缘具5个齿。唇基背面轻度隆起，中央纵脊不明显，前缘中度隆起。额叶窄，遮盖触角窝的一半。额脊短，到达复眼前缘水平。触角12节，柄节几乎到达头后缘，触角棒3节。复眼位于头侧缘中点稍前处，占据侧缘的1/5。侧面观，前胸背板轻度隆起。中胸背板和并胸腹节背面平直，向后缓坡形降低，缺前中胸背板缝和后胸沟。并胸腹节刺很短，粗壮，近三角形，长度约与基部宽相等；斜面近平直。并胸腹节侧叶短，顶端圆。腹柄前面小柄长，约与腹柄结等长；腹柄结近三角形，前面和后面轻度隆起，顶角窄圆；腹面近平直，前下角具小齿。后腹柄与腹柄结等高，前上角圆，后面近平直，腹面浅凹。背面观，前胸背板最宽，前缘中度隆起，肩角宽圆，侧面强烈隆起，缺前中胸背板缝。中胸中度收缩，侧缘浅凹。并胸腹节近长方形，侧缘近平直；并胸腹节刺短粗，近三角形，指向后侧方。腹柄近梯形，向后变宽，长大于宽，侧缘轻度隆起。后腹柄近梯形，向后变窄，约为腹柄宽的1.4倍，前侧角宽圆。后腹部长卵形。上颚具纵条纹。头部背面具密集纵皱纹，后部和侧面的界面具密集刻点，头侧缘后下部光滑发亮。唇基具密集纵皱纹。胸部具网纹；并胸腹节背面具粗网纹，后胸侧板具纵皱纹。并胸腹节侧面、腹柄和后腹柄具细网纹，后腹柄背面较光滑。后腹部光滑。身体背面具丰富亚直立、亚倾斜短毛和丰富倾斜绒毛被，胸部立毛亚倾斜；柄节和胫节具密集倾斜绒毛被。身体黑棕色，头部和后腹部黑色，上颚、柄节端部和基部、鞭节基部几节、转节、腿节端部和基部、胫节端部和跗节棕黄色。

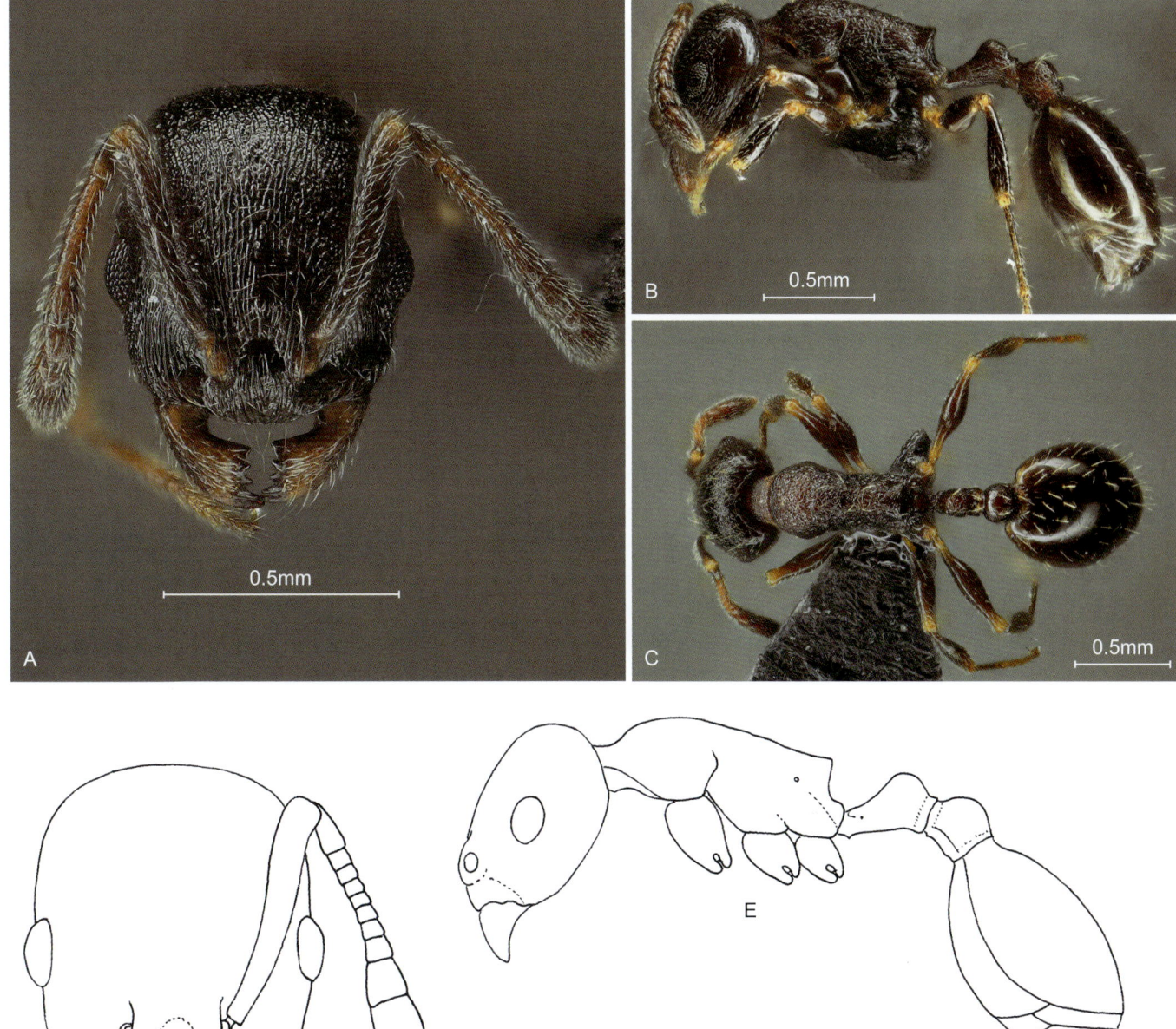

惊蛰切胸蚁 *Temnothorax jingzhe*

A, D. 工蚁头部正面观；B, E. 工蚁身体侧面观；C, F. 工蚁身体背面观（A-C. 引自Qian和Xu, 2024）
A, D. Head of worker in full-face view; B, E. Body of worker in lateral view; C, F. Body of worker in dorsal view (A-C. cited from Qian & Xu, 2024)

正模： 工蚁，中国云南省沧源县勐来乡窝坎大山（23.332309°N，99.19831°E），2450m，2012.Ⅳ.19，李春良采于苔藓常绿阔叶林树上，No. A12-1144。

词源： 该新种以中国二十四节气中的"惊蛰"命名。

立春切胸蚁
Temnothorax lichun Qian & Xu, 2024

Temnothorax lichun Qian & Xu, 2024, European Journal of Taxonomy, 936: 35, fig. 23 (w.) CHINA (Yunnan).

Holotype worker: TL 3.5, HL 0.78, HW 0.70, CI 91, SL 0.62, SI 88, ED 0.17, PW 0.46, WL 0.96, PL 0.37, PH 0.22, DPW 0.20. In full-face view, head roughly rectangular, longer than broad, posterior and lateral margins weakly convex, posterior corners broadly rounded. Mandibles triangular, masticatory margin with 5 teeth. Clypeal dorsum weakly convex, with median carina, anterior margin angularly concave in the center. Frontal lobes narrow, concealing half of antennal sockets. Frontal carinae relatively long, reaching to the level of midpoint of eyes. Antennae 12-segmented, scapes failing to reach posterior head margin by half diameter of scape, antennal clubs 3-segmented. Eyes located at midpoint of lateral head margin, occupying 1/4 of lateral margin. In lateral view, pronotum almost straight, weakly convex anteriorly, promesonotal suture absent. Mesonotum weakly convex and gently sloping down posteriorly, metanotal groove very shallowly impressed. Propodeal dorsum weakly concave, weakly prominent anteriorly and posteriorly; propodeal spines long, slightly longer than declivity, weakly down-curved; declivity weakly concave; propodeal lobes broad and rounded apically. Petiole with short anterior peduncle, about 1/2 length of petiolar node; petiolar node roughly trapezoidal and narrowing dorsally, anterior margin almost straight, dorsal and posterior margins weakly convex, anterodorsal and posterodrsal corners broadly rounded; ventral margin straight anteriorly, weakly concave posteriorly, anteroventral corner largely acutely toothed. Postpetiole about as high as petiole node, roughly trapezoidal and narrowing dorsally, dorsal margin weakly convex, anterodorsal and posterodorsal corners broadly rounded, ventral margin deeply concave. In dorsal view, pronotum broadest, anterior and lateral margins weakly convex, humeral corners narrowly rounded, promesonotal suture absent. Mesothorax weakly constricted, lateral margins weakly convex, metanotal groove shallowly impressed. Propodeum roughly rectangular, lateral margins slightly convex; spines long, weakly in-curved. Petiole roughly trapezoidal and widening posteriorly, longer than breadth, lateral margins weakly convex. Postpetiole roughly trapezoidal and narrowing posteriorly, about 1.2 times as broad as petiole, anterolateral corners narrowly rounded, lateral margins weakly convex. Gaster elongate oval. Mandibles longitudinally striate. Head dorsum loosely longitudinally rugose anteriorly, gradually reticulate posteriorly and laterally. Clypeus smooth, each side with 3 longitudinal rugae. Mesosoma reticulate. Petiole and postpetiole finely reticulate, ventral face of petiole densely punctate. Gaster smooth and shiny. Body dorsum with abundant erect to suberect apically blunt short hairs and abundant decumbent pubescence. Scapes and tibiae with dense decumbent pubescence. Head and gaster black; mesosoma, petiole and postpetiole reddish brown; mandibles, antennae and legs blackish brown.

Holotype: Worker, China: Yunnan Province, Malipo County, Malipo Town, Chongtou Village (23.101083°N, 104.724083°E), 1050 m, 2010.Ⅳ.2, collected by Liang Wang from a canopy sample of conifer-brodleaf mixed forest, No. A10-1994.

Etymology: The specific epithet refers to "lichun", one of the Twenty-four Solar Terms of China.

正模工蚁： 正面观，头部近长方形，长大于宽；后缘和侧缘轻度隆起，后角宽圆。上颚三角形，咀嚼缘具5个齿。唇基背面轻度隆起，具中央纵脊，前缘中央角状凹陷。额叶窄，遮盖触角窝的一半。额脊较长，到达复眼中点水平。触角12节，柄节未到达头后缘，柄节末端与头后缘距离为柄节直径的一半，触角棒3节。复眼位于头侧缘中点处，占据侧缘的1/4。侧面观，前胸背板近平直，前部轻度隆起，缺前中胸背板缝。中胸背板轻度隆起，向后缓坡形降低，后胸沟非常浅。并胸腹节背面轻度凹陷，前部和后部轻度隆起；并胸腹节刺长，稍长于斜面，轻度下弯；斜面浅凹。并胸腹节侧叶宽，顶端圆。腹柄前面具短的小柄，约为腹柄结长的1/2；腹柄结近梯形，向上变窄，前面近平直，背面和后面轻度隆起，前上角和后上角宽圆；腹面前部平直，后部浅凹，前下角大，尖齿状。后腹柄约与腹柄结等高，近梯形，向上变窄，背面轻度隆起，前上角和后上角宽圆，腹面深凹。背面观，前胸背板最宽，前缘和侧缘轻度隆起，肩角窄圆，缺前中胸背板缝。中胸轻度收缩，侧缘轻度隆起，后胸沟浅凹。并胸腹节近长方形，侧缘轻度隆起；并胸腹节刺长，轻度内弯。腹柄近梯形，向后变宽，长大于宽，侧缘轻度隆起。后腹柄近梯形，向后变窄，约为腹柄宽的1.2倍，前侧角窄圆，侧缘轻度隆起。后腹部长卵形。上颚具纵条纹。头部背面前部具松散的纵皱纹，后部和侧面逐渐呈网纹。唇基光滑，每侧具3条纵皱纹。胸部具网纹；腹柄和后腹柄具细网纹，腹柄腹面具密集刻点；后腹部光滑发亮。身体背面具丰富直立、亚直立短钝毛和丰富倾斜绒毛被；柄节和胫节具密集倾斜绒毛被。头部和后腹部黑色，胸部、腹柄和后腹柄红棕色，上颚、触角和足黑棕色。

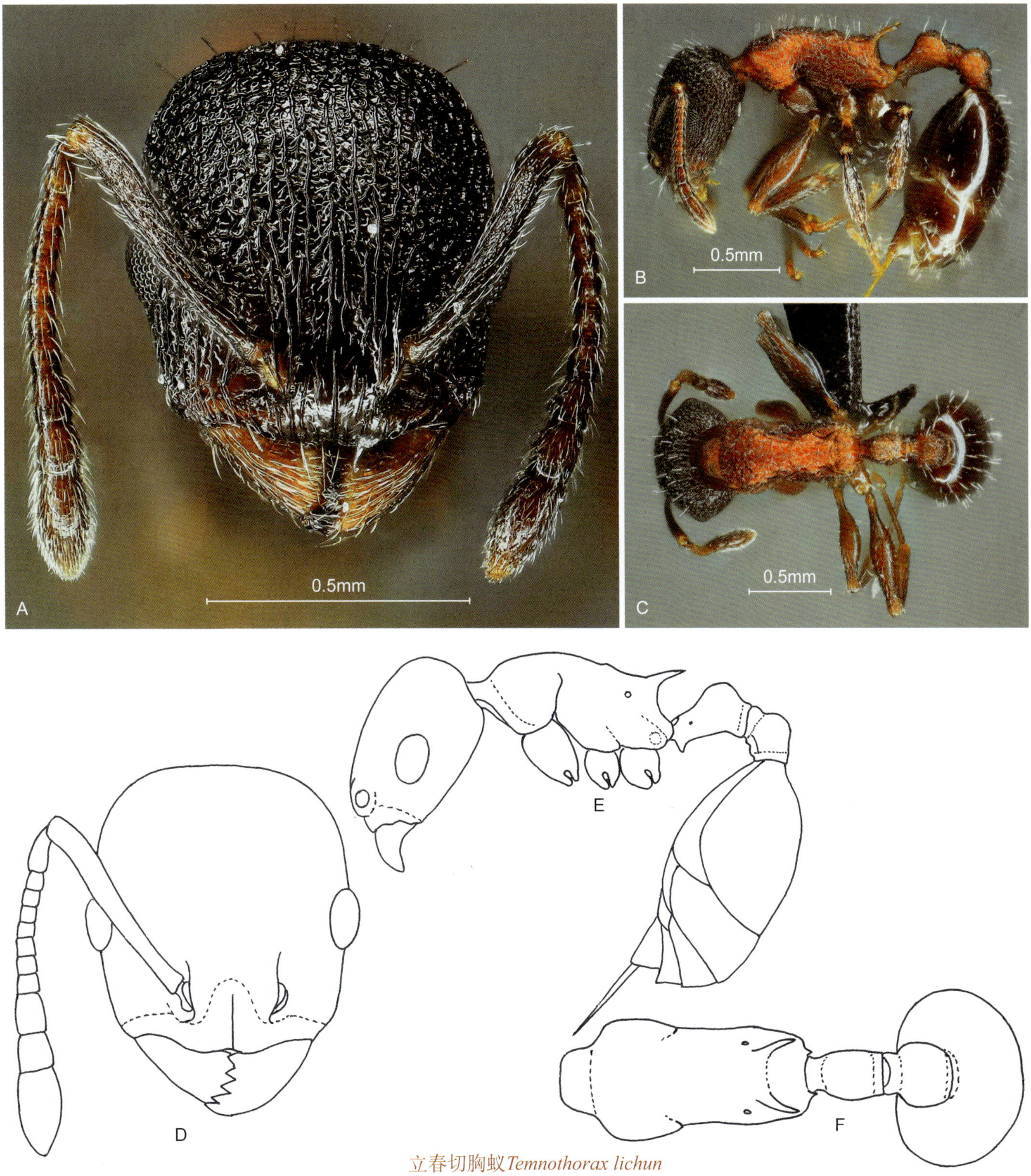

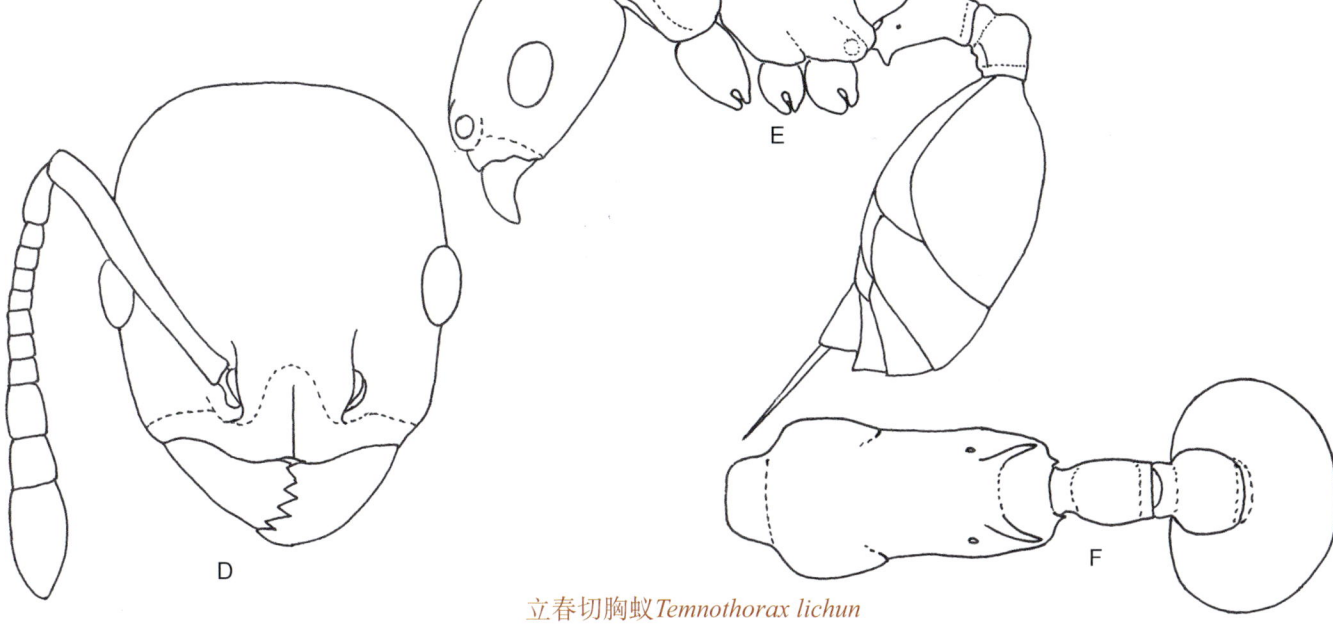

立春切胸蚁 *Temnothorax lichun*
A, D. 工蚁头部正面观; B, E. 工蚁身体侧面观; C, F. 工蚁身体背面观(A-C. 引自Qian和Xu, 2024)
A, D. Head of worker in full-face view; B, E. Body of worker in lateral view; C, F. Body of worker in dorsal view (A-C. cited from Qian & Xu, 2024)

正模:工蚁,中国云南省麻栗坡县麻栗坡镇冲头村(23.101083°N,104.724083°E),1050m,2010.Ⅳ.2,王亮采于针阔混交林树冠样中,No. A10-1994。

词源:该新种以中国二十四节气中的"立春"命名。

立冬切胸蚁
Temnothorax lidong Qian & Xu, 2024

Temnothorax lidong Qian & Xu, 2024, European Journal of Taxonomy, 936: 37, fig. 25 (w.) CHINA (Sichuan).

Holotype worker: TL 2.3, HL 0.57, HW 0.45, CI 79, SL 0.5, SI 111, ED 0.13, PW 0.33, WL 0.66, PL 0.26, PH 0.17, DPW 0.13. In full-face view, head roughly rectangular, longer than broad, posterior and lateral margins weakly convex, posterior corners broadly rounded. Mandibles triangular, masticatory margin with 5 teeth. Clypeal dorsum weakly convex, without median carina, anterior margin moderately convex. Frontal lobes narrow, concealing half of antennal sockets. Frontal carinae short, reaching to the level of anterior eye margins. Antennae 12-segmented, scapes just reaching to posterior head margin, club 3-segmented. Eyes located at midpoint of lateral head margin, occupying 1/5 of lateral margin. In lateral view, pronotum moderately convex, promesonotal suture distinct laterally. Mesonotal dorsum gently sloping down posteriorly, weakly convex anteriorly and posteriorly, weakly impressed in the middle. Metanotal groove absent. Propodeal dorsum slightly convex and gently sloping down posteriorly; spines very long, about as long as declivity, pointed posterodorsally and slightly down-curved; declivity weakly concave; propodeal lobes short and rounded apically. Petiole with short anterior peduncle, about 1/2 length of petiolar node; petiolar node roughly conical, anterior and posterior margins weakly convex, dorsum narrowly rounded; ventral margin weakly concave, anteroventral corner bluntly angled. Postpetiole as high as petiolar node, anterior margin strong convex, dorsal margin weakly convex, posterior margin almost straight; ventral margin weakly concave, anteroventral corner acutely toothed. In dorsal view, pronotum broadest, anterior and lateral margins moderately convex, humeral corners broadly rounded. Promesonotal suture and metanotal groove absent. Mesothorax weakly constricted, lateral margins weakly concave. Propodeum roughly rectangular, lateral margins slightly concave; spines stout and long, pointed posterolaterally and weakly in-curved. Petiole roughly trapezoidal, widening posteriorly, longer than broad, lateral margins almost straight. Postpetiole roughly square, about 1.7 times as broad as petiole, anterolateral corners narrowly rounded, lateral margins moderately convex. Gaster elongate oval. Mandibles longitudinally striate. Head dorsum longitudinally rugose, becoming smooth posteriorly and on the central strip; genae and clypeus longitudinally rugose. Mesosoma finely reticulate; promesonotal dorsum relatively smooth, lower part of metapleura longitudinally rugose. Propodeal sides, petiole and postpetiole densely punctate, interface appearing as micro-reticulation. Gaster smooth and shiny. Body dorsum with abundant erect to suberect apically blunt short hairs and abundant decumbent pubescence. Scapes and tibiae with dense decumbent pubescence. Body color black; mandibles, antennae and legs brownish yellow. **Paratype workers:** TL 2.3-2.5, HL 0.55-0.60, HW 0.45-0.48, CI 79-82, SL 0.40-0.50, SI 89-105, ED 0.10-0.15, PW 0.30-0.35, WL 0.60-0.66, PL 0.20-0.25, PH 0.15-0.20, DPW 0.10-0.11 ($n=2$). As holotype, in some individuals body size slightly different, propodeal spines slightly shorter and apex not pointed.

Holotype: Worker, China: Sichuan Province, Panzhihua City, Tongde Town, Longtang Village (26.742033°N, 101.571572°E), 1726 m, 2018.Ⅶ.22, collected by Zheng-hui Xu on the ground in monsoon evergreen broadleaf forest, No. C18-230.

Etymology: The specific epithet refers to "lidong", one of the Twenty-four Solar Terms of China.

正模工蚁： 正面观，头部近长方形，长大于宽；后缘和侧缘轻度隆起，后角宽圆。上颚三角形，咀嚼缘具5个齿。唇基背面轻度隆起，缺中央纵脊，前缘中度隆起。额叶窄，遮盖触角窝的一半。额脊短，到达复眼前缘水平。触角12节，柄节刚到达头后缘，触角棒3节。复眼位于头侧缘中点处，占据侧缘的1/5。侧面观，前胸背板中度隆起，前中胸背板缝仅在侧面明显。中胸背板背面向后缓坡形降低，前部和后部轻度隆起，中部轻度凹陷。缺后胸沟。并胸腹节背面轻微隆起，向后缓坡形降低；并胸腹节刺很长，约与斜面等长，指向后上方，轻微下弯；斜面浅凹。并胸腹节侧叶短，顶端圆。腹柄前面小柄短，约为腹柄结长的1/2；腹柄结近锥形，前面和后面轻度隆起，背面窄圆；腹面浅凹，前下角钝角状。后腹柄与腹柄结等高，前面强烈隆起，背面轻度隆起，后面近平直；腹面浅凹，前下角尖齿状。背面观，前胸背板最宽，前缘和侧缘轻度隆起，肩角宽圆。缺前中胸背板缝和后胸沟。中胸轻度收缩，侧缘浅凹。并胸腹节近长方形，侧缘轻微凹陷；并胸腹节刺粗长，指向后侧方，轻度内弯。腹柄近梯形，向后变宽，长大于宽，侧缘近平直。后腹柄近方形，约为腹柄宽的1.7倍，前侧角窄圆，侧面轻度隆起。后腹部长卵形。上颚具纵条纹。头部背面具纵皱纹，后部和中央纵带光滑，颊区和唇基具纵皱纹。胸部具细网纹；前中胸背板背面较光滑，后胸侧板下部纵皱纹。并胸腹节侧面、腹柄和后腹柄具密集刻点，界面呈微网纹。后腹部光滑发亮。身体背面具丰富直立、亚直立短钝毛和丰富倾斜绒毛被；柄节和胫节具密集倾斜绒毛被。身体黑色，上颚、触角和足

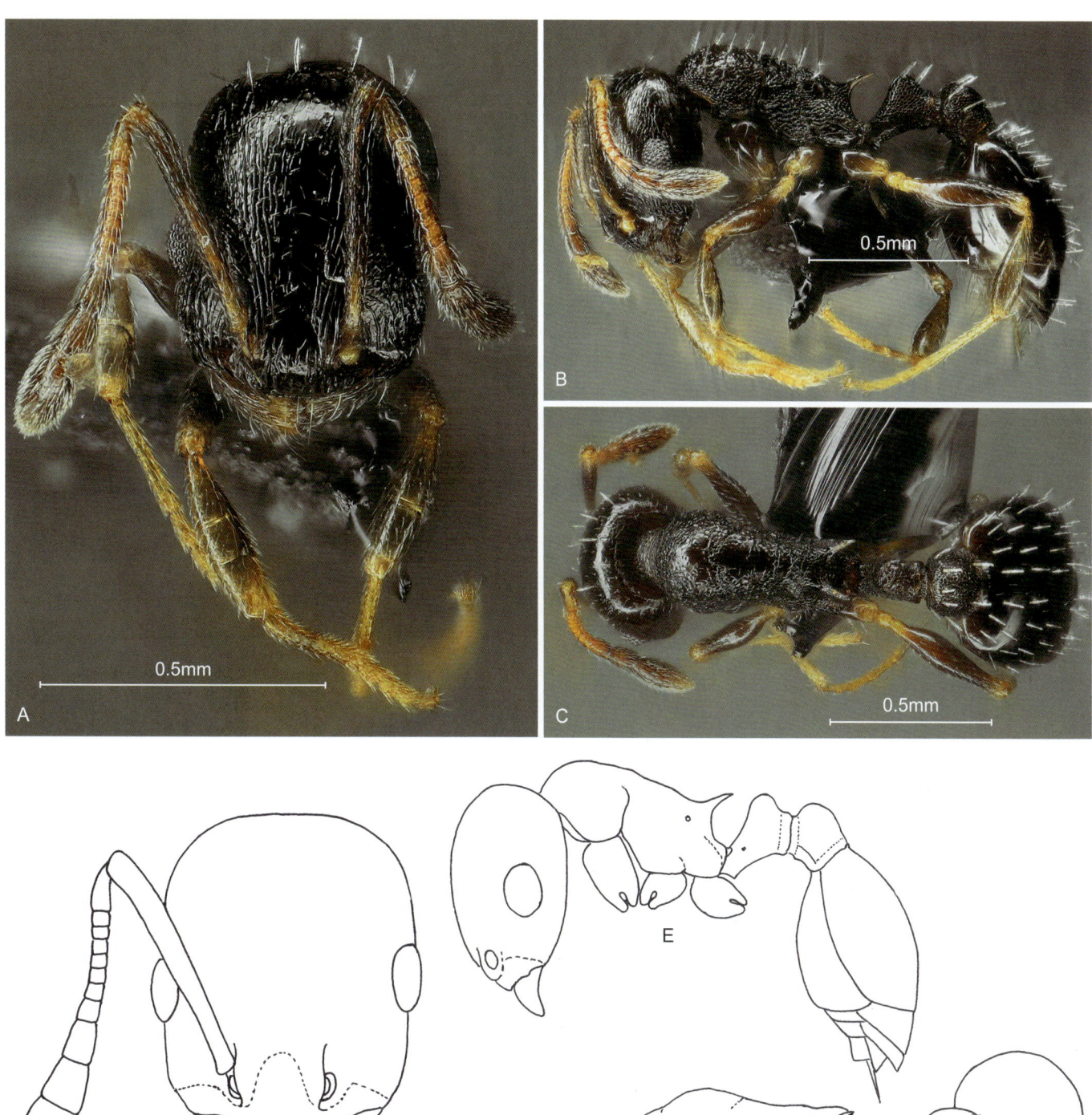

立冬切胸蚁 *Temnothorax lidong*

A, D. 工蚁头部正面观；B, E. 工蚁身体侧面观；C, F. 工蚁身体背面观（A-C. 引自Qian和Xu, 2024）
A, D. Head of worker in full-face view; B, E. Body of worker in lateral view; C, F. Body of worker in dorsal view (A-C. cited from Qian & Xu, 2024)

棕黄色。**副模工蚁：** 特征同正模，有的个体身体大小有轻微差异，并胸腹节刺稍短，末端不尖锐。

　正模： 工蚁，中国四川省攀枝花市同德镇龙塘村（26.742033°N，101.571572°E），1726m，2018.Ⅶ.22，徐正会采于季风常绿阔叶林地表，No. C18-230。

　词源： 该新种以中国二十四节气中的"立冬"命名。

103 立秋切胸蚁
Temnothorax liqiu Qian & Xu, 2024

Temnothorax liqiu Qian & Xu, 2024, European Journal of Taxonomy, 936: 40, fig. 27 (w.) CHINA (Sichuan).

Holotype worker: TL 3.0, HL 0.66, HW 0.55, CI 83, SL 0.62, SI 113, ED 0.15, PW 0.41, WL 0.83, PL 0.29, PH 0.20, DPW 0.16. In full-face view, head roughly rectangular, longer than broad, posterior and lateral margins weakly convex, posterior corners broadly rounded. Mandibles triangular, masticatory margin with 5 teeth. Clypeal dorsum weakly convex, with median carina, anterior margin moderately convex. Frontal lobes narrow, concealing half of antennal socket. Frontal carinae short, reaching to the level of anterior eye margin. Antennae 12-segmented, scapes surpassing posterior head margin by 1/7 of its length, club 3-segmented. Eyes located at midpoint of lateral head margin, occupying 1/4 of lateral margin. In lateral view, pronotum moderately convex, promesonotal suture absent. Mesonotum almost straight anteriorly and weakly convex posteriorly. Metanotal groove widely moderately impressed and distinct. Dorsum of propodeum straight and sloping down posteriorly; propodeal spines short and roughly triangular, about as long as their basal width, pointed posterodorsally; declivity weakly concave, shorter than dorsum; propodeal lobes small and rounded apically. Petiole with long anterior peduncle, about as long as petiolar node; petiolar node roughly trapezoidal and narrowed dorsally, anterior margin straight, dorsal margin weakly convex and gently sloping down posteriorly, posterior margin short and weakly convex, anterodorsal corner narrowly rounded, posterodorsal corner broadly rounded; ventral margin almost straight and weakly concave posteriorly, anteroventral corner shortly toothed. Postpetiole about as high as petiole, weakly inclined anteriorly, anterodorsal margin roundly convex, posterior margin straight, ventral margin almost straight. In dorsal view, pronotum broadest, anterior margin weakly convex, humeral corners broadly rounded, lateral margin strongly convex. Promesonotal suture absent. Mesothorax moderately constricted, lateral margins moderately concave. Metanotal groove widely impressed. Propodeum roughly square and weakly widening posteriorly, lateral margins almost straight; spines short, pointed posterolaterally. Petiole roughly trapezoidal and widening posteriorly, longer than broad, lateral margins weakly convex. Postpetiole roughly trapezoidal and narrowing posteriorly, about 1.4 times as broad as petiole, anterolateral corners narrowly rounded, lateral margins nearly straight. Gaster elongate oval. Mandibles longitudinally striate. Head dorsum longitudinally rugose anteriorly with rugae divergent posteriorly, and gradually reticulate posteriorly and laterally, interface densely punctate. Clypeus with three rugae on each side. Mesosomal dorsum reticulate; pronotal sides and mesopleura reticulate, metapleura longitudinally rugose; sides of propodeum, petiole and postpetiole finely reticulate. Gaster smooth and shiny. Body dorsum with abundant erect to suberect apically blunt short hairs and abundant decumbent pubescence, pubescence on head dorsum relatively denser; scapes and tibiae with dense subdecumbent to decumbent pubescence. Body color yellowish brown; head dorsum and first gastral segment blackish brown. **Paratype workers:** TL 2.3-3.2, HL 0.6-0.70, HW 0.45-0.60, CI 75-86, SL 0.50-0.55, SI 92-111, ED 0.10-0.11, PW 0.33-0.35, WL 0.60-0.80, PL 0.24-0.25, PH 0.15-0.18, DPW 0.10-0.13 (n=13). As holotype, in some individuals propodeal spines slightly longer than in holotype, body color lighter, metanotal groove deeply to shallowly impressed.

Holotype: Worker, China: Sichuan Province, Muli County, Xiamaidi Town, Mianbu Pass (27.687339°N, 101.221072°E), 3280 m, 2018.Ⅶ.26, collected by Zhao Huang from a soil sample of alpine conifer forest, No. C18-795.

Etymology: The specific epithet refers to "liqiu", one of the Twenty-four Solar Terms of China.

正模工蚁：正面观，头部近长方形，长大于宽；后缘和侧缘轻度隆起，后角宽圆。上颚三角形，咀嚼缘具5个齿。唇基背面轻度隆起，具中央纵脊，前缘中度隆起。额叶窄，遮盖触角窝的一半。额脊短，到达复眼前缘水平。触角12节，柄节的1/7超过头后缘，触角棒3节。复眼位于头侧缘中点处，占据侧缘的1/4。侧面观，前胸背板中度隆起，缺前中胸背板缝。中胸背板前部近平直，后部轻度隆起。后胸沟宽形中度凹陷。并胸腹节背面平直，向后坡形降低；并胸腹节刺短，近三角形，长度约等于其基部宽，指向后上方；斜面浅凹，短于背面。并胸腹节侧叶小，顶端圆。腹柄前面小柄长，约与腹柄结等长；腹柄结近梯形，向上变窄，前面平直；背面轻度隆起，向后缓坡形降低；后面短，轻度隆起；前上角窄圆，后上角宽圆；腹面近平直，后部浅凹，前下角短齿状。后腹柄约与腹柄等高，轻度前倾，前上缘圆形隆起，后面平直，腹面近平直。背面观，前胸最宽，前缘轻度隆起，肩角宽圆，侧缘强烈隆起。缺前中胸背板缝。中胸中度收缩，侧缘中度凹陷。后胸沟宽形凹陷。并胸腹节近方形，向后轻度变宽，侧缘近平直；并胸腹节刺短，指向后侧方。腹柄近梯形，向后变宽，长大于宽，侧缘轻度隆起。后腹柄近梯形，向后

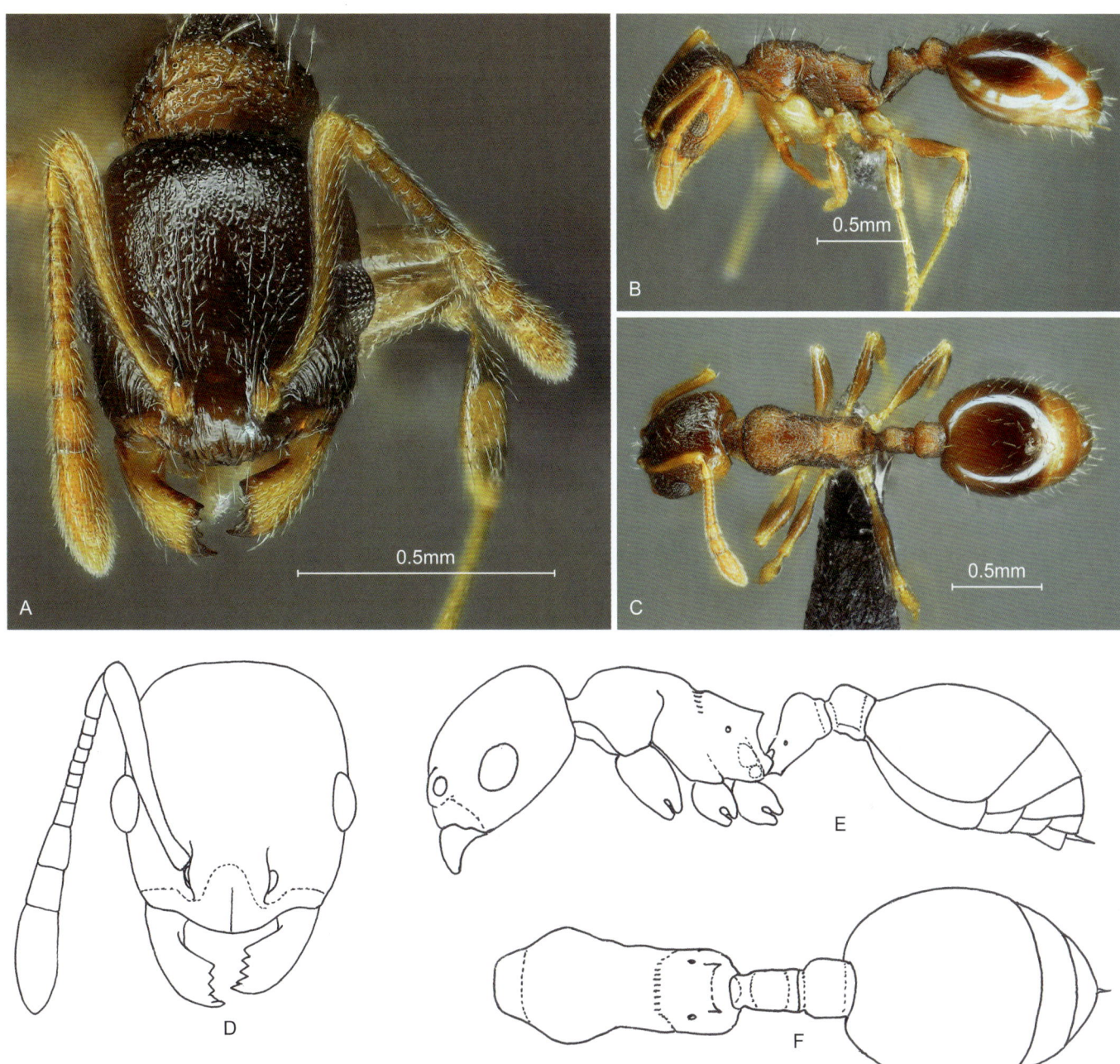

立秋切胸蚁 *Temnothorax liqiu*

A, D. 工蚁头部正面观；B, E. 工蚁身体侧面观；C, F. 工蚁身体背面观（A-C. 引自Qian和Xu, 2024）
A, D. Head of worker in full-face view; B, E. Body of worker in lateral view; C, F. Body of worker in dorsal view (A-C. cited from Qian & Xu, 2024)

变窄，约为腹柄宽的1.4倍，前侧角窄圆，侧缘近平直。后腹部长卵形。上颚具纵条纹。头部背面前部具向后发散的纵皱纹，后部和侧面逐渐呈网纹，界面具密集刻点。唇基每侧具3条皱纹。胸部背面具网纹；前胸背板两侧和中胸侧板具网纹，后胸侧板具纵皱纹；并胸腹节侧面、腹柄和后腹柄具细网纹。后腹部光滑发亮。身体背面具丰富直立、亚直立短钝毛和丰富倾斜绒毛被，头部背面绒毛被较密；柄节和胫节具密集亚倾斜、倾斜绒毛被。身体黄棕色，头部背面和后腹部第1节黑棕色。**副模工蚁：**特征同正模，部分个体并胸腹节刺稍长于正模，身体色较浅，后胸沟深凹至浅凹。

正模：工蚁，中国四川省木里县下麦地乡棉布垭口（27.687339°N，101.221072°E），3280m，2018.Ⅶ.26，黄钊采于高山针叶林土壤样中，No. C18-795。

词源：该新种以中国二十四节气中的"立秋"命名。

立夏切胸蚁
Temnothorax lixia Qian & Xu, 2024

Temnothorax lixia Qian & Xu, 2024, European Journal of Taxonomy, 936: 42, fig. 28 (w.) CHINA (Sichuan).

Holotype worker: TL 2.4, HL 0.62, HW 0.52, CI 84, SL 0.52, SI 100, ED 0.12, PW 0.38, WL 0.76, PL 0.25, PH 0.20, DPW 0.15. In full-face view, head roughly rectangular, longer than broad, posterior margin slightly convex, posterior corners broadly rounded, lateral margins weakly convex. Mandibles triangular, masticatory margin with 5 teeth. Clypeal dorsum weakly convex, without median carina, anterior margin straight in the middle. Frontal lobes narrow, concealing half of antennal sockets. Frontal carinae short, about as long as frontal lobes. Antennae 12-segmented, scapes almost reaching to posterior head margin, antennal clubs 3-segmented. Eyes located at midpoint of lateral head margin, occupying 1/4 of lateral margin. In lateral view, promesonotum weakly convex and gently sloping down posteriorly, promesonotal suture and metanotal groove absent. Propodeal dorsum almost straight and gently sloping down posteriorly, slightly convex anteriorly and medially; propodeal spines short and triangular, as long as their basal width, about 1/2 length of declivity; declivity slightly concave; propodeal lobes short, rounded apically. Petiole with short anterior peduncle, about 2/3 length of petiolar node; petiolar node roughly triangular, anterior margin straight, posterior margin moderately convex, top corner narrowly rounded; ventral margin almost straight anteriorly, weakly concave posteriorly, anteroventral corner minutely toothed. Postpetiole as high as petiolar node, anterodorsal margin rounded, posterior margin almost straight, ventral margin moderately concave. In dorsal view, pronotum broadest, anterior margin moderately convex, humeral corners very broadly rounded, lateral margins strongly convex. Promesonotal suture and metanotal groove absent. Mesothorax weakly constricted, lateral margins weakly concave. Propodeum roughly rectangular, lateral margins almost straight; spines short, roughly triangular, pointed posterolaterally. Petiole roughly trapezoidal and widening posteriorly, longer than broad, lateral margins slightly convex. Postpetiole roughly trapezoidal and narrowing posteriorly, about 1.4 times as broad as petiole, anterolateral corners narrowly rounded, lateral margins weakly convex. Gaster elongate oval. Mandibles longitudinally striate. Head dorsum densely longitudinally rugose, gradually reticulate posteriorly and laterally with interface densely punctate. Clypeus smooth, each side with 2-3 longitudinal rugae. Mesosoma reticulate; pronotal sides coarsely reticulate; mesopleura and upper part of metapleura finely reticulate; lower part of metapleura longitudinally rugose; propodeal sides, petiolar dorsum and postpetiolar dorsum densely punctate; petiolar sides and postpetiolar sides finely reticulate. Gaster smooth and shiny. Body dorsum with sparse erect to suberect apically blunt short hairs and sparse decumbent pubescence; hairs on head and gaster relatively abundant. Scapes and tibiae with dense decumbent pubescence. Body color black; mandibles, antennae and legs brownish yellow, scapes and femora blackish brown. **Paratype workers:** TL 2.3-2.5, HL 0.60-0.63, HW 0.50-0.55, CI 83-88, SL 0.48-0.50, SI 86-100, ED 0.08-0.10, PW 0.35-0.36, WL 0.70-0.75, PL 0.25-0.30, PH 0.18-0.20, DPW 0.10-0.13 (*n*=3). As holotype.

Holotype: Worker, China: Sichuan Province, Kangding County, Pengta Town, Tongling Village (30.501231°N, 102.300811°E), 2450 m, 2005.Ⅷ.31, collected by Zheng-hui Xu in *Pinus yunnanensis* forest using Winkler leaf litter extraction, No. A05-945.

Etymology: The specific epithet refers to "lixia", one of the Twenty-four Solar Terms of China.

正模工蚁： 正面观，头部近长方形，长大于宽；后缘和侧缘轻度隆起，后角宽圆。上颚三角形，咀嚼缘具5个齿。唇基背面轻度隆起，缺中央纵脊，前缘中部平直。额叶窄，遮盖触角窝的一半。额脊短，约与额叶等长。触角12节，柄节几乎到达头后缘，触角棒3节。复眼位于头侧缘中点处，占据侧缘的1/4。侧面观，前中胸背板轻度隆起，向后缓坡形降低，缺前中胸背板缝和后胸沟。并胸腹节背面近平直，向后缓坡形降低，前部和中部轻微隆起；并胸腹节刺短，三角形，长度与其基部宽相等，约为斜面长的1/2；斜面轻微凹陷。并胸腹节侧叶短，顶端圆。腹柄前面小柄短，约为腹柄结长的2/3；腹柄结近三角形，前面平直，后面中度隆起，顶角窄圆；腹面前部近平直，后部浅凹，前下角小齿状。后腹柄与腹柄结等高，前上缘圆，后缘近平直，腹面中度凹陷。背面观，前胸背板最宽，前缘中度隆起，肩角极宽圆，侧缘强烈隆起。缺前中胸背板缝和后胸沟。中胸轻度收缩，侧缘浅凹。并胸腹节近长方形，侧缘近平直；并胸腹节刺短，近三角形，指向后侧方。腹柄近梯形，向后变宽，长大于宽，侧缘轻微隆起。后腹柄近梯形，向后变窄，约为腹柄宽的1.4倍，前侧角窄圆，侧面轻度隆起。后腹部长卵形。上颚具纵条纹。头部背面具密集纵皱纹，后部和侧面逐渐呈网纹，界面具密集刻点。唇基光滑，每侧具2～3条纵皱纹。胸部具网纹；前胸背板侧面具粗网纹；中胸侧板和后胸侧板上部具细网纹；后胸侧板下部具纵皱纹；并胸腹节侧面、腹柄背面和后腹

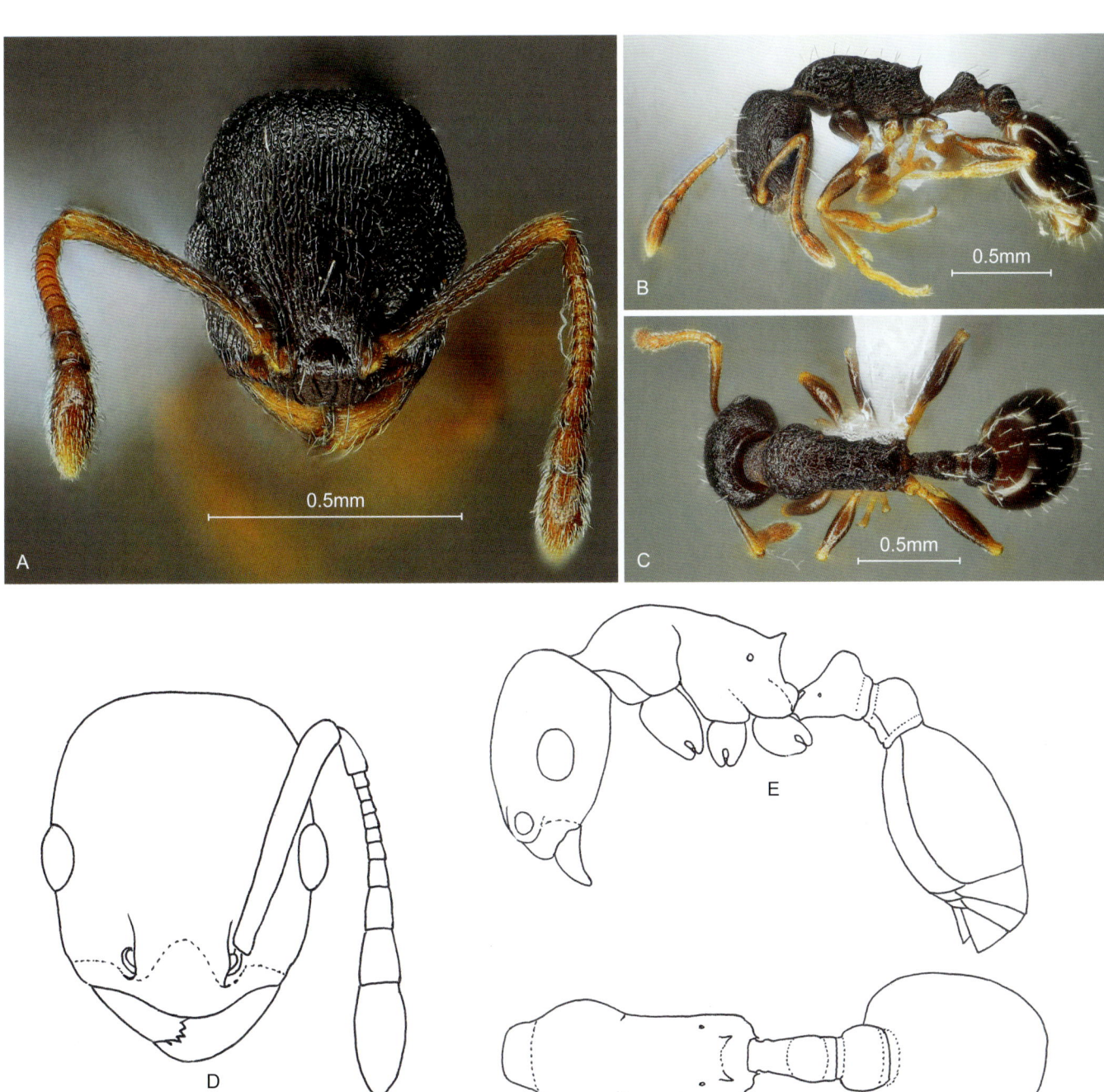

立夏切胸蚁 *Temnothorax lixia*

A, D. 工蚁头部正面观；B, E. 工蚁身体侧面观；C, F. 工蚁身体背面观（A-C. 引自Qian和Xu, 2024）
A, D. Head of worker in full-face view; B, E. Body of worker in lateral view; C, F. Body of worker in dorsal view (A-C. cited from Qian & Xu, 2024)

柄背面具密集刻点；腹柄侧面和后腹柄侧面具细网纹。后腹部光滑发亮。身体背面具稀疏直立、亚直立短钝毛和稀疏倾斜绒毛被，头部和后腹部立毛较丰富；柄节和胫节具密集倾斜绒毛被。身体黑色，上颚、触角和足棕黄色，柄节和腿节黑棕色。**副模工蚁：**特征同正模。

正模：工蚁，中国四川省康定县捧塔乡铜陵村（30.501231°N，102.300811°E），2450m，2005.Ⅷ.31，徐正会采用Winkler地被物分离法采于云南松林中，No. A05-945。

词源：该新种以中国二十四节气中的"立夏"命名。

105 芒种切胸蚁
Temnothorax mangzhong Qian & Xu, 2024

Temnothorax mangzhong Qian & Xu, 2024, European Journal of Taxonomy, 936: 44, fig. 29 (w.) CHINA (Yunnan).

Holotype worker: TL 3.7, HL 0.79, HW 0.64, CI 82, SL 0.72, SI 113, ED 0.16, PW 0.49, WL 1.14, PL 0.42, PH 0.25, DPW 0.21. In full-face view, head roughly rectangular, longer than broad, posterior and lateral margins weakly convex, posterior corners broadly rounded. Mandibles triangular, masticatory margin with 5 teeth. Clypeal dorsum weakly convex, with median carina, anterior margin moderately convex. Frontal lobes narrow, concealing half of antennal sockets. Frontal carinae very short, about as long as frontal lobes. Antennae 12-segmented, scapes just reaching to posterior head margin, antennal clubs 3-segmented. Eyes located at midpoint of lateral head margin, occupying 1/4 of lateral margin. In lateral view, pronotum weakly convex, promesonotal suture absent. Mesonotum nearly straight and gently sloping down posteriorly, metanotal groove widely shallowly impressed. Propodeal dorsum almost straight, posterodorsal corner rightly angled; declivity weakly concave, propodeal lobes broad ant truncated apically. Petiole with short anterior peduncle, about 1/2 length of petiolar node; petiolar node elongate and roundly convex dorsally, anterior margin almost straight, posterior margin weakly convex; ventral margin almost straight anteriorly, weakly concave posteriorly, anteroventral corner lowly minutely toothed. Postpetiole as high as petiolar node, anterodorsal margin moderately convex, posterior margin almost straight, ventral margin weakly convex. In dorsal view, pronotum broadest, anterior margin moderately convex, humeral corners broadly rounded, lateral margins strongly convex, promesonotal suture absent. Mesothorax moderately constricted, lateral margins moderately concave, metanotal groove shallowly impressed. Propodeum roughly rectangular, lateral margins slightly convex; propodeal spines acutely toothed. Petiole longer than broad, widening posteriorly, spiracles prominent, lateral margins of the node moderately convex, posterior margins strongly convex. Postpetiole about 1.3 times as broad as petiole, narrowing posteriorly, anterolateral corners broadly rounded, lateral margins weakly convex. Gaster elongate oval. Mandibles longitudinally striate. Head dorsum densely longitudinally rugose, densely finely reticulate posteriorly and laterally with interface densely punctate. Clypeus longitudinally rugose. Mesosoma densely finely reticulate, lower potion of metapleura longitudinally rugose. Petiole and postpetiole densely finely punctate, postpetiolar dorsum smooth. Gaster smooth and shiny. Body dorsum with abundant erect to suberect short hairs and abundant decumbent pubescence, hairs and pubescence denser on head. Scapes and tibiae with abundant subdecumbent hairs and dense decumbent pubescence. Body color reddish brown; head dorsum and gaster black; mandibles, scapes, trochanters and tarsi brownish yellow.

Holotype: Worker, China: Yunnan Province, Tengchong County, Shangying Town, Dahaoping (24.979244°N, 98.694517°E), 2000 m, 1999.Ⅳ.28, collected by Lei Fu from a canopy sample of semi-moist evergreen broadleaf forest, No. A99-100.

Etymology: The specific epithet refers to "mangzhong", one of the Twenty-four Solar Terms of China.

正模工蚁： 正面观，头部近长方形，长大于宽；后缘和侧缘轻度隆起，后角宽圆。上颚三角形，咀嚼缘具5个齿。唇基背面轻度隆起，具中央纵脊，前缘中度隆起。额叶窄，遮盖触角窝的一半。额脊很短，约与额叶等长。触角12节，柄节刚到达头后缘，触角棒3节。复眼位于头侧缘中点处，占据侧缘的1/4。侧面观，前胸背板轻度隆起，缺前中胸背板缝。中胸背板近平直，向后缓坡形降低。后胸沟宽形浅凹。并胸腹节背面近平直，后上角直角形；斜面浅凹。并胸腹节侧叶宽，末端平截。腹柄前面小柄短，约为腹柄结长的1/2；腹柄结伸长，背面圆形隆起，前面近平直，后面轻度隆起；腹面前部近平直，后部浅凹，前下角小齿状。后腹柄与腹柄结等高，前上缘中度隆起，后缘近平直，腹面轻度隆起。背面观，前胸背板最宽，前缘中度隆起，肩角宽圆，侧缘强烈隆起。缺前中胸背板缝。中胸中度收缩，侧缘中度凹陷。后胸沟浅凹。并胸腹节近长方形，侧缘轻微隆起；并胸腹节刺尖齿状。腹柄长大于宽，向后变宽，气门突起，腹柄结侧缘中度隆起，后缘强烈隆起。后腹柄约为腹柄宽的1.3倍，向后变窄，前侧角宽圆，侧面轻度隆起。后腹部长卵形。上颚具纵条纹。头部背面具密集纵皱纹，后部和侧面呈细密网纹，界面具密集刻点。唇基具纵皱纹。胸部具细密网纹，后胸侧板下部具纵皱纹。腹柄和后腹柄具细密刻点，后腹柄背面光滑。后腹部光滑发亮。身体背面具丰富直立、亚直立短毛和丰富倾斜绒毛被，头部的立毛和绒毛被较密集；柄节和胫节具丰富亚倾斜毛和密集倾斜绒毛被。身体红棕色，头部背面和后腹部黑色；上颚、柄节、转节和跗节棕黄色。

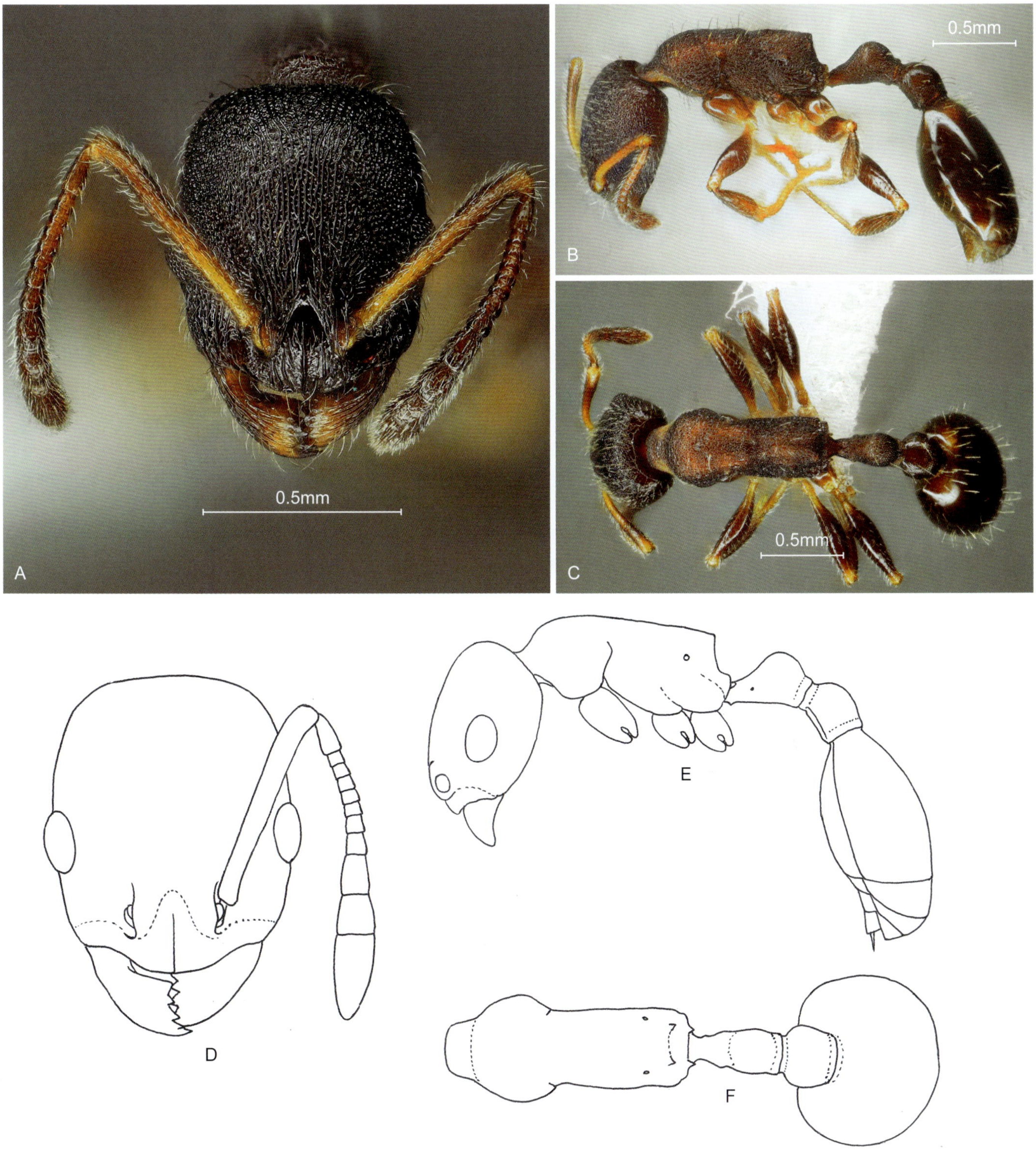

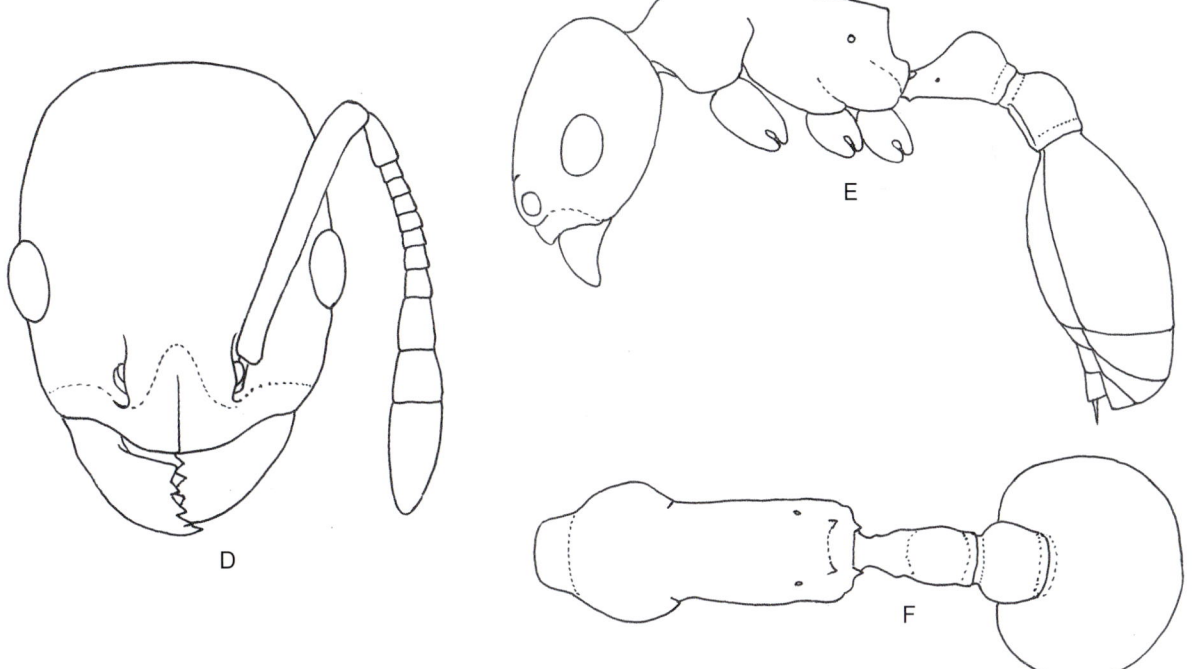

芒种切胸蚁 *Temnothorax mangzhong*

A, D. 工蚁头部正面观；B, E. 工蚁身体侧面观；C, F. 工蚁身体背面观（A-C. 引自Qian和Xu, 2024）
A, D. Head of worker in full-face view; B, E. Body of worker in lateral view; C, F. Body of worker in dorsal view (A-C. cited from Qian & Xu, 2024)

正模：工蚁，中国云南省腾冲县上营乡大蒿坪（24.979244°N，98.694517°E），2000m，1999.Ⅳ.28，付磊采于半湿润常绿阔叶林树冠样中，No. A99-100。

词源：该新种以中国二十四节气中的"芒种"命名。

清明切胸蚁
Temnothorax qingming Qian & Xu, 2024

Temnothorax qingming Qian & Xu, 2024, European Journal of Taxonomy, 936: 46, fig. 30 (w.) CHINA (Tibet).

Holotype worker: TL 2.1, HL 0.52, HW 0.40, CI 76, SL 0.44, SI 110, ED 0.10, PW 0.30, WL 0.60, PL 0.22, PH 0.16, DPW 0.12. In full-face view, head roughly rectangular, longer than broad, posterior and lateral margins weakly convex, posterior corners broadly rounded. Mandibles triangular, masticatory margin with 5 teeth. Clypeal dorsum weakly convex, with median carina, anterior margin moderately convex. Frontal lobes relatively broad, concealing antennal sockets. Frontal carinae short, reaching to the level of anterior eye margins. Antennae 12-segmented, scapes almost reaching to posterior head margin. Eyes located at midpoint of lateral head margin, occupying 1/4 of lateral margin. In lateral view, pronotum slightly convex, anterodorsal corner narrowly rounded. Promesonotal suture and metanotal groove absent. Mesonotum weakly convex and gently sloping down posteriorly. Propodeal dorsum straight and sloping down posteriorly; propodeal spines short and sharp, about as long as their basal width, about 1/2 length of declivity; declivity weakly concave; propodeal lobes broad, truncated apically. Petiole with short anterior peduncle, about 1/2 length of petiolar node; petiolar node roughly triangular, anterior and posterior margins slightly convex, dorsum narrowly rounded; ventral margin almost straight, weakly concave posteriorly, anteroventral corner acutely toothed. Postpetiole about as high as petiolar node, anterodorsal margin rounded, posterior margin straight, ventral margin deeply notched. In dorsal view, pronotum broadest, anterior and lateral margins moderately convex, humeral corners broadly rounded. Promesonotal suture and metanotal groove absent. Mesothorax moderately constricted, lateral margins moderately concave. Propodeum roughly rectangular, lateral margins almost straight, spines shortly toothed. Petiole roughly trapezoidal and widening posteriorly, longer than broad, lateral margins almost straight. Postpetiole roughly rectangular, about 1.5 times as broad as petiole, anterolateral corners narrowly rounded, lateral margins almost straight. Gaster elongate oval. Mandibles longitudinally striate. Head smooth and shiny, inner side of frontal carinae with 3 longitudinal rugae. Central part of clypeus smooth, each side with 3 longitudinal rugae; lateral part obliquely striate. Mesosoma, petiole and postpetiole uniformly densely punctate, dorsum of petiolar node and postpetiolar dorsum smooth. Gaster smooth and shiny. Body dorsum with sparse erect to suberect apically blunt short hairs and abundant decumbent pubescence. Scapes and tibiae with dense decumbent pubescence. Body color blackish brown; head dorsum and gaster black; tarsi yellowish brown.

Holotype: Worker, China: Tibet Autonomous Region, Linzhi County, Lulang Town, Dongjiu Village (29.950200°N, 94.795417°E), 2560 m, 2007.Ⅸ.19, collected by Long-guan Chen beneath stone in *Pinus yunnanensis* forest, No. A07-56.

Etymology: The specific epithet refers to "qingming", one of the Twenty-four Solar Terms of China.

正模工蚁： 正面观，头部近长方形，长大于宽；后缘和侧缘轻度隆起，后角宽圆。上颚三角形，咀嚼缘具5个齿。唇基背面轻度隆起，具中央纵脊，前缘中度隆起。额叶较宽，遮盖触角窝。额脊短，到达复眼前缘水平。触角12节，柄节几乎到达头后缘。复眼位于头侧缘中点处，占据侧缘的1/4。侧面观，前胸背板轻微隆起，前上角窄圆。缺前中胸背板缝和后胸沟。中胸背板轻度隆起，向后缓坡形降低。并胸腹节背面近平直，向后坡形降低；并胸腹节刺短而尖，长度约与其基部宽相等，约为斜面长度的1/2；斜面浅凹。并胸腹节侧叶宽，末端平截。腹柄前面小柄短，约为腹柄结长的1/2；腹柄结近三角形，前面和后面轻微隆起，背面窄圆；腹面近平直，后部浅凹，前下角尖齿状。后腹柄约与腹柄结等高，前上缘圆，后缘平直，腹面有深切口。背面观，前胸背板最宽，前缘和侧缘中度隆起，肩角宽圆。缺前中胸背板缝和后胸沟。中胸中度收缩，侧缘中度凹陷。并胸腹节近长方形，侧缘近平直；并胸腹节刺短齿状。腹柄近梯形，向后变宽，长大于宽，侧缘近平直。后腹柄近长方形，约为腹柄宽的1.5倍，前侧角窄圆，侧面近平直。后腹部长卵形。上颚具纵条纹。头部光滑发亮，额脊内侧具3条纵皱纹。唇基中部光滑，每侧具3条纵皱纹；两侧具倾斜条纹。胸部、腹柄和后腹柄具均匀一致的密集刻点，腹柄结背面和后腹柄背面光滑。后腹部光滑发亮。身体背面具稀疏直立、亚直立短钝毛和丰富倾斜绒毛被；柄节和胫节具密集倾斜绒毛被。身体黑棕色，头部背面和后腹部黑色，跗节黄棕色。

正模： 工蚁，中国西藏自治区林芝县鲁朗镇东久村（29.950200°N，94.795417°E），2560m，2007.Ⅸ.19，陈龙官采于云南松林石下，No. A07-56。

词源： 该新种以中国二十四节气中的"清明"命名。

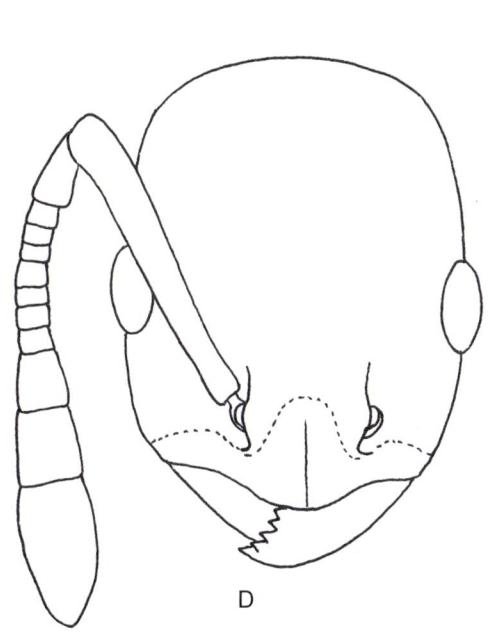
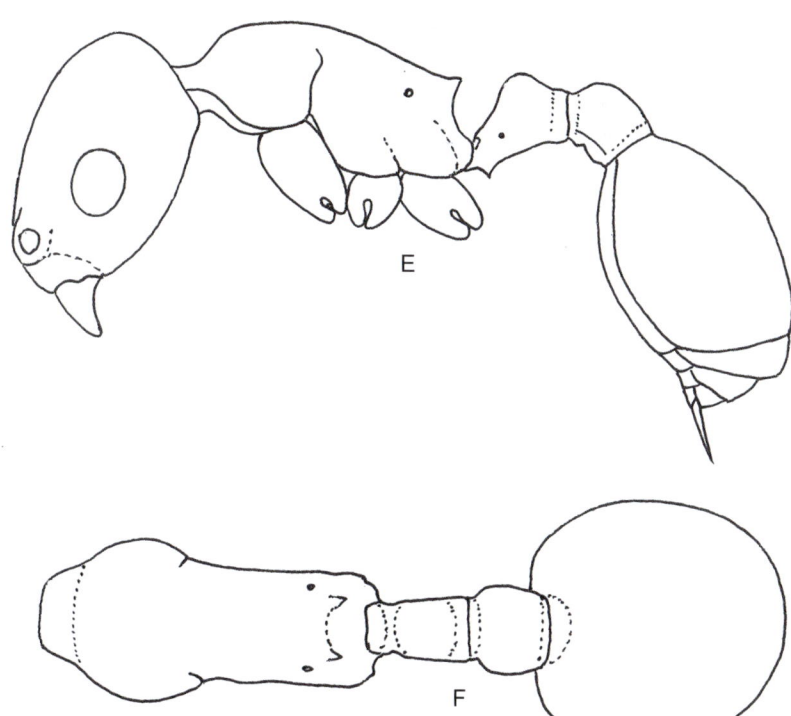

清明切胸蚁 *Temnothorax qingming*

A, D. 工蚁头部正面观；B, E. 工蚁身体侧面观；C, F. 工蚁身体背面观（A-C. 引自 Qian 和 Xu, 2024）
A, D. Head of worker in full-face view; B, E. Body of worker in lateral view; C, F. Body of worker in dorsal view (A-C. cited from Qian & Xu, 2024)

秋切胸蚁
Temnothorax qiu Qian & Xu, 2024

Temnothorax qiu Qian & Xu, 2024, European Journal of Taxonomy, 936: 48, fig. 32 (w.) CHINA (Sichuan).

Holotype worker: TL 2.4, HL 0.56, HW 0.44, CI 78, SL 0.51, SI 116, ED 0.12, PW 0.33, WL 0.63, PL 0.25, PH 0.20, DPW 0.13. In full-face view, head roughly rectangular, longer than broad, posterior and lateral margins weaky convex, posterior corners broadly rounded. Mandibles triangular, masticatory margin with 5 teeth. Clypeal dorsum weakly convex, with median carina, anterior margin moderately convex. Frontal lobes narrow, concealing half of antennal sockets. Frontal carinae short, reaching to the level of anterior eye margins. Antennae 12-segmented, scapes slightly surpassing posterior head margin. Eyes located slightly before midpoint of lateral head margin, occupying 1/4 of lateral margin. In lateral view, pronotum weakly convex, promesonotal suture present but not impressed. Mesonotum and propodeal dorsum slightly convex and gently sloping down posteriorly, metanotal groove absent. Propodeal spines moderate long, slightly shorter than declivity and about 2 times as long as their basal width; declivity weakly concave; propodeal lobes short and rounded apically. Petiole with short anterior peduncle, about 1/2 length of petiolar node; petiolar node roughly triangular, anterior margin slightly concave, posterior margin weakly convex, top corner narrowly rounded; ventral margin straight, weakly concave posteriorly, anteroventral corner bluntly angled. Postpetiole slightly lower than petiolar node, anterior margin almost straight, anterodorsal corner narrowly rounded, dorsal margin weakly convex, ventral margin moderately concave. In dorsal view, pronotum broadest, anterior margin moderately convex, humeral corners broadly rounded, lateral margins strongly convex, promesonotal suture present. Mesothorax weakly constricted, lateral margins weakly concave. Propodeum roughly rectangular, lateral margins slightly convex; spines moderate long and sharp, pointed posterolaterally. Petiole roughly trapezoidal and widening posteriorly, longer than broad, lateral margins almost straight. Postpetiole roughly spherical, about 1.6 times as broad as petiole, anterolateral corners broadly rounded, lateral margins moderately convex. Gaster elongate oval. Mandibles longitudinally striate. Head dorsum longitudinally rugose, with interface densely punctate posteriorly and laterally; lateral head margins below eyes relatively smooth. Clypeus smooth, each side with 4-5 short rugae. Mesosoma densely finely reticulate; propodeum, petiole and postpetiole densely punctate. Gaster smooth and shiny. Body dorsum with abundant erect to suberect apically blunt short hairs and abundant decumbent pubescence. Scapes and tibiae with dense decumbent pubescence. Body color black; antennae and legs blackish brown; mandibles and tarsi brownish yellow. **Paratype workers:** TL 2.3-2.5, HL 0.55-0.69, HW 0.48-0.54, CI 82-87, SL 0.48-0.55, SI 100-104, ED 0.11-0.13, PW 0.36-0.41, WL 0.63-0.74, PL 0.22-0.24, PH 0.2-0.22, DPW 0.13-0.16 ($n=13$). As holotype, in some individuals body color blackish brown, anterior margin of petiolar node nearly straight.

Holotype: Worker, China: Sichuan Province, Yanyuan County, Boda Town, Hebian Village (27.375331°N, 101.288239°E), 2280 m, 2018.Ⅶ.25, collected by Zhao Huang from a soil sample of *Pinus yunnanensis* forest, No. C18-561.

Etymology: The specific epithet refers to "qiu", one of the four seasons.

正模工蚁： 正面观，头部近长方形，长大于宽；后缘和侧缘轻度隆起，后角宽圆。上颚三角形，咀嚼缘具5个齿。唇基背面轻度隆起，具中央纵脊，前缘中度隆起。额叶窄，遮盖触角窝的一半。额脊短，到达复眼前缘水平。触角12节，柄节稍超过头后缘。复眼位于头侧缘中点稍前处，占据侧缘的1/4。侧面观，前胸背板轻度隆起。前中胸背板缝存在但不凹陷。中胸背板和并胸腹节背面轻微隆起，向后缓坡形降低，缺后胸沟；并胸腹节刺中等长，稍短于斜面，长约为其基部宽的2倍；斜面浅凹。并胸腹节侧叶短，顶端圆。腹柄前面小柄短，约为腹柄结长的1/2；腹柄结近三角形，前面轻微凹陷，后面轻度隆起，顶角窄圆；腹面平直，后部浅凹，前下角钝角状。后腹柄稍低于腹柄结，前面近平直，前上角窄圆，背面轻度隆起，腹面中度凹陷。背面观，前胸背板最宽，前缘中度隆起，肩角宽圆，侧缘强烈隆起。前中胸背板缝存在。中胸轻度收缩，侧缘浅凹。并胸腹节近长方形，侧缘轻微隆起；并胸腹节刺中等长，尖锐，指向后侧方。腹柄近梯形，向后变宽，长大于宽，侧缘近平直。后腹柄近球形，约为腹柄宽的1.6倍，前侧角宽圆，侧缘中度隆起。后腹部长卵形。上颚具纵条纹。头部背面具纵皱纹，后部和侧面皱纹间具密集刻点，头部侧面复眼下方较光滑。唇基光滑，每侧具4～5条短皱纹。胸部具细密网纹；并胸腹节、腹柄和后腹柄具密集刻点。后腹部光滑发亮。身体背面具丰富直立、亚直立短钝毛和丰富倾斜绒毛被；柄节和胫节具密集倾斜绒毛被。身体黑色，触角和足黑棕色，上颚和跗节棕黄色。**副模工蚁：** 特征同正模，部分个体身体黑棕色，腹柄结前缘近平直。

正模： 工蚁，中国四川省盐源县博大乡河边村（27.375331°N，101.288239°E），2280m，2018.Ⅶ.25，

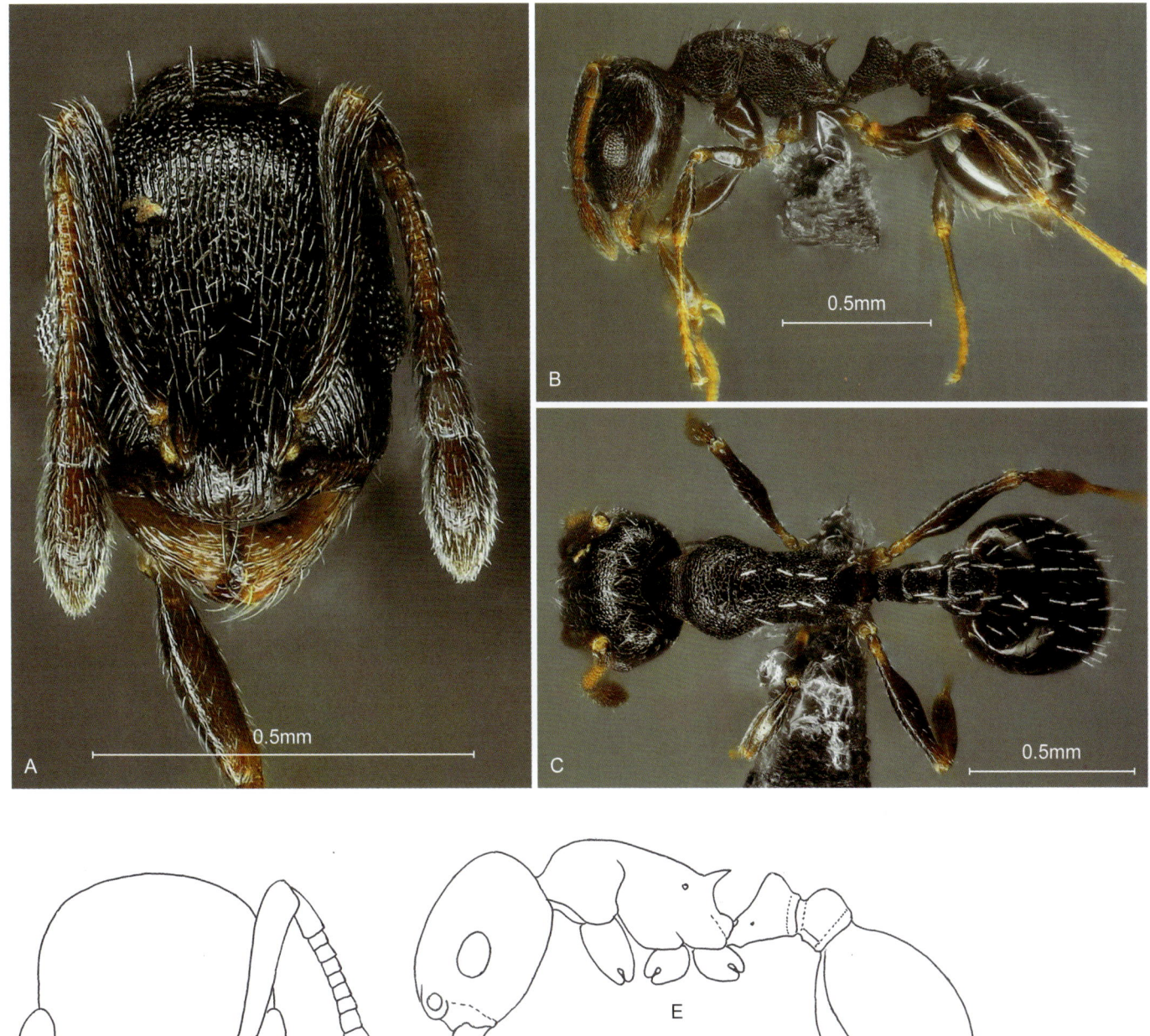

秋切胸蚁 *Temnothorax qiu*

A, D. 工蚁头部正面观；B, E. 工蚁身体侧面观；C, F. 工蚁身体背面观（A-C. 引自Qian和Xu, 2024）
A, D. Head of worker in full-face view; B, E. Body of worker in lateral view; C, F. Body of worker in dorsal view (A-C. cited from Qian & Xu, 2024)

黄钊采于云南松林土壤样中，No. C18-561。

词源： 该新种以四季中的"秋"命名。

108 秋分切胸蚁
Temnothorax qiufen Qian & Xu, 2024

Temnothorax qiufen Qian & Xu, 2024, European Journal of Taxonomy, 936: 50, fig. 33 (w.) CHINA (Sichuan).

Holotype worker: TL 2.7, HL 0.63, HW 0.51, CI 81, SL 0.60, SI 118, ED 0.13, PW 0.36, WL 0.80, PL 0.29, PH 0.22, DPW 0.16. In full-face view, head roughly rectangular, longer than broad, posterior margin almost straight, posterior corners narrowly rounded, lateral margins weakly convex. Mandibles triangular, masticatory margin with 5 teeth. Clypeal dorsum weakly convex, with median carina, anterior margin moderately convex. Frontal lobes broad, concealing antennal sockets. Frontal carinae short, reaching to the level of anterior eye margins. Antennae 12-segmented, scapes just reaching to posterior head margin. Eyes located at midpoint of lateral head margin, occupying 1/4 of lateral margin. In lateral view, pronotum slightly convex, anterodorsal corner narrowly rounded, promesonotal suture absent. Mesonotum weakly convex and gently sloping down posteriorly, metanotal groove very shallowly impressed. Propodeal dorsum straight and gently sloping down posteriorly; propodeal spines short and acute, roughly triangular, about as long as their basal width and about 1/2 length of declivity; declivity weakly concave; propodeal lobes short and rounded apically. Petiole with moderate long anterior peduncle, about 1/2 length of petiolar node; petiolar node roughly trapezoidal, anterior margin almost straight, posterior margin weakly convex, top margin slight convex; ventral margin straight, anteroventral corner largely acutely toothed. Postpetiole about as high as petiolar node, anterodorsal margin rounded, posterior margin straight, ventral margin moderately concave. In dorsal view, pronotum broadest, humeral corners broadly rounded, lateral margin moderately convex. Mesothorax weakly constricted, lateral margins weakly concave. Metanotal groove very shallowly impressed. Propodeum roughly trapezoidal, lateral margins almost straight; spines short and triangular, pointed posterolaterally. Petiole roughly trapezoidal and widening posteriorly, longer than broad, lateral margins weakly convex. Postpetiole roughly trapezoidal and narrowing posteriorly, about 1.6 times as broad as petiole, anterolateral corners narrowly rounded, lateral margins weakly convex. Gaster elongate oval. Mandibles longitudinally striate. Head dorsum loosely longitudinally rugose with the longitudinal central strip relatively smooth, and reticulate laterally; genae longitudinally rugose. Clypeus with longitudinal rugosity. Mesosomal dorsum coarsely reticulate, pronotum coarsely rugose, mesonotal dorsum loosely rugose; mesopleura and metapleura reticulate; propodeum, petiole and postpetiole densely punctate. Gaster smooth and shiny. Body dorsum with sparse erect to suberect apically blunt short hairs and sparse decumbent pubescence, hairs on head and gaster relatively abundant. Scapes and tibiae with dense decumbent pubescence. Body color blackish brown, mandibles, antennae and legs brown to brownish yellow. **Paratype worker:** TL 2.6, HL 0.63, HW 0.50, CI 80, SL 0.50, SI 100, ED 0.13, PW 0.35, WL 0.65, PL 0.25, PH 0.2, DPW 0.13 (*n*=1). As holotype.

Holotype: Worker, China: Sichuan Province, Ganzi County, Gonglong Town, Dingdalong Village (31.696508°N, 99.781661°E), 3480 m, 2018.Ⅷ.20, collected by Zhong-ping Xiong on the ground in *Betula albo-sinensis* forest, No. A18-972.

Etymology: The specific epithet refers to "qiufen", one of the Twenty-four Solar Terms of China.

正模工蚁： 正面观，头部近长方形，长大于宽；后缘近平直，后角窄圆，侧缘轻度隆起。上颚三角形，咀嚼缘具5个齿。唇基背面轻度隆起，具中央纵脊，前缘中度隆起。额叶宽，遮盖触角窝。额脊短，到达复眼前缘水平。触角12节，柄节刚到达头后缘。复眼位于头侧缘中点处，占据侧缘的1/4。侧面观，前胸背板轻微隆起，前上角窄圆，缺前中胸背板缝。中胸背板轻微隆起，向后缓坡形降低。后胸沟很浅。并胸腹节背面平直，向后缓坡形降低；并胸腹节刺短而尖，近三角形，长度约与其基部宽相等，约为斜面长的1/2；斜面浅凹。并胸腹节侧叶短，顶端圆。腹柄前面小柄中等长，约为腹柄结长的1/2；腹柄结近梯形，前面近平直，后面轻度隆起，背面轻微隆起；腹面平直，前下角大，尖齿状。后腹柄约与腹柄结等高，前上缘圆，后缘平直，腹面中度凹陷。背面观，前胸背板最宽，肩角宽圆，侧缘中度隆起。中胸轻度收缩，侧缘浅凹。后胸沟浅凹。并胸腹节近梯形，侧缘近平直；并胸腹节刺短，三角形，指向后侧方。腹柄近梯形，向后变宽，长大于宽，侧缘轻度隆起。后腹柄近梯形，向后变窄，约为腹柄宽的1.6倍，前侧角窄圆，侧面轻度隆起。后腹部长卵形。上颚具纵条纹。头部背面具疏松的纵皱纹，中央纵带较光滑，侧面具网纹；颊区具纵皱纹。唇基具纵皱纹。胸部背面具粗网纹，前胸背板具粗皱纹，中胸背板背面具稀疏皱纹；中胸侧板和后胸侧板具网纹，并胸腹节、腹柄和后腹柄具密集刻点。后腹部光滑发亮。身体背面具稀疏直立、亚直立短钝毛和稀疏倾斜绒毛被，头部和后腹部立毛较丰富；柄节和胫节具密集倾斜绒毛被。身体黑棕色，上颚、触角和足棕色至棕黄色。**副模工蚁：** 特征同正模。

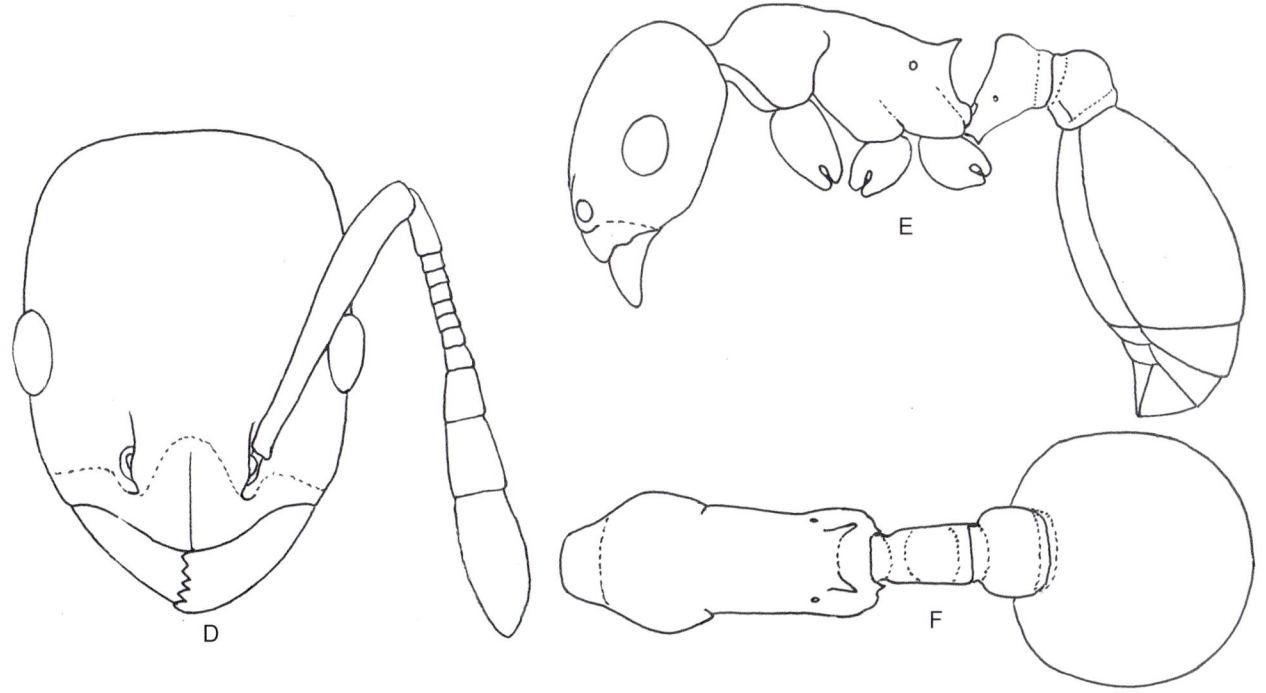

秋分切胸蚁 *Temnothorax qiufen*

A, D. 工蚁头部正面观；B, E. 工蚁身体侧面观；C, F. 工蚁身体背面观（A-C. 引自Qian和Xu, 2024）
A, D. Head of worker in full-face view; B, E. Body of worker in lateral view; C, F. Body of worker in dorsal view (A-C. cited from Qian & Xu, 2024)

正模：工蚁，中国四川省甘孜县贡隆乡丁达隆村（31.696508°N，99.781661°E），3480m，2018.Ⅷ.20，熊忠平采于白桦林地表，No. A18-972。

词源：该新种以中国二十四节气中的"秋分"命名。

霜降切胸蚁
Temnothorax shuangjiang Qian & Xu, 2024

Temnothorax shuangjiang Qian & Xu, 2024, European Journal of Taxonomy, 936: 52, fig. 34 (w.) CHINA (Yunnan).

Holotype worker: TL 3.5, HL 0.80, HW 0.72, CI 91, SL 0.59, SI 82, ED 0.18, PW 0.49, WL 1.05, PL 0.38, PH 0.28, DPW 0.20. In full-face view, head roughly rectangular, longer than broad, posterior margin almost straight, posterior corners narrowly rounded, lateral margins weakly convex. Mandibles triangular, masticatory margin with 5 teeth. Clypeal dorsum weakly convex, with waved median carina, anterior margin moderately convex. Frontal lobes broad, concealing antennal sockets. Frontal carinae short, reaching to the level of anterior eye margins. Antennae 12-segmented, scapes failing to reach posterior head margin by 1/7 of its length, antennal clubs 3-segmented. Eyes located slightly before midpoint of lateral head margin, occupying 1/4 of lateral margin. In lateral view, promesonotum weakly convex and weakly arched, anterodorsal corner of pronotum bluntly angled, promesonotal suture and metanotal groove absent. Propodeal dorsum almost straight and gently sloping down posteriorly; propodeal spines long and slender, as long as declivity, pointed posterodorsally, its dorsal margin weakly convex, ventral margin straight; declivity straight; propodeal lobes relatively large and triangular. Petiole with anterior peduncle, about 2/3 length of petiolar node; petiolar node roughly trapezoidal and narrowing dorsally, anterior margin almost straight, dorsal and posterior margins weakly convex, anterodorsal corner narrowly rounded, posterodorsal corner broadly rounded; ventral margin straight anteriorly, weakly concave posteriorly, anteroventral corner bluntly angled. Postpetiole about as high as petiolar node, anterodorsal margin rounded, posterior margin weakly concave, ventral margin angularly concave. In dorsal view, pronotum broadest, anterior and lateral margin moderately convex and rounded, humeral corners rounded. Promesonotal suture and metanotal groove absent. Mesothorax weakly constricted, lateral margins weakly concave. Propodeum roughly rectangular, lateral margins slightly concave, spines moderately long and straight, pointed posterolaterally. Petiole roughly trapezoidal and widening posteriorly, longer than broad, lateral margins weakly convex. Postpetiole roughly trapezoidal and narrowing posteriorly, about 1.4 times as broad as petiole, anterolateral corners broadly rounded, lateral margins weakly convex. Gaster elongate oval. Mandibles longitudinally striate. Head dorsum longitudinally rugose and reticulate posteriorly and laterally, sides below eyes densely finely reticulate. Clypeus loosely irregularly rugose. Mososoma coarsely reticulate. Petiole and postpetiole densely finely reticulate, anterior face of postpetiole smooth. Gaster smooth and shiny. Body dorsum with sparse erect to suberect apically blunt short hairs and sparse decumbent pubescence. Scapes and tibiae with dense decumbent pubescence. Body color black, mandibles, antennae and legs blackish brown.

Holotype: Worker, China: Yunnan Province, Huaping County, Xinzhuang Town, Tianxing Forest Farm (26.636450°N, 101.149017°E), 2030 m, 2013.Ⅲ.15, collected by Shun-rong Pu from a soil sample of *Pinus yunnanensis* forest, No. A13-198.

Etymology: The specific epithet refers to "shuangjiang", one of the Twenty-four Solar Terms of China.

正模工蚁：正面观，头部近长方形，长大于宽；后缘近平直，后角窄圆，侧缘轻度隆起。上颚三角形，咀嚼缘具5个齿。唇基背面轻度隆起，具波形中央纵脊，前缘中度隆起。额叶宽，遮盖触角窝。额脊短，到达复眼前缘水平。触角12节，柄节未到达头后缘，末端至头后缘距离为柄节长的1/7，触角棒3节。复眼位于头侧缘中点稍前处，占据侧缘的1/4。侧面观，前中胸背板轻度隆起呈弱弓形，前胸背板前上角钝角状，缺前中胸背板缝和后胸沟。并胸腹节背面近平直，向后缓坡形降低；并胸腹节刺细长，与斜面等长，指向后上方，背面轻度隆起，腹面平直；斜面平直。并胸腹节侧叶较大，三角形。腹柄前面具小柄，约为腹柄结长的2/3；腹柄结近梯形，向上变窄，前面近平直，背面和后面轻度隆起，前上角窄圆，后上角宽圆；腹面前部平直，后部浅凹，前下角钝角状。后腹柄约与腹柄结等高，前上缘圆，后缘浅凹，腹面角状凹陷。背面观，前胸背板最宽，前缘和侧缘中度隆起，肩角圆。缺前中胸背板缝和后胸沟。中胸轻度收缩，侧缘浅凹。并胸腹节近长方形，侧缘轻微凹陷；并胸腹节刺中等长，直，指向后侧方。腹柄近梯形，向后变宽，长大于宽，侧缘轻度隆起。后腹柄近梯形，向后变窄，约为腹柄宽的1.4倍，前侧角宽圆，侧缘轻度隆起。后腹部长卵形。上颚具纵条纹。头部背面具纵皱纹，后部和侧面具网纹，复眼下方具细密网纹。唇基具疏松不规则皱纹。胸部具粗糙网纹；腹柄和后腹柄具细密网纹，后腹柄前面光滑。后腹部光滑发亮。身体背面具稀疏直立、亚直立短钝毛和稀疏倾斜绒毛被；柄节和胫节具密集倾斜绒毛被。身体黑色，上颚、触角和足黑棕色。

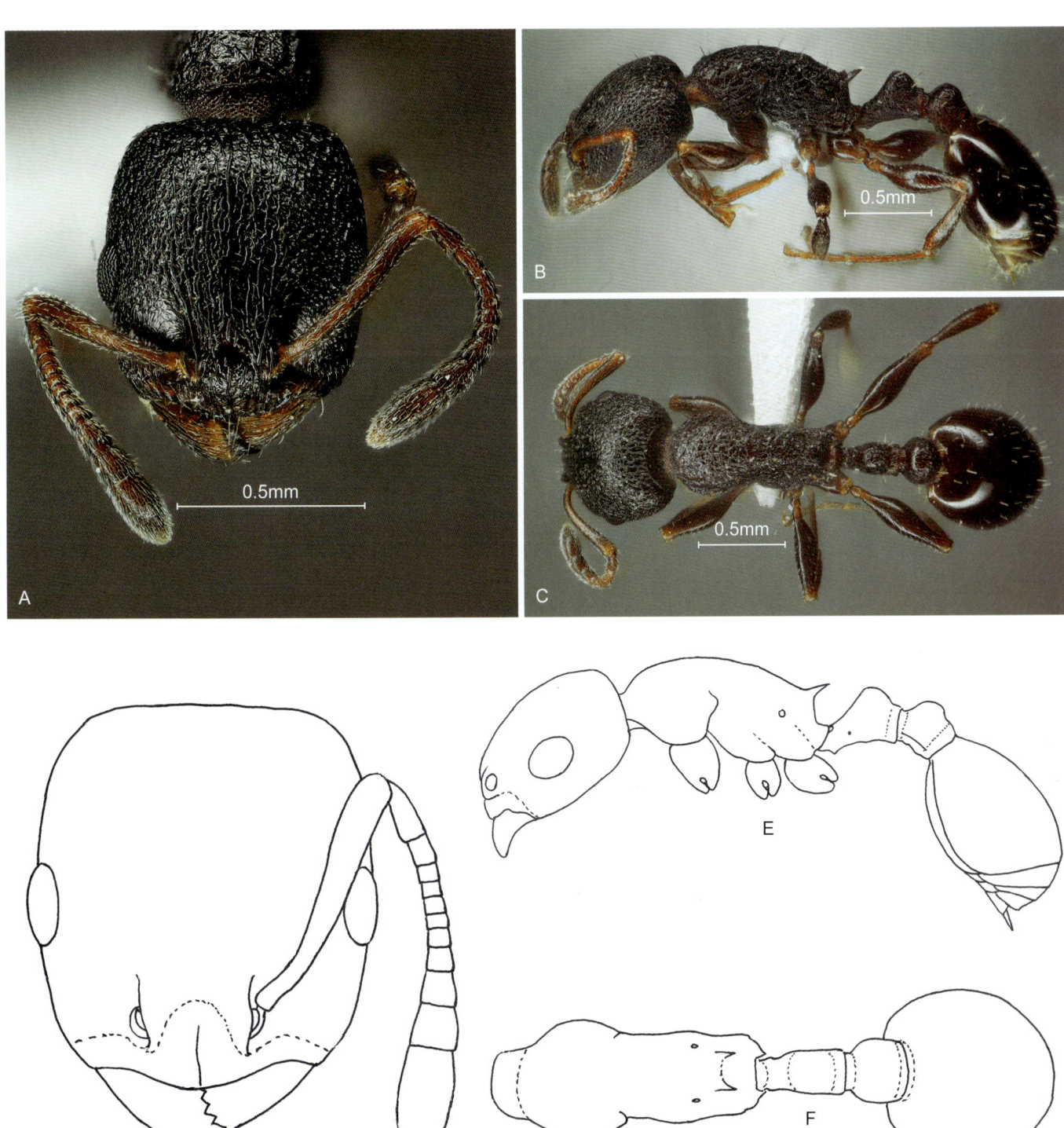

霜降切胸蚁 *Temnothorax shuangjiang*

A, D. 工蚁头部正面观；B, E. 工蚁身体侧面观；C, F. 工蚁身体背面观（A-C. 引自Qian和Xu, 2024）
A, D. Head of worker in full-face view; B, E. Body of worker in lateral view; C, F. Body of worker in dorsal view (A-C. cited from Qian & Xu, 2024)

正模：工蚁，中国云南省华坪县新庄乡天星林场（26.636450°N，101.149017°E），2030m，2013.Ⅲ.15，普顺荣采于云南松林土壤样中，No. A13-198。

词源：该新种以中国二十四节气中的"霜降"命名。

110 夏切胸蚁
Temnothorax xia Qian & Xu, 2024

Temnothorax xia Qian & Xu, 2024, European Journal of Taxonomy, 936: 54, fig. 35 (w.) CHINA (Yunnan).

Holotype worker: TL 2.9, HL 0.66, HW 0.55, CI 83, SL 0.58, SI 106, ED 0.12, PW 0.41, WL 0.79, PL 0.37, PH 0.19, DPW 0.17. In full-face view, head roughly rectangular, longer than broad, posterior margin straight, posterior corners narrowly rounded, lateral margins weakly convex. Mandibles triangular, masticatory margin with 5 teeth. Clypeal dorsum weakly convex, with median carina, anterior margin moderately convex. Frontal lobes narrow, concealing half of antennal socket. Frontal carinae short, reaching to the level of anterior eye margin. Antennae 12-segmented, scapes reaching to posterior head margin, club 3-segmented. Eyes located at midpoint of lateral head margin, occupying 1/5 of lateral margin. In lateral view, promesonotum relatively higher and weakly convex. Dorsa of posterior part of mesonotum and propodeum almost straight and gently sloping down posteriorly. Promesonotal suture and metanotal groove absent. Propodeal spines short and weakly down-curved, about 1.5 times as long as their basal width; declivity almost straight; propodeal lobes triangular, with apex bluntly angled. Petiole with moderately long anterior peduncle, about 2/3 length of petiolar node; petiolar node roughly semicircular, dorsal margin rounded, anterior and posterior margins sloping down; ventral margin straight for most of its length, concave posteriorly, anteroventral corner acutely toothed. Postpetiole as high as petiole, dorsum roundly convex, ventral margin short and almost straight. In dorsal view, pronotum broadest, anterior margin moderately convex, humeral corners broadly rounded, lateral margins strongly convex. Promesonotal suture and metanotal groove absent. Mesothorax moderately constricted and narrowing posteriorly, lateral margins moderately concave. Propodeum roughly rectangular, lateral margins almost straight; spines relatively short, pointed posterolaterally and weakly in-curved. Petiole roughly trapezoidal and widening posteriorly, longer than broad, lateral margins weakly convex. Postpetiole roughly trapezoidal and narrowing posteriorly, about 1.4 times as broad as petiole, anterolateral corners strongly convex. Gaster elongate oval. Mandibles longitudinally striate. Head dorsum longitudinally rugose, gradually reticulate laterally, with interface densely punctate. Clypeus with several short rugae on each side. Mesosomal dorsum coarsely reticulate; sides of pronotum obliquely costulate; mesopleura and metapleura coarsely reticulate; propodeal sides densely punctate. Petiole and postpetiole finely rugulose, dorsum of petiolar peduncle densely punctate. Gaster smooth and shiny. Body dorsum with sparse erect to suberect apically blunt short hairs and sparse depressed pubescence. Scapes and tibiae with dense decumbent pubescence. Body color blackish brown; gastral dorsum black, mandibles, antennae and legs yellowish brown. **Paratype workers:** TL 2.4-2.8, HL 0.63-0.70, HW 0.50-0.55, CI 79-80, SL 0.50-0.51, SI 91-100, ED 0.10-0.15, PW 0.33-0.35, WL 0.65-0.70, PL 0.25-0.26, PH 0.15-0.20, DPW 0.10-0.15 (n=5). As holotype, in some individuals promesonotum weakly rugose, sides of pronotum obliquely indistinctly costulate, mesopleura and metapleura coarsely weakly reticulate.

Holotype: Worker, China: Yunnan Province, Anning County, Yiliujie Town, Monande Village (24.543797°N, 102.336396°E), 2000 m, 1991.Ⅶ.17, collected by Zheng-hui Xu on the ground in conifer-broadleaf mixed forest, No. A91-353.

Etymology: The specific epithet refers to "xia", one of the four seasons.

正模工蚁： 正面观，头部近长方形，长大于宽；后缘平直，后角窄圆；侧缘轻度隆起。上颚三角形，咀嚼缘具5个齿。唇基背面轻度隆起，具中央纵脊，前缘轻度隆起。额叶窄，遮盖触角窝的一半。额脊短，到达复眼前缘水平。触角12节，柄节到达头后缘，触角棒3节。复眼位于头侧缘中点处，占据侧缘的1/5。侧面观，前中胸背板较高，轻度隆起；中胸背板后部和并胸腹节背面几乎平直，向后缓坡形降低。缺前中胸背板缝和后胸沟。并胸腹节刺短，轻度下弯，长度约为其基部宽的1.5倍；斜面几乎平直。并胸腹节侧叶三角形，末端钝角状。腹柄前面小柄中等长，约为腹柄结长的2/3；腹柄结近半圆形，背面圆，前面和后面坡形；腹面大部平直，后部凹陷，前下角尖齿状。后腹柄与腹柄等高，背面圆形隆起，腹面短，几乎平直。背面观，前胸背板最宽，前缘中度隆起，肩角宽圆，侧缘强烈隆起。缺前中胸背板缝和后胸沟。中胸中度收缩，向后变窄，侧缘中度凹陷。并胸腹节近长方形，侧缘几乎平直；并胸腹节刺较短，指向后侧方，轻度内弯。腹柄近梯形，向后变宽，长大于宽，侧缘轻度隆起。后腹柄近梯形，向后变窄，约为腹柄宽的1.4倍，前侧角强烈隆起。后腹部长卵形。上颚具纵条纹。头部背面具纵皱纹，侧面逐渐呈网纹，界面具密集刻点。唇基每侧具几条短皱纹。胸部背面具粗网纹；前胸背板两侧具倾斜脊纹，中胸侧板和后胸侧板具粗网纹，并胸腹节侧面具密集刻点。腹柄和后腹柄具细皱纹，腹柄的小柄背面具密集刻点。后腹部光滑发亮。身体背面具稀疏直立、亚直立短钝毛和稀疏平伏绒毛被；柄节和胫节具密集倾斜绒毛被。身体黑棕色，后腹

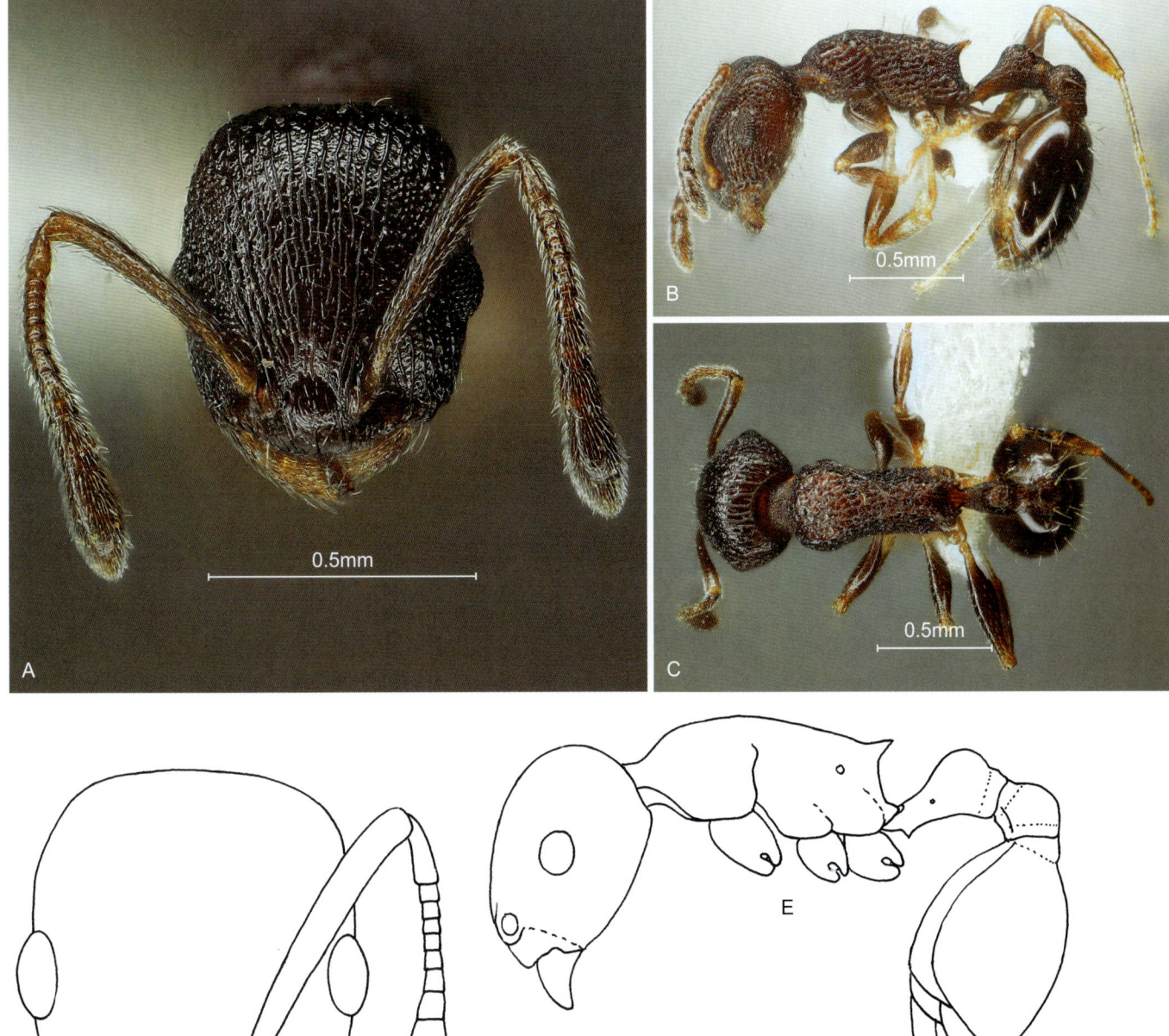

夏切胸蚁 *Temnothorax xia*

A, D. 工蚁头部正面观；B, E. 工蚁身体侧面观；C, F. 工蚁身体背面观（A-C. 引自Qian和Xu, 2024）
A, D. Head of worker in full-face view; B, E. Body of worker in lateral view; C, F. Body of worker in dorsal view (A-C. cited from Qian & Xu, 2024)

部背面黑色，上颚、触角和足黄棕色。**副模工蚁**：特征同正模，部分个体前中胸背板具弱皱纹，前胸背板侧面具不明显的倾斜脊纹，中胸侧板和后胸侧板具弱的粗网纹。

正模：工蚁，中国云南省安宁县一六街乡磨南德村（24.543797°N，102.336396°E），2000m，1991.Ⅶ.17，徐正会采于针阔混交林地表，No. A91-353。

词源：该新种以四季中的"夏"命名。

小寒切胸蚁
Temnothorax xiaohan Qian & Xu, 2024

111

Temnothorax xiaohan Qian & Xu, 2024, European Journal of Taxonomy, 936: 57, fig. 36 (w.) CHINA (Tibet).

Holotype worker: TL 3.0, HL 0.68, HW 0.58, CI 85, SL 0.66, SI 114, ED 0.15, PW 0.38, WL 0.86, PL 0.33, PH 0.20, DPW 0.15. In full-face view, head roughly rectangular, longer than broad, posterior and lateral margins weakly convex, posterior corners broadly rounded. Mandibles triangular, masticatory margin with 5 teeth. Clypeal dorsum weakly convex, without median carina, anterior margin moderately convex. Frontal lobes broad, concealing most of antennal sockets. Antennae 12-segmented, scapes surpassing posterior head margin, club 3-segmented. Eyes located at midpoint of lateral head margin, occupying 1/5 of lateral margin. In lateral view, pronotum weakly convex, promesonotal suture absent. Mesonotum slightly convex and gently sloping down posteriorly, metanotal groove absent. Propodeal dorsum straight; propodeal spines very short, about as long as their basal width, and about 1/2 length of declivity; declivity weakly concave; propodeal lobes very short and rounded apically. Petiole with long anterior peduncle, about 2/3 length of petiolar node; petiolar node roughly triangular, anterior margin almost straight, posterior margin weakly convex, top corner narrowly rounded; ventral margin straight for most of its length, weakly concave posteriorly, anteroventral corner very bluntly angled. Postpetiole slightly lower than petiolar node, anterodorsal margin rounded, posterior margin straight, ventral margin moderately concave. In dorsal view, pronotum broadest, lateral margins strongly covex, humeral corners broadly rounded. Promesonotal suture absent. Mesothorax moderately constricted, lateral margins deeply concave. Metanotal groove absent. Propodeum roughly rectangular, lateral margins slightly convex; spines short, weakly in-curved. Petiole roughly rectangulay, longer than broad, lateral margins of the node weakly convex. Postpetiole roughly trapezoidal and narrowing posteriorly, about 1.4 times as broad as petiole, anterolateral corners bluntly angled, lateral margins almost straight. Gaster elongate oval. Mandibles longitudinally striate. Head dorsum loosely longitudinally rugose and gradually reticulate posteriorly and laterally, interface coarsely reticulate. Clypeus very loosely longitudinally rugose, interface smooth. Mesosoma reticulate; pronotal sides and lower portion of metapleura longitudinally rugose. Petiole and postpetiole finely reticulate. Gaster smooth and shiny. Body dorsum with abundant suberect to subdecumbent short hairs and abundant decumbent pubescence, hairs on head dorsum relatively denser and erect. Scapes with abundant subdecumbent hairs and dense decumbent pubescence, tibiae with dense decumbent pubescence. Body color brownish yellow, head and gaster light black. **Paratype workers:** TL 2.6-3.3, HL 0.55-0.65, HW 0.45-0.62, CI 82-97, SL 0.50-0.64, SI 100-111, ED 0.10-0.15, PW 0.3-0.43, WL 0.65-0.93, PL 0.25-0.34, PH 0.15-0.21, DPW 0.10-0.14. (*n*=9). As holotype, in some individuals propodeal spines shorter than holotype and body color ligther.

Holotype: Worker, China: Tibet Autonomous Region, Linzhi County, Pailong Town, Zhaqu Village (29.928517°N, 95.156467°E), 2740 m, 2007.Ⅸ.27, collected by Zheng-hui Xu on the ground in subalpine moist evergreen broadleaf forest of the Yarlung Tsangpo Grand Canyon, No. A07-543.

Etymology: The specific epithet refers to "xiaohan", one of the Twenty-four Solar Terms of China.

正模工蚁： 正面观，头部近长方形，长大于宽；后缘和侧缘轻度隆起，后角宽圆。上颚三角形，咀嚼缘具5个齿。唇基背面轻度隆起，缺中央纵脊，前缘中度隆起。额叶宽，遮盖触角窝大部。触角12节，柄节超过头后缘，触角棒3节。复眼位于头侧缘中点处，占据侧缘的1/5。侧面观，前胸背板轻度隆起，缺前中胸背板缝。中胸背板轻度隆起，向后缓坡形降低。缺后胸沟。并胸腹节背面平直；并胸腹节刺很短，长度约与其基部宽相等，约为斜面长的1/2；斜面浅凹。并胸腹节侧叶很短，顶端圆。腹柄前面小柄长，约为腹柄结长的2/3；腹柄结近三角形，前面近平直，后面轻度隆起，顶角窄圆；腹面大部分平直，后部浅凹，前下角钝角状。后腹柄稍低于腹柄结，前上缘圆，后缘平直，腹面中度凹陷。背面观，前胸背板最宽，侧缘强烈隆起，肩角宽圆。缺前中胸背板缝。中胸中度收缩，侧缘深凹。缺后胸沟。并胸腹节近长方形，侧缘轻微隆起；并胸腹节刺短，轻度内弯。腹柄近长方形，长大于宽，腹柄结侧缘轻度隆起。后腹柄近梯形，向后变窄，约为腹柄宽的1.4倍，前侧角钝角状，侧缘几乎平直。后腹部长卵形。上颚具纵条纹。头部背面具疏松的纵皱纹，后部和两侧逐渐变为网纹，界面具粗网纹。唇基具很疏松的纵皱纹，界面光滑。胸部具网纹，前胸背板两侧和后胸侧板下部具纵皱纹。腹柄和后腹柄具细网纹。后腹部光滑发亮。身体背面具丰富亚直立、亚倾斜短毛和丰富倾斜绒毛被，头部背面立毛较密，直立；柄节具丰富亚倾斜毛和密集倾斜绒毛被，胫节具密集倾斜绒毛被。身体棕黄色，头部和后腹部浅黑色。**副模工蚁：** 特征同正模，部分个体并胸腹

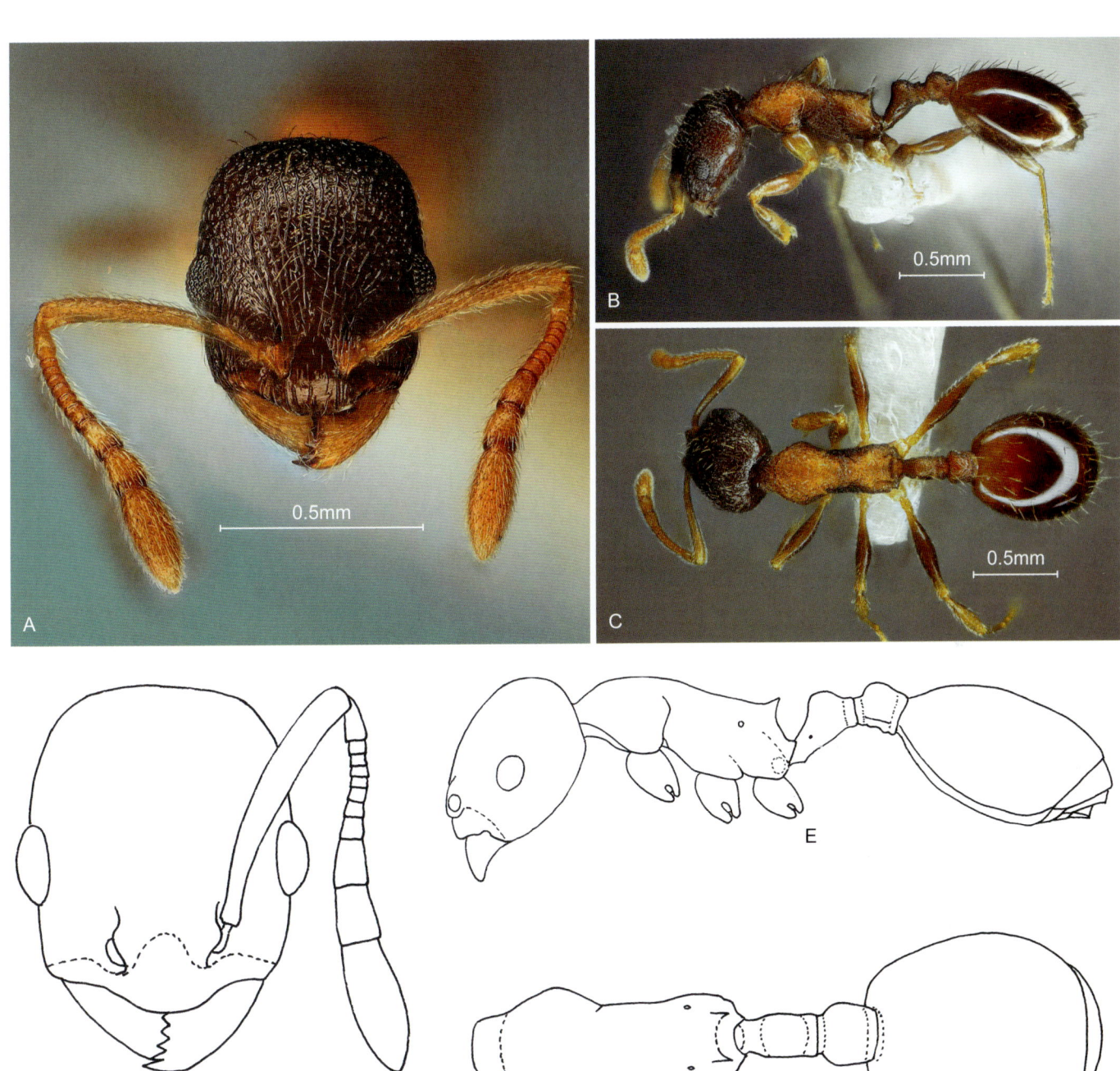

小寒切胸蚁 *Temnothorax xiaohan*

A, D. 工蚁头部正面观；B, E. 工蚁身体侧面观；C, F. 工蚁身体背面观（A-C. 引自Qian和Xu, 2024）
A, D. Head of worker in full-face view; B, E. Body of worker in lateral view; C, F. Body of worker in dorsal view (A-C. cited from Qian & Xu, 2024)

节刺短于正模，身体颜色较浅。

正模： 工蚁，中国西藏自治区林芝县排龙乡扎曲村（29.928517°N，95.156467°E），2740m，2007.Ⅸ.27，徐正会采于雅鲁藏布大峡谷的中山湿性常绿阔叶林地表，No. A07-543。

词源： 该新种以中国二十四节气中的"小寒"命名。

112 小满切胸蚁
Temnothorax xiaoman Qian & Xu, 2024

Temnothorax xiaoman Qian & Xu, 2024, European Journal of Taxonomy, 936: 59, fig. 37 (w.) CHINA (Shaanxi).

Holotype worker: TL 2.9, HL 0.66, HW 0.56, CI 85, SL 0.54, SI 96, ED 0.14, PW 0.42, WL 0.87, PL 0.32, PH 0.24, DPW 0.17. In full-face view, head roughly rectangular, longer than broad, posterior margin slightly convex, posterior corners narrowly rounded, lateral margins weakly convex. Mandibles triangular, masticatory margin with 5 teeth. Clypeal dorsum weakly convex, with median carina, anterior margin weakly convex. Frontal lobes broad, concealing antennal sockets. Frontal carinae very short, about half length of frontal lobe. Antennae 12-segmented, scapes failing to reach posterior head margin, club 3-segmented. Eyes located at midpoint of lateral head margin, occupying 1/5 of lateral margin. In lateral view, pronotum moderately convex, promesonotal suture absent. Dorsa of mesonotum and propodeum straight and gently sloping down posteriorly. Metanotal groove absent. Propodeal spines short, about as long as their basal width and pointed posterodorsally; declivity straight. Propodeal lobes roughly triangular, rounded apically. Anterior peduncle of petiole very short, about 1/2 length of petiolar node; petiolar node roughly triangular, anterior margin straight, posterior margin weakly convex, top corner blunt angled; ventral margin weakly concave posteriorly, anteroventral corner toothed. Postpetiole as high as petiole, anterodorsal margin roundly convex, posterior margin almost straight; ventral margin short and almost straight, anteroventral corner rightly angled. In dorsal view, pronotum broadest, anterior margin moderately convex, humeral corners broadly rounded, lateral margin weakly convex. Promesonotal suture absent. Mesothorax moderately constricted and narrowing posteriorly, lateral margins strongly concave. Metanotal groove absent. Propodeum roughly square, lateral margins weakly concave, spines short, pointed posterolaterally. Petiole elongate trapezoidal and widening posteriorly, longer than broad; lateral margins with a prominence each side anteriorly, and weakly convex posteriorly. Postpetiole roughly trapozoidal, about 1.4 times as broad as petiole, narrowing posteriorly, anterior margin moderately convex, lateral margins strongly convex, posterior margin short and nearly straight. Gaster elongate oval. Mandibles longitudinally striate. Head dorsum longitudinally rugose with interweaved fine rugulae between the main rugae, and gradually reticulate posteriorly and laterally, interface abundantly punctate. Clypeus with three longitudinal rugae on each side, interface smooth. Dorsum of mesosoma coarsely reticulate, dorsum of mesonotum smooth in the center. Pronotal sides coarsely reticulate anteriorly and longitudinally costulate posteriorly. Mesopleura, metapleura and propodeal sides densely reticulate. Petiole and postpetiole finely reticulate, anterior face of petiolar node finely punctate, postpetiolar dorsum smooth. Gaster smooth and shiny. Head dorsum with dense erect apically blunt short hairs and dense decumbent pubescence; mesosoma, petiole and postpetiole with sparse suberect apically blunt short hairs and sparse decumbent pubescence; gaster with abundant suberect apically blunt short hairs and abundant decumbent pubescence. Scapes with dense decumbent pubescence; tibiae with abundant decumbent pubescence. Body color blackish brown; mandibles, antennae and legs yellowish brown.

Holotype: Worker, China: Shaanxi Province, Taibai County, Qinling Mountains, Huangbaiyuan Scenic Area (33.813848°N, 107.526568°E), 1400 m, 1997.Ⅵ.27, collected by Cong Wei on the ground in conifer-broadleaf mixed forest, No. A97-4109.

Etymology: The specific epithet refers to "xiaoman", one of the Twenty-four Solar Terms of China.

正模工蚁： 正面观，头部近长方形，长大于宽；后缘轻微隆起，后角窄圆；侧缘轻度隆起。上颚三角形，咀嚼缘具5个齿。唇基背面轻度隆起，具中央纵脊，前缘轻度隆起。额叶宽，遮盖触角窝。额脊很短，约为额叶长的一半。触角12节，柄节未到达头后缘，触角棒3节。复眼位于头侧缘中点处，占据侧缘的1/5。侧面观，前胸背板中度隆起，缺前中胸背板缝。中胸背板和并胸腹节背面平直，向后缓坡形降低，缺后胸沟；并胸腹节刺短，长度约与其基部宽相等，指向后上方；斜面平直。并胸腹节侧叶近三角形，顶端圆。腹柄前面小柄很短，约为腹柄结长的1/2；腹柄结近三角形，前面平直，后面轻度隆起，顶角钝角状；腹面后部浅凹，前下角齿状。后腹柄与腹柄等高，前上缘圆形隆起，后缘几乎平直；腹面短，几乎平直，前下角直角形。背面观，前胸背板最宽，前缘中度隆起，肩角宽圆，侧缘轻度隆起。缺前中胸背板缝。中胸中度收缩，向后变窄，侧缘强烈凹陷。缺后胸沟。并胸腹节近方形，侧缘轻度凹陷；并胸腹节刺短，指向后侧方。腹柄长梯形，向后变宽，长大于宽；侧缘前部每侧具1个突起，后部轻度隆起。后腹柄近梯形，约为腹柄宽的1.4倍，向后变窄，前缘中度隆起，侧缘强烈隆起，后缘短，近平直。后腹部长卵形。上颚具纵条纹。头部背面具纵皱纹，纵皱纹间具交织的细皱纹，后部和侧面逐渐变为网纹，界面具丰富刻点。唇基每侧具3条纵皱纹，界面光滑。胸部背面具粗网纹，中胸背板背面中央光滑；前胸背板侧面前部具粗网纹，

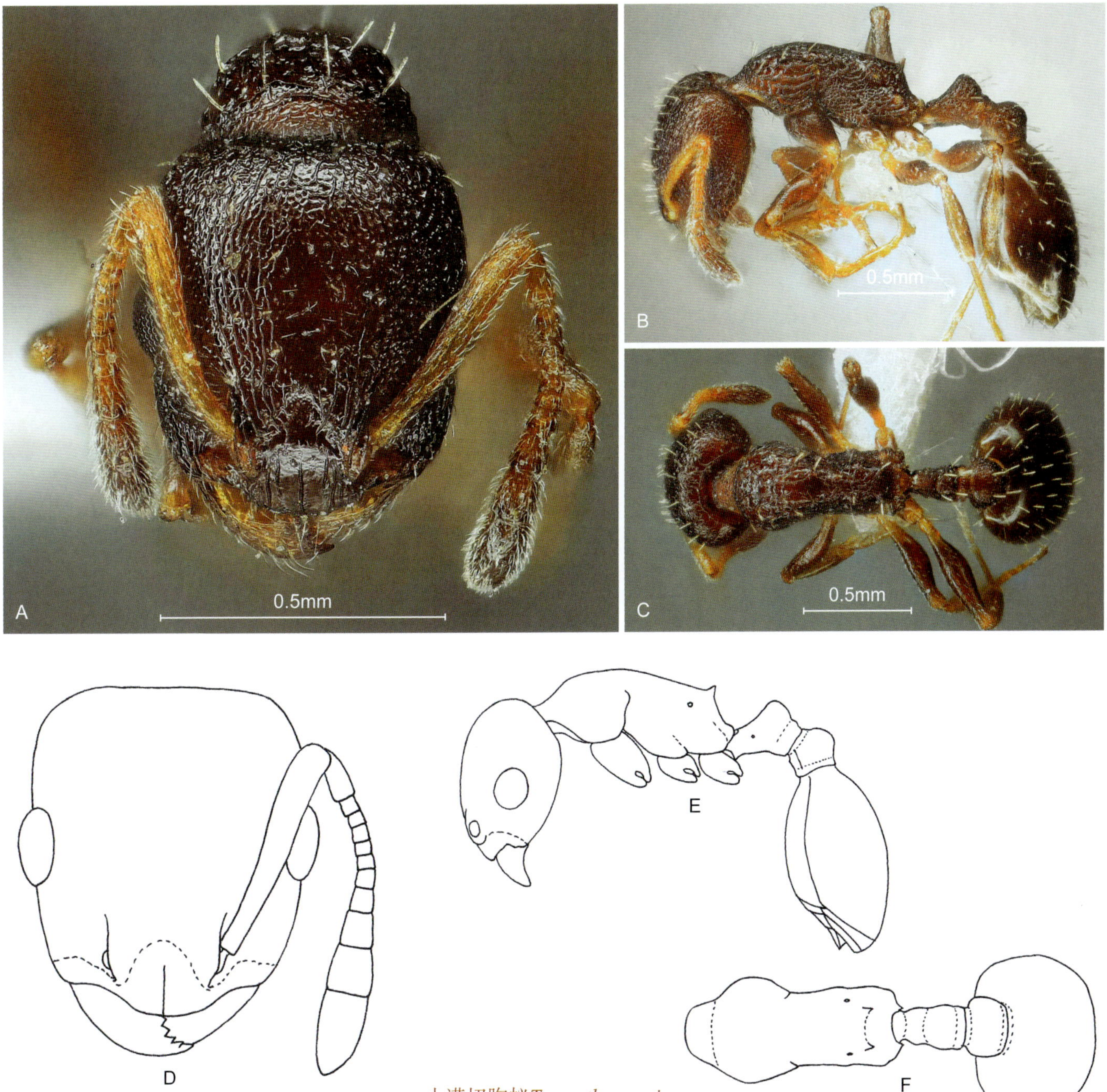

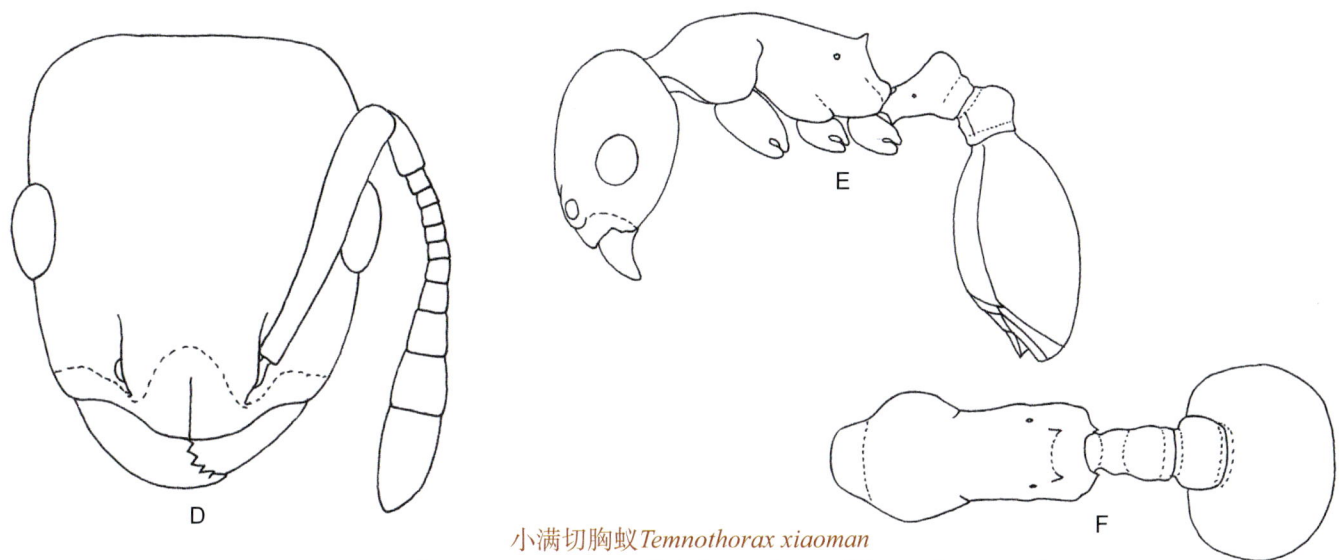

小满切胸蚁 *Temnothorax xiaoman*

A, D. 工蚁头部正面观；B, E. 工蚁身体侧面观；C, F. 工蚁身体背面观（A-C. 引自Qian和Xu, 2024）
A, D. Head of worker in full-face view; B, E. Body of worker in lateral view; C, F. Body of worker in dorsal view (A-C. cited from Qian & Xu, 2024)

后部具纵脊纹。中胸侧板、后胸侧板和并胸腹节侧面具密集网纹。腹柄和后腹柄具细网纹，腹柄结前面具细刻点，后腹柄背面光滑。后腹部光滑发亮。头部背面具密集直立短钝毛和密集倾斜绒毛被；胸部、腹柄和后腹柄具稀疏亚直立短钝毛和稀疏倾斜绒毛被；后腹部具丰富亚直立短钝毛和丰富倾斜绒毛被；柄节具密集倾斜绒毛被；胫节具丰富倾斜绒毛被。身体黑棕色，上颚、触角和足黄棕色。

正模：工蚁，中国陕西省太白县秦岭黄柏塬景区（33.813848°N，107.526568°E），1400m，1997.Ⅵ.27，魏琮采于针阔混交林地表，No. A97-4109。

词源：该新种以中国二十四节气中的"小满"命名。

113 小暑切胸蚁
Temnothorax xiaoshu Qian & Xu, 2024

Temnothorax xiaoshu Qian & Xu, 2024, European Journal of Taxonomy, 936: 61, fig. 38 (w.) CHINA (Yunnan).

Holotype worker: TL 3.3, HL 0.77, HW 0.64, CI 84, SL 0.74, SI 114, ED 0.16, PW 0.46, WL 0.93, PL 0.34, PH 0.22, DPW 0.18. In full-face view, head roughly rectangular, longer than broad, posterior and lateral margins weakly convex, posterior corners broadly rounded. Mandibles triangular, masticatory margin with 5 teeth. Clypeal dorsum weakly convex, with median carina, anterior margin moderately convex. Frontal lobes narrow, concealing half of antennal socket. Frontal carinae short, reaching to the level of anterior eye margin. Antennae 12-segmented, scapes surpassing posterior head margin by 1/12 of its length, club 3-segmented. Eyes located at midpoint of lateral head margin, occupying 1/4 of lateral margin. In lateral view, promesonotum moderately convex and gently sloping down posteriorly, anterior margin of pronotum marginated, promesonotal suture absent. Metanotal groove widely moderately impressed and distinct. Propodeal dorsum almost straight and gently sloping down posteriorly; propodeal spines short and stout, about as long as their basal width, blunt apically; declivity slightly concave, about 1/2 length of dorsum; propodeal lobes large and rounded apically. Petiole with anterior peduncle, about 3/4 length of petiolar node; petiolar node roughly triangular, anterior margin almost straight, posterior margin weakly convex, top corner narrowly rounded; ventral margin almost straight, weakly concave posteriorly, anteroventral tooth indistinct. Postpetiole as high as petiole, anterodorsal margin roundly convex, posterior margin short and almost straight, ventral margin short and weakly concave. In dorsal view, pronotum broadest, anterior margin strongly convex and weakly marginated, humeral corners broadly rounded, lateral margin strongly convex, promesonotal suture absent. Mesothorax moderately constricted and narrow, lateral margins moderately concave. Metanotal groove widely impressed. Propodeum roughly trapezoidal and weakly widening posteriorly, lateral margins weakly convex; spines short and weakly in-curved; propodeal lobes acutely toothed. Petiole roughly trapezoidal and widening posteriorly, longer than broad, lateral margins weakly convex, the node roughly circular. Postpetiole roughly trapezoidal and narrowing posteriorly, about 1.3 times as broad as petiole, anterolateral corners narrowly rounded, lateral margins nearly straight. Gaster elongate oval. Mandibles longitudinally striate. Head coarsely reticulate, anterior 1/5 longitudinally rugose and divergent posteriorly. Clypeus with 3-4 rugae on each side, interface smooth. Mesosoma coarsely reticulate, lower part of metapleuron with 2 longitudinal costulae. Petiole finely reticulate, postpetiole densely punctate. Gaster smooth and shiny. Body dorsum with abundant suberect to subdecumbent hairs and abundant decumbent pubescence, head dorsum with dense erect short hairs and dense decumbent pubescence. Scapes and tibiae with dense subdecumbent hairs and dense decumbent pubescence. Body color brownish yellow; first gastral segment blackish brown. **Paratype worker:** TL 3.3, HL 0.75, HW 0.60, CI 80, SL 0.60, SI 100, ED 0.10, PW 0.40, WL 0.85, PL 0.25, PH 0.20, DPW 0.13 ($n=1$). As holotype.

Holotype: Worker, China: Yunnan Province, Lushui City, Pianma Town, Pianma Village (25.993789°N, 98.660772°E), 2500 m, 1999.Ⅳ.25, collected by Zheng-hui Xu inside dead branch in subalpine moist evergreen broadleaf forest, No. A99-24.

Etymology: The specific epithet refers to "xiaoshu", one of the Twenty-four Solar Terms of China.

正模工蚁：正面观，头部近长方形，长大于宽；后缘和侧缘轻度隆起，后角宽圆。上颚三角形，咀嚼缘具5个齿。唇基背面轻度隆起，具中央纵脊，前缘中度隆起。额叶窄，遮盖触角窝的一半。额脊短，到达复眼前缘水平。触角12节，柄节的1/12超过头后缘，触角棒3节。复眼位于头侧缘中点处，占据侧缘的1/4。侧面观，前中胸背板中度隆起，向后缓坡形降低，前胸背板前缘具边缘，缺前中胸背板缝。后胸沟宽形中度凹陷，明显。并胸腹节背面近平直，向后缓坡形降低；并胸腹节刺短粗，长度约等于其基部宽，顶端钝；斜面轻微凹陷，约为背面长的1/2。并胸腹节侧叶大，末端圆。腹柄前面具小柄，约为腹柄结长的3/4；腹柄结近三角形，前面近平直，后面轻度隆起，顶角窄圆；腹面近平直，后部浅凹，前下角的齿不明显。后腹柄与腹柄等高，前上缘圆形隆起，后缘短，近平直；腹面短，浅凹。背面观，前胸背板最宽，前缘强烈隆起，具弱边缘；肩角宽圆，侧缘强烈隆起。缺前中胸背板缝。中胸中度收缩，狭窄，侧缘中度凹陷。后胸沟宽形凹陷。并胸腹节近梯形，向后轻度变宽，侧缘轻度隆起；并胸腹节刺短，轻度内弯。并胸腹节侧叶尖齿状。腹柄近梯形，向后变宽，长大于宽，侧缘轻度隆起，腹柄结近圆形。后腹柄近梯形，向后变窄，约为腹柄宽的1.3倍，前侧角窄圆，侧面近平直。后腹部长卵形。上颚具纵条纹。头部具粗网纹，前部1/5具纵皱纹，向后发散。唇基每侧具3~4条皱纹，界面光滑。胸部具粗网纹，后胸侧板下部具2条纵脊纹。腹柄具细网纹，后腹柄具密集刻点。后腹部光滑发亮。身体背面具丰富亚直立、亚倾斜毛和丰富倾斜绒毛被；

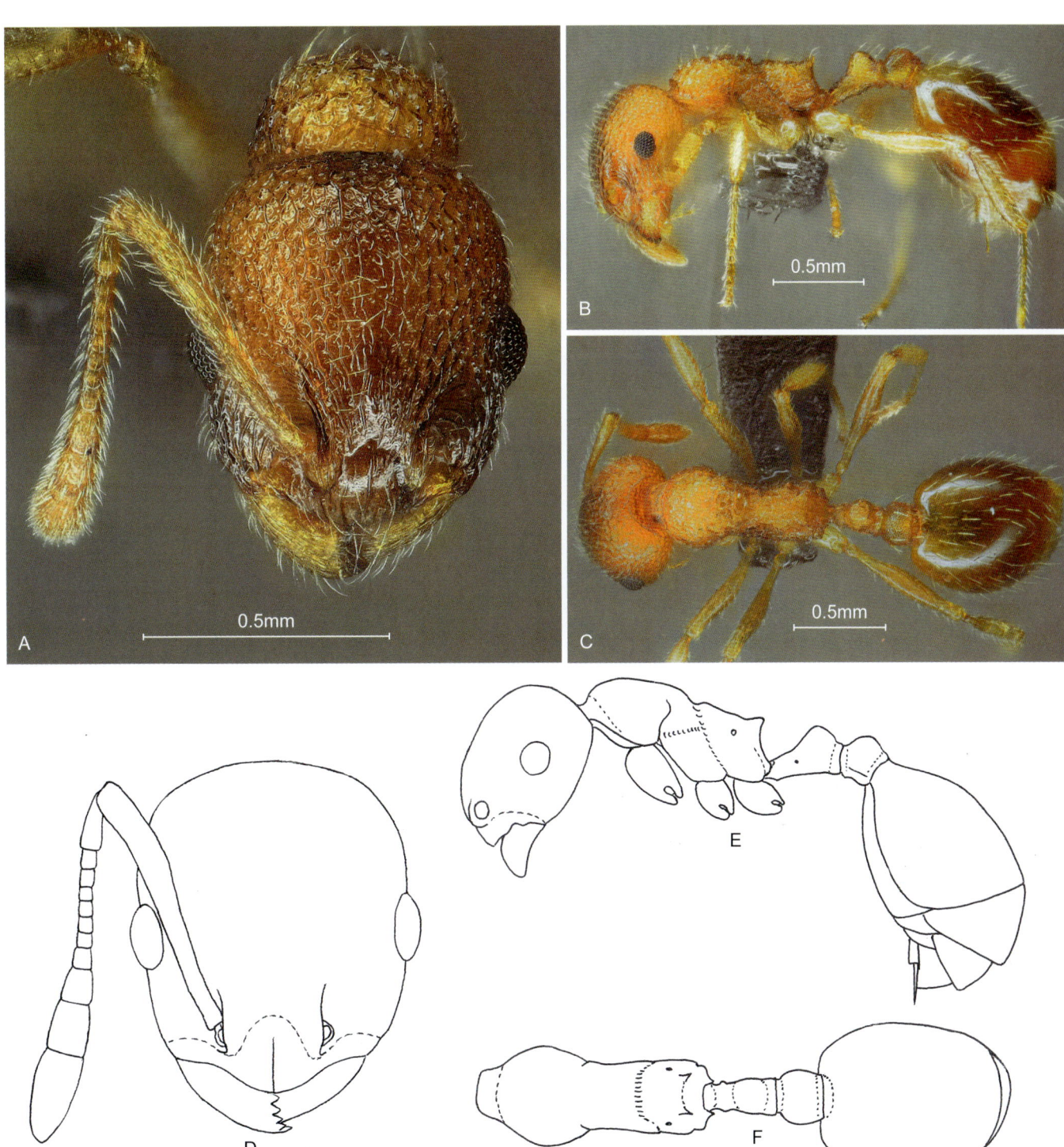

小暑切胸蚁 Temnothorax xiaoshu

A, D. 工蚁头部正面观；B, E. 工蚁身体侧面观；C, F. 工蚁身体背面观（A-C. 引自Qian和Xu, 2024）
A, D. Head of worker in full-face view; B, E. Body of worker in lateral view; C, F. Body of worker in dorsal view (A-C. cited from Qian & Xu, 2024)

头部背面具密集直立短毛和密集倾斜绒毛被；柄节和胫节具密集亚倾斜毛和密集倾斜绒毛被。身体棕黄色，后腹部第1节黑棕色。**副模工蚁：** 特征同正模。

正模： 工蚁，中国云南省泸水市片马镇片马村（25.993789°N，98.660772°E），2500m，1999.Ⅳ.25，徐正会采于中山湿性常绿阔叶林朽木内，No. A99-24。

词源： 该新种以中国二十四节气中的"小暑"命名。

小雪切胸蚁
Temnothorax xiaoxue Qian & Xu, 2024

Temnothorax xiaoxue Qian & Xu, 2024, European Journal of Taxonomy, 936: 63, fig. 39 (w.) CHINA (Yunnan).

Holotype worker: TL 2.5, HL 0.58, HW 0.48, CI 82, SL 0.59, SI 123, ED 0.14, PW 0.37, WL 0.74, PL 0.28, PH 0.25, DPW 0.16. In full-face view, head roughly rectangular, longer than broad, posterior and lateral margins weakly convex, posterior corners narrowly rounded. Mandibles triangular, masticatory margin with 5 teeth. Clypeal dorsum weakly convex, without median carina, anterior margin moderately convex. Frontal lobes narrow, concealing half of antennal sockets. Frontal carinae relatively long, reaching to the level of midpoint of eyes. Antennae 12-segmented, scapes surpassing posterior head margin by 1/9 of its length, antennal clubs 3-segmented. Eyes located at midpoint of lateral head margin, occupying 1/4 of lateral margin. In lateral view, promesonotum weakly convex and weakly arched, promesonotal suture and metanotal groove absent. Propodeal dorsum straight and gently sloping down posteriorly; propodeal spines moderately long and straight, slightly shorter than declivity, pointed posterodorsally; declivity almost straight; propodeal lobes triangular, bluntly angled apically. Petiole with short anterior peduncle, about 2/3 length of petiolar node; petiolar node roughly trapezoidal and narrowing dorsally, anterior and dorsal margins straight, posterior margin weakly convex, anterodorsal corner bluntly angled, posterodorsal corner narrowly rounded; ventral margin slightly concave anteriorly and posteriorly, slightly convex in the middle, anteroventral corner acutely toothed. Postpetiole as high as petiolar node, dorsum rounded, ventral margin weakly concave. In dorsal view, pronotum broadest, anterior margin moderately convex, humeral corners broadly rounded, lateral margins strongly convex. Promesonotal suture and metanotal groove absent. Mesothorax moderately constricted, lateral margins weakly concave. Propodeum roughly rectangular, lateral margins slightly convex; spines moderately long and blunt apically, slightly in-curved. Petiole roughly trapezoidal and widening posteriorly, longer than broad, lateral margins weakly convex. Postpetiole roughly trapezoidal and narrowing posteriorly, about 1.6 times as broad as petiole, anterolateral corners broadly rounded, lateral margins weakly convex. Gaster elongate oval. Mandibles longitudinally striate. Head dorsum loosely longitudinally rugose and reticulate laterally. Clypeus smooth, each side with 3-4 longitudinal rugae. Mesosomal dorsum reticulate; mesosomal sides longitudinally rugose, interface finely reticulate. Petiole and postpetiole finely reticulate. Gaster smooth and shiny. Body dorsum with sparse erect to suberect apically blunt short hairs and sparse decumbent pubescence. Scapes with abundant subdecumbent hairs and dense decumbent pubescence, tibiae with dense decumbent pubescence. Body color blackish brown; mandibles, antennae and legs brownish yellow. **Paratype worker:** TL 3.0, HL 0.65, HW 0.55, CI 85, SL 0.55, SI 100, ED 0.15, PW 0.40, WL 0.75, PL 0.30, PH 0.2, DPW 0.15 (*n*=1). As holotype.

Holotype: Worker, China: Yunnan Province, Kunming City, Xishan Mountain, Taihua Temple (24.967782°N, 102.636006°E), 1950 m, 2001.Ⅴ.2, collected by Ji-ling Zhang from a ground sample of conifer-broadleaf mixed forest, No. A00436.

Etymology: The specific epithet refers to "xiaoxue", one of the Twenty-four Solar Terms of China.

正模工蚁： 正面观，头部近长方形，长大于宽；后缘和侧缘轻度隆起，后角窄圆。上颚三角形，咀嚼缘具5个齿。唇基背面轻度隆起，缺中央纵脊，前缘轻度隆起。额叶窄，遮盖触角窝的一半。额脊较长，到达复眼中点水平。触角12节，柄节的1/9超过头后缘，触角棒3节。复眼位于头侧缘中点处，占据侧缘的1/4。侧面观，前中胸背板轻度隆起，呈弱弓形，缺前中胸背板缝和后胸沟。并胸腹节背面平直，向后缓坡形降低；并胸腹节刺直，中等长，稍短于斜面，指向后上方；斜面近平直。并胸腹节侧叶三角形，末端钝角状。腹柄前面小柄短，约为腹柄结长的2/3；腹柄结近梯形，向上变窄，前面和背面平直，后面轻度隆起，前上角钝角状，后上角窄圆；腹面前部和后部浅凹，中部轻微隆起，前下角尖齿状。后腹柄与腹柄结等高，背面圆，腹面浅凹。背面观，前胸背板最宽，前缘中度隆起，肩角宽圆，侧缘强烈隆起。缺前中胸背板缝和后胸沟。中胸中度收缩，侧缘浅凹。并胸腹节近长方形，侧缘轻微隆起；并胸腹节刺中等长，末端钝，轻微内弯。腹柄近梯形，向后变宽，长大于宽，侧缘轻度隆起。后腹柄近梯形，向后变窄，约为腹柄宽的1.6倍，前侧角宽圆，侧面轻度隆起。后腹部长卵形。上颚具纵条纹。头部背面具疏松纵皱纹，侧面呈网纹。唇基光滑，每侧具3~4条纵皱纹。胸部背面具网纹；胸部侧面具纵皱纹，界面具细网纹。腹柄和后腹柄具细网纹。后腹部光滑发亮。身体背面具稀疏直立、亚直立短钝毛和稀疏倾斜绒毛被；柄节具丰富亚倾斜毛和密集倾斜绒毛被，胫节具密集倾斜绒毛被。身体黑棕色，上颚、触角和足棕黄色。**副模工蚁：** 特征同正模。

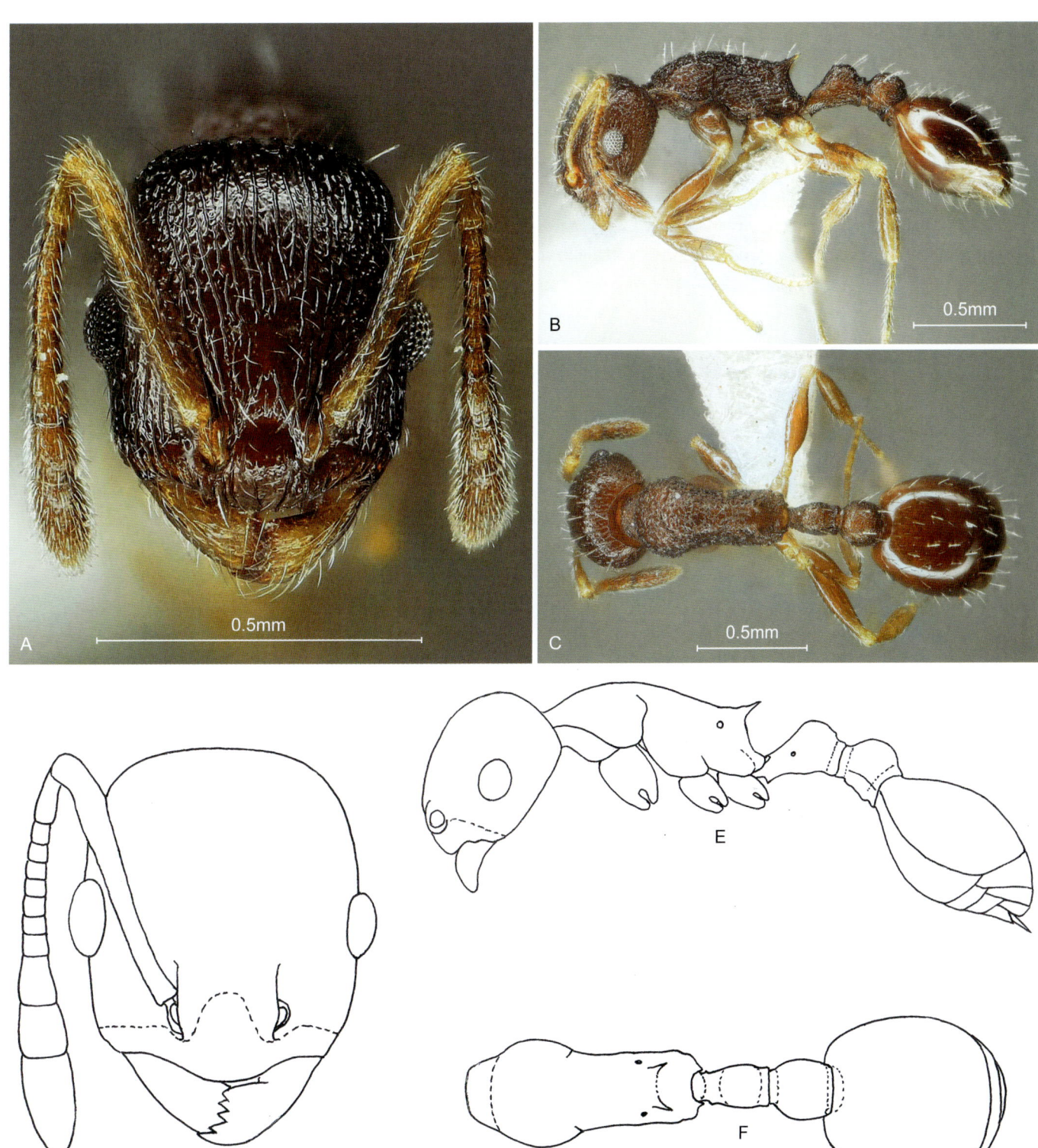

小雪切胸蚁 *Temnothorax xiaoxue*

A, D. 工蚁头部正面观；B, E. 工蚁身体侧面观；C, F. 工蚁身体背面观（A-C. 引自Qian和Xu, 2024）
A, D. Head of worker in full-face view; B, E. Body of worker in lateral view; C, F. Body of worker in dorsal view (A-C. cited from Qian & Xu, 2024)

正模：工蚁，中国云南省昆明市西山太华寺（24.967782°N，102.636006°E），1950m，2001.Ⅴ.2，张继玲采于针阔混交林地表样中，No. A00436。

词源：该新种以中国二十四节气中的"小雪"命名。

115 夏至切胸蚁
Temnothorax xiazhi Qian & Xu, 2024

Temnothorax xiazhi Qian & Xu, 2024, European Journal of Taxonomy, 936: 65, fig. 41 (w.) CHINA (Yunnan).

Holotype worker: TL 2.9, HL 0.69, HW 0.58, CI 84, SL 0.60, SI 104, ED 0.13, PW 0.40, WL 0.83, PL 0.33, PH 0.22, DPW 0.15. In full-face view, head roughly rectangular, longer than broad, posterior margin almost straight, posterior corners narrowly rounded, lateral margins weakly convex. Mandibles triangular, masticatory margin with 5 teeth. Clypeal dorsum weakly convex, with median carina, anterior margin moderately convex. Frontal lobes broad, concealing antennal sockets. Frontal carinae short, reaching to the level of anterior eye margins. Antennae 12-segmented, scapes just reaching to posterior head margin, club 3-segmented. Eyes located at midpoint of lateral head margin, occupying 1/4 of lateral margin. In lateral view, pronotum slightly convex, anterodorsal corner narrowly rounded, promesonotal suture absent. Mesonotum weakly convex and gently sloping down posteriorly, metanotal groove very shallowly impressed. Propodeal dorsum straight and gently sloping down posteriorly; propodeal spines short and acute, roughly triangular, about as long as their basal width and about 1/2 length of declivity; declivity weakly concave; propodeal lobes short and rounded apically. Petiole with long anterior peduncle, about as long as petiolar node; petiolar node roughly triangular, anterior margin almost straight, posterior margin weakly convex, top corner bluntly angled; ventral margin straight, anteroventral corner largely acutely toothed. Postpetiole about as high as petiolar node, anterodorsal margin rounded, posterior margin straight, ventral margin moderately concave. In dorsal view, pronotum broadest, humeral corners broadly rounded, lateral margin moderately convex, promesonotal suture absent. Mesothorax weakly constricted, lateral margins weakly concave. Metanotal groove very shallowly impreassed. Propodeum roughly trapezoidal and weakly widening posteriorly, lateral margins almost straight; spines short and triangular, pointed posterolaterally. Petiole roughly trapezoidal and widening posteriorly, longer than broad, lateral margins weakly convex. Postpetiole roughly trapezoidal and narrowing posteriorly, about 1.5 times as broad as petiole, anterolateral corners narrowly rounded, lateral margins weakly convex. Gaster elongate oval. Mandibles longitudinally striate. Head dorsum densely longitudinally rugose, and gradually reticulate posteriorly and laterally, interface between reticulation densely punctate. Clypeus smooth in the center, longitudinally rugose posteriorly and laterally. Mesosoma and petiolar node finely reticulate; mesopleura and metapleura reticulate; anterior face of pronotum, sides of propodeum, petiole except the node and postpetiole densely punctate with interface micro-reticulate. Gaster smooth and shiny. Body dorsum with sparse erect to suberect apically blunt short hairs and sparse decumbent pubescence, hairs on head and gaster relatively abundant. Scapes and tibiae with dense decumbent pubescence. Body color blackish brown, head dorsum and gaster black; mandibles, antennae and legs blackish brown to brownish yellow. **Paratype workers:** TL 2.7-2.8, HL 0.63-0.65, HW 0.45-0.5, CI 71-77, SL 0.50-0.51, SI 100-111, ED 0.10-0.11, PW 0.34-0.35, WL 0.75-0.83, PL 0.24-0.25, PH 0.15-0.20, DPW 0.10-0.15 (n=5). As holotype, in some individuals body size slightly different.

Holotype: Worker, China: Yunnan Province, Yulong County, Baisha Town, Yulong Snow Mountain (27.134850°N, 100.259831°E), 3000 m, 2004.Ⅹ.20, collected by Zheng-hui Xu from a ground sample of *Pinus yunnanensis* forest, No. A04-1198.

Etymology: The specific epithet refers to "xiazhi", one of the Twenty-four Solar Terms of China.

正模工蚁： 正面观，头部近长方形，长大于宽；后缘近平直，后角窄圆，侧缘轻度隆起。上颚三角形，咀嚼缘具5个齿。唇基背面轻度隆起，具中央纵脊，前缘中度隆起。额叶宽，遮盖触角窝。额脊短，到达复眼前缘水平。触角12节，柄节刚到达头后缘，触角棒3节。复眼位于头侧缘中点处，占据侧缘的1/4。侧面观，前胸背板轻微隆起，前上角窄圆，缺前中胸背板缝。中胸背板轻度隆起，向后缓坡形降低。后胸沟很浅。并胸腹节背面平直，向后缓坡形降低；并胸腹节刺短而尖，近三角形，长度约等于其基部宽，约为斜面长的1/2；斜面浅凹。并胸腹节侧叶短，末端圆。腹柄前面小柄长，约与腹柄结等长；腹柄结近三角形，前面近平直，后面轻度隆起，顶角钝角状；腹面平直，前下角大，尖齿状。后腹柄约与腹柄结等高，前上缘圆，后缘平直，腹面中度凹陷。背面观，前胸背板最宽，肩角宽圆，侧缘中度隆起。缺前中胸背板缝。中胸轻度收缩，侧缘浅凹。后胸沟浅凹。并胸腹节近梯形，向后轻度变宽，侧缘近平直；并胸腹节刺短，三角形，指向后侧方。腹柄近梯形，向后变宽，长大于宽，侧缘轻度隆起。后腹柄近梯形，向后变窄，约为腹柄宽的1.5倍，前侧角窄圆，侧面轻度隆起。后腹部长卵形。上颚具纵条纹。头部背面具密集纵皱纹，后部和侧面逐渐呈网纹，网纹界面具密集刻点。唇基中央光滑，后部和两侧具纵皱纹。胸部和腹柄结具细网纹；中胸侧板和后胸侧板具网纹；前胸背板前面、并胸腹节侧面、腹柄下部和后腹柄具密集刻点，界面

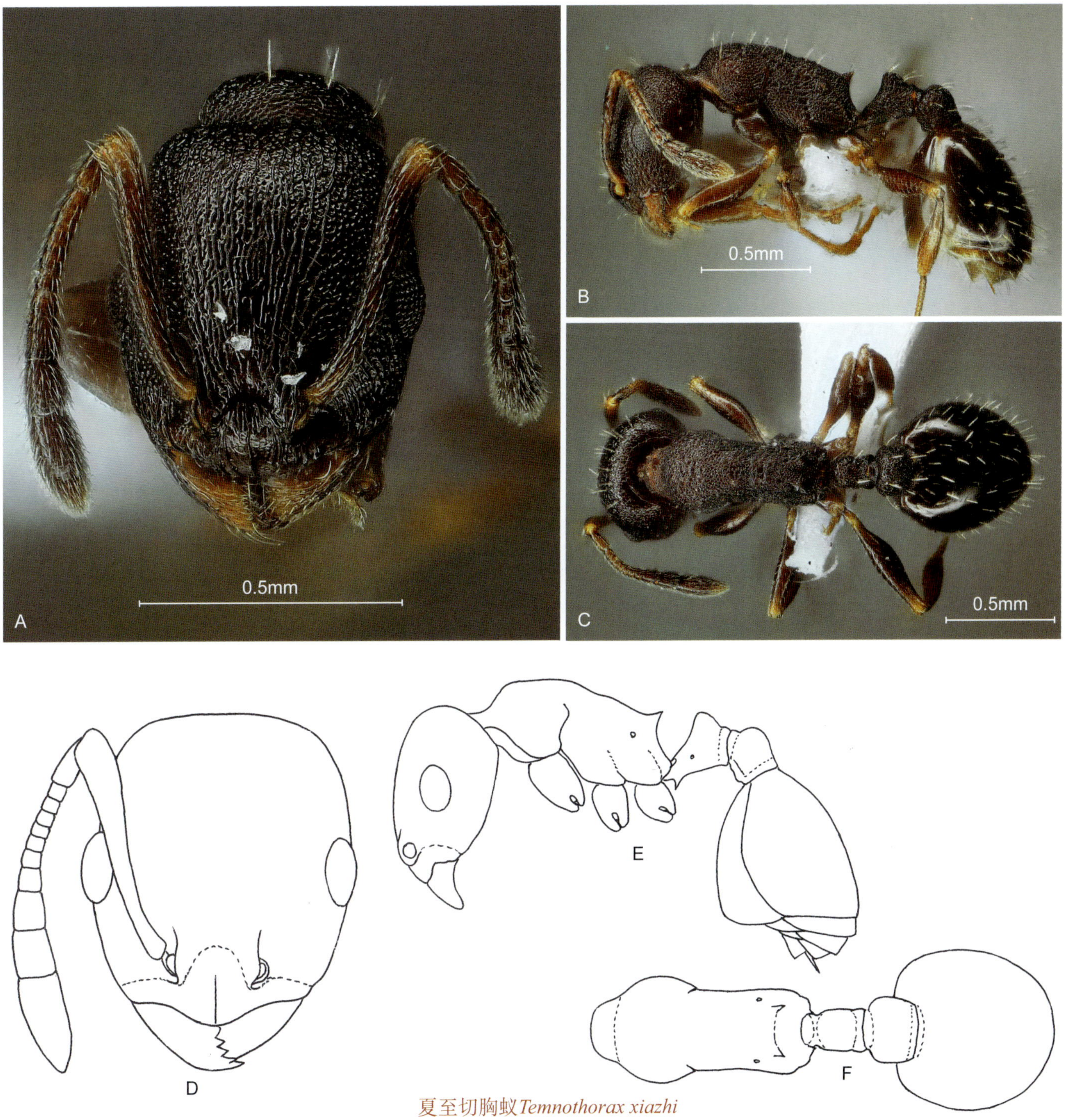

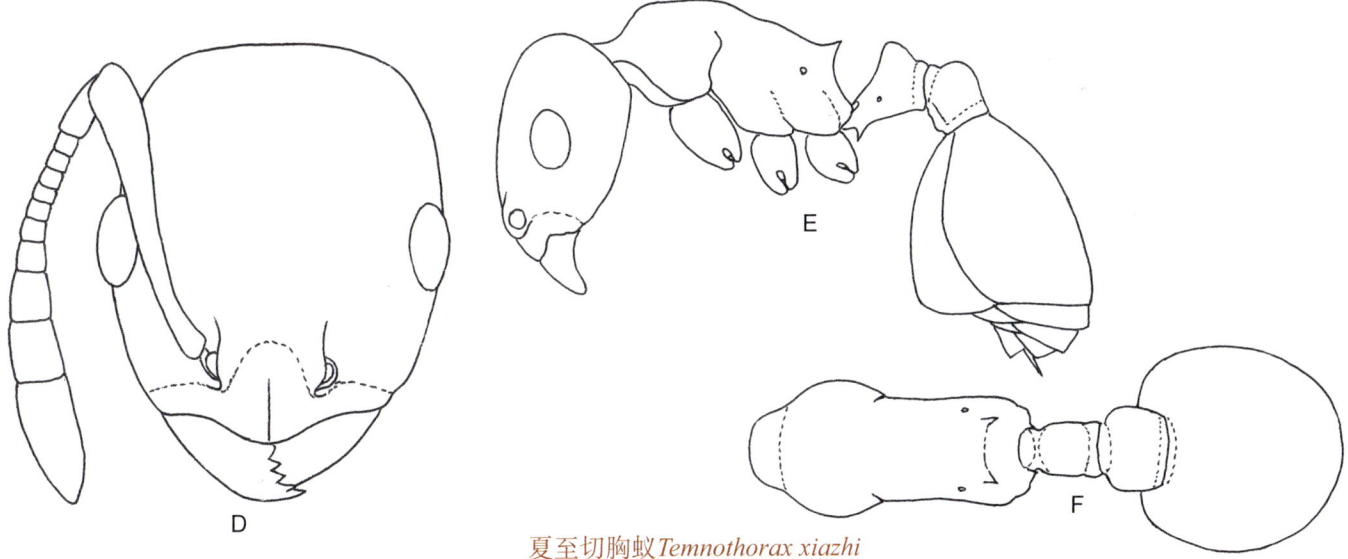

夏至切胸蚁 Temnothorax xiazhi

A, D. 工蚁头部正面观；B, E. 工蚁身体侧面观；C, F. 工蚁身体背面观（A-C. 引自Qian和Xu, 2024）
A, D. Head of worker in full-face view; B, E. Body of worker in lateral view; C, F. Body of worker in dorsal view (A-C. cited from Qian & Xu, 2024)

呈微网纹。后腹部光滑发亮。身体背面具稀疏直立、亚直立短钝毛和稀疏倾斜绒毛被，头部和后腹部立毛较丰富；柄节和胫节具密集倾斜绒毛被。身体黑棕色，头部背面和后腹部黑色，上颚、触角和足黑棕色至棕黄色。**副模工蚁**：特征同正模，部分个体身体大小稍有差异。

正模：工蚁，中国云南省玉龙县白沙乡玉龙雪山（27.134850°N，100.259831°E），3000m，2004.X.20，徐正会采于云南松林地表样中，No. A04-1198。

词源：该新种以中国二十四节气中的"夏至"命名。

116 雨水切胸蚁
Temnothorax yushui Qian & Xu, 2024

Temnothorax yushui Qian & Xu, 2024, European Journal of Taxonomy, 936: 67, fig. 43 (w.) CHINA (Sichuan).

Holotype worker: TL 2.8, HL 0.58, HW 0.50, CI 86, SL 0.50, SI 100, ED 0.13, PW 0.36, WL 0.71, PL 0.29, PH 0.16, DPW 0.13. In full-face view, head roughly rectangular, longer than broad, posterior margin moderately convex, posterior corners broadly rounded, lateral margins weakly convex. Mandibles triangular, masticatory margin with 5 teeth. Clypeal dorsum weakly convex, with median carina, anterior margin of clypeus moderately convex. Frontal lobes narrow, concealing half of antennal socket. Frontal carinae short, reaching to the midpoint level of eye. Antennae 12-segmented, scapes almost reaching posterior head margin, club 3-segmented. Eyes located slightly before midpoint of lateral head margin, occupying 1/4 of lateral margin. In lateral view, pronotum moderately convex, promesonotal suture absent. Anterior part of mesonotum almost straight, posterior part weakly convex and sloping down posteriorly. Metanotal groove widely moderately impressed. Propodeal dorsum almost straight and gently sloping down posteriorly; propodeal spines bluntly angled; declivity weakly concave, shorter than dorsum; propodeal lobes small and bluntly angled apically. Petiole with anterior peduncle, about 2/3 length of petiolar node; petiolar node roughly triangular, anterior margin straight, posterior margin weakly convex, top corner narrowly rounded; ventral margin almost straight, weakly concave posteriorly, anteroventral corner with a tiny tooth. Postpetiole as high as petiole, anterodorsal margin roundly convex, posterior margin short and straight, ventral margin short and almost straight. In dorsal view, pronotum broadest, anterior margin moderately convex, humeral corners broadly rounded, lateral margins strongly convex. Promesonotal suture absent. Mesothorax moderately constricted and narrow, lateral margins almost straight. Metanotal groove widely impressed. Propodeum roughly rectangular, lateral margins almost straight; propodeal spines very short and acutely toothed. Petiole roughly rectangular, longer than broad, lateral margins weakly convex. Postpetiole roughly trapezoidal and narrowing posteriorly, about 1.5 times as broad as petiole, anterolateral corners bluntly angled, lateral margins nearly straight. Gaster elongate oval. Mandibles longitudinally striate. Head dorsum and genae longitudinally rugose with interface smooth before posterior margins of eyes, gradually reticulate posteriorly and laterally with interface densely punctate. Clypeus with 4-5 rugae on each side, interface smooth. Mesosomal dorsum reticulate; pronotal sides and mesoplura reticulate; metapleura longitudinally rugose, the rugae become coarse ventrally; propodeal sides, petiole and postpetiole finely reticulate with interface densely punctate, postpetiolar dorsum smooth. Gaster smooth and shiny. Body dorsum with abundant erect to suberect short hairs and abundant decumbent pubescence, pubescence become denser on head dorsum, hairs become denser and subdecumbent on gastral dorsum. Scapes with sparse subdecumbent hairs and dense decumbent pubescence; tibiae with dense decumbent pubescence, but without standing hairs. Body color brownish yellow; mandibles, antennae and legs light yellow; dorsum of first gastral segment and eyes light black. **Paratype workers:** TL 2.5-2.6, HL 0.57-0.70, HW 0.48-0.55, CI 79-85, SL 0.50-0.56, SI 91-117, ED 0.10-0.12, PW 0.33-0.40, WL 0.70-0.75, PL 0.23-0.31, PH 0.17-0.20, DPW 0.10-0.14 (*n*=15). As holotype, in some individuals metanotal groove deeply to shallowly impressed, propodeal spines bluntly angled or slightly longer and with sharp apices.

Holotype: Worker, China: Sichuan Province, Muli County, Liziping Town, Lizigou Village (28.055228°N, 101.183367°E), 2766 m, 2018.Ⅶ.30, collected by Yu-cheng He from a soil sample of conifer-broadleaf mixed forest, No. C18-1446.

Etymology: The specific epithet refers to "yushui", one of the Twenty-four Solar Terms of China.

正模工蚁： 正面观，头部近长方形，长大于宽；后缘中度隆起，后角宽圆；侧缘轻度隆起。上颚三角形，咀嚼缘具5个齿。唇基背面轻度隆起，具中央纵脊，前缘中度隆起。额叶窄，遮盖触角窝的一半。额脊短，到达复眼中点水平。触角12节，柄节几乎到达头后缘，触角棒3节。复眼位于头侧缘中点稍前处，占据侧缘的1/4。侧面观，前胸背板中度隆起，缺前中胸背板缝。中胸背板前部近平直，后部轻度隆起，向后坡形降低。后胸沟宽形中度凹陷。并胸腹节背面近平直，向后缓坡形降低；并胸腹节刺钝角状；斜面浅凹，短于背面。并胸腹节侧叶小，末端钝角状。腹柄前面具小柄，约为腹柄结长的2/3；腹柄结近三角形，前面平直，后面轻度隆起，顶角窄圆；腹面近平直，后部浅凹，前下角具小齿。后腹柄与腹柄等高，前上缘圆形隆起，后缘短而直，腹面短，近平直。背面观，前胸背板最宽，前缘中度隆起，肩角宽圆，侧缘强烈隆起。缺前中胸背板缝。中胸中度收缩，狭窄，侧缘近平直。后胸沟宽形凹陷。并胸腹节近长方形，侧缘几乎平直；并胸腹节刺很短，尖齿状。腹柄近长方形，长大于宽，侧缘轻度隆起。

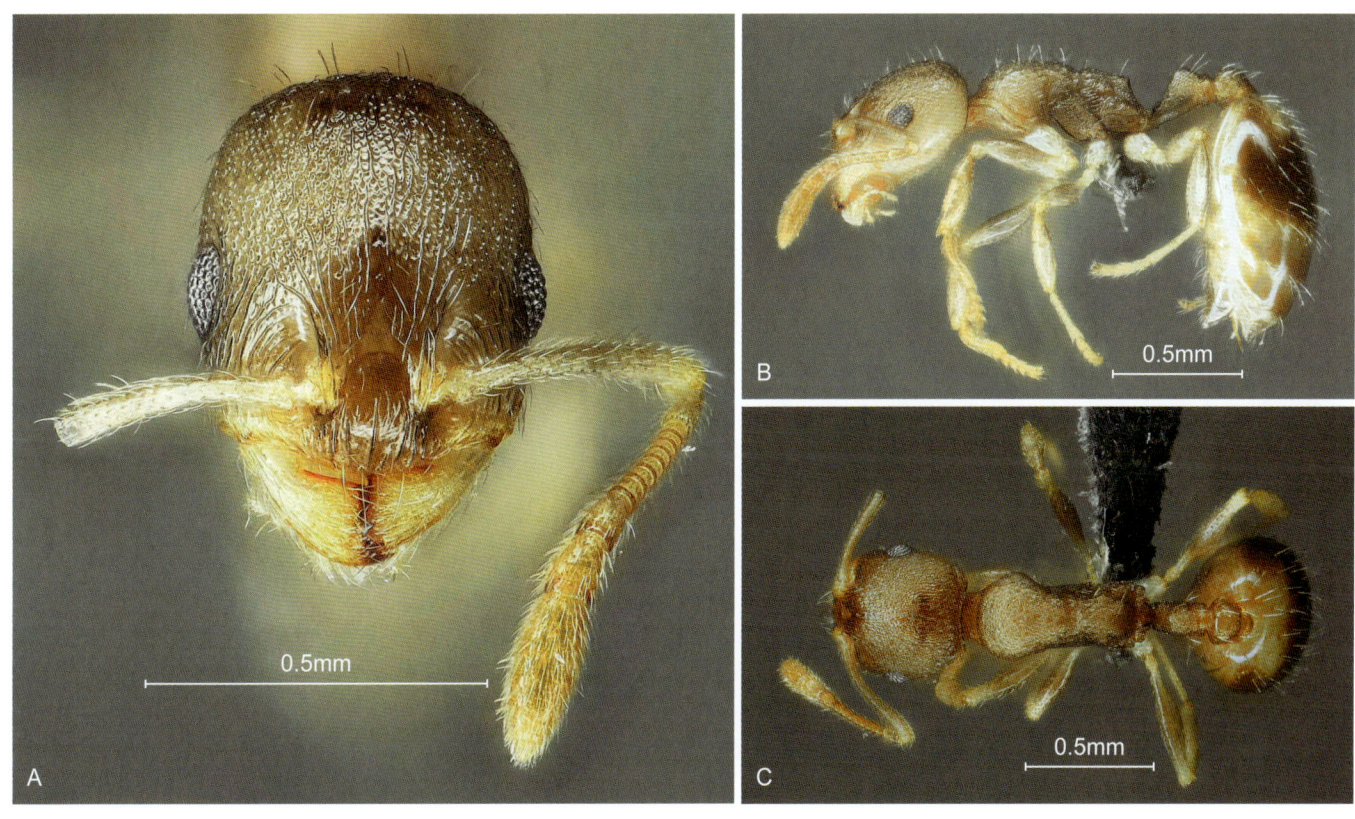

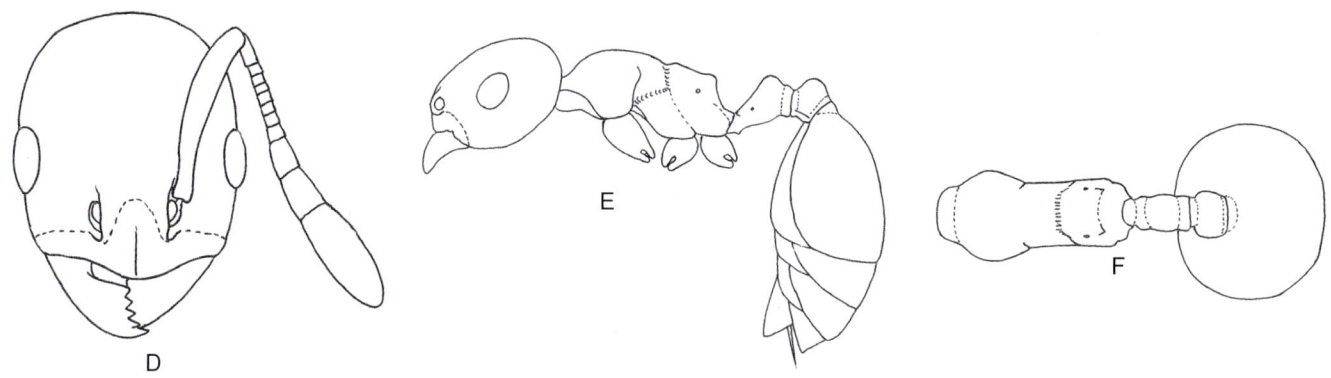

雨水切胸蚁 *Temnothorax yushui*

A, D. 工蚁头部正面观；B, E. 工蚁身体侧面观；C, F. 工蚁身体背面观（A-C. 引自Qian和Xu, 2024）
A, D. Head of worker in full-face view; B, E. Body of worker in lateral view; C, F. Body of worker in dorsal view (A-C. cited from Qian & Xu, 2024)

后腹柄近梯形，向后变窄，约为腹柄宽的1.5倍，前侧角钝角状，侧缘近平直。后腹部长卵形。上颚具纵条纹。头部背面和颊区具纵皱纹，复眼后缘之前界面光滑；后部和侧面逐渐变成网纹，界面具密集刻点。唇基每侧具4~5条皱纹，界面光滑。胸部背面、前胸背板侧面和中胸侧板具网纹；后胸侧板具纵皱纹，下部皱纹变粗；并胸腹节侧面、腹柄和后腹柄具细网纹，界面具密集刻点，后腹柄背面光滑。后腹部光滑发亮。身体背面具丰富直立、亚直立短毛和丰富倾斜绒毛被；头部背面绒毛被较密；后腹部背面立毛较密，亚倾斜；柄节具稀疏亚倾斜毛和密集倾斜绒毛被；胫节具密集倾斜绒毛被，缺立毛。身体棕黄色，上颚、触角和足浅黄色，后腹部第1节背面和复眼浅黑色。**副模工蚁**：特征同正模，部分个体后胸沟深凹至浅凹，并胸腹节刺钝角状或稍长且末端尖锐。

正模：工蚁，中国四川省木里县李子坪乡李子沟村（28.055228°N, 101.183367°E），2766m, 2018.Ⅶ.30, 和玉成采于针阔混交林土壤样中, No. C18-1446。

词源：该新种以中国二十四节气中的"雨水"命名。

心头铺道蚁
Tetramorium cardiocarenum Xu & Zheng, 1994

Tetramorium cardiocarenum Xu & Zheng, 1994, Entomotaxonomia, 16(4): 286, figs. 2, 9 (w.) CHINA (Yunnan).

Holotype worker: TL 2.7, HL 0.71, HW 0.63, CI 88, SL 0.50, SI 80, PW 0.43, AL 0.75, ED 0.14. In full-face view, head cardioform, occipital margin strongly concave, the head narrowing anteriorly. Mandibles striate. Anterior margin of clypeus roundly convex. Frontal carinae short, end in front of the level of the eyes. Pronotal corners rounded in dorsal view. Metanotal groove conspicuous, concave in lateral view. Propodeal spines long and acute, straight. Metapleural lobes small, rounded. Petiole with ventral convex, the node rectangular, the anterior and posterior faces parallel, the dorsal surface straight, in dorsal view, the node broader than long. Median portion of clypeus and dorsum of head with numerous parallel and fine longitudinal rugae. Spaces between the rugae with fine punctulations. Scrobe area densely reticulate-rugulose, lateral surfaces of head puncto-striatus. Alitrunk and pedicel segments finely and densely reticulate-rugulose. Dorsum of alitrunk with fine longitudinal rugae. Gaster smooth and shining. Dorsal surface of head and body with sparse, short and blunt suberect hairs. Dorsal surface of the antennal scapes with very short curved subdecumbent hairs. Decumbent hairs are present on the dorsal surfaces of the hind tibiae. Body in colour light yellowish brown, with the dorsal surfaces of head and alitrunk, and the gaster blackish yellowish brown. **Paratype workers:** TL 2.4-2.7, HL 0.70-0.75, HW 0.61-0.68, CI 88-91, SL 0.49-0.55, SI 76-84, PW 0.40-0.46, AL 0.70-0.79, ED 0.14-0.15 (*n*=10). As holotype, but in some specimens hairs are completely absent from dorsal surfaces of head and alitrunk.

Holotype: Worker, China: Yunnan Province, Dali County, Xizhou Town, Hudiequan Spring (25.75°N, 100.15°E), 2000 m, 1991.X.11, collected by Zheng-hui Xu on the ground in *Pinus yunnanensis* forest, No. A91-963.

Etymology: The species name *cardiocarenum* combines Greek *cardi-* (heart) + word root *caren* (head) + suffix *-um* (neutral form), it refers to the heart-shaped head in full-face view.

正模工蚁： 正面观，头部心形，向前变窄，后缘强烈凹陷。上颚具条纹。唇基前缘圆形隆起。额脊短，达到复眼之前。背面观前胸背板肩角圆。后胸沟明显，侧面观凹陷。并胸腹节刺长而直，尖锐。后侧叶小，末端圆。腹柄腹面隆起；腹柄结长方形，前面和后面平行，背面平直，背面观宽大于长。唇基中部和头部背面具众多平行的细纵皱纹，界面具细刻点；触角沟具密集网纹，头部侧面具条纹和刻点。胸部、腹柄和后腹柄具细密网纹，胸部背面具细纵皱纹。后腹部光滑发亮。头部和身体背面具稀疏亚直立短钝毛；触角柄节背面具很短的弯曲亚倾斜绒毛被；后足胫节背面具倾斜绒毛被。身体淡黄棕色，头胸部背面和后腹部暗黄棕色。**副模工蚁：** 特征同正模，但一些标本头胸部背面缺立毛。

正模： 工蚁，中国云南省大理县喜洲镇蝴蝶泉（25.75°N，100.15°E），2000m，1991.X.11，徐正会采于云南松林地表，No. A91-963。

词源： 该新种的种名"*cardiocarenum*"由希腊语"*cardi-*"（心）+词根"*caren*"（头）+后缀"*-um*"（中性形式）组成，指正面观头部心形。

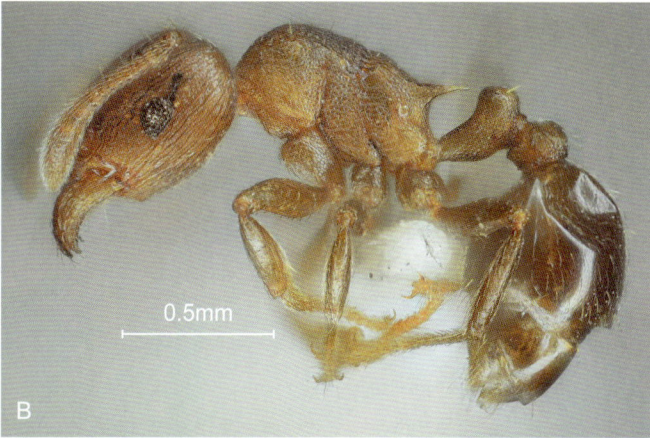

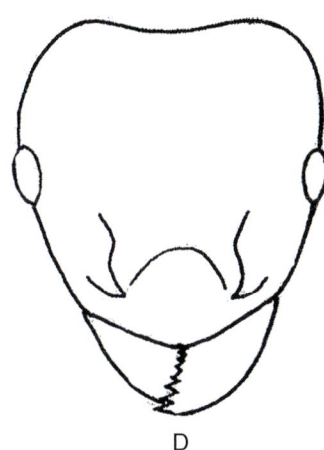

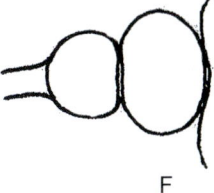

心头铺道蚁 *Tetramorium cardiocarenum*

A, D. 工蚁头部正面观；B, E. 工蚁身体侧面观；C. 工蚁身体背面观；F. 工蚁腹柄和后腹柄背面观（D-F. 引自Xu和Zheng, 1994）
A, D. Head of worker in full-face view; B, E. Body of worker in lateral view; C. Body of worker in dorsal view; F. Petiole and postpetiole of worker in dorsal view (D-F. cited from Xu & Zheng, 1994)

圆叶铺道蚁
Tetramorium cyclolobium Xu & Zheng, 1994

Tetramorium cyclolobium Xu & Zheng, 1994, Entomotaxonomia, 16(4): 288, figs. 4, 10 (w.) CHINA (Guangxi).

Holotype worker: TL 1.9, HL 0.56, HW 0.53, CI 93, SL 0.35, SI 67, PW 0.43, AL 0.58, ED 0.14. In full-face view, mandibles weekly striate. Anterior margin of clypeus entire and straight, with a narrow flange. Frontal carinae long and strong, extending back well beyond the eyes, and then curved down, forming the upper and poisterior margins of the scrobes. Scrobe deeply depressed, with a fine longitudinal carina. Occipital margin evenly convex. Alitrunk in dorsal view short and broad, the pronotal corners angled. Propodeal spines short and acute, straight. Metapleural lobes broad, rounded apically. Petiole in lateral view with the node anteroposteriorly compressed, the anterior face of the node and the dorsum of the peduncle are connected by an arched surface, in dorsal view, the node transverse. Median portion of clypeus with several fine longitudinal rugae and a strong longitudinal carina. Head, alitrunk and pedicel segments densely reticulate-rugulose. Dorsum of head with conspicuous median longitudinal carina. Gaster unsculptured, the first tergite with very fine basal striates. Dorsal surfaces of head and body uniformly clothed with a dense mat of pale trifid hairs, head also with a few elongate simple hairs which longer than the trifid ones. Lateral surfaces of alitrunk without hairs. Numerous short suberect hairs are present on the dorsal surfaces of scapes. Hind tibiae with dense trifid hairs on their dorsal surfaces. Body in color yellowish brown, dorsum of head and gaster blackish brown. **Paratype workers:** TL 1.9-2.1, HL 0.55-0.59, HW 0.53-0.56, CI 93-98, SL 0.34-0.38, SI 63-67, PW 0.40-0.44, AL 0.55-0.60, ED 0.13-0.14 (n=8). As holotype, but in some specimens striates on mandibles are conspicuous.

Holotype: Worker, China: Guangxi Zhuang Autonomous Region, Guilin City, Guilin (25.283333°N, 110.266667°E), 260 m, 1992.Ⅷ.16, collected by Zheng-hui Xu on the ground in broadleaf forest, No. A92-301.

Etymology: The species name *cyclolobium* combines Greek *cycl-* (round) + word root *lob* (lobe) + suffix *-um* (neutral form), it refers to the apically rounded metapleural lobes.

正模工蚁： 正面观，上颚具弱条纹。唇基前缘完整，平直，具窄的凸缘。额脊长而发达，向后延伸超过复眼，然后向下弯曲，形成触角沟的上缘和后缘。触角沟深凹，具1条细纵脊。头后缘均匀隆起。背面观胸部短而宽，前胸背板肩角角状。并胸腹节刺直，短而尖。后侧叶宽，顶端圆。侧面观腹柄结前后压扁，腹柄结前面和小柄背面之间由1个弓形面连接；背面观腹柄结横形。唇基中部具几条细纵皱纹和1条发达的中央纵脊。头部、胸部、腹柄和后腹柄具密集网纹。头部背面具明显的中央纵脊。后腹部无刻纹，第1节背板具很细的基纵脊。头部和身体背面被1层稠密且均匀一致的浅色三叉毛，头部还有少量伸长的简单毛，明显长于三叉毛；胸部侧面缺立毛。柄节背面具众多亚直立短毛，后足胫节背面具稠密的三叉毛。身体黄棕色，头部背面和后腹部黑棕色。**副模工蚁：** 特征同正模，但一些标本上颚条纹明显。

正模： 工蚁，中国广西壮族自治区桂林市桂林（25.283333°N，110.266667°E），260m，1992.Ⅷ.16，徐正会采于阔叶林地表，No. A92-301。

词源： 该新种的种名"*cyclolobium*"由希腊语"*cycl-*"（圆的）+ 词根"*lob*"（叶，裂片）+ 后缀"*-um*"（中性形式）组成，指后侧叶顶端钝圆。

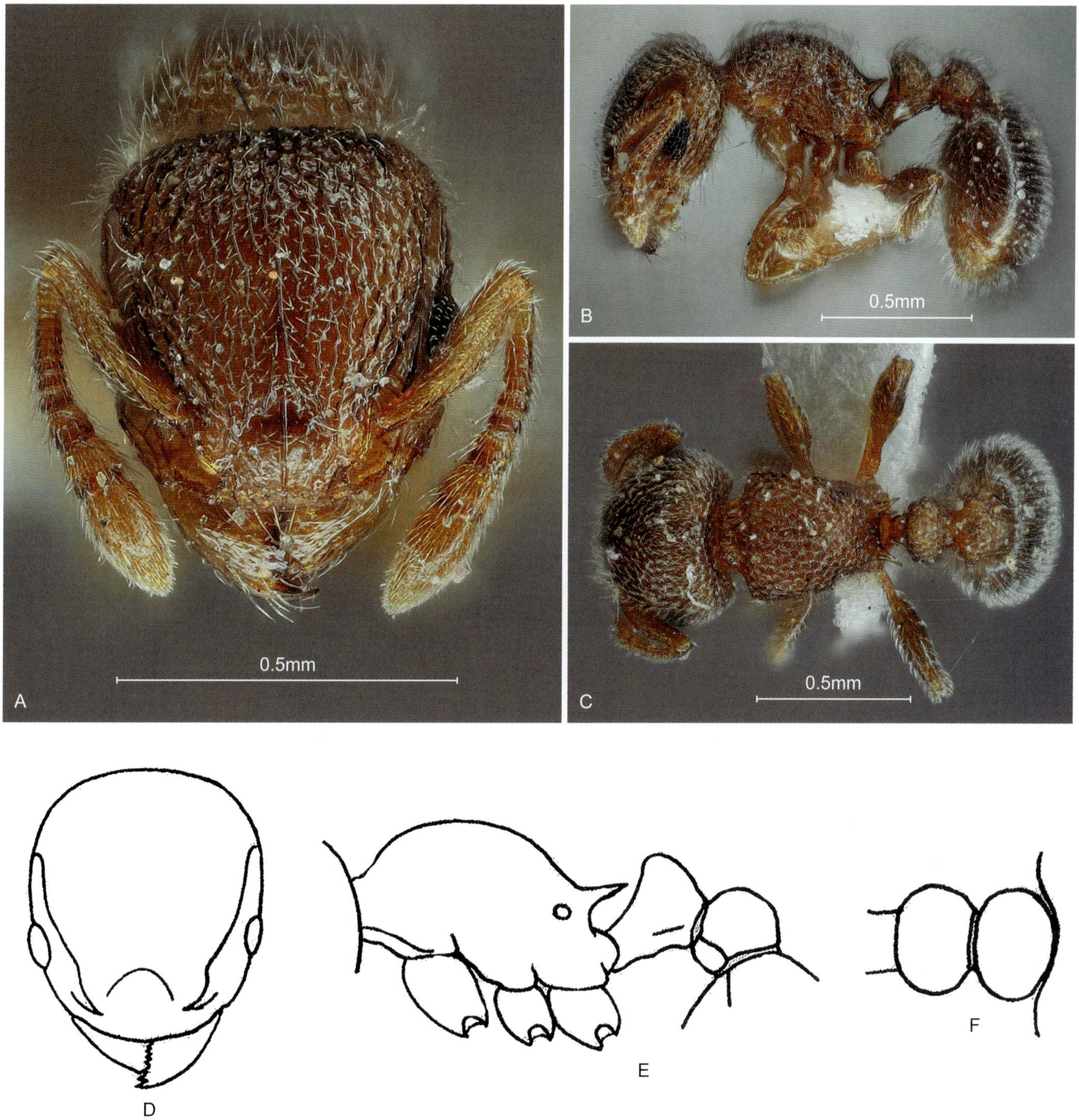

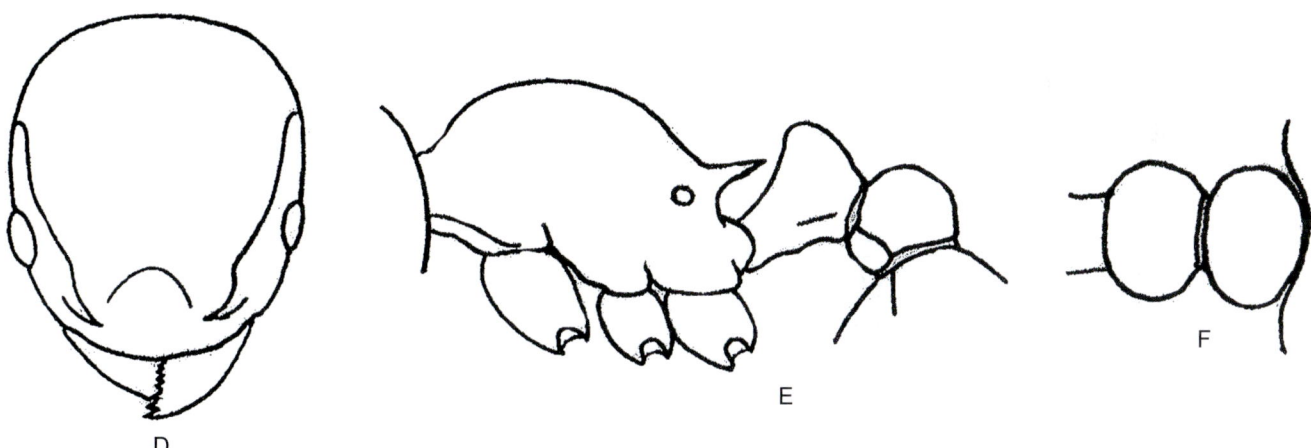

圆叶铺道蚁 *Tetramorium cyclolobium*

A, D. 工蚁头部正面观；B, E. 工蚁身体侧面观；C. 工蚁身体背面观；F. 工蚁腹柄和后腹柄背面观（D-F. 引自Xu和Zheng，1994）

A, D. Head of worker in full-face view; B, E. Body of worker in lateral view; C. Body of worker in dorsal view; F. Petiole and postpetiole of worker in dorsal view (D-F. cited from Xu & Zheng, 1994)

119 玉龙铺道蚁
Tetramorium yulongense Xu & Zheng, 1994

Tetramorium yulongense Xu & Zheng, 1994, Entomotaxonomia, 16(4): 288, figs. 6, 12 (w.) CHINA (Yunnan).

Holotype worker: TL 2.9, HL 0.79, HW 0.71, CI 90, SL 0.60, SI 84, PW 0.50, AL 0.93, ED 0.15. In full-face view, mandibles striate. Median portion of clypeus roundly and transversely convex, anterior margin slightly convex. Frontal carinae short, terminating at the level of the eyes. Scrobes absent. Occipital margin straight. In dorsal view, the pronotal corners bluntly angled. Metanotal groove depressed on the lateral surfaces. Propodeal spines short, elevated, acute apically. Metapleural lobes rounded. Petiole in lateral view with a small anteroventral tooth-like process, the node with its anterior and posterior faces norrowing above, dorsal surface feebly convex, the posterodorsal angle rounded. In dorsal view, the node broader than long. Median portion of clypeus with several fine longitudinal rugae, median longitudinal carina distinct. Dorsum of head with numerous fine longitudinal rugae, dense punctuations are present between the rugae. Dosum of alitrunk coarsely reticulate-rugulose, spaces between the reticula with fine punctulations. The lateral surfaces of alitrunk and the pedicel segments densely punctulate, punctulations on pedicel segments are smaller. The median portions of dorsum of the nodes smooth. Gaster unsculptured, shining. Dorsal surfaces of head and body with abundant acute erect or suberect hairs. Numerous short subdecumbent hairs are present on the dorsal surfaces of scapes and hind tibiae. Head and alitrunk blackish brown; tarsi yellowish brown; pedicel segments and gaster black. **Paratype workers:** TL 2.7-3.0, HL 0.70-0.83, HW 0.63-0.71, CI 85-93, SL 0.50-0.60, SI 77-84, PW 0.44-0.50, AL 0.80-0.93, ED 0.13-0.16 (*n*=8). As holotype.

Holotype: Worker, China: Yunnan Province, Lijiang County, Dayan Town, Heilongtan Park (26.916667°N, 100.283333°E), 2400 m, 1991.X.8, collected by Zheng-hui Xu on the ground in conifer-broadleaf mixed forest, No. A91-930.

Etymology: The new species is named after the famous snow mountain "Yulong", near the type locality.

正模工蚁： 正面观，上颚具条纹。唇基中部圆形横向隆起，前缘轻度隆起。额脊短，到达复眼水平。缺触角沟。头后缘平直。背面观，前胸背板肩角钝角状。后胸沟在侧面凹陷。并胸腹节刺短，向后升高，顶端尖锐。后侧叶顶端圆。侧面观腹柄前下角具1个齿状突起；腹柄结前面和后面向上变窄，背面轻微隆起，后上角圆。背面观，腹柄结宽大于长。唇基中部具几条细纵皱纹，中央纵脊明显。头部背面具众多细纵皱纹，皱纹间具密集刻点。胸部背面具粗网纹，网纹间具细刻点。胸部侧面、腹柄和后腹柄具密集刻点，腹柄和后腹柄的刻点较小，腹柄结和后腹柄结的背面中部光滑。后腹部无刻点，光亮。头部和身体背面具丰富直立、亚直立尖毛；柄节和后足胫节的背面具众多亚倾斜短毛。头部和胸部黑棕色，跗节黄棕色，腹柄、后腹柄和后腹部黑色。**副模工蚁：** 特征同正模。

正模： 工蚁，中国云南省丽江县大研镇黑龙潭公园（26.916667°N，100.283333°E），2400m，1991.X.8，徐正会采于针阔混交林地表，No. A91-930。

词源： 该新种以模式产地附近的著名雪山"玉龙"命名。

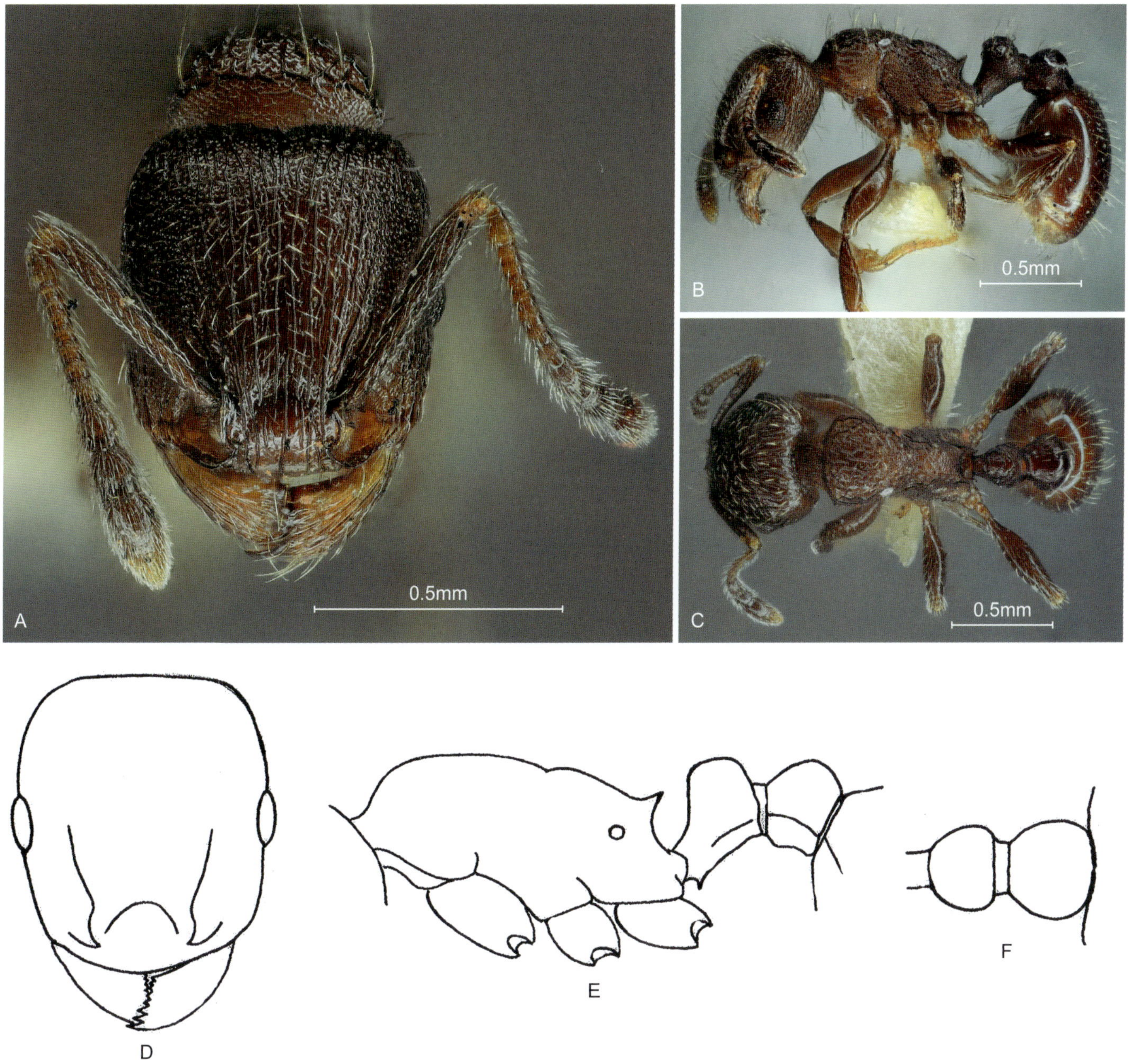

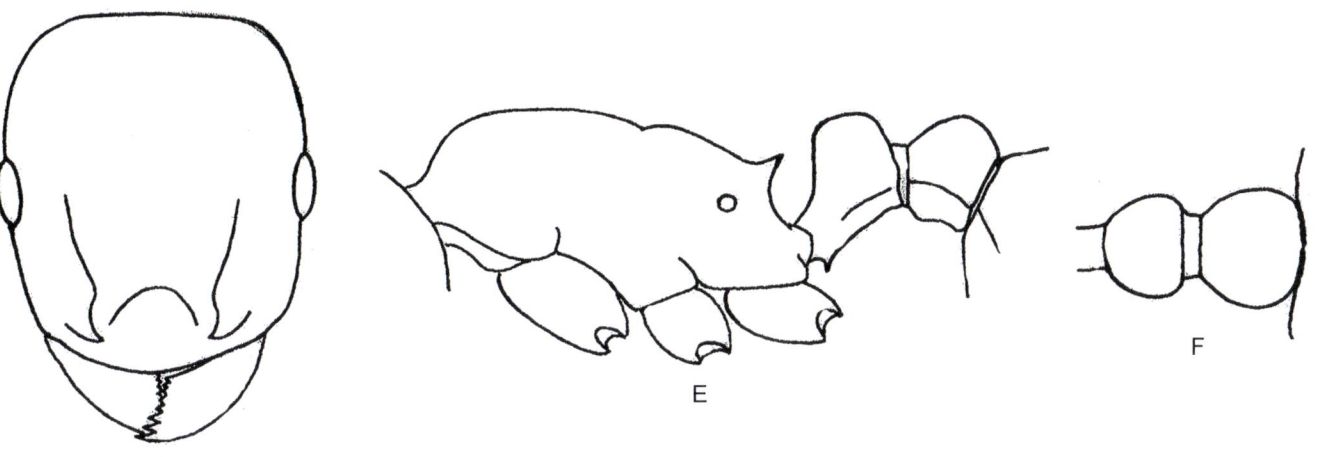

玉龙铺道蚁 Tetramorium yulongense

A, D. 工蚁头部正面观；B, E. 工蚁身体侧面观；C. 工蚁身体背面观；F. 工蚁腹柄和后腹柄背面观（D-F. 引自Xu和Zheng, 1994）

A, D. Head of worker in full-face view; B, E. Body of worker in lateral view; C. Body of worker in dorsal view; F. Petiole and postpetiole of worker in dorsal view (D-F. cited from Xu & Zheng, 1994)

西藏犁沟蚁
Vombisidris tibeta Xu & Yu, 2012

Vombisidris tibeta Xu & Yu, 2012, Sociobiology, 59 (3): 1500, figs. 10-13 (w.) CHINA (Tibet).

Holotype worker: TL 3.9, HL 0.83, HW 0.68, CI 82, SL 0.65, SI 96, ED 0.23, PW 0.53, AL 1.13. In full-face view, head roughly rectangular, longer than broad. Occipital margin weakly concave in the middle, occipital corners rounded. Sides weakly convex. Mandibles triangular, masticatory margin with 3 apical teeth, a long diastema, and 2 blunt basal teeth. Anterior clypeal margin strongly convex, posteriorly extended portion very broad, about 2.5 times as broad as frontal lobes. Antennae 12-segmented, apices of scapes just reached to occipital corners, antennal clubs 3-segmented. Frontal carinae fine and long, extended backward and close to the occipital corners. Eyes large, situated slightly before the midpoints of the sides. In lateral view, subocular groove complete, running from mandibular insertion to the lateroccipital margin. Promesonotum weakly convex, promesonotal suture vestigial on the dorsum. Metanotal groove shallowly notched. Propodeal dorsum straight and weakly slope down backward. Propodeal spines strong and long, slightly curved down backward, about 1.8 times as long as propodeal dorsum. Declivity weakly concave, about as long as propodeal dorsum. Propodeal spiracle small and circular, high up on the side. Propodeal lobes moderately developed, rounded at apices. Petiolar node elongate and dome-like, both anterior and posterior faces gently slope down, without distinct dorsal face; anterior peduncle very short, spiracle situated at about the mid-length of the peduncle; ventral face weakly convex about in the middle, and weakly concave afterwards; anteroventral corner acutely toothed. Postpetiolar node evenly convex, ventral face nearly straight. In dorsal view, sides of pronotum roundly convex. Sides of mesonotum without prominence. Propodeal spines weakly curved inward. Sides of petiole nearly straight, slightly widened backward; petiolar node longer than broad, length : width = 2 : 1. Postpetiole wider than petiole, sides weakly convex; postpetiolar node wider than long, width : length = 1.2 : 1. Mandibles smooth and shining, sparsely punctured. Head and alitrunk coarsely reticulate. Clypeus and sides of alitrunk longitudinally striate. Propodeal declivity longitudinally striate and densely finely punctured. Petiole and postpetiole finely reticulate and densely finely punctured. Gaster smooth and shining, basal costulae distinct, about 1/2 length of the postpetiole. Dorsal surfaces of head and body with sparse suberect to subdecumbent tapered hairs and abundant decumbent pubescence. Scapes with abundant suberect to subdecumbent tapered hairs and dense decumbent pubescence. Tibiae with sparse subdecumbent tapered hairs and abundant decumbent pubescence. Color brownish yellow, middle portion of gaster black, legs yellow. **Paratype workers:** TL 4.0-4.1, HL 0.85-0.88, HW 0.70-0.73, CI 82-83, SL 0.63-0.65, SI 89-90, ED 0.24, PW 0.53-0.58, AL 1.15-1.25 (*n*=2). As holotype, but metanotal groove even more shallow, sting extruding in one worker.

Holotype: Worker, China: Tibet Autonomous Region, Medog County, Damu Town, Damu Village, 1200 m, 2011.Ⅶ.20, collected by Xia Liu from a canopy sample in the valley tropical rainforest, No.A11-3928.

Etymology: The new species is named after the type locality "Tibet".

正模工蚁：正面观，头部近长方形，长大于宽；后缘中央轻度凹陷，后角圆；侧缘轻度隆起。上颚三角形，咀嚼缘具3个端齿、1个长的齿间隙和2个钝基齿。唇基前缘强烈隆起，后延部分很宽，约为额叶宽的2.5倍。触角12节，柄节刚到达头后角，触角棒3节。额脊细长，向后延伸并接近头后角。复眼大，位于头侧缘中点稍前处。侧面观，眼下沟完整，从上颚插入部延伸至后头侧缘。前中胸背板轻度隆起，前中胸背板缝在背面退化。后胸沟浅凹。并胸腹节背面平直，向后轻度坡形降低；并胸腹节刺长而粗，向后轻度下弯，约为并胸腹节背面长的1.8倍；斜面轻度凹陷，约与背面等长；气门小而圆，位于侧面上部。后侧叶中等发达，顶端圆。腹柄结伸长，圆丘状，前面和后面均呈缓坡形，缺明显的背面；前面小柄很短，气门约位于中部；腹面约在中部轻度隆起，向后轻度凹陷，前下角尖齿状。后腹柄结均匀隆起，腹面近平直。背面观，前胸背板侧缘圆形隆起。中胸背板侧缘无突起。并胸腹节刺轻度内弯。腹柄侧面近平直，向后轻微变宽；腹柄结长大于宽，长：宽=2：1。后腹柄宽于腹柄，侧面轻度隆起；后腹柄结宽大于长，宽：长=1.2：1。上颚光滑发亮，具稀疏刻点。头部和胸部具粗网状，唇基和胸部侧面具纵条纹，并胸腹节斜面具纵条纹和细密刻点。腹柄和后腹柄具细网纹和细密刻点；后腹部光滑发亮，基纵脊明显，约为后腹柄长的1/2。头部和身体背面具稀疏亚直立、亚倾斜尖毛和丰富倾斜绒毛被；柄节具丰富亚直立、亚倾斜尖毛和密集倾斜绒毛被；胫节具稀疏亚倾斜尖毛和丰富倾斜绒毛被。身体棕黄色，后腹部中部黑色，足黄色。**副模工蚁：**特征同正模，但是后胸沟更浅，1头工蚁的螯针伸出。

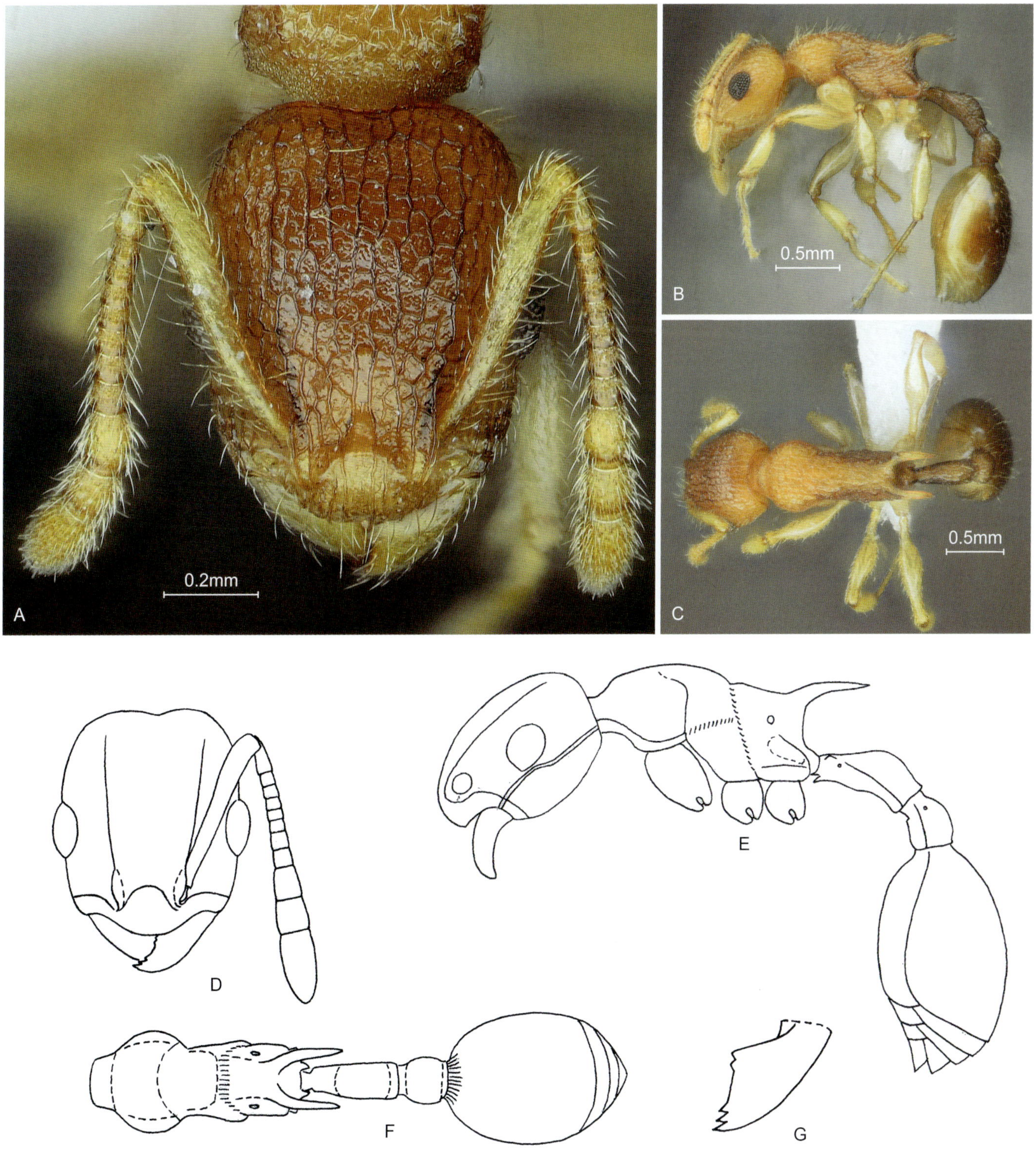

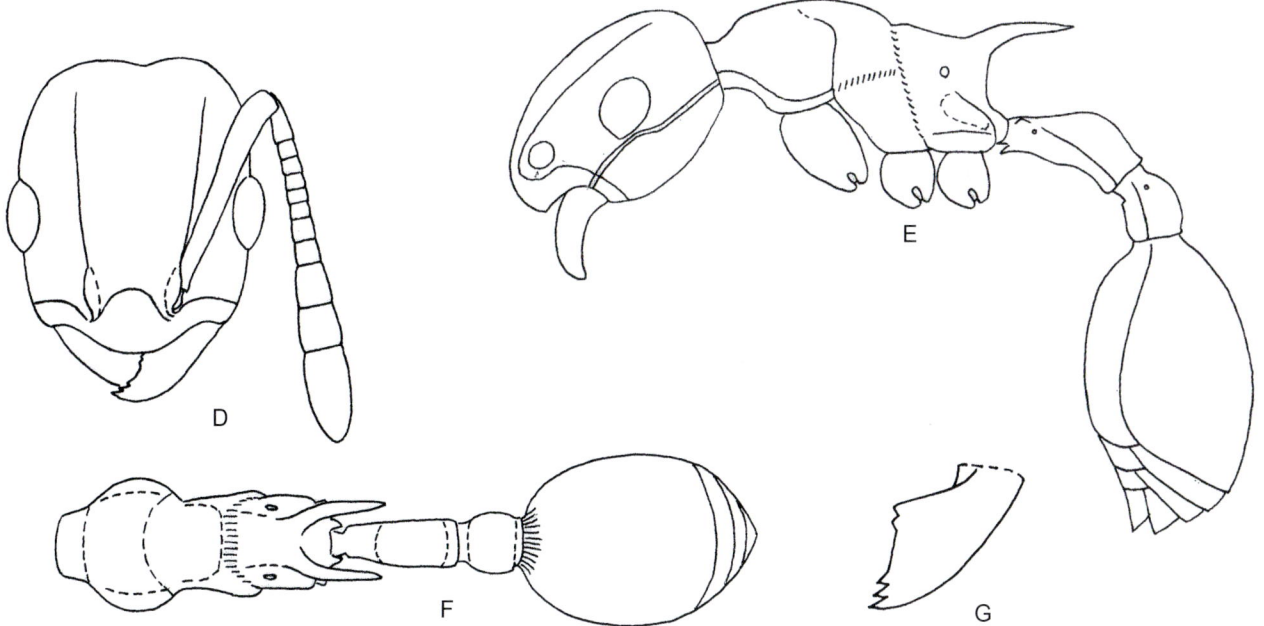

西藏犁沟蚁 *Vombisidris tibeta*

A, D. 工蚁头部正面观；B, E. 工蚁身体侧面观；C, F. 工蚁身体背面观；G. 工蚁上颚背面观（D-G. 引自Xu和Yu, 2012）
A, D. Head of worker in full-face view; B, E. Body of worker in lateral view; C, F. Body of worker in dorsal view; G. Mandible of worker in dorsal view (D-G. cited from Xu & Yu, 2012)

正模： 工蚁，中国西藏自治区墨脱县达木乡达木村，1200m，2011.Ⅶ.20，刘霞采于沟谷热带雨林树冠样中，No. A11-3928。

词源： 该新种以模式产地"西藏"命名。

凹头臭蚁
Dolichoderus incisus Xu, 1995

Dolichoderus incisus Xu, 1995a, Journal of Southwest Forestry College, 15(1): 37, figs. 9, 10 (w.) CHINA (Yunnan).

Holotype worker: TL 3.6, HL 1.08, HW 0.93, CI 86, SL 0.98, SI 105, PW 0.60, AL 1.38, ED 0.24. In full-face view, head roughly rectangular, distinctly longer than broad, sides convex, posterior margin broadly concave, posterior corners bluntly angled. Eyes large, located on midline of head. Mandibles with 11 teeth, the apical two teeth larger. Dorsum of clypeus weakly convex, without longitudinal central carina, anterior margin concave. Frontal carinae rised vertically as low ridges. Antennae 12-segmented, scapes surpassing posterior corners by 1/4 of their length; flagella gradually incrassate toward apex, the middle segments longer than broad. In lateral view, pronotum high, dorsum flat, marginated anteriorly. Promesonotal suture distinct and depressed. Mesonotum sloping down posteriorly, prominant at basal 1/3, dorsum weakly longitudinally depressed. Metanotal groove widely deeply concave. Propodeum low, dorsum flat, convex anteriorly, posterodorsal corner protruding and acutely toothed. In dorsal view, propodeum rectangular, posterior margin straigh; declivity strongly concave. Petiolar node relatively thicker, inclined anteriorly, anterior and posterior margins parallel, top margin rounded. In dorsal view, the node transverse, both anterior and posterior margins convex, dorsal margin straight. Anterior face of gaster vertically depressed. Mandibles smooth and shiny. Dorsa of head and alitrunk densely finely punctate, opaque; mesopleura coarsely longitudinally rugose and finely punctate, opaque; metapleura weakly punctate, relatively opaque. Petiole and gaster finely reticulate, relatively opaque. Dorsal face of head and body with abundant erect to suberect hairs and dense decumbent pubescence, pubescence on alitrunk sides sparse. Dorsa of antennal scapes and hind tibiae with abundant erect to suberect hairs and dense decumbent pubescence. Body color black; mandibles, antennae and legs reddish brown; coxae, profemora and tibiae black. **Paratype workers:** TL 3.6-4.4, HL 0.95-1.08, HW 0.83-0.98, CI 85-93, SL 0.90-1.00, SI 103-109, PW 0.53-0.63, AL 1.23-1.38, ED 0.23-0.26 ($n = 6$). As holotype.

Holotype: Worker, China: Yunnan Province, Lincang County, Lincang (23.8°N, 100.0°E), 1550 m, 1991.X.13, collected by Zheng-hui Xu on the ground in *Pinus kesiya* var. *langbianensis* forest, No. A91-1077.

Etymology: The species name *incisus* combines Latin *incis-* (cut into) + suffix *-us* (positive form), it refers to the concave posterior margin of head.

正模工蚁： 正面观，头部近长方形，长明显大于宽；两侧隆起，后缘宽形凹陷，头后角突出。复眼大，位于头中部。上颚具11个齿，端部2个齿较大。唇基轻度隆起，无中央纵脊，前缘凹陷。额脊垂直升起成低的脊状。触角12节，柄节约1/4超过头后角；鞭节向顶端逐渐加粗，中间各节长大于宽。侧面观前胸背板高，背面平坦，前部具边缘；前中胸背板缝明显，凹陷。中胸背板向后降低成坡形，基部1/3处隆起，背面轻度纵向凹陷；后胸沟宽而深凹。并胸腹节低，背面前部隆起，中部和后部平，后端突出，成尖齿状；背面观长方形，后缘直；端面强烈内凹。腹柄结较厚，前倾，前、后面平行，顶端圆形突起。背面观横形，前、后面均隆起；背缘直。后腹部前面纵向凹陷。上颚光滑发亮；头、胸部背面具密集细刻点，暗；中胸侧板具粗糙纵皱纹和细刻点，暗；后胸侧板刻点较弱，较暗。腹柄和后腹部具网状细刻纹，较暗。头和体背面具丰富直立、亚直立毛和密集倾斜绒毛被；胸部侧面绒毛被稀疏；触角柄节和后足胫节背面具丰富直立、亚直立毛和密集倾斜绒毛被。体黑色，上颚、触角和足红棕色，基节、前足腿节和胫节黑色。**副模工蚁：** 特征同正模。

正模： 工蚁，中国云南省临沧县临沧（23.8°N，100.0°E），1550m，1991.X.13，徐正会采于思茅松林地表，No. A91-1077。

词源： 该新种的种名"*incisus*"由拉丁语"*incis-*"（切入）+后缀"*-us*"（阳性形式）组成，指头部后缘凹陷。

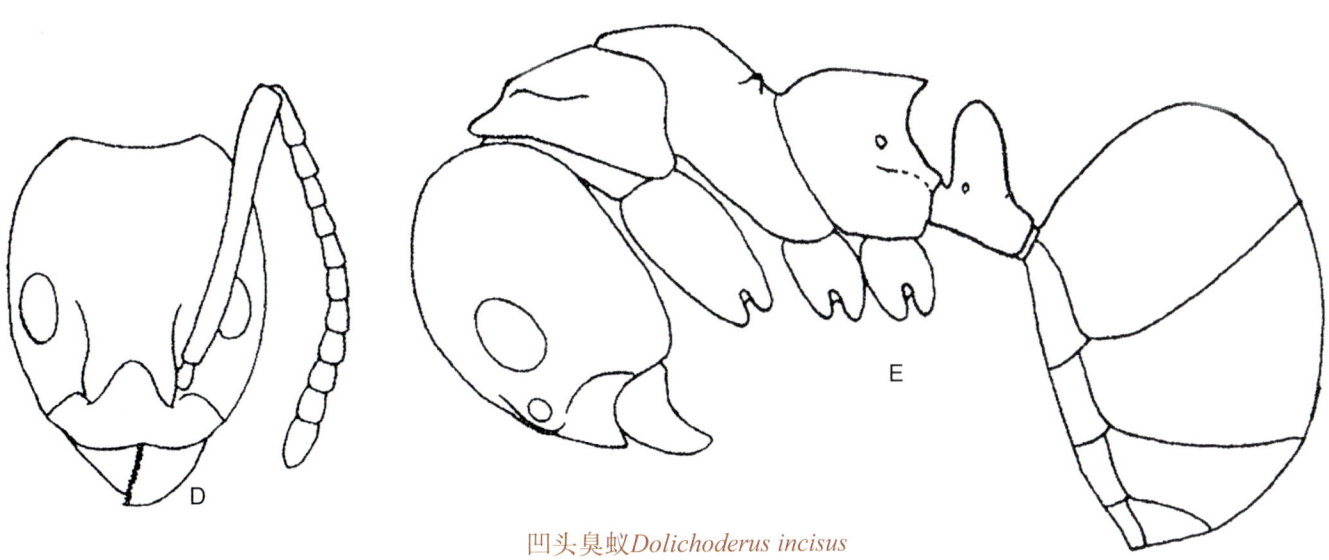

凹头臭蚁 *Dolichoderus incisus*

A, D. 工蚁头部正面观；B, E. 工蚁身体侧面观；C. 工蚁身体背面观（D-E. 引自 Xu, 1995a）
A, D. Head of worker in full-face view; B, E. Body of worker in lateral view; C. Body of worker in dorsal view (D-E. cited from Xu, 1995a)

鞍背臭蚁
Dolichoderus sagmanotus Xu, 2001

Dolichoderus sagmanotus Xu, 2001b, Acta Zootaxonomica Sinica, 26(3): 356, figs. 4-6 (w.) CHINA (Yunnan).

Holotype worker: TL 3.7, HL 0.90, HW 0.80, CI 89, SL 0.80, SI 100, PW 0.57, AL 1.17, ED 0.27. In full-face view, head longer than broad, narrowed forward. Occipital margin straight, occipital corners rounded, sides convex. Anterior margin of clypeus straight. Masticatory margin of mandible with 15 teeth. Scape of antenna surpassed occipital corner by about 1/8 of its length. Eyes large and convex. In lateral view, pronotum weakly convex, without margin. Promesonotal suture shallowly impressed. Mesonotum evenly convex. Metanotal groove wide and deep. Dorsum of propodeum strongly convex, longer than declivity, posterodorsal corner formed an acute angle. Declivity concave and margined along the lateral and dorsal margins. Node of petiole thick and inclined forward, anterior face short and vertical, posterior face long and slope down, dorsal face roundly convex. In dorsal view, posterior margin of propodeum complete and weakly convex. In front view, upper margin of petiolar node weakly convex. Mandibles with dense micro-punctures and sparse large punctures. Clypeus densely and longitudinally striate. Head and thorax foveolate. Foveolae of head relatively sparse, diameter of foveola about equal to distance between foveolae, interface with dense micro-punctures. Foveolae of thorax dense, interface formed coarse rugo-reticulation. Lower portion of lateral side of pronotum finely punctured. Lower portion of lateral side of mesothorax, lateral side of metathorax, and declivity of propodeum smooth and shining. Node of petiole finely and densely punctured. Gaster smooth and shining. Dorsum of head and body with sparse erect or suberect hairs and sparse decumbent pubescence, hairs on head shorter and relatively abundant. Scapes of antennae and tibiae of legs with sparse suberect hairs and sparse decumbent pubescence. Head, thorax and petiole reddish brown. Mandibles, antennae and legs yellowish brown. Eyes and gaster black.

Holotype: Worker, China: Yunnan Province, Mengla County, Shangyong Town, Manzhuang Village, 900 m, 1998.Ⅳ.10, collected by Tai-yong Liu in semi-evergreen monsoon forest, No. A98-429.

Etymology: The species name *sagmanotus* combines Greek *sagm-* (saddle) + word root *not* (back) + suffix *-us* (positive form), it refers to the saddle-shaped dorsum of mesosoma.

正模工蚁： 正面观，头部长大于宽，向前变窄；后缘平直，后角圆；侧缘隆起。唇基前缘平直。上颚咀嚼缘具15个齿。触角柄节约1/8超过头后角。复眼大而凸起。侧面观，前胸背板轻度隆起，无隆起的边缘。前中胸背板缝浅凹。中胸背板均匀隆起。后胸沟宽而深。并胸腹节背面强烈隆起，长于斜面，后上角成锐角状；斜面凹陷，两侧和背缘具边缘。腹柄结厚，前倾，前面短而垂直，后面长而呈坡形，背面圆形隆起。背面观，并胸腹节后缘完整，轻度隆起。前面观，腹柄结背缘轻度隆起。上颚具密集微刻点和稀疏大刻点。唇基具密集纵条纹。头部和胸部具凹坑；头部凹坑较稀疏，凹坑直径约等于凹坑间距，界面具密集微刻点。胸部凹坑密集，界面形成粗网纹。前胸背板侧面下部具细刻点；中胸侧面下部、后胸侧板和并胸腹节斜面光滑发亮。腹柄结具细密刻点。后腹部光滑发亮。头部和身体背面具稀疏直立、亚直立毛和稀疏倾斜绒毛被，头部立毛较短较丰富；触角柄节和足胫节具稀疏亚直立毛和稀疏倾斜绒毛被。头部、胸部和腹柄红棕色，上颚、触角和足黄棕色，复眼和后腹部黑色。

正模： 工蚁，中国云南省勐腊县尚勇乡曼庄村，900m，1998.Ⅳ.10，柳太勇采于半常绿季雨林中，No. A98-429。

词源： 该新种的种名"*sagmanotus*"由希腊语"*sagm-*"（马鞍）+词根"*not*"（背）+后缀"*-us*"（阳性形式）组成，指胸部背面呈马鞍形。

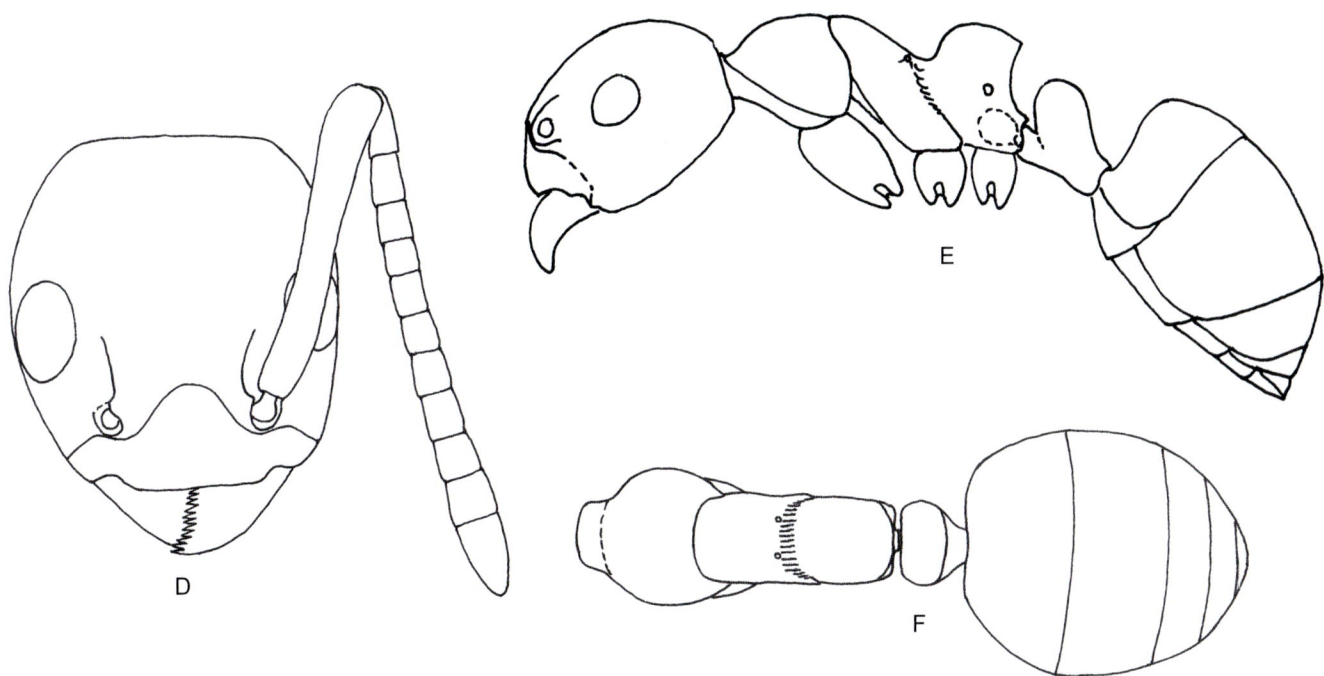

鞍背臭蚁 *Dolichoderus sagmanotus*
A, D. 工蚁头部正面观；B, E. 工蚁身体侧面观；C, F. 工蚁身体背面观（D-F. 引自Xu, 2001b）
A, D. Head of worker in full-face view; B, E. Body of worker in lateral view; C, F. Body of worker in dorsal view (D-F. cited from Xu, 2001b)

鳞结臭蚁
Dolichoderus squamanodus Xu, 2001

Dolichoderus squamanodus Xu, 2001b, Acta Zootaxonomica Sinica, 26(3): 357, figs. 22-24 (w.) CHINA (Yunnan).

Holotype worker: TL 3.5, HL 0.97, HW 0.93, CI 97, SL 0.80, SI 86, PW 0.67, AL 1.23, ED 0.27. In full-face view, head as broad as long, distinctly narrowed forward. Occipital margin nearly straight, occipital corners rounded, sides evenly convex. Anterior margin of clypeus shallowly concave in the middle. Mandible multi-dentate on inner and masticatory margins, inner margin with 7 teeth, masticatory margin with 11 ones. Scape of antenna surpassed occipital corner by about 1/7 of its length, segments 2-8 longer than broad. Dorsum of pronotum plane, anterior border marginate, the shoulder corners strongly convex, promesonotal suture distinct and shallowly impressed. In lateral view mesonotum convex and lowered down posteriorly, metanotal groove deeply notched. Propodeum elevated and formed a horizontal plane, almost as high as pronotum, anterior face nearly vertical, posterodorsal corner protruding in an acute angle, declivity roundly concave and about 1.4 times as long as dorsum. In dorsal view, posterior margin of propodeum roundly convex. In lateral view, petiolar node relatively thin and erect, scale-like, both anterior and posterior faces weakly convex, in front view, upper margin straight. Mandibles with dense micro-punctures. Head finely and weakly punctured, clypeus sparsely and longitudinally striate. Thorax densely foveolate, interface formed coarse rugo-reticulation, dorsum of pronotum coarsely and longitudinally rugose. Lateral side of pronotum densely and finely punctured. Lateral side of metathorax and declivity smooth and shining. Petiolar node and gaster with dense micro-punctures, relatively shining. Dorsum of body with abundant erect hairs and decumbent pubescence, scapes and tibiae with sparse erect hairs and abundant decumbent pubescence. Head and alitrunk black, petiolar node and gaster brownish black. Mandibles, clypeus, antennae and legs yellowish brown. **Paratype workers:** TL 2.7-3.7, HL 0.78-0.93, HW 0.80-0.93, CI 100-102, SL 0.60-0.80, SI 75-86, PW 0.53-0.67, AL 0.90-1.20, ED 0.20-0.27 (*n*=5). As holotype, but in some individuals with head, petiolar node and gaster very weakly punctured and relatively shining or smooth and shining.

Holotype: Worker, China: Yunnan Province, Mengla County, Mengla Town, Bubang Village, 760 m, 1996.Ⅲ.11, collected by Yong-chao Du in seasonal rain forest, No. A96-590.

Etymology: The species name *squamanodus* combines Latin *squam-* (scale) + word root *nod* (knot) + suffix *-us* (positive form), it refers to the scale-shaped petiolar node.

正模工蚁： 正面观，头部长宽相等，向前明显变窄；后缘近平直，后角圆；侧缘均匀隆起。唇基前缘中央浅凹。上颚内缘和咀嚼缘多齿，内缘具7个齿，咀嚼缘具11个齿。触角柄节约1/7超过头后角，第2～8节长大于宽。前胸背板背面平坦，前缘具隆起的边缘，肩角强烈隆起；前中胸背板缝明显，浅凹。侧面观，中胸背板隆起，向后降低，后胸沟深切。并胸腹节升高，形成1个平面，几乎与前胸背板等高，前面近垂直，后上角突出呈锐角状；斜面圆形凹陷，约为背面长的1.4倍。背面观，并胸腹节后缘圆形隆起。侧面观，腹柄结较薄，直立，呈鳞片状，前面和后面轻度隆起；前面观背缘平直。上颚具密集微刻点。头部具细弱刻点，唇基具稀疏纵条纹。胸部具密集凹坑，界面形成粗网纹；前胸背板背面具粗糙纵皱纹，侧面具细密刻点；后胸侧板和并胸腹节斜面光滑发亮。腹柄结和后腹部具密集微刻点，较光亮。身体背面具丰富直立毛和倾斜绒毛被；柄节和胫节具稀疏直立毛和丰富倾斜绒毛被。头部和胸部黑色，腹柄结和后腹部棕黑色，上颚、唇基、触角和足黄棕色。**副模工蚁：** 特征同正模，但是一些个体的头部、腹柄结和后腹部具很弱的刻点且较光亮，或者光滑发亮。

正模： 工蚁，中国云南省勐腊县勐腊镇补蚌村，760m，1996.Ⅲ.11，杜永超采于季节性雨林中，No. A96-590。

词源： 该新种的种名"*squamanodus*"由拉丁语"*squam-*"（鳞片）+词根"*nod*"（结）+后缀"*-us*"（阳性形式）组成，指腹柄结呈鳞片状。

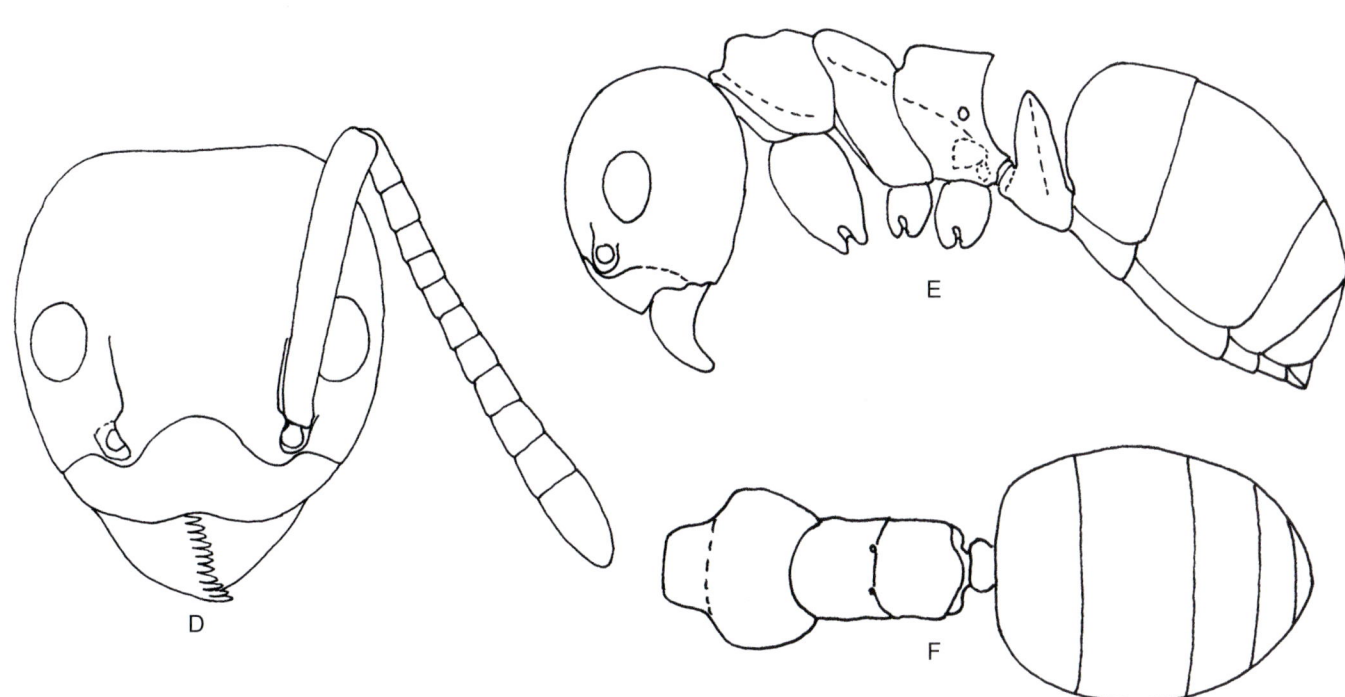

鳞结臭蚁 *Dolichoderus squamanodus*

A, D. 工蚁头部正面观；B, E. 工蚁身体侧面观；C, F. 工蚁身体背面观（D-F. 引自Xu, 2001b）
A, D. Head of worker in full-face view; B, E. Body of worker in lateral view; C, F. Body of worker in dorsal view (D-F. cited from Xu, 2001b)

尖齿刺结蚁
Lepisiota acuta Xu, 1994

Lepisiota acuta Xu, 1994b, Journal of Southwest Forestry College, 14(4): 234, figs. 4-6 (w.) CHINA (Yunnan).

Holotype worker: TL 2.8, HL 0.63, HW 0.60, CI 96, SL 0.68, SI 113, PW 0.43, AL 0.85, ED 0.20. In full-face view, head longer than broad, narrowing anteriorly; lateral margins convex, posterior corners rounded, posterior margin straight. Mandibles triangular, masticatory margin with 5 teeth. Clypeus convex in the center, without longitudinal central carina, anterior margin bluntly angled. Frontal carinae short and parallel to each other. Frontal area small and triangular. Antennae 11 segmented, scapes surpassing posterior head corner by 1/3 of their length, flagella incrassate toward apex. Eyes moderately large, located slightly behind midline of head, with 3 ocelli. In lateral view, pronotum high, dorsum flat. Promesonotal suture distinct, depressed on the sides. Mesonotum sloping down posteriorly, strongly constricted in the middle; spiracle prominent and located on the dorsum. Metanotal groove deeply impressed. Propodeum low, dorsum flat; propodeal spines extending posteriorly and acutely toothed; spiracles large, located at ventral base of the spines. In dorsal view, propodeum nearly crescent-shaped, posterior margin concave, propodeal spines bluntly toothed. Petiolar node high, anterior margin vertical, posterior margin sloping; upper margin with a pair of back-curved spines. In front view, petiolar node narrowing dorsally, upper margin roundly concave; spines slender, as long as the distance between their bases, laterodorsally pointed. Base of gaster protruding forward and overhanging the posterior part of petiole. Mandibles smooth. Head and prothorax densely finely reticulate, relatively shiny; mesothorax, metathorax and propodeum densely coarsely reticulate, relatively opaque; upper part of mesopleura longitudinally rugose. Petiole and gaster smooth and shiny. Dorsa of head and body with sparse erect to suberect hairs and sparse decumbent pubescence, hairs on gaster abundant. Dorsal faces of antennal scapes and hind tibiae with dense decumbent pubescence, standing hairs absent. Body color black, mandibles, antennae and tarsi brown. **Paratype workers:** TL 2.1-2.8, HL 0.58-0.65, HW 0.54-0.60, CI 90-96, SL 0.63-0.68, SI 113-118, PW 0.39-0.43, AL 0.80-0.90, ED 0.16-0.20 (n = 6). As holotype.

Holotype: Worker, China: Yunnan Province, Anning County, Wenquan Town, Yangjiaocun Village (24.9°N, 102.4°E), 1850 m, 1991.Ⅹ.1, collected by Zheng-hui Xu on the ground in conifer-broadleaf mixed forest, No. A91-900.

Etymology: The species name *acuta* combines Latin *acut-* (sharp) + suffix *-a* (feminine form), it refers to the acutely toothed propodeal spines.

正模工蚁： 正面观，头长大于宽，两侧隆起，后部宽于前部；后角圆突，后缘直。上颚长三角形，咀嚼缘具5个齿。唇基中央突起，无中央纵脊，前缘钝角状突出。额脊短，互相平行。额区小，三角形。触角11节，柄节约1/3超出头后角，鞭节向端部变粗。复眼中等大小，位于头中线稍后处，单眼3个。侧面观前胸背板高，背面平坦，前中胸背板缝明显，在侧面凹陷。中胸背板向后降低，中部强烈收缩，气门位于背面，突出，后胸沟深凹。并胸腹节低，背面平，并胸腹节刺后延成尖齿突，气门大，位于齿突的腹面；背面观，并胸腹节近新月形，后缘凹陷，并胸腹节刺钝齿状。腹柄结高，前面垂直，后面坡形，背缘具1对后弯的刺突；前面观，腹柄结向上变窄，背缘弧形凹陷，刺突细长，其长约等于2个刺基间距，指向侧上方。后腹部基部向前突出，悬覆于腹柄后部上方。上颚光滑。头部和前胸具密集网状细刻纹，较光亮；中后胸和并胸腹节具密集网状粗刻纹，较暗；中胸侧板上部具纵皱纹；腹柄和后腹部光滑发亮。头和体背面具稀疏直立、亚直立毛和稀疏倾斜绒毛被，腹部毛被丰富。触角柄节和后足胫节背面具密集倾斜绒毛被，缺立毛。体黑色，上颚、触角和跗节褐色。**副模工蚁：** 特征同正模。

正模： 工蚁，中国云南省安宁县温泉镇羊角村（24.9°N，102.4°E），1850m，1991.Ⅹ.1，徐正会采于针阔混交林地表，No. A91-900。

词源： 该新种的种名"*acuta*"由拉丁语"*acut-*"（尖锐的）+后缀"*-a*"（阴性形式）组成，指该种的并胸腹节刺呈尖齿状。

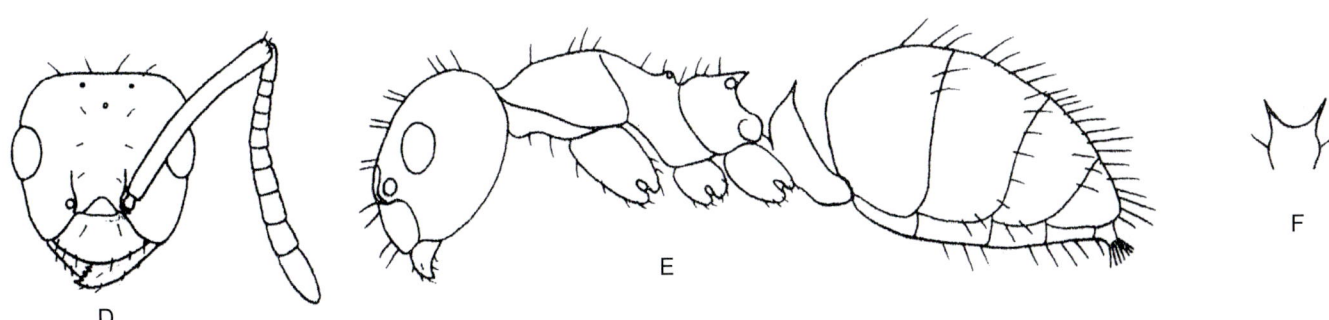

尖齿刺结蚁 *Lepisiota acuta*

A, D. 工蚁头部正面观；B, E. 工蚁身体侧面观；C. 工蚁身体背面观；F. 工蚁腹柄结前面观（D-F. 引自Xu, 1994b）
A, D. Head of worker in full-face view; B, E. Body of worker in lateral view; C. Body of worker in dorsal view; F. Petiolar node of worker in front view (D-F. cited from Xu, 1994b)

网纹刺结蚁
Lepisiota reticulata Xu, 1994

Lepisiota reticulata Xu, 1994b, Journal of Southwest Forestry College, 14(4): 234, figs. 7-9 (w.) CHINA (Guangxi).

Holotype worker: TL 2.4, HL 0.60, HW 0.55, CI 92, SL 0.60, SI 109, PW 0.38, AL 0.80, ED 0.16. In full-face view, head longer than broad, slightly narrowing anteriorly; lateral and posterior margins weakly convex, posterior corners rounded. Mandibles triangular, with 5 teeth. Clypeus convex in the center, without longitudinal central carina, anterior margin roundly convex. Frontal area triangle. Frontal carinae short, weakly diverging posteriorly. Antennae 11 segmented, scapes surpassing posterior head corners by about 1/3 of their length, flagella incrassate toward apex. Eyes moderately large, located slightly behind midline of head; with 3 small ocelli. In lateral view, pronotum high, dorsum flat, promesonotal suture obvious. Mesonotum sloping down posteriorly, strongly constricted in the middle; spiracles prominent, located on the dorsum. Metanotal groove deeply impressed. Propodeum low, dorsum flat; spines extending posteriorly, acutely toothed and weakly elevated; spiracles located on the ventral bases of the spines. In dorsal view, propodeum crescent-shaped, posterior margin concave, spines bluntly toothed. Petiolar node high, anterior margin vertical, posterior margin concave, upper margin with a pair of back-curved spines, pointing laterodorsally. In front viewed, spines slightly shorter than the distance between their bases, apart from each other; upper margin of petiolar node weakly concave, petiole with cylindrical posterior peduncle. Base of gaster protruding forward and overhanging the posterior part of petiole. Mandibles smooth. Head and alitrunk uniformly, densely and coarsely reticulate, opaque; clypeus weakly reticulate, relatively shiny; upper parts of mesopleura with few longitudinal costulae. Petiole and gaster smooth and shiny. Dorsa of head and body with abundant erect to suberect hairs and sparse decumbent pubescence, hairs on gaster dense. Dorsal faces of antennal scapes and hind tibiae with dense subdecumbent to decumbent pubescence, without standing hairs. Head, alitrunk and petiole reddish brown, gaster black; antennal apices, femora and tibiae of middle and hind legs blackish brown. **Paratype workers:** TL 2.0-2.7, HL 0.55-0.63, HW 0.50-0.58, CI 91-93, SL 0.58-0.63, SI 109-115, PW 0.35-0.40, AL 0.73-0.83, ED 0.16-0.18 ($n = 6$). As holotype.

Holotype: Worker, China: Guangxi Zhuang Autonomous Region, Lipu County, Lipu (24.5°N, 110.4°E), 300 m, 1992.Ⅵ.14, collected by Zheng-hui Xu on the ground in broadleaf forest, No. A92-243.

Etymology: The species name *reticulata* combines Latin *reticul-* (net) + suffix *-ata* (feminine form, with), it refers to the reticulate sculpture of head and alitrunk.

正模工蚁： 正面观，头长大于宽，两侧轻度隆起，后部稍宽于前部；头后角圆，头后缘轻度隆起。上颚三角形，具5个齿。唇基中部突起，无中央纵脊；前缘圆形突出。额区三角形。额脊短，向后轻度分歧。触角11节，柄节约1/3超出头后角，鞭节向顶端变粗。复眼中等大小，位于头中线稍后处；单眼小，3个。侧面观前胸背板高，背面平坦，前中胸背板缝明显。中胸背板向后降低，中部强烈收缩，气门位于背面，突起，后胸沟深凹。并胸腹节低，背面平，并胸腹节刺后延成尖齿状，轻度升高，气门位于齿突腹面。背面观并胸腹节新月形，后缘凹陷，并胸腹节刺钝齿状。腹柄结高，前面垂直，后面凹陷，背缘具1对后弯的刺突，指向侧上方；前面观刺突稍短于2个刺基间距，左右离开，背缘轻度弧形凹陷，腹柄后部形成圆柱形小柄。后腹部基部向前突出，悬覆于腹柄后部上方。上颚光滑。头胸部具均匀一致的密集粗糙网状刻纹，暗；唇基刻纹弱，较光亮；中胸背板两侧具少数纵皱纹。腹柄和后腹部光滑发亮。头和体背面具丰富直立、亚直立毛和稀疏倾斜绒毛被，腹部毛被密集；触角柄节和后足胫节背面具密集亚倾斜、倾斜绒毛被，缺立毛。头胸部和腹柄红棕色，腹部黑色，触角端部及中、后足腿节和胫节黑褐色。

副模工蚁： 特征同正模。

正模： 工蚁，中国广西壮族自治区荔浦县荔浦（24.5°N，110.4°E），300m，1992.Ⅵ.14，徐正会采于阔叶林地表，No. A92-243。

词源： 该新种的种名"*reticulata*"由拉丁语"*reticul-*"（网）+后缀"*-ata*"（阴性形式，具有）组成，指该种的头部和胸部具有网状的刻纹。

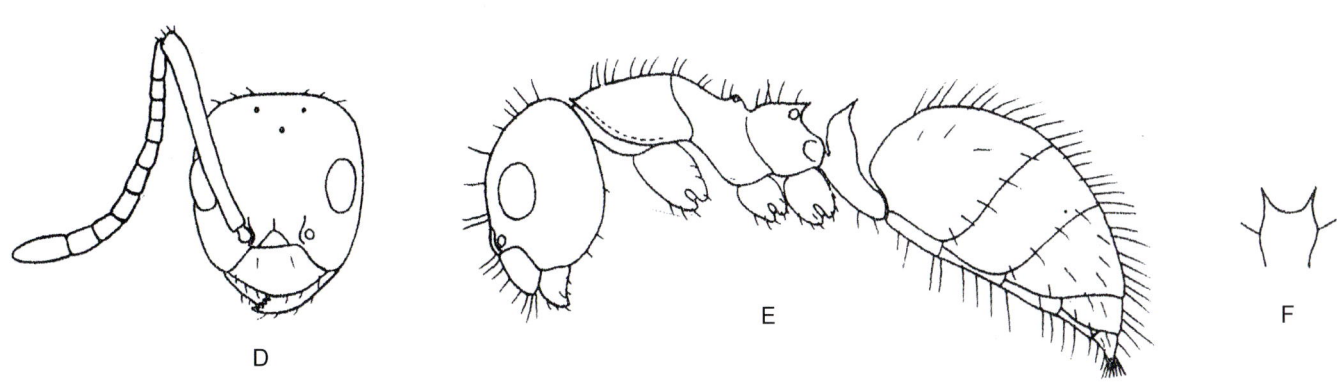

网纹刺结蚁 Lepisiota reticulata

A, D. 工蚁头部正面观；B, E. 工蚁身体侧面观；C. 工蚁身体背面观；F. 工蚁腹柄结前面观（D-F. 引自Xu, 1994b）
A, D. Head of worker in full-face view; B, E. Body of worker in lateral view; C. Body of worker in dorsal view; F. Petiolar node of worker in front view (D-F. cited from Xu, 1994b)

楔结长齿蚁
Myrmoteras cuneonodum Xu, 1998

Myrmoteras cuneonodum Xu, 1998c, Entomologia Sinica, 5(2): 125, figs. 12, 13 (w.) CHINA (Yunnan).

Holotype worker: TL 5.0, HL 1.10, HW 1.03, CI 94, SL 1.10, SI 106, PW 0.60, AL 1.40, ED 0.63, ML 1.53, MI 139. In full-face view, head nearly triangular, occipital corners extruding, nearly in right angle. Frontal sulcus distinct, narrow and deep, from middle ocellus to the level of antennal foveae. Mandibles long and slender, filiform, inner margin with 11 long teeth, becoming shorter successively from apex to base; with 2 small denticles between the apical 1st and 2nd teeth, the posterior one is smaller than the anterior one, between apical 2nd and 3rd, and 3rd and 4th teeth, each with 1 small denticle. Anterior margin of labrum straight, without a pair of long trigger hairs. Anterior margin of clypeus roundly concave, sides extruding into sharp angles. Transverse sulcus behind clypeus distinct. Antennal scapes surpass occipital corners by about 1/2 of their length, flagella filiform. Eyes long elliptic, with 3 ocelli. In lateral view, mesothorax constricted and cylindrical, metanotal spiracles protruding. Pronotum slightly convex. Dorsum and declivity of propodeum very weakly convex, nearly straight, declivity shorter than dorsum. In lateral view, petiolar node cuneiform, narrowering upward, anterior face nearly straight, upper portion of posterior face slightly convex. Head and mandibles smooth and shining; central dorsum of head finely rugose, the rugae branching backward; genae below eyes with short oblique rugae. Pronotum, anterior part of mesonotum and mesopleura smooth and shining; cervicum and propodeum transversely rugose; dorsum of the cylindrical portion of mesothorax granulate, the sides roughly longitudinally rugose. Petiolar node and gaster smooth and shining. Dorsa of head and body with sparse erect hairs, pubescence absent, hairs on gaster abundant. Antennal scapes and hind tibiae with rich suberect hairs. Body in colour reddish brown; gaster dark reddish brown; mandibles, antennae and legs brownish yellow.

Holotype: Worker, China: Yunnan Province, Jinghong County, Mengyang Town, Sanchahe River, 950 m, 1997.Ⅱ.28, collected by Yu-chu Lai from a ground sample of seasonal rain forest, No. A97-79.

Etymology: The species name *cuneonodum* combines Latin *cune-* (wedge-shaped) + word root *nod* (knot) + suffix *-um* (neutral form), it refers to the cuneiform petiolar node.

正模工蚁： 正面观，头部近三角形，后角突出，近直角形。额沟明显，狭窄而深，从中单眼延伸至触角窝水平。上颚细长，呈线形，内缘具11个长齿，从端部到基部依次变短；端部第1、2齿之间具2个小齿，后面的小齿小于前面的小齿；在端部第2、3齿之间和第3、4齿之间各具1个小齿。上唇前缘平直，缺1对长的触发毛。唇基前缘圆形凹陷，两侧突出成锐角。唇基后方横沟明显。触角柄节约1/2超过头后角，鞭节丝状。复眼长椭圆形，具3个单眼。侧面观，中胸收缩呈圆柱形，后胸气门突出。前胸背板轻度隆起。并胸腹节的背部和斜面轻度隆起，近平直，斜面短于背面。侧面观，腹柄结楔形，向上变窄，前面近平直，后面上部轻微隆起。头部和上颚光滑发亮；头部背面中央有细皱纹，皱纹向后分支；复眼下方颊区具短的倾斜皱纹。前胸背板、中胸背板前部和中胸侧板光滑发亮；颈部和并胸腹节具横皱纹；中胸圆柱形部分的背面具粒纹，侧面具粗糙纵皱纹。腹柄结和后腹部光滑发亮。头部和身体背面具稀疏直立毛，缺绒毛被；后腹部立毛丰富；触角柄节和后足胫节具丰富亚直立毛。身体红棕色，后腹部暗红棕色，上颚、触角和足棕黄色。

正模： 工蚁，中国云南省景洪县勐养镇三岔河，950m，1997.Ⅱ.28，赖玉初采于季节性雨林地表样中，No. A97-79。

词源： 该新种的种名"*cuneonodum*"由拉丁语"*cune-*"（楔形）+词根"*nod*"（结）+后缀"*-um*"（中性形式）组成，指该种的腹柄结呈楔形。

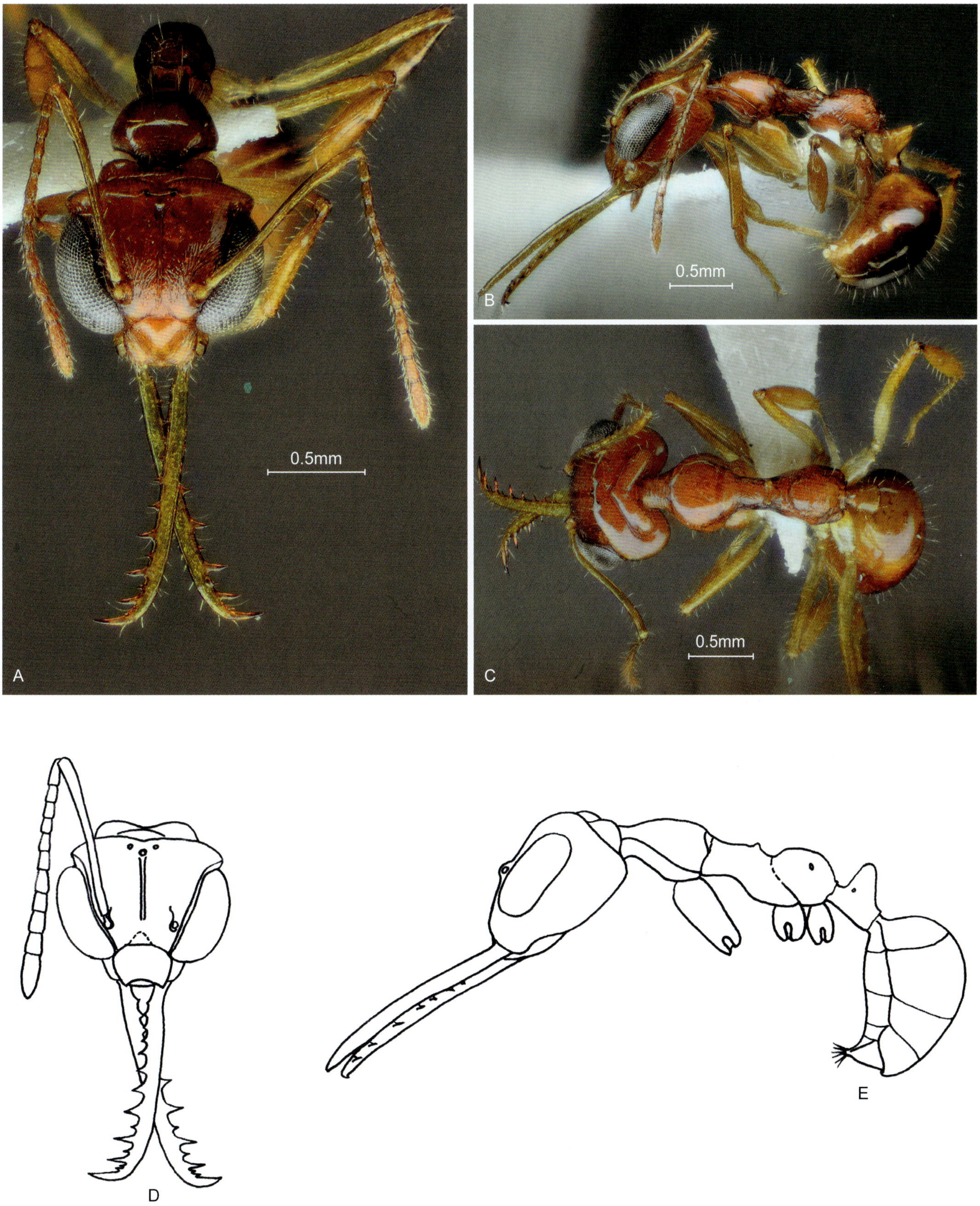

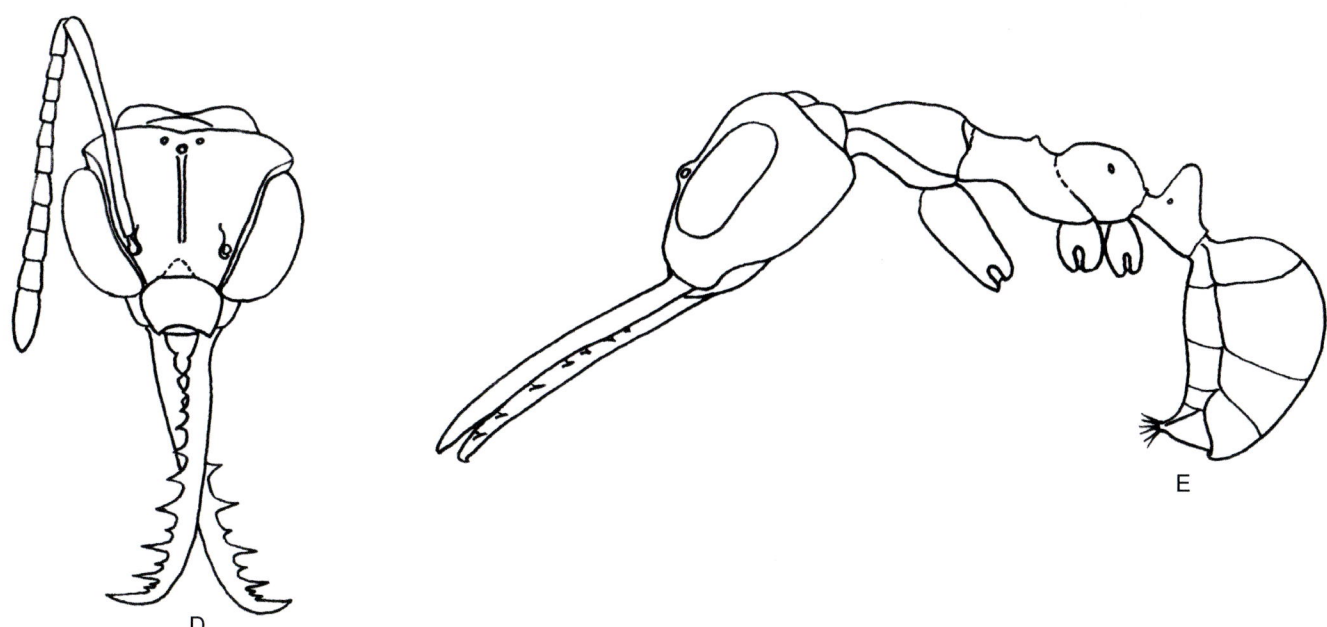

楔结长齿蚁 *Myrmoteras cuneonodum*

A, D. 工蚁头部正面观；B, E. 工蚁身体侧面观；C. 工蚁身体背面观（D-E. 引自 Xu, 1998c）

A, D. Head of worker in full-face view; B, E. Body of worker in lateral view; C. Body of worker in dorsal view (D-E. cited from Xu, 1998c)

127 巴卡多刺蚁
Polyrhachis bakana Xu, 1998

Polyrhachis bakana Xu, 1998b, Zoological Research, 19(3): 244, figs. 7-10 (w.) CHINA (Yunnan).

Holotype worker: TL 8.0, HL 2.20, HW 1.60, CI 73, SL 3.00, SI 188, PW 1.25, AL 3.15, ED 0.50. In full-face view, head longer than broad, sides nearly straight and parallel, narrowered near the bases of mandibles; occiput subtriangular, occipital margin strongly prominent in the middle, lateroposterior margins of occiput nearly straight. Eyes prominent and well behind midline of the head. Mandibles with 5 teeth. Clypeus longitudinally convex in the middle, but without carina; anteror margin slightly roundly convex. Antennal scapes compressed, surpassing occipital margin by 3/5 of its length. In lateral view, the place where pronotum meets mesonotum is the highest, slightly lowering down anteriorly and posteriorly, mesonotum slightly concave. Pronotum and mesonotum rounded laterally, but posterior 1/5 of mesonotal dorsum weakly marginate. Promesonotal suture distinct and slightly depressed; metanotal groove visible, only depressed on the sides. Pronotum and propodeum with long spines; pronotal spines straight and anteriorly pointed, slightly shorter than propodeal ones; propodeal spines dorsoposteriorly pointed, wide at bases. Dorsum of propodeum straight, longitudinally concave in the middle, and marginate laterally; declivity concave. In dorsal view, pronotal spines lateroanteriorly pointed, propodeal ones lateroposteriorly pointed and slightly curved outwards. Middle and hind tibiae compressed. In lateral view, anterior, posterior, and dorsal faces of petiolar node strongly convex; petiolar spines slender and curved backwards, about 1/2 length of propodeal ones; in dorsal view, the spines lateroposteriorly pointed. Mandibles finely longitudinally striate and sparsely punctate, subopaque. Head, alitrunk and petiole uniformly, closely and coarsely punctate, opaque. Gaster uniformly, closely and finely punctate, opaque. Whole body and appendages covered with sparse short depressed pubescence, distance between them about 2-3 times as long as one pubescence. Setae absent from whole body except mouthparts and apex of gaster. Color black; insertions of antennae, palpi and apex of gaster yellowish brown; eyes grayish brown. **Paratype workers:** TL 7.3-8.1, HL 1.80-2.20, HW 1.30-1.60, CI 68-75, SL 2.70-3.00, SI 188-214, PW 1.10-1.25, AL 2.75-3.15, ED 0.45-0.50 (*n*=5). As holotype, but in some individuals, propodeal spines only slightly divergent in dorsal view.

Holotype: Worker, China: Yunnan Province, Mengla County, Menglun Town, Bakaxiaozhai (21.9°N, 101.1°E), 840 m, 1996.Ⅲ.8, collected by Zhong-wen Yang from a nest containing 676 workers inside decayed bamboo on the ground in seasonal rain forest of Xishuangbanna National Nature Reserve, No. A96-347.

Etymology: The new species is named after the type locality "Bakaxiaozhai".

正模工蚁： 正面观，头长大于宽，侧缘近平直且互相平行，在上颚基部附近变窄；头后部近三角，后缘中部强烈突起，侧后缘近平直。复眼突出，位于头中线后方。上颚具5个齿。唇基中部纵向隆起，缺中央纵脊，前缘轻度圆形隆起。触角柄节压扁，3/5超过头后缘。侧面观，前胸背板与中胸背板交界处最高，向前向后均轻度降低，中胸背板轻微凹陷。前胸背板和中胸背板两侧圆，但是中胸背板背面后部1/5具弱边缘。前中胸背板缝明显，轻微凹陷；后胸沟可见，仅在侧面凹陷。前胸背板和并胸腹节具长刺；前胸背板刺直，指向前方，稍短于并胸腹节刺；并胸腹节刺指向后上方，基部宽。并胸腹节背面平直，中间纵向凹陷，两侧具隆起的边缘；斜面凹陷。背面观，前胸背板刺指向侧前方；并胸腹节刺指向侧后方，轻微外弯。中足和后足胫节压扁。侧面观，腹柄结的前面、后面和背面强烈隆起；腹柄结刺细长，后弯，约为并胸腹节刺长的1/2；背面观，腹柄结刺指向侧后方。上颚具细纵条纹和稀疏刻点，较暗。头部、胸部和腹柄具均匀一致的密集粗刻点，暗。后腹部具均匀一致的密集细刻点，暗。全身和附肢具稀疏平伏短绒毛被，绒毛间距约为绒毛长度的2～3倍；除口器和后腹部末端外，全身缺立毛。身体黑色，触角插入部、下颚须、下唇须和后腹部末端黄棕色，复眼灰棕色。**副模工蚁：** 特征同正模，但是背面观一些个体的并胸腹节刺仅轻度分歧。

正模： 工蚁，中国云南省勐腊县勐仑镇巴卡小寨（21.9°N，101.1°E），840m，1996.Ⅲ.8，杨忠文采于西双版纳国家级自然保护区季节性雨林地表1个包含676头工蚁的朽竹内巢中，No. A96-347。

词源： 该新种以模式产地"巴卡小寨"命名。

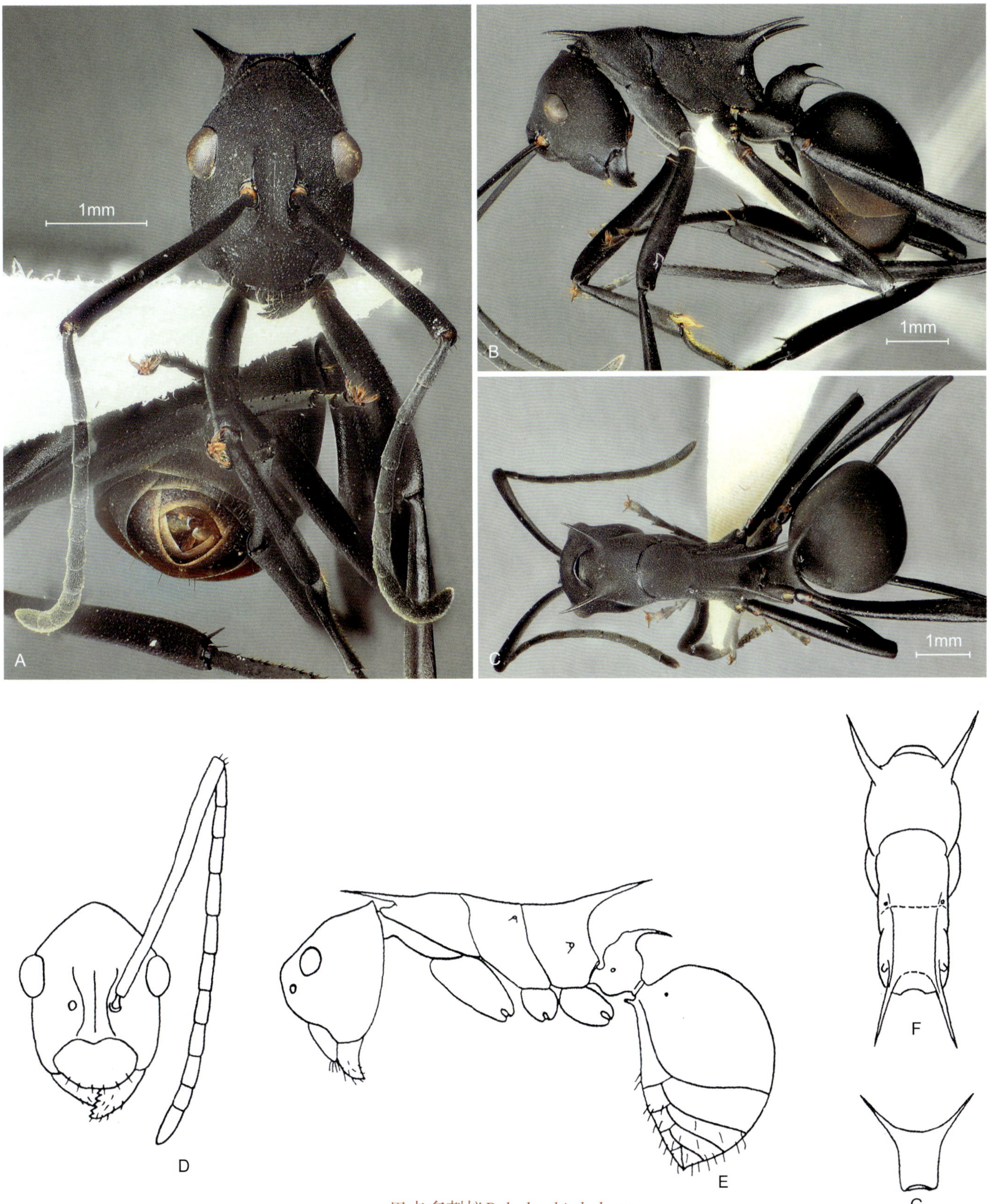

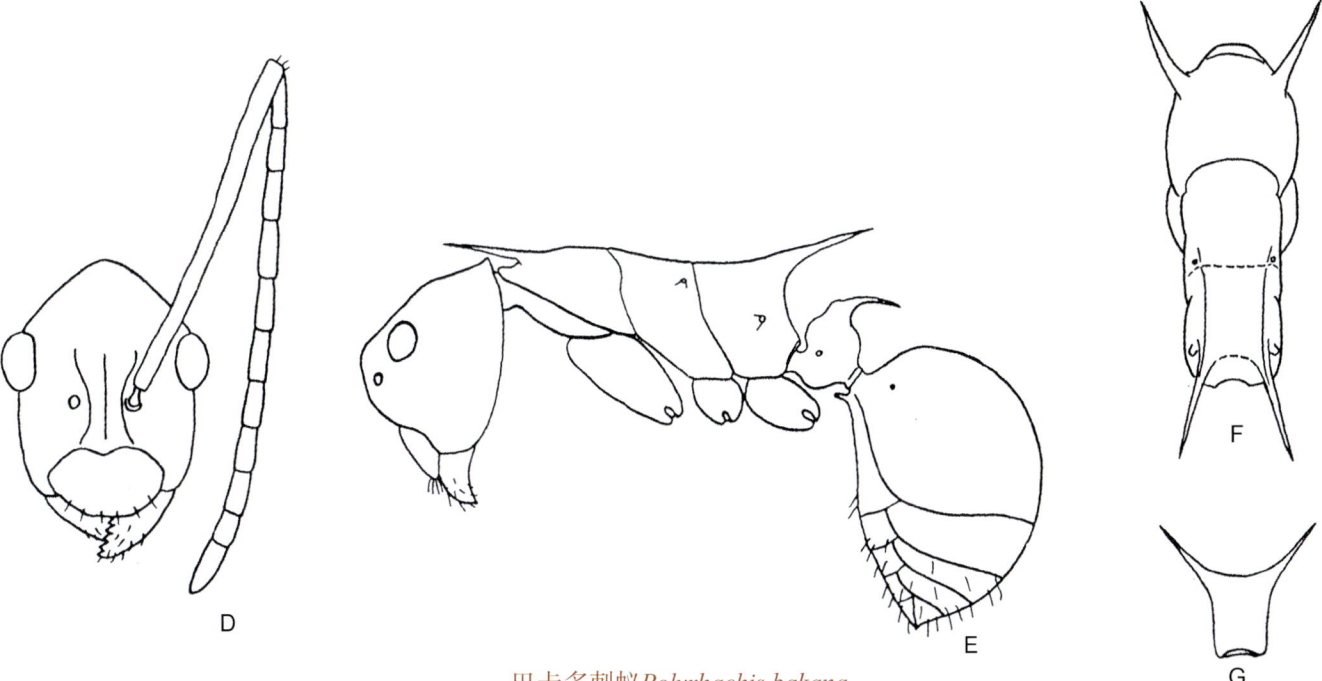

巴卡多刺蚁 *Polyrhachis bakana*

A, D. 工蚁头部正面观; B, E. 工蚁身体侧面观; C. 工蚁身体背面观; F. 工蚁胸部背面观; G. 工蚁腹柄结前面观 (D-G. 引自 Xu, 1998b)

A, D. Head of worker in full-face view; B, E. Body of worker in lateral view; C. Body of worker in dorsal view; F. Alitrunk of worker in dorsal view; G. Petiolar node of worker in front view (D-G. cited from Xu, 1998b)

短胸多刺蚁
Polyrhachis brevicorpa Xu, 2002

Polyrhachis brevicorpa Xu, 2002c, Acta Entomologica Sinica, 45(4): 528, figs. 34-37 (w.) CHINA (Yunnan).

Holotype worker: TL 5.0, HL1.50, HW1.45, CI 97, SL 1.85, SI 129, PW 1.10, AL 1.80, ED 0.40, MTL 2.25. In full-face view, head about as broad as long, roughly triangular. Occipital margin roundly convex, occipital corners rounded. Sides weakly convex. Mandibles with 5 teeth. Clypeus evenly convex, without longitudinal central carina, anterior margin with a pair of small denticles and straight between the denticles. Antennae long, scape surpassing occipital corner by about 2/3 of its length. Eyes convex, placed at occipital corners. In lateral view, dorsum of alitrunk strongly convex, promesonotal suture distinct, metanotal groove absent. Propodeum unarmed, dorsum and declivity straight and separated by a blunt conex. In dorsal view, shoulders of pronotum rounded. In front view, petiolar node with a pair of short acute lateral teeth, median teeth elongate and slender, longer than the lateral ones. In lateral view, petiolar node triangular, median teeth bent backward. Mandibles with dense micro-punctures, less shiny. Head, alitrunk, petiole and gaster uniformly, densely and finely punctured, dull. Whole body with abundant depressed short pubescence. Dorsum of head with sparse erect hairs, mesonotum with a pair of erect hairs, apex of gaster with abundant suberect hairs. Occiput, pronotum, propodeum, petiole and dorsum of first gastral segment without hairs. Scapes with dense depressed pubescence, femora and tibiae with abundant depressed pubescence, both scapes, femora and tibiae without erect hairs. Whole body black, eyes yellowish brown, legs black, hairs grayish white. **Paratype workers:** TL 5.3-5.8, HL 1.55-1.70, HW 1.50-1.65, CI 97, SL 1.80-1.85, SI 120-127, PW 1.15-1.35, AL 1.85-2.10, ED 0.45-0.50, MTL 2.20-2.65 (*n*=4). As holotype.

Holotype: Worker, China: Yunnan Province, Jinghong County, Dadugang Town, Guanping Village, 1120 m, 1997.Ⅷ.7, collected by Zheng-hui Xu from a canopy sample of mountain rain forest, No. A97-977.

Etymology: The species name *brevicorpa* combines Latin *brev-* (short) + word root *corp* (body) + suffix *-a* (feminine form), it refers to the relatively shorter alitrunk of the species.

正模工蚁： 正面观，头部长宽约相等，近三角形；后缘圆形隆起，后角圆，侧缘轻度隆起。上颚具5个齿。唇基均匀隆起，缺中央纵脊，前缘具1对小齿，小齿之间平直。触角长，柄节约2/3超过头后角。复眼隆起，位于头后角处。侧面观，胸部背面强烈隆起，前中胸背板缝明显，缺后胸沟。并胸腹节缺刺，背面和斜面平直，二者之间弱弓形轻微隆起。背面观，前胸背板肩角圆。前面观，腹柄结1对短的尖锐侧齿；中齿伸长，细长，长于侧齿。侧面观，腹柄结三角形，中齿后弯。上颚具密集微刻点，不太光亮。头部、胸部、腹柄和后腹部具均匀一致的细密刻点，暗。全身具丰富平伏短绒毛被。头部背面具稀疏直立毛，中胸背板具1对直立毛，后腹部末端具丰富亚直立毛。头后部、前胸背板、并胸腹节、腹柄和后腹部第1节背面缺立毛。柄节具密集平伏绒毛被；腿节和胫节具丰富平伏绒毛被；柄节、腿节和胫节均无立毛。身体黑色，复眼黄棕色，足黑色，立毛灰白色。**副模工蚁：** 特征同正模。

正模： 工蚁，中国云南省景洪县大渡岗乡关坪村，1120m，1997.Ⅷ.7，徐正会采于山地雨林树冠样中，No. A97-977。

词源： 该新种的种名"*brevicorpa*"由拉丁语"*brev-*"（短的）+词根"*corp*"（身体）+后缀"*-a*"（阴性形式）组成，指该种的胸部相对较短。

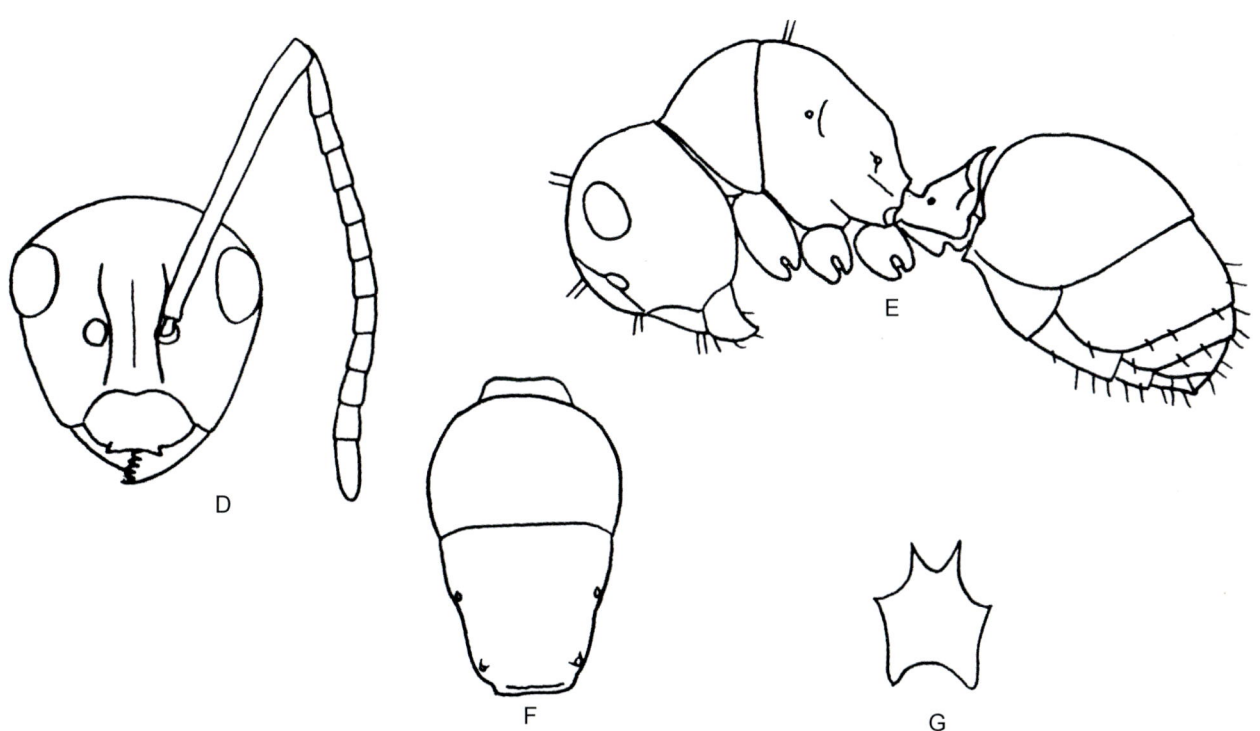

短胸多刺蚁 *Polyrhachis brevicorpa*

A, D. 工蚁头部正面观；B, E. 工蚁身体侧面观；C. 工蚁身体背面观；F. 工蚁胸部背面观；G. 工蚁腹柄结前面观（D-G. 引自 Xu, 2002c）

A, D. Head of worker in full-face view; B, E. Body of worker in lateral view; C. Body of worker in dorsal view; F. Alitrunk of worker in dorsal view; G. Petiolar node of worker in front view (D-G. cited from Xu, 2002c)

方肩多刺蚁
Polyrhachis cornihumera Xu, 2002

Polyrhachis cornihumera Xu, 2002c, Acta Entomologica Sinica, 45(4): 523, figs. 1-4 (w.) CHINA (Yunnan).

Holotype worker: TL 5.9, HL 1.80, HW 1.85, CI 103, SL 1.85, SI 100, PW 1.40, AL 2.00, ED 0.45, MTL 2.45. In full-fcae view, head slightly broader than long, roughly triangular, narrowed forward. Occipital margin weakly convex, occipital corners rounded. Sides of head evenly convex. Mandible with 5 teeth. Clypeus evenly convex, without longitudinal central carina, anterior margin with a pair of small denticles and weakly emarginate between the denticles. Antennae long, scape surpassing occipital corner by about 1/2 of its length. Eyes convex, close to occipital corners. In lateral view, dorsum of alitrunk strongly convex, promesonotal suture distinct, metanotal groove absent. Propodeum with a pair of small teeth, dorsum and declivity relatively straight. In dorsal view, shoulders of pronotum form a pair of right angles. In front view, petiolar node with a pair of long slender lateral spines, laterodorsally pointed and slightly curved inward, without median teeth, dorsal margin of the node straight. In lateral view, petiolar node nearly triangular, lateral spines bent backward. Mandibles sparsely and finely punctured. Head, alitrunk and petiole densely and finely punctured, relatively dull. Gaster with weak micro-reticulation, less shiny. Whole body with abundant depressed short pubescence. Anterior part of head with sparse suberect hairs, apex of gaster with abundant suberect hairs. Occiput, alitrunk and dorsum of first gastral segment without hairs. Scapes with dense depressed short pubescence, without hairs. Femora and tibiae with abundant depressed pubecence, without hairs. Body black, with blue metallic luster, eyes blackish brown, legs black. Hairs grayish white. **Paratype workers:** TL 5.9-6.1, HL 1.75-1.85, HW 1.85-1.95, CI 105-106, SL 1.85-1.90, SI 97-100, PW 1.40-1.55, AL 1.85-2.10, ED 0.45-0.50, MTL 2.35-2.55 (*n*=2). As holotype.

Holotype: Worker, China: Yunnan Province, Jinghong County, Mengyang Town, Sanchahe River, 1100 m, 1997.Ⅲ.1, collected by Se-ping Dai from a canopy sample in mountain rain forest, No. A97-112.

Etymology: The species name *cornihumera* combines Latin *corn-* (horn) + word root *humer* (shoulder) + suffix *-a* (feminine form), it refers to the rightly angled shoulders of pronotum.

正模工蚁： 正面观，头部宽稍大于长，近三角形，向前变窄；后缘轻度隆起，后角圆；侧缘均匀隆起。上颚具5个齿。唇基均匀隆起，缺中央纵脊，前缘具1对小齿，小齿之间轻度凹陷。触角长，柄节约1/2超过头后角。复眼隆起，接近头后角。侧面观，胸部背面强烈隆起，前中胸背板缝明显，缺后胸沟。并胸腹节具1对小齿，背面和斜面相对平直。背面观，前胸背板肩角形成1对直角。前面观，腹柄结具1对细长的侧刺，指向侧上方，轻微内弯；腹柄结背缘平直，缺中齿。侧面观，腹柄结近三角形，侧刺后弯。上颚具稀疏细刻点。头部、胸部和腹柄具细密刻点，较暗。后腹部具弱的微网纹，不太光亮。全身具丰富平伏短绒毛被。头前部具稀疏亚直立毛，后腹部末端具丰富亚直立毛。头后部、胸部和后腹部第1节背面缺立毛。柄节具密集平伏短绒毛被，缺立毛。腿节和胫节具丰富平伏绒毛被，缺立毛。身体黑色，有蓝色金属光泽；复眼黑棕色，足黑色，立毛灰白色。**副模工蚁：** 特征同正模。

正模： 工蚁，中国云南省景洪县勐养镇三岔河，1100m，1997.Ⅲ.1，代色平采于山地雨林树冠样中，No. A97-112。

词源： 该新种的种名"*cornihumera*"由拉丁语"*corn-*"（角）+词根"*humer*"（肩）+后缀"*-a*"（阴性形式）组成，指该种的前胸背板具有直角形的肩膀。

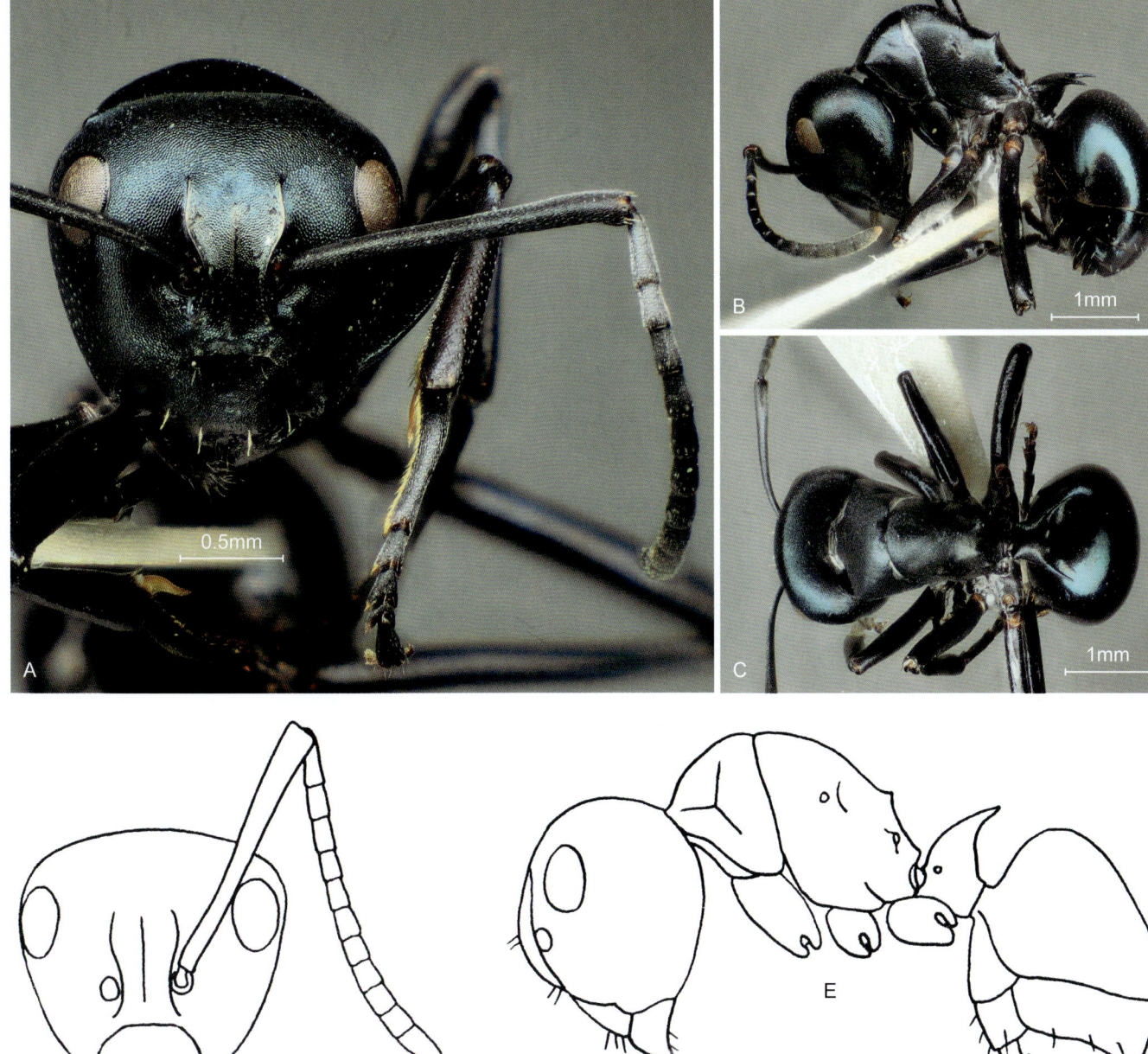

方肩多刺蚁 *Polyrhachis cornihumera*

A, D. 工蚁头部正面观；B, E. 工蚁身体侧面观；C. 工蚁身体背面观；F. 工蚁胸部背面观；G. 工蚁腹柄结前面观（D-G. 引自 Xu, 2002c）

A, D. Head of worker in full-face view; B, E. Body of worker in lateral view; C. Body of worker in dorsal view; F. Alitrunk of worker in dorsal view; G. Petiolar node of worker in front view (D-G. cited from Xu, 2002c)

驼背多刺蚁
Polyrhachis cyphonota Xu, 1998

Polyrhachis cyphonota Xu, 1998b, Zoological Research, 19(3): 243, figs. 1-4 (w.) CHINA (Yunnan, Guizhou).

Holotype worker: TL 5.7, HL 1.67, HW 1.63, Cl 98, SL 1.87, SI 114, PW 1.30, AL 2.07, ED 0.47. In full-face view, head subtriangular, narrowering anteriorly; occipital margin roundly convex, occipital corners rounded; sides evenly convex; ventral surface of head marginate laterally along whole length. Eyes close to occipital corners. Mandibles with 5 teeth. Clypeus without longitudinal central carina; anterior margin weakly emarginate in the center, with a pair of small anterolaterally pointed denticles. Antennal scapes surpassing occipital corners by 3/5 of its length. In lateral view, dorsum of alitrunk strongly convex, rounded laterally. Shoulders of pronotum rounded. Promesonotal suture obvious and depressed, metanotal groove absent. Propodeum with a pair of small blunt denticles, declivity nearly vertical. In lateral view, petiolar node narrowering upwards, anterior and posterior faces strongly convex; in front view, petiolar node with 4 subequal teeth, the inner pair closer and slightly posteriorly curved. Gaster ovate. Mandibles with close microreticulation, subopaque. Head, alitrunk, petiolar node, and gaster with close superficial microreticulation, more shining; lateral surfaces of meso- and metathorax, and base of petiole with close and coarse reticulation, opaque. Whole insect and its appendages covered with sparse depressed pubescence, distance between them about 2-3 times as long as one pubescence; flagella and tarsi with abundant pubescence. Setae restricted at mandibles, clypeus and the apical 4 segments of gaster, vertex of head and mesonotum each with a pair of erect setae; pronotum, propodeum, dorsum of first gastral segment, scapes, and tibiae without setae. Color black; trochanters, famora and tibiae red. **Paratype workers:** TL 4.7-5.7, HL 1.60-1.67, HW 1.50-1.65, Cl 94-100, SL 1.80-1.95, SI 112-127, PW 1.20-1.30, AL 2.00-2.10, ED 0.40-0.47 (*n*=6). As holotype.

Holotype: Worker, China: Yunnan Province, Wenshan County, Kaihua Town, Xihua Park (23.3°N, 104.2°E), 1320 m, 1991.VIII.30, collected by Zheng-hui Xu on the shrub in mountain area, No. A91-437.

Etymology: The species name *cyphonota* combines Greek *cyph-* (humpbacked) + word root *not* (back) + suffix *-a* (feminine form), it refers to the strongly convex and humpbacked dorsum of alitrunk.

正模工蚁： 正面观，头部近三角形，向前变窄；后缘圆形隆起，后角圆；侧缘均匀隆起，腹面两侧全长具隆起的边缘。复眼接近头后角。上颚具5个齿。唇基缺中央纵脊，前缘中央轻度凹陷，前侧角具1对小齿。触角柄节3/5超过头后角。侧面观，胸部背面强烈隆起，两侧圆。前胸背板肩角圆。前中胸背板缝明显凹陷，缺后胸沟。并胸腹节具1对小钝齿，斜面近垂直。侧面观，腹柄结向上变窄，前面和后面强烈隆起；前面观，腹柄结具4个近等长的齿；中间的1对齿较接近，轻微后弯。后腹部卵圆形。上颚具密集微网纹，较暗。头部、胸部、腹柄结和后腹部具密集的肤浅微网纹，较光亮；中胸侧面、后胸侧面和腹柄基部具密集的粗网纹，暗。全身和附肢具稀疏平伏绒毛被，绒毛间距为绒毛长的2～3倍；鞭节和跗节具丰富绒毛被。立毛仅分布于上颚、唇基和后腹部的端部4节，头顶和中胸背板各具1对直立毛；前胸背板、并胸腹节、后腹部第1节背面、柄节和胫节缺立毛。身体黑色，转节、腿节和胫节红色。**副模工蚁：** 特征同正模。

正模： 工蚁，中国云南省文山县开化镇西华公园（23.3°N，104.2°E），1320m，1991.VIII.30，徐正会采于山地灌丛上，No. A91-437。

词源： 该新种的种名"*cyphonota*"由希腊语"*cyph-*"（驼背的）+词根"*not*"（背）+后缀"*-a*"（阴性形式）组成，指该种的胸部背面强烈隆起而驼背。

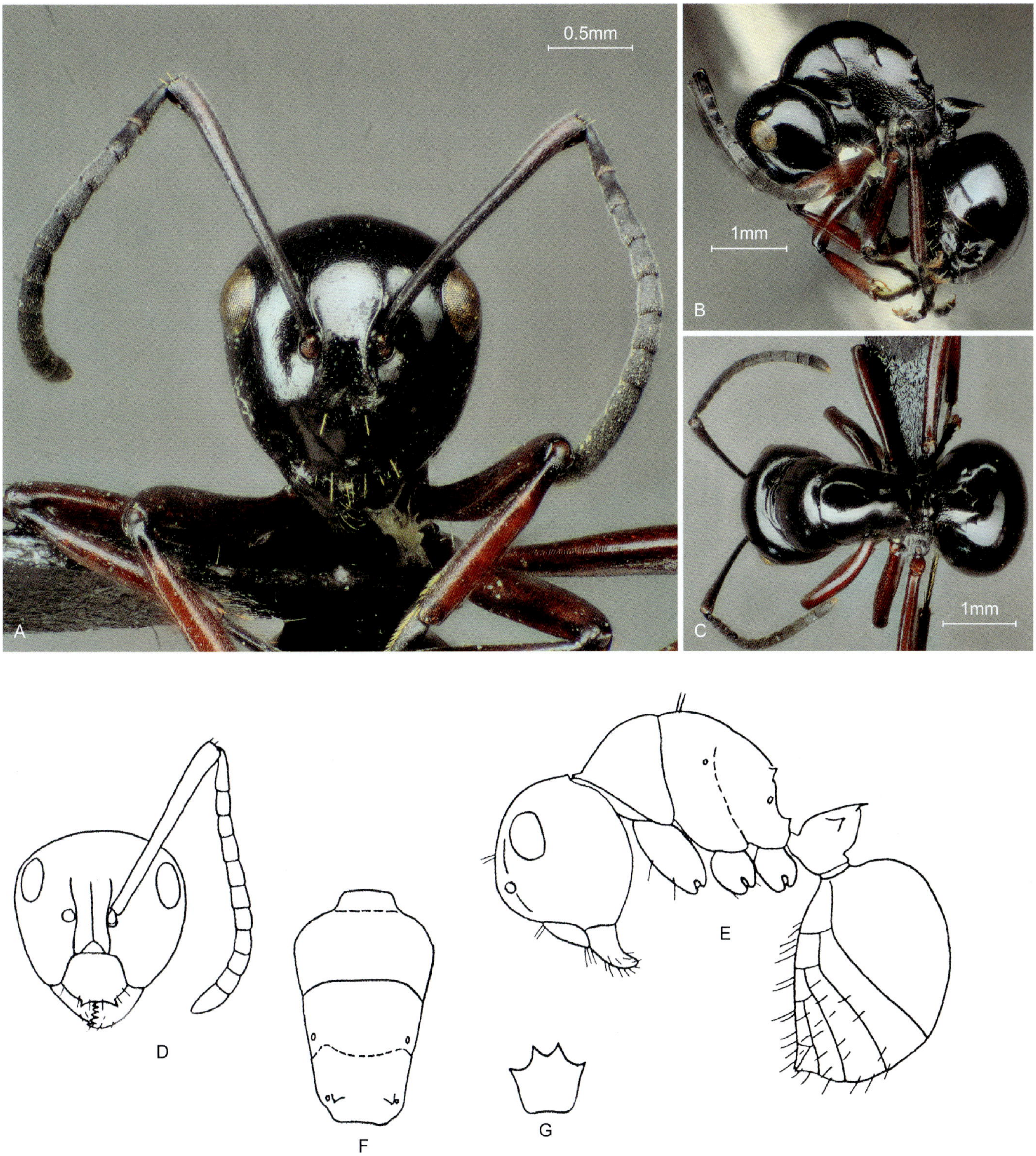

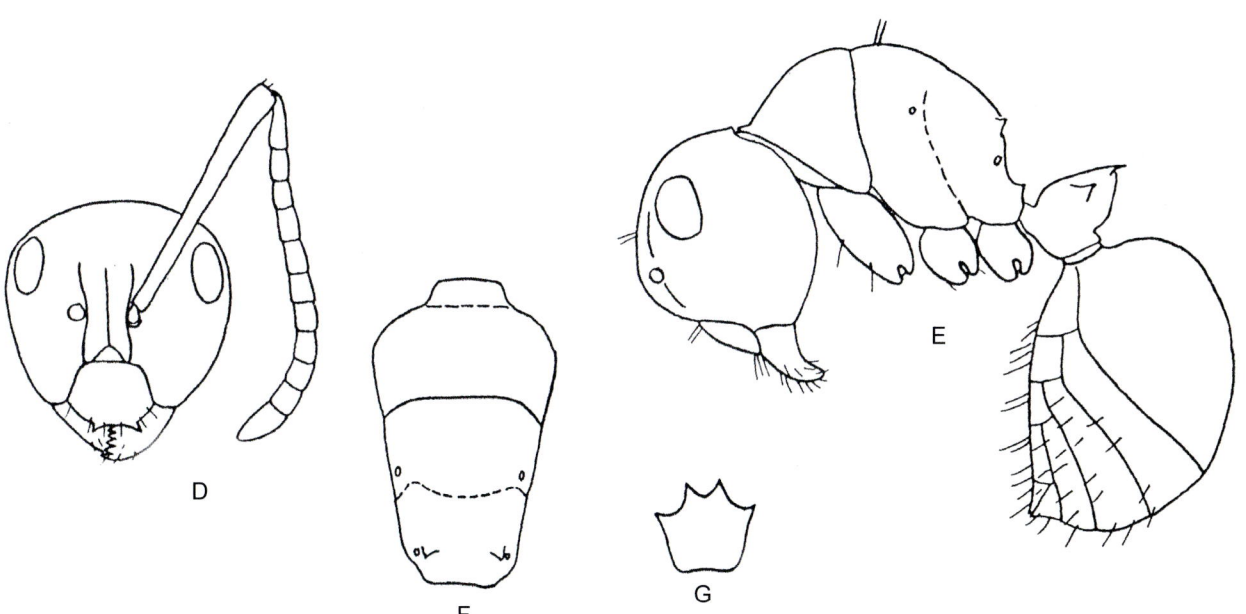

驼背多刺蚁 *Polyrhachis cyphonota*

A, D. 工蚁头部正面观；B, E. 工蚁身体侧面观；C. 工蚁身体背面观；F. 工蚁胸部背面观；G. 工蚁腹柄结前面观（D-G. 引自 Xu, 1998b）

A, D. Head of worker in full-face view; B, E. Body of worker in lateral view; C. Body of worker in dorsal view; F. Alitrunk of worker in dorsal view; G. Petiolar node of worker in front view (D-G. cited from Xu, 1998b)

齿肩多刺蚁
Polyrhachis dentihumera Xu, 2002

Polyrhachis dentihumera Xu, 2002c, Acta Entomologica Sinica, 45(4): 524, figs. 5-8 (w.q.) CHINA (Yunnan).

Holotype worker: TL 5.2, HL 1.55, HW1.65, CI 106, SL 1.80, SI 109, PW 1.20, AL 1.80, ED 0.40, MTL 2.35. In full-face view, head slightly broader than long, roughly triangular. Occipital margin weakly convex, occipital corners rounded. Sides evenly convex. Mandible with 5 teeth. Clypeus weakly convex, without longitudinal central carina, anterior margin with a pair of small denticles and weakly emarginate between the denticles. Antennae long, scape surpassing occipital corner by about 3/5 of its length. Eyes close to occipital corners. In lateral view, dorsum of alitrunk strongly convex, promesonotal suture distinct, metanotal groove absent. Propodeum unarmed, dorsum and declivity formed a single slope surface. In dorsal view, shoulders of pronotum form a pair of acute teeth. In front view, petiolar node with a pair of short lateral spines, laterodorsally pointed and slightly bent upward, without median teeth, dorsal margin nearly straight, with a notch in the center. In lateral view, petiolar node triangular, lateral spines bent backward. Mandibles finely and longitudinally striate, relatively dull. Head, alitrunk, petiole and gaster uniformly, densely and finely punctured, dull. Scapes and legs superficially, densely and finely punctured, less shiny. Whole body with abundant depressed short pubescence. Clypeus and vertex with sparse suberect hairs, ventral surface and apex of gaster with abundant suberect hairs; occiput, dorsum of alitrunk, petiole and dorsum of first gastral segment without hairs. Scapes, femora and tibiae with sparse depressed pubescence, without hairs. Whole body black, eyes brown, legs black, hairs white. **Paratype workers:** TL 5.0-6.2, HL 1.50-1.80, HW1.65-2.15, CI 110-119, SL 1.70-2.15, SI 100-105, PW 1.20-1.45, AL 1.70-2.30, ED 0.40-0.50, MTL 2.30-2.85 (*n*=5). As holotype, but in some individuals occiput and mesonotum with a pair of erect hairs separately.

Holotype: Worker, China: Yunnan Province, Mengla County, Mengla Town, Bubang Village, 790 m, 1997.XII.13, collected by Gang Hu in seasonal rain forest, No. A97-3140.

Etymology: The species name *dentihumera* combines Latin *dent-* (tooth) + word root *humer* (shoulder) + suffix *-a* (feminine form), it refers to the acutely toothed shoulders of pronotum.

正模工蚁： 正面观，头部宽稍大于长，近三角形；后缘轻度隆起，后角圆；侧缘均匀隆起。上颚具5个齿。唇基轻度隆起，缺中央纵脊，前缘具1对小齿，小齿之间轻度凹陷。触角长，柄节约3/5超过头后角。复眼接近头后角。侧面观，胸部背面强烈隆起，前中胸背板缝明显，缺后胸沟。并胸腹节无刺，背面和斜面形成1个单一的坡面。背面观，前胸背板肩角形成1对尖齿。前面观，腹柄结具1对短的侧刺，指向侧上方，轻微内弯；缺中齿，背缘近平直，中央具1个切口。侧面观，腹柄结三角形，侧刺后弯。上颚具细纵条纹，较暗。头部、胸部、腹柄和后腹部具均匀一致的细密刻点，暗。柄节和足具肤浅的细密刻点，不太光亮。全身具丰富的平伏短绒毛被；唇基和头顶具稀疏亚直立毛；头部腹面和后腹部末端具丰富亚直立毛；头后部、胸部背面、腹柄和后腹部第1节背面缺立毛；柄节、腿节和胫节具稀疏平伏短绒毛被，缺立毛。全身黑色，复眼棕色，足黑色，立毛白色。**副模工蚁：** 特征同正模，但是一些个体的头后部和中胸背板各具1对立毛。

正模： 工蚁，中国云南省勐腊县勐腊镇补蚌村，790m，1997.XII.13，胡刚采于季节性雨林中，No. A97-3140。

词源： 该新种的种名"*dentihumera*"由拉丁语"*dent-*"（齿）+词根"*humer*"（肩）+后缀"*-a*"（阴性形式）组成，指该种的前胸背板具有尖齿状的肩膀。

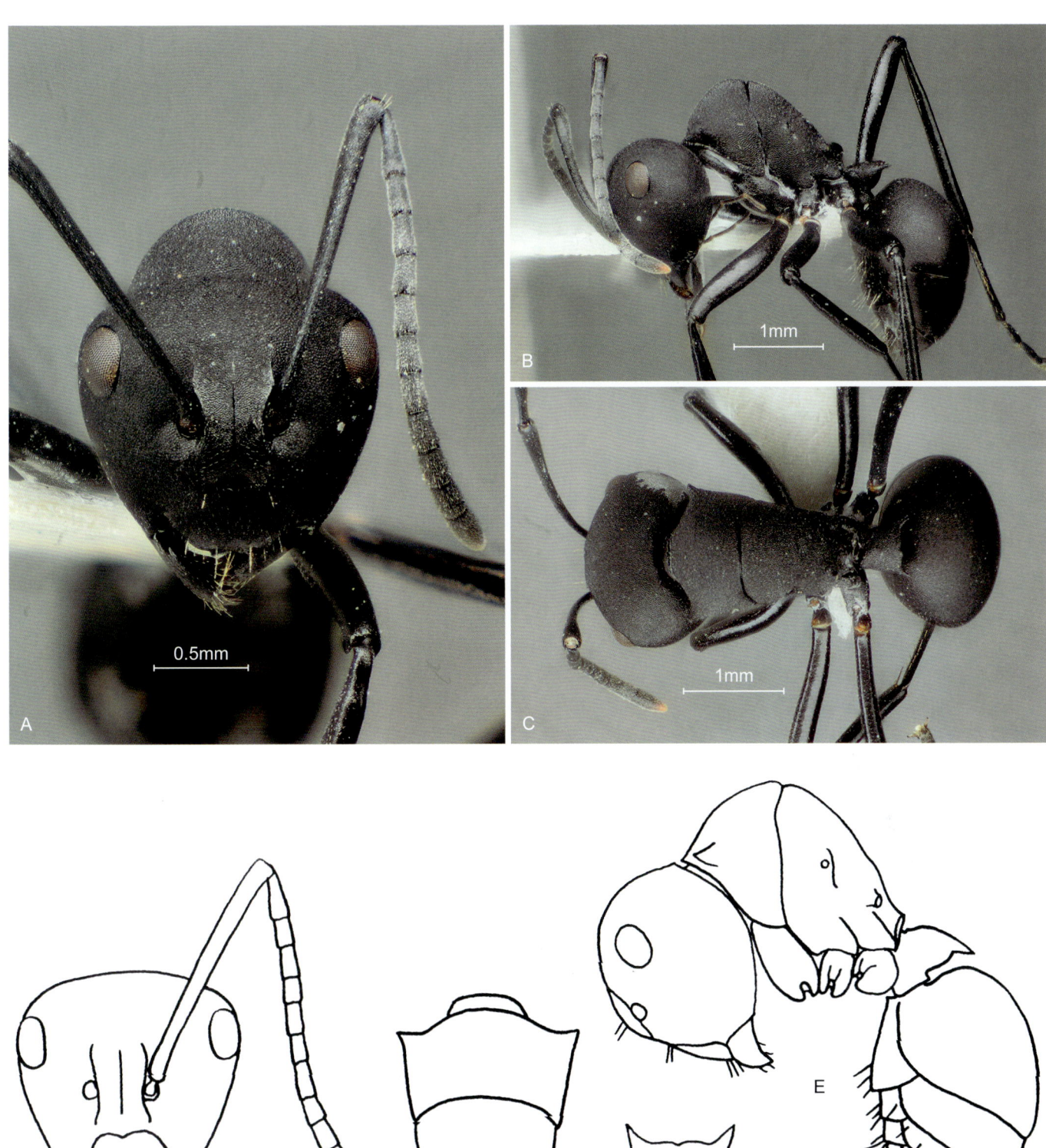

齿肩多刺蚁 *Polyrhachis dentihumera*

A, D. 工蚁头部正面观；B, E. 工蚁身体侧面观；C. 工蚁身体背面观；F. 工蚁胸部背面观；G. 工蚁腹柄结前面观（D-G. 引自 Xu, 2002c）

A, D. Head of worker in full-face view; B, E. Body of worker in lateral view; C. Body of worker in dorsal view; F. Alitrunk of worker in dorsal view; G. Petiolar node of worker in front view (D-G. cited from Xu, 2002c)

圆肩多刺蚁
Polyrhachis orbihumera Xu, 2002

Polyrhachis orbihumera Xu, 2002c, Acta Entomologica Sinica, 45(4): 526, figs. 26-29 (w.q.) CHINA (Yunnan).

Holotype worker: TL 5.1, HL 1.50, HW 1.50, CI 100, SL 1.75, SI 117, PW 1.15, AL 1.80, ED 0.45, MTL 2.00. In full-face view, head about as broad as long, roughly triangular. Occipital margin roundly convex, occipital corners rounded. Sides weakly convex. Mandibles with 5 teeth. Clypeus weakly convex, without longitudinal central carina, anterior margin with a pair of small denticles and weakly emarginate between the denticles. Antennae long, scape surpassing the occipital corner by about 3/5 of its length. Eyes placed at the occipital corners. In lateral view, dorsum of alitrunk strongly convex, promesonotal suture distinct, metanotal groove absent. Propodeum unarmed, dorsum and declivity weakly depressed and separated by a blunt convex. In dorsal view, shoulders of pronotum rounded. In front view, petiolar node with lateral teeth short and rightly angled, median teeth elongate and acute, distinctly longer than the lateral ones. In lateral view, petiolar node triangular, median teeth slightly bent backward. Mandibles densely and finely striate and sparsely punctured, relatively dull. Head, alitrunk and petiole uniformly, densely and superficially punctured, less shiny, gaster with similar punctures but weaker and shiny. Whole body with sparse depressed short pubescence. Anterior part of head with sparse erect hairs, mesonotum with a pair of erect hairs, ventral face and apex of gaster with abundant suberect hairs. Occiput, pronotum, propodeum, petiole and dorsum of first gastral segment without erect hairs. Scapes, femora and tibiae with abundant depressed short pubescence, without erect hairs. Body black, femora and tarsi brownish red, tibiae blackish brown. Hairs light yellow. **Paratype workers:** TL 4.8-5.8, HL 1.40-1.70, HW 1.40-1.75, CI 100-103, SL 1.65-1.95, SI 111-120, PW 1.00-1.35, AL 1.75-2.15, ED 0.45-0.50, MTL 1.95-2.35 (*n*=5). As holotype, but in some individuals with tibiae black, or mesonotum without erect hairs.

Holotype: Worker, China: Yunnan Province, Jinghong County, Puwen Town, Songshanling, 1270 m, 1998.Ⅲ.4, collected by Yun-feng He from a canopy sample of warm pine forest, No. A98-59.

Etymology: The species name *orbihumera* combines Latin *orb-* (round) + word root *humer* (shoulder) + suffix *-a* (feminine form), it refers to the rounded shoulders of pronotum.

正模工蚁： 正面观，头部长宽约相等，近三角形；后缘圆形隆起，后角圆；侧缘轻度隆起。上颚具5个齿。唇基轻度隆起，缺中央纵脊，前缘具1对小齿，小齿之间轻度凹陷。触角长，柄节约3/5超过头后角。复眼位于头后角处。侧面观，胸部背面强烈隆起，前中胸背板缝明显，缺后胸沟。并胸腹节无刺，背部和斜面轻微凹陷，二者交界处轻度隆起。背面观，前胸背板肩角圆。前面观，腹柄结具短的直角形侧齿，中齿伸长且尖锐，明显长于侧齿。侧面观，腹柄结三角形，中齿轻微后弯。上颚具细密条纹和稀疏刻点，较暗。头部、胸部和腹柄具均匀一致的密集肤浅刻点，不太光亮；后腹部具相似的刻点，但是刻点较弱，光亮。全身具稀疏平伏短绒毛被；头前部具稀疏直立毛，中胸背板具1对直立毛，腹面和后腹部末端具丰富亚直立毛；头后部、前胸背板、并胸腹节、腹柄和后腹部第1节背面缺立毛；柄节、腿节和胫节具丰富平伏短绒毛被，缺立毛。身体黑色，腿节和跗节棕红色，胫节黑棕色，立毛浅黄色。**副模工蚁：** 特征同正模，但是一些个体的胫节黑色，或者中胸背板缺立毛。

正模： 工蚁，中国云南省景洪县普文镇松山岭，1270m，1998.Ⅲ.4，何云峰采于暖性松林树冠样中，No. A98-59。

词源： 该新种的种名"*orbihumera*"由拉丁语"*orb-*"（圆的）+词根"*humer*"（肩）+后缀"*-a*"（阴性形式）组成，指该种的前胸背板具有圆的肩膀。

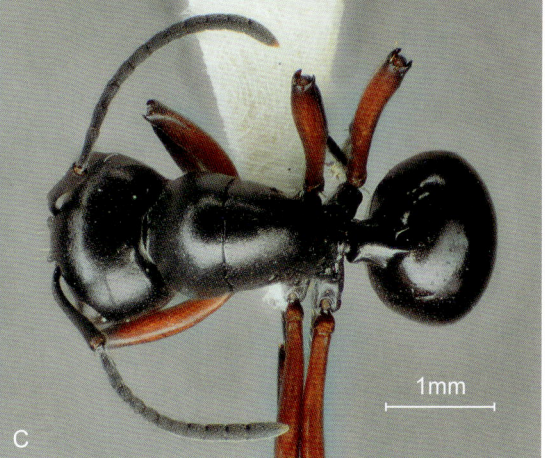
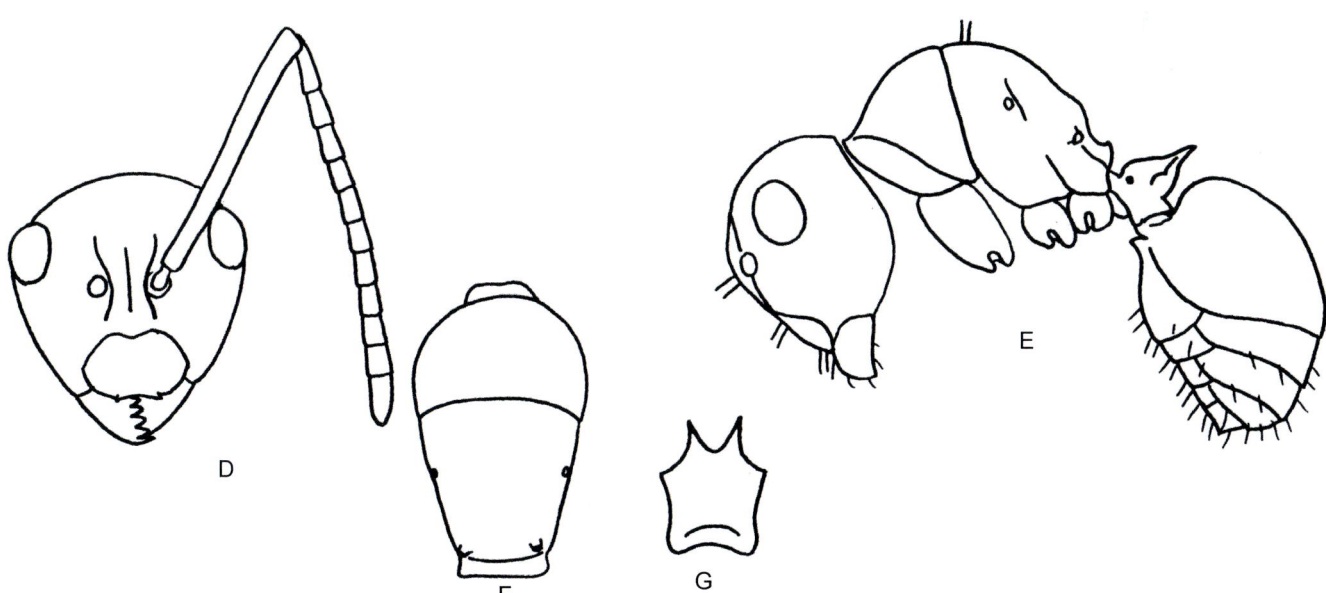

圆肩多刺蚁 *Polyrhachis orbihumera*

A, D. 工蚁头部正面观；B, E. 工蚁身体侧面观；C. 工蚁身体背面观；F. 工蚁胸部背面观；G. 工蚁腹柄结前面观（D-G. 引自 Xu, 2002c）

A, D. Head of worker in full-face view; B, E. Body of worker in lateral view; C. Body of worker in dorsal view; F. Alitrunk of worker in dorsal view; G. Petiolar node of worker in front view (D-G. cited from Xu, 2002c)

圆顶多刺蚁
Polyrhachis rotoccipita Xu, 2002

Polyrhachis rotoccipita Xu, 2002c, Acta Entomologica Sinica, 45(4): 528, figs. 30-33 (w.) CHINA (Yunnan).

Holotype worker: TL 5.3, HL 1.40, HW1.35, CI 96, SL 1.85, SI 137, PW 1.05, AL 1.75, ED 0.40, MTL 2.25. In full-face view, head slightly longer than broad, roughly triangular. Occipital margin roundly convex, occipital corners rounded. Sides evenly convex. Mandibles with 5 teeth. Clypeus evenly convex, without longitudinal central carina, anterior margin with a pair of small denticles and straight between the denticles. Antennae long, scape surpassing occipital corner by about 2/3 of its length. Eyes placed at occipital corners. In lateral view, dorsum of alitrunk strongly convex, promesonotal suture distinct, metanotal groove absent. Propodeum unarmed, dorsum and declivity separated by a roundly blunt convex. In dorsal view, shoulders of pronotum rounded. In front view, lateral teeth of petiolar node short and acute, median teeth elongate and slender, spine-like and longer than lateral ones. In lateral view, petiolar node triangular, with median teeth bent backward. Mandibles finely punctured, relatively dull. Head, alitrunk, petiole and gaster with weak micro-reticulation, relatively shiny. Whole body with abundant depressed short pubescence. Anterior portion of head, ventral face and apex of gaster with spares suberect hairs, mesonotum with a pair of erect hairs. Occiput, pronotum, propodeum, petiole and dorsum of first gastral segment without erect hairs. Scapes, femora and tibiae without erect hairs. Whole body black, with blackish blue metallic luster, eyes yellowish brown, legs black. Hairs grayish white. **Paratype workers:** TL 5.2-5.6, HL 1.40-1.65, HW 1.35-1.55, CI 93-97, SL 1.85-2.15, SI 135-146, PW 1.05-1.30, AL 1.75-2.10, ED 0.40-0.45, MTL 2.25-2.60 (*n*=5). As holotype, but in some individuals mesonotum without erect hairs.

Holotype: Worker, China: Yunnan Province, Mengla County, Menglun Town, Shihuishan Mountain, 830 m, 1997.Ⅸ.16, collected by Zheng-hui Xu from a canopy sample of karst monsoon forest, No. A97-2836.

Etymology: The species name *rotoccipita* combines Latin *rot-* (round) + word root *occipit* (back of head) + suffix *-a* (feminine form), it refers to the head with round posterior margin.

正模工蚁： 正面观，头部长稍大于宽，近三角形；后缘圆形隆起，后角圆；侧缘均匀隆起。上颚具5个齿。唇基均匀隆起，缺中央纵脊，前缘具1对小齿，小齿之间平直。触角长，柄节约2/3超过头后角。复眼位于头后角处。侧面观，胸部背面强烈隆起，前中胸背板缝明显，缺后胸沟。并胸腹节无刺，背部和斜面交界处圆形隆起。背面观，前胸背板肩角圆。前面观，腹柄结侧齿短而尖；中齿伸长，细长，刺状，长于侧齿。侧面观，腹柄结三角形，中齿后弯。上颚具细刻点，较暗。头部、胸部、腹柄和后腹部具弱的微网纹，较光亮。全身具丰富平伏短绒毛被。头前部、腹面和后腹部末端具稀疏亚直立毛，中胸背板具1对直立毛。头后部、前胸背板、并胸腹节、腹柄和后腹部第1节背面缺立毛。柄节、腿节和胫节缺立毛。全身黑色，有暗蓝色金属光泽；复眼黄棕色，足黑色，立毛灰白色。**副模工蚁：** 特征同正模，但是一些个体中胸背板缺立毛。

正模： 工蚁，中国云南省勐腊县勐仑镇石灰山，830m，1997.Ⅸ.16，徐正会采于石灰岩季雨林树冠样中，No. A97-2836。

词源： 该新种的种名"*rotoccipita*"由拉丁语"*rot-*"（圆的）+词根"*occipit*"（头后部）+后缀"*-a*"（阴性形式）组成，指该种的头部有圆形的后缘。

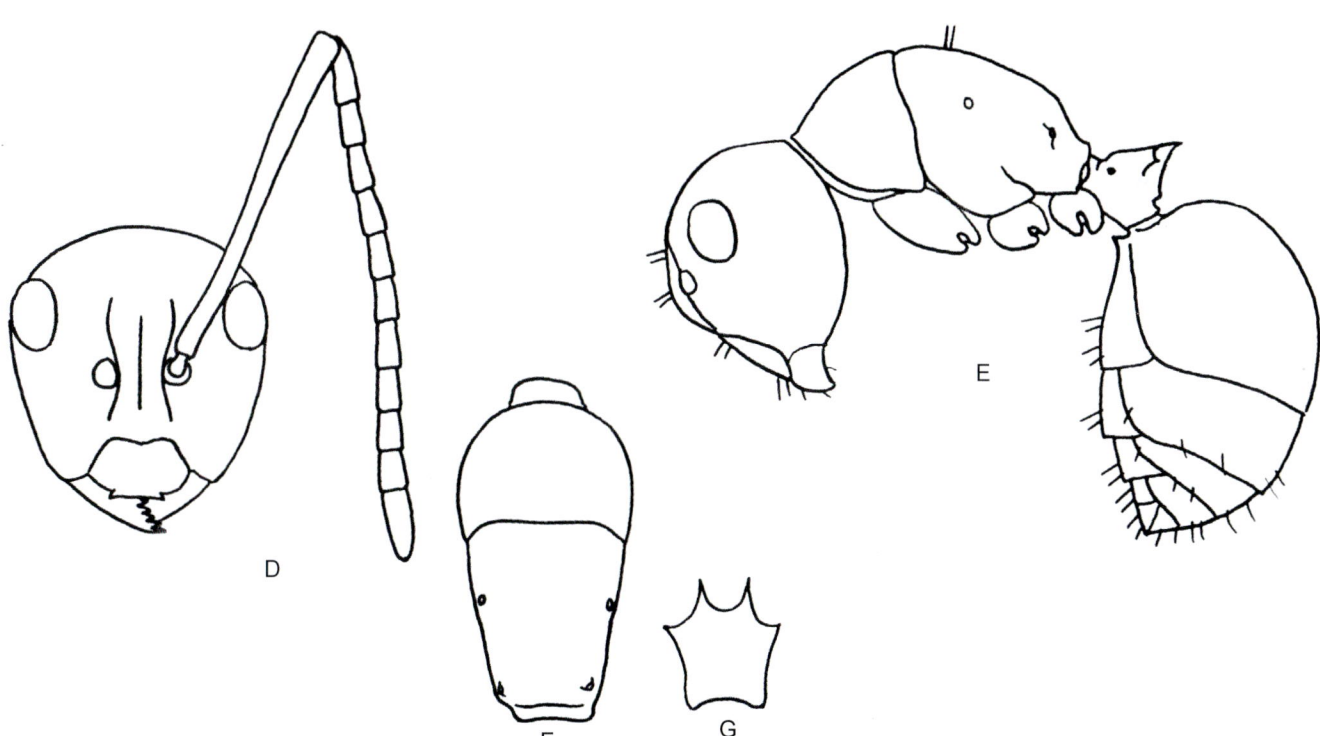

圆顶多刺蚁 *Polyrhachis rotoccipita*

A, D. 工蚁头部正面观；B, E. 工蚁身体侧面观；C. 工蚁身体背面观；F. 工蚁胸部背面观；G. 工蚁腹柄结前面观（D-G. 引自 Xu, 2002c）

A, D. Head of worker in full-face view; B, E. Body of worker in lateral view; C. Body of worker in dorsal view; F. Alitrunk of worker in dorsal view; G. Petiolar node of worker in front view (D-G. cited from Xu, 2002c)

134 大眼前结蚁
Prenolepis magnocula Xu, 1995

Prenolepis magnocula Xu, 1995b, Zoological Research, 16(4): 339, figs. 4-6 (w.) CHINA (Yunnan).

Holotype worker: TL 3.3, HL 0.75, HW 0.60, CI 80, SL 1.08, SI 179, PW 0.45, AL 1.03, ED 0.23. In full-face view, head longer than broad, distinctly broader posteriorly than in front. Occipital margin slightly convex, occipital corners rounded. Eyes large and prominent, placed behind the midlength of the head. Mandibles long triangular, armed with 6 teeth on the masticatory margin, the apical, the fourth and the basal ones larger. Clypeus broader than long, strongly convex in the centre, anterior margin roundly prominant and complete. Antenna filiform, with 12 segments, scape extending beyond the occipital margin of the head by more than half its length, flagellum slightly incrassate towards apex. Alitrunk distinctly constricted in the middle of mesothorax, pronotum and declivity slightly convex, propodeum roundly convex. Promesonotal suture and metanotal groove conspicuous. Legs long, basitarsus of hind leg very long, about 1.3 times the length of the rest 4 tarsal joints together. Petiolar node low and transverse, in lateral view, subtriangular, inclined forwards, narrowing upwards, dorsal margin straight. In lateral view, the first gastral segment strongly anteriorly convex, overhanging the pedicel, its anterior face broadly concave, in dorsal view, anterior margin of gaster straight. Mandibles smooth, head and alitrunk smooth and shining. Gaster with very weak and fine reticulations, less shining. Dorsum of head and body covered with abundant erect or suberect long hairs, hairs on the gaster denser. Head with sparse decumbent pubescence, pubescence on alitrunk and gaster quite rare. Dorsa of antennal scapes and hind tibiae with abundant subdecumbent hairs, hairs longer than the width of scape, and as long as the width of tibia. Head blackish brown, mandibles, antennae, alitrunk, legs and pedicel yellowish brown, gaster black. **Paratype workers:** TL 2.5-3.3, HL 0.63-0.75, HW 0.55-0.65, CI 80-90, SL 0.95-1.15, SI 175-183, PW 0.39-0.49, AL 0.88-1.10, ED 0.19-0.23 (*n*=8). As holotype, but in some individuals mandibles, antennae, alitrunk, legs and pedicel brownish yellow.

Holotype: Worker, China: Yunnan Province, Wenshan County, Kaihua Town, Xihua Park (23.3°N, 104.2°E), 1320 m, 1991.Ⅷ.30, collected by Zheng-hui Xu on the ground in conifer-broadleaf mixed forest, No. A91-445.

Etymology: The species name *magnocula* combines Latin *magn-* (large) + word root *ocul* (eye) + suffix *-a* (feminine form), it refers to the large eyes.

正模工蚁： 正面观，头部长大于宽，后部明显宽于前部；后缘轻度隆起，后角圆。复眼大而突起，位于头中线之后。上颚长三角形，咀嚼缘具6个齿，端齿、第4齿和基齿较大。唇基宽大于长，中部强烈隆起，前缘圆形隆起，完整。触角丝状，12节，柄节一半以上超过头后缘，鞭节向端部轻度变粗。胸部在中胸中部明显收缩，前胸背板和并胸腹节斜面轻度隆起，并胸腹节圆形隆起。前中胸背板缝和后胸沟明显。足长，后足基跗节很长，约为其余4个跗节合长的1.3倍。腹柄结低，横形，侧面观近三角形，前倾，向上变窄，背缘平直。侧面观，后腹部第1节向前强烈隆起，悬覆于腹柄上方，前面宽凹；背面观，后腹部前缘平直。上颚光滑。头部和胸部光滑发亮。后腹部具很弱的细网纹，不太光亮。头部和身体背面具丰富的直立、亚直立长毛；后腹部的立毛较密集；头部具稀疏倾斜绒毛被；胸部和后腹部的绒毛很稀少；触角柄节和后足胫节背面具丰富亚倾斜毛，立毛长于柄节宽度，与胫节宽度相等。头部黑棕色，上颚、触角、胸部、足和腹柄黄棕色，后腹部黑色。**副模工蚁：** 特征同正模，但是一些个体的上颚、触角、胸部、足和腹柄棕黄色。

正模： 工蚁，中国云南省文山县开化镇西华公园（23.3°N，104.2°E），1320m，1991.Ⅷ.30，徐正会采于针阔混交林地表，No. A91-445。

词源： 该新种的种名"*magnocula*"由拉丁语"*magn-*"（大的）+词根"*ocul*"（眼睛）+后缀"*-a*"（阴性形式）组成，指该种的复眼大。

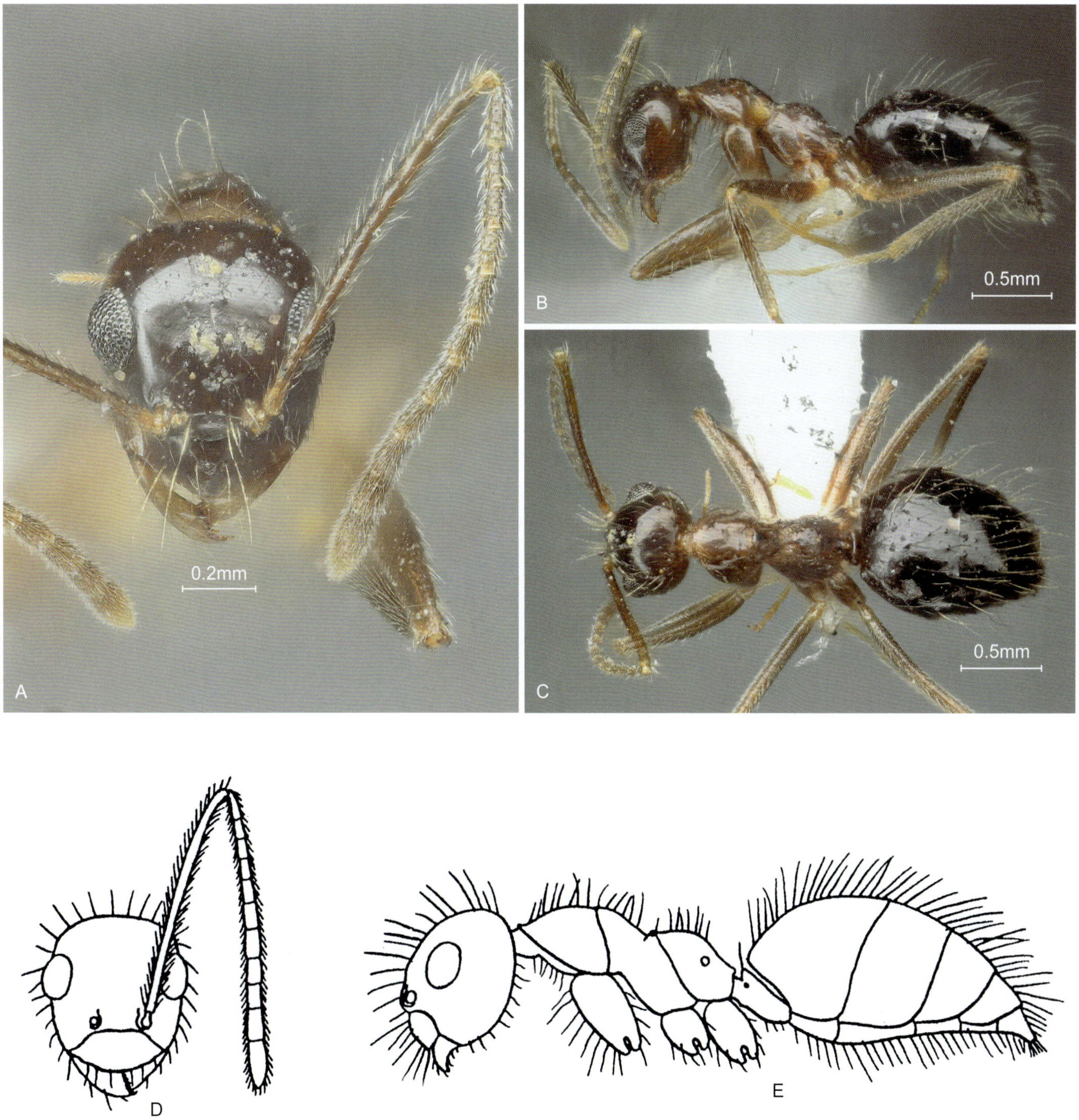

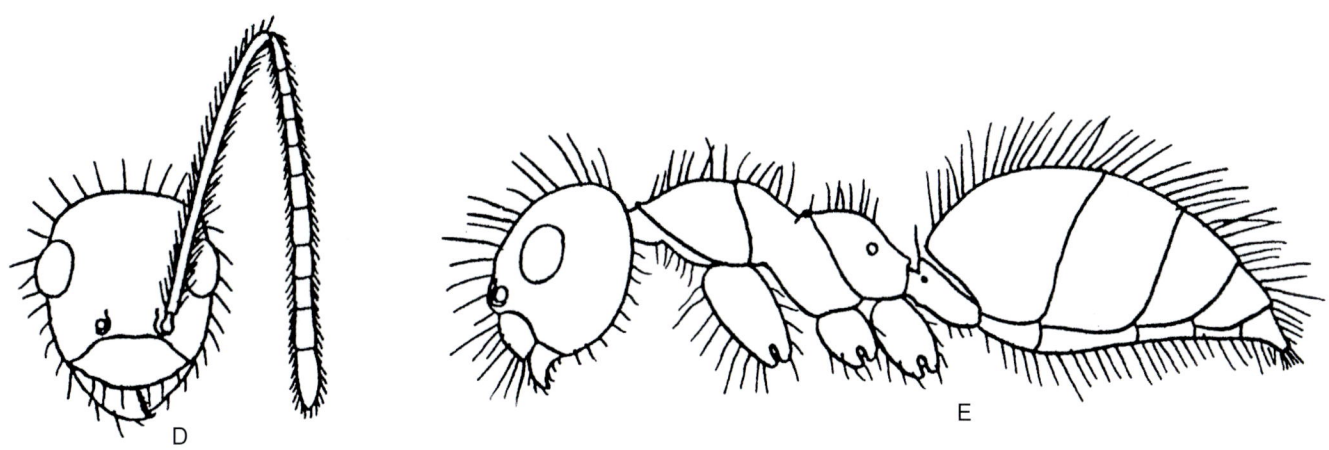

大眼前结蚁 Prenolepis magnocula

A, D. 工蚁头部正面观；B, E. 工蚁身体侧面观；C. 工蚁身体背面观（D-E. 引自 Xu, 1995b）
A, D. Head of worker in full-face view; B, E. Body of worker in lateral view; C. Body of worker in dorsal view (D-E. cited from Xu, 1995b)

135 黑角前结蚁
Prenolepis nigriflagella Xu, 1995

Prenolepis nigriflagella Xu, 1995b, Zoological Research, 16(4): 338, figs. 1-3 (w.) CHINA (Yunnan).

Holotype worker: TL 5.3, HL 1.25, HW 1.25, CI 100, SL 1.70, SI 136, PW 0.78, AL 1.85, ED 0.25. In full-face view, head subtriangular, as broad as long, much broader posteriorly than in front. Occipital margin straight, occipital comers rounded, cheeks convex. Eyes moderately large, situated above the midlength of the head. Mandibles armed with 6 teeth on the masticatory margin, the apical and the fourth ones larger. Clypeus large and transverse, convex in the centre, anterior margin roundly prominent, incised in the middle. Antenna long, filiform, with 12 segments, scape extending beyond the occipital margin of the head by more than half its length, flagellum slightly incrassate towards apex. Alitrunk strongly constricted in the middle of mesothorax, pronotum and propodpum strongly roundly convex, declivity nearly flat. Promesonotal suture and metanotal groove conspicuous. Legs long, basitarsus of hind leg very long, about 1.5 times the length of the rest 4 tarsal joints together. Petiolar node low and transverse, inclined forwards, subtriangular in lateral view, narrowing upwards, the dorsal margin straight. In lateral view, the first gastral segment angularly and anteriorly convex, overhanging the pedicel, its anterior face broadly concave, in dorsal view anterior margin of gaster nearly straight. Mandibles smooth, head, alitrunk and gaster smooth and shining. Dorsum of head and body covered with numerous erect or suberect long hairs and sparse subdecumbent pubescence. Dorsa of antennal scapes and hind tibiae with abundant suberect hairs, hairs much longer than the width of scape and tibia. Head and body in colour orange yellow, gaster black, antennal segments 3-12 blackish brown. **Paratype workers:** TL 4.9-5.8, HL 1.23-1.33, HW 1.20-1,33, CI 98-102, SL 1.65-1.80, SI 129-142, PW 0.75-0.85, AL 1.75-2.00, ED 0.25-0.28 (*n*=8). As holotype, but in some individuals antennal segments 3-12 black, or mandible with 7 teeth.

Holotype: Worker, China: Yunnan Province, Lincang County, Lincang (23.8°N, 100.0°E), 1550 m, 1991.X.13, collected by Zheng-hui Xu on the ground in *Pinus kesiya* var. *langbianensis* forest, No. A91-1103.

Etymology: The species name *nigriflagella* combines Latin *nigr-* (black) + word root *flagell* (flagellum) + suffix *-a* (feminine form), it refers to the apical antennal segments blackish brown.

正模工蚁： 正面观，头部近三角形，长宽相等，后部显著宽于前部；后缘平直，后角圆，颊区隆起。复眼中等大，位于头中线之后。上颚咀嚼缘具6个齿，端齿和第4齿较大。唇基大，横形，中部隆起，前缘圆形隆起，中央有切口。触角长，丝状，12节，触角柄节一半以上超过头后缘，鞭节向端部轻度变粗。胸部在中胸中部强烈收缩，前胸背板和并胸腹节强烈圆形隆起，斜面近平坦。前中胸背板缝和后胸沟明显。足长，后足基跗节很长，约为其余4个跗节合长的1.5倍。腹柄结低，横形，前倾，侧面观近三角形，向上变窄，背缘平直。侧面观，后腹部第1节向前角状突出，悬覆于腹柄上方，前面宽凹；背面观，后腹部前缘近平直。上颚光滑。头部、胸部和后腹部光滑发亮。头部和身体背面具密集直立、亚直立长毛和稀疏倾斜绒毛被。触角柄节和后足胫节背面具丰富亚直立毛，立毛长度远大于柄节和胫节宽度。头部和身体橙黄色，后腹部黑色，触角第3~12节黑棕色。**副模工蚁：** 特征同正模，但是一些个体的触角第3~12节黑色，或者上颚具7个齿。

正模： 工蚁，中国云南省临沧县临沧（23.8°N，100.0°E），1550m，1991.X.13，徐正会采于思茅松林地表，No. A91-1103。

词源： 该新种的种名"*nigriflagella*"由拉丁语"*nigr-*"（黑色的）+词根"*flagell*"（鞭节）+后缀"*-a*"（阴性形式）组成，指该种的触角端部各节呈黑棕色。

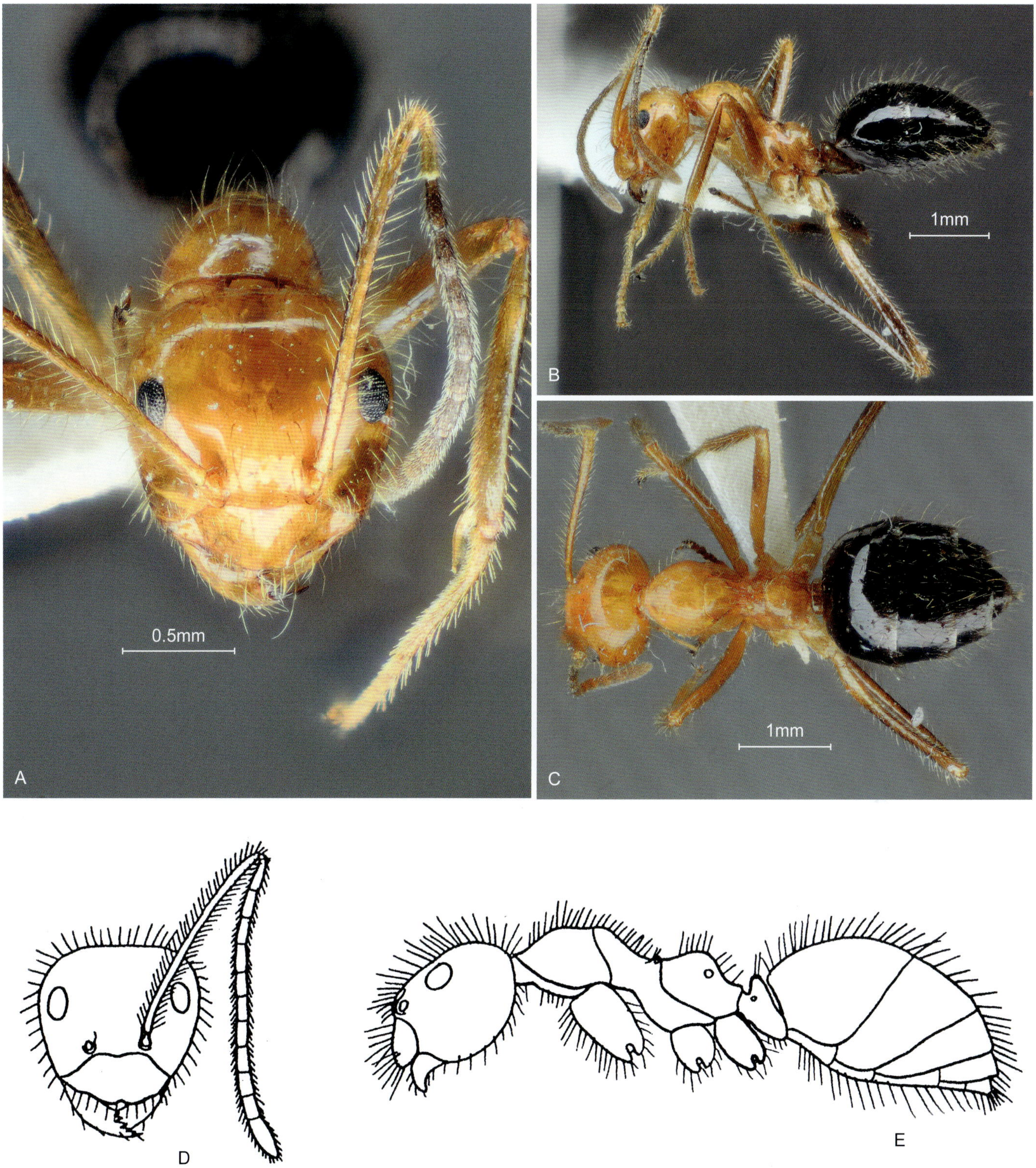

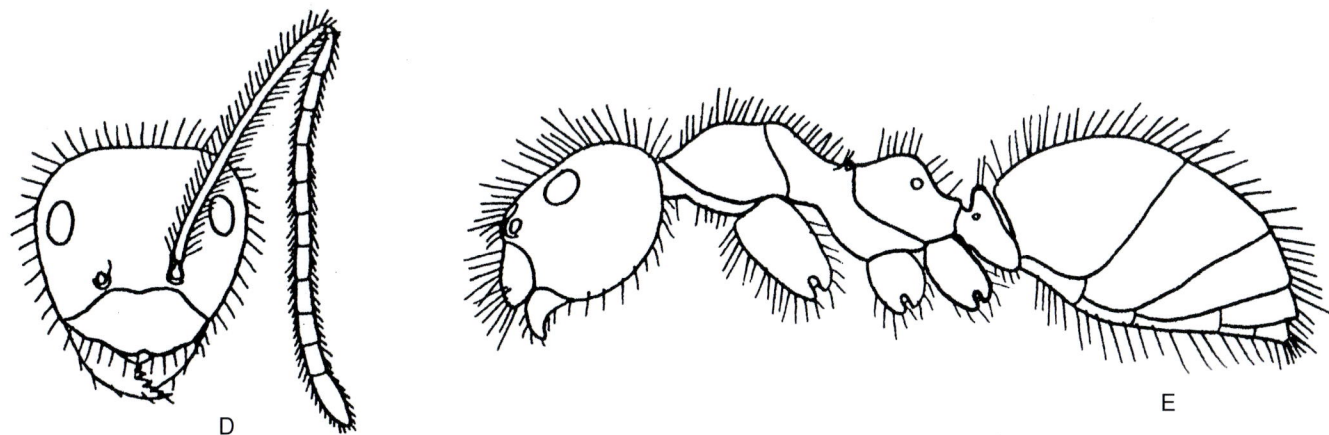

黑角前结蚁 *Prenolepis nigriflagella*

A, D. 工蚁头部正面观；B, E. 工蚁身体侧面观；C. 工蚁身体背面观（D-E. 引自 Xu, 1995b）

A, D. Head of worker in full-face view; B, E. Body of worker in lateral view; C. Body of worker in dorsal view (D-E. cited from Xu, 1995b)

双齿唇拟毛蚁
Pseudolasius bidenticlypeus Xu, 1997

Pseudolasius bidenticlypeus Xu, 1997, Zoological Research, 18(1): 3, figs. 14-16 (w.) CHINA (Guizhou).

Holotype major worker: TL 4.5, HL 1.45, HW 1.40, CI 97, SL 1.13, SI 80, PW 0.88, AL 1.63, ED 0.20. In full-face view, head subrectangular, about as broad as long, narrowed anteriorly at cheek, dorsum has a longitudinal central furrow, sides lightly convex, occipital margin shallowly and angularly excised, occipital corner bluntly rounded. Mandible long triangular, apical margin oblique, has 6 teeth, the besal 2 combined, the 4th smaller. Clypeus transverse, longitudinally convex in the middle, but without carina, anterior margin convex, with one blunt denticle at each side. Frontal carinae straight, parallel, apices reaching to the eye level. Eyes small, lightly convex, placed a little anterior to the midline of head, with 10 facets in their maximum diameter. Antenna has 12 segments, distance from apex of scape to occipital corner about 1/8 length of scapes, flagellum slightly incrassate toward apex, segments 3-9 longer than broad. In lateral view, promesonotum high, roundly convex, promesonotal suture distinct and depressed. Metanotal groove deep and wide, spiracle extruding. Propodeum low, dorsum short, evenly convex, rounded into declivity, the latter long and straight, about 1.5 times as long as dorsum, spiracle elliptic. Petiolar node erect, triangular, narrowing upward; in front view, upper margin rounded, shallowly emarginate in the middle. The lst gastral segment convex anteriorly, depressed anteroventrally so as to receipt the petiolar node. Mandible smooth and shining. Head finely and densely punctate, more shining. Alitrunk coarsely and densely punctate, dull. Gaster smoother. Head and body covered with abundant erect or suberect hairs and dense decumbent pubescence. Antennal scape with rich suberect hairs and dense subdecumbent pubescence, the longest hairs about as long as diameter of scape. Hind tibiae with rich suberect hairs and dense depressed pubescence. Body in color brownish yellow, dorsum of head and alitrunk brown, gastral dorsum blackish brown. Apical margin of mandible black. Hairs and pubescence light yellow. **Paratype median workers:** TL 3.4-4.2, HL 1.03-1.28, HW 0.98-1.23, CI 92-102, SL 0.93-1.08, SI 84-95, PW 0.70-0.83, AL 1.30-1.50, ED 0.15-0.20 ($n=7$). Body smaller, occipital margin narrowly and slightly depressed in the middle, Eye with 9 facets in the maximum diameter. Antennal scape surpassing occipital corner by 1/7 of its length. For the rest as in holotype. **Paratype minor workers:** TL 2.8-3.5, HL 0.83-1.10, HW 0.75-1.03, CI 91-97, SL 0.75-0.98, SI 94-100, PW 0.55-0.73, AL 1.00-1.38, ED 0.13-0.16 ($n=7$). Body very small, head longer than broad, occipital margin very narrowly and shallowly depressed in the middle, dorsum without longitudinal central furrow. Eye with 8 facets in the maximum diameter. Upper margin of petiolar node widely and shallowly emarginate in the middle. For the rest as in holotype.

Holotype: Major worker, China: Guizhou Province, Guiyang City, Qianling Park (26.6°N, 106.7°E), 1080 m, 1991.IX.14, collected by Zheng-hui Xu from a soil nest in conifer-broadleaf mixed forest, No. A91-778.

Etymology: The species name *bidenticlypeus* combines Latin *bi-* (two) + *dent-* (tooth) + word root *clype* (lip base) + suffix *-us* (positive form), it refers to the anterior margin of clypeus with two denticles.

正模大型工蚁：正面观，头部近方形，长宽约相等，颊区向前变窄，背面具中央纵沟；侧缘轻度隆起；后缘角状凹陷，后角钝圆。上颚长三角形，咀嚼缘倾斜，具6个齿，基部2个齿联合，第4齿较小。唇基横形，中部纵向隆起，缺中央纵脊；前缘隆起，每侧具1个钝的小齿。额脊直，互相平行，末端到达复眼水平。复眼小，轻度隆起，位于头中线稍前处，最大直径上具10个小眼。触角12节，柄节末端至头后角间距约为柄节长的1/8；鞭节向端部轻度变粗，第3～9节长大于宽。侧面观，前中胸背板高，圆形隆起，前中胸背板缝明显凹陷。后胸沟深且宽，气门突起。并胸腹节低，背面短，均匀隆起，圆形进入斜面；斜面长，平直，约为背面长的1.5倍；气门椭圆形。腹柄结直立，三角形，向上变窄；前面观背缘圆，中央浅凹。后腹部第1节向前隆起，前下方凹陷以接纳腹柄结。上颚光滑发亮。头部具细密刻点，较光亮。胸部具密集粗糙刻点，暗。后腹部较光滑。头部和身体具丰富直立、亚直立毛和密集倾斜绒毛被；触角柄节具丰富亚直立毛和密集倾斜绒毛被，最长的立毛长度约与柄节直径相等；后足胫节具丰富亚直立毛和密集平伏绒毛被。身体棕黄色，头部背面和胸部背面棕色，后腹部背面黑棕色，上颚咀嚼缘黑色，立毛和绒毛浅黄色。**副模中型工蚁：**身体较小，头后缘中央狭窄浅凹，复眼最大直径上具9个小眼，触角柄节1/7超过头后角。其余特征同正模。**副模小型工蚁：**身体很小；头部长大于宽，后缘中央轻微狭窄凹陷，背面缺中央纵沟；复眼最大直径上具8个小眼；腹柄结背缘中央宽形浅凹；其余特征同正模。

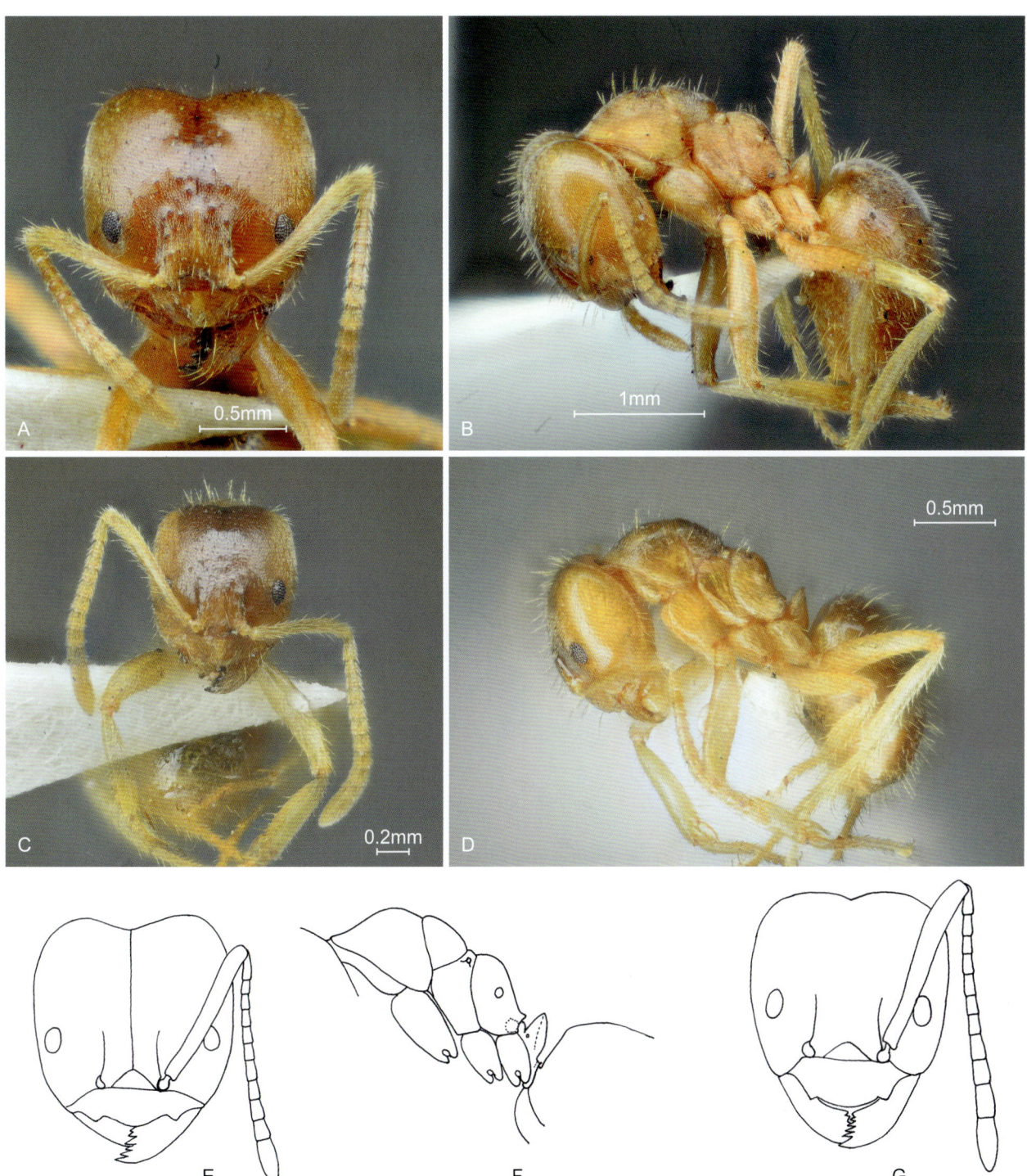

双齿唇拟毛蚁 Pseudolasius bidenticlypeus

A, E. 大型工蚁头部正面观；B, F. 大型工蚁身体侧面观；C, G. 小型工蚁头部正面观；D. 小型工蚁身体侧面观（E-G. 引自Xu, 1997）

A, E. Head of major worker in full-face view; B, F. Body of major worker in lateral view; C, G. Head of minor worker in full-face view; D. Body of minor worker in lateral view (E-G. cited from Xu, 1997)

正模：大型工蚁，中国贵州省贵阳市黔灵公园（26.6°N，106.7°E），1080m，1991.IX.14，徐正会采于针阔混交林土壤巢中，No. A91-778。

词源：该新种的种名"*bidenticlypeus*"由拉丁语"*bi-*"（二，双）+"*dent-*"（齿）+词根"*clype*"（唇基）+后缀"*-us*"（阳性形式）组成，指该种的唇基前缘具2个小齿。

参考文献

References

长有德, 贺达汉, 1998. 宁夏荒漠地区蚂蚁种类及分布[J]. 宁夏农学院学报, 19(4): 12–15. [CHANG Y D, HE D H, 1998. The ants in the desert regions of Ningxia and their distributions [J]. Journal of Ningxia Agricultural College, 19(4): 12–15.]

长有德, 贺达汉, 2002. 中国西北地区蚂蚁区系特征[J]. 动物学报, 48(3): 322–332. [CHANG Y D, HE D H, 2002. Ant fauna in the northwestern regions of China [J]. Acta Zoologica Sinica, 48: 322–332.]

陈守坚, 1962. 世界上最古老的生物防治: 黄柑蚁 Oecophylla smaragdina Fabr.在柑橘园中的放饲及其利用价值[J]. 昆虫学报, 11(4): 401–408. [CHEN S J, 1962. The earliest biological control method in the world: The liberation and breeding of the yellow citrus ant (*Oecophylla smaragdina* Fabr.) in citrus orchard and its significance in practice [J]. Acta Entomologica Sinica, 11(4): 401–408.]

陈志林, 于忠明, 周善义, 等, 2021. 广西花坪蚂蚁图鉴[M]. 桂林: 广西师范大学出版社. [CHEN Z L, YU Z M, ZHOU S Y, et al., 2021. Illustrated handbook of ants of Huaping, Guangxi [M]. Guilin: Guangxi Normal University Press].

程量, 叶勤, 杨毅, 1992. 斯里兰卡切叶蚁: 中国新记录[J]. 昆虫分类学报, 14(4): 244. [CHENG L, YE Q, YANG Y, 1992. *Atopomyrmex srilankensis*: a new record from China [J]. Entomotaxonomia, 14(4): 244.]

黄建华, 周善义, 2006. 中国蚁科昆虫名录: 切叶蚁亚科(I) [J]. 广西师范大学学报(自然科学版), 24(3): 87–94. [HUANG J H, ZHOU S Y, 2006. A checklist of family Formicidae of China: Myrmicinae (Part I) (Insecta: Hymenoptera) [J]. Journal of Guangxi Normal University (Natural Science Edition), 24(3): 87–94.]

黄建华, 周善义, 2007a. 中国蚁科昆虫名录: 切叶蚁亚科(Ⅱ)[J]. 广西师范大学学报(自然科学版), 25(1): 91–99. [HUANG J H, ZHOU S Y, 2007a. A checklist of family Formicidae of China – Myrmicinae (Part II) (Insecta: Hymenoptera) [J]. Journal of Guangxi Normal University (Natural Science Edition), 25(1): 91–99.]

黄建华, 周善义, 2007b. 中国蚁科昆虫名录: 切叶蚁亚科(Ⅲ)[J]. 广西师范大学学报(自然科学版), 25(3):

88–96. [HUANG J H, ZHOU S Y, 2007b. A checklist of family Formicidae of China – Myrmicinae (Part III) (Insecta: Hymenoptera) [J]. Journal of Guangxi Normal University (Natural Science Edition), 25(3): 88–96.]

黄人鑫, 欧阳彤, 吴卫, 等, 2004. 新疆蚁科昆虫42种中国新记录 [J]. 昆虫分类学报, 26(2): 156–160. [HUANG R X, OUYANG T, WU W, et al., 2004. Forty two new record species of family Formicidae (Hymenoptera: Formicoidea) from Xinjiang, China [J]. Entomotaxonomia, 26(2): 156–160.]

李参, 陈益, 1992 中国大头蚁属 (*Pheidole*) 两新种 (膜翅目: 蚁科) [J]. 浙江农业大学学报, 18(3): 55–57. [LI S, CHEN Y, 1992. Two new species of the ant genus *Pheidole* of China (Hymenoptera: Formicidae). Journal of Zhejiang Agricultural University, 18(3): 55–57.]

李淑萍, 刘福林, 康洁, 等, 2005. 河南省膜翅目蚁科昆虫名录 [J]. 河南农业科学 (5): 33–36. [LI S P, LIU F L, KANG J, et al., 2005. Hymenoptera Formicidae insects name record in Henan Province [J]. Journal of Henan Agricultural Sciences (5): 33–36.]

刘福林, 李淑萍, 王玉玲, 等, 2005. 河南商丘地区蚂蚁调查研究 [J]. 河南师范大学学报 (自然科学版), 33(4): 172–174. [LIU F L, LI S P, WANG, Y L, et al., 2005. The research of Henan Province Shangqiu District ant [J]. Journal of Henan Normal University (Natural Science), 33(4): 172–174.]

刘红, 袁兴中, 陈鹏, 1995. 吉林省东部山区蚂蚁资源研究 [J]. 山地研究, 13(3): 160–164. [LIU H, YUAN X Z, CHEN P, 1995. The study on ants resources in the mountainous region of eastern Jilin Province [J]. Mountain Research, 13(3): 160–164.]

骆春璇, 梁翠君, 陈志林, 2021. 中国新记录属种: 岩支蚁 *Cladomyrma scopulosa* Eguchi & Bui (膜翅目: 蚁科) 记述 [J]. 广西师范大学学报 (自然科学版), 39(1): 98–101. [LUO C X, LIANG C J, CHEN Z L, 2021. *Cladomyrma scopulosa* Eguchi & Bui (Hymenoptera: Formicidae), a new record genus and species in China [J]. Journal of Guangxi Normal University (Natural Science Edition), 39(1): 98–101.]

马丽滨, 柳青, 王波, 等, 2022. 云南树栖型蚂蚁高清图鉴 [M]. 郑州: 河南科学技术出版社. [MA L B, LIU Q, WANG B, et al., 2022. Arboreal ants of Yunnan: A field guide with high-resolution photographs [M]. Zhengzhou: Henan Science and Technology Press.]

马永林, 辛明, 宋伶英, 等, 2008. 宁夏蚁科昆虫种类及分布调查 [J]. 农业科学研究, 21(1): 35–38. [MA Y L, XIN M, SONG L Y, et al., 2008. A survey of ants (Hymenoptera: Formicidae) species and distribution in Ningxia [J]. Journal of Agricultural Sciences, 21(1): 35–38.]

梅象信, 晏增, 霍宝民, 等, 2019. 河南蚁科昆虫 [M]. 郑州: 黄河水利出版社. [MEI X X, YAN Z, HUO B M, et al., 2019. Formicidae insects of Henan Province [M]. Zhengzhou: Yellow River Water Conservancy Press]

冉浩, 周善义, 2011. 中国蚁科昆虫名录: 蚁型亚科群 (膜翅目: 蚁科)(Ⅰ) [J]. 广西师范大学学报 (自然科学版), 29(3): 65–73. [RAN H, ZHOU S Y, 2011. Checklist of Chinese Ants: the Formicomorph subfamilies (Hymenoptera: Formicidae) (I) [J]. Journal of Guangxi Normal University (Natural Science Edition), 29(3): 65–73.]

冉浩, 周善义, 2012. 中国蚁科昆虫名录: 蚁型亚科群 (膜翅目: 蚁科)(Ⅱ) [J]. 广西师范大学学报 (自然科学版), 30(4): 81–91. [RAN H, ZHOU S Y, 2012. Checklist of Chinese Ants: Formicomorph Subfamilies (Hymenoptera: Formicidae) (II) [J]. Journal of Guangxi Normal University (Natural Science Edition, 30(4): 81–91.]

冉浩, 周善义, 2013. 中国蚁科昆虫名录: 蚁型亚科群 (膜翅目: 蚁科)(Ⅲ) [J]. 广西师范大学学报 (自然科学版), 31(1): 104–111. [RAN H, ZHOU S Y, 2013. Checklist of Chinese Ants: Formicomorph Subfamilies

(Hymenoptera: Formicidae) (III) [J]. Journal of Guangxi Normal University (Natural Science Edition), 31(1): 104−111.]

寺山守, 1990. 台湾产针蚁亚科目录(膜翅目: 蚁科)[J]. 桐朋学园女子部研究纪要, 4: 25−50. [TERAYAMA M, 1990. A list of Ponerinae of Taiwan (Hymenoptera; Formicidae) [J]. Bulletin of the Toho Gakuen, 4: 25−50.]

唐觉, 李参, 黄恩友, 等, 1995. 中国经济昆虫志第四十七册: 膜翅目蚁科(一) [M]. 北京: 科学出版社. [TANG J, LI S, HUANG E Y, et al., 1995. Economic insect fauna of China. Fasc. 47. Hymenoptera: Formicidae (1) [M]. Beijing: Science Press].

王思忠, 徐鹏, 刘毅, 2011. 成都市园林树木蚂蚁种类调查[J]. 安徽农业科学, 39(9): 5045−5046, 5048. [WANG S Z, XU P, LIU Y, 2011. Investigation on ant species for garden tree in Chengdu [J]. Journal of Anhui Agricultural Science, 39(9): 5045−5046, 5048.]

王维, 沈作奎, 赵玉宏, 2009. 湖北省蚁科昆虫分类研究(昆虫纲: 膜翅目: 蚁科) [M]. 武汉: 中国地质大学出版社. [WANG W, SHEN Z K, ZHAO Y H, 2009. A taxonomic study on the family Formicidae from Hubei Province [M]. Wuhan: China University of Geosciences Press].

魏琮, 周善义, 刘铭汤, 1999. 红蚁属一中国新记录种记述 [J]. 昆虫分类学报, 21(1): 60. [WEI C, ZHOU S Y, LIU M T, 1999. A new record species of the genus *Myrmica* Latreille (Hymenoptera: Formicidae) from Shaanxi, China [J]. Entomotaxonomia, 21(1): 60.]

吴坚, 王常禄, 1995. 中国蚂蚁 [M]. 北京: 中国林业出版社. [WU J, WANG C L, 1995. The ants of China [M]. Beijing: China Forestry Publishing House].

吴志成, 1991. 蚂蚁与类风湿性关节炎 [M]. 南京: 江苏科学技术出版社. [WU Z C, 1991. Ants and rheumatoid arthritis [M]. Nanjing: Jiangsu Science and Technology Press].

夏永娟, 郑哲民, 1995. 蚁科一中国新记录属及一新种记述: 膜翅目 [J]. 昆虫分类学报, 17(3): 219−221. [XIA Y J, ZHENG Z M, 1995. A new record genus and a new species of Formicidae (Hymenoptera) from China [J]. Entomotaxonomia, 17(3): 219−221.]

夏永娟, 郑哲民, 1997a. 新疆蚁科昆虫调查[J]. 陕西师范大学学报(自然科学版), 25(2): 64−66. [XIA Y J, ZHENG Z M, 1997a. A survey of Formicidae from Xinjiang [J]. Journal of Shaanxi Normal University (Natural Science Edition), 25(2): 64−66.]

夏永娟, 郑哲民, 1997b. 新疆蚁属一新种记述(膜翅目:蚁科) [J]. 湖北大学学报(自然科学版), 19(4): 391−392. [XIA Y J, ZHENG Z M, 1997b. A new species of the genus *Formica* from Xinjiang [J]. Journal of Hubei University (Natural Science), 19(4): 391−392.]

徐正会, 1994a. 中国西南地区短猛蚁属分类研究(膜翅目: 蚁科: 猛蚁亚科) [J]. 西南林学院学报, 14(3): 181−185. [XU Z H, 1994a. A taxonomic study of the ant genus *Brachyponera* Emery in Southwestern China (Hymenoptera: Formicidae: Ponerinae) [J]. Journal of Southwest Forestry College, 14(3): 181−185.]

徐正会, 1994b. 中国西南地区刺结蚁属分类研究(膜翅目: 蚁科: 蚁亚科) [J]. 西南林学院学报, 14(4): 232−237. [XU Z H, 1994b. A taxonomic study of the ant genus *Lepisiota* Santschi from Southwestern China (Hymenoptera: Formicidae: Formicinae). Journal of Southwest Forestry College, 14(4): 232−237.]

徐正会, 1995a. 中国臭蚁属分类研究(膜翅目: 蚁科: 臭蚁亚科) [J]. 西南林学院学报, 15(1): 33−39. [XU Z H, 1995a. A taxonomic study of the ant genus *Dolichoderus* Lund in China (Hymenoptera: Formicidae: Dolichoderinae) [J]. Journal of Southwest Forestry College, 15(1): 33−39.]

徐正会, 2002a. 西双版纳自然保护区蚁科昆虫生物多样性研究 [M]. 昆明: 云南科技出版社. [XU Z H, 2002a. A study on the biodiversity of Formicidae ants of Xishuangbanna Nature Reserve [M]. Kunming: Yunnan Science and Technology Press.]

徐正会, 褚姣娇, 张成林, 等, 2011. 藏东南工布自然保护区的蚂蚁种类及分布格局 [J]. 四川动物, 30(1): 118–123. [XU Z H, CHU J J, ZHANG C L, et al., 2011 Ant species and distribution pattern in Gongbo Nature Reserve in Southeastern Tibet [J]. Sichuan Journal of Zoology, 30(1): 118–123.]

徐正会, 胡刚, 于新文, 1999a. 西双版纳热带雨林蚂蚁群落生物量和生态功能研究 [J]. 动物学研究, 20(6): 441–445. [XU Z H, HU G, YU X W, 1999a. Biomass and ecological function of ant communities in the tropical rain forest of Xishuangbanna, China [J]. Zoological Research, 20(6): 441–445.]

徐正会, 姜明, 杨桂良, 等, 2022. 高黎贡山蚂蚁图鉴 [M]. 北京: 中国林业出版社. [XU Z H, JIANG M, YANG G L, et al. 2022. Pictorial book of ants of Mt. Gaoligong [M]. Beijing: China Forestry Publishing House.]

徐正会, 曾光, 柳太勇, 等, 1999b. 西双版纳地区不同植被亚型蚁科昆虫群落研究 [J]. 动物学研究, 20(2): 118–125. [XU Z H, ZENG G, LIU T Y, et al., 1999b. A study on communities of Formicidae ants in different subtypes of vegetation in Xishuangbanna District of China [J]. Zoological Research, 20(2): 118–125.]

曾玲, 陆永跃, 何晓芳, 等, 2005. 入侵中国大陆的红火蚁的鉴定及发生为害调查 [J]. 昆虫知识, 42(2): 144–148, 230–231. [ZENG L, LU Y Y, HE X F, et al., 2005. Identification of red imported fire ant *Solenopsis invicta* to invade mainland China and infestation in Wuchuan, Guangdong [J]. Chinese Bulletin of Entomology, 42(2): 144–148, 230–231.]

张明伟, 王真才, 叶淑琴, 等, 1997. 辽宁地区蚂蚁的初步研究 [J]. 森林病虫通讯 (1): 20–23. [ZHANG M W, WANG Z C, YE S Q, et al., 1997. Preliminary study on ants in Liaoning Province [J]. Forest Pest and Disease (1): 20–23.]

张玮, 郑哲民, 2002. 四川省蚂蚁区系研究 (膜翅目: 蚁科) [J]. 昆虫分类学报, 24(3): 216–222. [ZHANG W, ZHENG Z M, 2002. Studies of ant (Hymenoptera: Formicidae) fauna in Sichuan Province [J]. Entomotaxonomia, 24(3): 216–222.]

张玮, 周善义, 2016. 南岭国家森林公园蚁科昆虫种类调查 [J]. 惠州学院学报, 36(3): 27–30. [ZHANG W, ZHOU S Y, 2016. An investigation on Formicidae species of Nanling National Park [J]. Journal of Huizhou University, 36(3): 27–30.]

周樑鎰, 寺山守, 1991. 台湾昆虫名录 (膜翅目: 蚁科) [J]. 中华昆虫, 11(1): 75–84. [CHOU L Y, TERAYAMA M, 1991. Name lists of insects in Taiwan – Hymenoptera: Apocrita: Formicidae [J]. Chinese Journal of Entomology, 11(1): 75–84.]

周善义, 2001. 广西蚂蚁 [M]. 桂林: 广西师范大学出版社. [ZHOU S Y, 2001. Ants of Guangxi [M]. Guilin: Guangxi Normal University Press].

周善义, 陈志林, 杨宇, 等, 2020. 中国习见蚂蚁生态图鉴 [M]. 郑州: 河南科学技术出版社. [ZHOU S Y, CHEN Z L, YANG Y, et al., 2020. Ecological illustrated book of common ant species from China [M]. Zhengzhou: Henan Science and Technology Press.]

AntMaps, 2024. AntMaps [Z/OL]. [2024-03-09]. https://antmaps.org/.

AntWiki, 2024. AntWiki [Z/OL]. [2024-03-09]. https://www.antwiki.org/.

BARONI URBANI C, DE ANDRADE M L, 2007. The ant tribe Dacetini: limits and constituent genera, with descriptions of new species (Hymenoptera, Formicidae) [J]. Annali del Museo Civico di Storia Naturale "Giacomo Doria", 99: 1-191.

BIHN J H, VERHAAGH M, 2007. A review of the genus *Mystrium* (Hymenoptera: Formicidae) in the Indo-Australian region [J]. Zootaxa, 1642: 1-12.

BOLTON B, 1975. A revision of the ant genus *Leptogenys* Roger (Hymenoptera: Formicidae) in the Ethiopian region with a review of the Malagasy species [J]. Bulletin of the British Museum (Natural History), Entomology, 31: 235-305.

BOLTON B, 1976. The ant tribe Tetramoriini (Hymenoptera: Formicidae). Constituent genera, review of smaller genera and revision of *Triglyphothrix* Forel [J]. Bulletin of the British Museum (Natural History), Entomology, 34: 281-379.

BOLTON B, 1994. Identification guide to the ant genera of the world [M]. Cambridge, Mass.: Harvard University Press.

BOLTON B, 1995. A new general catalogue of the ants of the world [M]. Cambridge, Mass.: Harvard University Press.

BOLTON B, 2007. Taxonomy of the dolichoderine ant genus *Technomyrmex* Mayr (Hymenoptera: Formicidae) based on the worker caste [J]. Contributions of the American Entomological Institute, 35(1): 1-149.

BOLTON B, 2024. An Online Catalog of the Ants of the World [Z/OL]. [2024-03-09]. http://www.antcat.org/.

BRADY S G, SCHULTZ T R, FISHER B L, et al., 2006. Evaluating alternative hypotheses for the early evolution and diversification of ants [J]. Proceedings of the National Academy of Sciences U.S.A., 103: 18172-18177.

BRADY S G, FISHER B L, SCHULTZ T R, et al., 2014. The rise of army ants and their relatives: diversification of specialized predatory doryline ants [J]. BMC Evolutionary Biology, 14(93): 1-14.

BRASSARD F, LEONG C, CHAN H, et al., 2020. A new subterranean species and an updated checklist of *Strumigenys* (Hymenoptera, Formicidae) from Macao SAR, China, with a key to species of the Greater Bay Area [J]. ZooKeys, 970: 63-116.

BROWN W L JR, 1958. Contributions toward a reclassification of the Formicidae. II. Tribe Ectatommini (Hymenoptera) [J]. Bulletin of the Museum of Comparative Zoology, 118: 173-362.

BROWN W L JR, 1978. Contributions toward a reclassification of the Formicidae. Part VI. Ponerinae, tribe Ponerini, subtribe Odontomachiti. Section B. Genus *Anochetus* and bibliography [J]. Studia Entomologica, 20: 549-638.

CAMACHO G P, FRANCO W, BRANSTETTER M G, et al., 2022. UCE phylogenomics resolves major relationships among ectaheteromorph ants (Hymenoptera: Formicidae: Ectatomminae, Heteroponerinae): a new classification for the subfamilies and the description of a new genus [J]. Insect Systematics and Diversity, 6 (1): 5: 1-20.

CHEN Z L, LIANG C J, DU C C, 2022. Revision of Chinese species of the ant genus *Parasyscia* Emery, 1882 (Hymenoptera: Formicidae: Dorylinae) [J]. Zootaxa, 5196 (3): 301-330.

CHEN Z L, YE D D, LU C W, et al., 2011. New species of the ant genus *Pheidole* (Hymenoptera: Formicidae) from Hainan Province, China [J]. Sociobiology, 58: 1-7.

COLLINGWOOD C A, 1962. Some ants (Hym. Formicidae) from northeast Asia [J]. Entomologisk Tidskrift, 83: 215-230.

COLLINGWOOD C A, 1970. Formicidae (Hymenoptera: Aculeata) from Nepal [J]. Khumbu Himal. Ergebnisse des Forschungsunternehmens Nepal Himalaya, 3: 371-387.

COLLINGWOOD C A, 1982. Himalayan ants of the genus *Lasius* (Hymenoptera: Formicidae) [J]. Systematic Entomology, 7: 283-296.

DARWIN C, 1859. On the origin of species by means of natural selection, or the preservation of favoured races in the struggle for life [M]. London: John Murray, Albemarle Street.

DLUSSKY G M, 1965. Ants of the genus *Formica* L. of Mongolia and northeast Tibet (Hymenoptera, Formicidae) [J]. Annales Zoologici (Warsaw), 23: 15-43.

DONISTHORPE H, 1929. The Formicidae (Hymenoptera) taken by Major R. W. G. Hingston, M.C., I.M.S. (ret.), on the Mount Everest Expedition, 1924 [J]. Annals and Magazine of Natural History, 4(10): 444-449.

DONISTHORPE H, 1947. New species of ants from China and Mauritius [J]. Annals and Magazine of Natural History, 13(11): 283-286.

DUBOIS M B, 1998. A revision of the ant genus *Stenamma* in the Palaearctic and Oriental regions (Hymenoptera: Formicidae: Myrmicinae) [J]. Sociobiology, 32: 193-403.

EIDMANN H, 1941. Zur Ökologie und Zoogeographie der Ameisenfauna von Westchina und Tibet. Wissenschaftliche

Ergebnisse der 2. Brooke Dolan-Expedition 1934-1935 [J]. Zeitschrift für Morphologie und Ökologie der Tiere, 38: 1-43.

ELMES G W, RADCHENKO A G, 1998. Ants of the genus *Myrmica* from Taiwan (Hymenoptera: Formicidae) [J]. Chinese Journal of Entomology, 18: 217-224.

EMERY C, 1889. Formiche di Birmania e del Tenasserim raccolte da Leonardo Fea (1885-87). Dummy reference [J]. Annali del Museo Civico di Storia Naturale, 27: 485-520.

EMERY C, 1893. Formicides de l'Archipel Malais [J]. Revue Suisse de Zoologie, 1: 187-229.

EMERY C, 1895. Viaggio di Leonardo Fea in Birmania e regioni vicine. LXIII. Formiche di Birmania del Tenasserim e dei Monti Carin raccolte da L. Fea. Parte II [J]. Annali del Museo Civico di Storia Naturale, 34: 450-483.

EMERY C, 1911. Hymenoptera. Fam. Formicidae. Subfam. Ponerinae [M]. [S.L.]: Genera Insectorum, 118: 1-125.

EMERY C, 1925. Revision des espèces paléarctiques du genre *Tapinoma* [J]. Revue Suisse de Zoologie, 32: 45-64.

FELLOWES J R, 2003. Ant genera on Hainan Island, China [J]. ANeT Newsletter, 6: 14-18.

FERNÁNDEZ F, 2010. A new species of *Carebara* from the Philippines with notes and comments on the systematics of the *Carebara* genus group (Hymenoptera: Formicidae: Myrmicinae) [J]. Caldasia, 32: 191-203.

FOREL A, 1879. Études myrmécologiques en 1879 (deuxième partie [1re partie en 1878]) [J]. Bulletin de la Société Vaudoise des Sciences Naturelles, 16: 53-128.

FOREL A, 1902. Variétés myrmécologiques [J]. Annales de la Société Entomologique de Belgique, 46: 284-296.

FOREL A, 1922. Glanures myrmécologiques en 1922 [J]. Revue Suisse de Zoologie, 30: 87-102.

GRIEBENOW Z, 2024. Systematic revision of the ant subfamily Leptanillinae (Hymenoptera, Formicidae) [J]. ZooKeys, 1189: 83-184.

GUÉNARD B, DUNN R R, 2012. A checklist of the ants of China [J]. Zootaxa, 3358: 1-77.

GUÉNARD B, BLANCHARD B, LIU C, et al., 2013. Rediscovery of the rare ant genus *Bannapone* (Hymenoptera: Formicidae: Amblyoponinae) and description of the worker caste [J]. Zootaxa, 3734(3): 371-379.

HAMER M T, LEE R H, GUÉNARD B, 2023a. First record of the genus *Temnothorax* Mayr, 1861 (Formicidae: Myrmicinae) in Hong Kong, with descriptions of two new species [J]. European Journal of Taxonomy, 879: 116-135.

HAMER M T, PIERCE M P, GUÉNARD B, 2023b. The Amblyoponinae (Formicidae) of Hong Kong [J]. Asian Myrmecology, 16: 1-37.

HÖLLDOBLER B, WILSON E O, 1990. The Ants [M]. Cambridge, Mass.: The Belknap Press of Harvard University Press.

HSU F C, ESTEVES F A, CHOU L S, et al., 2017a. A new species of *Stigmatomma* from Taiwan (Hymenoptera, Formicidae, Amblyoponinae) [J]. ZooKeys, 705: 81-94.

HSU P W, HSU F C, HSIAO Y, et al., 2017b. Taxonomic notes on the genus *Protanilla* (Hymenoptera: Formicidae: Leptanillinae) from Taiwan [J]. Zootaxa, 4268: 117-130.

HUANG J H, CHEN B, ZHOU S Y, 2004. A new species of the ant genus *Leptothorax* Mayr (Hymenoptera, Formicidae) from Hunan, China [J]. Acta Zootaxonomica Sinica, 29: 766-768.

HUANG J H, HUANG Y, ZHOU S Y, 2008. A new species of the genus *Myrmecina* Curtis, 1829 (Hymenoptera, Formicidae) from Hunan Province, China [J]. Acta Zootaxonomica Sinica, 33: 275-278.

HUANG J H, ZHOU S Y, 2006. *Vombisidris* Bolton (Hymenoptera, Formicidae), a new record genus in China, with description of a new species [J]. Acta Zootaxonomica Sinica, 31: 206-207.

HUANG Y Y, ZHONG Y, 2023. Two new species in the ant genus *Aphaenogaster* Mayr (Hymenoptera: Formicidae: Myrmicinae) from China [J]. Entomotaxonomia, 45 (4): 1-9.

HUXLEY J, 1940. The new systematics [M]. England: Oxford University Press.

JAITRONG W, XU Z H, KACHONPISITSAK S, 2022. A new species of the ant *Platythyrea clypeata* species group (Hymenoptera, Formicidae, Ponerinae) from continental Asia [J]. ZooKeys, 1115: 151-168.

KOHOUT R J, 1994. *Polyrhachis lama*, a new ant from the Tibetan plateau (Formicidae: Formicinae) [J]. Memoirs of the Queensland Museum, 35: 137-138.

LAPOLLA J S, 2004. *Acropyga* (Hymenoptera: Formicidae) of the world [J]. Contributions of the American Entomological Institute, 33(3): 1-130.

LATTKE J E, 2004. A taxonomic revision and phylogenetic analysis of the ant genus *Gnamptogenys* Roger in Southeast Asia and Australasia (Hymenoptera: Formicidae: Ponerinae)[J]. University of California Publications in Entomology, 122: 1-266.

LENGYEL S, GOVE A D, LATIMER A M, et al., 2009. Ants sow the seeds of global diversification in flowering plants [J]. PLoS ONE, 4(5): 1-6.

LEONG C M, GUÉNARD B, SHIAO S F, et al., 2019. Taxonomic revision of the genus *Ponera* Latreille, 1804 (Hymenoptera: Formicidae) of Taiwan and Japan, with a key to East Asian species [J]. Zootaxa, 4594: 1-86.

LEONG C M, SHIAO S F, LIU J J, et al., 2017. Records of *Odontoponera denticulata* (Smith, 1858) (Hymenoptera: Formicidae) from Taiwan, with a note on sculptural variation in workers [J]. Japanese Journal of Systematic Entomology, 23: 21-27.

LEONG C M, TSAI W H, TERAYAMA M, et al., 2018. Description of a new species of the genus *Anochetus* Mayr (Hymenoptera: Formicidae) from Orchid Island, Taiwan [J]. Journal of Asia-Pacific Entomology, 21: 124-129.

LEONG C M, YAMANE S, GUÉNARD B., 2018. Lost in the city: discovery of the rare ant genus *Leptanilla* (Hymenoptera: Formicidae) in Macau with description of *Leptanilla macauen*sis sp. nov. [J]. Asian Myrmecology, 10: 1-16.

LIN C C, WU W J, 1996. Revision of the ant genus *Strumigenys* Fr. Smith (Hymenoptera: Formicidae) of Taiwan [J]. Chinese Journal of Entomology, 16: 137-152.

LIN C C, WU W J, 1998. The ant tribe Myrmecinini (Hymenoptera: Formicidae) of Taiwan[J]. Chinese Journal of Entomology, 18: 83-100.

LIN C C, WU W J, 2001. Three new species of *Strumigenys* Fr. Smith (Hymenoptera: Formicidae) with a key to Taiwanese species [J]. Formosan Entomologist, 21: 159-170.

LINNAEUS C, 1758. Systema naturae per regna tria naturae, secundum classes, ordines, genera, species, cum characteribus, differentiis, synonymis, locis. Tomus I. Editio decima, reformata [M]. Holmiae [= Stockholm]: L. Salvii.

LIU C, GUÉNARD B, HITA GARCIA F, et al., 2015a. New records of ant species from Yunnan, China [J]. ZooKeys, 477: 17-78.

LIU C, HITA GARCIA F, PENG Y Q, et al., 2015b. *Aenictus yangi* sp. n. : a new species of the *A. ceylonicus* species group (Hymenoptera, Formicidae, Dorylinae) from Yunnan, China [J]. Journal of Hymenoptera Research, 42: 33-45.

LIU C, FISCHER G, LIU Q, et al., 2022. Updating the taxonomy of the ant genus *Myrmecina* (Hymenoptera, Formicidae) in China with descriptions of three new species [J]. Zootaxa, 5182 (2): 152-164.

LIU X, XU Z H, 2011. Three new species of the ant genus *Stenamma* (Hymenoptera: Formicidae) from Himalaya and the Hengduan Mountains with a revised key to the known species of the Palaearctic and Oriental regions [J]. Sociobiology, 58: 733-747.

LIU X, XU Z H, HITA GARCIA F, 2021. Taxonomic review of the ant genus *Lordomyrma* Emery, 1897 (Hymenoptera, Formicidae) from China, with description of two new species and an identification key to the known species of the world [J]. Asian Myrmecology, 14: 1-33.

LIU Z Y, ZHONG Y, HUANG Y Y, et al., 2024. A new ant species of the genus *Carebara* Westwood, 1840 (Hymenoptera, Formicidae, Myrmicinae) with a key to Chinese species [J]. ZooKeys, 1190: 1-37.

LUO Y Y, GUÉNARD B, 2016. Descriptions of a new species and the gyne in the rarely collected arboreal genera *Paratopula* and *Rotastruma* (Hymenopytera: Formicidae) from Hong Kong, with a discussion on their ecology [J]. Asian Myrmecology, 8: 1-16.

MAN P, RAN H, CHEN Z L, et al. 2017. The northern-most record of Leptanillinae in China with description of *Protanilla beijingensis* sp. nov. (Hymenoptera: Formicidae) [J]. Asian Myrmecology, 9: 1-12.

MATSUMURA S, UCHIDA T, 1926. Die Hymenopteren-Fauna von den Riukiu-Inseln [J]. Insecta Matsumurana, 1: 32-52.

MAYR E, 1942. Systematics and the Origin of Species [M]. New York: Columbia University Press.

MAYR G, 1866. Diagnosen neuer und wenig gekannter Formiciden [J]. Verhandlungen der k. k. Zoologisch-Botanischen Gesellschaft in Wien, 16: 885-908.

MAYR G, 1870. Neue Formiciden [J]. Verhandlungen der Kaiserlich-Königlichen Zoologisch-Botanischen Gesellschaft in Wien, 20: 939-996.

MAYR G, 1889. Insecta in itinare Cl. Przewalski in Asia centrali novissime lecta. 17. Formiciden aus Tibet [J]. Trudy Russkago Entomologicheskago Obshchestva, 24: 278-280.

MENOZZI C, 1939. Formiche dell'Himalaya e del Karakorum raccolte dalla Spedizione italiana comandata da S. A. R. il Duca di Spoleto (1929) [J]. Atti della Società Italiana di Scienze Naturali e del Museo Civico di Storia Naturale di Milano, 78: 285-345.

References

MOREAU C S, BELL C D, VILA R, et al., 2006. Phylogeny of the ants: Diversification in the age of Angiosperms [J]. Science, 312: 101-104.

MULLIS K B, 1990. The unusual origins of the polymerase chain reaction [J]. Scientific American, 262: 36-41.

OGATA K, TERAYAMA M, MASUKO K, 1995. The ant genus *Leptanilla*: discovery of the worker-associated male of *L. japonica*, and a description of a new species from Taiwan (Hymenoptera: Formicidae: Leptanillinae) [J]. Systematic Entomology, 20: 27-34.

OKIDO H, OGATA K, HOSOISHI S, 2020. Taxonomic revision of the ant genus *Myrmecina* in Southeast Asia (Hymenoptera: Formicidae) [J]. Bulletin of the Kyushu University Museum, 17: 1-108.

PIERCE M P, LEONG C M, GUÉNARD B, 2019. A new species and new record of the cryptobiotic ant genus *Ponera* Latreille, 1804 (Hymenoptera, Formicidae) from Hong Kong [J]. ZooKeys, 867: 9-21.

QIAN Y H, XU Z H, MAN P, et al., 2024. Three new species of the ant genus *Leptanilla* (Hymenoptera: Formicidae) from China, with a key to the world species [J]. Myrmecological News, 34: 21-44.

QIAN Y H, XU Z H, 2024. Taxonomy of the ant genera *Leptothorax* Mayr, 1855 and *Temnothorax* Mayr, 1861 (Hymenoptera: Formicidae) of China with descriptions of twenty-eight new species and a key to the known Chinese species [J]. European Journal of Taxonomy, 936: 1-97.

RADCHENKO A G, ELMES G W, 2010. *Myrmica* Ants (Hymenoptera: Formicidae) of the Old World. Fauna Mundi 3 [M]. Warsaw: Natura Optima Dux Foundation.

RUZSKY M, 1905. The ants of Russia. (Formicariae Imperii Rossici). Systematics, geography and data on the biology of Russian ants. Part I [J]. Trudy Obshchestva Estestvoispytatelei pri Imperatorskom Kazanskom Universitete, 38(4-6): 1-800.

RUZSKY M, 1914. Myrmekologische Notizen[J]. Archiv für Naturgeschichte, 79(9): 58-63.

RUZSKY M, 1915. On the ants of Tibet and the southern Gobi. On material collected on the expedition of Colonel P. K. Kozlov [J]. Ezhegodnik Zoologicheskago Muzeya, 20: 418-444.

SANTSCHI F, 1925. Contribution à la faune myrmécologique de la Chine [J]. Bulletin de la Société Vaudoise des Sciences Naturelles, 56: 81-96.

SANTSCHI F, 1928. Nouvelles fourmis de Chine et du Turkestan Russe [J]. Bulletin et Annales de la Société Entomologique de Belgique, 68: 31-46.

SCHMIDT C A, SHATTUCK S O, 2014. The higher classification of the ant subfamily Ponerinae (Hymenoptera: Formicidae), with a review of ponerine ecology and behavior [J]. Zootaxa, 3817 (1): 1-242.

SEIFERT B, 2003. The ant genus *Cardiocondyla* (Insecta: Hymenoptera: Formicidae) - a taxonomic revision of the *C. elegans, C. bulgarica, C. batesii, C. nuda, C. shuckardi, C. stambuloffii, C. wroughtonii, C. emeryi*, and *C. minutior* species groups [J]. Annalen des Naturhistorischen Museums in Wien. B, Botanik, Zoologie, 104: 203-338.

SEIFERT B, 2020. A taxonomic revision of the Palaearctic members of the subgenus *Lasius* s.str. (Hymenoptera, Formicidae) [J]. Soil Organisms, 92 (1): 15-86.

SEIFERT B, 2023a. A revision of the Palaearctic species of the ant genus *Cardiocondyla* Emery 1869 (Hymenoptera: Formicidae). Zootaxa, 5274 (1): 1-64.

SEIFERT B, 2023b. Two new species of *Formicoxenus* Mayr 1855 and *Leptothorax* Mayr 1855 from Tibet (Hymenoptera: Formicidae) [J]. Soil Organisms, 95 (2): 129-142.

SMITH F, 1858. Catalogue of hymenopterous insects in the collection of the British Museum. Part VI. Formicidae [M]. London: British Museum.

STAAB M, 2014. A new species of the *Aenictus wroughtonii* group (Hymenoptera, Formicidae) from South-East Asia [J]. ZooKeys, 391: 65-73.

STAAB M, HITA GARCIA F, LIU C, et al., 2018. Systematics of the ant genus *Proceratium* Roger (Hymenoptera, Formicidae, Proceratiinae) in China, with descriptions of three new species based on micro-CT enhanced next-generation-morphology [J]. ZooKeys, 770: 137-192.

STITZ H, 1923. Hymenoptera, VII. Formicidae [J]. Beiträge zur Kenntnis der Land- und Süsswasserfauna Deutsch-Südwestafrikas, 2: 143-167.

STITZ H, 1934. Schwedisch-chinesische wissenschaftliche Expedition nach den nordwestlichen Provinzen Chinas, unter Leitung von Dr. Sven Hedin und Prof. Sü Ping-chang. Insekten gesammelt vom schwedischen Arzt der Expedition Dr.

David Hummel 1927-1930. 25. Hymenoptera [J]. Arkiv för Zoologi, 27A(11): 1-9.

TANG K L, GUÉNARD B, 2023. Further additions to the knowledge of *Strumigenys* (Formicidae: Myrmicinae) within South East Asia, with the descriptions of 20 new species [J]. European Journal of Taxonomy, 907: 1-144.

TANG K L, GUÉNARD B, 2024. Further additions to the knowledge of *Strumigenys* (Formicidae: Myrmicinae) within South East Asia, with the description of 20 species – corrigendum [J]. European Journal of Taxonomy, 917: 194-197.

TANG K L, PIERCE M P, GUÉNARD B, 2019. Review of the genus *Strumigenys* (Hymenoptera, Formicidae, Myrmicinae) in Hong Kong with the description of three new species and the addition of five native and four introduced species records [J]. ZooKeys, 831: 1-48.

TERANISHI C, 1940. Works of Cho Teranishi. Memorial Volume [M]. Osaka: Kansai Entomological Society.

TERAYAMA M, 2009. A synopsis of the family Formicidae of Taiwan [J]. Research Bulletin of Kanto Gakuen University, 17: 81-266.

THOMAS J A, SETTELE J, 2004. Evolutionary biology: butterfly mimics of ants [J]. Nature, 432: 283-284.

VIEHMEYER H, 1912. Ameisen aus Deutsch Neuguinea gesammelt von Dr. O. Schlaginhaufen. Nebst einem Verzeichnisse der papuanischen Arten [J]. Abhandlungen und Berichte des Königlichen Zoologischen und Anthropologische-Ethnographischen Museums zu Dresden, 14: 1-26.

VIEHMEYER H, 1922. Neue Ameisen [J]. Archiv für Naturgeschichte, 88(7): 203-220.

WARD P S, FISHER B L, 2016. Tales of dracula ants: the evolutionary history of the ant subfamily Amblyoponinae (Hymenoptera: Formicidae) [J]. Systematic Entomology, 41: 683-693.

WEBER N A, 1947. A revision of the North American ants of the genus *Myrmica* Latreille with a synopsis of the Palearctic species. I [J]. Annals of the Entomological Society of America, 40: 437-474.

WEBER N A, 1950. A revision of the North American ants of the genus *Myrmica* Latreille with a synopsis of the Palearctic species. III [J]. Annals of the Entomological Society of America, 43: 189-226.

WEI C, XU Z H, HE H, 2001a. A new species of the ant genus *Strongylognathus* Mayr (Hymenoptera: Formicidae) from Shaanxi, China [J]. Entomotaxonomia, 23: 68-70.

WEI C, ZHOU S Y, HE H, et al., 2001b. A taxonomic study of the genus *Myrmica* Latreille from China (Hymenoptera: Formicidae) [J]. Acta Zootaxonomica Sinica, 26: 560-564.

WHEELER W M, 1909. Ants of Formosa and the Philippines [J]. Bulletin of the American Museum of Natural History, 26: 333-345.

WHEELER W M, 1930. A list of the known Chinese ants [J]. Peking Natural History Bulletin, 5: 53-81.

WHEELER W M, 1933. New ants from China and Japan [J]. Psyche (Cambridge), 40: 65-67.

WILLIAMS J L, LAPOLLA J S, 2016. Taxonomic revision and phylogeny of the ant genus *Prenolepis* (Hymenoptera: Formicidae) [J]. Zootaxa, 4200: 201-258.

WILSON E O, 1955. A monographic revision of the ant genus *Lasius* [J]. Bulletin of the Museum of Comparative Zoology, 113: 1-201.

WILSON E O, 1964. The true army ants of the Indo-Australian area (Hymenoptera: Formicidae: Dorylinae) [J]. Pacific Insects, 6: 427-483.

WONG T L, GUÉNARD B, 2020. Review of ants from the genus *Polyrhachis* Smith (Hymenoptera: Formicidae: Formicinae) in Hong Kong and Macau, with notes on their natural history [J]. Asian Myrmecology, 13: 1-70.

WU C F, 1941. Catalogus Insectorum Sinensium. Volume VI [M]. Peiping [= Beijing]: Yenching University.

XU Z H, 1995b. Two new species of the ant genus *Prenolepis* from Yunnan China (Hymenoptera: Formicidae) [J]. Zoological Research, 16: 337-341.

XU Z H, 1995c. A taxonomic study of the genus *Pachycondyla* Smith in China (Hyminoptera: Formicidae: Ponerinae). Pp. 103-112 in: Lian Z M (chief editor) 1995. Entomological Research. First issue. A collection of research papers commemorating 40 years of teaching by Professor Zheng Zhemin [C]. Xi'an: Shaanxi Normal University Press.

XU Z H, 1996. A taxonomic study of the ant genus *Pachycondyla* from China (Hymenoptera: Formicidae: Ponerinae) [J]. Zoological Research, 17: 211-216.

XU Z H, 1997. A taxonomic study of the ant genus *Pseudolasius* Emery in China (Hymenoptera: Formicidae) [J]. Zoological Research, 18: 1-6.

XU Z H, 1998a. Two new species of the genera *Mystrium* and *Cryptopone* from Yunnan, China (Hymenoptera: Formicidae) [J].

Zoological Research, 19: 160-164.

XU Z H, 1998b. Two new species of the ant genus *Polyrhachis* Smith from Yunnan, China (Hymenoptera: Formicidae) [J]. Zoological Research, 19: 242-246.

XU Z H, 1998c. Two new record genera and three new species of Formicidae (Hymenoptera) from China [J]. Entomologia Sinica, 5: 121-127.

XU Z H. 1999. Systematic studies on the ant genera of *Carebara*, *Rhopalomastix* and *Kartidris* in China (Hymenoptera: Formicidae: Myrmicinae) [J]. Acta Biologica Plateau Sinica, 14: 129-136.

XU Z H, 2000a. Two new genera of ant subfamilies Dorylinae and Ponerinae (Hymenoptera: Formicidae) from Yunnan, China [J]. Zoological Research, 21: 297-302.

XU Z H, 2000b. A systematic study of the ant genus *Proceratium* Roger from China (Hymenoptera: Formicidae) [J]. Acta Zootaxonomica Sinica, 25: 434-437.

XU Z H, 2000c. Five new species and one new record of the ant genus *Leptogenys* Roger (Hymenoptera: Formicidae) from Yunnan Province, China [J]. Entomologia Sinica, 7: 117-126.

XU Z H, 2000d. A new species of the ant genus *Epitritus* Emery (Hymenoptera: Formicidae) from China [J]. Entomotaxonomia, 22: 297-300.

XU Z H, 2001a. A systematic study on the ant genus *Ponera* Latreille (Hymenoptera: Formicidae) of China [J]. Entomotaxonomia, 23: 51-60.

XU Z H, 2001b. Two new species of the ant genus *Dolichoderus* Lund from Yunnan, China (Hymenoptera: Formicidae) [J]. Acta Zootaxonomica Sinica, 26: 355-360.

XU Z H, 2001c. Four new species of the ant genus *Ponera* Latreille (Hymenoptera: Formicidae) from Yunnan, China [J]. Entomotaxonomia, 23: 217-226.

XU Z H, 2001d. A systematic study on the ant genus *Amblyopone* Erichson from China (Hymenoptera: Formicidae) [J]. Acta Zootaxonomica Sinica, 26: 551-556.

XU Z H, 2002b. A systematic study on the ant subfamily Leptanillinae of China (Hymenoptera: Formicidae) [J]. Acta Entomologica Sinica, 45: 115-120.

XU Z H, 2002c. A systematic study on the ant subgenus *Cyrtomyrma* Forel of the genus *Polyrhachis* Smith of China (Hymenoptera: Formicidae) [J]. Acta Entomologica Sinica, 45: 522-530.

XU Z H, 2003. A systematic study on Chinese species of the ant genus *Oligomyrmex* Mayr (Hymenoptera: Formicidae) [J]. Acta Zootaxonomica Sinica, 28: 310-322.

XU Z H, 2006. Three new species of the ant genera *Amblyopone* Erichson, 1842 and *Proceratium* Roger, 1863 (Hymenoptera: Formicidae) from Yunnan, China [J]. Myrmecologische Nachrichten, 8: 151-155.

XU Z H, 2012a. A newly recorded genus and species, *Harpagoxenus sublaevis*, from China with a key to the known species of *Harpagoxenus* of the world (Hymenoptera: Formicidae) [J]. Sociobiology, 59: 19-25.

XU Z H, 2012b. *Gaoligongidris planodorsa*, a new genus and species of the ant subfamily Myrmicinae from China with a key to the genera of Stenammini of the world (Hymenoptera: Formicidae) [J]. Sociobiology, 59: 331-342.

XU Z H, 2012c. *Furcotanilla*, a new genus of the ant subfamily Leptanillinae from China with descriptions of two new species of *Protanilla* and *P. rafflesi* Taylor (Hymenoptera: Formicidae) [J]. Sociobiology, 59: 477-491.

XU Z H, BURWELL C J, NAKAMURA A, 2014a. Two new species of the proceratiine ant genus *Discothyrea* Roger from Yunnan, China, with a key to the known Oriental species [J]. Asian Myrmecology, 6: 33-41.

XU Z H, BURWELL C J, NAKAMURA A, 2014b. A new species of the ponerine ant genus *Myopias* Roger from Yunnan, China, with a key to the known Oriental species [J]. Sociobiology, 61 (2): 164-170.

XU Z H, CHAI Z Q, 2004. Systematic study on the ant genus *Tetraponera* F. Smith (Hymenoptera: Formicidae) of China [J]. Acta Zootaxonomica Sinica, 29: 63-76.

XU Z H, CHU J J, 2012. Four new species of the amblyoponine ant genus *Amblyopone* (Hymenoptera: Formicidae) from southwestern China with a key to the known Asian species [J]. Sociobiology, 59 (4):1175-1196.

XU Z H, HE Q J, 2011. Description of *Myopopone castanea* (Smith) (Hymenoptera: Formicidae) from Himalaya region [J]. Entomotaxonomia, 33 (3): 231-235.

XU Z H, HE Q J, 2015. Taxonomic review of the ponerine ant genus *Leptogenys* Roger, 1861 (Hymenoptera: Formicidae) with a key to the Oriental species [J]. Myrmecological News, 21: 137-161.

XU Z H, LIU X, 2012. Three new species of the ant genus *Myopias* (Hymenoptera: Formicidae) from China with a key to the known Chinese species [J]. Sociobiology, 59: 819-834.

XU Z H, WANG W H, 2004. The third species of the ant genus *Perissomyrmex* Smith (Hymenoptera: Formicidae) in the world [J]. Entomotaxonomia, 26: 217-221.

XU Z H, XU G L, 2011. A new species of the genus *Paratopula* Wheeler (Hymenoptera: Formicidae) from Tibet [J]. Acta Zootaxonomica Sinica, 36: 595-597.

XU Z H, YU N N, 2012. *Vombisidris tibeta*, a new myrmicine ant species from Tibet, China with a key to the known species of *Vombisidris* Bolton of the world [J]. Sociobiology, 59 (3): 1495-1507.

XU Z H, ZENG G, 2000. Discovery of the worker caste of *Platythyrea clypeata* Forel and a new species of *Probolomyrmex* Mayr in Yunnan, China (Hymenoptera: Formicidae) [J]. Entomologia Sinica, 7: 213-217.

XU Z H, ZHANG C L, 2012. Review of the myrmicine ant genus *Perissomyrmex* M.R. Smith, 1947 (Hymenoptera: Formicidae) with description of a new species from Tibet, China [J]. Myrmecological News, 17: 147-154.

XU Z H, ZHANG J L, 2002a. Two new species of the ant subfamily Leptanillinae from Yunnan, China (Hymenoptera: Formicidae) [J]. Acta Zootaxonomica Sinica, 27: 139-144.

XU Z H, ZHANG W, 1996. A new species of the genus *Gnamptogenys* (Hymenoptera: Formicidae: Ponerinae) from southwestern China [J]. Entomotaxonomia, 18: 55-58.

XU Z H, ZHANG Z Y, 2002b. Systematics of Chinese species of the ant genus *Pristomyrmex* Mayr (Hymenoptera: Formicidae) [J]. Entomologia Sinica, 9(4): 69-72.

XU Z H, ZHENG Z M, 1994. New species and new record species of the genus *Tetramorium* Mayr (Hymenoptera: Formicidae) from southwestern China [J]. Entomotaxonomia, 16: 285-290.

XU Z H, ZHENG Z M, 1995. Two new species of the ant genera *Recurvidris* Bolton and *Kartidris* Bolton (Hymenoptera: Formicidae: Myrmicinae) from southwestern China [J]. Entomotaxonomia, 17: 143-146.

XU Z H, ZHOU, X G, 2004. Systematic study on the ant genus *Pyramica* Roger (Hymenoptera, Formicidae) of China [J]. Acta Zootaxonomica Sinica, 29: 440-450.

XU Z H, ZHOU X Y, 2015. Species grouping and key to known species of the ant genus *Echinopla* Smith (Hymenoptera: Formicidae) with reports of Chinese species [J]. Asian Myrmecology, 7: 19-36.

YAMANE S, LEONG C M, LIN C C, 2018. Taiwanese species of the ant genus *Technomyrmex* (Formicidae: Dolichoderinae) [J]. Zootaxa, 4410: 35-56.

YANO M, 1911. The genus *Polyrhachis* of Japan [In Japanese] [J]. Dobutsugaku Zasshi (Zoological Magazine), 23: 249-256.

YASUMATSU K, 1941a. On the ants of the genus *Dolichoderus* of Angaran element from the Far East (Hymenoptera, Formicidae) [J]. Kontyû, 14: 177-183.

YASUMATSU K, 1941b. Ants collected by Mr. H. Takahasi in Hingan (Hsingan) North Province, North Manchuria (Hymenoptera, Formicidae) [J]. Transactions of the Natural History Society of Formosa, 31: 182-185.

YASUMATSU K, 1962. Notes on synonymies of five ants widely spread in the Orient (Hym.: Formicidae) [J]. Mushi, 36: 93-97.

YOSHIMURA M, FISHER B L, 2012. A revision of male ants of the Malagasy Amblyoponinae (Hymenoptera: Formicidae) with resurrections of the genera *Stigmatomma* and *Xymmer* [J]. PLoS ONE, 7 (3): e33325.

YOSHIMURA M, FISHER B L, 2014. A revision of the ant genus *Mystrium* in the Malagasy region with description of six new species and remarks on *Amblyopone* and *Stigmatomma* (Hymenoptera, Formicidae, Amblyoponinae) [J]. ZooKeys, 394: 1-99.

ZHOU S Y, XU Z H, 2003. Taxonomic study on Chinese members of the ant genus *Strumigenys* F. Smith (Hymenoptera: Formicidae) from the mainland of China [J]. Acta Zootaxonomica Sinica, 28: 737-740.

附 录
Appendix

模式标本馆藏在西南林业大学的蚂蚁新种名录
Checklist of New Species of Ants with Type Specimens Housed in Southwest Forestry University

依据Bolton（1995）分类系统，列出了模式标本馆藏在西南林业大学的136个蚂蚁新种的名录，这些新种隶属于10亚科39属。亚科参照系统发育顺序排序，亚科内各属按照属名拉丁字母顺序排序，属内物种按照种名拉丁字母顺序排序。依据AntWiki（2024）和AntMaps（2024）的地理分布信息，提供了每个物种在我国各省、自治区、直辖市、特别行政区的分布以及在国外各国家和地区的地理分布，以方便读者和研究人员参考。

1 钝猛蚁亚科 Amblyoponinae

1）钝猛蚁属 *Amblyopone* Erichson, 1842

（1）阿佤钝猛蚁 *Amblyopone awa* Xu & Chu, 2012
　　地理分布：中国（云南、西藏），印度。
（2）细齿钝猛蚁 *Amblyopone crenata* Xu, 2001
　　地理分布：中国（云南），越南，泰国。
（3）康巴钝猛蚁 *Amblyopone kangba* Xu & Chu, 2012
　　地理分布：中国（西藏、云南），不丹。
（4）梅里钝猛蚁 *Amblyopone meiliana* Xu & Chu, 2012
　　地理分布：中国（云南）。
（5）八齿钝猛蚁 *Amblyopone octodentata* Xu, 2006
　　地理分布：中国（西藏、云南）。
（6）三叶钝猛蚁 *Amblyopone triloba* Xu, 2001

According to the classification system of Bolton (1995), a checklist is provided for the 136 new species of ants with type specimens housed in Southwest Forestry University. These new species belongs to 10 subfamilies and 39 genera of Formicidae. Subfamilies are listed refering phylogenetic order. Genera of the subfamilies are listed in Latin alphabetical order by genus name. Species of the genera are listed in Latin alphabetical order by species name. According to the information of geographical distribution provided by AntWiki (2024) and AntMaps (2024), we provided the distribution information of each species in our provinces, autonomous regions, municipalities and special administrative regions, and the distribution information in foreign countries and regions, for the convenience of readers and researchers reference.

1 Amblyoponinae

1) *Amblyopone* Erichson, 1842

(1) *Amblyopone awa* Xu & Chu, 2012
　　Distribution: China (Tibet, Yunnan), India.
(2) *Amblyopone crenata* Xu, 2001
　　Distribution: China (Yunnan) , Thailand, Vietnam.
(3) *Amblyopone kangba* Xu & Chu, 2012
　　Distribution: China (Tibet, Yunnan), Bhutan.
(4) *Amblyopone meiliana* Xu & Chu, 2012
　　Distribution: China (Yunnan).
(5) *Amblyopone octodentata* Xu, 2006
　　Distribution: China (Tibet, Yunnan).
(6) *Amblyopone triloba* Xu, 2001

地理分布：中国（云南）。

（7）卓玛钝猛蚁 *Amblyopone zoma* Xu & Chu, 2012

地理分布：中国（西藏）。

2）版纳猛蚁属 *Bannapone* Xu, 2000

（8）木兰版纳猛蚁 *Bannapone mulanae* Xu, 2000

地理分布：中国（云南）。

3）迷猛蚁属 *Mystrium* Roger, 1862

（9）小眼迷猛蚁 *Mystrium oculatum* Xu, 1998

地理分布：中国（云南）。

2　刺猛蚁亚科 Ectatomminae

4）曲颊猛蚁属 *Gnamptogenys* Roger, 1863

（10）版纳曲颊猛蚁 *Gnamptogenys bannana* Xu & Zhang, 1996

地理分布：中国（云南）。

3　卷尾猛蚁亚科 Proceratiinae

5）盘猛蚁属 *Discothyrea* Roger, 1863

（11）版纳盘猛蚁 *Discothyrea banna* Xu, Burwell & Nakamura, 2014

地理分布：中国（云南、江西）。

（12）滇盘猛蚁 *Discothyrea diana* Xu, Burwell & Nakamura, 2014

地理分布：中国（云南）。

6）卷尾猛蚁属 *Proceratium* Roger, 1863

（13）布氏卷尾猛蚁 *Proceratium bruelheidei* Staab, Xu & Hita Garcia, 2018

地理分布：中国（江西，浙江）。

（14）克平卷尾猛蚁 *Proceratium kepingmai* Staab, Xu & Hita Garcia, 2018

地理分布：中国（江西，浙江）。

（15）龙门卷尾猛蚁 *Proceratium longmenense* Xu, 2006

地理分布：中国（云南）。

（16）怒江卷尾猛蚁 *Proceratium nujiangense* Xu, 2006

地理分布：中国（云南）。

（17）昌平卷尾猛蚁 *Proceratium shohei* Staab, Xu & Hita Garcia, 2018

地理分布：中国（云南）。

Distribution: China (Yunnan).

(7) *Amblyopone zoma* Xu & Chu, 2012
Distribution: China (Tibet).

2) *Bannapone* Xu, 2000

(8) *Bannapone mulanae* Xu, 2000
Distribution: China (Yunnan).

3) *Mystrium* Roger, 1862

(9) *Mystrium oculatum* Xu, 1998
Distribution: China (Yunnan).

2 Ectatomminae

4) *Gnamptogenys* Roger, 1863

(10) *Gnamptogenys bannana* Xu & Zhang, 1996
Distribution: China (Yunnan).

3 Proceratiinae

5) *Discothyrea* Roger, 1863

(11) *Discothyrea banna* Xu, Burwell & Nakamura, 2014
Distribution: China (Jiangxi, Yunnan).

(12) *Discothyrea diana* Xu, Burwell & Nakamura, 2014
Distribution: China (Yunnan).

6) *Proceratium* Roger, 1863

(13) *Proceratium bruelheidei* Staab, Xu & Hita Garcia, 2018
Distribution: China (Jiangxi, Zhejiang).

(14) *Proceratium kepingmai* Staab, Xu & Hita Garcia, 2018
Distribution: China (Jiangxi, Zhejiang).

(15) *Proceratium longmenense* Xu, 2006
Distribution: China (Yunnan).

(16) *Proceratium nujiangense* Xu, 2006
Distribution: China (Yunnan).

(17) *Proceratium shohei* Staab, Xu & Hita Garcia, 2018
Distribution: China (Yunnan).

（18）赵氏卷尾猛蚁 *Proceratium zhaoi* Xu, 2000
地理分布：中国（云南）。

4 猛蚁亚科 Ponerinae

7）短猛蚁属 *Brachyponera* Emery, 1900

（19）短背短猛蚁 *Brachyponera brevidorsa* Xu, 1994
地理分布：中国（云南、贵州、广西）。

8）隐猛蚁属 *Cryptopone* Emery, 1892

（20）直唇隐猛蚁 *Cryptopone recticlypea* Xu, 1998
地理分布：中国（云南）。

9）埃猛蚁属 *Emeryopone* Forel, 1912

（21）黑色埃猛蚁 *Emeryopone melaina* Xu, 1998
地理分布：中国（云南）。

10）细颚猛蚁属 *Leptogenys* Roger, 1861

（22）黄帝细颚猛蚁 *Leptogenys huangdii* Xu, 2000
地理分布：中国（云南）。

（23）老子细颚猛蚁 *Leptogenys laozii* Xu, 2000
地理分布：中国（云南、浙江）。

（24）孟子细颚猛蚁 *Leptogenys mengzii* Xu, 2000
地理分布：中国（云南、西藏）。

（25）盘古细颚猛蚁 *Leptogenys pangui* Xu, 2000
地理分布：中国（云南），越南，马来西亚，新加坡。

（26）孙子细颚猛蚁 *Leptogenys sunzii* Xu & He, 2015
地理分布：中国（云南）。

（27）炎帝细颚猛蚁 *Leptogenys yandii* Xu & He, 2015
地理分布：中国（西藏）。

（28）庄子细颚猛蚁 *Leptogenys zhuangzii* Xu, 2000
地理分布：中国（云南）。

11）小眼猛蚁属 *Myopias* Roger, 1861

（29）锥头小眼猛蚁 *Myopias conicara* Xu, 1998
地理分布：中国（云南、西藏），越南。

（30）傣小眼猛蚁 *Myopias daia* Xu, Burwell & Nakamura, 2014
地理分布：中国（云南）。

（31）哈尼小眼猛蚁 *Myopias hania* Xu & Liu, 2012
地理分布：中国（云南）。

（32）珞巴小眼猛蚁 *Myopias luoba* Xu & Liu, 2012

(18) *Proceratium zhaoi* Xu, 2000
Distribution: China (Yunnan).

4 Ponerinae

7) *Brachyponera* Emery, 1900

(19) *Brachyponera brevidorsa* Xu, 1994
Distribution: China (Guangxi, Guizhou, Yunnan).

8) *Cryptopone* Emery, 1892

(20) *Cryptopone recticlypea* Xu, 1998
Distribution: China (Yunnan).

9) *Emeryopone* Forel, 1912

(21) *Emeryopone melaina* Xu, 1998
Distribution: China (Yunnan).

10) *Leptogenys* Roger, 1861

(22) *Leptogenys huangdii* Xu, 2000
Distribution: China (Yunnan).

(23) *Leptogenys laozii* Xu, 2000
Distribution: China (Yunnan, Zhejiang).

(24) *Leptogenys mengzii* Xu, 2000
Distribution: China (Tibet, Yunnan).

(25) *Leptogenys pangui* Xu, 2000
Distribution: China (Yunnan), Malaysia, Singapore, Vietnam.

(26) *Leptogenys sunzii* Xu & He, 2015
Distribution: China (Yunnan).

(27) *Leptogenys yandii* Xu & He, 2015
Distribution: China (Tibet).

(28) *Leptogenys zhuangzii* Xu, 2000
Distribution: China (Yunnan).

11) *Myopias* Roger, 1861

(29) *Myopias conicara* Xu, 1998
Distribution: China (Tibet, Yunnan), Vietnam.

(30) *Myopias daia* Xu, Burwell & Nakamura, 2014
Distribution: China (Yunnan).

(31) *Myopias hania* Xu & Liu, 2012
Distribution: China (Yunnan).

(32) *Myopias luoba* Xu & Liu, 2012

地理分布：中国（西藏）。

(33) 门巴小眼猛蚁 *Myopias menba* Xu & Liu, 2012

地理分布：中国（西藏）。

12）厚结猛蚁属 *Pachycondyla* Smith, 1858

(34) 片突厚结猛蚁 *Pachycondyla lobocarena* Xu, 1995

地理分布：中国（云南、四川、广东），越南。

(35) 郑氏厚结猛蚁 *Pachycondyla zhengi* Xu, 1995

地理分布：中国（云南、西藏）。

13）猛蚁属 *Ponera* Latreille, 1804

(36) 巴卡猛蚁 *Ponera baka* Xu, 2001

地理分布：中国（云南、西藏）。

(37) 坝湾猛蚁 *Ponera bawana* Xu, 2001

地理分布：中国（云南）。

(38) 二齿猛蚁 *Ponera diodonta* Xu, 2001

地理分布：中国（云南、西藏）。

(39) 龙林猛蚁 *Ponera longlina* Xu, 2001

地理分布：中国（云南、西藏）。

(40) 勐腊猛蚁 *Ponera menglana* Xu, 2001

地理分布：中国（云南、西藏），马来西亚，新加坡。

(41) 南贡山猛蚁 *Ponera nangongshana* Xu, 2001

地理分布：中国（云南）。

(42) 五齿猛蚁 *Ponera pentodontos* Xu, 2001

地理分布：中国（云南）。

(43) 片马猛蚁 *Ponera pianmana* Xu, 2001

地理分布：中国（云南、西藏、四川）。

(44) 黄色猛蚁 *Ponera xantha* Xu, 2001

地理分布：中国（云南）。

14）小盲猛蚁属 *Probolomyrmex* Mayr, 1901

(45) 长柄小盲猛蚁 *Probolomyrmex longiscapus* Xu & Zeng, 2000

地理分布：中国（云南），越南，老挝。

5 行军蚁亚科 Dorylinae

15）云行军蚁属 *Yunodorylus* Xu, 2000

(46) 六刺云行军蚁 *Yunodorylus sexspinus* Xu, 2000

地理分布：中国（云南），泰国。

Distribution: China (Tibet).

(33) *Myopias menba* Xu & Liu, 2012

Distribution: China (Tibet).

12) *Pachycondyla* Smith, 1858

(34) *Pachycondyla lobocarena* Xu, 1995

Distribution: China (Guangdong, Sichuan, Yunnan), Vietnam.

(35) *Pachycondyla zhengi* Xu, 1995

Distribution: China (Tibet, Yunnan).

13) *Ponera* Latreille, 1804

(36) *Ponera baka* Xu, 2001

Distribution: China (Tibet, Yunnan).

(37) *Ponera bawana* Xu, 2001

Distribution: China (Yunnan).

(38) *Ponera diodonta* Xu, 2001

Distribution: China (Tibet, Yunnan).

(39) *Ponera longlina* Xu, 2001

Distribution: China (Tibet, Yunnan).

(40) *Ponera menglana* Xu, 2001

Distribution: China (Tibet, Yunnan), Malaysia, Singapore.

(41) *Ponera nangongshana* Xu, 2001

Distribution: China (Yunnan).

(42) *Ponera pentodontos* Xu, 2001

Distribution: China (Yunnan).

(43) *Ponera pianmana* Xu, 2001

Distribution: China (Sichuan, Tibet, Yunnan).

(44) *Ponera xantha* Xu, 2001

Distribution: China (Yunnan).

14) *Probolomyrmex* Mayr, 1901

(45) *Probolomyrmex longiscapus* Xu & Zeng, 2000

Distribution: China (Yunnan), Laos, Vietnam.

5 Dorylinae

15) *Yunodorylus* Xu, 2000

(46) *Yunodorylus sexspinus* Xu, 2000

Distribution: China (Yunnan), Thailand.

6　细蚁亚科 Leptanillinae

16）细蚁属 *Leptanilla* Emery, 1870

（47）北京细蚁 *Leptanilla beijingensis* Qian, Xu, Man & Liu, 2024

地理分布：中国（北京）。

（48）德宏细蚁 *Leptanilla dehongensis* Qian, Xu, Man & Liu, 2024

地理分布：中国（云南）。

（49）昆明细蚁 *Leptanilla kunmingensis* Xu & Zhang, 2002

地理分布：中国（云南）。

（50）秦岭细蚁 *Leptanilla qinlingensis* Qian, Xu, Man & Liu, 2024

地理分布：中国（陕西）。

（51）云南细蚁 *Leptanilla yunnanensis* Xu, 2002

地理分布：中国（云南）。

17）原细蚁属 *Protanilla* Taylor, 1990

（52）北京原细蚁 *Protanilla beijingensis* Man, Ran, Chen & Xu, 2017

地理分布：中国（北京）。

（53）双色原细蚁 *Protanilla bicolor* Xu, 2002

地理分布：中国（云南）。

（54）单色原细蚁 *Protanilla concolor* Xu, 2002

地理分布：中国（云南）。

（55）叉颚原细蚁 *Protanilla furcomandibula* Xu & Zhang, 2002

地理分布：中国（云南）。

（56）耿马原细蚁 *Protanilla gengma* Xu, 2012

地理分布：中国（云南），印度。

（57）西藏原细蚁 *Protanilla tibeta* Xu, 2012

地理分布：中国（西藏）。

7　伪切叶蚁亚科 Pseudomyrmecinae

18）细长蚁属 *Tetraponera* Smith, 1852

（58）无缘细长蚁 *Tetraponera amargina* Xu & Chai, 2004

地理分布：中国（云南、浙江）。

6 Leptanillinae

16) *Leptanilla* Emery, 1870

(47) *Leptanilla beijingensis* Qian, Xu, Man & Liu, 2024
Distribution: China (Beijing).

(48) *Leptanilla dehongensis* Qian, Xu, Man & Liu, 2024
Distribution: China (Yunnan).

(49) *Leptanilla kunmingensis* Xu & Zhang, 2002
Distribution: China (Yunnan).

(50) *Leptanilla qinlingensis* Qian, Xu, Man & Liu, 2024
Distribution: China (Shaanxi).

(51) *Leptanilla yunnanensis* Xu, 2002
Distribution: China (Yunnan).

17) *Protanilla* Taylor, 1990

(52) *Protanilla beijingensis* Man, Ran, Chen & Xu, 2017
Distribution: China (Beijing).

(53) *Protanilla bicolor* Xu, 2002
Distribution: China (Yunnan).

(54) *Protanilla concolor* Xu, 2002
Distribution: China (Yunnan).

(55) *Protanilla furcomandibula* Xu & Zhang, 2002
Distribution: China (Yunnan).

(56) *Protanilla gengma* Xu, 2012
Distribution: China (Yunnan), India.

(57) *Protanilla tibeta* Xu, 2012
Distribution: China (Tibet).

7 Pseudomyrmecinae

18) *Tetraponera* Smith, 1852

(58) *Tetraponera amargina* Xu & Chai, 2004
Distribution: China (Yunnan, Zhejiang).

（59）凹唇细长蚁 *Tetraponera concava* Xu & Chai, 2004
地理分布：中国（云南），泰国。

（60）隆背细长蚁 *Tetraponera convexa* Xu & Chai, 2004
地理分布：中国（云南、浙江）。

（61）叉唇细长蚁 *Tetraponera furcata* Xu & Chai, 2004
地理分布：中国（云南）。

（62）尖唇细长蚁 *Tetraponera protensa* Xu & Chai, 2004
地理分布：中国（云南）。

8　切叶蚁亚科 Myrmicinae

19）圆鳞蚁属 *Epitritus* Emery, 1869

（63）大禹圆鳞蚁 *Epitritus dayui* Xu, 2000
地理分布：中国（云南）。

20）高黎贡蚁属 *Gaoligongidris* Xu, 2012

（64）平背高黎贡蚁 *Gaoligongidris planodorsa* Xu, 2012
地理分布：中国（云南、四川）。

21）无刺蚁属 *Kartidris* Bolton, 1991

（65）阿诗玛无刺蚁 *Kartidris ashima* Xu & Zheng, 1995
地理分布：中国（云南、四川）。

（66）疏毛无刺蚁 *Kartidris sparsipila* Xu, 1999
地理分布：中国（云南）。

22）弯蚁属 *Lordomyrma* Emery, 1897

（67）景颇弯蚁 *Lordomyrma jingpo* Liu, Xu & Hita Garcia, 2021
地理分布：中国（云南）。

（68）尼玛弯蚁 *Lordomyrma nima* Liu, Xu & Hita Garcia, 2021
地理分布：中国（西藏）。

23）稀切叶蚁属 *Oligomyrmex* Mayr, 1867

（69）尖刺稀切叶蚁 *Oligomyrmex acutispinus* Xu, 2003
地理分布：中国（云南、四川）。

（70）高结稀切叶蚁 *Oligomyrmex altinodus* Xu, 2003
地理分布：中国（云南、西藏、江西、海南）。

（71）双角稀切叶蚁 *Oligomyrmex bihornatus* Xu, 2003
地理分布：中国（云南）。

（72）弯刺稀切叶蚁 *Oligomyrmex curvispinus* Xu, 2003
地理分布：中国（云南）。

(59) *Tetraponera concava* Xu & Chai, 2004
Distribution: China (Yunnan), Thailand.

(60) *Tetraponera convexa* Xu & Chai, 2004
Distribution: China (Yunnan, Zhejiang).

(61) *Tetraponera furcata* Xu & Chai, 2004
Distribution: China (Yunnan).

(62) *Tetraponera protensa* Xu & Chai, 2004
Distribution: China (Yunnan).

8 Myrmicinae

19) *Epitritus* Emery, 1869

(63) *Epitritus dayui* Xu, 2000
Distribution: China (Yunnan).

20) *Gaoligongidris* Xu, 2012

(64) *Gaoligongidris planodorsa* Xu, 2012
Distribution: China (Sichuan, Yunnan).

21) *Kartidris* Bolton, 1991

(65) *Kartidris ashima* Xu & Zheng, 1995
Distribution: China (Sichuan, Yunnan).

(66) *Kartidris sparsipila* Xu, 1999
Distribution: China (Yunnan).

22) *Lordomyrma* Emery, 1897

(67) *Lordomyrma jingpo* Liu, Xu & Hita Garcia, 2021
Distribution: China (Yunnan).

(68) *Lordomyrma nima* Liu, Xu & Hita Garcia, 2021
Distribution: China (Tibet).

23) *Oligomyrmex* Mayr, 1867

(69) *Oligomyrmex acutispinus* Xu, 2003
Distribution: China (Sichuan, Yunnan).

(70) *Oligomyrmex altinodus* Xu, 2003
Distribution: China (Hainan, Jiangxi, Tibet, Yunnan).

(71) *Oligomyrmex bihornatus* Xu, 2003
Distribution: China (Yunnan).

(72) *Oligomyrmex curvispinus* Xu, 2003
Distribution: China (Yunnan).

（73）钝齿稀切叶蚁 *Oligomyrmex obtusidentus* Xu, 2003

地理分布：中国（云南、西藏、四川、湖南），印度。

（74）直背稀切叶蚁 *Oligomyrmex rectidorsus* Xu, 2003

地理分布：中国（云南、西藏、四川、河北、河南、湖南）。

（75）纹头稀切叶蚁 *Oligomyrmex reticapitus* Xu, 2003

地理分布：中国（云南、西藏、四川、广西、海南）。

（76）条纹稀切叶蚁 *Oligomyrmex striatus* Xu, 2003

地理分布：中国（云南、四川）。

24）华丽蚁属 *Paratopula* Wheeler, 1919

（77）郑氏华丽蚁 *Paratopula zhengi* Xu & Xu, 2011

地理分布：中国（西藏）。

25）奇蚁属 *Perissomyrmex* Smith, 1947

（78）裂唇奇蚁 *Perissomyrmex fissus* Xu & Wang, 2004

地理分布：中国（云南）。

（79）墨脱奇蚁 *Perissomyrmex medogensis* Xu & Zhang, 2012

地理分布：中国（西藏）。

26）棱胸蚁属 *Pristomyrmex* Mayr, 1886

（80）弯钩棱胸蚁 *Pristomyrmex hamatus* Xu & Zhang, 2002

地理分布：中国（云南）。

27）塔蚁属 *Pyramica* Roger, 1862

（81）哀牢山塔蚁 *Pyramica ailaoshana* Xu & Zhou, 2004

地理分布：中国（云南）。

（82）弄巴塔蚁 *Pyramica nongba* Xu & Zhou, 2004

地理分布：中国（云南）。

（83）杨氏塔蚁 *Pyramica yangi* Xu & Zhou, 2004

地理分布：中国（云南）。

28）角腹蚁属 *Recurvidris* Bolton, 1992

（84）女娲角腹蚁 *Recurvidris nuwa* Xu & Zheng, 1995

地理分布：中国（贵州、云南、江西）。

29）棒角蚁属 *Rhopalomastix* Forel, 1900

（85）暗首棒角蚁 *Rhopalomastix umbracapita* Xu, 1999

---

(73) *Oligomyrmex obtusidentus* Xu, 2003

Distribution: China (Hunan, Sichuan, Tibet, Yunnan), India.

(74) *Oligomyrmex rectidorsus* Xu, 2003

Distribution: China (Hebei, Henan, Hunan, Sichuan, Tibet, Yunnan).

(75) *Oligomyrmex reticapitus* Xu, 2003

Distribution: China (Guangxi, Hainan, Sichuan, Tibet, Yunnan).

(76) *Oligomyrmex striatus* Xu, 2003

Distribution: China (Sichuan, Yunnan).

24) *Paratopula* Wheeler, 1919

(77) *Paratopula zhengi* Xu & Xu, 2011

Distribution: China (Tibet).

25) *Perissomyrmex* Smith, 1947

(78) *Perissomyrmex fissus* Xu & Wang, 2004

Distribution: China (Yunnan).

(79) *Perissomyrmex medogensis* Xu & Zhang, 2012

Distribution: China (Tibet).

26) *Pristomyrmex* Mayr, 1886

(80) *Pristomyrmex hamatus* Xu & Zhang, 2002

Distribution: China (Yunnan).

27) *Pyramica* Roger, 1862

(81) *Pyramica ailaoshana* Xu & Zhou, 2004

Distribution: China (Yunnan).

(82) *Pyramica nongba* Xu & Zhou, 2004

Distribution: China (Yunnan).

(83) *Pyramica yangi* Xu & Zhou, 2004

Distribution: China (Yunnan).

28) *Recurvidris* Bolton, 1992

(84) *Recurvidris nuwa* Xu & Zheng, 1995

Distribution: China (Guizhou, Jiangxi, Yunnan).

29) *Rhopalomastix* Forel, 1900

(85) *Rhopalomastix umbracapita* Xu, 1999

地理分布：中国（云南、广西）。

30）窄结蚁属 *Stenamma* Westwood, 1839

（86）哀牢窄结蚁 *Stenamma ailaoense* Liu & Xu, 2011, 2011

地理分布：中国（云南、四川）。

（87）乌蒙窄结蚁 *Stenamma wumengense* Liu & Xu, 2011

地理分布：中国（云南、四川）。

（88）雅鲁藏布窄结蚁 *Stenamma yaluzangbum* Liu & Xu, 2011

地理分布：中国（西藏、四川）。

31）切胸蚁属 *Temnothorax* Mayr, 1855

（89）白露切胸蚁 *Temnothorax bailu* Qian & Xu, 2024

地理分布：中国（西藏、四川、云南）。

（90）春切胸蚁 *Temnothorax chun* Qian & Xu, 2024

地理分布：中国（云南）。

（91）春分切胸蚁 *Temnothorax chunfen* Qian & Xu, 2024

地理分布：中国（云南）。

（92）处暑切胸蚁 *Temnothorax chushu* Qian & Xu, 2024

地理分布：中国（云南、四川）。

（93）大寒切胸蚁 *Temnothorax dahan* Qian & Xu, 2024

地理分布：中国（云南）。

（94）大暑切胸蚁 *Temnothorax dashu* Qian & Xu, 2024

地理分布：中国（云南、西藏）。

（95）大雪切胸蚁 *Temnothorax daxue* Qian & Xu, 2024

地理分布：中国（云南）。

（96）冬切胸蚁 *Temnothorax dong* Qian & Xu, 2024

地理分布：中国（云南）。

（97）冬至切胸蚁 *Temnothorax dongzhi* Qian & Xu, 2024

地理分布：中国（四川）。

（98）谷雨切胸蚁 *Temnothorax guyu* Qian & Xu, 2024

地理分布：中国（云南、四川）。

（99）寒露切胸蚁 *Temnothorax hanlu* Qian & Xu, 2024

地理分布：中国（云南）。

（100）惊蛰切胸蚁 *Temnothorax jingzhe* Qian & Xu, 2024

地理分布：中国（云南）。

Distribution: China (Guangxi, Yunnan).

30) *Stenamma* Westwood, 1839

(86) *Stenamma ailaoense* Liu & Xu, 2011, 2011
Distribution: China (Sichuan, Yunnan).

(87) *Stenamma wumengense* Liu & Xu, 2011
Distribution: China (Sichuan, Yunnan).

(88) *Stenamma yaluzangbum* Liu & Xu, 2011
Distribution: China (Sichuan, Tibet).

31) *Temnothorax* Mayr, 1855

(89) *Temnothorax bailu* Qian & Xu, 2024
Distribution: China (Sichuan, Tibet, Yunnan).

(90) *Temnothorax chun* Qian & Xu, 2024
Distribution: China (Yunnan).

(91) *Temnothorax chunfen* Qian & Xu, 2024
Distribution: China (Yunnan).

(92) *Temnothorax chushu* Qian & Xu, 2024
Distribution: China (Sichuan, Yunnan).

(93) *Temnothorax dahan* Qian & Xu, 2024
Distribution: China (Yunnan).

(94) *Temnothorax dashu* Qian & Xu, 2024
Distribution: China (Tibet, Yunnan).

(95) *Temnothorax daxue* Qian & Xu, 2024
Distribution: China (Yunnan).

(96) *Temnothorax dong* Qian & Xu, 2024
Distribution: China (Yunnan).

(97) *Temnothorax dongzhi* Qian & Xu, 2024
Distribution: China (Sichuan).

(98) *Temnothorax guyu* Qian & Xu, 2024
Distribution: China (Sichuan, Yunnan).

(99) *Temnothorax hanlu* Qian & Xu, 2024
Distribution: China (Yunnan).

(100) *Temnothorax jingzhe* Qian & Xu, 2024
Distribution: China (Yunnan).

（101）立春切胸蚁 *Temnothorax lichun* Qian & Xu, 2024

地理分布：中国（云南）。

（102）立冬切胸蚁 *Temnothorax lidong* Qian & Xu, 2024

地理分布：中国（四川、云南）。

（103）立秋切胸蚁 *Temnothorax liqiu* Qian & Xu, 2024

地理分布：中国（四川、云南）。

（104）立夏切胸蚁 *Temnothorax lixia* Qian & Xu, 2024

地理分布：中国（四川）。

（105）芒种切胸蚁 *Temnothorax mangzhong* Qian & Xu, 2024

地理分布：中国（云南）。

（106）清明切胸蚁 *Temnothorax qingming* Qian & Xu, 2024

地理分布：中国（西藏）。

（107）秋切胸蚁 *Temnothorax qiu* Qian & Xu, 2024

地理分布：中国（四川、云南）。

（108）秋分切胸蚁 *Temnothorax qiufen* Qian & Xu, 2024

地理分布：中国（四川、青海）。

（109）霜降切胸蚁 *Temnothorax shuangjiang* Qian & Xu, 2024

地理分布：中国（云南）。

（110）夏切胸蚁 *Temnothorax xia* Qian & Xu, 2024

地理分布：中国（云南、西藏）。

（111）小寒切胸蚁 *Temnothorax xiaohan* Qian & Xu, 2024

地理分布：中国（西藏）。

（112）小满切胸蚁 *Temnothorax xiaoman* Qian & Xu, 2024

地理分布：中国（陕西）。

（113）小暑切胸蚁 *Temnothorax xiaoshu* Qian & Xu, 2024

地理分布：中国（云南）。

（114）小雪切胸蚁 *Temnothorax xiaoxue* Qian & Xu, 2024

地理分布：中国（云南）。

（115）夏至切胸蚁 *Temnothorax xiazhi* Qian & Xu, 2024

(101) *Temnothorax lichun* Qian & Xu, 2024

Distribution: China (Yunnan).

(102) *Temnothorax lidong* Qian & Xu, 2024

Distribution: China (Sichuan, Yunnan).

(103) *Temnothorax liqiu* Qian & Xu, 2024

Distribution: China (Sichuan, Yunnan).

(104) *Temnothorax lixia* Qian & Xu, 2024

Distribution: China (Sichuan).

(105) *Temnothorax mangzhong* Qian & Xu, 2024

Distribution: China (Yunnan).

(106) *Temnothorax qingming* Qian & Xu, 2024

Distribution: China (Tibet).

(107) *Temnothorax qiu* Qian & Xu, 2024

Distribution: China (Sichuan, Yunnan).

(108) *Temnothorax qiufen* Qian & Xu, 2024

Distribution: China (Qinghai, Sichuan).

(109) *Temnothorax shuangjiang* Qian & Xu, 2024

Distribution: China (Yunnan).

(110) *Temnothorax xia* Qian & Xu, 2024

Distribution: China (Tibet, Yunnan).

(111) *Temnothorax xiaohan* Qian & Xu, 2024

Distribution: China (Tibet).

(112) *Temnothorax xiaoman* Qian & Xu, 2024

Distribution: China (Shaanxi).

(113) *Temnothorax xiaoshu* Qian & Xu, 2024

Distribution: China (Yunnan).

(114) *Temnothorax xiaoxue* Qian & Xu, 2024

Distribution: China (Yunnan).

(115) *Temnothorax xiazhi* Qian & Xu, 2024

地理分布：中国（云南）。

（116）雨水切胸蚁 *Temnothorax yushui* Qian & Xu, 2024

地理分布：中国（四川、云南）。

32）铺道蚁属 *Tetramorium* **Mayr, 1855**

（117）心头铺道蚁 *Tetramorium cardiocarenum* Xu & Zheng, 1994

地理分布：中国（云南、四川、贵州、广西）。

（118）圆叶铺道蚁 *Tetramorium cyclolobium* Xu & Zheng, 1994

地理分布：中国（广西、云南、四川）。

（119）玉龙铺道蚁 *Tetramorium yulongense* Xu & Zheng, 1994

地理分布：中国（云南）。

33）犁沟蚁属 *Vombisidris* **Bolton, 1991**

（120）西藏犁沟蚁 *Vombisidris tibeta* Xu & Yu, 2012

地理分布：中国（西藏、云南）。

9　臭蚁亚科 Dolichoderinae

34）臭蚁属 *Dolichoderus* **Mayr, 1855**

（121）凹头臭蚁 *Dolichoderus incisus* Xu, 1995

地理分布：中国（云南、浙江），越南。

（122）鞍背臭蚁 *Dolichoderus sagmanotus* Xu, 2001

地理分布：中国（云南、广西）、

（123）鳞结臭蚁 *Dolichoderus squamanodus* Xu, 2001

地理分布：中国（云南）。

10　蚁亚科 Formicinae

35）刺结蚁属 *Lepisiota* **Santschi, 1926**

（124）尖齿刺结蚁 *Lepisiota acuta* Xu, 1994

地理分布：中国（云南、四川）。

（125）网纹刺结蚁 *Lepisiota reticulata* Xu, 1994

地理分布：中国（广西、贵州、云南）。

36）长齿蚁属 *Myrmoteras* **Forel, 1893**

（126）楔结长齿蚁 *Myrmoteras cuneonodum* Xu, 1998

地理分布：中国（云南）。

37）多刺蚁属 *Polyrhachis* **Smith, 1857**

（127）巴卡多刺蚁 *Polyrhachis bakana* Xu, 1998

地理分布：中国（云南）。

Distribution: China (Yunnan).

(116) *Temnothorax yushui* Qian & Xu, 2024
Distribution: China (Sichuan, Yunnan).

32) *Tetramorium* **Mayr, 1855**

(117) *Tetramorium cardiocarenum* Xu & Zheng, 1994
Distribution: China (Guangxi, Guizhou, Sichuan, Yunnan).

(118) *Tetramorium cyclolobium* Xu & Zheng, 1994
Distribution: China (Guangxi, Sichuan, Yunnan).

(119) *Tetramorium yulongense* Xu & Zheng, 1994
Distribution: China (Yunnan).

33) *Vombisidris* **Bolton, 1991**

(120) *Vombisidris tibeta* Xu & Yu, 2012
Distribution: China (Tibet, Yunnan).

9 Dolichoderinae

34) *Dolichoderus* **Mayr, 1855**

(121) *Dolichoderus incisus* Xu, 1995
Distribution: China (Yunnan, Zhejiang), Vietnam.

(122) *Dolichoderus sagmanotus* Xu, 2001
Distribution: China (Guangxi, Yunnan).

(123) *Dolichoderus squamanodus* Xu, 2001
Distribution: China (Yunnan).

10 Formicinae

35) *Lepisiota* **Santschi, 1926**

(124) *Lepisiota acuta* Xu, 1994
Distribution: China (Sichuan, Yunnan).

(125) *Lepisiota reticulata* Xu, 1994
Distribution: China (Guangxi, Guizhou, Yunnan).

36) *Myrmoteras* **Forel, 1893**

(126) *Myrmoteras cuneonodum* Xu, 1998
Distribution: China (Yunnan).

37) *Polyrhachis* **Smith, 1857**

(127) *Polyrhachis bakana* Xu, 1998
Distribution: China (Yunnan).

（128）短胸多刺蚁 *Polyrhachis brevicorpa* Xu, 2002

地理分布：中国（云南）。

（129）方肩多刺蚁 *Polyrhachis cornihumera* Xu, 2002

地理分布：中国（云南、广西）。

（130）驼背多刺蚁 *Polyrhachis cyphonota* Xu, 1998

地理分布：中国（云南、贵州、广西、江西）。

（131）齿肩多刺蚁 *Polyrhachis dentihumera* Xu, 2002

地理分布：中国（云南）。

（132）圆肩多刺蚁 *Polyrhachis orbihumera* Xu, 2002

地理分布：中国（云南）。

（133）圆顶多刺蚁 *Polyrhachis rotoccipita* Xu, 2002

地理分布：中国（云南）。

38）前结蚁属 *Prenolepis* Mayr, 1861

（134）大眼前结蚁 *Prenolepis magnocula* Xu, 1995

地理分布：中国（云南）。

（135）黑角前结蚁 *Prenolepis nigriflagella* Xu, 1995

地理分布：中国（云南）。

39）拟毛蚁属 *Pseudolasius* Emery, 1886

（136）双齿唇拟毛蚁 *Pseudolasius bidenticlypeus* Xu, 1997

地理分布：中国（贵州、云南）。

(128) *Polyrhachis brevicorpa* Xu, 2002

Distribution: China (Yunnan).

(129) *Polyrhachis cornihumera* Xu, 2002

Distribution: China (Guangxi, Yunnan).

(130) *Polyrhachis cyphonota* Xu, 1998

Distribution: China (Guangxi, Guizhou, Jiangxi, Yunnan).

(131) *Polyrhachis dentihumera* Xu, 2002

Distribution: China (Yunnan).

(132) *Polyrhachis orbihumera* Xu, 2002

Distribution: China (Yunnan).

(133) *Polyrhachis rotoccipita* Xu, 2002

Distribution: China (Yunnan).

38) *Prenolepis* Mayr, 1861

(134) *Prenolepis magnocula* Xu, 1995

Distribution: China (Yunnan).

(135) *Prenolepis nigriflagella* Xu, 1995

Distribution: China (Yunnan).

39) *Pseudolasius* Emery, 1886

(136) *Pseudolasius bidenticlypeus* Xu, 1997

Distribution: China (Guizhou, Yunnan).

中文名索引
Index of Chinese Names

A

阿诗玛无刺蚁	164
阿佤钝猛蚁	036
哀牢山塔蚁	196
哀牢窄结蚁	206
鞍背臭蚁	278
暗首棒角蚁	204
凹唇细长蚁	152
凹头臭蚁	276

B

八齿钝猛蚁	044
巴卡多刺蚁	288
巴卡猛蚁	106
坝湾猛蚁	108
白露切胸蚁	212
版纳盘猛蚁	056
版纳曲颊猛蚁	054
北京细蚁	128
北京原细蚁	138
布氏卷尾猛蚁	060

C

叉唇细长蚁	156
叉颚原细蚁	144
昌平卷尾猛蚁	068
长柄小盲猛蚁	124
齿肩多刺蚁	296
处暑切胸蚁	218
春分切胸蚁	216
春切胸蚁	214

D

大寒切胸蚁	220
大暑切胸蚁	222
大雪切胸蚁	224
大眼前结蚁	302
大禹圆鳞蚁	160
傣小眼猛蚁	094
单色原细蚁	142
德宏细蚁	130
滇盘猛蚁	058
冬切胸蚁	226
冬至切胸蚁	228
短背短猛蚁	072
短胸多刺蚁	290
钝齿稀切叶蚁	180

E

二齿猛蚁	110

F

方肩多刺蚁	292

G

高结稀切叶蚁	174
耿马原细蚁	146
谷雨切胸蚁	230

H

哈尼小眼猛蚁	096
寒露切胸蚁	232
黑角前结蚁	304
黑色埃猛蚁	076
黄帝细颚猛蚁	078
黄色猛蚁	122

J

尖齿刺结蚁	282
尖唇细长蚁	158
尖刺稀切叶蚁	172
惊蛰切胸蚁	234
景颇弯蚁	168

K

康巴钝猛蚁	040

K
克平卷尾猛蚁	062
昆明细蚁	132

L
老子细颚猛蚁	080
立春切胸蚁	236
立冬切胸蚁	238
立秋切胸蚁	240
立夏切胸蚁	242
裂唇奇蚁	190
鳞结臭蚁	280
六刺云行军蚁	126
龙林猛蚁	112
龙门卷尾猛蚁	064
隆背细长蚁	154
珞巴小眼猛蚁	098

M
芒种切胸蚁	244
梅里钝猛蚁	042
门巴小眼猛蚁	100
勐腊猛蚁	114
孟子细颚猛蚁	082
墨脱奇蚁	192
木兰版纳猛蚁	050

N
南贡山猛蚁	116
尼玛弯蚁	170
弄巴塔蚁	198
怒江卷尾猛蚁	066
女娲角腹蚁	202

P
盘古细颚猛蚁	084
片马猛蚁	120
片突厚结猛蚁	102
平背高黎贡蚁	162

Q
秦岭细蚁	134
清明切胸蚁	246
秋分切胸蚁	250
秋切胸蚁	248

S
三叶钝猛蚁	046
疏毛无刺蚁	166
双齿唇拟毛蚁	306
双角稀切叶蚁	176
双色原细蚁	140
霜降切胸蚁	252
孙子细颚猛蚁	086

T
条纹稀切叶蚁	186
驼背多刺蚁	294

W
弯刺稀切叶蚁	178
弯钩棱胸蚁	194
网纹刺结蚁	284
纹头稀切叶蚁	184
乌蒙窄结蚁	208
无缘细长蚁	150
五齿猛蚁	118

X
西藏犁沟蚁	274
西藏原细蚁	148
细齿钝猛蚁	038
夏切胸蚁	254
夏至切胸蚁	264
小寒切胸蚁	256
小满切胸蚁	258
小暑切胸蚁	260
小雪切胸蚁	262
小眼迷猛蚁	052
楔结长齿蚁	286
心头铺道蚁	268

Y
雅鲁藏布窄结蚁	210
炎帝细颚猛蚁	088
杨氏塔蚁	200
雨水切胸蚁	266
玉龙铺道蚁	272
圆顶多刺蚁	300
圆肩多刺蚁	298
圆叶铺道蚁	270
云南细蚁	136

Z
赵氏卷尾猛蚁	070
郑氏厚结猛蚁	104
郑氏华丽蚁	188
直背稀切叶蚁	182
直唇隐猛蚁	074
庄子细颚猛蚁	090
锥头小眼猛蚁	092
卓玛钝猛蚁	048

学名索引
Index of Scientific Names

A

Amblyopone awa Xu & Chu, 2012	036
Amblyopone crenata Xu, 2001	038
Amblyopone kangba Xu & Chu, 2012	040
Amblyopone meiliana Xu & Chu, 2012	042
Amblyopone octodentata Xu, 2006	044
Amblyopone triloba Xu, 2001	046
Amblyopone zoma Xu & Chu, 2012	048

B

Bannapone mulanae Xu, 2000	050
Brachyponera brevidorsa Xu, 1994	072

C

Cryptopone recticlypea Xu, 1998	074

D

Discothyrea banna Xu, Burwell & Nakamura, 2014	056
Discothyrea diana Xu, Burwell & Nakamura, 2014	058
Dolichoderus incisus Xu, 1995	276
Dolichoderus sagmanotus Xu, 2001	278
Dolichoderus squamanodus Xu, 2001	280

E

Emeryopone melaina Xu, 1998	076
Epitritus dayui Xu, 2000	160

G

Gaoligongidris planodorsa Xu, 2012	162
Gnamptogenys bannana Xu & Zhang, 1996	054

K

Kartidris ashima Xu & Zheng, 1995	164
Kartidris sparsipila Xu, 1999	166

L

Lepisiota acuta Xu, 1994	282
Lepisiota reticulata Xu, 1994	284
Leptanilla beijingensis Qian, Xu, Man & Liu, 2024	128
Leptanilla dehongensis Qian, Xu, Man & Liu, 2024	130
Leptanilla kunmingensis Xu & Zhang, 2002	132
Leptanilla qinlingensis Qian, Xu, Man & Liu, 2024	134
Leptanilla yunnanensis Xu, 2002	136
Leptogenys huangdii Xu, 2000	078
Leptogenys laozii Xu, 2000	080
Leptogenys mengzii Xu, 2000	082

Leptogenys pangui Xu, 2000 — 084
Leptogenys sunzii Xu & He, 2015 — 086
Leptogenys yandii Xu & He, 2015 — 088
Leptogenys zhuangzii Xu, 2000 — 090
Lordomyrma jingpo Liu, Xu & Hita Garcia, 2021 — 168
Lordomyrma nima Liu, Xu & Hita Garcia, 2021 — 170

M

Myopias conicara Xu, 1998 — 092
Myopias daia Xu, Burwell & Nakamura, 2014 — 094
Myopias hania Xu & Liu, 2012 — 096
Myopias luoba Xu & Liu, 2012 — 098
Myopias menba Xu & Liu, 2012 — 100
Myrmoteras cuneonodum Xu, 1998 — 286
Mystrium oculatum Xu, 1998 — 052

O

Oligomyrmex acutispinus Xu, 2003 — 172
Oligomyrmex altinodus Xu, 2003 — 174
Oligomyrmex bihornatus Xu, 2003 — 176
Oligomyrmex curvispinus Xu, 2003 — 178
Oligomyrmex obtusidentus Xu, 2003 — 180
Oligomyrmex rectidorsus Xu, 2003 — 182
Oligomyrmex reticapitus Xu, 2003 — 184
Oligomyrmex striatus Xu, 2003 — 186

P

Pachycondyla lobocarena Xu, 1995 — 102
Pachycondyla zhengi Xu, 1995 — 104
Paratopula zhengi Xu & Xu, 2011 — 188
Perissomyrmex fissus Xu & Wang, 2004 — 190
Perissomyrmex medogensis Xu & Zhang, 2012 — 192
Polyrhachis bakana Xu, 1998 — 288
Polyrhachis brevicorpa Xu, 2002 — 290
Polyrhachis cornihumera Xu, 2002 — 292
Polyrhachis cyphonota Xu, 1998 — 294
Polyrhachis dentihumera Xu, 2002 — 296
Polyrhachis orbihumera Xu, 2002 — 298
Polyrhachis rotoccipita Xu, 2002 — 300
Ponera baka Xu, 2001 — 106
Ponera bawana Xu, 2001 — 108
Ponera diodonta Xu, 2001 — 110
Ponera longlina Xu, 2001 — 112
Ponera menglana Xu, 2001 — 114
Ponera nangongshana Xu, 2001 — 116
Ponera pentodontos Xu, 2001 — 118
Ponera pianmana Xu, 2001 — 120
Ponera xantha Xu, 2001 — 122
Prenolepis magnocula Xu, 1995 — 302
Prenolepis nigriflagella Xu, 1995 — 304
Pristomyrmex hamatus Xu & Zhang, 2002 — 194
Probolomyrmex longiscapus Xu & Zeng, 2000 — 124
Proceratium bruelheidei Staab, Xu & Hita Garcia, 2018 — 060
Proceratium kepingmai Staab, Xu & Hita Garcia, 2018 — 062
Proceratium longmenense Xu, 2006 — 064
Proceratium nujiangense Xu, 2006 — 066
Proceratium shohei Staab, Xu & Hita Garcia, 2018 — 068
Proceratium zhaoi Xu, 2000 — 070
Protanilla beijingensis Man, Ran, Chen & Xu, 2017 — 138
Protanilla bicolor Xu, 2002 — 140
Protanilla concolor Xu, 2002 — 142
Protanilla furcomandibula Xu & Zhang, 2002 — 144
Protanilla gengma Xu, 2012 — 146
Protanilla tibeta Xu, 2012 — 148
Pseudolasius bidenticlypeus Xu, 1997 — 306
Pyramica ailaoshana Xu & Zhou, 2004 — 196
Pyramica nongba Xu & Zhou, 2004 — 198
Pyramica yangi Xu & Zhou, 2004 — 200

R

Recurvidris nuwa Xu & Zheng, 1995 — 202
Rhopalomastix umbracapita Xu, 1999 — 204

S

Stenamma ailaoense Liu & Xu, 2011 — 206
Stenamma wumengense Liu & Xu, 2011 — 208
Stenamma yaluzangbum Liu & Xu, 2011 — 210

T

Temnothorax bailu Qian & Xu, 2024 — 212
Temnothorax chun Qian & Xu, 2024 — 214
Temnothorax chunfen Qian & Xu, 2024 — 216
Temnothorax chushu Qian & Xu, 2024 — 218
Temnothorax dahan Qian & Xu, 2024 — 220
Temnothorax dashu Qian & Xu, 2024 — 222
Temnothorax daxue Qian & Xu, 2024 — 224
Temnothorax dong Qian & Xu, 2024 — 226
Temnothorax dongzhi Qian & Xu, 2024 — 228
Temnothorax guyu Qian & Xu, 2024 — 230
Temnothorax hanlu Qian & Xu, 2024 — 232
Temnothorax jingzhe Qian & Xu, 2024 — 234
Temnothorax lichun Qian & Xu, 2024 — 236
Temnothorax lidong Qian & Xu, 2024 — 238
Temnothorax liqiu Qian & Xu, 2024 — 240
Temnothorax lixia Qian & Xu, 2024 — 242
Temnothorax mangzhong Qian & Xu, 2024 — 244
Temnothorax qingming Qian & Xu, 2024 — 246
Temnothorax qiu Qian & Xu, 2024 — 248
Temnothorax qiufen Qian & Xu, 2024 — 250
Temnothorax shuangjiang Qian & Xu, 2024 — 252
Temnothorax xia Qian & Xu, 2024 — 254
Temnothorax xiaohan Qian & Xu, 2024 — 256
Temnothorax xiaoman Qian & Xu, 2024 — 258
Temnothorax xiaoshu Qian & Xu, 2024 — 260
Temnothorax xiaoxue Qian & Xu, 2024 — 262
Temnothorax xiazhi Qian & Xu, 2024 — 264
Temnothorax yushui Qian & Xu, 2024 — 266
Tetramorium cardiocarenum Xu & Zheng, 1994 — 268
Tetramorium cyclolobium Xu & Zheng, 1994 — 270
Tetramorium yulongense Xu & Zheng, 1994 — 272
Tetraponera amargina Xu & Chai, 2004 — 150
Tetraponera concava Xu & Chai, 2004 — 152
Tetraponera convexa Xu & Chai, 2004 — 154
Tetraponera furcata Xu & Chai, 2004 — 156
Tetraponera protensa Xu & Chai, 2004 — 158

V

Vombisidris tibeta Xu & Yu, 2012 — 274

Y

Yunodorylus sexspinus Xu, 2000 — 126